Biological Control of Plant-parasitic Nematodes, 2nd Edition

Soil Ecosystem Management in Sustainable Agriculture

_____

This book is dedicated to the memory of two good friends, Brian Kerry and Gregor Yeates. Both contributed prolifically to our knowledge of nematode ecology, but unfortunately did not live to see the final manuscript. Brian kindly wrote the Foreword to my previous book, but sadly, he suffered serious health problems at the end of his career and passed away in 2011. He made a major contribution to the field of biological control of nematodes; the MBE he was awarded in 2008 fittingly recognized his services to nematology and to science in general, both in the United Kingdom and internationally. Gregor's death in 2012 was also an untimely loss, as he was still very active in 'retirement'. When Gregor commenced his career, nematologists were mainly interested in plant parasites, but through his pioneering efforts, they now take a much wider view of the nematode community and its role in the soil ecosystem.

Biological Control of Plant-parasitic Nematodes, 2nd Edition

Soil Ecosystem Management in Sustainable Agriculture

_____

Graham R. Stirling

Biological Crop Protection Pty Ltd, Brisbane, Australia

www.cabi.org

CABI is a trading name of CAB International

CABI	CABI
Nosworthy Way	38 Chauncey Street
Wallingford	Suite 1002
Oxfordshire OX10 8DE	Boston, MA 02111
UK	USA
Tel: +44 (0)1491 832111	Tel: +1 800 552 3083 (toll free)
Fax: +44 (0)1491 833508	Tel: +1 617 395 4051
E-mail: info@cabi.org	E-mail: cabi-nao@cabi.org
Website: www.cabi.org	

A catalogue record for this book is available from the British Library, London, UK.

Library of Congress Cataloging-in-Publication Data

Stirling, Graham R.
  Biological control of plant-parasitic nematodes : soil ecosystem manage-ment in sustainable agriculture / by Graham R. Stirling. -- 2nd ed.
    p. cm.
  Includes bibliographical references and index.
  ISBN 978-1-78064-415-8 (alk. paper)
1. Plant nematodes--Biological control. 2. Soil management. I. Title.

  SB998.N4S85 2014
  632'.6257--dc23

                            2013040394

ISBN-13: 978 1 78064 415 8

Commissioning editor: Rachel Cutts
Editorial assistant: Alexandra Lainsbury
Production editor: Laura Tsitlidze

Typeset by SPi, Pondicherry, India
Printed and bound in the UK by CPI Group (UK) Ltd, Croydon, CR0 4YY.

Contents

**SECTION V   NATURAL SUPPRESSION
AND INUNDATIVE BIOLOGICAL CONTROL**

SECTION VI   SUMMARY, CONCLUSIONS, PRACTICAL GUIDELINES AND FUTURE RESEARCH

Foreword

Graham Stirling's previous book on biological control of plant-parasitic nematodes reflected the state of the science at that time. In the intervening 20 years, our perception of processes regarding the regulation and suppression of nematodes in agricultural ecosystems has become much more holistic. It has been enhanced by advances in our knowledge of the role of soil biota in ecosystem processes, and by research on cropping systems that has been conducted under the auspices of agricultural sustainability initiatives. The current book recognizes that soil organisms function in an intricately connected and interacting food web; that they compete for, or exploit each other as, food resources; and that their activity and diversity is strongly influenced by the ways in which crops and soil are managed.

One might expect a dichotomy of interests regarding approaches to biological control of soil organisms. On the one hand, there is the commercial interest in the identification of specific parasites or predators that have potential for production in mass culture and application to field sites. On the other hand, there is the growing understanding that a plethora of well-adapted organisms occurs in the soil and that, through appropriate soil stewardship, an environment can be developed in which pest species are suppressed by their endemic natural enemies. The two interests seem at odds in that stewardship promoting the development of a suite of naturally occurring antagonists may limit the commercial value of introduced organisms by reducing the required frequency or biomass of applications. This book provides important information that allows a rational synthesis of both perspectives.

Inundative releases of organisms into a diverse and complex system, which has all micro-sites occupied and exploited, are usually only effective in the short term. The introduced organisms are unlikely to become established unless they possess invasive characteristics that render them more successful than the incumbents. Such an approach is more likely to be successful in systems that have been biologically damaged by previous management, or by lack of resources. For example, inundative releases may be very useful for regulating pest species during the multi-year transition period from a conducive to a suppressive soil environment. Also, in cases where there is no incentive for long-term stewardship of leased or rented land, or in situations where a rapid solution to a pest problem is necessary, inundative releases will be appropriate.

The organisms of the soil differ in physiology and behaviour, and contribute to ecosystem services in a multitude of ways. They occupy, and are variously adapted to, the array of soil micro-niches that is created by distribution of resources, the irregularities of the physical and

chemical components of the soil matrix, and gradients of water, temperature and gaseous diffusion. Consequently, the communities of organisms inhabiting each micro-site may differ in
physiological and behavioural characteristics. The integral effect of the dynamic interactions at
each micro-site represents the magnitude of the service performed by the resident taxa. General
and specific suppressiveness to plant-parasitic nematodes is one of the services provided by
the soil biota, and examples of situations where it has been enhanced through management are
well documented in extremely readable passages in this book.

Over the years between Dr Stirling's two books, his perceptions of how soil biological
systems function have evolved based on his observation and vast practical experience. This
book is centred on stewardship of the intrinsic suppressiveness of soils that results from their
physical, chemical and biological complexity. Dr Stirling explains that plant residues, root exudates and other organic inputs fuel the biological components of the system, and that the organic
resources that would normally drive undisturbed systems have become severely depleted by
modern agricultural practices.

Dr Stirling emphasizes that certain agricultural activities, including applications of chemicals, excessive tillage and prolonged fallow periods between crops, are major deterrents to
intrinsic biological suppression or regulation of pest species. Soil tillage disrupts soil microhabitats and has abrasive effects on larger-bodied organisms. Most soil organisms inhabit the
thin films of water that surround soil particles. Agricultural chemicals applied to the soil dissolve
in the water films and their concentrations change with fluxes in water volume. Soil organisms
differ in their tolerance to these chemical stresses, but some will be adversely affected. For example,
large-bodied nematodes of the nematode orders Dorylaimida and Mononchida include
important generalist and specialist predators of other soil nematodes. They are frequently abundant under undisturbed conditions but often they are not detectable in conventionally managed
agricultural soils. The effects of agricultural practices are recorded, perhaps indelibly, in the biological, chemical and physical state of the soil; the effects are not easily reversed.

Once a soil has been converted to intensive agricultural production and has lost its natural
suppressiveness, pest species become more abundant. One convenient and effective solution is
to apply a pesticide. However, many of the pesticides used to control nematodes, and those
that have been used in the past, are broad-spectrum biocides that further reduce levels of
organisms which might otherwise contribute to natural soil suppressiveness. Thus, stepping
onto the pesticide treadmill as a management response to plant damage by plant-parasitic
nematodes and other soil pests has long-term consequences: natural suppressive mechanisms
are depleted to such an extent that it becomes difficult to discontinue chemical-intensive management. It may be achieved by activating the soil biological community through inputs of
organic matter, and eliminating constraining factors (e.g. preventing soil compaction, minimizing tillage, reducing the length of fallows, limiting chemical inputs) but, as Dr Stirling
explains, adopting these practices does not result in instantaneous success. Organisms that are
missing from the system or are below detectable levels may take many years to return and
re-establish. There is a management learning curve during the transition to biologically intensive systems, and the risk of pest-induced losses is greater during the transition. Organic inputs
to increase soil carbon and fuel the soil food web will not have the instantaneous effect of a
biocide. It may take years of input and continuous stewardship before resource levels are sufficient and the necessary successional changes have occurred in soil communities. Dr Stirling
provides examples of systems in which the transition has been successfully accomplished.

A factor often not considered in cropping system design studies is that modern high-
yielding crop cultivars that have been selected for high-input agriculture may lack the root
mass and morphology for exploitation of organic nutrient sources. These cultivars, when
grown under the conditions for which they are selected, do not need an active soil biota. So,
perhaps there is a need to select cultivars with different morphological characteristics for bio-
intensive systems. Root exudation and rhizodeposition are major drivers of the soil food web.
A larger root system may enhance the priming of biological activity in the soil food web.

This book is about stewardship of the soil ecosystem through management strategies that are continued across years. Here factors and decisions associated with land tenure become important. Management-intensive stewardship is more likely to be applied to land that is owned than to land that is rented and in which there may not be long-term interest, a metaphorical tragedy of the commons. At a global level, agriculture is a net nutrient-export enterprise; there is an urgent need to consider the global cycling of carbon, nitrogen and other nutrients. These minerals are harvested from the soil via the crops that are produced, transported to centres of human population, and then excreted into rivers and oceans. How can the nutrients be returned to the land so that they are cycled rather than permanently exported? How can organic wastes that accumulate around cities and livestock enterprises be used to enhance suppressive services of the soil food web that have been diminished by the constant removal of harvested product?

In the final chapters of this book, Dr Stirling provides a valuable 'how to' synthesis for the transition to, and stewardship of, biologically suppressive soils with the goal of minimizing damage due to nematode pests. Hopefully the approaches he has suggested will encourage scientists and land managers to tackle a primary cause of nematode problems in agriculture: practices, or lack thereof, that result in diminution of the soil biota.

Howard Ferris
Department of Entomology and Nematology
University of California, Davis

Preface

My first book on biological control of plant-parasitic nematodes was published in 1991, at a time when biological control was first being seriously considered as an option for managing nematode pests. Several widely used nematicides had been removed from the market in the previous 15 years, and nematologists were seeking alternatives that had little or no impact on human health or the environment. Although research on natural enemies of nematodes was in its infancy, the number of scientists working on biological control was increasing, and their research programmes were being supported by a surge of activity in related areas such as soil biology and ecology.

In the last 22 years, the level of interest in biological control of nematodes has continued to increase, and an ever-expanding group of researchers has addressed numerous issues of relevance to that topic. Some of the key areas of investigation include: the ecology of nematophagous fungi; molecular taxonomy of parasites and predators; regulatory forces within soil food webs and their effects on the nematode community; interactions between bacteria in the genus *Pasteuria* and various plant-parasitic nematodes; host specificity of *Pasteuria* spp.; the impact of endophytes, mycorrhizal fungi and plant growth-promoting rhizobacteria on nematodes; nematode-suppressive soils; induced systemic resistance to soilborne pathogens; the role of organic matter in stimulating natural enemies of nematodes; molecular ecology of soil and rhizosphere organisms; and biotic contributions to soil health and sustainable agriculture. Given the amount of new knowledge that has been generated during that period, it is clearly time to review progress and update Stirling (1991).

In considering whether I should attempt to write another book, my initial thoughts were that the field of biological control of nematodes was now so large that it would be impossible for one person to do justice to the topic. However, if biological control is to ever become a key component of future nematode management programmes, there is a need to integrate the large body of existing knowledge into a package that is useful to a diverse audience: fellow scientists who may focus on only one or two biological control agents; students wishing to undertake studies in this fascinating but challenging field; nematologists working in areas other than biological control; plant pathologists, agronomists, horticulturalists, soil scientists, extension personnel, pest management consultants, and others who constantly face soil-related problems but are not conversant with nematode management; and farmers seeking sustainable methods of reducing losses from nematode pests.

Given the breadth of my experience, I decided that I had as much chance as anyone of tackling this task. I was raised on a farm, my family is still farming, and my first degree was in agricultural science; so I have an abiding interest in ensuring that our farming systems continue to become more productive and sustainable. My career commenced in the temperate region of southern Australia, working mainly on grapevines and citrus, but a mid-career move to Queensland enabled me to broaden my experience and work with a number of tropical and subtropical crops, including rice, pineapple, sugarcane and ginger, and a range of vegetable crops. I have also been involved in research for the Australian grains industry, which continues to flourish in some of the poorest soils and most demanding climates in the world. Thus, I have been fortunate to have worked on a wide range of nematode problems in diverse environments. Project and consultancy work in Asia and the Pacific has also given me some awareness of the problems faced by agricultural professionals and farmers in developing countries. Although I have never specialized entirely on biological control, I have worked with egg-parasitic fungi, nematode-trapping fungi and *Pasteuria*; collaborated with companies interested in developing commercially acceptable biological products for nematode control; and contributed to our understanding of nematode-suppressive soils. Thus, I have broad interests in biological control and believe that it has the potential to play an important role in nematode management.

For 13 years, I was a member of a research group working on yield decline of sugarcane, and my interaction with members of that team (agronomists, soil scientists, soil microbiologists, plant pathologists, agricultural engineers, extension specialists and economists) taught me that problems often considered to be caused by nematodes and soilborne pathogens are actually the result of inappropriate farming systems. Thus, the solution is often to modify the way a crop is grown rather than focus on controlling selected root pathogens. Biological suppressiveness to nematodes can be enhanced by modifying the farming system: a lesson I have learnt through first-hand experience.

Unusually for a research scientist, I have spent about one-third of my career operating a small science-based company that undertakes research and provides diagnostic services in nematodes and soilborne diseases. I am, therefore, aware of the multi-faceted nature of running a business and the need for financial viability. Thus, I have some empathy for land managers, who must face economic realities when dealing with soilborne disease problems. My main clients are grower/government-funded research corporations wishing to see practical outcomes from their research investments, and this explains why a recurring theme in this book is that agricultural research must be relevant to the farming community; that outcomes must be communicated to those who can benefit from the work; and that research is not complete until outcomes are tested, adapted and validated in the field.

One thing I have learnt during my career is that plant-parasitic nematodes are rarely the only cause of suboptimal crop performance. If a poor-growth problem is soil-related, it will generally have multiple causes, and so it is important to provide holistic solutions rather than a temporary fix that just focuses on the nematode component. Biological control has been a continuing interest, but from my perspective, it is only one of many tools that can be used by farmers to improve soil health and limit losses from plant-parasitic nematodes.

Someone with my background will obviously have a different view of biological control than the specialists in many different fields whose contributions are essential for the development and implementation of biological methods of managing nematodes. This book deliberately takes a broad view of the topic and attempts to integrate our knowledge of soil health and sustainable agriculture with our understanding of nematode ecology and the suppressive services that are provided by the soil food web. Hopefully, the end result will be useful to anyone with an interest in agriculture, soil biology or ecology. Since biological control should be a component of all integrated nematode management programmes, the book focuses on the tools available to farmers (e.g. organic matter management strategies, rotation schemes, tillage practices, and water and nutrient inputs) and considers how they might be used to enhance the activity of natural enemies and reduce losses from nematode pests. The mass production and release of biological control agents is also covered, but is not seen as the universal panacea that it is often considered to be.

I do not claim that this book provides a detailed coverage of every subject that is discussed. A voluminous amount of literature is available on ecological and agricultural issues that are pertinent to biological control (e.g. conservation agriculture, sustainable farming systems, soil organic matter, precision agriculture, the rhizosphere, soil health, the soil biological community and regulatory mechanisms within the soil food web) and if additional information on topics of this nature is required, it should be obtained from the cited references. In the same way, review articles and chapters in multi-authored books must be consulted for detailed information on specific aspects of biological control, as they have been written by experts in the particular subject area. However, many such reviews have a relatively narrow focus, whereas this book intentionally takes a broader view and considers biological control of nematodes from an ecological and farming systems perspective. Hopefully it will motivate readers to start thinking holistically about how the nematode community and a diverse range of natural enemies can be manipulated to achieve effective and sustainable systems of suppressing nematode pests.

Although my original intention was to revise my 1991 book, it soon became apparent that a complete rewrite was required. The previous book summarizes much of the early work on biological control of nematodes and remains relevant, but this book is deliberately different. It has a wider focus; the content has been substantially rearranged; and it concentrates on research that has been carried out over the last 25 years. The book has been subdivided into six sections, and after an initial introductory chapter, the second section covers the soil environment and the organisms that live in soil, and how they are influenced by plants and farming systems. The third and fourth sections deal with the natural enemies of nematodes (parasitic and predatory fungi, invertebrate predators, bacterial parasites and viruses), and a diverse range of fungal and bacterial symbionts that have the capacity to interfere in some way with nematode development. Methods of reducing populations of plant-parasitic nematodes through biological means are discussed in the fifth section, with a particular focus on a concept referred to as 'integrated soil biology management'. The final section summarizes the main points made in the book, and offers some suggestions on priorities for future research. It also includes a chapter that encourages advisors and practitioners to think about the biological status of their soils, and provides guidelines on how soil biological processes can be utilized to reduce losses from nematode pests.

Graham Stirling
August 2013

Acknowledgements

Many people helped me complete this book, but Howard Ferris, Rob McSorley, Andreas Westphal, Sara Sánchez-Moreno, Kathy Ophel-Keller and Mike Bell deserve a special mention. They acted as referees and made many incisive comments on my first draft. Although I take total responsibility for the final version, their contributions are very much appreciated.

My association with Howard Ferris (University of California, Davis) began at UC Riverside in 1975, when I took a one-semester student project and worked with a nematode–grapevine model he had developed. At the time, Howard was compiling the results of a survey which showed that root-knot nematode populations in California peach orchards were unexpectedly low (Ferris et al., 1976). I decided that it would be interesting to find out why, and my PhD studies with Ron Mankau led to the discovery of *Dactylella oviparasitica* (now *Brachyphoris oviparasitica*) and its role in suppressing the nematode. Since then, Howard's wide-ranging contributions in the field of nematode ecology made him an obvious choice to not only comment on an early draft of the book, but also write the Foreword.

I first came across Rob McSorley (University of Florida) when we competed against each other in a student's presentation session at a meeting of the Society of Nematologists. Rob won the prize, but we remained friends! He spent a sabbatical year with me in Brisbane during the 1980s, and is widely known for his work on nematode ecology and management. Andreas Westphal (Institute for Plant Protection in Field Crops and Grassland, Germany) was another whose comments I sought, as he has worked in many quite different environments in the United States and Europe, and has also made a major contribution to our understanding of nematode-suppressive soils. I have never met Sara Sánchez-Moreno (National Institute for Agricultural and Food Research and Technology, Spain), but her ecological papers are substantive and she has experience with the agriculture in California and southern Europe.

My two Australian colleagues are not nematologists, and so they were able to view what I had written from a different perspective. Kathy Ophel-Keller (South Australian Research and Development Institute) leads a research group that established the world's first commercial DNA-based testing service for soilborne diseases (Ophel-Keller et al., 2008). Thus, she has a good knowledge of molecular technologies, and understands the complexities involved in managing beneficial organisms and multiple pathogens in soil. Mike Bell (Queensland Alliance for Agriculture and Food Innovation) provided an agronomic viewpoint. He has worked with tropical and subtropical crops (particularly sugarcane, cereals, peanuts and soybean) for many years, and recognizes that the soil ecosystem must be managed effectively if agriculture is to be productive and sustainable in the long term.

Many other people also contributed by making comments or responding to my queries, and I would like to thank them sincerely. They included Robin Giblin-Davis, Keith Davies, Richard Sikora, Mario Tenuta, Chris Hayward, Alan McKay, Megan Ryan, Nikki Seymour, Richard Humber, Roger Shivas and Emily Rames. Also, a special thank you to Professor David Guest for his support, as the book was compiled while I was an Honorary Associate in the Department of Plant and Food Sciences, Faculty of Agriculture and Environment, The University of Sydney.

I also wish to thank the book production staff at CABI (Rachel Cutts, Alexandra Lainsbury, Laura Tsitildze and their colleagues) for their contribution. They responded quickly to my emails, were always courteous and efficient, and made many helpful suggestions.

My lovely wife, Marcelle, deserves special thanks for her constant support. As a fellow scientist, she understood the time commitment involved in preparing a book of this nature. While I was ensconced in 'writing mode', she maintained a home, kept our business operating and was also able to use her artistic talent to prepare Figures 2.6, 3.5, 5.1, 5.2, 5.3, 6.2, 6.4, 8.1, 14.1, 14.2, 14.3 and 14.4. Marcelle, I will be eternally grateful for your help and encouragement.

Finally, I wish to acknowledge the contributions of the nematologists, soil biologists and soil ecologists whose work is cited in this book. Your collective efforts have expanded our knowledge of plant-parasitic nematodes and their natural enemies to the point where biological control now has the capacity to play a crucial role in integrated nematode management programmes.

Graham Stirling
August 2013

Section I

Setting the Scene

_____

1

Ecosystem Services and the Concept of 'Integrated Soil Biology Management'

Plant-parasitic nematodes are important pests of most of the world's crops. Estimates of crop loss usually range from 5% to 15%, but higher losses sometimes occur, and there are situations where nematodes are a major factor limiting the production of a particular crop. Although numerous methods are available to either reduce populations of plant-feeding nematodes or enable crops to better tolerate the damage they cause, biological control is generally perceived as playing little or no role in current nematode management programmes. This book will demonstrate that the regulatory mechanisms forming the basis of biological control are actually operating in many agricultural systems (although usually at sub-optimal levels), and that their effectiveness can be enhanced through appropriate management. It argues that modern agriculture must not only be highly productive, but also provide a full range of ecosystem services, including pest and disease suppression; and that this is achievable by incorporating practices into the farming system that increase and sequester carbon, enhance soil biological activity, minimize soil disturbance, improve soil health and ensure long-term sustainability.

Agriculture from an Ecological Perspective

Plants are the lifeblood of all agricultural systems, as they capture energy from the sun and convert it into biomass through the process of photosynthesis. Some of that biomass is then harvested and used to feed and clothe the human population, satisfy the needs of livestock, or provide feedstock for biofuel production. However, what is often forgotten is that plants have many other important roles within agroecosystems: their roots host symbionts that transfer mineral nutrients to the plant, convert atmospheric nitrogen to ammonia, and enhance disease resistance and tolerance mechanisms; plant biomass is the primary source of energy for a decomposer community that breaks down organic matter and recycles nutrients into forms available to plants; and roots, rhizodeposits and plant residues support a complex community of organisms that influence numerous soil physical and chemical properties and also interfere with the pests and pathogens that obtain resources from roots.

The primary role of agriculture is to supply human needs for food and fibre, and in ecological terms this is referred to as a provisioning service (Power, 2010). However, it is clear from the previous paragraph and from Fig. 1.1 that agriculture also provides a range of other important ecosystem services: soil aggregate formation; nutrient cycling; immobilization of nutrients; nitrogen fixation; enhancement of air and water quality; carbon sequestration; detoxification of pollutants, pest control and disease suppression. These supporting and regulating services are provided through the soil biota and are particularly important from an agricultural sustainability and environmental perspective (Powlson *et al.*, 2011a). However, because they are difficult to value in a monetary sense, they are not valued by the agricultural community to the same extent as food and fibre production.

One of the aims of this book is to discuss ways in which soil and plant management practices can be modified so that ecosystem processes within agricultural systems continue to supply provisioning services, but also provide a range of services that support future provisioning. If that goal can be accomplished, one of the outcomes will be a greater level of suppressiveness to plant-parasitic nematodes and other soilborne pathogens.

Biotic Interactions within the Soil Food Web

The community of organisms that live and interact in soil is referred to as the soil food web. The organisms that occupy this food web are dependent on each other for sources of carbon and energy, while the whole biological community is sustained by the photosynthetic activity of plants. Since plants are the primary producers, they form the first trophic level in the soil food web. Primary consumers (bacteria, fungi, plant-feeding nematodes and root-grazing insects) form the second trophic level, as they either feed on living roots or decompose detritus that was originally derived from plants. These organisms then become food and energy sources for higher trophic levels (e.g. bacteria are consumed by nematodes and protozoa; fungal hyphae and spores are eaten by fungivorous nematodes and springtails; and plant-feeding and free-living nematodes are parasitized by fungi or consumed by nematode or arthropod predators). Thus, the soil food web consists of a vast array of interacting organisms that transfer energy from plants to primary and secondary consumers.

Bacteria and fungi are by far the most important component of the soil food web, but many other organisms are also present,

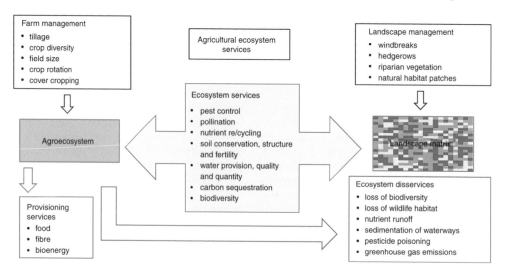

Fig. 1.1. Impacts of farm management and landscape management on the flow of ecosystem services and disservices to and from agroecosystems. (From Power, 2010, with permission.)

including protozoa, nematodes and enchytraeids; a range of microarthropods (e.g. mites, springtails and symphylans); and larger fauna such as termites, millipedes, earthworms, burrowing reptiles and rodents. The main function of the soil food web is to decompose plant material, and mineralize and store nutrients. However, during the decomposition process, soil organisms compete for resources and interact with each other through many different mechanisms, including parasitism, predation, competition and antibiosis. Thus, regulatory forces operating within the soil food web determine the size, composition and activity of the soil biological community and prevent the uncontrolled proliferation of opportunistic organisms. This phenomenon, which is sometimes referred to as 'biological buffering' or the 'balance of nature', affects all organisms, including nematode pests, and is the basis of biological control.

Biological Control of Plant-parasitic Nematodes

The term 'biological control' has different meanings to different people. To most farmers, the general public and many pest management consultants, biological control simply means replacing toxic pesticides with safe biological products, a tactic that is often referred to as 'inundative' biological control. To some scientists, particularly entomologists and weed scientists, the term refers to what might be termed 'inoculative' biological control, where a parasite or predator is deliberately introduced from elsewhere, becomes permanently established in its new environment, and eventually reduces pest populations to acceptable levels. A third form of biological control is not as widely recognized, but is particularly important in soil. It is usually referred to as 'conservation' biological control, and involves either providing the resources that endemic natural enemies need to improve their effectiveness, or reducing the factors that are preventing them from being effective parasites or predators. All three forms of biological control apply to nematode pests and are considered in this book.

The essence of biological control is that it encompasses any ecologically based strategy that ultimately results in a reduction in pest populations, or in the capacity of a pest to cause damage. Such effects are manifested through the actions of living organisms and can occur naturally, but may also be achieved by manipulating the soil food web or introducing one or more antagonists. However, when this concept is applied to plant-parasitic nematodes, there is room for debate about whether particular practices should be included under the umbrella term 'biological control'. For example, nematode populations can be reduced with nematode-resistant crops, and since plants are living organisms, some would consider that resistant plants are 'biological' control agents. In the same way, organic amendments may act against nematodes by producing nematicidal by-products or by stimulating natural enemies, and so it is questionable whether the control mechanisms are chemical or biological (see Chapter 9). In yet another example, practices that improve the health of nematode-infested crops through biological processes, but may not reduce nematode populations, are sometimes designated as biological 'controls'. Since there will always be different views on what is meant by the term, this book focuses on a more important point: that the management practices used in agriculture must be ecologically sound and foster a soil food web capable of keeping populations of plant-parasitic nematodes at levels that do not cause economic damage.

One issue that arises throughout this book is whether the effects of natural enemies in reducing nematode populations should be referred to as 'control', 'suppression' or 'regulation'. I have used the latter term to describe situations where populations of a nematode and its parasites or predators respond to each other in a density-dependent manner. Jaffee (1993) described those situations succinctly: 'when hosts are plentiful, parasites multiply; when parasites are plentiful, hosts are suppressed and parasites then decline'. Although such a model is reasonable when dealing with a single pest and a relatively specific parasite or predator, it is difficult to apply this concept to a complex environment such as soil. Nematodes, for example, are preyed upon by numerous fungi, other nematodes and a wide range of microarthropods, and may also have specific bacterial and fungal parasites. These natural enemies all have different abundances; they are distributed in various micro-niches; they respond to ambient conditions at different rates; they are often in competition; and they may also complement each other. In such a situation, it is difficult to describe these community effects in terms of density-dependent regulation. Thus, I have chosen to refer to this complex soil biotic environment as being suppressive to the pest species.

Due to their widespread use, terms such as 'biological control', 'biocontrol' and 'biocontrol agent' will always be with us. However, the term 'control' has connotations of humans being the dominant force, reducing pest populations to low levels with chemicals or other tactics. Suppression seems to be a softer word, and is often used to describe situations where pest populations are reduced through biological processes. It is, therefore, commonly used in this book. Thus, a suppressive soil is one that, due to a multitude of managed or unmanaged biotic and abiotic factors, is not conducive to high pest population levels or exponential multiplication of the pest.

Sustainable Agriculture

The earth's landscape has been dramatically transformed in the thousands of years since humans began growing crops, but it is still able to produce enough food and fibre to meet the needs of an ever-increasing human population. However, the dependence of modern cropping systems on non-renewable fossil fuels and their role in degrading natural resources and the quality of air, water and soil raises questions about the long-term sustainability of agriculture. Human survival is dependent on the resilience and regenerating capacity of the thin layer of soil covering the earth's surface, and these attributes are the essence of sustainable agriculture. Not only must our soils produce adequate amounts of food, but they must be farmed in a way that

protects their integrity, remediates historical damage and ensures they remain productive for future generations. Thus, the essence of sustainable agriculture is:

> the management and utilization of the agricultural ecosystem in a way that maintains its biological diversity, productivity, regeneration capacity, vitality and ability to function, so that it can fulfil – today, and in the future – significant ecological, economic and social functions at the local, national and global level, and does not harm other ecosystems (Lewandowski *et al.*, 1999).

Agricultural sustainability became a major issue for discussion in the latter part of the 20th century due to concerns about the long-term viability of current food production systems (Gliessman, 1984; Pesek, 1989; Edwards *et al.*, 1990; Harwood, 1990; Lal, 1994; Meerman *et al.*, 1996). The prevailing system, characterized by large-scale farms; rapid technological innovation; large capital investments in production and management technology; single crops grown continuously over many seasons; genetically uniform high-yielding crops; extensive use of pesticides and fertilizers; high external energy inputs; dependency on agribusiness; and replacement of farm labour with machinery; has delivered tremendous gains in food production over the last 60 years, but there are questions about the long-term viability of what has been variously termed 'conventional farming', 'modern agriculture' or 'industrial farming' (Gold, 2007). The main concerns are the negative effects on the world's ecosystems due to soil degradation, water pollution, overuse of surface and ground water, and loss of wetlands and wildlife habitat. Thus, sustainability has become an integral component of the agricultural policies of many countries. Although there are many conflicting views on what elements are, or are not, acceptable and appropriate in a sustainable farming system (Gold, 2007), increasing numbers of farmers are now embarking on their own paths towards sustainability. However, sustainable agriculture should not be seen as a prescribed set of practices. Instead, it is a concept that encourages food producers to reflect on agriculture from an ecological perspective, to think about the long-term implications of their management practices, and then to take steps to redesign or restructure inappropriate farming systems.

Soil Health

Soil, the vital natural resource that sustains agricultural production, is non-renewable and easily ruined by mismanagement. It is continually subject to water and wind erosion; it can be further degraded by compaction and loss of organic matter; and is rendered unproductive by salinization and desertification. Maintaining healthy soil is, therefore, a key component of sustainable agriculture (Jenny, 1984; Doran et al., 1994, 1996; Glanz, 1995; Doran and Safley, 1997). Since soil organisms play a critical role in many important processes associated with soil health (e.g. maintaining soil structure and fertility, cycling nutrients and minimizing pest and disease outbreaks), they are intimately linked with the issue of sustainability. Thus, an important component of sustainable agriculture is the development of soil conservation practices, soil fertility programmes and pest management practices that protect and nurture the organisms responsible for the soil's long-term stability and productivity.

Although biological attributes (e.g. microbial biomass carbon, soil respiration rate, microbial activity) are commonly used as indicators of soil health, soil organic matter status is arguably the most important parameter determining whether a soil has the capacity to sustain plant productivity and maintain other important soil functions (Powlson et al., 2011a). In fact, the effects of organic matter on soil properties are so far-reaching that they are out of proportion to the relatively small amounts of carbon (0.1–4% w/w) usually found in agricultural soils. The positive impacts of soil organic matter on soil physical, chemical and biological properties are discussed in detail in Chapters 2 and 3, but the role of management in influencing levels of soil organic matter, and its flow-on effects to the soil biological community, is a recurring theme throughout this book.

The Rise of Conservation Agriculture

One of the most important agricultural developments in the last 30 years has been the rise of conservation agriculture (Baker et al., 2006). Defined as an agricultural management system that combines minimum soil disturbance with permanent soil cover and crop rotation, it is the major practical outcome of the recent move towards more sustainable systems of agricultural production. Conservation tillage (a collective term that encompasses a number of commonly used terms, including no-till, direct drilling, minimum tillage and reduced tillage) is a key component of conservation agriculture, and is now practised on more than 95 million hectares of land worldwide (Hobbs et al., 2008). The benefits from conservation tillage are summarized by Hobbs et al. (2008) and discussed more fully in Chapter 4, but include reduced soil compaction, enhanced water infiltration, better moisture retention, improved soil structure, fewer soil losses due to erosion, increased soil organic carbon, higher soil microbial biomass, lower labour costs and fewer inputs of fossil fuels. Conservation tillage, therefore, improves soil health through its effects on soil physical, chemical and biological fertility, and is usually more profitable than farming systems based on conventional tillage. When combined with rotational or cover cropping and retention of plant residues on the soil surface, it provides a farming system that can be used to produce most of the world's major food and fibre crops, and can be adopted in widely different environments. Given its many advantages and the high rate of adoption in countries such as the United States, Brazil, Argentina, Canada and Australia, conservation agriculture will eventually become the world's dominant agricultural production system.

Biological Control of Nematodes: Current Status and the Way Forward

Soil fumigants and nematicides were just one of many technological developments during the 1950s and 1960s that resulted in a huge surge in world food production. The sometimes spectacular responses obtained with these chemicals demonstrated that plant-parasitic nematodes were a major constraint to crop production, stimulated scientific and commercial interest in nematodes, and heralded the development of nematology as a discipline. Not surprisingly, biological control was not seen as a high priority during this period. Nematologists concentrated on improving their understanding of the biology, ecology, physiology and taxonomy of this relatively unknown group of pests, while applied research was directed towards maximizing the effectiveness of fumigants and nematicides. Thus, by the time the health and environmental problems associated with the use of nematicides were recognized in the 1980s (Thomason, 1987), research on biological control of plant-parasitic nematodes was still in its infancy and there were 'no widely accepted examples of the contrived use of an antagonist to control a plant-parasitic nematode' (Stirling, 1991).

The need to find alternatives to chemical nematicides and decisions in the 1990s to phase out the use of methyl bromide because of its ozone-depleting properties (Ristaino and Thomas, 1997) provided compelling reasons to add biological control to the range of tools available to manage nematode pests. Mainstream research programmes in nematology were modified to include biological

control, and scientists with an interest in the natural enemies of nematodes were employed by many research agencies. Thus, biological control became one of the main growth areas in nematology, with biological control and resistance being the only two subject areas where the number of papers published in the *Journal of Nematology* more than doubled in the years since 1979 (Fig. 1.2).

Although an infusion of resources into biological control has increased our understanding of the way nematodes and their antagonists interact in soil, it is often argued that this change in the focus of nematological research has produced relatively little in terms of practical outcomes. Biological control is rarely recognized as a component of current nematode management programmes, and so it is reasonable to ask why. I would argue that this situation has developed because biological control is usually viewed in a very limited sense as replacing a chemical pesticide with a biological alternative (i.e. replacing one 'silver bullet' with another). There is certainly evidence to suggest that such an approach is sometimes useful, as several biological products with activity against nematodes are now available in the marketplace (see Chapter 12). However, the use of inoculants in mainstream nematode management programmes will always be hampered by economic limitations and biological constraints.

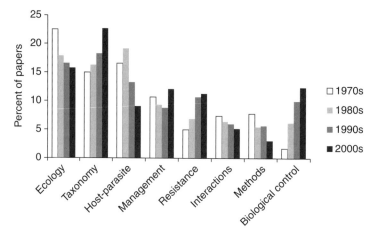

Fig. 1.2. Changes in the level of activity within various areas of nematology over four decades, as determined by the percentage of papers published in *Journal of Nematology*. (Figure prepared from data presented by McSorley, 2011a, and published with permission.)

In many agricultural situations, it is not feasible to mass produce an antagonist, transport it to the required location, and incorporate it into soil at the application rates required to achieve useful levels of nematode control; while the buffering effects of the microbial community inevitably operate against the biological control agent once it is introduced into soil. The alternative approach of manipulating the farming system, the environment or the host plant to achieve a soil biological community capable of suppressing nematode pests is, therefore, the major focus of this book. It is based on the proposition that wherever agriculture is practised, the requisite components of the soil food web will already be present and adapted to local conditions. The challenge is to redesign management practices to enhance their activity.

Integrated Soil Biology Management

The main reason that plant-parasitic nematodes are important crop pests is that much of the land currently used for agriculture has been exploited for many years and is physically, chemically and biologically degraded. The regulatory mechanisms that normally suppress nematode populations no longer operate as they should, while a sub-optimal physical and chemical environment means that plants are unable to tolerate the damage nematodes cause to their root systems. Thus, excessive fertilizer and pesticide inputs are required to maintain production, further weakening an already depleted soil biological community. Given that carbon levels in agricultural soils are almost always more than 50% lower than undisturbed soils in their natural state; that soil organic matter plays a central role in improving a soil's physical, chemical and biological properties; and that improvements in soil health result from interactions between occupants of the detritus-based food web; the ultimate solution is to focus on building an active and diverse biological community by increasing carbon inputs, minimizing carbon losses and reducing management impacts on soil organisms. The advantage of this approach is that it targets the primary cause of the nematode problem (management-induced diminution of the soil biota and shortcomings in the farming system) rather than the secondary effect (the presence of a nematode community dominated by plant parasites). Improved soil health and enhanced sustainability will be the most important outcomes of such an approach, but biological mechanisms that suppress nematodes will also be activated. Other outcomes are likely to be broad-spectrum suppression of other root pathogens and an increased capacity of crops to tolerate the effects of these pathogens.

The important point in the preceding paragraph is that the management practices required to enhance soil biological activity and diversity are the foundation on which effective systems of minimizing losses from plant-parasitic nematodes can be built. This argument is based on the premise that once crop rotation and cover cropping become integral components of the farming system, cropping frequency is increased; bare fallows are eliminated; tillage is minimized; crop residues are retained on the soil surface; levels of soil organic matter are raised; compaction and other sub-optimal soil constraints are removed; inputs of nutrients are optimized; and pesticides are used judiciously, then biological control mechanisms will begin to operate effectively, and will provide enhanced levels of nematode suppression. Evidence that this occurs is presented throughout this book.

The advantage of this approach is that the focus is no longer on the pest. Instead, the focus is on fostering the biological resources that must be in place if the agroecosystem is to provide the full range of ecosystem services depicted in Fig 1.1. It is, therefore, a wider concept than integrated pest management (IPM), an approach that is commonly used to minimize damage caused by insect pests, weeds, plant pathogens and nematodes. It has been termed 'integrated soil biology management' in this book, and it aims to build an active and diverse soil biological community that will not only suppress nematode pests but also improve the soil's physical properties, enhance biological nutrient cycling and provide the many other services provided by the soil biota. As discussed in Chapter 11, some of the tactics used in IPM approaches to managing nematodes are incompatible with the concept of integrated soil biology management, or are of questionable value (see Table 11.2). The two approaches are, therefore, quite different. Integrated soil biology management has a pest-management component, but is really an integrated approach to managing soil biological resources with a view to promoting sustainable and productive agriculture. Thus, it has similarities to some of the approaches discussed in a workshop on soil management sponsored by the Food and Agriculture Organization of the United Nations (FAO) in 2002 (FAO, 2003).

In some cropping systems and environments, serious nematode pests of the principal crop may not be present, and so when sustainable soil and crop management systems are being developed, it is not necessary to include tactics that target particularly damaging nematode species. Improving soil health and enhancing suppressiveness to soil-borne pests may be all that is required to minimize losses from plant-parasitic nematodes. However, in situations where a virulent nematode is capable of causing crop losses at low population densities, management tactics directed at that pest (e.g. a resistant/tolerant cultivar or rootstock, or a non-host rotation crop) will be an important component of any management system that is devised.

One potential problem with management systems that aim to enhance suppressiveness to plant-parasitic nematodes is that in many agricultural soils, the predatory component of the soil food web may already have been eliminated or severely depleted by many years of mismanagement. In such situations, remediation measures are likely to be needed, and it may take many years to restore a fully functional soil biological community. The challenges involved in tackling these issues are discussed in Chapter 11.

Transferring Ecological Knowledge into Practical Outcomes

The key message from this introductory chapter is that agroecosystems should be managed with two principal aims in mind: to sustain long-term plant productivity, and to maintain a full range of ecosystem services. Thus, any practice that is integrated into the system in an attempt to reduce losses from plant-parasitic nematodes must be considered holistically, and should recognize the following:

*   The primary goal of management must be to establish a farming system that protects the soil resource from degradation; enhances rather than diminishes soil carbon levels; increases soil microbial activity; enhances biological diversity; and is productive and sustainable in the long term. Any practice that is used to manage nematodes must be compatible with these goals.
*   Nematode management practices must have a sound ecological basis. Plant-parasitic nematodes are only one component of a complex biological community, and tactics used to reduce their population densities should not diminish the capacity of the wider community to fulfil its functions.
*   Reducing populations of plant-parasitic nematodes to low levels may not always be necessary. In a well-functioning agroecosystem, the soil health benefits associated with improved physical and chemical fertility may increase damage thresholds to the point where the crop suffers minimal losses, despite the presence of plant-feeding nematodes.

The primary purpose of this book is to consider how our knowledge of the soil ecosystem can be used to reduce losses from plant-parasitic nematodes. It commences with three chapters that explore the links between the soil biological community, soil health and sustainable agriculture (Chapters 2–4). The natural enemies of nematodes and the symbionts that have the capacity to kill nematodes or interfere in some way with their development are then discussed (Chapters 5–8), and examples demonstrating that these antagonists can suppress populations of plant-parasitic nematodes are given in Chapters 9 and 10. The management practices required to conserve and enhance natural enemies of nematodes are considered in Chapter 11, while recent progress with inundative biocontrol is reviewed in Chapter 12. The main points made in the book are summarized in Chapter 13, and the many important issues that require further research are also discussed. Chapter 14 is a practical guide to improving soil health. It is written for land managers and their direct advisors in the hope that it will encourage them to improve the sustainability of their farming systems and introduce practices that enhance the suppressiveness of their soils to plant-parasitic nematodes.

Given the broad range of topics that are covered (the soil environment, the soil food web, the nematode community and soil ecosystem management within sustainable agriculture), it is important to appreciate that the segments of this book are not designed to stand alone. The soil biological community is complex; organisms within the soil food web interact with plants and with each other; both plants and the soil biota are affected by the soil environment; and options for managing agroecosystems depend on the soil resource, climate, the principal crop and many other factors. Thus, to gain a holistic view of the main messages, it is necessary to read the entire book.

Section II

The Soil Environment, Soil Ecology, Soil Health and Sustainable Agriculture

_____

2

The Soil Environment and the Soil–Root Interface

Any discussion of biological control of plant-parasitic nematodes would be incomplete without some consideration of the soil environment. Plant-parasitic nematodes spend most of their lives in soil, but since they also develop an intimate relationship with roots during the feeding process, it is perhaps more correct to consider them as occupying the soil–root interface rather than the bulk soil mass. Eggs of some plant-parasitic nematodes hatch in response to substances that diffuse from roots, and feeding takes place in non-suberized areas of the root system near root tips and root hairs, and in regions where lateral roots emerge. The bodies of ectoparasitic nematodes remain in the thin layer of soil no more than 2 mm thick surrounding the root; adult females of some sedentary endoparasites protrude into this zone; and the eggs of many species are aggregated on the root surface. Such distinctions about the habitat of plant-parasitic nematodes may seem trivial but they are vitally important when population regulation and biological control are considered. Essentially, biological control is concerned with interactions between pests and antagonists and it is important to define the battlefield where this 'biological warfare' occurs. During fallow periods between host crops, this battlefield is localized microsites within the bulk soil mass. However, once a crop is planted, plant-parasitic nematodes aggregate near roots, and so biological control agents must be effective at the soil–root interface.

In introductory remarks to his book on soil biology, Bardgett (2005) argued that the first step in understanding the factors that control the abundance and activity of soil organisms and cause spatial and temporal variability within soil biological communities is to understand the physical and chemical matrix in which they live. However, that is not as simple as it may seem, because soil physical and chemical properties not only affect soil biological activity, but are also modified by the organisms that reside in soil. Thus, soil is a complex and dynamic medium, and its final state is the result of numerous interactions between its mineral fraction, organic matter and the soil biota. Plants, microorganisms and the soil fauna play a key role in soil formation and influence many important soil properties. Consequently, the first part of this chapter concentrates on the biologically inert components of soil, the process of soil formation, and the soil physical and chemical parameters that influence soil organisms. It then discusses the effects of organic matter on soil properties and the critical role of plant roots in harbouring soil organisms and influencing soil properties. Further information on these matters can be found in various texts on soil science and soil biology, including Coleman and Crossley

(2003), Davet (2004), Bardgett (2005), Sylvia *et al.* (2005), Lavelle and Spain (2005), Uphoff *et al.* (2006), White (2006), van Elsas *et al.* (2007a), Brady and Weil (2008), Chesworth (2008) and Tan (2009).

The Process of Soil Formation and the Composition of Soil

The soil mineral fraction

Soil is the result of weathering processes that have operated over millions of years to break down the parent rock into smaller and smaller particles. Thus, the mineralogical composition of the parent material influences the types of soils that are formed and the character of the vegetation they support. The original parent material, temperature, the amount of precipitation in the region where the soil developed and the time involved in the soil-formation process all affect soil texture (the proportion of sand, silt and clay), and this in turn has a major impact on many important soil properties. The coarser sand and silt fractions are comparatively inert, because the surface area of the particles is small relative to their weight, and their capacity to retain cations is negligible. Therefore, sand and silt particles do not readily retain water, nutrients and humic substances, and do not support a high proportion of the indigenous microbial population. Their main role is to enhance the porosity of fine-textured soils, improve the propensity of a soil to drain water and facilitate water and gas diffusion.

From a biological perspective, clays are by far the most important inert constituent of soil for several reasons:

- The negative charge on the surface of clay particles confers a capacity to buffer pH changes. H^+ ions can be exchanged for other cations, and this helps organisms that are present in the liquid phase to withstand sudden changes in pH.
- Reversible electrostatic bonds that are formed between clays and positively charged particles confer the ability to retain and exchange cations. This property (known as the cation exchange capacity) is of vital importance to plant growth and has direct and indirect effects on soil organisms.

- Clay soils are capable of storing large quantities of water. Thus, in climates where rainfall is variable or limited, moisture conditions favour biological activity for much longer in soils with moderate to high clay contents than in coarse-textured soils.
- Clays not only retain mineral ions, but also organic molecules. In clay soils, organic substances are often held within microaggregates and are, therefore, temporarily protected from degradation by microorganisms and soil enzymes, ensuring the slow release of some of the soil's nutrient reserves.

Soil organic matter

Although the mineral fraction is by far the dominant component of soil, the organic fraction (which usually comprises only 1–4% of the weight of most agricultural soils), is the key factor influencing the soil biota and a major determinant of soil properties. However, most routine analytical methods for measuring soil organic matter (SOM) actually determine the content of soil organic carbon (SOC). Conversion factors that account for the proportion of SOM that is not carbon are then applied to obtain SOM contents. These conversion factors range from 1.7 to 2.0, but a factor of 1.72 is typically used (i.e. SOM = SOC × 1.72). Since no single conversion factor is appropriate for all soils, it is now more common to report results in terms of SOC (Baldock and Skjemstad, 1999).

Soil organic carbon contents vary enormously between soils and also within a specific soil type (Fig. 2.1), due largely to differences in the major soil-forming factors (climate, soil mineral composition, topography, vegetation and soil biota), and the way a soil is managed by humans. In natural systems, the amount of soil organic carbon tends towards an equilibrium that is determined by the rates of carbon inputs and losses, but in agricultural soils,

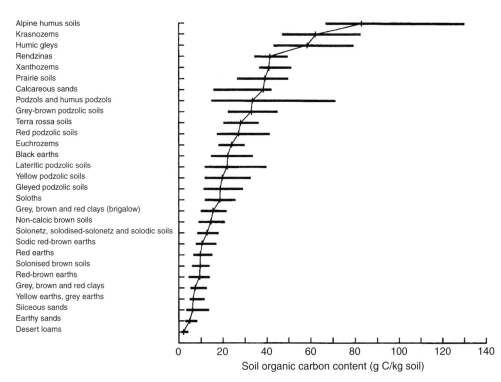

Alpine humus soils
Krasnozems
Humic gleys
Rendzinas
Xanthozems
Prairie soils
Calcareous sands
Podzols and humus podzols
Grey-brown podzolic soils
Terra rossa soils
Red podzolic soils
Euchrozems
Black earths
Lateritic podzolic soils
Yellow podzolic soils
Gleyed podzolic soils
Soloths
Grey, brown and red clays (brigalow)
Non-calcic brown soils
Solonetz, solodised-solonetz and solodic soils
Sodic red-brown earths
Red earths
Solonised brown soils
Red-brown earths
Grey, brown and red clays
Yellow earths, grey earths
Siiceous sands
Earthy sands
Desert loams

0   20   40   60   80   100   120   140

Soil organic carbon content (g C/kg soil)

Fig. 2.1. Variability in soil organic carbon between major soil groups in Australia, and within specific soil types. Values (median and quartile ranges) are expressed in SI units but can be converted to the percentage values traditionally used to measure soil organic carbon by dividing by 10. (Modified from Baldock and Skjemstad, 1999, with permission.)

continual changes in management and cropping practices result in levels of soil organic carbon that are in a continual state of flux. However, if management practices remain consistent for long enough (20–50 years?), it is possible for an agricultural soil to attain an average equilibrium value that reflects the rotation or suite of management practices imposed.

Soil organic matter is derived from plants, the primary producers in all ecosystems, as they convert energy from the sun and carbon from the atmosphere into organic molecules and living tissues. Thus, the primary organic inputs into soil are leaves and twigs that fall to the ground; dead roots; cells that are sloughed from roots; carbohydrates, proteins and other substances that are exuded from roots; and carcasses and wastes from all the animals that feed on plants. This detritus is then subject to microbial decomposition. Labile compounds (e.g. sugars and proteins) are the first to

decompose, and then the more resilient materials (which largely consist of cellulose and lignin) are converted into a complex range of polysaccharides, polypeptides and organic acids. As this process continues, detritus is progressively transformed into complex and relatively recalcitrant molecules known as humus. Thus, soil organic matter consists of a number of components that vary in physical size, chemical composition, degree of association with soil minerals and extent of decomposition. These components are depicted in Fig. 2.2 and comprise: (i) living biomass (microorganisms, larger soil organisms and intact plant and animal tissues); (ii) organic materials in the soil solution; (iii) particulates (roots and other recognizable plant residues); (iv) humus (a largely amorphous and colloidal mixture of organic substances that are no longer identifiable as plant or animal tissues); and (v) inert organic materials such as charcoal.

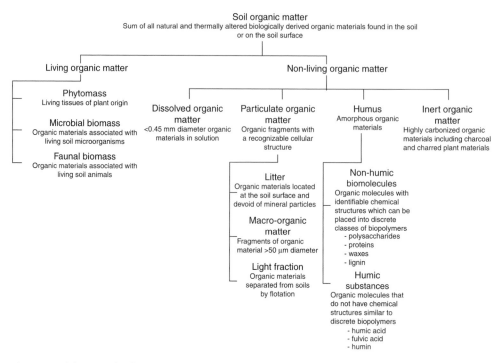

Fig. 2.2. A definition of soil organic matter and descriptions of its various components. (From Baldock and Skjemstad, 1999, with permission.)

Inputs of organic matter are particularly important in soil profile development. Litter deposited on the soil surface, organic materials derived from roots and the biological activity associated with roots and detritus produce soils with a layered structure. The litter layer and the organic horizon immediately beneath it are the most biologically active layers in the soil profile, as they are largely composed of fresh plant material and organic matter in various stages of decomposition. The uppermost mineral layer (A horizon) is also vitally important, as it contains humified organic material and organic matter derived from above. Collectively, these surface layers, which are usually no more than 10–15 cm deep, are the most functionally important zones in the soil profile, as they determine a soil's capacity to respond to environmental stresses and carry out ecosystem functions.

Soil organic matter is composed of several carbon fractions that vary in their turnover times or rate of decomposition. The labile fraction, which turns over in less than 5 years, is the primary food source for soil microorganisms, and since fungal hyphae and bacterial mucilages play a major role in binding soil particles into microaggregates, labile carbon is particularly important in influencing soil physical properties. It is also important from a chemical perspective, because nutrient cycling in soil is largely associated with the organisms involved in decomposing labile compounds from plant residues and root exudates.

The humic fraction is also a significant contributor to soil chemical and physical properties. This fraction is a mixture of large, amorphous, colloidal, dark-coloured polymers with aromatic, ring-type structures (e.g. fulvic acids, humic acids and humins) and comprises about 50–80% of soil organic matter. These substances are remarkably resistant to microbial degradation, with fulvic acid surviving unchanged in soil for decades and more recalcitrant materials having half-lives that are measured in centuries. Nevertheless, humus is subject to limited but continual microbial attack and so recurring additions of plant residues are required to maintain levels of soil organic matter.

One of the key attributes of humic substances is that they interact with proteins to form complexes that protect nitrogen and other essential nutrients from rapid mineralization and loss from soil. They also form stable complexes with clay particles, and during this process, nitrogen-containing organic compounds are entrapped between and within the clay matrix and are, therefore, protected to some extent from microbial degradation. Another important attribute of humus is its high cation exchange capacity. Cations are held in an available form and can be readily exchanged for other cations, a process that helps maintain the constant supply of nutrients required for optimum plant growth. Since the C:N ratio of humus (10:1) is much narrower than the original plant materials from which it was derived, it also stores considerable quantities of nitrogen. Humus also contains about 0.5% P, 0.5% S and smaller amounts of K, Ca, Mg and micronutrients, and is, therefore, an important source of all the nutrients required by plants and soil organisms. Since it has the capacity to absorb substantial quantities of water, humus also plays a role in supplying water to plants and in creating a habitable environment for soil organisms.

Impact of Organic Matter on Soil Properties

There are many reasons why organic matter has a major impact on many important soil properties: it plays a critical role in creating and stabilizing soil structure, which in turn produces good tilth, adequate drainage and the capacity to resist erosion; it increases soil water availability by enhancing water infiltration and increasing a soil's water-holding capacity; it is the principal long-term nutrient store and the primary short-term source of nearly all nutrients required by plants and soil organisms; and it also provides a food source for the organisms that cycle carbon and protect plants from disease. These effects are discussed in detail in comprehensive texts

on soil organic matter (e.g. Wolf and Snyder, 2003; Magdoff and Weil, 2004a), and no more than a summary is given here.

Figure 2.3 shows the many ways in which soil organic matter influences soil properties and plant productivity, and indicates its subsequent effects on the environment. What is apparent from this figure is that there is a complex chain of multiple benefits from organic matter. When organic matter is added to soil, there are both direct and indirect effects on soil physical, chemical and biological properties, with one effect leading to another. One of its primary effects is to maintain or increase soil biological activity and diversity; and it is for this reason that organic matter is mentioned incessantly throughout this book.

Organic matter and soil physical fertility

Sand, silt and clay particles are the basic building blocks of soil, and the way these particles are arranged into larger units or aggregates determines its structure, and the nature of the pore spaces between soil particles. Soil structural development involves physical, chemical and biological factors, but is heavily dependent on soil organisms and their interactions with organic matter (Degens, 1997). Roots and microbes produce polysaccharide glues that bind soil particles together into aggregates. A variety of organic compounds contribute to this binding process, while fungal hyphae tend to enmesh mineral particles with organic materials (Beare et al., 1997; Ritz and Young, 2004; and Fig. 2.4). These binding forces tend to stabilize aggregates, so that they are not readily dispersed when exposed to rainfall impact or to stresses imposed by cultivation or trafficking. Earthworms and other burrowing animals contribute to structural development by ingesting soil particles, macerating organic materials and creating a network of macro- and micropores that play a vital role in many soil functions.

The connectivity and size distribution of the labyrinth of large and small pores between particles, aggregates and peds is described as soil structure. The pore spaces within soil

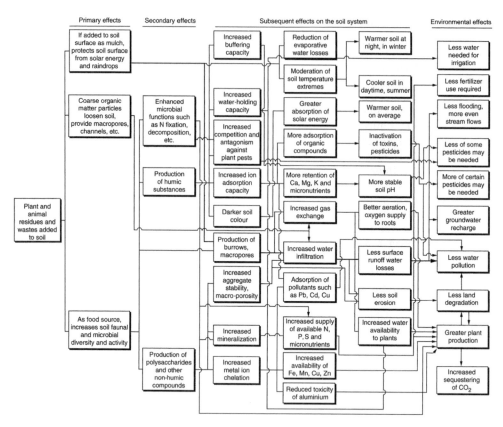

Fig. 2.3. An illustration of how addition of organic materials to a soil results in a complex web of direct and indirect changes that alter the soil's properties, behaviour and environmental impacts. (From Weil and Magdoff, 2004, with permission.)

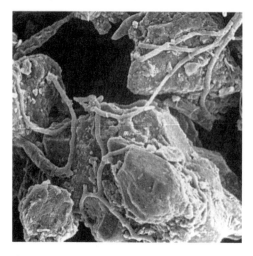

Fig. 2.4. A network of mycelial filaments clustering around mineral aggregates and organic particles in soil. (From Davet, 2004, with permission.)

provide a residence for roots and the soil biota; regulate the movement of air, water and nutrients; and reduce the probability that plant growth and biological activity will be adversely affected by anaerobic conditions. Good soil structure is a key attribute of a fertile soil and is governed to a large extent by organic matter. It is manifested in a number of ways:

- *Reduced bulk density*. One of the benefits of good soil structure is that it lowers bulk density (the weight of solids per unit volume of soil). A reduction in bulk density is associated with better aeration and improved diffusion of nutrients. Root growth is increasingly restricted as bulk density increases.

- *Better infiltration of water*. Well-structured soils have enough macropores to allow rainfall and irrigation water to rapidly infiltrate the soil and then drain readily. Any impairment of this process will have serious consequences, as loss of topsoil and nutrients in runoff diminishes the most fertile layer in the soil profile and also contaminates the wider environment. Conversely, impaired drainage reduces aeration and limits the availability of oxygen to the microbial community.

- *Improved storage of water and air.* Retention of water after initial drainage is dependent on the small pores and capillaries within the soil matrix, and the presence of humus and clay particles that readily retain water due to their large surface areas. However, a capacity to store large amounts of water often means low air storage, and so a balance between air and water is important for overall soil fertility. Well-structured soils will have both high plant-available water contents and high air-filled porosity.

- *Better root growth.* Plant roots grow readily in soils where there is little resistance to penetration, but growth is impeded as soil strength increases. Thus soil strength (which is often referred to as penetration resistance because it is measured with a penetrometer) must be as low as possible, and should never exceed the capacity of roots to grow through soil at times when the water content is adequate for plant growth. In soils with good physical fertility, root growth is usually not restricted at penetration resistances of <1 MPa (Cass, 1999).

- *Better structural stability.* One important aspect of soil physical fertility is the ability of a soil to resist adverse changes in its structural arrangement when external stresses are imposed. These structural changes can result from either slaking or dispersion. Slaking refers to the rapid disintegration of large aggregates when placed in water, and is largely due to lack of strong organic bonding between particles and micro-aggregates. Dispersion occurs when clay structures that bind fine aggregates to silt and sand break down, causing individual clay particles to move into soil pore spaces, and block water and air flow. Structural stability is enhanced by high levels of biological activity, and

this occurs when there is good root growth, numerous fungal hyphae and a general open and porous structure. Such conditions are predominantly, but not exclusively, associated with high contents of organic matter. Development of crusts and the presence of impenetrable surface or sub-surface layers are examples of structural stability problems. They are usually associated with excessive tillage and a lack of organic inputs.

Organic matter and soil chemical fertility

For many years, the nutrient management practices used in agriculture focused on ensuring that nutrient inputs were sufficient to maximize plant productivity in a given environment. However, the need to reduce costs and increase nutrient use efficiency, together with concerns about the long-term availability of nutrients and the loss of nutrients such as nitrogen and phosphorus into ground and surface waters, have prompted interest in improving the way nutrients are managed. Since organic matter plays a central role in storing and cycling nutrients, practices such as cover cropping, crop rotation, tillage and inputs of inorganic and organic amendments will play an important role in achieving this goal (Seiter and Horwarth, 2004). The main effects of organic matter on soil chemical fertility are summarized as follows:

- *Source of nutrients.* Most of the elements in the atmosphere and in soil are taken up by plants. Thus, soil organic matter, which is derived primarily from plants, is an important source of all the macronutrients (N, P, Ca, K, Mg and S) and micronutrients (B, Cl, Cu, Fe, Mn, Mo, Zn and Si) required by crops. Soil organic matter contains nearly all the nitrogen and much of the phosphorus and sulfur found in soils and is the principal long-term storage medium and the primary short-term source of these and other nutrients (Weil and Magdoff, 2004). Even when inorganic N fertilizer is applied, numerous studies (e.g. Omay *et al.*, 1998) have shown that most of the nitrogen taken up by a crop comes from mineralization of soil organic pools rather than directly from the current year's fertilizer.

- *Storage and release of nutrients.* During the decomposition process, the nutrients present in plant tissue are transferred to heterotrophic organisms in the soil food web, as they rely directly or indirectly on organic matter as a food source. Since these nutrients remain in an organic form and are effectively immobilized within microorganisms, they are unavailable to plants. However, losses to groundwater due to leaching are minimized. Thus, immobilization of nutrients in microbial biomass acts as a storage mechanism, locking up nutrients that might otherwise be lost to the environment. These nutrients are eventually released into the soil solution in an inorganic form through a process known as mineralization. Nitrogen, for example, is secreted as plant-available mineral nitrogen (mainly NH_4^+) by the fungi and bacteria involved in decomposing organic matter, and also by the nematodes, protozoa and other organisms that feed on them. These immobilization and mineralization processes occur simultaneously and are major factors governing the supply of nutrients to plants. Together with the capacity of humus and other pools of organic carbon to store nutrients and exchange cations, organic matter decomposition clearly plays a major role in determining whether plant nutrient demands are satisfied. Because the activities of soil organisms involved in the degradation process are temperature dependent, the processes driving decomposition and nutrient availability proceed more rapidly in warm or tropical climates than in cool temperate regions.
- *Buffering effects.* Humic substances have a capacity to resist changes in soil pH, and along with clays, provide most of the buffering capacity of soils. Interactions with common cations in alkaline soils (Ca and Mg) and in acid soils (Al and Fe) result in the formation of humates that are not readily dispersed or solubilized, thereby restricting pH changes resulting from the acidifying effect of fertilizers.

Organic matter and soil biological fertility

Soil biological fertility refers to processes mediated by soil organisms that directly or indirectly improve plant growth. Many of those processes were discussed in preceding sections and involve interactions between soil organisms, plant roots and decomposing organic matter that result in modifications to a soil's physical or chemical properties. Organic matter also has other effects on soil biological fertility, but since they are discussed throughout this book, they will only be mentioned briefly here:

- *Enhancement of microbial biomass and biological diversity.* Since above- and belowground carbon inputs are the energy sources that drive the soil biological community, soil microbial biomass is usually positively correlated with the total organic carbon content of soil. Although the strength of this relationship is affected by climate, soil properties (e.g. parent material, mineralogy and texture) and a range of manageable factors (e.g. soil and residue management practices, plant species, nutrient inputs and the application of amendments), the major factors influencing soil microbial biomass are the amount and quality of organic matter inputs (Gonzales-Quiñones *et al.*, 2011). Provided the soil is not disturbed by tillage, inputs of organic matter also increase biological diversity (Adl *et al.*, 2006).
- *Suppression of soilborne pathogens.* Organisms that parasitize or prey on plant pests, or are antagonistic to them in some other way, are an important component of the soil food web. They play a crucial role in regulating populations of root-feeding insects and plant-parasitic nematodes, and are, therefore, critically important in maintaining plant productivity. Many of these organisms are influenced by organic matter, either because they are facultative saprophytes or parasites, or because they use organisms in the detritus food web as an alternative food source.

In concluding this brief summary of the way in which organisms associated with organic matter contribute to biological soil fertility, it is important to remember that organic matter has multi-faceted roles in soils. These roles have been subdivided here into three categories, but the fact that they do not occur in isolation should never be forgotten. Physical, chemical and biological properties influence each other (Fig. 2.5), and this must be kept in mind when considering the impact of organic matter on soil properties and overall soil fertility.

The Soil Environment and Its Impact on Nematodes and Other Soil Organisms

Soils vary tremendously across a landscape. Major and quite subtle changes can also occur within relatively small areas, such as in an agricultural field. This variability influences the physical matrix of the soil, its hydraulic behaviour and many other soil properties. Since the microbes, nematodes and other organisms that live in soil are also affected by changes in the soil environment, the impact of key environmental factors on the soil biota is considered here.

Physical structure

The composition of the soil mineral fraction and the manner in which soil particles are aggregated determines the degree to which pores are filled with air or water, and the physical nature of the living space available to soil organisms. A typical soil aggregate contains sand, silt and clay particles bound together by organic substances and microorganisms, and the pore space provides a number of niche habitats (Fig. 2.6). Bacteria either occupy microsites less than 5 μm in diameter, attach themselves to fragments of organic matter or live in water films, while fungi grow over the surface of soil particles or are confined to contiguous pores that are at least as wide as the diameter of their hyphae. Most protozoans are small enough to occupy the cavities depicted in Fig. 2.6, but as faunal size increases, the physical matrix imposes limitations

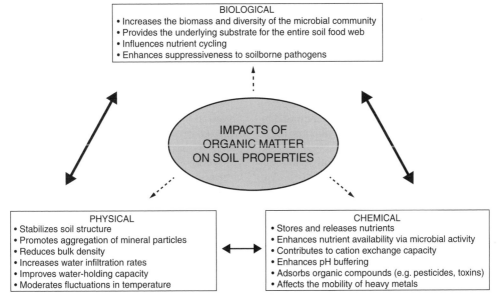

Fig. 2.5. Impacts of organic matter on various soil properties. The solid arrows indicate that each soil property can be affected by other properties.

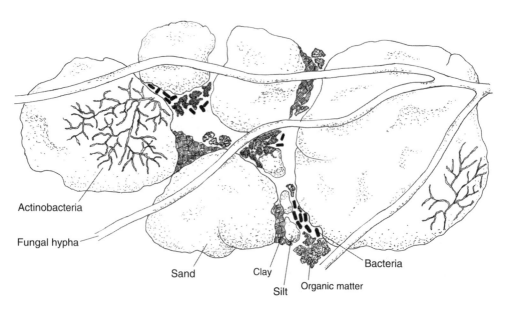

Fig. 2.6. A diagrammatic representation of a typical soil aggregate, showing sand, silt and clay particles held together by organic matter and soil microorganisms. Each aggregate contains a diverse range of niche habitats. Bacteria are associated with organic matter, while fungal hyphae and the branching filaments of actinobacteria grow over surfaces.

on movement. The mesofauna (e.g. rotifers, nematodes and symphylans) cannot exert enough pressure to force a passage between soil particles and aggregates, and consequently are confined to the soil's existing pore system. Soil animals (the macrofauna) live in much larger spaces and in some cases (e.g. earthworms) living space is created through burrowing activities or by ingesting soil.

Most vermiform nematodes range from 15 to 60 µm in diameter and the fraction of the pore space they can occupy is determined by the cross-sectional diameter of the pore necks through which they are just able to pass. Many nematodes are small enough to enter some of the pore spaces depicted in Fig. 2.6, but they tend to be confined to somewhat larger pores (e.g. 30–60 µm diameter) because, at the narrower end of the range, usage is restricted by tortuosity and blind ends (Jones, 1978). Thus, smaller pores provide bacteria and fungi with a safe refuge from the microbivorous nematodes that prey on them, and this is just one example of how soil structure can affect the nature and intensity of trophic interactions within soil.

Soil water

Water is the most important constituent of the soil environment, as it carries the dissolved nutrients essential for plant growth and satisfies the transpiration requirements of growing plants. The availability of water also affects the survival, movement and activity of the soil microflora and microfauna, and has important effects on soil temperature and aeration. The effects of soil moisture result in part from the aquatic nature of many soil organisms and the impact of water on their capacity to move. There is a relationship between matric potential and the diameter at which soil pores empty, and so the size and volume of water-filled pores available for the movement of soil organisms is strongly influenced by soil moisture. The largest pores drain first, and by the time matric potential has decreased to −0.05 MPa, all pores greater than about 6 µm in diameter will have emptied (Papendick and Campbell, 1981). This means that the movement of organisms that live in water films (e.g. bacteria, protozoa and nematodes) is progressively limited as the water-filled pores or pore necks become too small to permit their passage (Wallace, 1963; Jones, 1978; Griffin, 1981).

In addition to influencing the dispersion of soil organisms, changes in soil water content have profound effects on their activity. Adequate moisture is essential for optimum growth, and the contribution of some microorganisms to total microbial respiration begins to decline at water potentials of between −0.3 and −0.6 MPa (Wilson and Griffin, 1975). Soil organisms differ in their capacity to survive in situations of even greater moisture stress, but most can tolerate moisture potentials to −10 MPa (Harris, 1981). In dry soils, soil air is always maintained at or very near a relative humidity of 100% (Papendick and Campbell, 1981) and this is perhaps the main reason that organisms can survive relatively low soil moisture potentials.

The impact of soil water on the soil biota is perhaps best illustrated with specific examples. Depending on soil particle size, nematodes are best able to move when pores are full of water or when the soil moisture potential corresponds to the stage when pores are draining of water (Wallace, 1968). As the soil drains to potentials of −0.01 MPa and −0.02 MPa, pore necks about 30 μm and 15 μm in diameter, respectively, are emptied (Papendick and Campbell, 1981). Pores of this size are too small for some nematodes, and also, surface tension forces in the remaining water film pull the nematode against the soil particles. Thus, nematodes are immobilized at relatively slight potentials (Jones, 1978). Egg hatch will occur at soil moistures that inhibit movement, but it decreases rapidly as soil dries to moisture potentials of about −0.3 MPa (Baxter and Blake, 1969a).

The response of nematodes to decreasing soil moisture contrasts sharply with that of some other soil organisms, which tend to remain active at low soil moisture potentials. Because of their filamentous habit, fungi can continue to ramify in relatively dry soils. They are not confined to water-filled pores because their hyphae can bridge air spaces and can spread along the walls of drained pores. The growth of most fungi is not restricted until the water potential drops to between −3 and −6 MPa, and some fungi can germinate and grow at potentials as low as −20 to −40 MPa (Cook and Papendick, 1972; Griffin, 1972; Harris, 1981). Bacteria are less tolerant of soil moisture stress than fungi, mainly because of their inability to move towards or over their substrates in the absence of free water. Bacterial movement is restricted to water-filled pores large enough to permit passage of the bacterium, and pore necks of this size (2–3 μm in diameter) drain at potentials of about −0.1 MPa. In general, bacterial activity continues to decline at matric potentials below those that have a marked effect on mobility, and becomes negligible for most purposes near −0.15 MPa (Griffin, 1981). The factors involved are not known with certainty, but the decline is probably not attributable to a direct effect of potentials on respiration and growth. For many species, these activities continue at slow rates in pure culture at osmotic potentials as low as −10 MPa (Griffin, 1981; Harris, 1981).

Aeration

Since higher plants and most of the soil microflora and microfauna are aerobic organisms, the rate of their metabolic processes is influenced by oxygen supply. Soils with good structure have sufficient air-filled pores to allow adequate gas exchange with the atmosphere. When soil pores are blocked, either with water or because of inadequate structure, sealed microsites can develop, and when active metabolism takes place around roots or decomposing organic residues, oxygen becomes depleted and levels of carbon dioxide increase. As a result of this process, all soils have a heterogeneous distribution of oxygen, and even a so-called aerobic soil can develop anaerobic pockets. This may markedly affect the distribution and activity of soil organisms.

The limited amount of work done with nematodes suggests that the activity of most species is rarely limited by lack of oxygen under the conditions normally found near the surface of most agricultural soils. Provided moist soils drain rapidly and oxygen concentration increases to above 10%, hatch of *Meloidogyne javanica* eggs and migration of juveniles proceeds normally (Baxter and Blake, 1969b). Oxygen concentrations of 2% and 7% did not affect survival of *M. incognita* and *Tylenchulus semipenetrans* but were detrimental to *Xiphinema americanum* and *Paratrichodorus minor* (van Gundy *et al.*, 1962), suggesting that some species may be more susceptible to reduced aeration than others. None of these four species survives exposure to 0% oxygen concentration for 10 days, but such extreme conditions are rarely encountered in agricultural soils, other than those used for rice production.

Laboratory studies on the effects of low oxygen concentrations on the growth of fungi have produced differing results, depending on the methods used. Linear growth rates on agar generally do not decline until oxygen concentrations are lowered to about 4%, whereas the dry weight of mycelium produced in agitated liquid cultures declines in a linear fashion as oxygen concentration decreases below 21%. Griffin (1972) suspected that sensitivity to oxygen in liquid culture was associated with problems of diffusion within the fungal pellets. He believed linear growth more accurately measured the direct response of fungi to changes in oxygen concentration, and argued that soil fungi were relatively insensitive to severe reductions in oxygen concentration. If Griffin's arguments are accepted, fungal growth is unlikely to be affected by changes in the concentration of oxygen within gas-filled pores that are in contact with the atmosphere. However, in water-filled pores and pores not in contact with the atmosphere, some fungal species may be affected when oxygen concentrations decline below 4%.

Most bacteria are aerobes and they dominate the bacterial community in surface soils. However, when the soil oxygen concentration is reduced to less than 1% (which can occur after heavy rainfall or when large amounts of readily decomposable organic matter are added to soil), the microbial population shifts from being predominantly aerobic to anaerobic. One important soil process that proceeds under anaerobic conditions is denitrification. Provided appropriate amounts of organic matter are available as an energy source, a wide range of relatively common soil bacteria (e.g. *Pseudomonas*, *Alcaligenes*, *Flavobacterium*,

and *Bacillus*) reduce nitrate to dinitrogen and/or nitrous oxide, thereby reducing nitrogen availability to plants. Other bacterially mediated processes that are activated under anaerobic conditions are the mineralization of organic compounds to methane and the reduction of sulfate to hydrogen sulfide (Tate, 2000; Sylvia *et al.*, 2005)

Although the effects of various gaseous environments on the activity of particular organisms have been studied under controlled conditions in the laboratory, it is difficult to use results of such studies to predict behaviour in soil, as it is structurally heterogeneous and oxygen concentration may differ at a microsite level. The atmosphere in macropores occupied by the larger animals such as nematodes may be largely aerobic, but there may be much less oxygen within aggregates, where many microorganisms reside. Respiration by microorganisms, combined with slow gaseous diffusion through water-filled micropores, creates heterogeneous conditions within aggregates, which may be aerobic on the surface and anaerobic in the centre. Thus, at the level of a soil aggregate, aerobic organisms may be able to function normally in close proximity to microsites where anaerobic processes such as denitrification are occurring.

pH

Most agricultural soils fall within a pH range of 5.5–7.5, but there are many situations where soil pH is outside that range. Alkaline soils are relatively common in arid and semi-arid regions, while acidic soils constitute a major portion of the world's soil resources, particularly in the tropics and subtropics. However, the removal of plant and animal produce from the land, the application of ammonium-based fertilizers, and the leaching of excess nitrate and accompanying cations such as calcium, are all acidifying processes. Soil acidification is, therefore, a major degradation problem in agriculture. Most soil microorganisms prefer a near-neutral pH of 6–7 and will grow within the pH range 4–9, but are increasingly stressed at the outer limits of this range. In acid soils, for example, elements that are toxic to microbes (e.g. aluminium, Al^{3+}) may be released, while microbial growth can be affected by a reduction in the availability of essential nutrients such as phosphorus. Most fungi are acid-tolerant and are commonly found in acidic soils, whereas bacteria, particularly the actinobacteria, are more sensitive to acidity and become a relatively less important component of the microbial community as soils acidify. In fact, soil pH is a relatively good predictor of bacterial community composition and extracellular enzyme activities (Fierer *et al.*, 2009). With regard to soil animals, earthworms are intolerant of acidity (Edwards and Bohlen, 1996), whereas their close relatives, the enchytraeids, can increase to high population densities in acid soils (Cole *et al.*, 2002).

Although there is no doubt that pH affects the activity and diversity of soil organisms, this does not mean that it is easy to determine whether extremes of soil pH affect the provision of the ecological services for which these organisms are responsible. Soil is a heterogeneous entity, and acid-limited microbes, for example, may function normally in microsites of moderate pH, despite the fact that they are surrounded by a soil with a pH that would restrict their activity. Also, the majority of soil microorganisms have never been cultured and their ecological role is unknown. It is possible that some of them are responsible for important functions when soil pH is outside the normal range. For these reasons, the occurrence of particular processes in the field (e.g. nitrification, degradation of particular compounds) is likely to be a better indicator of microbial capabilities under acid and alkaline conditions than extrapolations from laboratory studies of pH effects on particular microbes.

Soil temperature

Within the normal range of temperatures that occur in agricultural soils, the rates of various chemical and biological processes double for every 10°C increase in temperature. Thus, soil temperature is one of the most important environmental variables of biological significance. Soils freeze in some climates and rise to more than 50°C in others, while at a local level, soil temperatures fluctuate diurnally and also vary with depth and season of the year. Soil organisms are generally well adapted to the environments in which they occur, and are, therefore, able to cope with temperature variability and extremes of temperature. Most have temperature optima for growth and activity of 15–35°C, but since some activity will occur at temperatures well outside this range, biological activity in most agricultural soils is rarely limited completely by temperature.

The Soil–Root Interface

It should be clear from the preceding discussion that organic inputs from plants drive the biological processes that ultimately influence numerous soil properties. Some of these inputs come from above ground, but the contribution from roots is also important, particularly in agricultural soils, where potential sources of organic matter (stems, leaves, fruits and seeds) are often removed as harvested product. Although root biomass varies with plant and soil type, the quantities present in soil are significant, with root biomass in cropped soils averaging 2 t/ha (Jackson *et al.*, 1996). Since roots usually grow to depths of 1–1.5 m and root length density in upper layers of the soil profile often exceeds 100 m of root/L of soil (Gregory, 2006), it is not surprising that roots have a major impact on the soil environment. Soil pH and the soil atmosphere is modified when roots utilize oxygen, and carbon dioxide is respired by root cells and microorganisms; moisture gradients become established near roots when plants transpire; and the soil's nutrient status is continually modified when nitrogen, phosphorus, potassium and other elements are removed by plants. Thus, the physical and chemical environment at the soil–root interface differs markedly from bulk soil, and this has an impact on the myriad of biological processes that occur in that region.

The zone of soil that is influenced by roots is known as the rhizosphere, while the actual root surface is termed the rhizoplane.

Since organic materials derived from roots (exudates, sloughed cells and decaying roots) act as a food source for the soil microbial community, populations of bacteria, archaea and fungi are highest on the rhizoplane and decline markedly with increasing distance from the root surface, a phenomenon known as the rhizosphere effect. Comparative estimates of populations in rhizosphere soil (R) and in non-rhizosphere soil (S), expressed as the R:S ratio, have shown that populations of most organisms are usually 5–20 and sometimes more than 100 times higher in the rhizosphere than in the surrounding soil (Curl and Truelove, 1986). However, it has long been recognized that such estimates do not provide a true picture of the rhizosphere effect, as the cultural methods often used to estimate microbial populations recover only a small percentage of the bacteria and fungi that can be seen on the rhizoplane under a microscope (Goodman *et al.*, 1998; Bowen and Rovira, 1999). Culture-independent molecular technologies can now be used to fingerprint microbial communities on the root surface, and studies using such methods are revealing the true extent of biodiversity within the rhizosphere (Sørensen *et al.*, 2009; Buée *et al.*, 2009 a, b).

The soil–root interface is the region where plants interact with the soil biological community, and where organisms in that community interact with each other. Since root-feeding nematodes and their antagonists are a component of that community, the rhizosphere must be considered in any essay on biological control of plant-parasitic nematodes. Therefore, the next part of this chapter considers why the rhizosphere is a hotspot of biological activity, and then focuses on the tripartite interactions between plants, pathogens and beneficial organisms that occur in this zone. A complete review of rhizosphere biology is beyond the scope of this book, but further information can be obtained from various sources, including Curl and Truelove (1986); Foster (1986); Lynch (1990); Bowen and Rovira (1999); Singh *et al.* (2004); Watt *et al.* (2006); Bais *et al.* (2006); Mukerji *et al.* (2006); Cardon and Whitbeck (2007); Hinsinger *et al.* (2009); and Raaijmakers *et al.* (2009).

Roots and rhizodeposits: the energy source that sustains the soil biological community

Detritus from leaves, branches, bark and other above-ground materials is often considered the main source of carbon inputs into the soil. However, a sizeable proportion of total plant biomass occurs below ground, particularly as fine roots, and the high turnover rate of these roots means that they are the principal source of detritus at depths below the soil surface. Hundreds of studies over the last few decades have attempted to determine the fate of carbon that is fixed by plants during photosynthesis and a summation of the results of that work (Jones *et al.*, 2009) indicates that roughly 40% of photosynthates are partitioned below ground. In cereals and grasses, for example, total annual below-ground carbon inputs are in the order of 1.5–1.75 t/ha, and are even higher when plant growth is optimized through practices such as nitrogen fertilization (Kuzyakov and Domanski, 2000). Because of the technical difficulties inherent in the study of plant roots, there is still much to be learnt about the fate of this carbon. However, current evidence suggests that for cereal crops, roughly one-half of the carbon that is partitioned below ground is retained in root biomass; about one-third is released from the rhizosphere within a few days by root or microbial respiration; and most of the remaining fraction is incorporated into rhizosphere biomass and soil organic matter (Jones *et al.*, 2009).

The organic compounds that enrich the rhizosphere are known collectively as rhizodeposits, and they are derived from a variety of sources (Table 2.1). Some originate from the breakdown of sloughed root cap cells (also known as border cells), or from root hairs and cortical cells that die from abrasion or pathogen attack. Mucilage is actively secreted by root cap cells and the gelatinous layer that is formed around the root tip is one of the few clearly visible signs of organic carbon release from roots. Other rhizodeposits are lost passively from the root tip and from points where lateral roots emerge through the cortical tissue of primary roots. However, as pointed out by Dennis *et al.* (2010), roots are not the only source of rhizodeposits. Some carbon is transferred to microbial symbionts

Table 2.1. Origin and nature of organic materials in the rhizosphere.

Mucilage	Gelatinous material surrounding the root tip. Comprising mainly complex polysaccharides, it also contains proteins and some phospholipids. Mucilage is one of the few clearly visible signs of organic carbon excretion from roots
Mucigel	A complex of natural and modified mucilages, bacterial cells, metabolic by-products, mineral particles and colloidal organic matter
Exudates	Diffusible organic compounds that are lost passively by the root, and over which the root exerts little control
Lysates	Materials arising from the lysis of root and microbial cells
Volatiles	Volatile compounds that are released by root cells or microorganisms, diffuse through the soil and influence the activity of soil organisms at some distance from the point of release
Sloughed root cells	Cells that are continually sloughed off as roots grow. They include root hairs, senescent cells from the outermost region of the root and cells that are damaged by abrasion or pathogens
Border cells	A specific type of sloughed-off cell. These cells are detached in large numbers from the external layers of the root cap and remain alive for several days

Compiled from Rovira (1979), Curl and Truelove (1986) and Jones *et al.* (2009).

such as arbuscular mycorrhizae, while all exudates from roots are rapidly assimilated by rhizosphere microorganisms. Thus, many of the compounds found in the rhizosphere carbon pool are lysates and other materials of microbial origin (Fig. 2.7).

An extensive range of organic materials is released from roots, including various low molecular weight compounds (mainly amino acids, organic acids and sugars), and an array of secondary metabolites and compounds with high molecular weights such as mucilage and proteins (Table 2.2). Sugars and amino acids occur in the exudates of most plants and one of their most important roles is to provide readily available carbon and nitrogen sources for rhizosphere microbes. Other constituents of root exudates affect the biology of the rhizosphere in other ways. Organic acids modify rhizosphere pH, and because of their metal-chelating capacity, play an important role in the absorption and translocation of nutrient elements; while free fatty acids, sterols and vitamins are essential for the activity of some microorganisms and contribute to the nutritional requirements of rhizosphere-inhabiting heterotrophs. However, it is important to note that the quantity and composition of rhizodeposits is never constant, as it is influenced by a diverse range of factors, including the developmental stage of the plant, light intensity, soil compaction, drought, excessive soil moisture, nutrient

deficiency, the presence of rhizosphere microorganisms and various conditions associated with plant stress (Curl and Truelove, 1986; Bowen and Rovira, 1999; Badri and Vivanco, 2009; Gupta and Knox, 2010).

Although much of the early information on root exudation was obtained from plants grown hydroponically, more recent studies have shown that the losses of substrate from plants growing in soil are much larger than the minimal losses observed in solution culture. This suggests that the normal soil microflora is intimately involved in the exudation process, increasing carbon losses from roots by more than 100% (Barber and Martin, 1976; Prikryl and Vancura, 1980). The actual mechanism by which non-pathogenic, rhizosphere-inhabiting organisms increase root exudation is poorly understood, but their role in rapidly and continually removing exudates from the rhizosphere is almost certainly involved. Most of the sugars, amino acids and organic acids in root exudates are mineralized within a few hours (Ryan *et al.*, 2001) while mucilage has a half-life of approximately 3 days. Since internal carbon pools are continually replenished by the plant, the concentration of simple organic compounds inside the root is typically much greater than in the soil solution (Jones *et al.*, 2009), thus ensuring that high diffusion rates are maintained. The premature senescence of root cells by subclinical pathogens, and the production by

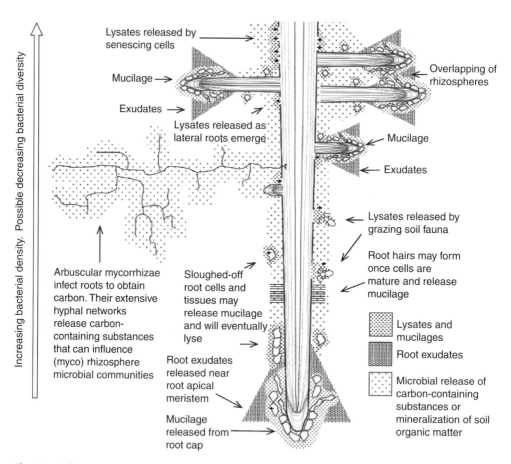

Fig. 2.7. A diagrammatic representation of the rhizosphere, showing that a diverse range of carbon pools are present in the root zone; and that they originate from roots, either directly as lysates, mucilages and exudates, or indirectly through the activities of microorganisms or the soil fauna. (From Dennis *et al.*, 2010, with permission.)

microorganisms of compounds that increase cell permeability, are other mechanisms by which the rhizosphere microbial community increases leakage of exudates from roots (Bowen and Rovira, 1999).

One point that should be apparent from the preceding discussion is that rhizodeposition comes at a significant carbon cost to the plant. Assuming that roots and microorganisms contribute equally to respiration in the rhizosphere, it has been estimated that about 27% of the carbon allocated to roots finds its way into rhizodeposits (Jones *et al.*, 2009). Although this loss of photosynthates could be considered an encumbrance to the plant, the fact that plants benefit in a number of ways from this process is now well recognized.

Carbon is transferred to mycorrhizal symbionts, for example, so that the plant can take advantage of its fungal partner's capacity to absorb water and mineral nutrients. Mucilage (the gelatinous layer surrounding the root tip) is produced to reduce frictional resistance as the root tip moves through soil, and because its high intrinsic affinity for water helps maintain the continuity of water flow towards the root (Jones *et al.*, 2009). Cells that detach themselves from the root cap are not a resource that is lost unnecessarily, as these border cells remain alive for several days and are thought to act as a decoy for nematodes and fungal pathogens, preventing them from reaching the growing root tip and causing infection (Hawes *et al.*, 1998, 2000).

Table 2.2. Organic compounds released by plant roots.

Sugars	arabinose, fructose, galactose, glucose, maltose, mannose, mucilages of various compositions, oligosaccharides, raffinose, rhamnose, ribose, deoxyribose, sucrose, xylose
Amino acids	α-alanine, β-alanine, γ-aminobutryric, α-aminoadipic, arginine, asparagines, aspartic, citrulline, cystathionine, cysteine, cystine, deoxymugineic, 3-epihydroxymugineic, glutamine, glutamic, glycine, histidine, homoserine, isoleucine, leucine, lysine, methionine, mugineic, ornithine, phenylanaline, praline, proline, serine, threonine, tryptophan, tyrosine, valine
Organic acids	acetic, aconitic, ascorbic, aldonic, benzoic, p-hydroxybenzoic, butyric, caffeic, citric, p-coumaric, erythronic, ferulic, formic, fumaric, glutaric, glycolic, glyoxilic, lactic, malic, malonic, oxalacetic, oxalic, piscidic, propionic, pyruvic, succinic, syringic, tartaric, tetronic, valeric, vanillic
Fatty acids	linoleic, linolenic, oleic, palmitic, stearic
Sterols	campesterol, cholesterol, sitosterol, stigmasterol
Growth factors and vitamins	p-amino benzoic acid, biotin, choline, N-methyl nicotinic acid, niacin, pantothenate, pathothenic, pyridoxine, riboflavin, thiamine
Enzymes	amylase, invertase, peroxidase, phenolase, acid/alkaline phosphatase, polygalacturonase, protease
Flavonones and purines/ nucleotides	adenine, flavonone, guanine, uridine/cytidine
Miscellaneous	auxins, scopoletin, hydrocyanic acid, glucosides, unidentified ninhydrin-positive compounds, unidentifiable soluble proteins, reducing compounds, ethanol, glycinebetaine, inositol, myo-inositol-like compounds, Al-induced polypeptides, dihydroquinone, sorgoleone, isothiocynates, inorganic ions and gaseous molecules (e.g. CO_2, H_2, H^+, OH^-, HCO_3^-), some alcohols, fatty acids and alklyl sulfides

From Dennis *et al.* (2010), with permission.

Rhizodeposition also benefits the plant indirectly through a myriad of biological interactions that are ultimately dependent on the release of organic carbon into the rhizosphere. These benefits are discussed elsewhere in this book and include improvements in soil physical and chemical properties, greater parasitism and predation on pests and pathogens, enhanced disease resistance, and improved nutrient cycling.

These observations suggest that root exudation is not simply a passive process that results in loss of carbon to the surrounding soil. Instead, plants use exudates to actively orchestrate their interactions with microorganisms, and in the process they generate a soil biological community that provides innumerable benefits. This point was made succinctly by Bonkowski *et al.* (2009), who noted that 'plant roots have been shown to be neither defenceless victims of root feeders nor passive recipients of nutrients, but instead, play

a much more active role in defending themselves and in attracting beneficial soil microorganisms and soil fauna'. Evidence for the role of root exudates in regulating soil microbial diversity was obtained by Broeckling *et al.* (2008), who showed that exudates from two quite different plant species (a legume and a brassica) influenced the composition of soil fungal communities in diverse soil types. Such observations support the point of view of Bonkowski *et al.* (2009), who argued that soil food webs are mainly fuelled by carbon inputs from roots and not by the introduction of organic matter derived from plant litter. This has important implications for managing agricultural soils because it suggests that lengthening the period when plants are present (e.g. by introducing cover crops and green manure crops, or short-duration crops that are terminated by herbicides), may do more to enhance soil biological fertility than retaining crop residues.

Microbial inhabitants of the soil and rhizosphere

The microorganisms that reside in the rhizosphere presumably originate from the surrounding soil and, therefore, are a component of the microbial community that is present in bulk soil. For most of the 20th century, much of our knowledge of that community was obtained by isolating organisms on culture media that either had a much higher nutrient status than the natural environment or were somewhat selective for certain groups of organisms. Since about 90% of the microbial cells observed by microscopy on soil particles or plant roots are not recovered by cultivation-based isolation procedures (Goodman *et al.*, 1998), microbiologists were working with only a relatively small subset of the soil and rhizosphere microbial community.

Culture-independent molecular technologies are now providing a much broader and less biased picture of the microorganisms that occur in these relatively obscure habitats. A study in which DNA pyrosequencing was used to analyse bacterial sequences from four quite different soils (Roesch *et al.*, 2007; Fulthorpe *et al.*, 2008) demonstrates how cultivation-independent methods are changing our perception of soil and rhizosphere biology. First, some of the most abundant bacteria in the soils examined were genera that had rarely been isolated from soil and had barely been studied previously. Thus, *Chitinophaga* was the most abundant or second most abundant culturable genus in all four soils, despite the fact that this group of chitinolytic, gliding bacteria had rarely been cited in the microbiological literature. Presumably the predominance of *Chitinophaga* was due to the fact that it uses fungal hyphae and insects as a source of chitin, and these resources are readily available in soil. Second, genera that have long been considered the dominant members of the soil bacterial community were either rarely recovered or were not detected in some soils. Of the nine bacteria considered to be significant soil inhabitants by Alexander (1977), six genera (*Agrobacterium, Alcaligenes, Arthrobacter, Micromonospora, Nocardia* and *Streptomyces*) were not abundant enough to be included in a list of the 23 common

genera obtained in these studies, while the other genera (*Flavobacterium, Pseudomonas* and *Bacillus*) were only abundant in some of the four soils. Third, Archaea represented a fairly significant proportion of total sequences. Their number and diversity was particularly high in agricultural soils, perhaps because nitrogen fertilizers were applied periodically.

With regard to the rhizosphere, cultivation-independent techniques are also showing that the rhizosphere bacterial community is more complex than was apparent from culture-based studies. Genera that are readily isolated from the rhizosphere using selective media (e.g. *Pseudomonas* and *Burkholderia*) are still identified using molecular methods, but a wider group of Proteobacteria, together with Actinobacteria and Firmicutes, are also commonly detected (Costa *et al.*, 2006a; Hawkes *et al.*, 2007; Buée *et al.*, 2009a; DeAngelis *et al.*, 2009; Mendes *et al.*, 2011). Since the Proteobacteria and Actinobacteria contain culturable members that cause plant disease, promote plant growth or produce antibiotics, they are the most-studied groups of rhizobacteria. However, DeAngelis *et al.* (2009) questioned the long-held assumption that easily cultivable Proteobacteria predominate in the rhizosphere, as this group was only one of many phyla in a subset of 147 taxa that responded to the presence of the root. Some less well documented rhizosphere colonizers, including Actinobacteria, Verrucomicrobia and Nitrospira, were well represented, and a reasonable number of non-culturable organisms that could not be allocated to a known taxonomic group were also present.

The microbial groups most commonly found in the rhizosphere using culture-independent methods are listed in Table 2.3. However, it must be recognized that the presence or absence of a particular group is rather unpredictable, as our knowledge of rhizosphere microbial communities remains incomplete. Also, the rhizosphere is a variable and dynamic environment, and community composition will be influenced by factors associated with the plant (e.g. plant species or cultivar, the stage of plant development, foliar treatments or the niche on the root surface being colonized); the soil environment (e.g. soil type, temperature, moisture, tillage

Table 2.3. The main groups of bacteria and archaea found in the rhizosphere using culture-independent methods[a].

Taxonomic group[b]	Notable members	Functions of some members (with relevance to the rhizosphere)[c]
α Proteobacteria	Rhizobium, Azospirillum, Beijerinckia, Rhodobacter, Agrobacterium	Nitrifiers, diazotrophs, methanotrophs, phototrophy, plant pathogens
β Proteobacteria	Nitrosomonas, Rhodocyclus, Ralstonia, Rubrivivax	Nitrifiers, diazotrophs, phototrophy, plant pathogens
	Burkholderia	Degradation of organic molecules, positive and negative effects on plant growth, opportunistic human pathogens
γ Proteobacteria	Azotobacter, Xanthomonas, Lysobacter, Escherichia, Vibrio, Serratia, Klebsiella	Nitrogen fixation, production of extracellular enzymes and antibiotics, predation on bacteria and fungi. Plant, human and animal pathogens
	Pseudomonas	Plant growth promotion, antibiotic production, plant pathogens
δ Proteobacteria	Desulfuromonas, Myxococcus, Bdellovibrio	Strictly anaerobic sulfur-reduction, degradation of complex polysaccharides, predation on other bacteria
Acidobacteria		Physiologically diverse, with many uncultured members
Actinobacteria	Streptomyces, Nocardia, Rhodococcus, Arthrobacter	Antibiotic production, degradation of complex polysaccharides, plant pathogens
Bacteroidetes	Cytophaga, Bacteroides	Degradation of complex polysaccharides, dominant member of human intestinal microflora
Firmicutes	Bacillus, Paenibacillus	Degradation of organic molecules, endophytes, pathogens of insects and mammals
Verrucomicrobia		Unknown. Few described species, but found in freshwater and soil environments
Unclassified		Unknown, as members cannot be allocated to any known taxonomic group
Crenarchaeota		Ammonia oxidizers

[a]Based on van Elsas et al. (2007b); Hawkes et al. (2007); Buée et al. (2009a); DeAngelis et al. (2009) and Mendes et al. (2011).
[b]The taxa listed are phyla, except for the four classes of the Phylum Proteobacteria.
[c]Most of the listed functions are not representative of the entire taxonomic group, as each group is functionally diverse.

practice, inputs of organic matter, or the presence of pesticides); the composition of the microbial community (e.g. the growth rates of individual members, their nutritional versatility and their capacity to produce antibiotics); and the presence of predators such as protozoans and bacterivorous nematodes.

Fungi are the other key group of rhizosphere inhabitants, with ergosterol-based measurements indicating that saprotrophic fungi make up about 40% of the microbial biomass in the rhizosphere of grassland plants, for example (Joergensen, 2000). Yeasts and sugar fungi utilize the sugars and other carbon compounds exuded from roots, and other fungal groups are specialized to degrade more complex substances (e.g. cellulose, hemicelluloses and lignin) that are found in all root cells. Fusarium, Penicillium, Trichoderma and Aspergillus, together with Mortierella and some other Mucorales, feature prominently in lists of fungi isolated from the rhizosphere (Curl and Truelove, 1986), but since these fungi are also common in soil, they are not necessarily specialized rhizosphere inhabitants. It is also likely that such lists are skewed towards fungi that grow rapidly, sporulate heavily or are readily isolated on selective media. Molecular profiling of soil and root samples provides a different picture, as an early study of the wheat rhizosphere using DNA-based methods that were still in their infancy revealed many species that are not commonly recognized as rhizosphere inhabitants (Smit *et al.*, 1999).

Arbuscular mycorrhizal fungi are intimately associated with the roots of many plants (see Chapter 8) and are, therefore, an important component of the rhizosphere mycoflora. About 200 species in the phylum Glomeromycota are recognized, with *Glomus* the most commonly encountered genus. Mycorrhizal species have traditionally been detected by microscopic observation of spores collected in the field or by planting 'trap' crops in soil samples, but PCR-based molecular tools are now used extensively to identify these fungi, assess their abundance and diversity, and study their spatial and temporal variation within different environments (see reviews by Lanfranco *et al.*, 1998; Redecker *et al.*, 2003; Reddy *et al.* 2005; Redecker and Raab, 2006). As could perhaps be expected, studies of this nature are indicating that when identification is based solely on morphology, the true diversity of the Glomeromycota is underestimated.

Microbial colonization of the rhizosphere

Contrary to earlier ideas developed from studies in solution or sand culture, it is now known that the entire surface of the root is not covered by microorganisms. After measuring microbial occupation of *Pinus radiata* roots grown in soil, Bowen and Theodorou (1973) concluded that there was great variability in the microbial colonization of roots and that the percentage of the surface covered was usually less than 10% when root segments were less than 3 weeks old. Often the cover of the root by microorganisms was extremely low, especially on segments 1–2 days old, and even when segments were 90 days old, the cover was only 37%. In a similar study, Rovira *et al.* (1974) found that bacteria covered only 4–10% of the root surface of eight plant species grown for 12 weeks in fertile grassland soil. Individual senescent cells and the junctions of epidermal cells were the preferred

sites for colonization, as they are major sources of lysates and root exudates.

Since root colonization is an active process, and not simply a chance encounter between soil microbes and a passing root (Kloepper and Beauchamp, 1992), an organism requires certain traits to colonize the rhizosphere effectively. One of the most important of those traits is motility (Lugtenberg *et al.*, 2001). Several studies have shown that non-motile mutants of *Pseudomonas fluorescens* are unable to colonize the root tip as readily as wild-type strains (de Weger *et al.*, 1987; Dekkers *et al.*, 1998; de Weert *et al.*, 2002); with the latter study indicating that amino acids and organic acids in root exudates were the major chemoattractants in the rhizosphere. Another important root colonization trait is a capacity to obtain the nutrients required for growth and multiplication (Dennis *et al.*, 2010). However, many other factors also play a role in the colonization process. Bacteria occupy water films around soil particles and invariably move further and are more evenly distributed in the rhizosphere when water percolates through the soil (Parke *et al.*, 1986; Bahme *et al.*, 1988; Davies and Whitbread, 1989; Liddell and Parke, 1989). They also live in fragments of organic matter, and can colonize the root tip when it comes into contact with bacterial colonies as it moves through soil. Bacteria may also be ingested by nematodes and defecated elsewhere in a viable form, or they may adhere to the external surfaces of soil fauna and be moved around the rhizosphere (Bonkowski *et al.*, 2009). In some cases, the interaction between roots and the soil fauna is quite precise, with the roots of some legumes emitting specific volatile signals to recruit bacterivorous nematodes, thereby improving their inoculation with rhizobia (Horiuchi *et al.*, 2005).

Once bacteria arrive at the root surface, the colonization process is a two-stage phenomenon involving attachment and then multiplication (Howie *et al.*, 1987). In the initial attachment phase, bacteria attach to and are then transported on the elongating root tip as it moves through soil (James *et al.*, 1985). Bacterial attachment to plant cells is mediated by proteinaceous filamentous extensions on bacteria known as fimbriae or pili, and by a number of cell surface polysaccharides (Weller, 1988). However, as the root elongates, the initial inoculum at the root tip is diluted and some bacteria are physically displaced by soil particles. Thus, the multiplication phase is vital to the colonization process, as the bacteria must multiply, spread locally, and compete with other indigenous microflora to adequately colonize roots.

A study of the dynamics of microbial colonization and succession processes along wheat roots in the period 2–4 weeks after planting showed that bacterial populations oscillated in a regular manner; that the location and pattern of the waves changed progressively from week to week; and that these wave-like fluctuations were not consistently correlated with total organic carbon concentrations or the location of lateral root formation (Semenov et al., 1999). These oscillations were initially thought to be associated with the death of microbes following substrate depletion and their subsequent regrowth on carbon recycled from dead bacteria. However, when soils were amended with organic matter (which at a microniche level is somewhat similar to providing a nutrient impulse from a growing root), fluctuations in bacterial populations were also affected by the regulatory activities of protozoan and nematode predators (Zelenev et al., 2004). Thus, microbial populations in the rhizosphere are spatially and temporally variable, not only because the root tip is a continually moving source of nutrients, but also because bacterial numbers and diversity are influenced by many other factors, including root age, degree of suberization, the quantity and quality of root exudates, and the activities of predators. Since different trophic groups will peak in activity at different times and at different locations on the root surface, and all of these groups can affect organisms that feed on roots, van Bruggen et al. (2006) argued that a complex biotic community is more likely to suppress root pathogens than a single trophic group or species.

Since fungal growth rates of up to 3 mm per day have been recorded in natural soils (Bowen and Rovira, 1976), one advantage fungi have over bacteria is that they can move relatively long distances to colonize roots. Also, fungi are not necessarily limited by a lack of nutrients in micro-niches within the soil profile, as they can obtain sustenance from a nutrient source (e.g. a fragment of organic matter or a rhizodeposit) and grow through low-nutrient niches to reach the root surface. Little is known about the rhizosphere colonization process, but presumably exudates from root cells stimulate spore germination or attract hyphae from afar, and then fungal adhesives are produced that enable hyphae to adhere to the root surface. Since these processes are the first component of the colonization process in mycorrhizal fungi (Buée et al., 2000), and also occur when the phylloplane is colonized by foliar plant pathogens (Tucker and Talbot, 2001), they are likely to occur in rhizosphere-inhabiting fungi. Given the competitive nature of the rhizosphere environment, fungi must also be able to ward off or avoid antagonists during the colonization process. Due to their filamentous nature, one way fungi can do this is to grow across soil voids and thereby obtain some spatial escape from competitors.

Communication within the rhizosphere

Plants and the organisms that live near roots have co-evolved, and are, therefore, dependent on each other for survival. Thus, it is not surprising that plants produce signals which guide beneficial organisms to roots and that microbes, in turn, elicit physiological changes in the plant that help them establish in appropriate niches within the root. The symbiotic associations between legumes and rhizobia, and between plants and mycorrhizae, are the most-studied examples of plant–microbe relationships, and recent research has shown that signalling by both parties plays a key role in establishing these relationships (Long, 2001: Kosuta et al., 2003; Phillips et al., 2003; Bais et al., 2004, 2006). Signalling molecules are also produced by plants in many other situations, and often involve secondary metabolites that are commonly found in root exudates. Examples include sugar-containing substrates that induce the soil diazotrophic community to fix nitrogen (Bürgmann et al., 2005) and unknown compounds that actively shape the soil fungal community (Broekling *et al.*, 2008). Signalling molecules are also almost certainly associated with the production of phytoalexins, which are produced by plants in response to pathogen attack (Bais *et al.*, 2006).

There are also many other examples of signalling being involved in interactions within the rhizosphere food web. Colonies of bacteria exchange signal molecules with nearby cells to coordinate their activities, as this allows them to act like a multicellular organism. This process, which is known as quorum sensing, is density dependent, and is only effective when populations of bacterial cells are high (Waters and Bassler, 2005). Plants and other microorganisms respond to these signals by disrupting bacterial communication before populations reach a critical level, a process that helps them defend themselves from potential pathogens and competitors (Hartmann *et al.*, 2009). Signalling processes also affect the soil fauna, as bacteria are able to sense the presence of predators such as amoebae by up-regulating antibiotic production (Bonkowski *et al.*, 2009). Thus, communication within the rhizosphere is not just limited to roots and closely associated microbial symbionts: it occurs between all inhabitants and is manifested at all levels of interaction (Bonkowski *et al.*, 2009).

Plant–microbe–faunal interactions in the rhizosphere

Since the rhizosphere is not completely occupied by microorganisms, it might be thought that rhizosphere inhabitants would have little effect on each other. However, despite the fact that microbial communities are spatially discrete, they tend to occupy the same favoured sites and so they interact in localized areas. In fact, Bowen (1979) visualized microbial communities on the root surface as a series of microislands that tended to interact when their 'territories' were crossed. From a practical point of view, such interactions are probably the most important phenomena in rhizosphere biology, as they determine the success or failure of microbes that are introduced as inoculants for biological control of root pests and pathogens.

Unfortunately, our knowledge of these interactions is limited because of the complexity of the habitat and the difficulties involved in studying microbial behaviour in the rhizosphere. Much of our information has been obtained from relatively simple experimental systems and may not translate to actual behaviour in natural soils.

Interactions between the mixed populations of microorganisms in the rhizosphere can either promote the growth of individuals within those populations through commensalism, or inhibit them by competition, antibiosis, hyperparasitism or predation. Two types of commensal relationships are commonly observed in the rhizosphere. The first involves microbial transformations that release previously unavailable nutrients for use by other microorganisms. This occurs, for example, when microorganisms produce organic acids or ammonify amino acids, causing pH changes that increase the availability of some major and minor nutrients to other soil organisms. A second example of commensalism involves the synthesis by one species of a growth factor that is essential for another organism. Interactions of this type often occur in the rhizosphere, as for example when thiamine, biotin, vitamin B_{12}, pantothenic acid and riboflavin are synthesized by some bacteria and used by others.

In the struggle among soil organisms for occupancy of a place in the rhizosphere or on an organic substrate, there will be competition for available nutrients, water, oxygen and space. However, substrate or nutrient supply appears to be the primary resource for which microorganisms actively compete (Curl and Truelove, 1986). Root exudates in the rhizosphere are somewhat like a culture medium that only supports the growth of certain species. Other species are suppressed, either because essential nutrients are depleted or because biostatic metabolites are produced. Thus, microorganisms genetically endowed with enzyme systems for rapid metabolism and high respiration rates are likely to prevail in the rhizosphere. The ability of some rhizosphere-inhabiting microorganisms to obtain a competitive advantage by sequestering iron (Leong, 1986) provides one of the best-documented examples of a mechanism by which competition for nutrients occurs in the rhizosphere.

Antagonism mediated by specific or non-specific metabolites of microbial origin, by lytic agents, enzymes, volatile compounds or other toxic substances, is known as antibiosis. Of the various agents responsible for antibiosis, most attention has been directed towards organic compounds of low molecular weight that are detrimental to other organisms at relatively low concentrations. Thousands of different antibiotic compounds are produced by soil organisms in the laboratory, and there is now sufficient evidence, obtained mainly by genetic methods, to indicate that some of these antibiotics have a function in nature (Fravel, 1988; Raaijmakers et al., 2002; Haas and Keel, 2003). The first clear-cut demonstration that a bacterially produced antibiotic could act against another rhizosphere inhabitant was obtained by Thomashow and Weller (1988), who identified a phenazine antibiotic, phenazine-1-carboxylic acid (PCA), as a biocontrol factor produced by *Pseudomonas fluorescens* strain 2-79. This strain, originally isolated from the wheat rhizosphere, was able to suppress take-all disease caused by the fungal pathogen *Gaeumannomyces graminis* var. *tritici*, whereas PCA-negative mutants were unable to inhibit the fungus *in vitro* or suppress the disease. Since that study was published, phenazine biosynthesis has been documented in numerous bacteria, and although the physiological role of phenazines remains uncertain, it is clear that they contribute to the competitiveness of strains that produce them in the environment (Mavrodi et al., 2006).

One valuable outcome of the work on suppression of root pathogens by *Pseudomonas* spp. is the observation that *in vitro* production of an antimicrobial compound does not necessarily mean that the antibiotic will be produced in the soil or rhizosphere, or that the microorganism producing the compound has biocontrol potential. The steps involved in obtaining evidence that biological control activity is associated with antibiotic production were discussed by Haas and Keel (2003) and are outlined here. However, because these steps have not been fully completed for many antagonist–pathogen interactions,

there is still some debate about the role of antibiosis in suppressing pathogens.

- The antibiotic of interest is produced by the biocontrol strain *in vitro*, can be identified chemically, and inhibits the pathogen *in vitro*.
- The structural and most important regulatory genes required to produce the antibiotic are identified by isolating non-producing mutants or over-producing strains, and characterizing the genes involved.
- The antibiotic can be detected in the rhizosphere.
- Antibiotic-negative mutants give reduced plant protection (in comparison to the wild-type strain) in assay systems consisting of sterile or natural soil, a host plant and the pathogen. However, this criterion is not necessarily definitive for biocontrol strains capable of producing multiple antibiotics, because mutants lacking one antibiotic may compensate for its loss by over-producing another, thereby maintaining total antibiotic production and biological control.
- The expression of antibiotic biosynthetic genes can be monitored *in situ* by fusing the antibiotic structural gene(s) to a reporter gene and monitoring its expression.
- Antibiotic production in the rhizosphere can be shown to play a role in sustaining the ecological fitness of the producer strain.

Given the enormous number of antibiotics synthesized by microorganisms and the fact many known antibiotic producers are abundant in the rhizosphere (e.g. *Streptomyces, Bacillus, Pseudomonas, Penicillium, Trichoderma* and *Aspergillus*); it is likely that microbial activity in the rhizosphere is influenced by antibiosis. However, as Haas and Keel (2003) pointed out, it should not be assumed that the rhizosphere community is simply composed of antibiotic-producing 'killers', antibiotic sensitive 'victims' and antibiotic-resistant 'heroes'. Organisms in the rhizosphere may avoid the effects of antibiotics because (i) antibiosis occurs mainly in localized microsites where the nutrients essential for antibiotic production are available; (ii) antibiotics are transitory because they are frequently bound to clay colloids and organic matter, or are quickly decomposed by microorganisms; and (iii) estimates of the amount of antibiotic associated with roots suggest that the concentrations likely to be present are usually insufficient to kill bacteria and fungi, and may not even be inhibitory. Therefore, antibiotics are most likely to act at sub-inhibitory concentrations through much more subtle mechanisms such as blockage of cell wall signalling, interference with adhesion processes, and repression of lytic enzymes and antibiotics that are produced by competitors (Haas and Keel, 2003).

In most soils, fungal propagules are restricted to some extent in their ability to germinate and grow. This phenomenon, which is known as soil fungistasis or mycostasis, occurs in situations where (i) viable fungal propagules that are not subject to constitutive dormancy do not germinate in soils under favourable temperature and moisture conditions; or (ii) growth of fungal hyphae is retarded or terminated by soil environmental conditions other than temperature or moisture (Watson and Ford, 1972). Although the causes of fungistasis have been the subject of continuing debate, studies during the 1970s suggested that both competition for nutrients and antibiosis were involved (Watson and Ford, 1972; Griffin and Roth, 1979; Balis and Kouyeas, 1979). Many fungal spores do not germinate unless an exogenous energy source of organic carbon and other nutrients is available, and microbial activity near spores can deplete these nutrients to such an extent that spore germination is inhibited. Demonstration that rhizosphere exudates and carbon substrates such as glucose can nullify this inhibition led some workers to conclude that nutrient deprivation is the major factor in fungistasis (Lockwood, 1977). However, there is also evidence that inhibitors of microbial origin, particularly volatiles such as ammonia, ethylene and formaldehyde, are also implicated in the fungistatic phenomenon (Balis and Kouyeas, 1979).

In more recent studies, treatments that affect soil organisms (e.g. microwave irradiation, nutrient inputs, heat treatments and changes in agricultural management practices) have been imposed in an attempt to determine which components of the microbial community

are primarily responsible for fungistasis. The results of De Boer *et al.* (2003) indicated that levels of fungistasis were highest when pseudomonads were a component of the soil bacterial community, and that soils dominated by bacilli were not fungistatic. Similar results were obtained by Wu *et al.* (2008), who found that retention of bacterial diversity was important in maintaining fungistasis. Other studies showed that soils with diverse communities of *Bacillus* and *Pseudomonas* were most likely to suppress hyphal extension in *Rhizoctonia solani* (Garbeva *et al.*, 2004, 2006). Collectively, these results suggest that that the structure, activity, and diversity of the microbial community all play a role in determining the level of soil fungistasis. Since De Boer *et al.* (2003) also showed that fungistasis was not induced by treatments that resulted in intense competition for nutrients, these authors considered that antibiotic production was primarily responsible for the phenomenon.

The question of whether the microorganisms involved in fungistasis act by limiting available carbon (the nutrient-deprivation hypothesis) or by producing antifungal compounds (the antibiosis hypothesis) has still not been definitively answered, but a review by Garbeva *et al.* (2011a) summarizes current knowledge.

• Fungistasis is not due to the total metabolic activity of the soil biomass, but is the result of the activities of specific components of the soil microbial community. Secondary metabolites produced by soil microorganisms are primarily responsible, but nutrient depletion by competing microbes is also involved.

• Interspecific interactions within the microbial community play a significant role in soil fungistasis by triggering the production of broad-spectrum antibiotics with inhibitory effects on fungi. These interactions may involve strains that had previously been assumed to play no role in fungal suppression (de Boer *et al.*, 2007).

• Some fungistatic interactions occur between spatially separated organisms because volatile compounds are involved that can diffuse through aqueous solutions and air spaces. Numerous soil bacteria produce fungus-inhibiting volatiles, and the diversity of these compounds and their concentration is often correlated with fungistatic activity.

- Fungi possess defence strategies that help protect them from the actions of potentially antagonistic microorganisms. Since external energy inputs are required to activate these defensive mechanisms, fungi have the capacity to resist the effects of fungistatic compounds in the nutrient-rich rhizosphere environment, but may succumb to them in bulk soil. Fungistasis is, therefore, beneficial to the fungus, because it limits spore germination and growth in situations where a host plant or other sources of nutrition are absent.

- Fungistasis should be seen as a component of a more general biostatic phenomenon that occurs in every soil and affects all soil microorganisms (Ho and Ko, 1982). Fungi, bacteria and actinobacteria produce inhibitors during competitive interactions, and those inhibitors not only affect their direct microbial competitors but also affect plants, other microorganisms and associated soil fauna. Although various terms have been used to describe the phenomenon (Ho and Ko, 1982, 1986), the simple term 'biostasis' is perhaps most appropriate.

The role of parasitism and predation in the rhizosphere also deserves some comment, as there are many opportunities for antagonistic interactions in an environment that is occupied by large numbers of different organisms. Bacteriophages (often referred to as phages) replicate within bacteria and then lyse bacterial cells, and given their roles in marine and freshwater environments, they are likely to be important predators of soil bacteria (Kimura et al., 2008). Little is known about bacteriophages in soil, but they do have the potential to affect bacteria that are introduced into soil to improve plant growth. Substantial increases in naturally occurring bacteriophages were observed following an increase in populations of its bacterial host (Ashelford et al., 2000), while a bacteriophage that was widely distributed in Irish soils was responsible for the decline of a growth-promoting

fluorescent pseudomonad that was introduced into soil on sugarbeet seed (Stephens et al., 1987). Members of the bacterial genus Bdellovibrio are also capable of multiplying in other bacteria and lysing their host cells. Protozoa and bacterial-feeding nematodes are the other important predators of bacteria. Since a single protozoan is capable of consuming millions of bacteria per day, protozoa play an important role in regulating bacterial populations and influencing the composition of the bacterial community. Bacterial-feeding nematodes have similar effects, as they compete with protozoa for the same food source.

Fungi and oomycetes that inhabit the rhizosphere are also subject to parasitism and predation by other soil organisms. Actinobacteria are recognized parasites of these organisms (Sutherland and Lockwood, 1984; Tu, 1988; El-Tarabily et al., 1997), but other bacteria also lyse fungal structures and degrade hyphae (Mitchell and Alexander, 1962; Mitchell and Hurwitz, 1965; Campbell and Ephgrave, 1983; Ordentlich et al., 1988). A number of cell wall-degrading enzymes are often considered to be responsible for these effects, but there is little direct evidence for their presence and activity in the rhizosphere (Whipps, 2001). Spores, sclerotia and hyphae may also be parasitized by fungi (reviewed by Lumsden, 1981; Whipps, 2001), with most observations linked to biocontrol of fungal pathogens by Trichoderma (Benitez et al., 2004; Harman et al., 2004). Predation on fungi by soil fauna is also common. Fungal-feeding nematodes are often a principal component of soil nematode communities, while many enchytraeids and microarthropods are mycophagous, as evidenced by the variety of fungal spores and mycelial fragments observed in their gut contents.

Given the complexity of the soil food web, the interactions discussed here are only a small subset of the plant–microbe–faunal interactions that occur in soil. All soil animals can be parasitized by microorganisms, and predatory interactions between protozoa, nematodes, tardigrades, mites, Collembola and other soil fauna are common. Since those types of interaction form the basis of biological control of nematodes, they are discussed in detail elsewhere in this book (see Chapters 5, 6 and 7).

Effects of rhizosphere inhabitants on plant growth

Three well-known groups of rhizosphere-inhabiting microorganisms (rhizobia, mycorrhizae and root pathogens) either establish symbiotic relationships that improve plant growth, or destroy roots and cause plant disease. However, the rhizosphere microflora also includes many other beneficial and deleterious elements with the potential to significantly influence plant growth and crop yields. Both bacteria and fungi may have these effects, but only plant growth-promoting rhizobacteria (PGPR) and deleterious rhizobacteria (DRB) have been widely studied (Schippers *et al.*, 1987; Schroth and Becker, 1990; Nehl *et al.*, 1996; Compant *et al.*, 2005, 2010; Saharan and Nehra, 2011).

PGPR are associated with most – if not all – plant species, and are found in all situations where crops are grown. Some of these bacteria inhabit the rhizosphere, while others establish endophytic populations in specific ecological niches within the root (Fig. 2.8 and also Hallmann *et al.*, 1997; Kloepper *et al.*, 1999; Compant *et al.*, 2010). However, regardless of their location, all are able to directly or indirectly promote plant growth. In many cases they suppress pathogens or other detrimental organisms by producing antibiotics, siderophores or lytic enzymes; by detoxifying or degrading pathogen virulence factors; or by inducing systemic resistance to pathogens (Compant *et al.*, 2005). Since these mechanisms of action are important in many biocontrol systems, they are considered in more detail in Chapters 8 and 12. Other growth-promotion mechanisms are also possible, including non-symbiotic nitrogen fixation, hormone production and phosphate solubilization. In many cases, multiple mechanisms are involved (Hayat *et al.*, 2010: Saharan and Nehra, 2011).

Deleterious rhizosphere bacteria are minor pathogens that are capable of inhibiting root and shoot growth. These bacteria occupy the root cortex but do not enter the vascular tissue, and although they sometimes produce no visually obvious symptoms, they may inhibit root hair development or cause root discoloration and necrosis. Deleterious rhizosphere bacteria act primarily by producing phytotoxins, but phytohormone production, competition for nutrients and inhibition of mycorrhizal function may also be involved

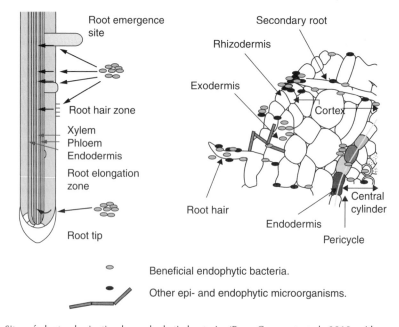

Fig. 2.8. Sites of plant colonization by endophytic bacteria. (From Compant *et al.*, 2010, with permission.)

(Nehl *et al.*, 1996). Since the same groups of bacteria (commonly isolates of *Pseudomonas* and *Bacillus*) and the same mechanisms of action are associated with plant growth promotion, the categorization of rhizobacteria as deleterious or beneficial to the plant is somewhat problematic. Nehl *et al.* (1996) discussed the reasons for this, and concluded that individual isolates could improve plant growth in some situations and be detrimental in others, with the effect varying with host genotype, environmental conditions and mycorrhizal status. Given this functional variability, and the fact that isolates which promote and depress plant growth can be obtained from the same root system (Sturz, 1995; Surette *et al.*, 2003), the challenges involved in consistently improving crop production by introducing microorganisms to the rhizosphere should not be underestimated.

Manipulating the rhizosphere community

The practice of inoculating legumes with rhizobia is already relatively common in agriculture and there is increasing recognition that mycorrhizae can contribute to sustainable agricultural production by reducing reliance on inorganic fertilizers. However, after more than 50 years of research, efforts to improve plant productivity by modifying non-symbiotic microbial processes have been ineffective or inconsistent (Rovira, 1991; Handelsman and Stabb, 1996). Nevertheless, our understanding of these plant–microbe interactions has developed to the point where there is optimism that we may eventually be able to influence rhizosphere processes and use them to our advantage. There are many possible ways in which the rhizosphere could be manipulated for the benefit of the plant (see the following list, which has been modified from Burns, 2010). Since research into many of these strategies is already under way, there are likely to be practical outcomes in some of these areas within the next decade.

- Using microbial inoculants or manipulating the indigenous rhizosphere community, for biocontrol purposes or for bioremediation.

- Inoculating planting material with endophytes.
- Controlling the release of plant nutrients and the enhancement of other plant growth-promoting processes.
- Exposing plants to microorganisms (or microbial metabolites) to induce a specific or general immune response to pathogens and predators.
- Planting crops that are compatible with the indigenous microflora, soil type and climate.
- Stimulating or designing rhizospheres that will support plant growth in soils with limited supplies of available nutrients, or will ameliorate the effects of stress factors such as salinity, pollution, erosion or a changing climate.
- Selecting or designing bacteria and fungi with rhizosphere 'competence'.
- Breeding plants that constitutively or inductively secrete signals that generate and support a beneficial rhizosphere.
- Designing self-sustaining soil biospheres for agricultural production in and around crowded cities.

To date, most attempts to manipulate rhizosphere microbial communities have involved the introduction of readily cultured bacteria into soil in an attempt to enhance plant growth or reduce the impact of root pathogens. Results of initial studies soon indicated the difficulties involved with this approach. Acea *et al.* (1988), for example, studied the survival and multiplication of several common soil bacteria (species of *Rhizobium, Agrobacterium, Pseudomonas, Micrococcus, Corynebacterium, Streptococcus* and *Bacillus*) after they were introduced into soil and found that: (i) some species initially increased in numbers but populations of all species declined within a few days, sometimes to very low levels; (ii) patterns of decline were similar in the two soils tested; (iii) the bacteria survived and multiplied in sterile soils, indicating that soil biological factors were associated with population decline in non-sterile soil; and (iv) populations of protozoa increased as populations of introduced bacteria declined, while microorganisms capable of lytic activity or antibiotic production were also present, suggesting that naturally occurring antagonists were responsible for the poor establishment.

Since bacteria survive poorly when introduced into soil, seed treatment is usually the preferred method of introduction. The process of seed germination releases an abundance of carbohydrates and amino acids in the form of seed exudates (Lynch, 1978; Subrahmanyam *et al.*, 1983), and as microorganisms on the seed surface can utilize these substrates, they are in an ideal position to colonize the root as it emerges. Numerous studies have shown that colonization does occur, as bacteria applied on or around seeds can later be isolated from the rhizosphere. However, a review of those studies (Kluepfel, 1993) indicates that populations typically increase exponentially for the first few days following inoculation and then decline steadily until detection limits are reached. Since inoculum density usually has little impact on rhizosphere colonization, the high initial population densities probably represent the carrying capacity of the root. Over a longer time period, the slow decline in bacterial density is probably due to changes in the quantity or composition of exudates as the plant ages, and to the cumulative effects of competitors and predators. Although the population density of immigrant bacteria is often limited, several studies cited by Kluepfel (1993) showed that their presence affected the indigenous microflora to some degree, at least temporarily.

Although it is not possible to markedly modify the rhizosphere microflora by introducing rhizobacteria to soil or applying them to seeds, this does not mean that this practice is not worth pursuing. It may, for example, be possible to establish disease-suppressive microorganisms in favourable niches within roots or on the rhizoplane, where they interact with a particular pathogen but are essentially 'insulated' from the rest of the rhizosphere community (Weller *et al.*, 2002). Also, many plant growth-promoting bacteria act against pathogens by inducing systemic resistance mechanisms within the plant (Whipps, 2001; Compant *et al.*, 2005; Saharan and Nehra, 2011), and it may require relatively minor changes in the structure of the rhizosphere community to achieve such effects.

The literature on establishing biocontrol fungi and oomycetes in the rhizosphere is almost as large as it is for bacteria. 'Rhizosphere competence', the term used to describe the capacity of an organism to grow and function in the rhizosphere, has been studied in non-pathogenic strains of *Fusarium oxysporum* and several species of *Pythium* and *Trichoderma* (Whipps, 2001), and in some fungal parasites of nematodes (Kerry, 2000). The nematode studies are discussed in Chapters 5 and 8, but the work on fungal biocontrol agents, most of which involves *Trichoderma*, will be mentioned briefly here. Early studies by Chao *et al.* (1986) indicated that *Trichoderma harzianum* and *T. koningii* had a limited capacity to grow in the rhizosphere, as both species were unable to colonize more than 2 cm from the seed on which inoculum had been placed. Ahmad and Baker (1987) confirmed that *T. harzianum* was not rhizosphere-competent and then produced mutants that could multiply at the root tip to population densities of 10^6 colony-forming units per gram of rhizosphere soil. Since that time, it has been generally accepted that many *Trichoderma* spp. can establish robust and long-lasting relationships with root surfaces. Some *Trichoderma* strains penetrate into the epidermis and a few cells below this level, but remarkably, they remain confined to this region and do not become plant pathogens (see review by Harman *et al.*, 2004).

Although most attempts to deliberately manage the rhizosphere have focused on inoculants, results from several recent studies suggest that it may be easier to change microbial diversity or function by manipulating the plant. It has long been known that relatively minor changes in a plant's genetic composition can alter the composition of the rhizosphere community (Neal *et al.*, 1973), and this observation has been built on in recent years (see reviews by Drinkwater and Snapp (2007) and Hawkes *et al.* (2007)). Thus it is now known, for example, that the bacterial endophyte communities of modern wheat cultivars are more diverse than those of ancient land races (Germida and Siciliano, 2001), and that plant–microbe interactions in the cotton rhizosphere are influenced by cultivar (Knox *et al.*, 2009). In rice, production of root cells and mucilage differ between cultivars and this has flow-on effects on bacteria–protozoa interactions in the rhizosphere and their subsequent effects on plant growth (Somasundaram et al., 2008). Some rice cultivars respond to these interactions by taking up more nutrients and increasing their biomass, whereas others do not, suggesting that specific plant traits which have co-evolved between ancestor rice species and soil microorganisms may have been lost during the selection of high-yielding cultivars under modern, high-input agricultural practice. Collectively, these results suggest that in addition to concentrating on above-ground performance, plant breeders should adopt a 'designer rhizosphere' approach in their breeding programmes and target the selection of plant–microbe consortia that are beneficial to crops (Somasundaram et al., 2008; Bonkowski et al., 2009). Practices such as crop rotation and the foliar application of herbicides or nutrients also influence soil microbial communities (Bowen and Rovira, 1999; Kremer et al., 2005; Gupta and Knox, 2010) and could possibly also be used to generate a rhizosphere with the desired biological characteristics.

Implications for Biological Control

It should be obvious from the material presented in this chapter that biological control of nematodes cannot be viewed in simplistic terms as merely involving interactions between nematodes and their natural enemies. Biological control takes place in a complex and dynamic environment where both nematodes and their antagonists are influenced by the physical and chemical environment, by soil organic matter, by the presence of plants and by the activities of the soil microflora and microfauna. Thus, this chapter concludes by considering the impact of the soil and rhizosphere environment on biological control of nematodes.

Impact of the soil environment

In their book on biological control of plant pathogens, Baker and Cook (1974) pointed out that 'environment controls the outcome of all host-pathogen-antagonist interactions'. This is especially true when the pathogen is a plant-parasitic nematode, because those interactions take place in a physical, chemical and biological environment that not only affects the behaviour and activity of the nematode and the organisms that regulate its populations, but also influences the capacity of the plant to tolerate nematode damage through flow-on effects to soil heath. The soil environment is also very dynamic, as it is continually being modified by inputs of organic matter, the presence of roots, the actions of soil microflora and fauna, seasonal and diurnal fluctuations in temperature, and constant changes in moisture. Thus, the establishment and maintenance of successful systems of biological control will ultimately depend on having a good understanding of the way in which physical, chemical and biological factors influence the relationship between a nematode pest, its host plant and its natural enemies. Since this point will be emphasized throughout this book, only a few examples are given here.

- Physical properties such as soil particle size will vary across a landscape or field, and this innate variability in soil texture has major effects on plants. The presence of clay, for example, will increase a soil's water- and nutrient-holding capacity and thereby improve its ability to support plant growth. Clays also influence a soil's biological properties. For example, their capacity to retain moisture helps provide an environment that is amenable to biological activity; their capacity to absorb antibiotics and extracellular enzymes will influence the outcome of interactions between microorganisms; and their effects on pore size will affect the movement of soil fauna, including plant-parasitic nematodes. Thus, clay content influences a soil's biological fertility, and will also affect a plant's capacity to tolerate nematode damage. In a sandy soil, for example, an increase in clay content from 5% to 15% can mean the difference between severe nematode damage and a healthy crop. Given this wide range of effects of clays, the clay content of a soil, or the type of clay, could be expected to influence the efficacy of many biological control systems, as has been observed for some fungal pathogens (Stutz et al., 1989; Alabouvette, 1999).

• Soil is not a uniform entity. Bulk soil and the soil in close proximity to roots contain numerous niche habitats where soil organisms interact. These habitats include relatively large voids between soil particles, microsites within aggregates and various microhabitats in the rhizosphere (e.g. the meristematic zone at the root tip, points where lysates accumulate as lateral roots emerge, and living or senescing cortical cells on the root surface). Plant-parasitic nematodes live in this zone and may be able to avoid relatively large predatory arthropods and nematodes by retreating into small pore spaces. However, this defence strategy is not likely to be effective against fungal and bacterial antagonists, which not only can occupy the same water films as their nematode prey, but are also able to access micropores within aggregates.

• Bacteria in the genus *Pasteuria* are well-recognized parasites of plant-parasitic nematodes (see Chapter 7), but their capacity to act as biological control agents is influenced by soil physical properties. A review of the impact of the soil environment on the host–parasite–soil relationship (Mateille et al., 2010) shows that *Pasteuria* endospores percolate quickly in coarse sandy soils and may be leached to a point where they are no longer present in the same part of the profile as the target nematode. They may also be adsorbed onto clay particles or become trapped in non-dispersible aggregates. Thus, the number of *Pasteuria* endospores available to infect nematodes is markedly influenced by textural properties such as macroporosity, microporosity and aggregate stability, and this probably explains why the parasite is an effective control agent in some situations but not in others.

• Water limits biological activity in many soil environments, and has major effects on the activity of most soil organisms, including zoosporic parasites of nematodes. Zoospores are only able to move in relatively large soil pores (10–20 μm in diameter), and since they are not filled with water unless the soil is very wet, nematode parasites such as *Nematophthora gynophila* and *Catenaria auxiliaris* are only likely to be infective during or immediately after heavy rainfall or irrigation. They will also be more active in wet than dry seasons (Kerry et al., 1980, 1982a, b; Stirling and Kerry, 1983).

• Soil temperature fluctuates diurnally, and also varies with depth and season of the year. A biological control agent must, therefore, be able to survive temperature extremes and also be effective when temperatures are ideal for the target nematode. Since all organisms have specific maximum, minimum and optimum temperatures for activity, interactions between nematodes and their antagonists will be affected by temperature. For example, the optimum temperatures for hatch of *Meloidogyne incognita* eggs and growth of the fungal egg parasite *Brachyphoris oviparasitica* were similar, but the growth of the fungus was not affected by low temperatures to the same extent as the development and hatch of eggs (Stirling, 1979b). Invasion of egg masses by the fungus, and penetration and growth in eggs took longer as temperatures decreased, but the slower rate of production, development and hatch of eggs more than compensated for the decreased growth of the fungus. Consequently, *B. oviparasitica* was a more efficient parasite as temperatures decreased from 27°C to 15°C. On the other hand, temperature had the opposite effect on the interaction between *M. javanica* and the bacterium *Pasteuria penetrans*, with pathogenesis being favoured by high temperatures (Stirling, 1981).

Multitrophic interactions in a complex environment

The soil in the vicinity of plant roots is one of the most biologically complex environments on earth. Many basic texts in soil biology provide estimates of the number of organisms in soil, and those estimates are always staggering. Bacterial populations are often greater than 10^9 cells/g soil, and 1 m^2 of most topsoils contains kilometres of fungal hyphae and many thousands of small animals ranging in size from protozoa to earthworms. However, as scientists explore the soil and rhizosphere habitat using molecular technologies, it is becoming apparent that the soil biological community is much more complex and diverse than those estimates suggest. Individuals in this community interact with each other, and during the process they influence plant growth, the soil biological community and their environment.

Although all parasites and predators of nematodes are components of this biological community, most biocontrol studies ignore the complexity of the soil environment and concentrate on tritrophic interactions that simply involve a plant, a target nematode and a predator. Experiments are often carried out in relatively sterile soils or potting mixes in highly controlled glasshouse environments, and given the difficulties involved in working with soil organisms, such an approach is understandable. However, it is questionable whether the results of such studies bear much relationship to the situation in the field. Those interested in biological control must, therefore, recognize that multitrophic interactions are the norm in the soil environment, and need to put more effort into determining how they influence populations of plant-parasitic nematodes and their damage thresholds. Such interactions are exceedingly complex, but they will eventually be studied using mathematical models that integrate our knowledge of soil organisms, molecular biology, soil ecology, soil health and crop agronomy. One area where such an approach would be particularly useful is to better understand the web of interactions involved in organic matter-mediated suppressiveness to nematodes. That knowledge could then be used to better define the situations where suppressiveness is effective enough to reduce damage caused by plant-parasitic nematodes (see Chapter 9 for further discussion).

The soil and rhizosphere as a source of antagonists

All plant-parasitic nematodes are intimately associated with plant roots and consequently spend most of their lives in the immediate vicinity of plant roots. Since plants recruit potentially useful microbes into the rhizosphere and profit from their activities (Hartman et al., 2009), the root surface and the soil within 1–2 mm of the root are likely to be rich sources of organisms with biological control potential. Molecular technologies are providing opportunities to study biological communities that inhabit this region, and will almost certainly play a role in finding new antagonists of nematodes. One area where they will be particularly useful is in locating endoparasites capable of attacking nematodes or their eggs. At present, such parasites are not easily isolated using traditional techniques because the affected carcase disintegrates too quickly.

Although most previous work on natural enemies of nematodes has focused on parasites and predators that are discussed later in this book, it is important that in future, the challenge of identifying organisms with biological control potential is not viewed too narrowly. For example, we are aware of endophytic organisms that promote plant growth or induce pathogen-resistance mechanisms in plants (see Chapter 8 for more detail), but given the limited amount of research in this area, it is likely that many other organisms with similar mechanisms of action are present in the rhizosphere. Biological control agents with new modes of action should also be sought. For example, much of the work on biological control of plant pathogens has been done with microorganisms that produce antibiotics, but there have been few similar studies with nematodes. We know that avermectins are produced by *Streptomyces avermitilis* (Burg et al., 1979), and since this group of metabolites are potent nematicides (Sasser et al., 1982; Garabedian and Van Gundy, 1983; Faske and Starr, 2006; Cabrera et al., 2009), it is possible that other soil or rhizosphere microorganisms produce compounds of this nature. The key question is whether these nematicidal substances are produced in soil at concentrations sufficient to kill nematodes or influence their behaviour. Viruses are another group of potentially useful antagonists that deserve more attention, as they are important parasites of insects (Miller and Ball, 1998) and are relatively common in soil (Kimura et al., 2008). The fact that novel viruses have recently been found in Caenorhabditis and Heterodera (Félix et al., 2011; Bekal et al., 2011) suggests that if serious attempts were made to find viruses in plant-parasitic nematodes, they would be successful.

Establishment of biological control agents in soil and the rhizosphere

Any microorganism applied to seed or planting material or introduced into soil for biocontrol purposes must be able to colonize those sections of root where the target nematode is most likely to be found. Typically, these areas are near the root tip, a region of intense microbial activity. The organisms likely to be successful colonists in this competitive environment will have at least some of the following traits: (i) flagella that enable bacterial cells to move along roots, or fungal spores that germinate quickly and produce rapidly growing hyphae; (ii) a capacity to sense chemical signals emanating from roots; (iii) a capacity to attach to the root surface; (iv) the ability to utilize carbohydrates and other substrates present in root exudates and root tip mucilage; (v) a capacity to produce metabolites that are detrimental to competitors; and (vi) tolerance to inhibitory substances produced by rhizosphere inhabitants.

The last two traits listed above are testimony to the fact that an introduced organism always faces competition from the indigenous microflora and fauna. These competitive forces are a normal part of the soil environment and will affect the establishment and proliferation of any introduced organisms, including antagonists of nematodes. Observations by Monfort et al. (2006), for example, clearly show the extent to which fungistatic forces operate against nematophagous fungi. Two of these fungi (Pochonia chlamydosporia and Purpureocillium lilacinum) grew readily when inoculum was sandwiched between sterilized soils, but did not grow in a sandwich of non-sterilized

soil. Furthermore, growth was markedly inhibited when the inoculum was separated from non-sterile soil by a membrane, suggesting that diffusible compounds were at least partly responsible for the fungistatic effect. Further examples (given in Chapter 12), together with mycological literature dating back more than 50 years, prompted Jaffee (1999) to comment that adding nematophagous fungi to soil seemed foolish, as the introduced fungi were likely to be eliminated by resident antagonists.

Given the difficulties involved in establishing introduced biological control agents in soil or rhizosphere, it is not surprising that this approach to controlling soilborne pathogens has been plagued with inconsistency. In cases of failure, it is important to know whether the introduced organism is present but ineffective, or whether it has failed to establish, and this can only be determined by monitoring it in the field. The need to determine the fate of genetically engineered organisms in the soil environment has resulted in the development of many different tracking methods for bacteria (Jansson et al., 2000) and fungi (Bae and Knudsen, 2000; Thornton, 2008), and PCR-based methods have already been used to monitor nematophagous fungi (Atkins et al., 2003b, 2005, 2009). Tracking tools of this nature should be used routinely to determine the fate of organisms that are introduced into soil for biocontrol purposes.

Manipulating the soil biological community

Situations where populations of plant-parasitic nematodes are maintained at levels that do not cause economic damage are usually not due to the actions of a single antagonist. They are more likely to be the result of multitrophic interactions within the soil biological community. Thus, one way of achieving biological control is to manage this community of organisms so that its suppressive potential is maintained or enhanced. We are currently a long way from being able to do this, as our knowledge of how to manipulate the rhizosphere for the benefit of the plant is in its infancy. Some of the options available were listed previously, but perhaps the initial starting point should be to develop crop plants that are better able to take up nutrients at low soil concentrations. Such a breeding programme might generate plants with greater root:shoot ratios, or root systems and mycorrhizal associations that are better able to scavenge for nutrients and water. Since such plants would not require a highly concentrated supply of nitrogen, for example, nutrient inputs could perhaps be dispersed more widely through the soil profile, thereby reducing their negative impacts on some members of the soil biological community. Another step in the path towards a designer rhizosphere would be to understand the genetics and temporal dynamics of rhizodeposition, and how it can be manipulated to achieve a desired rhizosphere community.

The role of organic matter

Perhaps the most important message from this chapter is that organic matter plays a key role in enhancing soil and plant health, and in building soil biological communities that suppress soilborne pathogens, including plant-parasitic nematodes. Organic matter influences numerous physical and chemical properties that impact on plant growth, but its biological effects are even more important. From the perspective of biological control of nematodes, the bacteria and fungi that decompose organic matter are consumed by bacterivorous and fungivorous nematodes, and since they then become a food source for generalist predators, continuous organic inputs are the key to maintaining a diverse range of natural enemies (see Chapter 6). The quantity and quality of soil organic matter also determines whether facultative predators such as the nematode-trapping fungi live as saprophytes or actively prey on nematodes (see Chapter 5). Thus, the role of soil organic matter in enhancing soil fertility, together with its role in fuelling natural systems of biological control, is so all-pervading that it is mentioned in almost every chapter in this book.

3

The Soil Food Web and the Soil Nematode Community

In a book on biological control of nematodes, it is appropriate that the soil food web is one of the first topics considered, as all the organisms that parasitize or prey on nematodes, or in some way influence their behaviour, are components of that food web. However, biological control is not just about nematodes and their natural enemies. As pointed out in Chapter 2, the soil biological community contains huge numbers of organisms, and they can influence biological control systems in many different ways. Perhaps the most obvious examples are the organisms that attack biocontrol agents (e.g. fungivorous nematodes that feed on nematophagous fungi; zoosporic fungi that parasitize predatory nematodes; protozoa and other predators that consume *Pasteuria* spores; and arthropods that prey on predatory nematodes). However, there are other less obvious examples. Some soil organisms have plant growth-promoting properties, while others affect plant growth by influencing soil physical and chemical fertility; and these direct and indirect effects influence the plant's capacity to tolerate nematode damage. Thus, in situations where biological activity and diversity is low and soil health is poor, biological control may never be feasible, because the plant suffers damage at very low nematode population densities. In more enriched and better-balanced biological environments, plants are more tolerant of nematode parasitism, and since damage thresholds are higher, natural enemies are more likely to provide the level of nematode suppression required for optimum plant growth and productivity.

Clearly, the organisms that live in soil play a critical role in biological control. However, the soil biological community is complex and taxonomically diverse, and most people, including many scientists with an interest in soil biology, find it difficult to fully comprehend the functions of even a small subset of the organisms that are present. Nevertheless, anyone with an interest in soil organisms must at some point consider the broader picture of how they interact with each other. This chapter, therefore, begins by providing a brief overview of the soil biota and their interactions within the soil food web. It then considers the nematode community, and the manner in which various groups of nematodes interact with each other, and with the rest of the soil biological community. Useful texts covering these topics include Wardle (2002), Coleman and Crossley (2003), Bardgett (2005), Thies and Grossman (2006) and Neher (2010).

Major Groups of Organisms in Soil

The organisms that live in soil span several orders of magnitude in terms of body size. The smallest, most numerous and most diverse groups are the bacteria and fungi, which are known collectively as the microbiota but are often termed 'microbes' or 'microorganisms'. The larger animal species vary in size from the microfauna (e.g. nematodes, protozoa, tardigrades and rotifers, with a body width less than 0.1 mm) to the mesofauna (e.g. enchytraeids, mites and springtails, with body widths of 0.1–2 mm), and the macrofauna (e.g. earthworms, termites and millipedes, with body widths greater than 2 mm). Since a comprehensive coverage of all these organisms is beyond the scope of this book, the following brief discussion is limited to the major groups of soil organisms and their roles within the soil food web. Further information can be found in texts on soil science (Lal, 2005), soil microbiology (Sylvia *et al.*, 2005), soil biology (Burges and Raw, 1967; Dindal, 1990; Abbott and Murphy, 2007; Dion, 2010) and soil ecology (Coleman and Crossley, 2003; Bardgett, 2005).

Microbiota

Bacteria are the most abundant component of the soil microbiota, as populations frequently exceed 10^8 individuals per gram dry weight of soil, population densities that are several orders of magnitude greater than fungi or algae. However, numbers alone do not necessarily reflect the relative importance of members of the microbiota. Fungi sometimes contribute much more to total soil biomass than bacteria, and collectively, the contribution of fungi and bacteria far exceeds that of plants, algae, protozoa, nematodes and other soil animals (Sylvia *et al.*, 2005).

Bacteria and fungi perform a huge range of ecosystem functions in soil, but their principal role is to decompose and recycle organic matter that is produced primarily by plants. Numerous extracellular enzymes are produced that break down the main constituents of plant material (large polymers such as cellulose, hemicellulose and lignin) into smaller units that can be transported across the microbial cell membrane. Fungi and actinobacteria play an important role in degrading these large molecular weight materials, but

once those compounds are fragmented, the entire microbial community participates in the decomposition process. The end result is that carbon, nitrogen, sulfur, phosphorus and other elemental constituents of plant tissue are biologically transformed (mineralized) from organic to inorganic forms.

The microbial component of the soil food web also provides other important ecosystem functions. Algae and cyanobacteria account for a significant proportion of microbial biomass in agricultural soils and they contribute by fixing atmospheric nitrogen and carbon dioxide; synthesizing and liberating plant growth-promoting substances; and solubilizing phosphorus. Bacteria exhibit enormous metabolic diversity, but their roles in nitrification, denitrification and nitrogen fixation are very important in agriculture. Some soil bacteria are phototrophs or methanotrophs; many produce antibiotics that impact on their competitors; some are capable of degrading xenobiotic compounds that are toxic to other soil organisms; and bacterially produced polysaccharides contribute to the soil formation process. Fungi enhance structural stability by binding soil particles together; some fungal species are plant pathogens; and others (the mycorrhizae) form mutualistic associations with plant roots that aid nutrient uptake. Both bacteria and fungi are a food source for microbe-feeding fauna, and are, therefore, of great significance to organisms at higher trophic levels in the soil food web.

Microfauna

The two most abundant groups of soil microfauna are the protozoa and nematodes, and collectively they make up a small but significant fraction of the total faunal biomass in agricultural soils. Soil protozoa are microscopic, unicellular, eukaryotic organisms and are commonly subdivided into four groups: flagellates, naked amoebae, testate amoebae and ciliates. They reside in water-filled pores and water films and feed primarily on bacteria, although fungal feeders, predators and saprophytes are known. Protozoa multiply rapidly when populations of their bacterial prey increase (e.g. following a rainfall event, or when organic matter is added to soil) and their feeding activities affect the composition and activity of the microbial community. In such situations, they also have major effects on the soil environment, as a large proportion of the ingested nutrients are excreted in mineral form and are available for uptake by plants (Bardgett and Griffiths, 1997).

Nematodes are the most numerous multicellular animals in soil, and are discussed in detail later in this chapter. Like protozoa, they move and feed in water films. The species in soil are predominantly microbivorous, but plant-parasitic, omnivorous and predatory species also occur. Bacterial-feeding nematodes ingest their prey though a tube-like or funnel-shaped stoma, while most fungal feeders use a delicate stylet to pierce fungal hyphae and spores. The subsequent effects of these nematode feeding activities (changes in the microbial community and nutrient excretion) are similar to those of protozoa.

Mesofauna

The predominant mesofauna in soil are the enchytraeids (Oligochaeta), and a variety of collembolans, mites and small insects that are collectively known as microarthropods. Enchytraeids are anatomically similar to earthworms, and although much smaller, they appear to play a similar role in soil. They ingest organic matter together with microorganisms associated with decaying litter and roots, but their main food source appears to be fungal mycelium.

The Acari (mites) and Collembola (springtails) are by far the most numerous and widespread of the microarthropod fauna in soil, and collectively they are the third most abundant group of animals, behind the protozoa and nematodes. The several hundred families of mites represented in soil include predatory species as well as detritus feeders and fungivores, whereas the Collembola are largely fungal feeders. Predation on nematodes has been observed in both groups. Mites and Collembola are most commonly found in the upper soil layers and both are favoured by situations where there is a discrete litter layer.

Since soil porosity decreases with increasing depth and mites and Collembola can only move in pores that are larger than themselves, the smaller species tend to become relatively more abundant as depth increases.

Macrofauna

The macrofauna include millipedes, centipedes, spiders, termites, ants, scorpions and earthworms, and because they are visible to the naked eye, they are the most familiar members of the soil fauna. Coleman and Crossley (2003) briefly discuss their feeding habits, which are very diverse and may involve the consumption of organic debris and associated microorganisms; predation on large and small arthropods; plant feeding; and opportunistic omnivory. Of all the animals in soil, earthworms have perhaps the greatest impact on the soil environment, as they are involved in soil formation, the creation of macropores, litter decomposition and the redistribution of organic matter within the soil profile. Epigeic species live in leaf litter on the soil surface; endogeic species tend to make horizontal burrows in the upper soil layer; while anecic species build permanent burrows deep into the mineral layer of soil. Since earthworms are highly sensitive to tillage, their populations and biomass are almost always much higher in non-tilled soils than in conventionally cultivated soils (Wardle, 1995).

Structure of the Soil Food Web

The soil food web refers to the community of organisms (plants, microbes and animals) that live in the soil and are dependent on each other for sources of carbon and energy. In ecological terms, these organisms are either producers (plants, algae and some bacteria) or consumers (herbivores, decomposers and predators). The producers (also known as autotrophs) use their photosynthetic capacity to convert carbon dioxide into organic compounds that are the primary energy source for the consumers (heterotrophs). Most of the organisms in soil are heterotrophic, and they obtain their carbon and energy from other organisms by either consuming them (as parasites and predators), or by feeding on their waste products, secretions or remains.

In terrestrial ecosystems, plants are the dominant primary producers and consequently form the base of the food web. Herbivores and detritivores form a second trophic level, as they, respectively, consume roots and organic matter derived from plants. In turn, they are consumed by predators (a third trophic level), which are then consumed by higher-order predators, as depicted in Fig. 3.1. Thus, the biomass that can be supported at any particular trophic level depends on the biomass of organisms in the trophic level immediately below it. Since consumption is an energy inefficient process, most of the energy contained in primary producers is lost as heat and does not flow from one trophic level to the next. Thus, food webs have a finite number of trophic levels. Studies reviewed by Wardle (2002), for example, suggest that

inputs of organic matter sometimes enhance organisms at three trophic levels (e.g. microbes, microbial-feeding nematodes and predators of nematodes).

Another way of looking at the structure of the soil food web is to aggregate soil organisms into functional groups that are based on their role within the soil ecosystem, and represent them diagrammatically. Given the diversity that exists within the soil biological community, this approach results in extremely complex diagrams in which numerous boxes are used to depict organisms within the food web, and multiple interconnecting arrows show the interactions between them. Omnivory, recycling of dead higher-level predators into organic matter, and interconnection of above- and below-ground food webs lead to further complexity. Nevertheless, relatively simple diagrams are useful in demonstrating key points about the structure of the soil food web. One such example (Fig. 3.2) shows that the organisms which obtain their sustenance from living

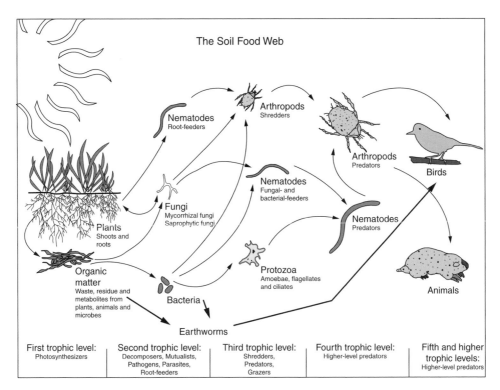

Fig. 3.1. A simplified version of the soil food web, emphasizing feeding relationships between soil organisms. (From Thies and Grossman, 2006, with permission.)

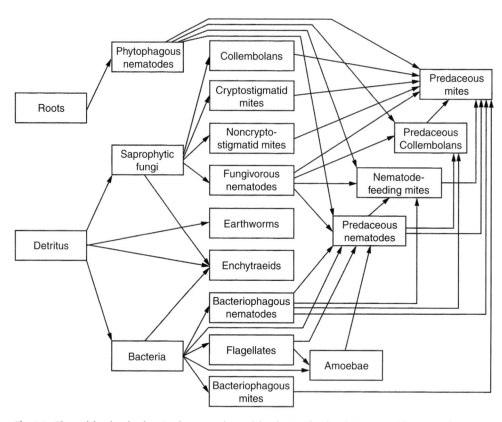

Fig. 3.2. The soil food web, showing how members of the detritus food web interact with nematode pests that are sustained by living roots. (From de Ruiter *et al.*, 1993, with permission.)

plant roots (e.g. phytophagous nematodes) are affected by the activities of organisms in the detritus food web. Another example (Fig 3.3) shows that the detritus food web can be divided into three major subsystems with different functional roles. The 'microfood-web' consists of the bacteria and fungi that decompose organic matter and the nematodes, protozoa and other animals that directly feed on them and each other. The 'litter transformers' (arthropods, earthworms and enchytraeids) either shred organic matter into fragments or consume plant debris and egest it as faecal pellets. Since shredding increases the surface area available to decomposers and faecal pellets are highly favourable for microbial growth, the activities of litter transformers increase decomposition rates. The 'ecosystem engineers' create physical structures that provide a modified habitat for the soil microflora and other soil fauna. Earthworms, for example, stimulate the

microflora by improving aeration and moisture conditions, secreting mucous and shredding litter.

Impact of Land Management on Energy Channels within the Soil Food Web

The disruptive processes involved in converting natural vegetation into farmland (deforestation, burning and cultivation) result in a substantial loss of soil organic matter. Ongoing tillage then enhances organic matter mineralization rates and disrupts fungal hyphae and large soil organisms, while seasonal production and the practice of growing a limited number of crops reduces the amount and complexity of the organic residues returned to soil. The end result is that the soils used for agriculture have fewer organisms, less biodiversity

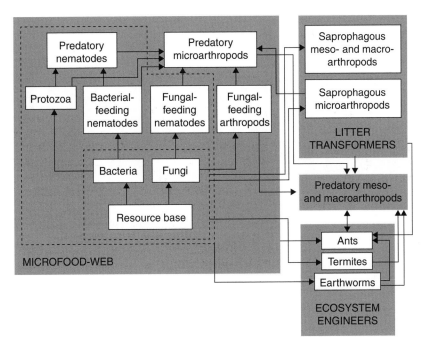

Fig. 3.3. The structure of the detritus or decomposer food web, showing only major groups of organisms and well-established linkages between them. (From Wardle, 2002, by permission of Princeton University Press.)

and shorter food chains than soils in natural ecosystems. The inputs and management practices involved in crop and pasture production (e.g. the use of fertilizers and pesticides, the type of tillage, the level of vehicle traffic, the diversity of crops grown and the intensity of grazing practices) have additional impacts that vary with different farming systems.

Early European studies (cited by Hendrix *et al.*, 1986) showed that cultivated agricultural soils had high decomposition rates, and their bacterial-based food webs were dominated by organisms with small body sizes, short generation times, high metabolic activity, omnivorous feeding habits and a capacity to disperse rapidly. Studies at sites in various regions of the USA confirmed those observations. Under the conventional tillage practices that were typically used in agriculture at the time, Hendrix *et al.* (1986) found that distribution of plant residue throughout the ploughed layer promoted bacterial activity and hence the abundance and activity of bacterivorous fauna. Thus, decomposition and nutrient mineralization proceeded rapidly in

tilled soil. In contrast, plant residue was localized on the soil surface in no-till systems, and this promoted fungal growth and increased the abundance and activity of fungivorous fauna in surface layers. Nutrients in the plant residue were immobilized and mineralization proceeded more slowly. This shift in food web structure due to tillage has been observed many times since that study (e.g. Moore, 1994; Wardle, 1995; Beare, 1997; Kladivko, 2001).

Studies in temperate grasslands (summarized by Bardgett, 2005) also show that management intensity affects the structure of soil food webs. Intensive pastures, characterized by high levels of fertilizer use, high grazing pressures and reduced soil organic matter content, invariably have bacterial-dominated food webs, with fungal biomass and fungal-to-bacterial biomass ratios being consistently and significantly greater in traditionally managed, unfertilized grasslands (Fig. 3.4). The role of soil management practices in altering food web structure is most apparent in intensive horticultural and vegetable production systems, where bacterial dominance sometimes

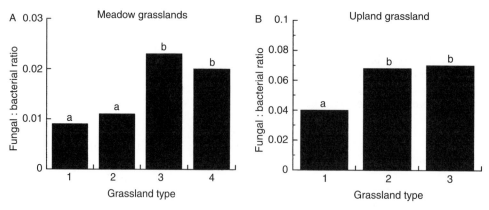

Fig. 3.4. Impact of management intensity on fungal-to-bacterial biomass ratios in meadow and upland grasslands in the United Kingdom. In both (A) and (B), management intensity decreases from left to right due to changes in fertilizer use, livestock density or the degree of pasture improvement practices. Bars with the same letter are not significantly different. (From Bardgett, 2005, p. 180, with permission, Oxford University Press. Original material from Bardgett and McAlister, 1999, with kind permission from Springer Science + Business Media B.V. and Grayston *et al.*, 2004, with permission.)

becomes so extreme that fungal energy channels are unimportant and fungivorous nematodes and larger soil fauna are rarely seen.

There are many reasons why agricultural soils have lost fungi, fungivorous arthropods and macrofauna, and why their soil food webs are often bacterial-dominant. Bardgett (2005) suggested that this change was most likely due to the combined effects of at least some of the following factors: (i) the dominance of fast-growing plant species that produce litter and rhizodeposits that encourage bacterial growth; (ii) reductions in the complexity and heterogeneity of organic matter inputs; (iii) reductions in habitat complexity within the soil environment; (iv) increases in soil physical disturbance due to traffic, grazing and tillage; (v) inhibitory effects of fertilizers on the growth of decomposer fungi; (vi) disruption of mycorrhizal mycelial networks by tillage; and (vii) suppression of mycorrhizal infection due to the addition of phosphate fertilizers. Nitrogen fertilizers, in particular, are almost certainly another contributing factor, as the addition of inorganic nitrogen influences microbial activity and rates of decomposition of organic matter (Fog, 1988; Hobbie, 2008; Treseder, 2008; Ramirez *et al.*, 2010). Also, some nitrogen sources are detrimental to some components of the soil microfauna (Tenuta and Ferris, 2004).

Given that the management practices used in agriculture change the composition of the soil biological community and result in the loss of some species, it is reasonable to ask whether these changes can be reversed. Currently available evidence suggests that this may be possible, but it will take many years. Bardgett (2005) cited studies showing that fungal-feeding mites had not returned 20 years after high-input sites were managed for nature conservation, and that the oribatid mite community increased very slowly following cessation of cultivation. Smith *et al.* (2003) found that it took nine years of traditional, low input management to detect recovery of the microbial community's fungal-to-bacterial biomass ratio at an intensively managed grassland site. In another study that compared agricultural sites with varying lengths of time under no-till management, species richness only approached that of undisturbed sites in fields that had not been tilled for 8–26 years (Adl *et al.*, 2006). A related question is whether soils can provide a full range of ecosystem services in the absence of some organisms. What is clear is that loss of symbionts such as mycorrhizal fungi and nitrogen-fixing bacteria will have major consequences for nutrient cycling, while reductions in key species and trophic groups (e.g. predators, shredders, fungivores, root feeders and bioturbators) is likely to have far-reaching and unpredictable consequences for ecosystem functioning (Wall *et al.*, 2010). Thus, the challenges involved in restoring a more balanced and fully functional biological community to agricultural soils should not be underestimated, as most soils used for agriculture have been under intensive management for at least 50 years.

Interactions within the Soil Food Web

Given the complexity of the soil food web, it is not surprising that the constituents interact in many different ways. These interactions can result in favourable, unfavourable or negligible effects on the organisms involved. Davet (2004) gave examples of the common types of relationships between soil organisms and the most important of them are described as follows.

Symbiosis is a close, almost obligate, relationship between two organisms that is beneficial to both. The two most commonly cited examples of symbiotic relationships in soil are those between plants and mycorrhizal fungi, and those between leguminous plants and bacteria in the genera *Rhizobium* and *Bradyrhizobium*.

Mutualism is similar to symbiosis in that two partners obtain mutual benefits from their relationship, but differs in that the partners also have the capacity to live independently. An example of a mutualistic association occurs in the gut of earthworms. Free-living microorganisms thrive in that environment due to favourable moisture and pH conditions and the presence of a readily available substrate (intestinal mucous) comprised of low-molecular weight carbon sources. In return, the microorganisms produce cellulases that break down organic matter entering the gut, releasing nutrients to the earthworm (Wardle, 2002).

Commensalism also involves a positive interaction between two organisms. However, in a commensal relationship, one organism benefits but the other is neither helped nor harmed. There are many examples of such relationships in soil, and they often involve one microorganism utilizing the secondary metabolic products of another microorganism.

The interactions most relevant to biological control are antagonism and competition. The word *antagonism* is sometimes used to

describe the situation where one organism inhibits another through antibiotic production. However, it is used in a more general sense in this book to cover all situations where one organism is detrimentally affected by another. Such a definition is commonly used in the literature on biological pest control to describe the general suppressive effects of a biological control agent, regardless of its mode of action. Antagonistic interactions are manifested in the following ways.

Antibiosis involves the inhibition of one organism by the metabolic product of another. It is usually associated with interactions where the adversary is killed, inhibited or repelled, but is not consumed. The metabolic products (usually soluble or volatile antibiotics) are produced in such small quantities by bacteria or fungi that it is difficult to prove conclusively that they are present in the natural environment. Nevertheless, they are known to play a role in interactions between various plant pathogens and the soil biota, with one well-studied example being inhibition of the take-all pathogen *Gaeumannomyces graminis* var. *tritici* by two antibiotics (2,4-diacetylphloroglucinol and phenazine-1-carboxylic acid) produced by fluorescent pseudomonads on wheat roots (Weller *et al.*, 2002).

Lysis is similar to antibiosis in that its effects are manifested at a distance from the organism responsible for lytic activity, but differs in that the adversary is exploited. It occurs when an organism produces extracellular enzymes (e.g. chitinases, cellulases and glucanases) that digest the cell wall or cuticle of another organism. Sometimes the process is accompanied by the production of toxins that immobilize or kill the prey. Bacteria, and more particularly actinobacteria, are significant producers of lytic enzymes and toxins, and important agents in the lysis of fungi.

Predation is generally characterized by the consumption or assimilation of one organism (the prey) by another, often larger, organism (the predator). It requires intimate contact between the two organisms and usually involves an active search for the prey by the predator. Protozoans, nematodes and microarthropods all have the capacity to consume other soil organisms, some feeding indiscriminately on a wide range of organisms and others having quite specific food preferences. With respect to nematodes, predators of bacteria and fungi are considered 'bacterivores' and 'fungivores', as this clearly specifies their food source and trophic position. Predators of organisms further along the food chain are referred to as 'top predators'.

Parasitism occurs when an organism (the parasite) lives in or on another organism (the host) and obtains all or part of its nutritional resources from that host. Bacteria and viruses are known to parasitize some soil organisms (e.g. protozoans and nematodes), but fungi are probably the most important parasitic organisms in soil. Numerous fungal parasites of arthropods and nematodes are known, and mycoparasitism (parasitism of one fungus by another) is also commonly observed.

Competition involves situations where the volume of habitable space, or the amount an essential substrate or nutrient, is insufficient to satisfy the needs of organisms requiring that resource. The organism most adept at accessing the limiting factor, making it inaccessible to others, or eliminating those trying to obtain it, will prosper relative to its competitors. Competition is a universal phenomenon within the soil food web, but becomes particularly intense when organisms in the same ecological niche are attempting to access the same scarce resource.

Regulation of Populations by Resource Supply and Predation

The mechanisms outlined depict the types of interactions that occur between organisms in the soil food web. However, outcomes from those interactions are not easy to predict. Environmental factors have marked effects on relationships between organisms, while the interactions between two organisms will be modified by the introduction of a third organism. Thus, the structure of the biological community and the population size of individual components is the result of environmental effects and multiple interactions that are almost impossible to fully comprehend.

Although the regulatory mechanisms that operate within the soil food web are complex, one simple way of looking at them is to consider whether component populations are regulated by 'bottom-up' or 'top-down' forces. 'Bottom-up' control occurs when a population is limited by resource inputs (i.e. food supply), whereas 'top-down' control involves regulation of populations via predation. Wardle (2002) reviewed the literature on the regulatory forces operating within the soil food web and concluded that as a whole, microbial biomass was 'bottom-up' controlled. His conclusion was based on the following evidence: (i) microbial biomass increases rapidly when readily available carbon (e.g. simple sugars) are added to soil; (ii) labile carbon compounds that originate in the rhizosphere of actively growing plants enhance microbial biomass; (iii) the quality of resource inputs (e.g. C:N ratio, or the presence of plant-derived secondary metabolites) influence microbial biomass; and (iv) fungi, which are often the main component of microbial biomass, strongly compete to gain access to organic substrates in soil.

When the effects of predation were considered in the same review, it was apparent that 'top-down' forces also operate on the microbial component of the soil food web. These forces are particularly important for bacteria, as they have few defences against predation and are readily ingested by protozoa and bacterial-feeding nematodes. Thus, several studies have shown that bacterial populations are reduced, or bacterial turnover rates increase, when populations of predators increase. In contrast, fungi appear to be less affected by predation, possibly because they have many morphological and chemical defence mechanisms, and also because their higher C:N ratio means that they are a poorer quality food source than bacteria.

With regard to the impact of regulatory forces on the soil fauna, evidence for 'bottom-up' control is provided by studies which show that faunal populations increase when basal resources are added to field soil or microcosms, either as amendments or via rhizodeposits from roots. In some cases, the effect of resource inputs is apparent at the secondary consumer level (e.g. microbe-feeding nematodes), while in other cases it extends to higher trophic levels (e.g. predatory nematodes, nematode-trapping fungi, springtails and spiders). The response of microbe-feeding nematodes to the addition or selective removal of predators provides evidence that the soil fauna is also regulated by 'top-down' processes.

Perhaps the main message to be taken from this discussion is that 'bottom-up' and 'top-down' regulatory processes determine the size and composition of the soil biological community. The interactions involved in these processes become more complex as diversity within the soil food web increases, and this means that multiple forces exert pressures on all components of the community, preventing the uncontrolled proliferation of particular populations. These regulatory forces have the effect of stabilizing the community that makes up the food web, a phenomenon known as homeostasis.

Impacts of the Soil Food Web on Ecosystem Processes: Storage and Cycling of Nutrients

Organisms that occupy the soil food web perform numerous functions within the soil ecosystem (Box 3.1), and most are relevant to the subject matter contained in this book. Interactions between soil organisms and organic matter and their role in influencing soil physical and chemical fertility were discussed in the previous chapter, as was the role of rhizosphere inhabitants in plant growth promotion. Regulatory mechanisms that operate within the soil food web are the essence of biological control and are, therefore, mentioned in several later chapters. Issues associated with the cycling of nutrients are also important, and since they are not covered elsewhere, they are briefly examined here.

Organic matter plays a central role in biological nutrient cycling, as it is a major reservoir of nutrients and the primary source of carbon, nitrogen, phosphorus, sulfur and other elements in agroecosystems. Some of those nutrients are held in roots, surface litter and microbial biomass, but most are contained within soil organic matter. The nitrogen content of soil organic matter, for example, is usually around 5%, and since most agricultural soils contain 10–40 g organic matter/kg of soil, this means that significant quantities (up to several tonnes per hectare) of this essential plant nutrient are retained within soil organic matter. The components of soil organic matter vary in physical size, chemical composition, degree of association with soil minerals and extent of decomposition, but are often considered to belong to either an active or a passive pool. The active pool is composed of recently produced plant residues, root exudates and microbial biomass, while the very stable humus fraction makes up the passive pool. One of the most important ecosystem services provided by the soil biological community is to break down these organic materials, and in the process mineralize the nutrients they contain, thereby making them available for use by plants and other soil organisms.

The nitrogen mineralization process provides a good indication of how nutrients are cycled by soil microorganisms. Amino acids and other relatively simple nitrogen-containing compounds found in root exudates are absorbed directly by microbial cells. More complex polymers that comprise the bulk of soil organic matter (e.g. proteins, nucleic acids and chitin) are too large to pass through microbial membranes, and so bacteria and fungi produce extracellular enzymes that break these compounds into smaller, water soluble subunits, which can then be utilized for growth. The fate of the nitrogen contained within organic matter depends on its C:N ratio (Tate, 2000). Heterotrophic bacteria have C:N ratios ranging from 4:1 to 6:1, while the C:N ratios of fungi generally vary between 10:1 and 12:1, and so there is a nitrogen deficit when soil microbes decompose organic materials with high C:N ratios (e.g. 60:1). In such situations, the microbes balance the carbon they have utilized by incorporating some soil nitrogen into their biomass. This process reduces nitrogen availability to plants and is known as immobilization. As the C:N ratio of the organic substrate declines, a point is reached where bacteria and fungi are limited by carbon rather than nitrogen. In this situation, they excrete plant-available nitrogen (NH_4^+) as a waste product, a process known as mineralization. Since microorganisms respire about 50% of their carbon uptake and cannot readily utilize some organic matter fractions (e.g. lignin), mineralization begins to occur at C:N ratios of about 30:1. It is, therefore, a normal process in most agricultural soils, as the C:N ratio is generally less than 20:1.

Box 3.1. The role of soil organisms in ecosystem processes

Soil microbes and the soil fauna play a crucial role in many ecosystem processes and are indispensable to plant productivity. Some of their more important functions are listed below.

Decomposition of organic matter

Soil organisms macerate and decompose plant and animal residues. Since this process has flow-on effects that impinge on most aspects of soil fertility and plant productivity, it is probably the most important function of the soil biological community.

Soil formation, and stabilization of aggregates

Soil microorganisms influence soil structure by producing polysaccharide glues that are important in the initial stages of aggregate formation. Networks of fungal hyphae bind mineral and organic particles together and contribute to the formation of macro-aggregates. The larger soil fauna act as ecosystem engineers, impacting on soil physical properties by moving or ingesting soil, or creating mounds, casts and burrows.

Production, uptake and transfer of nutrients

Many soil organisms impact directly on plant nutrition. For example, *Nitrosomonas* and *Nitrobacter* oxidize ammonium to nitrate; *Rhizobium* increases nitrogen supply in a symbiotic relationship with legume roots; free-living bacteria such as *Azotobacter* and *Clostridium* contribute to the nitrogen requirements of plants by fixing small amounts of nitrogen; arbuscular mycorrhizal fungi promote the uptake of phosphorus and other nutrients; siderophore-producing bacteria chelate iron, thereby increasing its availability; and both fungi and bacteria play a role in phosphorus nutrition by solubilizing mineral phosphates.

Storage and cycling of nutrients

When organic matter is decomposed by soil organisms, complex organic molecules are converted into simpler compounds and eventually into mineralized nutrients. During this process, some of the inorganic nutrients are immobilized (i.e. incorporated into organic molecules within living cells). Since mineralization and immobilization processes occur simultaneously, the supply of nutrients to plants and nutrient equilibrium levels in soil are governed by the soil biota.

Herbivory

Many soil organisms obtain sustenance from living roots. Most cause little damage, but some are important pests and pathogens.

Plant growth promotion and suppression of pests and pathogens

Rhizosphere inhabitants produce plant growth-promoting substances and endophytic micro-organisms induce resistance to pathogens. Interactions within the soil food web (e.g. competition, parasitism, predation and antibiosis) stabilize populations of all soil organisms, including those that attack plants.

Degradation of toxic compounds

Soil organisms help maintain a healthy environment by producing enzymes that break down pesticides, pollutants and other contaminants.

Although bacteria and fungi commence the process of mineralizing nitrogen and other essential plant nutrients, other components of the soil food web also contribute. Bacteria are consumed by nematodes and protozoa; fungal hyphae are pierced by stylet-bearing nematodes; plant-feeding and free-living nematodes are parasitized by fungi or eaten by predators, and these secondary consumers are eventually utilized by organisms at higher levels in the soil food web. During this process, nutrients are defecated and excreted, enhancing the rate at which nutrients are cycled.

One point that was made frequently by Wardle (2002) with regard to nutrient cycling is that the producer and decomposer subsystems of the soil food web are mutually dependent on each other. Plants (the producers) are the energy source that drives the system, but heterotrophic organisms that feed on living roots, utilize root exudates or decompose detritus, are ultimately responsible for governing the supply of nutrients required for plant productivity. Thus, each of the two components carries out processes that are required for the long-term maintenance of the other.

The Soil Nematode Community

Nematodes are the most numerous multicellular organisms on earth, and are exceeded in species diversity only by the arthropods (Bernard, 1992). About 14,000 free-living, invertebrate- and plant-associated nematode species are known, while a further group of about 12,000 species is parasitic on vertebrates. However, the soil environment is a major source of nematode diversity, with 42% of the described species being terrestrial (Hodda et al., 2009). This section provides a brief overview of the latter group of nematodes.

Morphological criteria are available to identify the nematodes that live in soil (e.g. Goodey, 1963; Bongers, 1988; Jairajpuri and Ahmad, 1992; Hunt, 1993; Siddiqi, 2000; Ahmad and Jairajpuri, 2010), but because taxonomic skills and high-resolution microscopy are required to assign a nematode to a particular genus or species, nematodes are generally not identified to this level in ecological studies. They are usually allocated to broader categories, but even at this level of categorization, the identification process is protracted and it is usually not possible to identify more than a small proportion of the nematodes within a community to species level. Advances in the molecular taxonomy of nematodes (van Megen et al., 2009)

are changing that situation, with DNA barcoding and other molecular technologies providing new tools to identify and quantify soil nematodes. Since greater numbers of nematodes can be assessed, taxonomic resolution is improved and results are obtained relatively quickly, such methods will soon be used routinely in soil ecological studies, not only to assess nematode assemblages but also other components of the soil food web (Neilson et al., 2009; Porazinska et al., 2009, 2010; Donn et al., 2011, 2012).

Although molecular technologies will enable us to fine-tune our understanding of the ecological role of various nematodes, our current understanding of their ecology is based on knowledge of their feeding habits and life histories. Thus, in the remainder of this chapter, nematodes are grouped according to their perceived functions within the soil ecosystem, and the role of each group is summarized. Since plant-parasitic nematodes are the subject of this book, the chapter concludes by outlining the various groups of nematodes that feed on plants and considers the factors that make some of them important pests.

Trophic groups within the soil nematode community

The various trophic groups that occur within the nematode community are defined by their feeding habits, which are usually inferred from the structure of the head region (Fig. 3.5). Plant parasites are armed with a hypodermic-like spear that is used to feed on root cells, while fungal-feeding nematodes penetrate fungal cell walls with a similar, but finer, spear. Some bacterial-feeding nematodes use their stoma and muscular oesophageal structures to ingest suspensions of bacteria and soil water, while others have specially developed lips that scrape bacteria from the surface of soil particles and organic matter. The mouth of most predatory nematodes is armed with a tooth that is used to capture and consume nematodes and other small animals, while omnivorous and some predatory nematodes have a spear with a wide aperture that is used to suck out the body contents of their prey.

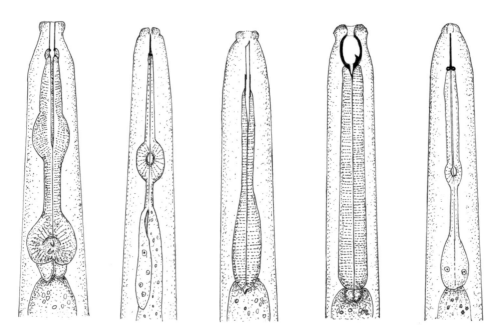

Fig. 3.5. Categorization of soil nematodes into different feeding groups on the basis of the structure of their mouth parts and oesophageal region. From left to right: bacterial-feeding rhabditid; fungal-feeding aphelenchid; omnivorous dorylaimid; predatory mononchid; and plant-feeding tylenchid.

The multiplicity of food sources and feeding habits within the nematode community (Box 3.2) gives some indication of the diverse roles of nematodes within the soil food web. Some feed directly on primary producers (higher plants and algae), but most use the fungi and bacteria associated with decomposing organic matter as a food source. Smaller numbers occupy higher trophic levels and they are predators of nematodes and other invertebrates, parasites of animals, or omnivores.

Although mouth parts provide an indication of a nematode's most likely food sources and its role within the soil food web, it is important to recognize that a variety of potential food sources are available in soil and we know little about what most nematodes actually eat in their natural habitat. The following examples illustrate that point:

- Some fine-tailed tylenchids (e.g. *Filenchus*) were previously classified as plant associates on the basis that they fed on epidermal cells and root hairs (Yeates *et al.*, 1993a). However, they have since been successfully cultured on fungi (Okada and Kadota, 2003), raising questions about the relative importance of roots and fungi in sustaining this relatively common group of soil nematodes.

- Some predatory mononchids can be raised in the laboratory on bacterial lawns (Yeates, 1987a; Allen-Morley and Coleman, 1989), but we do not know whether bacteria are an important nutrient source for these nematodes in soil. Predatory nematodes show preferences for certain prey in simple microcosms (Small and Grootaert, 1983), but what are the preferred or main food sources of individual species in the soil environment?

- Omnivorous nematodes are thought to feed on a wide range of foods, including fungi, algal cells, protozoa, nematodes and other invertebrates, but what does that mean at a genus or species level? Do certain nematodes really have a diverse range of food sources in nature, or is omnivory a convenient way of grouping nematodes with unknown feeding habits?

Box 3.2. Principal feeding habits of plant and soil nematodes, based on food resources utilized

The following feeding habits are found within the soil nematode community. Although examples of genera with particular feeding habits are given, it must be recognized that more than one feeding type may occur within a genus; that the food source of some nematodes may vary with circumstances; and that our knowledge of the feeding habits of some nematodes is inadequate.

Plant-feeding

- A stylet is used to feed on plant cells. Common feeding sites include the meristematic zone of root tips and cortical tissues of roots.
- Plant growth is reduced to varying degrees, depending on the nematode species, its feeding site and population density, and the level of stress to which the host plant is subjected.
- Examples: *Meloidogyne, Heterodera, Pratylenchus, Helicotylenchus, Paratrichodorus*

Plant-associated

- A stylet is used to feed on root hairs and epidermal cells, or on fungal hyphae and spores. However, the feeding habits of this group of nematodes are poorly documented.
- High population densities may occur in soil or the rhizosphere, but plant growth is not reduced.
- Examples: some species of *Tylenchus* sensu lato, *Dorylaimellus*

Hyphal-feeding

- Use a stylet to feed on fungal hyphae.
- Examples: *Aphelenchus, Aphelenchoides, Leptonchus, Diphtherophora*

Bacterial-feeding

- Feed on bacteria using a tubular or funnel-shaped stoma that is generally unarmed.
- Examples: *Rhabditis, Cephalobus, Alaimus*

Substrate ingestion

- During the process of feeding on bacteria or microfauna associated with detritus, organic substrates may be ingested through a broad stoma.
- Example: some Diplogasterida

Predation

- Animals are ingested through a broad stoma with teeth, or prey are killed by sucking their body contents through the narrow lumen of a stylet
- Examples: *Mononchus, Mononchoides, Nygolaimus, Seinura, Labronema*

Feeding on unicellular eukaryotes

- Some nematodes are thought to feed on diatoms and other algae, or to ingest fungal spores and yeast cells. However, this is difficult to confirm because intestinal contents of nematodes are hard to identify and many food sources lack marker structures.
- Examples: Chromadoridae

Animal parasitism

- Some dispersive or infective stages of animal parasites occur in soil and feed on fungi or bacteria.
- Examples: *Deladenus, Heterorhabditis*

Omnivory

- A variety of food sources is consumed, including cyanobacteria, fungi, microfauna, diatoms and algae.
- Examples: Some Dorylaimida, e.g. *Eudorylaimus*

Compiled from Yeates *et al.*,1993a, Yeates, 1998, 2010

There are no simple answers to the questions listed on page 60, as the opaque nature of a nematode's intestine, together with a lack of marker structures in various food sources, makes it difficult to search gut contents for clues. A range of molecular and other techniques have been used recently to identify the organisms ingested by different nematodes (e.g. Neilson *et al.*, 2006; Ladygina *et al.*, 2009; Treonis *et al.*, 2010b; Majdi *et al.*, 2012; Cabos *et al.*, 2013), and as such methods are further developed, they will provide a much clearer picture of the food preferences of nematodes in their natural environment.

A functional guild classification for soil nematodes

Although the trophic groups discussed in the previous section are useful, groupings that also consider life history provide a better indication of a nematode's role within the soil ecosystem. Such a classification system has been developed for nematodes and is based on the r-K scale used by ecologists to indicate the reproductive strategy favoured by an organism. R-strategist species, which are characterized by rapid development, early reproduction, small body size and short life spans, have been termed 'colonizers' and K-strategists, which exhibit the opposite characteristics, have been termed 'persisters'. When nematodes are considered in this manner, it is possible to categorize them according to where they fit on a colonizer-persister (c-p) scale of 1–5 (Box 3.3). By integrating feeding groups with life history characteristics, nematode families are then allocated to various functional guilds (Table 3.1).

The advantage of a functional guild classification is that nematodes become useful indicators of the biological processes that are occurring in soil (Ferris *et al.*, 2001). Thus, a predominance of bacterial scavengers in the Cephalobidae (Ba$_2$) and fungal feeders in the Aphelenchidae, Aphelenchoididae and Anguinidae (Fu$_2$) is indicative of food web under stress due to resource limitations, adverse environmental conditions or recent contamination. The presence of large numbers of nematodes in the Ba$_1$ guild (the Rhabditidae, Panagrolaimidae and Diplogasteridae) is indicative of enrichment. In such situations, new resources have become available to the soil food web due to disturbance, the addition of organic matter or a favourable shift in the environment, and these nematodes are taking advantage of the subsequent flush of microbial activity, and multiplying rapidly. An increase in the proportion of nematodes in c-p class 3 (the Ba$_3$ Prismatolaimidae, the Fu$_3$ Diptherophoridae and the Ca$_3$ Tripylidae) indicates that the food web is recovering from stress and is developing some rudimentary structure. The presence of nematodes in higher c-p classes (the Ca$_4$ Mononchidae, Om$_4$ Dorylaimidae, Fu$_4$ Leptonchidae, Ca$_5$ Discolaimidae and the Om$_5$ Thornematidae and Qudsianematidae) is indicative of even greater community structure, with more links in the soil food web, greater levels of predation and more multitrophic interactions.

Since nematodes are ubiquitous and respond to environmental change, they have been widely used as soil health indicators (Neher, 2001). The functional guild approach to classifying soil nematodes has proved particularly useful, and has formed the basis of many different studies in agricultural soils in recent years. Examples include: the impact of irrigation and cover cropping on nitrogen mineralization (Ferris *et al.*, 2004; DuPont *et al.*, 2009); succession patterns in the soil biological community during the decomposition of organic amendments (Wang *et al.*, 2004a); the effect of climate and plant species on the biological status of soils under pasture (Stirling and Lodge, 2005); the effect of changes in tillage practices and cropping patterns on soil food webs in a conventionally managed agricultural soil (Sánchez-Moreno *et al.*, 2006); and the impact of heavy metal pollutants on soil nematode assemblages (Nagy, 2009).

Ecological roles of free-living nematodes

It should be apparent from the preceding material that the soil nematode community is

---

Box 3.3. Assignment of free-living soil nematodes to a colonizer-persister (c-p) series on the basis of their r and K characteristics

---

Soil nematodes are categorized into a c-p series ranging from extreme r- to extreme K-strategists. 'Colonizer' nematodes are at the lower end of the scale and 'persister' nematodes at the high end. Five groups of nematodes with a range of characteristics can be distinguished:

c-p1

- Primarily bacterial-feeding nematodes with high metabolic activity, short generation times and a large proportion of the body occupied by gonads.
- Often termed 'enrichment opportunists' because large numbers of small eggs are produced and population growth is explosive under food-enriched conditions.
- Tolerant of pollutants and products of organic matter decomposition.
- Cease feeding and form dauer-larvae when microbial biomass and activity decreases.

c-p2

- Bacterial feeders, fungal feeders and a few predatory nematodes that respond more slowly than c-p1 nematodes to environmental enrichment, but still have short generation times and relatively high reproductive rates.
- Occur in all environments, including those with abundant and relatively scarce resources.
- Do not form dauer-larvae.
- Tolerant of pollutants and other disturbances.

c-p3

- Bacterial feeders, fungal feeders and some predatory nematodes with longer generation times than c-p2 nematodes, and greater sensitivity to disturbance.

c-p4

- Small dorylaimids and large non-dorylaimids with long generation times and a low ratio of gonad to body volume.
- Includes large carnivores, smaller omnivores and some bacterial feeders.
- Characterized by a permeable cuticle and high sensitivity to pollutants.
- The non-carnivorous nematodes are relatively sessile, whereas the carnivores actively seek prey.

c-p5

- Large dorylaimids with a long lifespan, low reproduction rates, low metabolic activity and slow movement. Includes the larger omnivores and predators.
- The gonads are small relative to body volume and produce a small number of large eggs.
- Very sensitive to pollutants and other disturbances.

Derived from Bongers and Bongers (1998), Ferris et al. (2001) and Ferris and Bongers (2009)

---

extremely diverse. Since components of that community interact with each other and with other soil organisms, it is hardly surprising that nematodes have a wide range of direct and indirect effects on ecosystem processes. The impact of plant-parasitic nematodes is discussed later in this chapter, while the following summarizes the main ecological effects of free-living species.

Microbial feeding

One of the most important ecosystem functions of bacterial-feeding nematodes is to maintain the metabolic activity of their prey. As nematodes graze on bacteria, they increase bacterial biomass and activity by keeping the prey population young and reproductively active (Fu *et al.*, 2005). When this process occurs in the rhizosphere, additional organic

Table 3.1. Functional guild classification matrix for soil nematodes, with example taxa[a] for each guild.

Feeding habit	Life history characteristics (colonizer-persister classification)				
	1	2	3	4	5
Plant-feeding	Pf_1	Pf_2 Tylenchidae	Pf_3 Pratylenchidae	Pf_4 Trichodoridae	Pf_5 Longidoridae
Hyphal-feeding	Fu_1	Fu_2 Aphelenchidae	Fu_3 Diphtherophoridae	Fu_4 Tylencholaimidae	Fu_5
Bacterial feeding	Ba_1 Panangrolaimidae	Ba_2 Cephalobidae	Ba_3 Teratocephalidae	Ba_4 Alaimidae	Ba_5
Animal predation	Ca_1 some Diplogasteridae	Ca_2 Seinuridae	Ca_3 some Tripylidae	Ca_4 Mononchidae	Ca_5 Discolaimidae
Unicellular feeding	Un_1 Monhysteridae	Un_2 Microlaimidae	Un_3 Achromadoridae	Un_4	Un_5
Omnivorous	Om_1 some Diplogasteridae	Om_2	Om_3	Om_4 Qudsianematidae	Om_5 Thornematidae

[a]In situations where example taxa are not listed, no nematode is known to occupy that guild. (From Ferris and Bongers, 2006, with permission.)

carbon is exuded from roots (Sundin *et al.*, 1990) and the subsequent effects on microbial turnover, carbon transfer and nutrient recycling ultimately affect root growth and plant performance (Bonkowski *et al.*, 2009). Bacterial-feeding nematodes also influence plant growth in other ways, as there is evidence to suggest that grazing-induced changes to the microbial community stimulate hormone production, and that this has flow-on effects to root architecture (Mao *et al.*, 2006, 2007).

Since fungal-feeding nematodes are ubiquitous in soil, nematode-grazing on fungal mycelia and spores almost certainly plays a role in many processes within the soil ecosystem. Issues related to nutrient cycling are covered later, but the effects of fungal-feeding nematodes on plant-pathogenic and beneficial fungi have also received some attention. There is a large body of literature on the potential of using *Aphelenchus* (usually *A. avenae*) and *Aphelenchoides* for biological control of fungal root pathogens (e.g. *Rhizoctonia solani* and *Fusarium solani*) and oomycetes (e.g. *Pythium* spp.) and these studies consistently show that fungal-feeding nematodes have the potential to suppress the diseases caused by these organisms (e.g. Rhoades and Linford, 1959; Barnes *et al.*, 1981; Lagerlöf, 2011). Since feeding by *Aphelenchoides* species also constrains the efficacy of the biocontrol fungus *Trichoderma harzianum* (Bae and Knudsen, 2001), complex interactions will obviously occur when nematodes feed on both the introduced fungus and the target pathogen (Jun and Kim, 2004). Fungal-feeding nematodes also utilize mycorrhizal fungi as a food source, and in some circumstances, their impact is large enough to reduce the plant growth response that is normally obtained with these fungi (Hussey and Roncadori, 1981).

Microbial transport

Since microorganisms are relatively immobile in soil, the soil fauna plays a crucial role in vectoring microbial symbionts in the soil and rhizosphere. Bacterial-feeding nematodes seem to be particularly important in transporting bacteria, as bacterial cells attach to cuticular mucilage and are carried externally (Bonkowski *et al.*, 2009), or are deposited in a viable form when the nematode defecates (Chantanao and Jensen, 1969). Thus, bacteria are dispersed to new areas and to new substrates as their principal predators migrate through soil. In some cases, this transportation process has important practical outcomes. One such example involves the initiation of a nitrogen-fixing symbiosis in the legume *Medicago truncatula*. Volatiles released by the plant attract bacterial-feeding nematodes, and in the process, roots are inoculated with rhizobia (Horiuchi *et al.*, 2005). Other studies with bacteria applied to seed for biocontrol purposes have shown that rhizosphere colonization is substantially increased by the presence of bacterial-feeding nematodes (Knox *et al.*, 2003, 2004).

Nutrient cycling

As mentioned earlier, nematodes are intimately involved in the nutrient mineralization processes performed by the soil food web. A landmark study in microcosms (Ingham *et al.*, 1985) demonstrated that plants initially took up more nitrogen and grew faster when grown in soil with bacteria and bacterial-feeding nematodes than in soil with bacteria alone. These responses were due to increased nitrogen mineralization by bacteria, excretion of NH_4^+ by nematodes and rhizosphere effects that promoted early root growth. Later work by Bouwman *et al.* (1994) confirmed the role of bacterial-feeding nematodes in mineralizing nitrogen, and led to more detailed studies on the mineralization process (Ferris *et al.*, 1997, 1998). Measurements on eight nematodes and six bacteria showed that the mean C:N ratios of bacterial-feeding nematodes and their prey were 5.89 and 4.12, respectively, supporting the contention that the feeding activities of these nematodes results in a surfeit of nitrogen. Based on observed nematode population levels in the field and previously determined effects of temperature on rates of metabolism, development and fecundity, it was estimated that the bacterial-feeding nematode community in the rhizosphere mineralized up to 1.01 µg N/g soil/day. Nitrogen was primarily mineralized in the NH_4^+ form, with mineralization rates varying from 0.0012 to 0.0058 µg N/nematode/day, depending on the nematode species. From the practical standpoint of plant nutrition and crop production, the importance of this node of the soil food web was greatest when C:N ratios of organic substrates fuelling the web were in an intermediate range (i.e. 22:1–32:1).

Ingham *et al.* (1985) noted that fungal-feeding nematodes had a more limited impact on plant growth and nitrogen uptake, as they excreted less NH_4^+ than bacterial feeders. One reason for this is that the C:N ratios of fungal-feeding nematodes and fungi are similar, ranging from 8:1 to 11:1 for nematodes and 8:1 to 10:1 for fungi (Chen and Ferris, 1999). Thus, only small amounts of excess nitrogen are generated during the feeding process. However, the extent of the nematode feeding effect is variable, presumably because individual fungi support different populations of fungal-feeding nematodes; environmental factors such as temperature influence nematode reproduction; fungi differ significantly in their capacity to mineralize nitrogen in the absence of nematodes; and fungus/nematode combinations vary in their N-mineralization potential (Chen and Ferris, 1999, 2000). Nevertheless, the contribution of fungal-feeding nematodes to nitrogen mineralization is generally small, and is probably most pronounced under conditions that optimize the growth rates of the fungi and nematodes present in soil (Okada and Ferris, 2001).

Regulation of populations

Since nematodes occupy several trophic levels within the soil food web, they are involved in predatory interactions that influence the population densities of many other soil organisms. The impact of microbial-feeding nematodes on their food sources has already been discussed, and the role of predatory nematodes in regulating populations of plant-parasitic nematodes is considered in Chapter 6. Predatory nematodes also influence populations of free-living nematodes, and that issue is covered briefly here.

Regulation of soil fauna by other soil fauna appears to be a common occurrence (Wardle, 2002) and has been shown to occur within the free-living nematode community. Mikola and Setälä (1998) established a microcosm experiment in which 20 species of bacteria and fungi formed the first trophic level, bacterial- and fungal-feeding nematodes comprised the second trophic level, and a predatory nematode (*Prionchulus punctatus*) the third level. Nematode populations were monitored over a 21-week period, and the results showed that the predator consistently and significantly reduced the biomass of the bacterivores and fungivores (Fig. 3.6). Towards the end of the experiment, the biomass of the predator exceeded the biomass of the prey, but neither of the microbivores was completely eliminated. In an even simpler microcosm, Allen-Morley and Coleman (1989) found that another mononchid (in this case *Mononchus tunbridgensis*) reduced populations of bacterial- and fungal-feeding nematodes. Although such results suggest that predatory nematodes regulate populations of other members of the nematode community, it has been difficult to demonstrate that this occurs in more complex systems such as soil.

Yeates and Wardle (1996) reviewed the literature on regulatory processes in typical

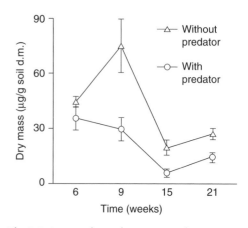

Fig. 3.6. Impact of a predatory nematode (*Prionchulus punctatus*) on biomass of microbivorous nematodes (mean ± SE) over a 21-week period in a laboratory microcosm. (From Mikola and Setälä, 1998, with permission.)

soil environments and noted that statistically significant relationships between predators and their assumed prey were often lacking. There are a number of possible reasons for this, including: (i) predatory nematodes are often facultative rather than obligate predators of nematodes, and so other food sources (bacteria, protozoa, enchytraeids, etc.) may be an important part of their diet in some circumstances; (ii) nematodes are likely to exhibit various forms of anti-predation behaviour (e.g. hiding in roots or soil aggregates), and so predators will be more effective against some prey than others; (iii) susceptibility to predation is likely to vary with soil texture, as some nematodes probably escape larger predators in some soils by occupying small voids that exist within the soil matrix; (iv) the reproductive capacity of most predators is relatively low, and so they may be unable to respond to the rapid increases in population that can occur with microbivorous nematodes; (v) predator–prey interactions are likely to show distinct annual cycles, and relationships between predator and prey may only become apparent when data are collected from a series of samples; and (vi) predators and prey vary in size and so it may be necessary to use predator biomass or consumption rather than numbers to assess predatory activity.

Plant-parasitic nematodes

Nematodes capable of feeding on plants are the most well-known component of the nematode community, as they damage the crops and pastures that provide humans with their food supply. However, from an ecological perspective, they are simply a group of nematodes that have evolved the capacity to use living plant tissue as a food source. Plant-feeding nematodes are always found in natural habitats but usually do not reach high population densities because a diverse range of plant species is present and not all are good hosts, and because populations are kept in check by the regulatory forces that operate in fully functional soil

food webs (van der Putten et al., 2006). Plant-parasitic nematodes multiply readily in agroecosystems, because susceptible hosts are grown repeatedly, and natural regulatory mechanisms no longer operate effectively.

The following material provides a brief overview of plant-parasitic nematodes, but concentrates on the most important pests, their economic impact and how they are managed. Further information on these issues can be found in numerous texts in plant nematology, including Evans et al. (1993), Barker et al. (1998), Chen et al. (2004), Luc et al. (2005) and Perry and Moens (2006). The chapter concludes with a brief discussion of why plant-parasitic nematodes are a difficult target for biological control agents.

Major groups of plant-feeding nematodes and their economic impact

Nematodes that feed on plants are grouped in different ways by various authors, but the categories outlined in Table 3.2 are frequently used. Categorization by feeding group is useful, because it provides an indication of how various groups of plant-parasitic nematodes interact with their hosts. However, the examples listed do not fully illustrate the breadth of the community of nematodes that depend on living plants for food. Only some of the genera known to feed on plants have been included, and many genera of plant-parasitic nematodes contain 20–100 species, each with different environmental requirements and host ranges.

Although large numbers of plant-parasitic nematodes have been described, most of them cause little or no damage to their hosts, or their effects are limited to certain crops or specific situations. Thus, from an agricultural perspective, it is perhaps more useful to concentrate on the nematodes most likely to cause economic losses. Thus, the most important pests of the world's major food and fibre crops, and the pastures and grasslands that support livestock production, are discussed in Box 3.4 and listed in Table 3.3. Although the list omits crops that

Table 3.2. Principal groups of plant-feeding nematodes.

Feeding group	Characteristics	Representative taxa
Sedentary endoparasites	The entire nematode penetrates root tissue and induces a sophisticated trophic system of nurse cells or syncytia The specialized feeding site enhances nutrient flow from the host, allowing females to become sedentary, obese in form and highly fecund Prior to establishing a permanent feeding site, a migratory stage is always present	*Globodera, Heterodera, Meloidogyne*
Sedentary semi-endoparasites	Only the anterior part of the nematode penetrates the root and establishes a permanent feeding site. The posterior portion remains in soil	*Rotylenchulus, Tylenchulus*
Migratory endoparasites		
Root-feeding	All life stages can be found within root tissue and also in soil Feeding occurs as nematodes migrate through and between root cells	*Hirschmanniella, Pratylenchus, Radopholus*
Parasitize tissues other than roots	Feed on aerial parts of plants, on bulbs and tubers, or on surface tissues of leaves and buds	*Anguina, Aphelenchoides, Ditylenchus*
Ectoparasites		
Produce symptoms on roots	A stylet is used to puncture plant cells and feed on the cytoplasm. The body remains in soil and does not enter the root Feeding depth is determined by the length of the stylet. Damage is usually limited to necrosis of cells penetrated by the stylet Nematodes that feed on meristematic cells in the root tip cause the most damage	*Belonolaimus, Criconemella, Criconemoides, Dolichodorus, Helicotylenchus, Hemicycliophora, Hoplolaimus, Mesocriconema, Paratrichodorus, Paratylenchus, Rotylenchus, Tylenchorhynchus, Xiphinema*
Plant associates	Feed superficially on epidermal cells and root hairs but have no impact on plant growth	Some members of the Tylenchidae and Atylenchidae

may be locally or regionally important; does not include high-value crops that may be economically important but grown on relatively small areas; and ignores some nematodes that cause crop loss in particular soil types, cropping systems or environments, it provides a good indication of the crops and nematodes that must be targeted if losses from plant-parasitic nematodes are to be markedly reduced.

It should be apparent from Table 3.3 that two major groups of nematodes are economically important worldwide. The sedentary endoparasites are universally seen as the most damaging nematode pests, as they not only affect most of the world's major crops, but also damage many crops that are not listed. Root-knot nematodes (*Meloidogyne* spp.) are widely distributed, particularly in the tropics and subtropics and in regions with warm to hot climates, and because many species have extremely wide host ranges, they are capable of attacking most crops. The cyst nematodes (*Heterodera* and *Globodera*) tend to be more restricted in their distribution, but their capacity to damage major food crops such as cereals, rice, potato and soybean makes them very important pests. The other group of major economic importance is the migratory endoparasites. *Pratylenchus* has a wide host range and causes losses in most agronomic and horticultural crops; *Radopholus* damages banana and many other crops in the tropics; and *Hirschmanniella* infests more than half of the world's rice fields.

It is difficult to estimate the crop losses caused by plant-parasitic nematodes, as their effects on yield are influenced by management practices, interactions with other pathogens and numerous environmental and biological factors. However, the general consensus among nematologists is that for most crops, yield losses usually range from 5% to 15% (Sasser and Freckman, 1987). Losses tend to be higher in developing countries and in

---

Box 3.4. Nematode pests of major economic importance to world agriculture

The plant-parasitic nematodes responsible for major losses in the world's most important crops and pastures are listed in Table 3.3. The decision to include a particular nematode was based on information collated by various authors and contained in three comprehensive books on nematode problems in temperate, tropical and subtropical crops (Evans *et al.*, 1993; Barker *et al.*, 1998; Luc *et al.*, 2005). Figures on losses due to nematodes on various crops are taken from Sasser and Freckman (1987). The crop loss figure listed for pulses is an average of the percentage loss for field bean, pigeon pea and cowpea. The separation of nematodes into categories of differing importance is somewhat arbitrary, but nematodes that are economically damaging in most areas where the crop is grown are considered of primary importance. Less damaging nematodes, or nematodes with a restricted distribution, are categorized as being of secondary importance. However, this does not mean nematodes that are not listed, particularly the ectoparasites, are unimportant. Factors such as climate, soil type, farming system and the host status of a crop influence the level of damage caused by nematodes, and this means that such nematodes may cause problems on particular crops, or may be important at a regional level.

Choice of crops and pastures is based on land coverage worldwide. Pastures occupy about 34 million km^2 of land, or about two-thirds of the area used for agricultural production. Some pastures are permanent and others are part of an arable rotation, and they include farming systems that vary enormously in management intensity (i.e. grasslands and rangelands that remain in a relatively natural state; semi-natural areas subject to minimal inputs; and highly managed pastures). Given this diversity, and the fact that little is known of the importance of nematodes in all but highly managed systems, pastures have not been included in Table 3.3. However, the available literature suggests that nematodes are most likely to reduce productivity in heavily fertilized, highly stocked or irrigated pastures with little sward diversity. For example, Cook and Yeates (1993) reported that *Heterodera* and *Meloidogyne* are widespread and damaging to ryegrasses in Europe and New Zealand; *Ditylenchus* is an important nematode pest of intensively managed lucerne and clovers, and *Meloidogyne*, *Heterodera* and *Pratylenchus* damage herbage legumes in situations where they are a dominant component of the sward. Nematodes probably cause little damage to pastures under less intensive management. For example, undisturbed perennial pastures are often used as a benchmark when monitoring the ecological condition of soils (Neher and Campbell, 1994; Stirling *et al.*, 2010), and there was little evidence that populations of plant-parasitic nematodes reached damaging levels in pastures that are grown extensively in Australia (Stirling and Lodge, 2005).

The 18 major crops listed in Table 3.3 were chosen using census data that provided global information on a crop's total harvested area. Collectively, these crops represent 85% of the world's total non-pasture cropland (Leff *et al.*, 2004). Crops in which no area is listed were considered 'minor' crops, but have been included because they are cultivated on a relatively large scale in some parts of the world.

Figures such as those in Table 3.3 are often used to estimate crop losses due to nematodes at a regional level. However, it must be remembered that these figures were derived from field trials conducted during the 1960s, 1970s and 1980s, and that there have been marked improvements in the way crops are managed since that time, particularly with regard to optimizing nutrient and water availability. Since nematode damage is exacerbated by nutrient and moisture stress (McSorley and Phillips, 1993), it is likely that in many modern farming systems, losses from nematodes are much lower today.

Continued

Box 3.4. Continued.

Table 3.3. Major world crops, total area harvested, estimated crop losses caused by plant-parasitic nematodes and the most important nematodes on those crops.

Crop	Area (km² × 10³)	% crop loss	Nematodes of primary importance	Other widespread and economically important nematodes
Wheat	4028	7.0	*Heterodera*	*Meloidogyne, Pratylenchus*
Maize	2271	10.2	–	*Meloidogyne, Pratylenchus*
Rice	1956	10.0	*Aphelenchoides, Hirschmanniella*	*Meloidogyne, Pratylenchus*
Barley	1580	6.3	*Heterodera*	*Meloidogyne, Pratylenchus*
Soybean	927	10.6	*Heterodera, Meloidogyne*	*Rotylenchulus, Pratylenchus*
Pulses[a]	794	13.1	*Heterodera, Meloidogyne*	*Rotylenchulus*
Cotton	534	10.7	*Meloidogyne, Rotylenchulus*	*Pratylenchus*
Potato	501	12.2	*Globodera, Meloidogyne*	*Pratylenchus*
Sorghum	501	6.9	*Pratylenchus*	
Millet	331	11.8	–	–
Sunflower	290	–	–	–
Rye	288	3.3	*Heterodera, Meloidogyne*	*Pratylenchus*
Rapeseed/canola	283	–	–	–
Sugarcane	265	15.3	*Meloidogyne, Pratylenchus*	*Paratrichodorus, Xiphinema*
Groundnut/peanut	247	12.0	*Meloidogyne, Pratylenchus*	–
Cassava	235	8.4	*Meloidogyne*	*Pratylenchus*
Sugarbeet	154	10.9	*Heterodera, Meloidogyne*	–
Oil palm	72	–	–	*Bursaphelenchus*
Oat	–	4.2	*Heterodera*	*Ditylenchus*
Coffee	–	15.0	*Meloidogyne, Pratylenchus*	–
Coconut	–	17.1	–	*Radopholus, Bursaphelenchus*
Sweet potato	–	10.2	*Meloidogyne, Rotylenchulus*	*Pratylenchus*
Grape	–	12.5	*Meloidogyne*	*Pratylenchus*
Olive	–	–	–	*Meloidogyne, Pratylenchus*
Banana	–	19.7	*Radopholus*	*Helicotylenchus, Pratylenchus*

[a]Includes bean, pea, chickpea, cowpea, pigeon pea, lentil and lupin

situations where control strategies are not used, while particularly damaging nematodes can cause total crop loss if they are not managed appropriately. Sedentary and migratory endoparasites cause most of these losses, and this is the reason that the biological and integrated management practices mentioned throughout this book generally target these nematodes.

Population dynamics and damage thresholds

Knowledge of current nematode population densities and their potential for future increase is critical for anticipating crop damage and making management decisions that minimize crop loss. Consequently, reviews of literature on population dynamics and damage thresholds are included in most texts on plant-parasitic nematodes (Ferris and Noling, 1987; McSorley and Phillips, 1993; McSorley, 1998; McSorley and Duncan, 2004), and they form the basis of this brief discussion.

Most plant-parasitic nematodes have relatively short life cycles (3–6 weeks under ideal conditions) and high fecundity. Therefore, population densities can increase quite rapidly from low initial levels. Although numbers may increase exponentially for a short period, food supply or some environmental factor will eventually limit population growth, and so a logistic model is commonly used to describe the manner in which nematode populations increase (Fig. 3.7). Knowledge of the multiplication rates of various nematodes and an understanding of the environmental and other factors that influence nematode multiplication on particular crops is essential when designing management systems that will maintain nematode populations at relatively low levels.

It is well recognized that plants can tolerate certain numbers of nematodes without suffering yield loss, but at some point, increasing population densities will progressively affect crop performance. Thus, it is important to understand the impact of population density on yield when deciding whether control measures should, or should not, be applied. The crop loss model depicted in Fig. 3.8 is often used to describe the relationship between nematode populations and crop yield, as it provides an indication of the nematode density below which yield is not affected (i.e. the tolerance limit). Damage functions relating yield to initial nematode population density (Pi) can be derived through field experimentation, and are preferable to those obtained in glasshouse studies. However, they must be interpreted carefully, as factors such as site location, soil type, season of the year, plant cultivar, temperature and nutrient levels affect the population density at which nematode damage becomes apparent.

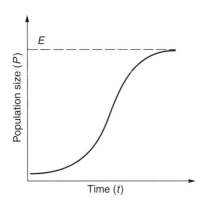

Fig 3.7. A logistic growth model of nematode population increase with time. E = equilibrium density. (From McSorley and Duncan, 2004, with permission.)

Models relating nematode population density to crop loss have proved useful in annual crops but are rarely used in perennial crops. Also, they are difficult to apply in situations where nematodes interact with other soilborne pathogens, or where plants are attacked by more than one nematode species.

Implications for Biological Control

The role of the soil food web and the soil environment

The main reason for discussing the soil food web early in this book is to make the point that plant-parasitic nematodes cannot be considered in isolation from other components of the soil biological community. Their root-feeding habit brings them into contact with a vast number of root and rhizosphere-associated microorganisms, and they also interact with numerous organisms in the detritus food web. Additionally, the activities of plant-parasitic nematodes and other soil organisms are influenced, directly and indirectly, by various soil physical and chemical properties, and by environmental factors such as temperature and moisture. These ecological realities must be recognized in any discussion of biological control.

The section on interactions within the soil food web is particularly important, because it demonstrates that biological control is a normal part of a properly functioning soil ecosystem. Numerous soil organisms interact with nematodes and with each other (Fig. 3.9) and in that process they contribute to the regulatory mechanisms that maintain the stability of the soil food web. Since plant-feeding nematodes become pests when these biological buffering processes are inadequate, biological control should be thought of as maintaining, restoring or enhancing the natural suppressive mechanisms that exist in all soils (Stirling, 2011a).

The roles of the soil biota within the soil ecosystem were outlined in Box 3.1, and from a biological control perspective, it would be easy to consider that their role in suppressing

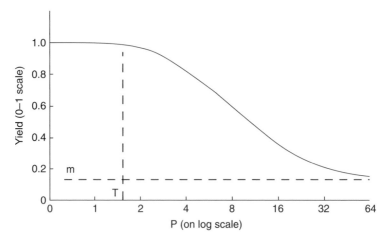

Fig. 3.8. General relationship between crop yield (Y) and nematode population (P), based on the relationship of Seinhorst (1965): $Y = m + (1-m) z^{P-T}$, where m = minimum yield and T = tolerance limit. (From McSorley and Phillips, 1993, with permission.)

pests and pathogens was the only function that is important. However, that is not the case. Soil organisms have many other important functions in soil (improving its physical structure; producing and cycling nutrients) that ultimately impinge on plant growth. Thus, plants in healthy, biologically active soils will yield much better than nutrient-poor plants growing in compacted or poorly aerated soils. Since healthy plants can tolerate many more plant-parasitic nematodes than plants under stress, improving the biological status of soils not only increases the number and diversity of antagonists acting against plant-parasitic nematodes, but also reduces the level of control that is required. It is for this reason that this book takes a holistic view of the soil biota and does not focus only on nematode pests and their natural enemies. The soil biological community has the capacity to reduce nematode populations, but it also has other important functions that in many cases will have even greater effects on plant productivity.

Major crops and nematode pests: their relevance to biological control

The inclusion of a section on major nematode pests and their role in reducing crop yields

may seem out of place in a book on biological control of nematodes. However, such information is relevant to this topic for a number of reasons. Categorizing nematodes as pests of primary and secondary importance is useful, because it indicates the nematodes that should be the main focus of biological control studies. Seven nematode genera (*Meloidogyne*, *Heterodera*, *Globodera*, *Rotylenchulus*, *Pratylenchus*, *Radopholus* and *Hirschmanniella*) are the main nematode pests worldwide, and are likely to be the principal target in most studies. Biological control strategies directed towards other nematodes will, therefore, have only a minor or localized impact. The fact that relatively few nematodes are economically important on some crops, and that problems at a regional level are sometimes due to a single nematode (e.g. *Globodera* on potato, *Meloidogyne* on many vegetable crops) is also important, because it means that research programmes on biological control can focus on a few major nematode pests, thereby increasing the chances of a successful outcome. However, the situation is not as clear-cut for crops with multiple nematode pests. Sugarcane and grapevines, for example, have two major nematode pests (*Meloidogyne* and *Pratylenchus*), but are also attacked by a community of nematodes that may include *Xiphinema*, *Paratrichodorus*, various criconematids and several other

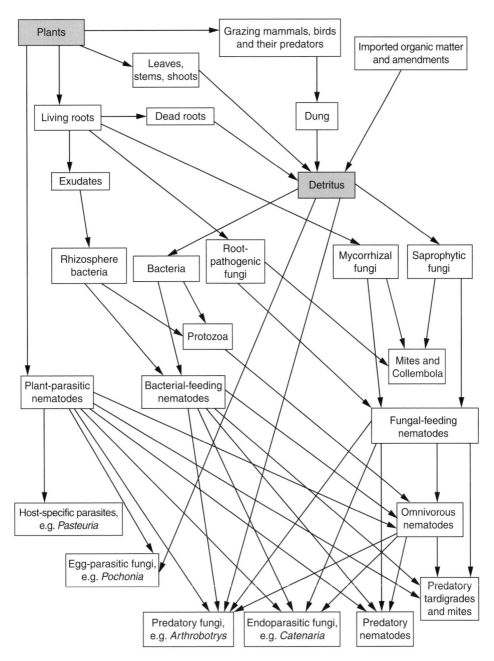

Fig. 3.9. Representation of a soil food web, showing the main interactions between plants, plant-parasitic nematodes, some other primary consumers and organisms in the detrital food web. (From Stirling, 2011a, with kind permission from Springer Science + Business Media B.V.)

plant-parasitic nematodes. Thus, it is questionable whether controlling the primary pests will have an impact, as other nematodes may simply take advantage of the resulting healthier root system and become dominant. In such situations, biological control strategies need to target the herbivorous nematode community as a whole.

One point that emerges from the data on the relative importance of various food and fibre crops is that about two-thirds of the world's cropland is dominated by five crops (wheat, maize, rice, barley and soybean). However, this issue is rarely recognized in the literature on biological control of nematodes. Statements that a particular organism 'has potential as a biological control agent' are often made, but the question of how such an agent might be deployed in major food-producing crops is never addressed. A related issue is the level of damage caused by nematodes on such crops, many of which are relatively low in value and are traded on the world market as commodities. Since crop losses caused by nematodes are often less than 10%, the inundative biocontrol strategies that could possibly be employed in such situations will always be constrained by economics. Thus, a strong theme that runs throughout this book is that developing biological products for nematode control will often not be a worthwhile strategy, as it is only feasible to use such products in high-value markets. Biological control is more likely to make a contribution to world agriculture if research efforts are focused on developing farming systems that improve soil health and enhance the regulatory mechanisms that exist in all soils.

Endoparasitic nematodes as a target for biological control agents

Sedentary endoparasites

One of the main impediments to successful biological control of root-knot and cyst nematodes is their endoparasitic habit. For most of their life cycles, sedentary endoparasites are completely surrounded by root tissue and are, therefore, protected from soilborne parasites and predators. Obligate parasites in the genus *Pasteuria* can infect the endoparasitic stages of *Meloidogyne*, *Heterodera* and *Globodera*, but in all these genera, infection occurs from spores that attach to second-stage juveniles before they invade the root. However, one difference between the root-knot and cyst nematodes is that female cyst nematodes are more vulnerable to attack by soilborne antagonists. Once

immature females rupture the root cortex and are exposed on the root surface, they can be destroyed in a few days by a range of fungal and oomycete parasites (see Chapter 5). Females of *Heterodera avenae* are a good example, as they are susceptible to parasitism by *Nematophthora gynophila* and *Pochonia chlamydosporia*, provided the attack occurs before the cuticle hardens to form a cyst (Kerry *et al.*, 1982a).

Prior to the 1970s, biological control studies directed at sedentary endoparasites commonly targeted second-stage juveniles. However, it is now recognized that this stage of the life cycle is an elusive target. For example, numbers of root-knot nematode juveniles fluctuate according to soil environmental conditions, and high populations may only be found when soil moisture conditions are favourable for egg hatch (i.e. following rainfall or irrigation). Juveniles of some cyst nematodes are even more transient, as they sometimes occur for only 1–2 months each year. Thus, second-stage juveniles are not consistently available to their antagonists, and when they are present, parasites and predators often have to cope with large numbers of individuals.

Another characteristic that makes the juveniles of sedentary endoparasites a difficult target for biological control agents is their capacity to escape parasitism and predation by quickly invading roots. *Meloidogyne* juveniles inoculated experimentally in pots or on agar surfaces can be found in roots within 24 h, while in the field they can move up to 50 cm in 3 days and establish themselves in roots (Prot and Netscher, 1979). Obviously an antagonist would have to be positioned near the root tip and show a high level of predacious or parasitic activity to be able to kill large numbers of juveniles before they find a safe haven within roots.

There are also other reasons why natural enemies that target second-stage juveniles of *Meloidogyne* may have little impact on populations of the nematode. First, in host plants where the galls induced by the nematode are reinfected by juveniles of succeeding generations, the offspring of female nematodes may multiply in a large gall without ever encountering soilborne parasites or predators. Second, root-knot nematode females often produce more than 1000 eggs, and only a small proportion of the second-stage juveniles hatching from those eggs have to establish feeding sites in roots in multiplication to occur. Thus, destroying large numbers of second-stage juveniles may do little to limit population increase or the level of nematode damage.

Since *Meloidogyne* females are protected by the root, and juveniles are mobile, dispersed and transient, eggs are the stage of the life cycle most vulnerable to attack by antagonists. Eggs are generally located on the root surface, and even under ideal conditions, they take at least 10 days to develop and hatch. Because root-knot nematode eggs are aggregated in an egg mass, an effective parasite or predator established in that vicinity could be expected to eliminate many of the eggs produced by an individual female. The gelatinous matrix surrounding the eggs is densely populated with bacteria (Papert and Kok, 2000) and may provide some protection from microbial attack (Orion *et al.*, 2001), but if that is the case, the protective function appears to be relatively limited. Fungi readily grow through the gelatinous matrix to parasitize eggs (Stirling *et al.*, 1979; Oka *et al.*, 1997), and eggs embedded in or removed from the gelatinous matrix respond similarly to suppressive agents in field soil (Stirling *et al.*, 2012). Roots probably provide more protection, as *Meloidogyne* eggs are often deposited within galled tissue, where they are out of reach of soilborne antagonists. However, factors such host plant species, temperature and initial nematode inoculum density affect the proportion of root-knot eggs deposited within and outside the gall, and will, therefore, influence the efficacy of antagonists that parasitize or prey on eggs.

The eggs of cyst nematodes might be expected to be less vulnerable to attack than those of root-knot nematodes, because the tough, leathery cyst that surrounds the eggs has a protective function. In fact, the cyst wall is such an effective barrier against the vagaries of the soil environment that cyst contents can remain viable in soil for many years. Nevertheless, the cyst does have natural openings (e.g. the vulva), and the cyst wall itself is not an impossible structure to penetrate. Thus, there have been many reports of fungal parasitism of cyst nematode females and eggs

(see Chapters 5 and 10), and there tends to be a progressive increase in the amount of fungal colonization as development proceeds to the brown cyst stage (Gintis *et al.*, 1983).

Sedentary semi-endoparasites such as *Tylenchulus* and *Rotylenchulus* differ from the root-knot and cyst nematodes in that their saccate females are partly exposed on the root surface throughout their lives. Since the immature stages of these nematodes spend a relatively long period in the rhizosphere before establishing a feeding site, and their eggs are deposited on the root surface, this group of nematodes could be expected to be reasonably amenable to biological control. Surveys in Florida and Spain (Walter and Kaplan, 1990; Gené *et al.*, 2005), indicate that a wide range of parasites and predators are associated with citrus nematode (*Tylenchulus semipenetrans*), but there have been no comprehensive studies on their role in regulating populations of the nematode. Nevertheless, the authors of the latter study observed that in one Spanish citrus orchard, 75% of the females and 10% of the eggs were parasitized by the fungus *Purpureocillium lilacinum*, and *Pasteuria* endospores were also observed on second-stage juveniles of *T. semipenetrans*. Nematode population densities in this orchard were eight times lower than in other surveyed orchards, and it was thought that microbial parasites were responsible.

Migratory endoparasites

Of all the groups of plant-parasitic nematodes, the migratory endoparasites are likely to be one of the most difficult to control with natural enemies. Nematodes such as *Pratylenchus* and *Radopholus* spend much of their lives in roots and tend to be found in soil only when their host plants are stressed, senescing or diseased, or when their hosts have died or been destroyed after harvest. Since eggs are often laid in root tissue and juveniles may hatch and develop to maturity without ever moving from roots, multiplication can sometimes proceed for several generations without nematodes being exposed to soilborne antagonists. Biological control agents that normally reside in soil could, therefore, be expected to have no more than a limited impact on these nematodes, as they are unable to access them during the pest's critical multiplication phase. For this reason, endophytic organisms are more likely to be useful suppressive agents (see Chapters 8 and 12 for further discussion).

Features that protect plant-parasitic nematodes from parasitism and predation

In addition to the features already discussed, plant-parasitic nematodes exhibit a number of characteristics that not only help them survive and flourish in the harsh and competitive soil environment, but also provide protection from antagonists. Since these features were discussed in detail by Stirling (1991), they are only considered briefly here.

The body of a nematode is protected by a multi-layered, proteinaceous cuticle that functions as a flexible skeleton and a barrier to undesirable elements in the soil environment. The developing embryo is also well protected, as it is surrounded by a triple-layered egg shell 1–2 μm thick composed mainly of chitin. Once embryological development is complete, antagonists face the problem of not only penetrating the egg shell, but also the cuticle of an unhatched juvenile that by this stage is capable of movement. This probably explains why eggs in the later stages of development often do not succumb to attack by fungal parasites.

The high reproductive capacity of most plant-parasitic nematodes not only helps make them significant pests, but also ensures that they are difficult to control. The life cycle of most important nematode pests takes only a few weeks at optimum temperatures, and each female has the capacity to produce hundreds, and in some cases more than 1000, progeny. On a susceptible crop under ideal growing conditions, nematode populations that are at virtually non-detectable levels at planting can increase to damaging levels in 6–12 weeks. This remarkable capacity for multiplication tends to negate the effects of antagonists, as high levels of parasitism and predation may simply remove surplus nematodes and do little to diminish final nematode numbers. Since the rate of nematode reproduction is strongly influenced by the host plant, biological control agents are more likely to be successful on crops that are not highly susceptible to nematodes.

Another important point that is relevant to the control of endoparasitic nematodes is that the population structure of the target nematode changes during the growing season as juveniles hatch from eggs; parasitic stages develop in roots; and adults begin to produce eggs. Typically, the population structure is pyramidal, with relatively few of the large numbers of eggs and juveniles at the base of the pyramid eventually developing into adults (Evans, 1969). Pests with this type of population structure are difficult to control with antagonists that target eggs or juveniles, because they tend to remove individuals that are surplus to requirements, and may not necessarily reduce the number of nematodes in roots.

4

Global Food Security, Soil Health and Sustainable Agriculture

The decision to commence this book with a broad overview of the many issues likely to affect the development of biological controls for plant-parasitic nematodes was deliberate. The complexity of the soil environment; the diversity of agricultural production systems; and the sheer number of economic and production-related issues that must be considered by today's food producers mean that numerous factors will impinge on any attempt to introduce alternative methods of managing nematodes. The soil environment; the soil biota; the role of organic matter; the biological interactions that occur at the root–soil interface; the soil food web; and the soil nematode community were discussed in Chapters 2 and 3. This chapter aims to cover some of the agricultural issues that affect our attempts to establish reliable systems of biological control. Land managers and farmers live in the real world, and so their management options are limited by climate, the inherent properties of the soil resource, economics, market requirements and many other factors. Alternative pest control strategies will only be adopted if they are feasible, cost-effective and consistently successful.

Global Food Security

Agriculture is a vibrant, innovative and successful industry. Despite a doubling of the global population in the last 50 years, food production has increased to an even greater extent, markedly decreasing the proportion of malnourished people in the world. New livestock and crop production technologies have enabled food to be produced in ways that would never have been contemplated by previous generations of farmers. However, world population is expected to reach 9 billion by 2050, and since steps are unlikely to be taken to regulate population growth, the level of innovation that characterized the latter part of the 20th century will have to continue unabated for many more years. Thus, the challenge facing agriculture is to meet the food requirements of a larger and more affluent population in an era when food producers are experiencing greater competition for land, water and energy (Godfray *et al.*, 2010; Gomiero *et al.*, 2011a).

Increases in crop production derive from three main sources: expansion of arable land; increases in cropping intensity (the frequency with which crops are harvested from a given area); and improvements in yield. Since the 1960s, the United Nations Food and Agriculture Organization (FAO) has shown that yield improvements made by far the greatest contribution to the increase in global food production, accounting for about 78% of the increase between 1961 and 1999. The remainder came from an expansion in the arable area (15%) and increased cropping intensity (7%) (Bruinsma, 2003). Future projections suggest that this situation is unlikely to change. There will be opportunities to expand the arable area in sub-Saharan Africa and Latin America, but in developing countries overall, 80% of increased crop production will have to come from intensification, higher yields, greater levels of multiple cropping and shorter fallow periods. In developed economies, where agricultural land is increasingly being planted to energy-producing biomass crops, virtually all the required increase in food production will come from yield improvements and intensification.

The need to produce more food on land that is already being used for food crops raises the question of whether land can be farmed more intensively without increasing the rate of soil degradation. Agricultural land is automatically degraded when nutrients are removed in harvested crops, but further degradation may occur through water and wind erosion, desertification, salinization and leaching of nutrients. Nevertheless, there are a number of reasons why agricultural intensification will not necessarily increase the rate of these processes: (i) evolving technologies in no-till/conservation agriculture can maintain year-round soil cover and increase soil organic matter, thereby reducing water and wind erosion while maintaining soil health; (ii) most irrigated agricultural land is relatively flat and is little affected by erosion, while abandonment of marginal land that is too steep for agriculture, together with practices such as contour banking, will reduce water erosion; (iii) agroforestry (the integration of cropping and or/livestock production with trees or shrubs) offers opportunities to reduce soil erosion, restore soil fertility and increase biodiversity; (iv) a shift towards raising livestock in more intensive systems will reduce grazing pressures on dryland pastures; (v) a range of intensification practices (increased fertilizer consumption, more efficient fertilizer use, the introduction of drought- and salt-tolerant crops, the use of grazing-tolerant pastures and the introduction of irrigation) will reduce erosion by increasing plant biomass, root growth and ground cover; and (vi) the cultivation of legumes in cropping and mixed crop–livestock farming systems will add nitrogen to soils and improve their stability and texture.

Attention to issues associated with soil degradation will always be an important priority for land managers, but the important message from the previous paragraph is that as food production becomes more intensive, many practices are available to minimize soil degradation, and they must be components of future soil management programmes. At the same time, future farming systems will have to address the environmental impact of agriculture. Agricultural activities, particularly those resulting in emissions to air and water, can have significant effects long distances from where those activities take place. Pesticides and nutrients can move into surface and ground water, while greenhouse gases can be emitted to the atmosphere, and so steps must be taken to minimize these negative impacts. Practices such as integrated pest management; optimization of water, nutrient and pesticide inputs through precision agriculture; and a whole range of practices to conserve soil and water (e.g. conservation tillage, cover cropping, controlled traffic, contour farming and mulching) will also have to be adopted more widely by farming communities. Food production must continue to increase in the 21st century, but it will have to be done in an environmentally sustainable manner, a process that has been termed 'sustainable intensification' (Royal Society, 2009).

Sustainable Farming Systems

Responsible and enduring stewardship of agricultural land is the essence of sustainable agriculture: land is managed in ways that maintain its long-term productivity, resilience and vitality while minimizing adverse environmental impacts. Although it has been variously defined (Hamblin, 1995; Lewandowski et al., 1999; Gliessman, 2007; Gold, 2007;

Pretty, 2008), sustainable agriculture is generally considered to: (i) replenish the resource base that sustains agricultural production, and then maintain it in a condition that does not compromise its use by future generations; (ii) integrate biological and ecological processes such as nutrient cycling, nitrogen fixation, soil regeneration, allellopathy, competition and the regulatory effects of pests' natural enemies, into food production systems; (iii) utilize ecological knowledge, the basics of agricultural science and the management skills and ingenuity of farmers to develop farming systems appropriate for the soil, climate and production goals; (iv) optimize the use of resources and minimize the use of non-renewable inputs; and (v) minimize the impact of pest management, crop nutrition, irrigation and other production practices on human health and the environment.

Sustainable agriculture is not a prescribed set of practices. It challenges land managers to think about the long-term implications of the practices they use, and to understand the interactions that occur within and between the many components of agricultural systems. A key goal is to view agriculture from an ecological perspective and balance the requirements for productivity and profitability with an understanding of nutrient and energy dynamics and the biological interactions that occur in agroecosystems. Any new practice or technology that improves productivity for farmers but does not cause undue harm to the soil biological environment is likely to enhance sustainability.

Sustainable agricultural intensification

Sustainable agricultural intensification is essentially about increasing productivity in ways that make better use of existing resources while minimizing environmental harm. There are many pathways to agricultural sustainability, and no single configuration of technologies, inputs and management practices is likely to be applicable in all situations

(Pretty, 2008). However, a number of key practices are consistently associated with sustainability (Goulding et al., 2008; Shennan, 2008; Wilkins, 2008; Kassam et al., 2009), and they are summarized next. Applied together, or in various combinations, these practices work synergistically to increase productivity and also to contribute important ecosystem services that enhance sustainability. However, it is important to recognize that there is no prescriptive list of sustainable practices: farmers have many options available to them, and their management practices must be chosen and adapted according to local production conditions and environmental constraints.

Reduced tillage

The negative impacts of mechanical tillage on soil carbon reserves and the increased susceptibility of cultivated soil to water and wind erosion have demonstrated that farming systems based on inversion tillage are not sustainable. Many options are available to reduce the depth, frequency and intensity of tillage operations, but no-till is associated with the least physical disturbance. It improves levels of soil organic matter, has profound direct and indirect effects on soil structure and aggregation, does not disrupt the soil biota, minimizes the consumption of fossil fuels and reduces labour requirements. Since a move to reduced tillage can increase compaction problems, particularly in farming systems with heavy equipment and random traffic, controlled traffic systems are usually an integral component of no-till agriculture.

Continual cropping and maintenance of a permanent cover of plant residues

This component of intensification minimizes the length of fallow periods and helps to ensure that the resources provided by roots and their exudates are continually available to the soil biological community. Continual cropping (within the limits imposed by the environment) and maintenance of a protective cover of organic matter on the soil surface mimics, to some extent, the way plants and soil interact in the natural environment.

Roots produced by previous crops are not disturbed, while the residues produced by the primary crops and any cover crops are left on the soil surface, where they moderate temperature fluctuations, conserve water and nutrients, protect the soil from erosion, minimize weeds and promote soil biological activity. Thus, crop residues are seen as a valuable resource in a sustainable farming system, rather than something that should be burnt or removed because it is a hindrance to future production. The extent of the benefits from residue retention will depend on the quantity and quality of the residues produced, and how they are manipulated.

Greater plant diversity

The practice of crop rotation plays a vital role in sustainable agriculture, as it is one of the simplest ways of minimizing losses from pests and pathogens that are often relatively crop-specific. However, there are many other options that can be used to increase biological diversity within agroecosystems, and enhance system resilience and sustainability. Examples include the maintenance of natural habitats in farming areas; integration of various forms of forestry with agriculture; the use of intercrop systems in which multiple crops are grown in mixed or structured arrangements; the planting of hedgerows and alley crops or the introduction of banker plants to encourage predators of pests; the retention or provision of windbreaks; and the introduction of legumes to fix nitrogen. In the long term, research aimed at replacing annual grain and oilseed crops with perennials (see Cox et al., 2006; Glover et al., 2007) will not only provide opportunities to increase crop diversity, but also help to make agriculture more sustainable.

Improved crop yield potential

The process of improving crops through plant breeding and genetic modification has played a major role in increasing world food production over the last 50 years. However, the effects on agricultural sustainability have been mixed. On the positive side, modern crop varieties with pest and disease resistance

prevent the world's major food crops from being regularly decimated by rusts, mildews and a range of insect and nematode pests, and do so in such an effective manner that fungicides, insecticides and nematicides are not widely used on many crops. However, there are concerns that plant breeders working in high-input systems have inadvertently selected plants with poor root systems; that higher external inputs are often needed to obtain improved yields; and that genetic uniformity and a narrowing of the genetic base may lead to decreased resilience in the face of environmental stress. Future plant-breeding programmes must, therefore, concentrate on producing well-adapted varieties that not only resist or tolerate the effects of important pests and pathogens, but also have a root structure and biomass capable of retrieving nutrients effectively. The cultivars available in future must also have a greater capacity to cope with common abiotic stresses such as heat, drought and salinity.

Optimized crop nutrition

In many modern farming systems, fertilizers and manures are applied excessively because the economic response in crop yield far outweighs the cost of the fertilizer. Consequently, enormous quantities of fertilizer are being wasted. For example, about 50%, and sometimes even more, of the synthetic nitrogen applied to crops is usually lost to the environment as gaseous emissions to the atmosphere, leaching to groundwater, and runoff to surface waters (Tomich et al., 2011). Improved nutrient-use efficiency must, therefore, be one of the cornerstones of sustainable intensification. This is not an impossible task, as many relatively simple practices are available to optimize crop nutrition. Some of the options include the use of soil analyses to determine nutrient requirements; application of lime to maintain the appropriate pH for optimum nutrient supply; use of leaf and sap analysis to match nutrient applications to the crop's requirements; inclusion of organic soil amendments in the nutrition programme; use of controlled-release fertilizers or nitrogenous products containing nitrification inhibitors; and the introduction of legumes to provide biologically fixed nitrogen. Ultimately, however, sustainable nutrient management must also take into account the nutrients immobilized and mineralized by the soil biota. Interactions between soil organisms and organic matter govern nutrient availability to plants, and greater efforts must be made at a research level to reliably predict the outcome of these interactions, so that nutrient applications can be adjusted accordingly.

Efficient water management

Although only about 18% of the world's cropped land is irrigated, this land is vitally important, as it produces about 45% of the global food supply (Morison *et al.*, 2008). However, irrigation water will always be a scarce resource due to the competing demands of agriculture and industry, and the requirements for human consumption. From an agricultural perspective, misuse of irrigation water results in soil health and environmental problems, while current and future water supplies will be depleted if irrigation use outpaces recharge rates. Thus, efficient water management is the key to sustainable irrigated agriculture. It can be achieved by improving irrigation infrastructure; by reducing evaporative losses; and by matching water inputs to the crop's requirements through the use of technologies that monitor soil moisture, environmental conditions and plant growth.

Site-specific management

Precision farming or site-specific management involves observing, measuring and then responding to intra-field variability so that agronomic practices and resource allocation are matched to soil and crop requirements. Nutrients, pesticides and other inputs are applied differentially using predefined maps based on soil or crop condition, or with sensors that control application as machinery traverses the field (Srinivasan, 2006). The capacity of precision agriculture to vary inputs based on variability in soil properties (e.g. soil texture, water-holding capacity, organic matter status), or biological factors (e.g. weed populations, insect populations,

disease occurrence, crop growth, harvestable yield) offers the potential to improve sustainability by maintaining or enhancing crop yields while reducing some of the environmental problems associated with nutrient and pest management.

Integrated pest management

Integrated pest management has been defined in various ways (Stirling, 1999), but is essentially about using our knowledge of pest and plant biology to prevent pests from causing economic damage. Pest populations are monitored; damage thresholds are determined; the impact of environmental factors on interactions between the pest and the plant are understood; a wide range of techniques (e.g. genetically resistant hosts, environmental modifications and biological control agents) are used to reduce pest populations to tolerable levels; and pesticides are only used as a last resort. IPM systems that reflect this philosophy will enhance sustainability because they are based on sound ecological principles.

Integrated crop and livestock production

One factor that impacts negatively on the sustainability of modern agriculture, particularly in developed countries, is the trend towards farm-level specialization. Crop production is becoming more specialized and crop and livestock enterprises are often separated, despite the clear soil health and environmental benefits associated with mixed crop–livestock systems. Farm livestock excrete some 50–95% of the nutrients they consume, and from an efficiency and sustainability perspective, there is a strong case for better integration of crop and livestock production, both at the individual farm and regional level (Wilkins, 2008; Kirkegaard *et al.*, 2010).

Soil Health

The quality or health of the soils used to produce crops and livestock is intimately linked to the issue of sustainable agriculture. Although some minor crops are grown hydroponically – and commercial facilities may be established to produce livestock, poultry and fish – food production is largely dependent on the thin layer of soil that covers the earth's surface. This non-renewable resource has a number of ecologically important functions: providing a suitable medium for plant growth; sustaining biological processes responsible for decomposing organic matter; cycling nutrients; maintaining soil structure; regulating pest populations; and detoxifying hazardous compounds (Powlson *et al.*, 2011a). It is important from an agricultural production, environmental and sustainability perspective that those functions are maintained indefinitely.

Although it is widely recognized that maintaining healthy soil is a vital component of sustainable agriculture (Doran *et al.*, 1994; 1996; Lal and Stewart, 1995; Doran and Safley, 1997; Rapport *et al.*, 1997; Gregorich and Carter, 1997; Kibblewhite *et al.*, 2008), the term 'soil health' has been the subject of fierce debate in the scientific literature (Sojka and Upchurch, 1999; Karlen *et al.*, 2001; 2003a,b; Letey *et al.*, 2003; Sojka *et al.*, 2003). First, arguments abound as to how soil health should be defined and whether 'soil quality' is a more appropriate term. Second, it has been particularly difficult to find a definition of soil quality/soil health that satisfies everyone, because soil performs multiple functions simultaneously. Thus, high soil quality for one function (e.g. crop production) does not guarantee high quality for another function (e.g. environmental protection), and vice versa. Third, attempts to develop soil-quality indices have been criticized on the basis that the process does nothing more than provide a highly generalized and non-specific assessment of the overall worth, value or condition of a soil. One of the major concerns is that assessment tools do not objectively and simultaneously consider both the potential positive and negative impacts of all indicators on production, sustainability and the environment. Thus, some highly valued parameters such as levels of soil organic matter and numbers of earthworms are almost always viewed positively, even though an increase in these parameters may sometimes result in negative outcomes. Finally, there are differences between those who evaluate soil health or quality on the basis of biodiversity, bioactivity or some other attribute believed to be reflective of 'natural' benchmark conditions, and those who argue that production agriculture is not a natural system, and that the debate should be about how soils are managed to achieve the required production and environmental outcomes.

The terms 'soil health' and 'soil quality' are often used synonymously, but the former term is used here because most farmers have at least some understanding of the concept. Soil is healthy if it is fit for a purpose, which in the case of agriculture is the production of a particular crop. However, agricultural land is a component of a larger ecosystem, so it must also provide functions that prevent degradation of neighbouring environments. The definition used by Kibblewhite *et al.* (2008) encompasses both of these important functions:

> a healthy agricultural soil is one that is capable of supporting the production of food and fibre, to a level and with a quality that is sufficient to meet human requirements, together with continued delivery of other ecosystem services that are essential for maintenance of the quality of life for humans and the conservation of biodiversity.

Management impacts on soil health and the role of conservation agriculture

Farmers and land managers are well aware of the many constraints that affect the productivity of the soils used for agriculture. Those constraints are too numerous to discuss here, but include soil compaction; poor structure; surface crusting; limited water infiltration; excessive leaching of nutrients; susceptibility to erosion; high weed pressure; poor nutrient retention; low water-holding capacity; nutrient deficiencies; sub-optimal pH; excessive salinity; low biological activity; limited biological diversity; and high levels of soilborne pathogens. Although most soils have only some of these problems, no soil could ever be considered completely healthy from a production perspective, while environmental issues such as off-site movement of nutrients and pesticides, and greenhouse gas emissions, are universal problems. Thus, one of the most important roles of a farm manager is to identify and prioritize the main factors causing soil-related problems, and then attempt to improve the health of the soil through management.

In a complex system such as soil where many factors interact, the most robust approach to soil health problems is to consider how a farming system could be modified to rectify existing problems and prevent them from recurring. For example, poor soil health is often associated with low levels of soil organic matter, and so tackling that issue in particular can lead to improvements in a whole range of soil physical, chemical and biological properties. Thus, the practices previously identified as the keys to sustainable agriculture are also the keys to improving soil health.

The three most important soil improvement practices (minimal soil disturbance, permanent plant residue cover and crop rotation) form the basis of conservation agriculture (Baker *et al.*, 2006; Hobbs *et al.*, 2008; Kassam *et al.*, 2009), a relatively recent agricultural management system that has been adopted widely in some parts of the world, particularly North America, Latin America and Australia. Conservation tillage (variously described as minimum tillage, reduced tillage, no-till or direct drill) and retention of crop residues are the primary components of conservation agriculture, and when used together, these practices reduce soil erosion and slow or reverse the precipitous decline in soil organic matter that has occurred in conventionally tilled agricultural soils over the last 100 years (Reeves, 1997; Uri, 1999; Paustian *et al.*, 2000; Franzluebbers, 2004). When combined with diversified crop rotations that include cover crops, mulch-producing crops and nitrogen-fixing legumes, many soil properties are affected (Table 4.1), soil health generally improves, and suppressiveness to root pathogens is often enhanced (Sturz *et al.*, 1997). However, as pointed out by many authors, including Blevins and Frye (1993) and Sojka *et al.* (2003), there are situations where the effects of conservation agriculture and increased levels of soil organic matter are not always positive. Examples include the impact of mulch cover on soil temperature, which can improve crop growth in a hot climate but slow early emergence and growth in temperate regions; decreased availability of plant-available

nitrogen due to immobilization; exacerbation of diseases caused by pathogens that survive on crop residues; difficulties associated with managing some weeds in the absence of tillage; herbicide carryover and runoff; the potential for weed populations to become resistant to herbicides; and the impact of soil organic matter and earthworm burrowing on porosity and macropore formation, which can increase the risk of nutrients and pesticides becoming groundwater contaminants.

The individual economic, soil health and other benefits listed in Table 4.1 will not be obtained in every situation, but collectively these benefits provide compelling reasons for farmers to minimize tillage and incorporate residue retention and crop rotation into their farming systems. However, perhaps the most persuasive reason for adopting the soil and crop management practices associated with conservation agriculture is that they enhance soil organic carbon pools, thereby reducing atmospheric CO_2 emissions associated with climate change (Powlson *et al.*, 2011b). Continuous surface cover and the increase in water-holding capacity associated with higher levels of soil organic matter also help mitigate the effects of any change in climate by increasing the tolerance of crops to higher temperatures and drought conditions (Lal, 2009: Lal *et al.*, 2011).

Other management practices to improve soil health

Although integrating conservation tillage, residue retention and crop rotation is the first step towards greater sustainability and improved soil health, further incremental improvements can be obtained by adopting a range of other practices.

Well-adapted, high-yielding varieties

Genome sequencing, DNA marker technologies, and phenotype analysis are just some of the many tools currently being used by plant breeders to improve the resistance of crops to pests and diseases, and to increase their tolerance to abiotic stresses. The additional biomass produced by higher-yielding

Table 4.1. The effects of the principal components of conservation agriculture on soil health and sustainability.

Effect	Mulch cover (crop residues, cover crops, green manures)	No tillage (minimal or no soil disturbance)	Crop rotation (includes legumes for nitrogen fixation)
Maintains a permanent residue cover on the soil surface	+	+	+
Reduces evaporative loss from upper soil layers	+	+	
Maintains the natural stratification of the soil profile	+	+	
Minimizes oxidation of soil organic matter		+	
Sequesters carbon and minimizes CO_2 loss	+	+	+
Minimizes compaction by intense rainfall	+		
Minimizes temperature fluctuations at the soil surface	+		
Maintains a supply of organic matter for the soil biota	+	+	+
Increases and maintains nitrogen levels in the root zone	+	+	+
Increases cation exchange capacity	+	+	+
Maximizes rainfall infiltration and minimizes runoff	+	+	
Minimizes erosion losses from water and wind	+	+	
Increases water-holding capacity	+	+	
Minimizes weeds	+	+	
Increases the rate of biomass production	+	+	+
Speeds the recuperation in soil porosity by the soil biota	+	+	+
Rebuilds damaged soil conditions and dynamics	+	+	+
Recycles nutrients	+	+	+
Reduces pests and diseases			+
Reduces labour input		+	
Reduces fuel-energy input		+	

Modified from Kassam *et al.* (2009).

crops should result in soil organic matter gains that will improve the health of agricultural soils. Well-adapted, disease-resistant varieties could also help to reduce the off-site impacts of agriculture, provided they have root systems that utilize applied nutrients more effectively than their susceptible counterparts.

Optimal nutrient management

In soils used for agriculture, nutrient levels must be maintained by replacing the nutrients removed in the harvested product. However, whenever industrially produced fertilizers and their organic alternatives are applied excessively, the nutrients they contain will either become environmental pollutants or be detrimental to some components of the soil biota. Thus, high nutrient-use efficiency is an important component of maintaining soil fertility, but is also essential for minimizing off-site impacts.

Efficient water management

The soil health and environmental problems associated with irrigation are widely recognized. Water tables rise when irrigation water is applied; salinity is a constant threat to irrigated agriculture; excessive inputs of water cause waterlogging and drainage problems; and salts, nutrients, herbicides and pesticides that are leached through the soil profile or transported by overland water flow will reduce the quality of downstream water. However, it is possible to avoid these negative impacts. Trickle irrigation; precision land levelling to improve surface irrigation; and monitoring soil water at multiple depths in the profile and then using the data to match irrigation inputs to plant uptake are just some of the practices that will markedly reduce deep percolation losses to groundwater. Deficit irrigation and partial root-zone drying are other management techniques that can be used to improve water use efficiency and minimize the off-site impacts of irrigation (Loveys *et al.*, 2004; Morison *et al.*, 2008).

Integrated pest management

Although IPM is widely promoted as a pest and disease-control strategy, the rates of adoption and the tactics employed vary considerably from industry to industry and from one pest to another. In some crops, pest populations are monitored and crop losses are minimized by integrating various cultural and biological controls, while in others, IPM involves little more than pesticide management. In situations where insecticides and fungicides are included in IPM programmes for above-ground pests, the possibility that residues could impact negatively on the soil biological community is rarely considered. Thus, the ultimate land management objective should be to develop a fully integrated system of managing the soil, the crops, and all pests and diseases. The IPM component would ideally be effective enough to control the key pests with minimal need for pesticides, while the crop and soil management component would aim to generate a healthy, biologically active soil capable of degrading any pesticide that might be required, thereby preventing it from becoming an environmental contaminant.

Variable-rate application and site-specific management

Intra- and inter-field variability in soil properties such as texture, depth, nutrient content and disease levels are the norm in an agricultural landscape. Variable-rate application techniques associated with precision agriculture provide the tools to optimize management in such situations. Soil variability across a field is mapped, a satellite positioning system (e.g. GPS) determines the location of farm equipment within the field, and variable-rate applicators can then apply fertilizers, pesticides and biological products in amounts that are appropriate for the crop's needs in a given location. In the same way, optical sensors are used to detect weeds and ensure that chemical or non-chemical weed controls are applied only where they are needed. Thus, variable-rate application minimizes the environmental footprint of farming, reduces costs and ensures that soil is not degraded by excessive external inputs.

A range of soil sensors is available in precision agriculture to measure various soil physical and chemical properties (Adamchuck *et al.*, 2004), while geo-referenced soil samples are widely used to determine nutrient requirements in fields that may vary in soil type, topography, cropping history or previous fertilizer inputs. Such samples can also be used to obtain an accurate base map of organic matter status. Although such information is a useful starting point for managing the soil biological community in a site-specific manner, the ultimate research objective should be to provide growers with data on the spatial and temporal variability of key soilborne pests and their natural enemies. High-throughput molecular methods of enumerating soil organisms are currently too expensive to be used in diagnostic services, but since this will change with improvements in sequencing methods and advances in bioinformatics, it will eventually be possible to integrate molecular diagnostics with the technologies available in precision agriculture.

Integrated crop and livestock production

The permanent nature of pastures and the continual presence of perennial plant species mean that soils under pasture are generally healthier than cropped soils. Pasture-based crop and livestock production systems (e.g. mixed farming systems and zero-grazing cut-and-carry systems) also require fewer external nutrient inputs than systems dominated by cropping, and so they tend to be more sustainable (Wilkins, 2008). Also, in landscapes that are subject to dryland salinity, the inclusion of deep-rooted perennials such as lucerne in a cropping rotation reduces deep drainage and prevents salinization (Bellotti, 2001). Thus, from soil health and sustainability perspectives, it makes sense to integrate livestock production and cropping.

Ecologically sound management systems: the pathway to healthy soils

The ultimate challenge of agricultural land management is to integrate the best available practices into a farming system that is not only productive and profitable, but also sustains the soil's productive capacity. The actual farming system that is chosen will depend on climatic factors, the basic properties of the soil being farmed, the resources available to implement change, and the commodity being produced. However, there is little doubt that major improvements are possible in all the world's current farming systems. The fact that the principles of conservation agriculture are being incorporated into a diverse range of farming systems in developed and developing countries around the world (Hobbs *et al.*, 2008; Kassam *et al.*, 2009) is testimony to the fact that progress is being made.

Soil-health benefits from conservation agriculture and precision farming: Australian examples

Conservation agriculture is widely practised in five countries: the United States, Brazil, Argentina, Canada and Australia, but the Australian situation is of particular interest. Australia has the poorest soils and one of the most variable climates in the world, and its successes with conservation agriculture suggest that the principles involved could be adopted by farmers producing almost any crop, in most regions of the world. Australia's major crops are grains (wheat and other cereals, pulses and oilseeds) and sugarcane, and the practices now used to produce those crops, and their impact on productivity and soil health, are summarized next. Further detail on grain-cropping systems can be obtained from various chapters in Tow *et al.* (2011); the sugarcane farming system is discussed by Garside *et al.* (2005); while issues associated with soil health and the soil biota are reviewed by Bell *et al.* (2007), Stirling (2008), Gupta and Knox (2010) and Gupta *et al.* (2011).

In Australia, grain is grown in a wide variety of climatic zones (dry subtropics to cool temperate and Mediterranean climates) and on vastly differing soil types (from heavy clays to coarse sands), and crops are almost always produced under rainfed conditions. Rainfall is generally low and highly variable, with most cropping regions receiving between 250 and 600 mm of rain per year. However, despite the limitations of soil and climate and the absence of government subsidies, Australian agriculture has achieved greater productivity growth than most other agricultural economies over the last 30 years (Mullen, 2007). This success has largely been achieved through the widespread adoption of conservation agriculture. Although management practices vary at a regional and local level, most leading farmers have made the change to no-till agriculture; crops are sown using equipment that incorporates improved disc-seeding technologies; in-field traffic is controlled using GPS guidance; rotational cropping or pasture leys are included in the farming system; legume crops provide nitrogen inputs; crop residues are retained on the soil surface; scanning technologies and variable-rate injection systems are used to optimize chemical application; optical-sensing devices ensure that herbicides are applied on weeds rather than on bare soil; while in-vehicle, aerial or remote sensing systems are available to provide information on environmental factors such as temperature and humidity, and the health status of the crop.

From the perspective of soil health, the introduction of these practices has generally had a positive effect. The move towards reduced-till and direct-drill systems, with associated stubble retention and traffic control, has improved most measures of physical structure (e.g. aggregate stability, the presence of stable macropores and shear strength) and also reduced compaction, thereby reversing the negative effects of conventional tillage on soil physical properties. Soil organic carbon levels have generally improved, particularly in the upper 10 cm of the profile, although studies in some environments have shown no significant change. The development of biologically mediated suppression of two of the most important soilborne diseases of wheat (Rhizoctonia bare patch and take-all) has also been observed in long-term experiments and some commercial fields (Roget, 1995; Roget *et al.*, 1999; Pankhurst *et al.*, 2002; Gupta *et al.*, 2011) and is associated with a build-up of organic carbon and microbial biomass under direct-drilling with stubble retention (Pankhurst *et al.*, 2002). There is also evidence that soil in the upper 25 cm of the soil profile is suppressive to root lesion nematode, *Pratylenchus thornei*, a major constraint to production in subtropical grain-growing areas (Stirling, 2011b). Another important soil-health benefit has been an increase in the capacity of soils to infiltrate and store water, and an improvement in the ability of roots to extract water from the soil (Turner, 2004). Therefore, crops are much more likely to reach their water-limited yield potential, with concomitant effects on productivity and the amount of organic matter returned to soil. Because there have also been negative effects in some situations (e.g. slower early-season growth under direct drill, nutrient stratification in surface soils and increases in diseases such as crown rot where pathogen inoculum survives on stubble), ongoing research and constant fine-tuning by farmers is required to continually improve and fully optimize the new system.

The Australian sugar industry is vastly different from the grains industry. Farms are much smaller (commonly 40–200 ha), the crop is grown largely as a monoculture, and inputs of fertilizer and pesticides are much higher. Also, the industry's location in the tropics and subtropics means that water is not a limitation: between 1200 and 4200 mm of rain is received each year, and in the drier areas, rainfall is supplemented by irrigation.

In the early 1990s, the Australian sugar industry was facing an uncertain future because productivity was declining due to a problem known as yield decline. At that time, sugarcane was grown on beds 1.5 m apart, machinery wheel spacing did not match crop row spacing, and the crop residues remaining after harvest were often burnt. After a plant and 3–4 ratoon crops, an expensive programme of ripping and cultivation was required to remove the old crop, alleviate compaction caused by farm machinery and

then replant the field to sugarcane. A multidisciplinary research team was established to develop solutions to the problem, and its initial studies showed that soils under long-term sugarcane monoculture were physically and chemically degraded. Results of later experiments indicated that biological constraints were also limiting productivity, as large yield responses were obtained when soil fumigants, fungicides and nematicides were applied; or pasture, another crop species, or bare fallow were included in the rotation (Garside and Bell, 2011a, b). Ultimately, the 12-year research programme (summarized by Garside *et al.*, 2005; Stirling, 2008) resulted in the development of a new farming system based on permanent raised beds, residue retention, minimum tillage, a leguminous rotation crop and controlled traffic using GPS guidance. This system is now being adopted by growers because it increases sugar yields, reduces costs and provides additional income from rotation crops such as soybean and peanut.

Although economic considerations (lower fuel and labour costs, and the replacement of fertilizer nitrogen with biologically fixed nitrogen from legumes) motivated growers to adopt the new sugarcane farming system, improvements in soil health were the main reason that yield increases of 20–30% were consistently obtained. Random trafficking of fields, often in wet conditions, by the heavy machinery used to plant, harvest and transport the crop meant that soil compaction was a major problem in the previous farming system. Soil physical properties improved markedly when beds were widened to accommodate controlled traffic. The introduction of a rotation crop reduced populations of fungal and nematode pathogens that were constraining crop production. A reduction in tillage increased earthworm populations, with consequential effects on macroporosity and water infiltration rates. In rainfed situations, improved water capture in periods of low rainfall contributed to yield increases, while improved percolation through macropores and retention of surface cover protected soils from erosion during intense tropical storms. There have also been signs of improvement in some chemical, biochemical and biological properties associated with soil health (Stirling *et al.*, 2010), and surface soils are now suppressive to *Pratylenchus zeae*, the main nematode pest of sugarcane (see Chapter 11 for details). Further improvements are expected to occur over time, but it is likely to take at least 15 years to fully realize the benefits from the new farming system (Stirling *et al.*, 2010, 2011b).

In summary, these Australian examples show that: (i) the principles of conservation agriculture are applicable in diverse environments and quite different farming systems; (ii) major changes in farming systems can be made relatively quickly in an environment where there is a strong level of agronomic research and good communication between scientists, extension personnel and farmers; and (iii) the economic and other benefits of conservation agriculture (e.g. reduced labour costs, much lower energy inputs, improved timeliness of operations and greater profitability) are so compelling that growers are generally willing to consider making changes to their farming system.

Although minimum tillage, residue retention and crop rotation interact together to improve a whole range of physical, chemical and biological properties associated with soil health, this does not mean that farming systems based on these practices are problem-free. Numerous issues are the subject of continuing research (e.g. nutrient stratification; management of herbicide-resistant weeds; overcoming soil structural problems during the transition to minimum till; alternative crops for inclusion in rotations), while growers may need to modify some management practices to fit the soil and climatic conditions on their farms. It is also recognized that soil organisms are a major determinant of a soil's productive capacity, and that further research is needed to fully harness the biological potential of soil.

Indicators of soil health

Soil health cannot be measured directly, because a soil's capacity to produce crops and also safeguard the environment is determined by numerous physical, chemical and biological properties, and the way they interact. However, the literature is replete with lists of measurable properties that collectively provide a broad indication of the health of a soil, and can be used to assess the impact of soil management practices on soil health (e.g. Doran and Parkin, 1994; Pankhurst *et al.*, 1997). A variety of physical, chemical and biological parameters are usually measured during the soil-health assessment process (e.g. Idowu *et al.*, 2009) and a subset of these indicators is then used to compile a relatively simple report that is designed to help growers make management decisions (e.g. Gugino *et al.*, 2009). Although such reports are useful, the practicality of measuring numerous parameters is often questioned, as certain parameters (e.g. total carbon and labile carbon) are often closely correlated with the physical and chemical properties assessed in soil-health tests.

Although it is recognized that most soil properties are ultimately determined by interactions between soil organic matter and the soil biota, biological measurements are rarely included in soil-health tests, as levels of organic matter and active carbon are often used as surrogates for biological activity and diversity. Hundreds of 'potentially useful' biological indicators have been proposed, but in many cases they simply reflect the discipline bias of the proponent. Also, most do not meet the criteria proposed by Doran and Zeiss (2000) as being required for any biological parameter that is to be used as an indicator of soil health: (i) sensitivity to variations in management; (ii) well-correlated with beneficial soil functions; (iii) useful for elucidating ecosystem processes; (iv) comprehensible and useful to land managers; and (v) easy and inexpensive to measure. This prompted Ritz *et al.* (2009) to look for biological indicators that not only provided information on important soil functions, but were also suitable for use in national-scale soil-monitoring programmes. A list of top-ranked indicators was developed (Box 4.1), but at the end of the process, the authors recognized that many of the selected indicators did not

Box 4.1. Biological indicators for use in monitoring soil health or quality

Ritz *et al.* (2009) ranked the plethora of biological methods that have been suggested for monitoring soil quality and produced a list of 21 biological indicators that were considered ecologically relevant. However, four of these could not be deployed in national-scale monitoring schemes, as further methodological development was required. The remaining indicators have been consolidated into eight groups, and the authors' comments on each of these groups are outlined as follows.

Soil microbial taxa and community structure using molecular techniques

Recent advances in molecular technologies have provided a range of methods that can be used to monitor various components of the soil biological community. Although terminal restriction length polymorphism (TRFLP) analysis has been used widely, the advent of faster, cheaper and higher-resolution sequencing technologies is providing many other options. Molecular methods are useful because high throughput is possible; information on biodiversity is obtained; results can be related to function; and DNA can be archived. However, further work is required to identify the most suitable primers, and to optimize the polymerase chain reaction (PCR), restriction and fingerprinting steps for particular groups of organisms. Also, these methods are extremely sensitive and discriminatory, and so it is not yet known how they are best applied in field situations, where spatial and temporal variability is the norm.

Soil microbial community structure and biomass from phospholipid fatty acids

Extracted lipids, in particular phospholipid fatty acids (PLFA), can be used as signature lipid biomarkers in studies of soil microbial communities. The total PLFA content is indicative of total viable biomass. Individual PLFAs (or suites thereof) can be related to community structure, as they are found predominantly, but not exclusively, in distinct microbial groups (e.g. fungi, bacteria, Gram-negative bacteria and actinobacteria). The main advantage of PLFA profiling is that it is semi-quantitative, does not rely on cultivability and provides wide coverage of the soil microbial community.

Soil respiration and carbon cycling from multiple substrate-induced respiration

Carbon cycling is fundamental to soil function, and the respiration of CO_2 from soils, arising from community-level biotic activity, is an intrinsic indicator of carbon cycling. Since measurement of this property in isolation does not provide discrimination, a multiple substrate-induced respiration (MSIR) approach is more useful, as it characterizes how a soil community responds when exposed to a range of carbon substrates of differing chemical status. As it is a laborious process to generate MSIR profiles, this method is constrained by difficulties involved in achieving high-throughput systems.

Biochemical processes from multi-enzyme profiling

Biochemical reactions in soils are mediated by enzymes produced by the soil biota. A plethora of individual enzymes can be profiled, relating to virtually any defined biochemical transformation. However, a multiple enzyme fluorometric approach is particularly useful, as sensitive measurements can be made on small samples; high-throughput assay systems are possible; and several ecological processes can be assessed in a single assay. Another advantage of this approach is that many different fluorescently labelled substrates are available to target carbon-transforming enzymes, and phosphatase and sulfatase activity.

Nematodes

The potential of nematodes as biological indicators has long been recognized, as they are abundant in soil and have a wide range of feeding habits. The total number of nematode taxa, the abundance of individual functional groups and a wide range of indices that reflect the composition of the nematode community are widely used as indicators. Since nematode extraction methods are laborious, and highly trained experts are required to identify nematodes (even to functional group level), ultimately nematode community analysis will be carried out using molecular techniques.

Continued

Box 4.1. Continued.

Microarthropods

Acari (mites) and Collembola (springtails) have been proposed as potential biological indicators because they are the dominant arthropods in many soils. Extraction methods are fairly straightforward, but the use of arthropods as indicators has been limited by the need for expert skills in identification, and by concerns about which metrics are most useful in ecological studies.

On-site visual recording of soil fauna and flora

Organisms that are readily visible (ants, earthworms and fungal fruiting bodies) are considered useful biological indicators because data can be collected relatively easily. However, consistent methodologies are required before such parameters can be used in on-site recording.

Pitfall traps for ground-dwelling and soil invertebrates

Pitfall traps are a well-established technique for assessing ground-dwelling and soil invertebrates, and are used widely in environmental surveillance. However, one disadvantage is that return visits are required to collect data.

---

satisfy all of the criteria considered essential by Doran and Zeiss (2000). Also, they did not know whether the indicators would prove to be reliable when used across diverse landscapes, and under environmental conditions that vary from season to season.

Soil microbial biomass is one of the simplest and most widely used means of estimating a soil's biological status, but it did not appear in the suite of top-ranked indicators identified by Ritz *et al.* (2009), largely because it was seen as a relatively gross measure that did not discriminate between various components of the soil biological community. Gonzales-Quiñones *et al.* (2011) generally agreed with that assessment, and argued that soil biomass data would be more useful if critical values were established at a regional level for specific soil type × land use combinations. They also indicated that the relationship between soil microbial biomass measurements and management practice would have to be better understood before farmers could use this parameter as a reliable indicator of soil health.

Although a paucity of reliable biological indicators limits the value of most soil-health assessments, an even greater problem is that soilborne pests and pathogens are usually ignored. A soil cannot be considered healthy if the crops that are grown in it suffer losses from soilborne diseases, and yet few of the data sets used to evaluate soil health attempt to quantify populations of nematode or insect pests; measure the inoculum density of particular pathogens; or assess the suppressiveness of the soil to key pests or pathogens. The latter characteristic is a particularly useful indicator of soil health, because a capacity to prevent soilborne pests from multiplying to destructive levels demonstrates that an active and diverse biological community is present, and that the regulatory functions within the soil food web are operating effectively. Unfortunately, however, there is not yet any simple way to assess suppressiveness to root diseases. Pathogen-specific bioassays are time-consuming and labour-intensive; multiple measurements are required to monitor characteristics that are related to suppressiveness (e.g. resilience in the face of a disturbance or stress event); and the various microbial parameters that have been assessed do not show consistent relationships to suppressiveness (van Bruggen and Semenov, 2000; Janvier *et al.*, 2007).

Ecological Knowledge, Biotic Interactions and Agricultural Management

Modern agriculture faces the twin challenges of being both highly productive and sustainable, and as Gliessman (2007) pointed out, this can only be achieved by applying ecological concepts and principles to the design and management of agroecosystems. Although there is general agreement on the need for an ecological approach to farming, discussion on what constitutes ecologically sound agriculture is often polarized by arguments about what type of farming system (e.g. traditional, organic, biodynamic, conventional, integrated, community-based, free-range, low input, etc.) is best able to provide the world's food needs with the least environmental impact (Trewavas, 2001, 2004; Badgley et al., 2007; Badgley and Perfecto, 2007; Cassman, 2007). However, when farming systems are viewed from an ecological perspective, they all contain both positive and negative elements. Thus, it is more enlightening to focus on the impact of management on the ecosystem services provided by the soil biota: production of food and fibre; maintenance of soil structure; storage and supply of nutrients; retention and release of water; and regulation of populations of soil organisms.

Management effects on the soil biota and the limiting role of the environment

It should be clear from this and preceding chapters that the provision of ecosystem services by the soil biota is intimately linked to the quantity and quality of soil organic matter. Management practices determine carbon inputs and losses, with flow-on effects to the soil biota, but levels of soil organic matter will also be influenced by climatic factors (particularly temperature and precipitation) and numerous soil properties (e.g. texture, depth and mineral composition). Thus, management plays an important role in determining the biological status of a soil, but the environment limits what is achievable.

The role of environmental factors in limiting carbon sequestration in soils was recognized by Ingram and Fernandes (2001), who used the term 'attainable' to describe the amount of carbon that could be sequestered in a particular situation, given the limitations of soil and climate. Since there is a close association between soil organic matter and the soil biota, this conceptual framework was extended by Gonzales-Quiñones et al. (2011) and used to explain the size of the soil microbial biomass pool. *Potential* soil microbial biomass was considered to be the maximum microbial population that could be sustained given otherwise non-limiting conditions, and was constrained by inherent site and soil characteristics (Fig. 4.1). For any particular land-use system, the *attainable* target value for soil microbial biomass was defined by factors controlling inputs of organic carbon to soil. Any management practice that increased carbon inputs (e.g. greater net primary production due to fertilization or irrigation) would tend to increase *attainable* soil microbial biomass towards the *potential*. For soil-monitoring purposes, two lower limits were added. *Critical* was the soil microbial biomass value below which biological soil function was lost irreversibly, while *constraining* described the situation where soil biological function still occurred but was at the lower limit of the desirable range of values (Fig. 4.1). Once appropriate *attainable* values were determined, management action could be initiated to move the *actual* value towards the desired target.

Although these concepts are relevant to soil carbon and microbial biomass, there is no reason why they could not be applied to other soil biological characteristics of interest (e.g. the level of nutrient cycling; the extent of biodiversity; the rate of a particular biochemical transformation; or the level of suppressiveness to a particular pest or pathogen). The advantage of visualizing soil biological attributes in this way is that it encourages scientists and land managers to consider the edaphic and climatic constraints that limit what is possible in a specific situation, and then think about how management might move levels of a particular attribute from current or *actual* levels towards an *attainable* target. The challenge for biological scientists is to provide farmers with some indication of what targets are achievable in a given soil type, land use and environment.

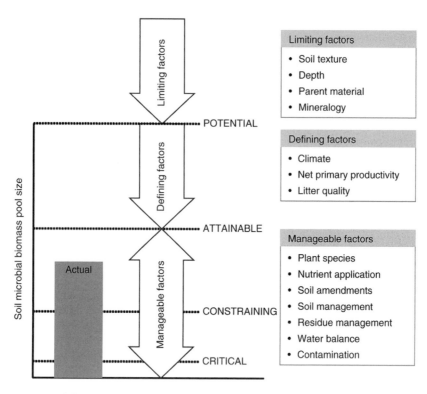

Fig. 4.1. Conceptual diagram showing the size of the soil microbial biomass pool in relation to target values and their defining factors. (From Gonzales-Quiñones *et al.*, 2011, with permission.)

Provision of ecosystem services by the soil biota and the role of management

The soil biological community is responsible for providing a range of ecosystem services, and when these natural services are lost due to biological simplification, they must be replaced by external inputs (Altieri, 1999; Shennan, 2008). Thus, insecticides, nematicides and fungicides are required when regulatory processes are no longer effective enough to suppress pests and pathogens. The essence of ecologically sound agriculture is to understand the biological interactions involved in the provision of a desired service and then use that knowledge to manipulate the biota through management, thereby minimizing or eliminating the need for external inputs. In the case of crop nutrition, this process might involve: (i) knowing the level of nutrients required to replace the nutrients removed in the harvested product; (ii) reducing leaching losses by increasing the soil's cation exchange capacity; (iii) minimizing denitrification losses through better drainage; (iv) including legumes in the rotation to supply biologically fixed nitrogen; (v) optimizing mycorrhizal symbioses with plant roots to enhance phosphorus and micronutrient uptake; (vi) understanding the mineralization and immobilization processes associated with soil organic matter, and then adjusting residue management practices so that naturally cycled nutrients are available when they are required by the crop; (vii) ensuring that nutrients are not applied in excess of crop requirements; and (viii) applying nutrients in forms that do not impact negatively on key components of the soil biota.

Given the number of issues listed above, it should be apparent that taking an ecological approach to nutrient management is a huge challenge. Biotic interactions are important mediators of nutrient availability, but we are not yet able to connect population and community ecology with fluxes of nutrients. Also, the interactions involved in biological nutrient cycling are complex and often multitrophic, and our understanding of the processes involved is too rudimentary to be used successfully in crop management (Plenchette *et al.*, 2005; Goulding *et al.*, 2008). Results from organic farming systems, where nutrient-cycling processes form the basis of crop nutrition, are testimony to the difficulties involved in taking such an approach to maintaining soil fertility: yields are generally lower than in conventional farming systems (Posner *et al.*, 2008); the biological community is not always enhanced by organic inputs (Shannon *et al.*, 2002); and high levels of mycorrhizal colonization do not necessarily overcome phosphorus deficiency (Kirchmann and Ryan, 2004). Thus, there is a need for research to better understand and manage the microbially mediated processes that impact on soil nutrient dynamics. A return to more diverse crop rotations and increased use of cover or catch crops, for example, is likely to improve soil health and provide some pest and disease control, but may have both positive and negative effects from a nutritional perspective. Fertilizer nitrogen requirements may be reduced and nutrient-use efficiency may improve, but more nitrates may be leached from the profile, an outcome which indicates that changes in practice do not always produce positive results (Goulding *et al.*, 2008).

A similar situation applies to another important service provided by the soil biota: regulation of soilborne pests and diseases. There is plenty of evidence to demonstrate that biological mechanisms of suppression operate in agricultural soils; that they act against a wide range of pests and pathogens; and that they are influenced by management (Baker and Cook, 1974; Stirling, 1991; Hoitink and Boehm, 1999; Kerry, 2000; Weller *et al.*, 2002; Stone *et al.*, 2004). Sometimes the suppressive agents are highly specialized antagonists, while in other cases suppression is associated with the wider biological community. However, we are not yet able to define the type of biological community and the level of biological activity required to maintain populations of specific pests and pathogens at levels that do not cause economic damage.

Integrated Soil Biology Management

The material presented in this chapter outlines the many issues that must be considered by anyone wishing to manage soil biotic interactions in such a way that crop productivity is maintained or improved; soil health is enhanced; off-site environmental impacts are minimized; natural processes are contributing to crop nutrition; and regulatory mechanisms are providing some pest and disease control. Clearly, this is an overwhelming task that involves what was referred to in Chapter 1 as 'integrated soil biology management': a management approach that not only requires crop-specific and site-specific knowledge, but a capacity to integrate various practices into a productive and sustainable farming system that provides a full range of ecosystem services. Although the complexities associated with this approach are obvious, it is only by considering the full gamut of issues that affect crop production (crops, soils, climate, water, organic matter, crop rotations, tillage practice, nutrients, weeds, pests, diseases and beneficial organisms) that farming systems can be improved. Improvements will always be incremental, but that is to be expected, given the inherent complexity of all farming systems.

The essence of integrated soil biology management is to manage agroecosystems at a farm and landscape level so that services such as soil formation, carbon sequestration, nutrient cycling, water regulation and pest and disease suppression are maximized and disservices such as nutrient runoff, sedimentation of waterways and greenhouse gas emissions are minimized. The issues involved were depicted previously in Fig. 1.1 and discussed by Power (2010) and Powlson *et al.* (2011a). Although the approach involves manipulating soil, crops and farm animals to achieve the desired effects, the term 'integrated soil biology management' is used because the outcomes are manifested through the soil biota.

Managing soil organic carbon is central to integrated soil biology management because the quantity and quality of soil organic inputs affect the activity and diversity of organisms within the soil food web, and they, in turn, influence numerous soil properties relevant to ecosystem function and crop growth. Since organic carbon levels in many cropped soils have declined to the point where agricultural productivity is being compromised, practices that promote carbon sequestration are actively promoted in the literature on sustainable farming systems, and soil organic carbon is always seen as a keystone indicator of soil health. There is no particular critical threshold for organic carbon that is applicable across different soils and environments (see review by Loveland and Webb, 2003), but it is clear that carbon levels in most agricultural soils are well below the maximum level attainable in a given soil type and location. Raising total carbon levels normally provides benefits, but as the above review indicated, the 'active' or 'fresh' fraction seems to be more important in determining services associated with soil structure and aggregation. Since these fractions support the soil biota, they are also the key to maintaining functions such as nutrient cycling and biological suppressiveness.

The organic carbon content of a soil is the result of a balance between inputs (plant roots, root exudates, plant residues and manures) and outputs (evolution of CO_2 due to respiration by the soil food web, leaching of soluble organic carbon compounds down the soil profile, and particulate losses from erosion). However, in the absence of substantial inputs of manure or other organic materials, it is difficult to influence these processes and increase the total organic carbon content of arable soils. Nevertheless, this is a worthwhile objective, as small changes in total carbon content can have disproportionally large effects on a range of soil properties (Powlson *et al.*, 2011a). Thus, from the perspective of

enhancing the activity and diversity of the soil biological community, farmers must focus on adopting practices that increase total carbon inputs, maintain labile carbon fractions in the rooting zone and reduce the rate of decomposition of organic matter. Such practices include rotational cropping, intercropping with perennials, green manuring, residue retention, mulching and minimum tillage.

Since nutrient cycling/recycling is one of the ecosystem services provided by soil organisms, plant nutrition is one of the key elements to be considered when attempts are made to manipulate the soil biological community. Crop residues and organic amendments are the only source of nitrogen and other plant nutrients in low input and organic farming systems, and their availability to the crop is mediated by the soil biota. However, in conventional farming systems, these natural decomposition and mineralization processes are largely overridden, with crop nutrient requirements being supplied as mineral fertilizers. The challenge of integrated soil biology management is to manage carbon inputs so that biological processes provide a greater proportion of the nutrients required by the crop. Instead of applying a single pulse of fertilizer at excessive rates (because leaching or other losses are expected later in the growing season), the aim would be to develop better ways of utilizing the nutrients gradually made available from soil reserves, crop residues and biological nitrogen fixation. The plants' additional growth requirements would be satisfied with judiciously placed and appropriately timed inputs of organic or synthetic fertilizers at critical periods.

As we learn more about the organisms that provide these nutrient-cycling services, and are more easily able to enumerate them using molecular methods, this type of management will become increasingly feasible. For example, in a series of studies cited by Ferris (2010), bacterial-feeding nematodes mineralized nitrogen throughout the growing season, but due to the different temperature adaptations of the nematodes involved, the nutritional service was mainly provided by rhabditids in spring and cephalobids in summer (Fig. 4.2). In another example, concentrations of mineral

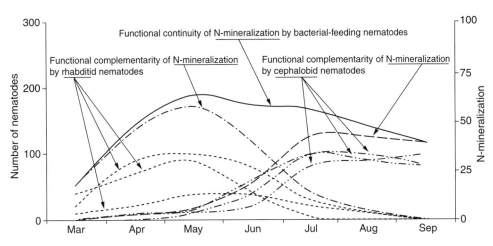

Fig. 4.2. Functional continuity of nitrogen mineralization (based on lifetime µg N per individual) for diverse assemblages of rhabditid and cephalobid nematodes adapted to temperature conditions in California during spring (March–May) and summer (June–September), respectively. (From Ferris, 2010, with permission.)

nitrogen available to a tomato crop were enhanced by manipulating cover cropping and irrigation practices to create conditions suitable for biological activity, particularly bacterivore nematodes (Ferris *et al.*, 2004). Since environmental conditions that favoured bacterivore nematodes probably also favoured other microbial grazers, including protozoa, the abundance of these nematodes was considered a useful indicator of overall grazing activity and nitrogen mineralization rates from soil fauna. Therefore, both these examples suggest that identifying the key organisms responsible for nutrient-cycling services and then manipulating their populations through management is a pathway to more sustainable nutrient usage.

One ecosystem service that has been lost in modern agriculture is the capacity of the system to suppress soilborne pests and diseases. Plant-parasitic nematodes and other soilborne pathogens normally do not cause problems in natural systems because they are kept under control by organisms that compete with them for the same food resource, and by parasites and predators at higher trophic levels in the soil food web. This interacting biological community remains active and diverse because it is continually maintained by carbon inputs from plants. The suppressive services provided

by this community have largely disappeared from agricultural soils because the soil food web is repeatedly disrupted by cultivation; beneficial organisms are disturbed or killed by pesticide and nutrient inputs; vital food reserves are depleted because carbon is exported as harvested product and not replaced; and decomposition processes that convert organic matter to CO_2 are accelerated by tillage. Integrated soil biology management means thinking holistically about all the practices used to produce crops; recognizing that the suppressive services provided by the soil food web are affected by management; and then redesigning the farming system so that those services begin to operate effectively. This issue is explored in more detail in Chapter 11.

Ecologically Based Management Systems and the Role of Farmers

It is easy to idealize that ecologically based decision making should play a greater role in agriculture. However, achieving this in practice will require mechanisms for dealing with the complexity that is inevitably associated with production systems that rely heavily on ecological processes. Shennan (2008) encapsulated the issues in this way: (i) in real farming systems, multiple variables interact in site-specific and farmer-specific ways; (ii) the outcomes of complex biotic interactions can be unpredictable and idiosyncratic; (iii) crop-specific and site-specific knowledge, together with an understanding of general system behaviour, is required to manage the whole system; and (iv) management complexity and perceived higher risks (relative to the continued use of chemical inputs) are a significant barrier to the wider adoption of ecological agriculture. It was also noted that discipline-based researchers are not well equipped to cope with this complexity, as they study only a few variables and try to control others in their attempts to understand the mechanisms driving ecological interactions.

Given the difficulties involved in managing ecologically based systems of agriculture, there is a need to consider how growers can assimilate the knowledge that is generated by research and use it to improve their farming systems. The traditional approach to agricultural extension has been to develop management recommendations based on mechanistic research and then extend them to farmers in different growing regions. Shennan (2008) argued that this research/extension model was no longer applicable and should be replaced by an interactive learning model involving farmers and researchers. This process would accommodate knowledge held by both parties and should, therefore, result in two outcomes: farmers would increase their understanding of ecological processes so they could better adapt management approaches to suit their own situation, while researchers would become more aware of the multiple variables that were interacting within a real farming system and could begin to consider how they might be responding to management. The new model would be accompanied by an increase in field-based adaptive research, with an emphasis on monitoring performance as adaptations were made.

Implications for Biological Control

The first section of this chapter highlights the fact that food security will be one of this century's key global challenges. More food must be produced from existing resources without damaging the environment, and this will largely be achieved through a process of sustainable intensification. Those involved in agriculture will continue to innovate, but proven technologies associated with conservation agriculture and site-specific crop management will form the basis of future farming systems. Thus, in the foreseeable future, farmers in many countries face the challenge of introducing practices such as reduced or zero tillage, cover cropping, residue retention, crop rotation and site-specific management into crop production systems where such practices are not yet used widely. Given the complexity of farming systems, together with the natural predisposition of farmers to resist change, and the fact that many governments stifle innovation by subsidizing agriculture, this will not be an easy task. Nevertheless, on-farm research sites must be established to demonstrate that farming systems can be modified, and agronomists and extension personnel will need to work with farmers to develop and then fine-tune the new systems. Nematologists must be involved in this process, because the task of reducing losses from nematode pests, and the problems involved in manipulating the free-living nematode community to improve nutrient cycling and regulatory processes, are inextricably linked to the development of crop and soil management systems that improve soil health, increase productivity and enhance sustainability.

Once farmers begin to use variable-rate-application equipment and other methods to optimize nutrient inputs, and crop moisture stress is reduced by minimizing evaporative losses, increasing the soil's water-holding capacity and improving irrigation management, nematologists will have to address the issue of whether plant-parasitic nematodes are still causing economic losses. Stress associated with inadequate water or nutrient uptake exacerbates damage caused by nematodes (Barker and Olthof, 1976; McSorley and Phillips, 1993), and so it is likely that problems caused by nematodes will become less severe when these stress factors are ameliorated or removed. Thus, estimates of the severity of losses caused by nematodes (see Box 3.4) may no longer be relevant, as they were obtained in farming systems that in many cases are markedly different to those operating today. Crop loss estimates must, therefore, be redetermined under conditions where crops are grown in a more sustainable manner, and water and nutrients are managed using the best available technologies. The critical question to be answered is whether the optimization of crop management results in situations where damage thresholds increase to a point where plant-parasitic nematodes become a relatively minor constraint to crop production. All available evidence suggests that nematodes will cause fewer problems when environmental stresses are minimized, opening up opportunities to manage them using strategies that focus on improving soil and plant health rather than eliminating the pest.

From an ecological perspective, the main benefit from introducing some or all of the practices associated with conservation agriculture will be to enhance levels of soil organic matter, particularly in surface layers. Given the role of soil organic matter and its associated biota in improving soil moisture relations and in storing and cycling nutrients, this improvement will have major flow-on effects to the crop loss/damage threshold issue discussed in the previous paragraph. However, it will also affect the natural processes that regulate populations of nematodes and other soil-borne pathogens. Organic matter-mediated disease suppression is a widely recognized phenomenon (Stone *et al.*, 2004), and so levels of suppressiveness should increase when soil is no longer tilled and growers begin to see organic matter as a valuable resource rather than something that interferes with farm operations. The challenge facing nematologists is to understand how the quantity, quality and timing of organic inputs, and how they are managed, influence the regulatory processes that stabilize populations of plant-parasitic nematodes. There are many questions to be answered:

- What level of soil organic carbon is required to achieve useful levels of suppressiveness?

- How does this vary with soil type and climate?
- Are particular forms or fractions of soil organic matter associated with suppressiveness?
- What practices must be implemented to maintain the continuity of regulatory services?
- What suppressive agents are involved and does their relative importance vary with crop, soil type and environment?
- What easily measured parameters can be used as indicators of suppressiveness?
- What is the impact of potentially disruptive treatments (e.g. tillage, nutrient inputs, pesticides) on suppressiveness?
- Are particular soil properties (e.g. clay content) associated with suppressiveness?
- What is the role of environmental factors such as moisture and temperature?
- Can organic inputs be manipulated to achieve higher or more constant levels of suppression?

These issues are discussed later in this book, but it will be apparent from the discussion in Chapter 9 that we have taken no more than a few small steps towards understanding how organic inputs influence natural regulatory processes in managed ecosystems. Organic amendments, for example, are often used as a nematode management tool, but outcomes are essentially unpredictable due to our lack of knowledge of the chemical and biological processes associated with the decomposition of organic matter.

Complexity is a theme that runs throughout this book. Biological complexity is inevitable when dealing with terrestrial ecosystems, as all six kingdoms of life (Bacteria, Archaea, Protista, Plantae, Fungi and Animalia) occur in soil and there is incredible diversity within each kingdom. Since it is simply impossible for one person to be familiar with the taxonomic and ecological attributes of all these organisms, soil biologists tend to specialize. Thus, if science is to ever come to grips with the complexity of the soil biota, interdisciplinary cooperation must occur, and biologists working with soil must have a broad understanding of soil ecology.

When the complexity associated with modern farming systems is added, it should be apparent that biological control of nema-todes cannot be considered in isolation from the many other economic, environmental, productivity and sustainability issues that affect the way crops are grown. It is for this reason that integrated soil biology man-agement, rather than biological control, is emphasized throughout this book. Biological control will only become a component of nematode management programmes if the interactions required to achieve it operate within farming systems that are both profit-able and sustainable. Thus, the first step in establishing effective methods of biologic-al control is to work towards developing farming systems that provide a better biotic and abiotic environment for growing plants.

Once those systems are in place, the stresses associated with poor soil health should be minimized, while the mechanisms regula-ting nematode populations should start to function as a result of interactions between plants, crop residues, the nematode commu-nity and the environment. In many situa-tions, no further steps will be necessary, other than the ongoing process of steward-ship, as nematodes will no longer be pests of major economic concern. However, there will also be situations where a particularly virulent nematode pest is attacking a crop, and in those cases, the level of nematode control may have to be increased by fine-tuning the new farming system or introduc-ing other management tactics.

Section III

Natural Enemies of Nematodes

———————————

5

Nematophagous Fungi and Oomycetes

Of all the natural enemies of nematodes, the nematophagous fungi are the most diverse. They are found in many different taxonomic groups within the fungal kingdom, and use a variety of mechanisms to capture and kill nematodes. Hundreds of species have been described, and this chapter makes no attempt to cover them all. Instead, it focuses on the most widely studied groups, discusses how they parasitize or prey on nematodes, and reflects on the ecological characteristics most likely to affect their capacity to suppress nematode populations. Further general information on this group of fungi can be found in Barron (1977), Gray (1987, 1988), Morgan-Jones and Rodriguez-Kabana (1988), Jansson and Nordbring-Hertz (1988), Siddiqui and Mahmood (1996), Chen and Dickson (2004a) and Nordbring-Hertz *et al.* (2006).

One issue that impacts on any discussion of the nematophagous fungi is the many taxonomic changes that have occurred in the last 15 years. DNA analysis is now used routinely in fungal systematics, and the results of such analyses have often challenged historical groupings of species at the genus level, which were previously based on morphology. A list of fungal names, together with commonly used synonyms (Table 5.1) will hopefully help overcome some of the current confusion. Phylogenetic studies have also resulted in changes at higher classification levels (Hibbett *et al.*, 2007), and as the following list shows, fungi that use nematodes as a nutrient source are widely distributed across the fungal kingdom.

- Ascomycota
    - Hypocreales (*Drechmeria, Harposporium, Hirsutella, Fusarium, Pochonia, Purpureocillium*).
    - Orbiliales (*Arthrobotrys, Brachyphoris, Dactylella, Dactylellina, Drechslerella, Duddingtonia, Gamsylella, Monacrosporium, Orbilia*).
- Basidiomycota (*Nematoctonus, Hohenbuehelia*)
- Blastocladiomycota (*Catenaria*)
- Zoopagomycotina (*Cystopage, Stylopage, Rhopalomyces*)
- Entomophthoromycota (*Meristacrum*)

Additionally, there are several nematophagous genera that are oomycetes, including *Haptoglossa, Myzocytiopsis, Nematophthora* and taxa of the Lagenidiaceae. The Oomycota belong to the kingdom Straminipila (Chromista), but are often thought of as fungi due to similarities in their mycelial growth habits and modes of nutrition.

Table 5.1. Names of nematophagous fungi and oomycetes commonly mentioned in this chapter, and some synonyms.

Current name	Synonyms
Arthrobotrys eudermata	*Monacrosporium eudermatum*
Arthrobotrys gephyropaga	*Monacrosporium gephyropagum, M. cionopagum, Gamsylella gephyropaga*
Arthrobotrys musiformis	
Arthrobotrys oligospora	
Arthrobotrys paucispora	*Geniculifera paucispora*
Arthrobotrys psychrophila	*Monacrosporium psychrophilum*
Arthrobotrys thaumasia	*Monacrosporium thaumasium*
Arthrobotrys superba	
Brachyphoris oviparasitica	*Dactylella oviparasitica*
Catenaria auxiliaris	*Tarichium auxiliare*
Dactylellina ellipsospora	*Monacrosporium ellipsosporum*
Dactylellina candidum	*Arthrobotrys haptotyla, Dactylella candida, Monacrosporium haptotylum, Dactylellina haptotyla*
Dactylellina lysipaga	*Monacrosporium lysipagum*
Drechmeria coniospora	*Meria coniospora*
Drechslerella dactyloides	*Arthrobotrys dactyloides*
Hirsutella rhossiliensis	
Haptocillium balanoides	*Cephalosporium balanoides, Verticillium balanoides*
Lecanicillium lecanii	*Verticillium lecanii*
Macrobiotophthora vermicola	*Entomophthora vermicola*
Myzocytiopsis vermicola	*Myzocytium vermicola*
Myzocytiopsis glutinospora	*Myzocytium glutinospora*
Nematophthora gynophila	
Plectosphaerella cucumerina	
Pochonia chlamydosporia (teleomorph = *Metacordyceps chlamydosporia*)	*Verticillium chlamydosporium, Cordyceps chlamydosporia*
Purpureocillium lilacinum	*Paecilomyces lilacinus*

Taxonomy, Infection Mechanisms, General Biology and Ecology

Nematode-trapping fungi in the order Orbiliales

Trapping structures

Several groups of fungi produce specialized trapping devices to capture nematodes, but genera in the order Orbiliales are by far the most common and widely studied. Usually referred to as either nematode-trapping or predatory fungi, these fungi sometimes produce conidial traps spontaneously, but more commonly respond to the presence of nematodes by producing mycelial traps that capture and kill their prey. The following five types of trapping device are recognized, and are depicted in Fig. 5.1.

ADHESIVE NETWORKS. This is the most common trapping mechanism. A lateral branch grows from the vegetative mycelium and then curves around and fuses with the parent hypha to form a loop. The trap may remain as a single ring, but in most species, more loops are formed to produce a two- or three-dimensional network of traps that are covered with adhesive material.

ADHESIVE KNOBS. The trapping device is a globose or subglobose knob that is sometimes sessile on the hypha but is more commonly borne on a slender, short, erect stalk of one to three cells. The trap is covered with adhesive. Nematodes are either caught by the knobs, or will break free with knobs still attached. In the latter case, the fungus is still able to infect the nematode.

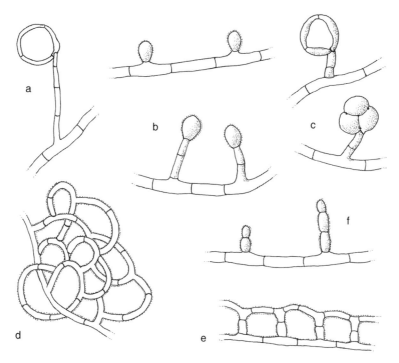

Fig. 5.1. Trapping devices used by nematode-trapping Orbiliales to capture nematodes. (a) non-constricting rings; (b) adhesive knobs; (c) constricting rings; (d) three-dimensional adhesive network; (e) two-dimensional adhesive net; (f) adhesive branches.

CONSTRICTING RINGS. This type of trap is the most sophisticated of all the trapping devices. An erect branch from the hypha curves over at the apex and follows a circular pathway until it fuses with the original hypha to form a three-celled ring on a short, stout stalk. The inside diameter of the rings is about 20 μm, and when a nematode moves into the ring, the cells ring inflate instantaneously and hold the victim securely. Assimilative hyphae then fill the body of the prey and consume its contents.

NON-CONSTRICTING RINGS. These traps are formed in the same way as constricting rings, but the cells do not inflate when a nematode moves into the ring. Also, the support stalk is longer and often breaks when a nematode is wedged in the ring. Fungi producing non-constricting rings will often produce adhesive knobs as well.

ADHESIVE BRANCHES. These short, erect trapping structures are only a few cells in height. They are covered with adhesive and arise from short laterals that grow from prostrate hyphae. Sometimes branches cannot be distinguished from unstalked knobs, and since one forms from another, the two trap types are often formed side by side (Scholler *et al.*, 1999).

Taxonomy

Traditionally, the nematode-trapping fungi were assigned to three genera (*Arthrobotrys, Dactylella* and *Monacrosporium*) on the basis of the morphology of their conidia and conidiophores (Fig 5.2). The shape, size and septation of conidia; the degree of branching of the conidiophores; and the presence of denticles, branches and other modifications at the tip of the conidiophore were the most important taxonomic characters. In *Arthrobotrys*, terminal heads of conidia were borne on denticles that formed on the swollen tips of erect conidiophores. Conidia usually had a single septum, but some species had none, or several septa. In *Dactylella*, single elongate, spindle-shaped

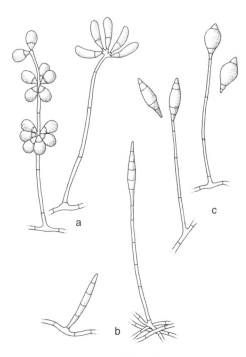

Fig. 5.2. Features of conidial and conidiophore morphology traditionally used to differentiate various genera of nematode-trapping fungi. (a) *Arthrobotrys*; (b) *Dactylella*; and (c) *Monacrosporium*.

to cylindrical conidia did not have an inflated central cell and were produced at the tips of simple conidiophores. In *Monacrosporium*, the conidiophores were simple, or formed a few short branches near the apex. Conidia were spindle-shaped and had two or more septa, but the central cell was the longest and widest cell. Cooke and Godfrey (1964), Van Oorshot (1985) and Rubner (1996) revised the taxonomy of this group of fungi, and their publications provide a good indication of the taxonomic status of the three genera at the time of publication.

Arthrobotrys, *Dactylella* and *Monacrosporium* reproduce asexually (i.e. they are anamorphs), and they have sexual (teleomorph) stages in the genus *Orbilia*, family Orbiliaceae (Pfister, 1997). *Orbilia* is characterized by small (usually less than 3 mm in diameter) variously coloured apothecial ascomata that are found on the surface of logs, fallen tree trunks and decayed bark in forested areas. When nematodes are added to

cultures derived from ascospores, typical nematode-trapping organs are commonly induced. Examples showing such teleomorph-anamorph connections within the Orbiliaceae can be found in Pfister (1994), Pfister and Liftik (1995), Yang and Liu (2005), Yu *et al.* (2006, 2007) and Qiao *et al.* (2012). However, not all anamorphs of *Orbilia* are nematophagous. For example, Webster *et al.* (1998) isolated *O. fimicoloides* (syn. *Dactylella oxyspora*) from dung pellets and found no evidence that its anamorphic stage produced trapping organs or captured nematodes. Other non-predatory examples are given by Yang and Liu (2005) and Yu *et al.* (2007).

Molecular studies of the Orbiliales commenced in the 1990s using internal transcribed spacer (ITS) and 18 S rDNA sequences, and the results indicated that trapping devices were more informative than other morphological structures in delimiting genera (Liou and Tzean, 1997; Ahrén *et al.*, 1998; Hagedorn and Scholler, 1999; Scholler *et al.*, 1999). Based on these studies, Scholler *et al.* (1999) divided the predatory taxa into four genera based on the type of traps produced: *Arthrobotrys* for adhesive networks, *Drechslerella* for constricting rings, *Dactylellina* for stalked adhesive knobs and non-constricting rings, and *Gamsylella* for adhesive columns and unstalked knobs. In a later study, Li *et al.* (2005) argued that *Gamsylella* did not warrant generic status, as unstalked knobs continue to grow and eventually form branches that can then fuse to form loops or two- or three-dimensional networks. They also used sequences of 28S rDNA, 5.8S rDNA and β-tubulin genes to show that the members of the group that formed loops were closely aligned to *Dactylellina*, while network-forming species were better grouped with *Arthrobotrys*. However, studies cited later in this chapter have shown that one of the species transferred from *Gamsylella* to *Arthrobotrys* (*G. gephyropaga*) is more predacious than typical *Arthrobotrys* spp. in soil, raising questions about its placement in that genus.

In addition to the predatory Orbiliales, there are also species that are known to attack other hosts such as copepods, collembolans, rotifers, mites and fungi. Rubner (1996) argued that the genus *Dactylella* should be reserved for non-nematophagous species, and this contention was supported by Chen *et al.* (2007a, b) who used ITS sequence alignment to show that anamorphic species of Orbiliales that do not form trapping devices should remain in the genus *Dactylella*. A new genus *Brachyphoris* was also established in a companion paper (Chen *et al.*, 2007c) to accommodate *Dactylella oviparasitica*, a species that produces very short conidiophores and is a parasite of nematode females and eggs. A pictorial view of the main characters now used to differentiate genera of nematode-trapping, parasitic and non-nematophagous Orbiliales is presented as Fig. 5.3.

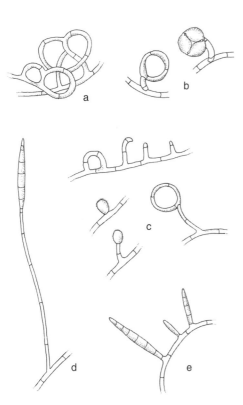

Fig. 5.3. Key taxonomic characters of some nematode-trapping, parasitic and non-nematophagous Orbiliales at a generic level. Trapping species are differentiated on the basis of their trapping structures into three genera: (a) *Arthrobotrys*; (b) *Drechslerella*; and (c) *Dactylellina*. Non-parasitic and parasitic genera are differentiated, respectively, on the basis of conidiophore length into (d) *Dactylella* and (e) *Brachyphoris*.

In summary, the names of many of the nematode-trapping fungi commonly used in biological control studies have changed in recent years, creating confusion for those who are not specialists in this group of fungi. Recent changes to the rules (the *International Code of Nomenclature for algae, fungi and plants*) that regulate the naming of fungi mean that from 1 January 2013, one fungus can only have one name. In cases where a fungus has anamorphic and teleomorphic names (as with many nematode-trapping fungi), the name chosen will be based on priority (first published), with exceptions granted for some names that are widely used (Hawksworth, 2012). Thus, further name changes are likely to occur in future, as lists of accepted names are developed.

Occurrence

Nematode-trapping fungi are relatively easy to detect in soil. A small quantity of soil (approximately 1 g) is sprinkled on a water agar plate, a suspension of nematodes is added as bait and within 1–2 weeks, trapped nematodes, trapping organs and conidia will be visible at low magnification under a microscope. Results of surveys carried out using this 'sprinkle-plate' method (summarized by Gray, 1987) have shown that nematode-trapping fungi are commonly found in agricultural, forest and garden soils, and are especially abundant in soils rich in organic matter. However, it is now clear that culture-based detection methods do not provide a full picture of the diversity of nematode-trapping fungi in soil, as initial studies with a PCR-based detection technique revealed many uncultured Orbiliales that were closely related to nematode-trapping fungi and fungal parasites of nematode eggs (Smith and Jaffee, 2009). Nevertheless, the sprinkle-plate method will remain useful in ecological studies, because it detected more nematode-trapping Orbiliales than the molecular assay in the study cited above. Presumably the molecular assay detected Orbiliales with abundant DNA in the sampled substrate, whereas sprinkle plates detected species that were present at low population densities but increased in response to the presence of nematodes.

The nematode-trapping fungi are so diverse that their biology and ecology cannot be discussed without taking species differences into consideration. Some species are highly dependent on nematodes as a food source; others exist mainly as saprophytes; while an intermediate group have the capacity to switch from one form of nutrition to the other. Since the relative importance of each mode of nutrition influences the way a nematode-trapping fungus interacts with nematodes, it is almost impossible to succinctly summarize the way this group of fungi behave in soil. Nematode-trapping fungi are nutritionally variable, and so their ecological role must be considered at an individual species level. Thus, information on the biology and ecology of particular species has been incorporated into relevant sections of this chapter.

Fungal and oomycete parasites of vermiform nematodes

Stylopage

Fungi previously referred to as zygomycetes use a variety of mechanisms to parasitize nematodes, rhizopods and rotifers. The most widely studied genus is *Stylopage*, which captures nematodes on sparse, irregularly branched, adhesive hyphae. Once a nematode has been captured, an appressorium-like outgrowth develops at the point of contact and haustoria or infection hyphae develop within the host (Saikawa, 2011). Common species such as *S. hadra* also produce erect conidiophores bearing one or a few large conidia (Fig. 5.4). *Stylopage* is commonly found in surveys for nematode-trapping fungi, but our knowledge of its biology and ecology is almost non-existent, largely because it has never been cultured in the absence of its host.

Catenaria

Catenaria anguillulae, a saprophytic fungus capable of parasitizing nematodes, nematode eggs, rotifers and tardigrades, is a member of a small family of zoosporic fungi formerly categorized as chytridiomycetes but now reclassified in a new phylum, Blastocladiomycota (Porter *et al.*, 2011). The zoospores have a single whiplash flagellum and they differentiate within zoosporangia inside the host body (Fig. 5.5), escape through the tip of a solitary exit tube and swim towards their nematode prey. On reaching a host, the spores encyst and germ tubes either enter the body through orifices or penetrate directly through the cuticle to initiate a new infection. Since the zoospores of *C. anguillulae* must quickly locate a host because they have limited energy reserves, the widespread occurrence of this fungus is undoubtedly due to its ability to utilize many common substrates in soil (Barron, 1977).

Those who have worked with *C. anguillulae* have generally considered it to be a relatively weak parasite of nematodes. Boosalis and Mankau (1965) believed that it was largely saprophytic on dead or injured microfauna; Sayre and Keeley (1969) concluded that it was not an effective control agent, as the level of parasitism achieved in liquid suspensions was not obtained in soil or sand; while Jaffee and Shaffer (1987) found that the fungus persisted in soil for 3 months but did not reduce numbers of *Xiphinema*. *C. anguillulae* was also a relatively poor parasite of root-knot nematode, as observations *in vitro*

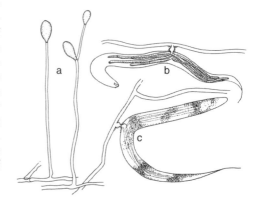

Fig. 5.4. *Stylopage hadra*. (a) Conidiophores and conidia; (b) recently captured nematode containing assimilative hyphae; (c) assimilative hyphae empty of contents and partly obscured by bacteria. (From Barron, 1977, with permission.)

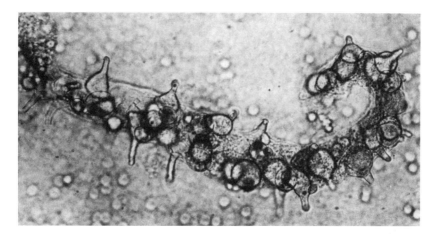

Fig. 5.5. Zoosporangia of *Catenaria anguillulae* inside an infected nematode, with exit tubes being produced to the exterior. (From Barron, 1977, with permission.)

indicated that zoospores did not readily encyst on the egg surface; juveniles ready to hatch were not killed; and those that had hatched were never attacked (Wyss *et al.*, 1992). However, these negative results contrast with observations that high levels of natural parasitism due to *C. anguillulae* occur in some agricultural fields in India (see references cited in Singh *et al.*, 2007). However, the results of a study purporting to show that *C. anguillulae* regulates populations of *Meloidogyne graminicola* on rice were not convincing. Levels of parasitism were recorded after nematodes retrieved from the field were incubated in cavity blocks for four days (Singh *et al.*, 2007), and so it is not clear whether the infections observed occurred in the field or developed during the incubation process.

C. *auxiliaris*, the only other species of *Catenaria* known to parasitize nematodes, is quite different from *C. anguillulae*, as it attacks saccate females of endoparasites rather than vermiform nematodes. In his description of the fungus, Tribe (1977) noted that it completely destroyed young females of *Heterodera schachtii*. However, if infection occurred at a later stage of development, females were largely destroyed but eggs were unharmed. *C. auxiliaris* produces rhizomycelium typical of *Catenaria*, but unlike *C. anguillulae*, each swollen segment develops into a precursor sporangium that either develops into a relatively round zoosporangium with about six exit tubes, or into a resting spore. Tribe (1977) considered that *C. auxiliaris* was related to *C. spinosa*, a parasite of midge eggs, and since the ribosomal DNA evidence of Porter *et al.* (2011) indicated that the latter species grouped separately from *C. anguillulae*, the taxonomy of *Catenaria* may need revision.

C. *auxiliaris* is mainly associated with cyst nematodes, being widespread on *H. schachtii* in Europe, (Tribe, 1977), but it also occurs on *H. avenae* in England and Australia (Kerry, 1975; Stirling and Kerry, 1983), and on *H. glycines* in the United States (Crump *et al.*, 1983). Since *C. auxiliaris* has never been cultured, it may have a more obligate relationship with nematodes than *C. anguillulae*. However, the above surveys suggest that levels of parasitism are usually relatively low. The fact that species of *Catenaria* require water for zoospore movement will limit their distribution, and this is almost certainly the reason that *C. auxiliaris* is restricted to moist, fine-textured soils in Australia (Stirling and Kerry, 1983).

A recent report indicates that *C. auxiliaris* has a wider host range than just the cyst nematodes, as it was found parasitizing the reniform nematode, *Rotylenchulus reniformis*, in Alabama (Castillo and Lawrence, 2011). Since an unidentified species of *Catenaria* has also been associated with entomopathogenic nematodes in Florida citrus orchards (Duncan *et al.*, 2007, 2013; Pathak *et al.*, 2012), it is to be hoped that these recent reports will stimulate interest in the biology of this under-researched group of fungi.

Nematoctonus, Hohenbuehelia and Pleurotus

When *Nematoctonus* was originally described by Drechsler (1941), the presence of clamp connections indicated that the fungus was a basidiomycete, but it was placed in the deuteromycetes because there was no evidence of sexual reproduction. However, *Nematoctonus* was later found to be the anamorph of *Hohenbuehelia* (Barron and Dierkes, 1977), and this observation enabled Thorn and Barron (1986) to connect many *Hohenbuehelia* species with their *Nematoctonus* anamorphs. Since *Hohenbuehelia* is a gilled mushroom, Thorn and Barron (1984) tested other gilled fungi for nematophagous activity and found that species in two other genera (*Pleurotus* and *Resupinatus*) were also able to attack and digest nematodes. These genera are closely related, and phylogenetic analyses support *Hohenbuehelia-Nematoctonus* as a monophyletic clade of the Pleurotaceae (Thorn *et al.*, 2000; Koziak *et al.*, 2007).

Studies with the oyster mushroom (*Pleurotus ostreatus*) showed that nematodes are captured by adhesive knobs (Saikawa and Wada, 1986). The captured nematodes are rapidly immobilized by a nematoxin (transdecenedioic acid), and hyphae then grow towards the orifices of the immobilized nematode and penetrate, colonize and digest the prey (Thorn and Barron, 1984; Barron and Thorn, 1987; Kwok *et al.*, 1992). Phylogenetic analyses support the idea that a common ancestor of *Pleurotus* and *Nematoctonus* was able to produce a nematoxin, but that this capacity appears to have been lost in *Nematoctonus* (Koziak *et al.*, 2007). Instead, *Nematoctonus* produces hourglass-shaped adhesive knobs on hyphae, adhesive knobs on conidia, or both types of trapping devices, and uses them to capture nematodes (Fig. 5.6).

Species of *Hohenbuehelia* are commonly associated with rotting wood (Barron, 1992), but are also found in soil. However, virtually nothing is known about their ecology in either habitat, raising questions about their importance in agricultural soils, where woody substrates are seldom available. It is possible that these lignolytic and cellulolytic fungi are involved in degrading recalcitrant crop residues, and that species of *Hohenbuehelia* only capture nematodes when nitrogen is limiting (see the discussion later in this chapter), but there is no experimental evidence to support either hypothesis. *Nematoctonus concurrens* and *N. haptocladus* both reduced nematode populations when conidia were added to sterile sand, but severe mycostatic and lytic effects limited their efficacy in soil (Giuma and Cooke, 1974). On the basis of this result, these species were considered poor biological control agents, but that may have been an unreasonable conclusion, given that the conidia were added to soil 7 days after it was amended with green cabbage leaf tissue at a rate equivalent to 50 t/ha. Considerable amounts of nitrogen would have been added with such an amendment, and since we now know that *Hohenbuehelia-Nematoctonus* is associated with high-carbon, low-nitrogen substrates, future work should concentrate on its predacious activity when associated with such substrates.

Drechmeria

Drechmeria coniospora is one of the most distinctive fungal endoparasites of nematodes, as it produces conical or teardrop-shaped conidia that have a small adhesive bud at the distal end. These spores generally adhere in clusters around the sensory organs of healthy nematodes, and after infection has occurred and the fungus has fully occupied the body of its host, reproductive hyphae emerge through the cuticle and groups of conidia are produced from the cells of the conidiophore (Fig. 5.7). Originally described by Drechsler (1941) as *Meria coniospora*, the genus *Drechmeria* was later erected to accommodate this species and a closely related species parasitic in ciliated protozoans (Gams and Jansson, 1985). More recent taxonomic studies placed *D. coniospora* in the family Clavicipitaceae, a fungal family in the perithecial ascomycete order Hypocreales that

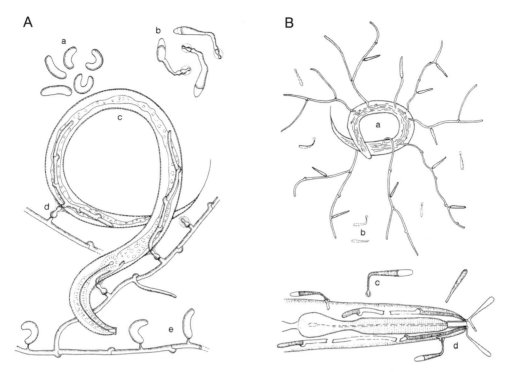

Fig. 5.6. Modes of infection by different species of *Nematoctonus*. (A) A predatory species showing (a) conidia; (b) germinating conidia producing an adhesive knob; (c) nematode caught by adhesive knobs on hyphae; (d) hourglass-shaped adhesive knob; and (e) conidia being produced on hypha. (B) An endoparasitic species showing (a) an infected host with conidiiferous hyphae; (b) and (c) an erect outgrowth with terminal adhesive bud produced by fallen conidia; and (d) infected host with conidia attached to the cuticle. (From Barron, 1977, with permission.)

contains many parasites of insects and nematodes (Gernandt and Stone, 1999).

Most research on *D. coniospora* has focused on host recognition, spore attachment processes and elucidation of infection mechanisms (e.g. Dijksterhuis *et al.*, 1990; Jansson and Friman, 1999), with much of this work having been done with the bacterivorous nematode *Panagrellus redivivus*, an excellent host of the parasite. Other bacterivores are also hosts, and the only ecological studies with *D. coniospora* have involved this group of nematodes (see discussion later in this chapter). Plant-parasitic nematodes seem to be poor hosts, as *Ditylenchus dipsaci*, *Tylenchorhynchus claytoni* and *Meloidogyne incognita* were the only species to be infected in laboratory tests with 12 nematode species in 10 different genera (Jansson *et al.*, 1985a, 1987). The observation that root-knot nematode was susceptible to *D. coniospora* led to

greenhouse experiments which showed that galling caused by this nematode was reduced when the fungus was added to potting mix as a conidial suspension or as infected nematodes (Jansson *et al.*, 1985b). However, despite these promising results, the fungus was never developed into a formulated product suitable for nematode control, largely because later work with a commercial partner indicated that it was difficult to mass produce (H-B. Jansson, pers. comm.).

Harposporium

The genus *Harposporium* contains *H. anguillulae*, the first fungus recorded as a parasite of nematodes, and many other species that are pathogens of nematodes, rotifers and tardigrades. Barron (1977), Esser and El-Gholi (1992) and Glockling (1993) provide brief overviews of these fungi, which attack bacterivorous nematodes that commonly occur in high numbers in situations where organic matter is decomposing (e.g. soil, leaf mould, rotten wood, dung and compost). Species of *Harposporium* produce a variety of unusually shaped conidia, but relatively long, narrow, curved or helicoid conidia are fairly common. Although the conidia of some species will adhere to the cuticle of the prey, infection generally occurs when spores are ingested by nematodes and lodge in various parts of the digestive tract. The conidia germinate, assimilative hyphae absorb the body contents of the host, and then conidiophores break out through the cuticle and produce phialides. Conidia are formed apically at the end of these phialides (Fig. 5.8). Taxonomic studies based on cultural data and large subunit nuclear ribosomal DNA sequences indicate that *Harposporium* is the anamorph of *Podocrella* and is closely related to other clavicipitaceous genera (Chaverri *et al.*, 2005).

Since stylet-bearing nematodes cannot ingest spores of *Harposporium*, species in this genus have seldom been considered in studies on biological control of plant-parasitic nematodes. However, it should not be forgotten that species of *Harposporium* could have indirect effects, as they will kill nematodes that are potential hosts for other nematophagous fungi. *Harposporium* spp. are well adapted to parasitizing bacterivorous nematodes, and when they occupy the same habitat as other predators, they will compete strongly for that food source.

Interestingly, there may be an association between *Harposporium* and *Hirsutella*, the next fungus to be discussed in this chapter.

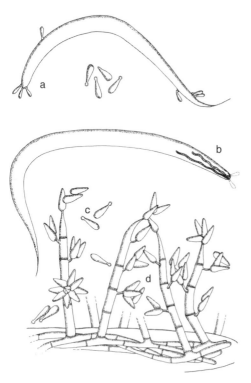

Fig. 5.7. *Drechmeria coniospora.* (a) adhesive conidia attached to the cuticle of a host nematode; (b) sinuous infection hyphae inside host body; (c) mature conidia showing adhesive bud; and (d) conidia on conidiophores emerging from a nematode cadaver. (From Barron, 1977, with permission.)

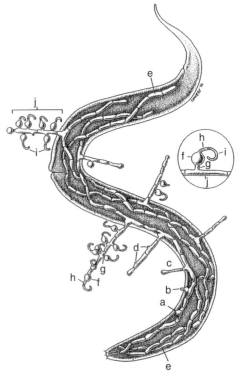

Fig. 5.8. A nematode parasitized by *Harposporium*, showing (a) appressorium; (b) papillate bud; (c) conidiophore growth; (d) lateral buds; (e) mycelia; (f) phialide; (g) pedicel; (h) sterigma; (i) conidia; and (j) mature conidiophore. (From Esser and El-Gholi, 1992, with permission.)

Hodge *et al.* (1997) found that two species of *Harposporium* (*H. anguillulae* and *H. cerberi*) both formed synanamorphs attributable to *Hirsutella* (i.e. they produced helicoid or sickle-shaped conidia typical of *Harposporium* and drop-shaped conidia typical of *Hirsutella*), leading the authors to examine the possibility that *Hirsutella*-type conidia might enhance the fitness of the fungus by enabling it to infect stylet-bearing nematodes. However, infection was not detected in the two nematodes tested (*Aphelenchus* sp. and *Meloidogyne hapla*). Therefore, the ecological significance of the two distinct types of conidia remained unclear, although it was suggested that it may be an adaptation to enable dispersal by different methods, or a mechanism that allows the fungus to attack different hosts.

Hirsutella

When nematodes succumb to parasitism by *Hirsutella rhossiliensis*, hyphae extend from the cadaver of the host, radiate into the soil and bear non-motile, adhesive spores on the tips of bottle-shaped phialides (Fig. 5.9). Spores adhere to passing vermiform nematodes and in the process are detached from the phialides. The spores then germinate; the fungus penetrates the nematode cuticle, occupies the body

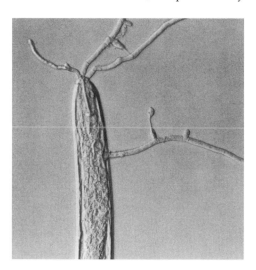

Fig. 5.9. Phialides and spores of *Hirsutella rhossiliensis* on hyphae growing from an infected nematode. Magnification 400×. (From Jaffee, 1992, with permission.)

cavity and kills the nematode. More external hyphae and spores are then produced, with spore production peaking after 3–4 days and then declining to zero when the body contents of the host are exhausted. About 100 spores are produced from infected juveniles of *Heterodera schachtii* in 12 days (Jaffee *et al.*, 1990).

Species in the genus *Hirsutella* are primarily reported as parasites of arthropods, and when Minter and Brady (1980) described *H. rhossiliensis*, it was the only species known to attack nematodes. Later, a second nematophagous species was isolated from second-stage juveniles of *Heterodera glycines* collected from a soybean field in Minnesota and named *H. minnesotensis* (Chen *et al.*, 2000). Both species have been studied extensively because of their potential as biological control agents, and much of that research is discussed later in this chapter and also in Chapters 10 and 12. The following brief overview covers ecological issues that are not considered elsewhere.

Both *H. rhossiliensis* and *H. minnesotensis* have broad host ranges that include plant-parasitic, entomopathogenic and free-living nematodes (Timper *et al.*, 1991; Tedford *et al.*, 1994; Liu and Chen, 2001b). Tests with 25 *H. rhossiliensis* isolates obtained from various sources showed that all isolates colonized *Heterodera schachtii*, *Meloidogyne javanica* and *Steinernema glaseri*, although isolates obtained from *Rotylenchus robustus* colonized nematodes more slowly than other isolates (Tedford *et al.*, 1994). However, when 93 isolates were screened by Liu and Chen (2001a), there was more variation in pathogenicity, with isolates obtained from bacterivorous nematodes not parasitizing second-stage juveniles of *H. glycines*. *H. minnesotensis* was also included in these tests, and it parasitized fewer nematodes than *H. rhossiliensis*.

The role of biotic factors in inhibiting the establishment of *Hirsutella* in soil is discussed in Chapter 12, but many abiotic factors also influence the activity of species in this genus. Jaffee *et al.* (1990) showed that soil texture was important, as *H. rhossiliensis* spores were transmitted more readily to nematodes in loamy sand than in coarse sand, presumably because the larger pore diameter in the coarser-textured soil limited nematode motility and the probability of contact between nematodes and spores. Furthermore, pores wider than 53 μm (the length of the phialide plus spore plus the diameter of the nematode) may allow nematodes to pass spores without contacting them (Fig. 5.10). Since studies with finer-textured soils indicated that parasitism tended to increase as sand content increased, *H. rhossiliensis* is likely to be most active in soils of intermediate texture and least effective as a biological control agent in coarse sands and heavy clays (Liu and Chen, 2009).

Studies on the impact of soil moisture have consistently shown that *Hirsutella* is not an effective parasite at high water potentials. Transmission of *H. rhossiliensis* to *Heterodera schachtii* and *Meloidogyne javanica* decreased linearly with increasing soil moisture, and this appeared to result from reduced sporulation of the fungus rather than reduced movement of the nematodes (Tedford *et al.*, 1992). Since the fungus does not sporulate when submerged, the authors assumed that soil pores had to drain to some extent before sporulation would occur. Assays in soil artificially infested with *H. rhossiliensis* showed that *Pratylenchus penetrans* was less likely to be parasitized as soil moisture increased (Timper and Brodie, 1993), while both the amount of *H. minnesotensis* DNA/g soil and the percentage of *Heterodera glycines* juveniles parasitized by the fungus decreased markedly at high soil water contents (Xiang *et al.*, 2010a).

As expected for a host–parasite interaction, temperature influences nematode–*Hirsutella*

interactions through effects on both the fungus and its host. *H. rhossiliensis* has an optimum temperature for growth on agar of 20–25°C (Velvis and Kamp, 1996), but will infect *Heterodera schachtii* at temperatures of 10–35°C. However, the fungus is most effective in reducing nematode activity at 20°C (Tedford *et al.*, 1995a). Soil temperatures around this optimum were thought to have been one of the reasons that *H. rhossiliensis* parasitized *Heterodera glycines* more heavily during mid-summer than in spring or autumn in a field trial in Minnesota (Chen and Reese, 1999). *H. minnesotensis* seems to have a lower optimum temperature, and this may explain why it is a prominent parasite of soybean cyst nematode in the cooler regions of north-eastern China (Xiang *et al.*, 2010a).

The possibility that soil pH may have affected the activity of *H. rhossiliensis* was first noted by Jaffee (1999), who found that when introduced as alginate pellets into two vineyards, the fungus performed well in only one of two vineyards, possibly because of pH differences. Soil pH was nearly neutral where the fungus performed poorly and quite acidic where it performed well. Follow-on studies in which soil from the neutral site was acidified and soil from the acidic site was neutralized showed that the activity of pelletized *H. rhossiliensis* was highest in acid soil and declined to near zero as pH increased to 6.5 (Jaffee and Zasoski, 2001). However, when the neutral soil was heated, fungal activity increased to levels equivalent to the acidified soil. This suggested

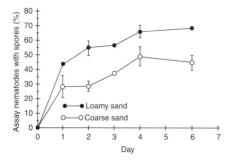 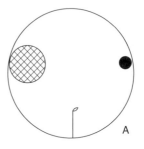

Fig. 5.10. Impact of soil texture on infection of assay nematodes (second-stage juveniles of *Heterodera schachtii*) by *Hirsutella rhossiliensis*. The cross-sectional diagrams on the right show the influence of pore diameter on the likelihood of a spore contacting a nematode host. A and B represent pores 150 μm and 40μm, respectively, and show a phialide with spore, second-stage juveniles of *H. schachtii* (solid circles), and a female of *Mesocriconema xenoplax* (circle with stripes). (From Jaffee *et al.*, 1990, with permission.)

that the pH effect may have been indirect, with soil organisms that interfered with the growth of *H. rhossiliensis* from alginate pellets being suppressed by low soil pH. Results of a Chinese survey and assays of soils in Minnesota also suggest that *Hirsutella* prefers low pH soils, as the occurrence of *H. rhossiliensis* and *H. minnesotensis* in the survey and levels of parasitism in the assays were highest in soils with the lowest pH (Ma *et al.*, 2005; Liu and Chen, 2009). However, the reasons for the pH effect are not clear, as the Chinese soils where the fungus was found had higher levels of organic matter, and presumably higher levels of biological activity, than soils where the fungus was not detected.

In conclusion, it is apparent from this brief discussion that many factors influence the activity of *Hirsutella* spp. in soil. Since *H. rhossiliensis* and *H. minnesotensis* are the most-studied endoparasites of nematodes, the range of abiotic factors affecting these species is a reminder of how little we know about the ecology of the many other fungal and oomycete antagonists of nematodes discussed in this chapter.

Nematophagous oomycetes: Myzocytiopsis, Haptoglossa, Nematophthora *and* Lagenidiaceae

The nematophagous oomycetes are generally considered relatively primitive organisms, producing simple holocarpic thalli within the host body and reproducing asexually by means of zoospores, aplanospores or chlamydospores, or sexually through oospore production. However, the methods used by these organisms to attack their motile hosts belie the primitive designation, as the infection mechanisms are complex, sophisticated and highly efficient. Since the taxonomy, biology, infection strategies and evolutionary phylogeny of these parasites has been comprehensively reviewed (Glockling and Beakes, 2000a; Beakes *et al.*, 2012), these issues are discussed only briefly here.

Most of the biflagellate parasites of nematodes and rotifers were initially placed in the genera *Myzocytium* and *Lagenidium*, but a taxonomic review by Dick (1997) separated

them from closely related algal parasites by moving them to a new genus, *Myzocytiopsis*. The genus *Chlamydomyzium* was also erected to accommodate chlamydospore-producing former members of the Lagenidiales. One species (*C. oviparasiticum*) that was originally described as an aggressive parasite of rotifer eggs, is also able to attack nematodes, and its development within *Rhabditis* has been described (Glockling and Beakes, 2006a). Other nematophagous oomycetes capable of attacking vermiform nematodes are found in two long-established genera, *Haptoglossa* and *Gonimochaete*.

Within *Myzocytiopsis*, species differ in the manner in which they infect nematodes. Aplanosporic species produce elongate, tapered aplanospores that germinate on release from the sporangium to form an adhesive tip that attaches to passing nematodes. However, most *Myzocytiopsis* species are zoosporic, with some species (e.g. *M. lenticularis*) producing zoospores that swim to the host and encyst on the cuticle, while others (e.g. *M. humicola*, *M. vermicola* and *M. glutinospora*) have only a short swimming phase before the zoospores lose their flagella and encyst in the environment. The encysted spores then develop an adhesive bud that attaches to the cuticle of a passing nematode (Fig. 5.11). Regardless of the mode of infection, infection thalli are produced within the host once the cuticle has been penetrated, and they eventually develop into sporangia in which asexual spores are formed. All *Myzocytiopsis* species also reproduce sexually, with the thick-walled oospore providing a means of survival in adverse conditions. An ultrastructural study of *M. vermicola* provides a detailed account of the manner in which one of these parasites infects nematodes (Glockling and Beakes, 2006b). The cited paper also shows how the parasite develops within *Rhabditis*, and details the processes of asexual and sexual reproduction.

Although *Haptoglossa* may be no more widespread than many other genera of nematophagous oomycetes, the infection cell that is used to rupture the host cuticle has proved to be of particular interest to scientists due to its complexity and level of sophistication. As mentioned earlier, the genus contains

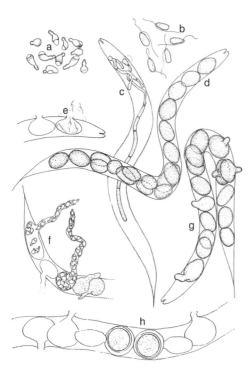

Fig. 5.11. *Myzocytiopsis humicola.* (a) encysted zoospores producing adhesive buds; (b) biflagellate zoospores; (c) young infection thalli inside host; (d) and (g) host filled with linear series of zoosporangial segments; (e) zoospores exiting a zoosporangium; (f) adhesive spores formed from zoospores that have encysted inside zoosporangia; (h) sexual state showing oogonia with oospores inside. (From Barron, 1977, with permission.)

both aplanosporic and zoosporic species, but regardless of the type of asexual spore, these spores form an infection spore that differentiates into an infection cell at maturity. This cell was termed a 'gun' cell by Robb and Barron (1982) because a harpoon-shaped projectile housed within the cell was propelled through the host cuticle when the tip was touched by a nematode or rotifer. Detailed ultrastructural studies (Beakes and Glockling, 1998; Glockling and Beakes, 2000b, 2002) have shown that the gun cell contains an elaborate inverted tubular system which functions as a retracted hypodermic syringe. When stimulated to fire by the touch of a passing nematode, the needlelike projectile breaches the host cuticle and the finger-like tube is ejected directly into the host cytoplasm, where it expands to form the infective sporidium inside the host (Fig. 5.12).

Little is known about this group of oomycetes from an ecological perspective. They are commonly found in wet terrestrial habitats rich in organic matter, but also occur in soils used for agriculture. For example, in one of the most comprehensive surveys for nematophagous fungi ever undertaken (1733 soil samples collected mainly from cultivated soils in Queensland, Australia), *Myzocytiopsis vermicola, Haptoglossa heterospora* and *Gonimochaete horridula* were recovered from 2.1%, 6.1% and 1.7% of the samples, respectively (McCulloch, 1977). The capacity of these parasites to suppress nematode populations is unknown, but our inability to grow most of them in pure culture suggests that they are relatively specialized parasites of nematodes. Most nematophagous oomycetes have been studied in bacterivorous nematodes, but *Myzocytiopsis glutinospora* is also known to attack the entomopathogenic nematode *Heterorhabditis marelatus*, increasing to high population densities (>10^3 propagules/g soil) when moth larvae killed by this nematode were added to microcosms (see the section on nematophagous fungi and entomopathogenic nematodes later in this chapter).

Two oomycete parasites of *Heterodera* females are commonly found in English cereal-growing soils, and both were described by Kerry and Crump (1980). Their role in suppressing populations of cereal cyst nematode is discussed in Chapter 10. A *Lagenidium*-like organism was only found in a few nematodes, whereas the other oomycete (*Nematophthora gynophila*) was much more common, killing large numbers of *Heterodera avenae* when the soil was wet during the period when white females were first exposed on the root surface (Kerry *et al.*, 1980). Recently infected nematodes contained broad hyphae that gave rise to sporangia, and their discharge tubes penetrated the disrupted cuticle and penetrated about 50 μm into the soil. Zoospores formed within the sporangia and once development was complete, they were rapidly released to initiate new infections. Thick-walled resting spores (oospores) were produced laterally on undifferentiated hyphal segments, and the infected nematode was eventually filled with a spore mass consisting of about 3000 of these spores. Although the parasite will grow saprophytically in axenic culture, growth is

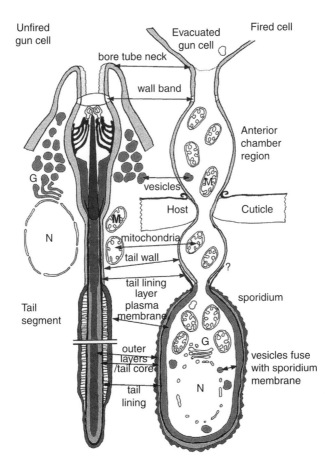

Fig. 5.12. Diagrammatic representation of an inverted tube in an unfired mature gun cell of *Haptoglossa heteromorpha* (left) and an everted tube of a fired cell with a sporidium (right). (From Glockling and Beakes, 2000b, with permission.)

extremely slow, hyphal tips are fragile, the fungus is nutritionally exacting and colonies are comparatively short-lived (Graff and Madelin, 1989). Thus, in nature, *N. gynophila* is almost certainly an obligate parasite capable of growing only within the body of its nematode host.

Cyst and egg parasites

Pochonia

Prior to 2001, many widely distributed parasites of higher plants, fungi, insects, nematodes and rotifers were lumped together in the anamorph genus *Verticillium*. Morphological observations and sequences of the ITS region and SSU and LSU rDNA then showed that this heterogeneous group of fungi could be subdivided into four distinct clades (Zare *et al.*, 2001a; Sung *et al.*, 2001). Clade A encompassed well-known plant pathogens, while clade B contained a diverse collection of entomogenous and fungicolous taxa that were transferred to the genus *Lecanicillium*. *L. lecanii* is a member of this clade, even though it is known to be nematophagous (see later in this chapter). Species in clade C were parasites of free-living nematodes and they were transferred to a new genus, *Haptocillium* (also mentioned later in this chapter). Clade D was dominated by species that mainly parasitize nematode cysts and eggs, and a new genus *Pochonia* was erected to accommodate these species (Gams and Zare, 2001; Zare *et al.*, 2001b).

Four nematophagous species were accommodated in *Pochonia* and a key prepared by Zare *et al.* (2001b) distinguishes them from each other on the basis of conidial shape and the position and abundance of dictyochlamydospores (a stalked, variable-shaped, hyaline, thick-walled, multicellular resting stage that is hereafter referred to as a chlamydospore). Based on whether conidia were arranged in chains or heads, variation within *P. chlamydosporia* and *P. suchlasporia* was also recognized at a sub-species level. However, the two varieties within each species could not be consistently separated on the basis of morphology and were only distinguished using parsimony analysis of sequences from the ITS region. The teleomorph of *P. chlamydosporia* was found to be *Metacordyceps chlamydosporia*, a parasite of mollusc eggs, indicating that this widely distributed species is a parasite of both snails and nematodes. Phialides, conidia and chlamydospores of *P. chlamydosporia* var. *chlamydosporia* are shown in Fig. 5.13.

Since *P. chlamydosporia* is an important pathogen of many economically important plant-parasitic nematodes, but particularly the cyst and root-knot nematodes (*Heterodera* spp. and *Meloidogyne* spp.), most studies of *Pochonia* have focused on this species (see reviews by Kerry and Hirsch (2011) and Manzanilla-Lopez *et al.* (2011a, 2013)). The following is a brief summary of what is known about its behaviour in soil and its interaction with plants, organic matter, nematodes and other soil organisms.

- Isolates of *P. chlamydosporia* vary markedly in pathogenicity (Irving and Kerry, 1986; Hidalgo-Díaz *et al.*, 2000; Moosavi *et al.*, 2010; Yang *et al.*, 2012a; Dallemole-Giaretta *et al.*, 2012). However, isolates tend to be most pathogenic to the host from which they were isolated, with isolates from root-knot nematode often not highly pathogenic to cyst nematodes and vice versa (de Leij and Kerry, 1991; Morton *et al.*, 2004; Mauchline *et al.*, 2004; Siddiqui *et al.*, 2009).
- There is considerable genetic diversity in isolates of *P. chlamydosporia* from different hosts and locations (Morton *et al.*, 2004). Isolates collected from a specific nematode

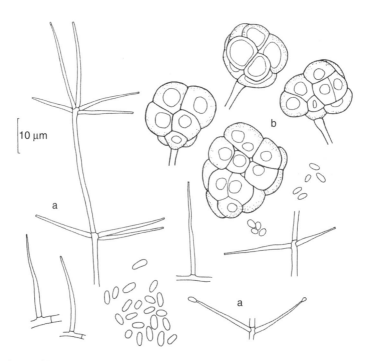

Fig. 5.13. *Pochonia chlamydosporia* var. *chlamydosporia* showing: (a) conidia and conidiophores; and (b) chlamydospores. (From Gams, 1988, with kind permission from Springer Science + Business Media B.V.)

host in a particular region tend to show much less diversity (Hidalgo-Díaz et al., 2000). For example, the dominant biotype of *P. chlamydosporia* var. *chlamydosporia* in tunnel houses used for tomato production in Portugal was identical at two widely separated sites, perhaps because the fungus population had responded to selection pressure during long-term cultivation of crops susceptible to root-knot nematode (Manzanilla- López et al., 2009).

• Chlamydospores are produced as a survival mechanism and do not germinate unless they are triggered by nutrients leaking from roots, or by the presence of organic matter (de Leij et al., 1993).

• *P. chlamydosporia* is a saprotroph in soil, a facultative parasite of invertebrates and a hyperparasite of other fungi. It has been observed growing on the surface of crop residues (Dallemole-Giaretta et al., 2011) and its population density increased markedly when peat was added to sand in a pot experiment (de Leij et al., 1993). In addition to being parasitic on nematodes, the fungus is able to parasitize mollusc eggs (Kerry, 1995) and *Phytophthora* oospores (Sneh et al., 1977; Sutherland and Papavizas, 1991). *P. chlamydosporia* coils around the hyphae of *Rhizoctonia solani* (Jacobs et al., 2003) and reduces disease symptoms caused by the take-all fungus (*Gaeumannomyces graminis*) on wheat (Monfort et al., 2005). However, the manner in which it interacts with soilborne fungal pathogens is poorly understood.

• Females and eggs of endoparasitic nematodes are embedded in roots or located on the root surface, and *P. chlamydosporia* can only gain access to those nematodes by colonizing the rhizosphere. A capacity to use root exudates as a nutrient source and then proliferate in the rhizosphere is, therefore, one of the main factors that determines whether an isolate of *P. chlamydosporia* will be an effective biological control agent (Bourne et al., 1994; Hirsch et al., 2001; Mauchline et al., 2002).

• The population density of *P. chlamydosporia* increases markedly when roots are infected by root-knot nematode, and this increase occurs at about the same time as egg masses are produced on the root surface (Kerry and Bourne, 1996; Bourne et al., 1996; Mauchline et al., 2002).

• Plant species differ in their ability to support the proliferation of *P. chlamydosporia* in the rhizosphere. For example, in a glasshouse test that compared the host status of various winter and summer crops, fungal population densities were highest following pearl millet and oil radish (Dallemole-Giaretta et al., 2011). However, there is no simple relationship between the abundance of *P. chlamydosporia* in the rhizosphere and the proportion of nematode eggs parasitized (Bourne and Kerry, 1999; Siddiqui et al., 2009; Atkins et al., 2009).

• The effectiveness of *P. chlamydosporia* as a biological control agent is influenced by nematode inoculum density, the host status of the plant and its tolerance to nematode attack. In the case of root-knot nematode, the fungus is most effective on plants that are poor hosts of the nematode; on hosts where egg masses are exposed on the root surface rather than being embedded in gall tissue; and in situations where populations of the host nematode are low. Since the fungus does not prevent juveniles from invading roots, it is most effective when plants are tolerant of nematode damage (Kerry, 2000).

• *P. chlamydosporia* is able to colonize roots endophytically under laboratory conditions (Bordallo et al., 2002; Lopez-Llorca et al., 2002), and also in washed sand (Manzanilla-Lopez et al., 2011b). The presence of the fungus in roots and on the root surface has been confirmed using real- time PCR and visualized using molecular beacon probes (Maciá-Vicente et al., 2009a; Escudero and Lopez-Llorca, 2012; Fig. 5.14). Endophytic colonization by *P. chlamydosporia* promotes growth of the host plant (Monfort et al., 2005; Maciá-Vicente et al., 2009b; Escudero and Lopez-Llorca, 2012).

• When *P. chlamydosporia* parasitizes a nematode egg, thin penetration hyphae invade the egg, but appressoria may also be produced on the egg surface (Escudero and Lopez-Llorca, 2012). The egg shell and juvenile cuticle are disrupted as infection proceeds (Morgan-Jones et al., 1983). Although various enzymes (proteases, esterases, lipases and chitinases)

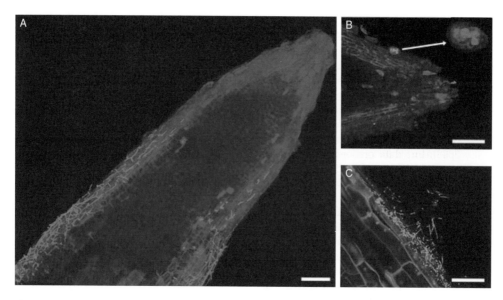

Fig. 5.14. Colonization of tomato roots by a GFP transformant of *Pochonia chlamydosporia* as viewed by laser confocal microscopy. (A) general view of the root apex showing colonization near the meristematic zone 3 days after inoculation; (B) close-up of the root cap showing a chlamydospore (magnified at the end of the arrow); and (C) longitudinal section of the root showing colonization of epidermal cells. Bar in all photographs = 75 μm. (From Escudero and Lopez-Llorca, 2012, with kind permission from Springer Science + Business Media B.V.)

are thought to be involved in the infection process (Morton *et al.*, 2004), enzyme activity *in vitro* was not correlated with either egg parasitism or rhizosphere colonization (Esteves *et al.*, 2009).

The material presented is only a brief summary of what is known about *P. chlamydosporia*, but it indicates that much has been learnt about the biology and ecology of this fungus over the last 30 years. *P. chlamydosporia* clearly has characteristics which suggest that it has potential for development as a biopesticide, as it is a parasite of the two major nematode pests worldwide; it can develop an intimate association with the target nematodes in the rhizosphere; it can survive host-free periods as chlamydospores or by living saprophytically; and it can be cultured on a wide range of substrates. It is, therefore, not surprising that the techniques developed to study the fungus and the knowledge that has been accumulated on host specificity and rhizosphere competence have mainly been used to identify isolates suitable for targeting particular nematodes, and also to monitor mass-produced isolates

following their introduction into soil. Examples of some of that work are presented in Chapter 12.

Despite our growing knowledge of *P. chlamydosporia*, there have been few serious attempts to understand how natural populations or introduced isolates behave in the soil environment, where multitrophic interactions are the norm (i.e. *P. chlamydosporia* interacts with plant roots, organic matter, nematodes, many potential food sources and other soil organisms). Given the complexity of that environment and the range of nutritional options available to the fungus, such studies will not be easy, but the molecular methodologies now available to monitor the fungus provide some of the tools required to address this issue. What is particularly important is to identify the main nutritional sources for the fungus in field soils (e.g. roots, root exudates, eggs of endoparasitic nematodes, fragments of soil organic matter, eggs of vermiform nematodes, other soil organisms); determine the key factors responsible for maintaining populations of the fungus; and understand the factors that cause the trophic switch from saprophytism to egg parasitism. Once these issues are understood, it may be possible to manipulate populations of the fungus by modifying crop or soil management practices.

P. chlamydosporia is often found in association with endoparasitic nematodes in agricultural soils (e.g. Kerry, 1975; Gintis *et al.*, 1983; Wang *et al.*, 2005; Sun *et al.*, 2006; Bent *et al.*, 2008), but data on levels of parasitic activity in the field are often not obtained. Nevertheless, there are situations where high levels of parasitism do occur in agricultural soils. The contribution of *P. chlamydosporia* to the decline of cereal cyst nematode populations in England is discussed in Chapter 10, while Mertens and Stirling (1993) provided data to show that root-knot nematode is sometimes heavily parasitized by the fungus in Australia. When tomato seedlings were planted in a kiwifruit orchard and retrieved when the first generation of egg masses was produced, 41% and 87% of the newly produced eggs were parasitized by fungi in spring and summer, respectively. *P. chlamydosporia* and *Purpureocillium lilacinum* were the predominant parasites and the population of the former fungus was about 5800 cfu/g soil. Since population densities and levels of parasitism of this magnitude are rarely obtained when the fungus is mass produced and applied to soil, this observation suggests that more effort should be put into determining the levels of parasitic activity that actually occur in soils used for agriculture, and understanding the factors that influence fungal population densities and levels of parasitism.

Based on what we know about the ecology of *P. chlamydosporia*, we could predict that a virulent isolate of the fungus would thrive on the endoparasitic nematodes that are continually available in the rhizosphere of an undisturbed perennial crop. Although its nematode host may be inactive in winter due to temperature constraints, the fungus should survive readily in orchards and vineyards during this period, as organic inputs from roots and the mown pasture that is normally grown between rows of trees and vines would provide a food source suitable for saprophytic growth. Experimental evidence is required to confirm these hypotheses, but if the above

factors were found to be important, it is then a matter of redesigning annual cropping systems to achieve an environment that encourages the high levels of parasitic activity observed in perennial cropping systems. Some of the options available to do this are discussed in Chapter 11.

Purpureocillium

The fungus known widely as *Paecilomyces lilacinus* is ubiquitous. It is commonly isolated from soil, decomposing organic matter, insects and nematodes; is a contaminant of laboratory air and medical materials; and is an increasingly important cause of infection in man and other vertebrates. From a nematological perspective, *P. lilacinus* is the most widely researched biological control agent of nematodes, and one strain (*P. lilacinus* strain 251) is registered as a bionematicide in many countries (Kiewnick, 2010). From a medical perspective, infections caused by *P. lilacinus* are most commonly manifested as keratitis and are generally found in patients with compromised immune systems or intraocular lens implants (Luangsa-ard *et al.*, 2011).

In a study aimed at determining whether *P. lilacinus* formed a monophyletic assemblage within the Hypocreales, and whether isolates pathogenic to vertebrates and invertebrates were in different clades, isolates from clinical specimens, soil, insects, nematodes and the general environment were compared using 18S rRNA gene, ITS and partial translation elongation factor (TEF) sequences (Luangsa-ard *et al.*, 2011). Since the results indicated that *P. lilacinus* was not related to *Paecilomyces* (as represented by the type species *P. variotii*), this species was transferred to the new genus *Purpureocillium*. Thus, species belonging to *Paecilomyces sensu stricto* are now characterized by their olive-coloured conidia, the presence of chlamydospores, and a capacity to grow at high temperatures (30–45°C); whereas *Purpureocillium* has lilac conidia, does not produce chlamydospores, and grows at lower temperatures (25–33°C). Furthermore, the molecular study showed that isolates of *Purpureocillium lilacinum* grouped into two clades. Saprobic species formed one small clade, while the major clade contained all isolates originating from clinical specimens,

hospital environments, insect larvae, nematodes and decaying vegetation. Some of the
clinical and nematode isolates shared the same
ITS and TEF sequences. Although harmful
and beneficial isolates of *Purpureocillium
lilacinum* could not be separated by the methods
used in this study, the authors did indicate
that genetic differences may exist at a sub-
species level and that they could possibly be
detected using higher-resolution genotyping
techniques.

There is a voluminous literature on
Purpureocillium lilacinum, and some of it can
be accessed by consulting general texts on
biological control of nematodes (e.g. Morgan-
Jones and Rodriguez-Kabana, 1988; Stirling,
1991; Chen and Dickson, 2004a) and two specific reviews (Cannayane and Sivakumar,
2001; Kiewnick, 2010). However, much of the
research on *P. lilacinum* has focused on its use
as an inundative biocontrol agent, and since
that work is discussed in Chapter 12, it will
not be considered here. Instead, this discussion will focus on the biological and ecological factors that influence the capacity of
P. lilacinum to parasitize nematodes or their
eggs in agricultural soils, and the possible
role of crop and soil management practices in
influencing parasitic activity.

P. lilacinum (Fig. 5.15) is the most-studied
fungal parasite of nematodes for several
reasons. First, it is widely distributed, particularly in tropical and subtropical soils.
Second, it is commonly found parasitizing
the eggs of root-knot nematodes, the world's
most damaging nematode pest, and so it is
seen as a potentially useful biological control agent in many cropping systems. Third,
the eggs of this nematode are aggregated in
an egg mass on the root surface, and so
parasitism is easy to observe and the fungus can be retrieved easily. Fourth, the fungus will grow on a wide range of substrates
and sporulates prolifically, and so inoculum is easy to produce for experimental
purposes.

Studies of the colonization and infection
process with a scanning electron microscope
(Dunn *et al.*, 1982) showed that hyphae of
P. lilacinum grew over the surface of the egg
and rapidly penetrated the egg shell, occasionally through the formation of appressoria

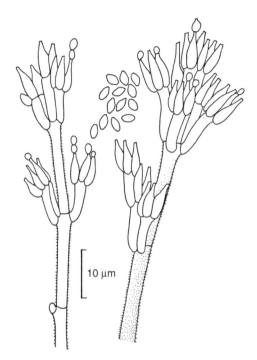

Fig. 5.15. Conidiophores and conidia of
Purpureocillium lilacinum. (From Domsch *et al.*,
1980, with permission.)

but more commonly through simple penetration by individual hyphae (Fig. 5.16). When
sectioned material was examined by transmission microscopy it was apparent that a
number of ultrastructural changes occurred
in the shells of parasitized eggs, and that the
cuticle of juveniles within these eggs was
also disrupted (Morgan-Jones *et al.*, 1984a).
Leucinotoxins, chitinases and proteases are
presumed to play a role in the infection process, as they are produced by *P. lilacinum* in
the laboratory (Khan *et al.*, 2003; Park *et al.*,
2004). However, evidence that they are produced in effective quantities at the site of
infection is lacking.

Given the importance of rhizosphere
competence in influencing the efficacy of
other egg-parasitic fungi (see earlier discussion of *Pochonia chlamydosporia*), the capacity
of *P. lilacinum* to colonize the rhizosphere has
been assessed (Rumbos and Kiewnick, 2006).
When the fungus was introduced as conidia
into a mixture of field soil and sand, comparisons of soil without plants and soil planted

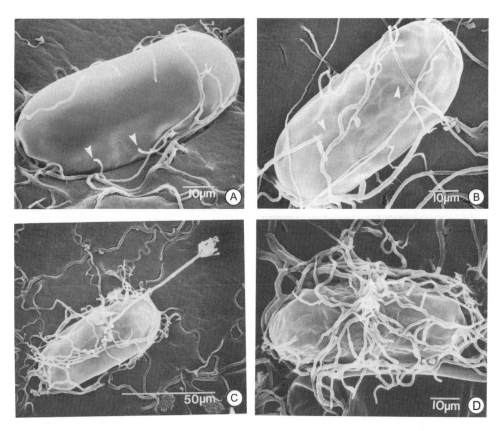

Fig. 5.16. Colonization and infection of *Meloidogyne incognita* eggs by *Purpureocillium lilacinum* on agar. (A) early and (B) later stages of colonization; (C) sporulation on an egg; and (D) a late stage of colonization, showing hyphal proliferation around the egg. (From Dunn *et al.*, 1982, with permission.)

with one of 12 plant species showed that the presence of plants had little impact on *P. lilacinum*, as its population density generally declined by 70–90% from initial levels, whether a plant was present or not. *P. lilacinum* was recoverable from rhizosphere soil, usually at the same population density as in bulk soil, and in cases where a rhizosphere effect was observed (e.g. barley, maize and aubergine), the effect was not consistent from experiment to experiment. The fungus was isolated in significant numbers from healthy barley roots, and endophytic root colonization was also detected in wheat, maize, banana and cabbage. However, in another study with the same isolate, the fungus was not detected in roots of any of the eight plants tested, including wheat and barley (Holland *et al.*, 2003). These

observations suggest that *P. lilacinum* can live endophytically in the roots of some plants; that it sometimes occurs in the rhizosphere; and that it is not strongly dependent on roots as a nutrient source. However, until more isolates and plant species are tested, the latter conclusion may be premature, as oilseed rape and sugarbeet strongly influenced the abundance of *P. lilacinum* in a laboratory test (Manzanilla-Lopez *et al.*, 2011b).

P. lilacinum appears to be a relatively good saprophyte, as it thrives on a wide range of organic materials, including plant tissue, oil cakes and industrial waste products (e.g. Mani and Anandam, 1989). It also grows and sporulates on numerous carbon sources, including substrates that are commonly found in soil (Sun and Liu, 2006), and persists better in the

presence than in the absence of organic matter (Rumbos *et al.*, 2008). Thus, the fungus has
at least four sources of nutrients available to
it in soil: living roots, root exudates, fragments of organic matter and nematodes.
What we do not know is how it responds to
the presence of these nutritional sources in
the natural soil environment. *P. lilacinum* is
widely distributed in agricultural soils (see
following paragraphs for results of surveys
in the United States and Australia), and is
by far the most common fungal species
associated with *Meloidogyne* females and
eggs in China (Sun *et al.*, 2006), but we know
little about the key factors that influence its
population density and levels of parasitic
activity.

In a survey of commercial tomato fields
in California, Gaspard *et al.* (1990a) detected
P. lilacinum in soil samples from most fields,
usually at population densities between 1×10^3
and 15×10^3 cfu/g soil. However, the presence of the fungus was not correlated with
the number of root-knot nematode juveniles
in the samples. Nevertheless, a bioassay with
nematode-infested and non-infested plants
clearly showed that *Meloidogyne incognita* was
an important nutritional source for *P. lilacinum*. Population densities were relatively low
in the absence of the nematode, but growing
a tomato plant infested with root-knot nematode increased the fungal population in several soils to levels $>10^5$ cfu/g soil. *Pochonia
chlamydosporia* was also present in many of
the soils, and in some cases the presence of a
nematode-infested tomato plant caused populations of both *P. lilacinum* and *P. chlamydosporia* to increase, while in other cases, one or
the other of the fungi was favoured.

Stirling and West (1991) used a bioassay with root-knot nematode to compare
levels of egg parasitism in 46 soil samples
representative of agricultural and virgin
soils in tropical and subtropical Australia,
and made three observations relevant to
naturally occurring egg parasitism. First,
Pupureocillium lilacinum and *Pochonia chlamydosporia* were the only fungi consistently
isolated from parasitized egg masses, and
both fungi were commonly obtained from
the same batch of parasitized eggs. Second,
egg parasitism was much more common in

soils where root-knot nematode was present than where the nematode was absent.
Third, high levels of parasitism were
observed in only two soils, and they were
planted to grapes. In both cases, almost all
of the eggs in 75% of the egg masses were
parasitized.

Based on the results of the two surveys
discussed, which were carried out in quite
different environments, it is apparent that
P. lilacinum is likely to be present in
most agricultural soils; that populations of
P. lilacinum can increase rapidly to high
densities when root-knot nematode is present; that *P. lilacinum* often occurs in association with *P. chlamydosporia*; and that
collectively, the two fungi sometimes kill a
large proportion of the eggs of root-knot
nematode. The observation that high levels
of parasitism were only observed in the
two soils planted to grapes (Stirling and
West, 1991) may also be important, because
it suggests that egg parasites are most
likely to be effective when they interact
with their nematode hosts in situations
where the host is permanently present.
Supportive evidence for that hypothesis
was obtained in a study discussed in the
previous section on *Pochonia*. When tomato
seedlings were planted into a vineyard and
a kiwifruit orchard in spring and summer,
the first cohort of root-knot nematode eggs
produced on tomato was heavily parasitized by both *Purpureocillium lilacinum*
and *Pochonia chlamydosporia* (Mertens and
Stirling, 1993).

Another conclusion that can be drawn
from these studies is that populations of
P. lilacinum and levels of parasitism increase as
the host nematode multiplies, and reach a
peak at the end of the growing season. In a
perennial crop, the fungus presumably over-
winters in eggs and also survives saprophytically on decomposing roots, and is, therefore,
in an ideal position to parasitize the first eggs
produced the following spring. However, in
an annual crop, that scenario is unlikely to
occur because *P. lilacinum* and other potentially useful organisms are either disrupted or
destroyed by the practices usually employed
prior to planting the next crop (e.g. aggressive
tillage with a rotary hoe, fallowing and soil

fumigation). Having established a community of beneficial organisms in the ecological niche where they are required (i.e. on the roots of the crop), it seems logical that this community should form the foundation on which the next year's nematode-control programme is built. Thus, future research on *P. lilacinum* and other egg parasites in annual crops should focus on developing cropping systems that preserve the high populations of beneficial fungi likely to be present at the end of each crop. Some of the options available to do this are discussed in Chapter 11.

Brachyphoris, Vermispora and the ARF fungus

The anamorphic genera *Brachyphoris* and *Vermispora* form two major clades within the *Dactylella* complex (Chen *et al.*, 2007a, c), and both clades contain parasites of nematodes. *Brachyphoris oviparasitica* (Fig. 5.17) is a parasite of *Meloidogyne* eggs and *Heterodera* females (Stirling and Mankau, 1978a, b; Westphal and Becker, 2001a), while *Vermispora fusarina* parasitizes eggs of *Heterodera pallida* (Burghouts and Gams, 1989). The nutritional habits of other species are poorly understood, but some capture and consume testaceous rhizopods, and others are known to parasitize oospores of *Pythium* and zygospores of *Cochlonema* (Chen *et al.*, 2007c).

Phylogenetic analyses of *B. oviparasitica* and its closest relatives (Yang *et al.*, 2012b)

have shown that they form a clade (referred to as the DO clade) containing an assemblage of fungi from several countries. Arkansas Fungus isolate ARF-L, an unnamed sterile fungus that parasitizes eggs, juveniles and young females of *Heterodera glycines* (Kim and Riggs, 1991: Kim *et al.*, 1998; Timper and Riggs, 1998; Timper *et al.*, 1999; Wang *et al.*, 2004d), was also found to be a member of that clade. Since *B. oviparasitica* and ARF-L share similar characteristics in terms of parasitism of sedentary endoparasitic nematodes, Yang *et al.* (2012b) suggested that other members of the DO clade may also have nematophagous capabilities.

From a biological control perspective, this group of fungi clearly deserves attention for two reasons. First, evidence presented in Chapter 10 shows that both *B. oviparasitica* and various isolates of the Arkansas Fungus are at least partly responsible for suppressiveness to root-knot and cyst nematodes in several different environments. Second, these fungi are widely distributed, as *B. oviparasitica* is relatively common in California soils (Yang *et al.*, 2012b); was also found in mushroom compost in South Africa (Koning *et al.*, 1996); and was detected in soils under sugarcane in Australia (Stirling *et al.*, 2011b). Also, members of the *Brachyphoris* clade, including *B. oviparasitica*, were present on soil, litter and wood from a natural reserve in California (Smith and Jaffee, 2009), while the isolates used by Yang *et al.* (2012b) in their

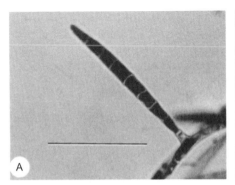
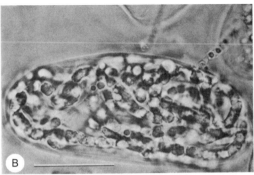

Fig. 5.17. (A) Mature conidium of *Brachyphoris oviparasitica* (stained with acid fuchsin); and (B) a *Meloidogyne* egg parasitized by the fungus. Bars = 25 μm. (From Stirling and Mankau, 1978a, reprinted with permission from *Mycologia* © Mycological Society of America; and from Stirling and Mankau, 1978b, with permission.)

phylogenetic studies were obtained from the United States, Canada, China and several European countries.

The fact that *Brachyphoris* was found using molecular techniques in many of these studies is a reminder that such techniques are particularly useful for detecting slow-growing fungi that are often difficult to isolate from an infected host. Some isolates of *B. oviparasitica*, for example, only grow at rates of 1–2 mm/day (Stirling and Mankau, 1978a) and tend to be overrun in culture media by faster-growing saprophytes. Molecular studies are also showing that *Brachyphoris* is much more widespread than previously recognized, and this finding raises questions about the ecological role of these fungi in soil. *Brachyphoris* is certainly able to parasitize eggs that are aggregated within cysts or egg masses, but given their widespread distribution, these fungi must also be able to utilize other substrates. Perhaps this group of fungi (and possibly other fungi) parasitize nematode eggs that are laid singly in soil or roots. Given the difficulties involved in studying parasites of individual eggs in soil, it is not surprising that this is a neglected area of research. Nevertheless, it warrants attention, because eggs of free-living and non-sedentary plant-parasitic nematodes provide a food source that will obviously be utilized by some soil organisms.

Studies of both *B. oviparasitica* and the Arkansas Fungus have shown that isolates vary enormously in their saprophytic and parasitic abilities. Three isolates obtained from *Meloidogyne* egg masses grew at different rates in culture (Stirling and Mankau, 1978a) and varied in pathogenicity (Stirling and Mankau, 1978b; 1979; Stirling *et al.*, 1979). Similar variation in pathogenicity was observed with the Arkansas Fungus (Kim *et al.*, 1998; Timper and Riggs, 1998; Timper *et al.*, 1999), with one isolate being an excellent saprophyte capable of decomposing wheat straw and soybean roots (Wang *et al.*, 2004c). When such variability is observed, researchers tend to select easily cultured isolates for further study, and will often argue the case for utilizing them in biological products for nematode control. However, such

isolates may not necessarily be the most effective parasites. Another concern is that studies with isolates from the saprophytic end of the nutritional spectrum may lead to inaccurate conclusions being drawn about the importance of saprophytism in the survival of the fungus in soil.

The latter point is important because observations in a field where *B. oviparasitica* was known to suppress *Heterodera schachtii* have shown that the fungus can survive in the absence of the host, but is highly dependent on the nematode for nutrition (Yang *et al.*, 2012b). Two years after a crop capable of hosting the nematode had been grown in the field, a sequence-selective qPCR assay rarely detected *B. oviparasitica* in old cysts that remained in the soil. However, when a host of the nematode (Swiss chard) was planted in the soil, *B. oviparasitica* was readily recovered from *H. schachtii* females. This result suggests that at least some isolates of *B. oviparasitica* are well adapted to parasitism, with the authors speculating that infection is initiated at the late juvenile stage or when the young female first breaks through the root surface and is exposed to the rhizosphere. The poor recovery of *B. oviparasitica* from old cysts also suggests that the fungus is not a good competitive saprophyte. After it has killed the host nematode and occupied the cadaver, it faces competition from other soil microorganisms and its population density declines. Resurgence will only occur when more host nematodes become available. The challenge of the future is to understand how to manipulate populations of the host nematode to favour parasites such as *B. oviparasitica*, and then utilize this knowledge in integrated nematode management programmes.

Other fungi

Since the biomass of fungi in soil often exceeds that of all other organisms, it is not surprising that regardless of the nematode or location, many different fungi will almost certainly be obtained in surveys for fungal parasites of nematodes or their eggs (see Table 10.1 in Chapter 10 for some examples). The results from two surveys are summarized here, simply to show the diversity of fungi normally associated with endoparasitic nematodes. In the first of those surveys, 62 fungal species were isolated from 4500 cysts of *Heterodera glycines* collected in 45 fields across southern Minnesota. Fungi were found in 55% of the cysts, with *Cylindrocarpon destructans*, *Fusarium solani*, *F. oxysporum*, *Pyrenochaeta terrestris* and *C. olidum* the most common species (Chen and Chen, 2002). In the second survey, 28 root and soil samples were collected from various crops in China, and isolations from about 1400 females and 14,000 eggs of *Meloidogyne* spp. yielded 455 fungal isolates in 24 genera. The predominant fungi were *Purpureocillium lilacinum*, *Fusarium* spp., *Pochonia chlamydosporia*, *Penicillium* spp., *Aspergillus* spp. and *Acremonium* spp. (Sun *et al.*, 2006). Having obtained data of this nature, nematode pathologists then face the long and tedious process of assessing pathogenicity on agar or in soil. Nevertheless, this process has now been undertaken by researchers working with various nematodes in many different environments, and it has provided us with an indication of the species most likely to be pathogenic to nematodes, and the species that are mainly saprophytes.

Although this approach is understandable and there is really no alternative, it raises the question of whether the results of pathogenicity tests in the laboratory or greenhouse are relevant in the soil environment, where fungi are almost certainly interacting with each other and with other soil organisms. In that environment, some fungi will primarily target nematodes; some will parasitize nematodes in certain situations and also survive saprophytically; while others may only become nematophagous after encountering a nematode that is affected in some way by other soil organisms. Thus, determining whether a fungus can parasitize a nematode in a highly contrived environment is only a small step towards understanding its behaviour in soil.

Since it is impossible to cover all the fungal associates of nematodes in a book of this nature, only well-studied species have been included in the first part of this chapter. However, this may mean that ecologically important parasites at the obligate end of the

nutritional spectrum have been omitted, as they are rarely studied due to the difficulties involved in growing them in culture. Also, other fungi will not have been included because they either parasitize nematodes with a limited distribution, or are only found in certain environments. Therefore, the following snippets of information are added in the hope that they will provide a more complete picture of the diverse range of fungi that are found in association with nematodes.

- *Fusarium oxysporum* and *F. solani* are ubiquitous soil fungi and are best known as plant pathogens. However, these species are also relatively good saprophytes, and some isolates have the capacity to parasitize nematodes (see Chapter 12). Endophytic strains of *F. oxysporum* are also known to reduce nematode multiplication in several plant species (see Chapters 8 and 12).
- In addition to the *Fusarium* endophytes mentioned above, several other endophytic fungi protect plants to some degree from damage caused by plant-parasitic nematodes. Arbuscular mycorrhizal fungi, *Piriformospora indica*, and grass endophytes in the genus *Neotyphodium* are the best-known examples, and they are discussed in Chapter 8.
- *Trichoderma* species are common soil inhabitants that are well recognized as biological control agents of soilborne plant pathogens. They do not appear to be naturally associated with nematodes, but in the course of studies with commercial *Trichoderma* products and laboratory-produced formulations containing *Trichoderma*, the fungus has been found to have activity against nematodes (see Chapter 12 for more information).
- *Cylindrocarpon destructans* causes a severe root rot problem in ginseng and is also a minor pathogen of a wide variety of plants (Seifert and Axelrood, 1998). Although commonly found in association with cyst nematodes, it is a relatively weak pathogen (Vovlas and Frisullo, 1983), and presumably survives largely on roots or lives as a saprotroph.

- *Plectosphaerella cucumerina* has been isolated from egg masses of *Meloidogyne hapla* (Yu and Coosemans, 1998), and is also a parasite of potato cyst nematode (Jacobs *et al.*, 2003). Although PCR, bait and culture methods are available to quantify *P. cucumerina* and monitor its activity in soil (Atkins *et al.*, 2003a), it is a poor competitor and has limited potential as a biological control agent (Jacobs *et al.*, 2003). That observation is supported by Bent *et al.* (2008), who used rRNA gene analysis to identify *P. cucumerina* in egg masses of *M. incognita* collected from a suppressive soil, and found that it showed a positive rather than negative association with nematode population density.

- *Haptocillium balanoides* (syn. *Cephalosporium balanoides*, *Verticillium balanoides*) is the best-known species in a genus established to accommodate a group of nematophagous species with conidia that attach to nematodes by means of an adhesive apical wall-thickening (Gams and Zare, 2001; Zare and Gams, 2001). Spore transmission and epidemiology studies showed that more than 10,000 spores of *H. balanoides* were produced in the cadavers of some infected nematodes on agar (Atkinson and Dürschner-Pelz, 1995), but because microorganisms will compete with the fungus for the cadaver, it is questionable whether this level of spore production occurs in soil. Since *H. balanoides* survives by continually parasitizing nematodes, Atkinson and Dürschner-Pelz (1995) developed a simple epidemiological model which suggested that a transient density-dependent epizootic may occur when high numbers of a host nematode (*Ditylenchus dipsaci*) enter *H. balanoides*-infested soil to overwinter.

- *Lecanicillium lecanii* (syn. *Verticillium lecanii*) is generally considered a parasite of insects and fungi, but has often been isolated from cysts and females of *Heterodera*. Attempts to use the fungus to control *H. glycines* produced inconsistent results (Meyer and Meyer, 1996; Meyer *et al.*, 1997), possibly because the strains introduced into soil were weak egg parasites (Meyer and Wergin, 1998) and also poor colonizers of the rhizosphere (Meyer *et al.*, 1998).

- Although *Verticillium leptobactrum* is associated with cysts of *Heterodera glycines* and parasitizes eggs of *Meloidogyne* in Alabama (Morgan-Jones and Rodriguez-Kábana, 1981; Godoy *et al.* 1982), it is more commonly isolated from decaying wood and forest soils (Gams, 1988). Recent greenhouse experiments showed that when isolates from Tunisia were introduced into soil, populations of *Meloidogyne* spp. on tomato were reduced and egg mortality increased (Regaieg *et al.*, 2011).

- *Rhopalomyces elegans* is a parasite of eggs and juvenile stages of bacterivorous nematodes and is generally found in rotting plant and animal debris, or in decomposing animal dung (Barron, 1973). The fungus is rarely encountered in agricultural soils, presumably because levels of organic matter are usually relatively low.

- A *Scytalidium*-like fungus was isolated from black-coloured egg masses of root-knot nematode and found to reduce the number of juveniles of *Meloidogyne javanica* hatching from egg masses in the laboratory (Oka *et al.*, 1997). The number of juveniles penetrating tomato roots was not reduced in a glasshouse test, but the fungus markedly reduced the number of nematodes in soil after one generation. However, species of *Scytalidium* are associated with diseases of plants and humans (Hay, 2002), and this may mean that nematophagous isolates have limited potential as biological control agents.

- Mushrooms are generally considered saprobes, but it is now known that many species use nematodes for nutritional purposes. One such example is *Stropharia rugosoannulata*, which produces cells with finger-like projections known as acanthocytes. In experiments on agar and in soil, these spiny cells rapidly immobilized and killed *Panagrellus redivivus*. Large numbers of acanthocytes were then produced on the degraded cadavers (Luo *et al.*, 2006).

- *Macrobiotophthora vermicola*, a member of the Entomophthorales, is characterized by forcibly discharged primary spores and the production of sticky but passive secondary spores that attach to passing nematodes. In a survey of soybean fields in Tennessee, *M. vermicola* was the most frequently encountered nematophagous fungus, but its host range appeared to be restricted to microbivorous diplogasterids, rhabditids and aphelenchids. Species of *Criconemella*, *Helicotylenchus*, *Heterodera* and *Meloidogyne* were not parasitized (Bernard and Arroyo, 1990).

Fungal–Nematode Interactions in Soil

Saprophytic and parasitic modes of nutrition

Given the taxonomic diversity of the nematophagous fungi present in soil, it is not surprising that they vary markedly in their reliance on nematodes as a food source. The endoparasitic fungi sit at one end of the nutritional spectrum, as they are specialized parasites of nematodes and have little or no saprophytic ability. At the other end of the spectrum there are fungi that may inadvertently consume nematodes or their eggs as they grow on decomposing organic materials or parasitize living plant tissue. Between those two extremes is an intermediate group containing the nematode-trapping fungi and also the egg parasites. In some circumstances these fungi will acquire nutrients from organic matter, whereas in other situations they will obtain sustenance from nematodes.

Early studies on the saprophytic and predacious behaviour of the nematode-trapping fungi (summarized by Cooke, 1968; Gray, 1987, 1988; and Stirling, 1991) showed that they are not a single group from a nutritional perspective. Nematode-trapping species vary markedly in their dependence on nematodes, with some predominantly predacious and others largely saprophytic. Ring-formers and species with adhesive branches and knobs are the most efficient predators, as traps form rapidly as conidia germinate or as hyphae grow through soil. Often termed spontaneous trap-formers, this group has low growth rates and poor saprophytic ability. In contrast, species forming adhesive network traps have rapid growth rates in culture but do not produce traps spontaneously. They appear to be relatively inefficient predators but are good saprophytes.

Studies cited earlier in this chapter have shown that isolates of egg-parasitic fungi (e.g. *Pochonia chlamydosporia*, *Purpureocillium lilacinum* and *Brachyphoris oviparasitica*) also vary in their capacity to parasitize nematode eggs. Thus, the important message from an ecological perspective is that the nutritional preferences of the nematophagous fungi vary widely. Species that are highly dependent on nematodes are more likely to have an impact on nematode populations than species that live largely as saprophytes.

Factors influencing the saprophytic and parasitic activity of nematophagous fungi in soil

The evolution of carnivory in the nematode-trapping fungi has always fascinated scientists (Ahren and Tunlid, 2003; Yang *et al.*, 2012c), but it is the factors that cause these fungi to switch from a saprophytic to a predacious mode of nutrition that are the key to understanding their behaviour as biological control agents. Three hypotheses have been proposed to explain trap induction in the nematode-trapping fungi and the process of carnivory: (i) it is a density-dependent response to increasing numbers of nematodes; (ii) it is a response to competition from the microflora for nutrients; or (iii) it is a mechanism that provides the fungus with nitrogen in low-nitrogen environments.

Density-dependent response as nematode populations increase

The hypothesis most commonly advanced to explain fungal carnivory is that nematophagous fungi are dependent on nematodes for food, and so they respond to an increase in nematode population density by increasing their predatory activity. First proposed by Linford *et al.* (1938) to explain the decline in populations of plant-parasitic nematodes that occurred when organic matter was added to soil, it is based on the premise that nematophagous fungi multiply in response to the microbivorous nematodes associated with decomposing organic matter, and then prey on all components of the nematode community, including plant parasites.

The problem with testing this hypothesis is that researchers usually add organic matter to soil and then look for relationships between numbers or activity of nematophagous fungi and nematode population density. However, because soil contains a diverse range of nematophagous fungi – and organic amendments not only increase populations of free-living nematodes but are also a nutrient source for fungi with some saprophytic capacity – it has proved difficult to show a relationship between nematode populations and predatory activity (Cooke, 1962a, b; Gray, 1987). Nevertheless, as mentioned earlier, the general conclusion from these early studies was that obligate parasites (e.g. endoparasitc fungi) and trapping fungi producing constricting rings and adhesive knobs showed a much stronger relationship to nematode population density than network-forming fungi that are now located in the genus *Arthrobotrys*. The results of more recent studies (discussed later in the section on the regulatory capacity of the nematophagous fungi) have generally supported that conclusion.

Competition from other soil organisms for nutrients

Trap induction by the nematophagous fungi is commonly studied on agar plates (e.g. Scholler and Rubner, 1994; Gao *et al.*, 2007), but it is questionable whether the results of such studies are relevant to the soil environment for two reasons: (i) fungal cultures on agar plates are not subject to the competitive forces that always exist in soil; and (ii) spores and mycelium are usually used to inoculate agar plates, whereas mycelium

growing from parasitized nematodes will often be the main source of inoculum in soil. Thus, Jaffee *et al.* (1992b) made an important contribution by comparing the behaviour of four nematode-trapping species in soil and on agar using parasitized nematodes as inoculum. This study indicated that trapping fungi produced few traps on agar unless they were exposed to nematodes, whereas abundant traps were produced when they grew into nematode-free saturation extracts of four soils (Fig. 5.18). Consequently, the authors concluded that nematophagous fungi (including nematode-trapping species that had previously been characterized as spontaneous and non-spontaneous trap-formers), did not require specialized conditions to induce traps. The endoparasite *Hirsutella rhossiliensis* was also included in this study and it produced infective structures (in this case spores on phialides) after hyphae grew from soil extract into the atmosphere. Thus, it was suggested that all the nematophagous species tested were primed for parasitism when they grew into soil from parasitized nematodes.

Galper *et al.* (1995) used a simple soil microcosm to show that at least some species of nematode-trapping fungi were also predacious when introduced into mineral soil as mycelium. Within 5 days of being added to soil, *Drechslerella dactyloides* (a constricting ring-former) and *Dactylellina candidum* (which produces non-constricting rings and knobs) both produced numerous traps and consistently reduced the number of juveniles of *Meloidogyne javanica* inoculated into the soil. Network-forming species were less effective predators, as they did not reduce nematode populations to the same extent and usually produced fewer traps in soil. However, later observations suggested that their poorer performance was a temporal effect, as it took 10–15 days rather than 5 days for this group of fungi to produce traps. These results, therefore, support the argument that predatory activity is normal behaviour for all groups of nematode-trapping fungi in soil.

Results of other studies also support the notion that nematode-trapping fungi produce traps when exposed to the microbial

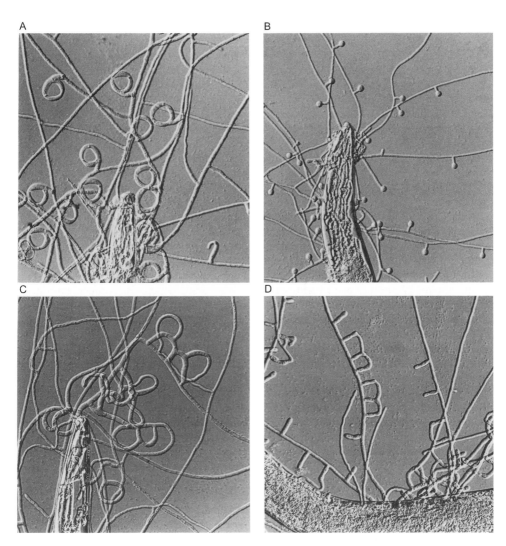

Fig. 5.18. Nematode-trapping fungi growing from parasitized nematodes (*Steinernema glaseri*) in soil extract. (A) constricting rings produced by *Drechslerella dactyloides*; (B) adhesive knobs produced by *Dactylellina ellipsospora*; (C) adhesive, three-dimensional networks produced by *Arthrobotrys oligospora* (the networks appear two-dimensional because they have been flattened by a coverslip); (D) adhesive branches and two-dimensional networks produced by *Arthrobotrys gephyropaga*. Magnification of all micrographs is 240×. (From Jaffee *et al.*, 1992b, with permission.)

community with which they always interact. Trap formation is influenced by bacteria (Rucker and Zachariah, 1987; Li *et al.*, 2011), and a study with several common rhizosphere bacteria isolated from vegetable-growing soils in Senegal showed that they increased trapping by *A. oligospora* and improved its capacity to suppress populations of root-knot nematode (Duponnois *et al.*, 1998).

When conidia of three fungi with different tendencies to form traps spontaneously were added to soil extracts in micro-well plates, all species produced conidial traps (Persmark and Nordbring-Hertz, 1997). Pre-incubation of soil extracts before inoculation facilitated trap formation, whereas removal of microorganisms from soil extracts by sterile filtration, dilution or heating decreased the production of conidial traps. Therefore, Persmark and Nordbring-Hertz (1997) concluded that the formation of conidial traps was a fungistatic phenomenon involving competition with other microorganisms for nutrients. In the soil environment, normal germination into germ tubes is inhibited, but the fungus is able to respond by producing conidial traps.

In conclusion, these results suggest that carnivory in nematophagous fungi has evolved as a means of overcoming the nutritional stresses imposed by competing soil microorganisms. When fresh organic materials become available in soil, nematode-trapping fungi and other soil microorganisms are stimulated into competition for the available substrates. Since the nematode-trapping fungi are in a unique position to use an additional substrate that is unavailable to other organisms, they form traps and use nematodes as a nutrient source. Once the decomposition process is complete, predacious activity declines. Fungi at the predacious end of the nutritional spectrum (e.g. endoparasites and spontaneous trap-formers) are always in a position to utilize nematodes as a food source, but it appears that even the more saprophytic groups (e.g. network-formers such as *Arthrobotrys oligospora*) will produce trapping devices when subject to microbial competition.

Competition for nitrogen in high-carbon, low-nitrogen environments

It has long been known that many fungi, including nematode-trappers and species of *Harposporium*, prey on nematodes in decomposing wood (Dixon, 1952). However, the link between wood and nematophagous fungi did not become apparent until much later, when *Hohenbuehelia*, a mushroom associated with rotting wood, was found to be the teleomorph of one of the nematode-trapping fungi (*Nematoctonus*); and several species of *Hohenbuehelia* and its close relative *Pleurotus* (including the oyster mushroom, *P. ostreatus*) were found to attack and consume nematodes (Barron and Dierkes, 1977; Thorn and Barron, 1984, 1986). The later discovery that *Orbilia* was the perfect stage of other nematode-trapping fungi (see references in the taxonomic

section of this chapter), and the fact that *Pleurotus*, *Hohenbuehelia* and *Orbilia* are commonly found on rotted logs, dead wood and bark in natural environments, provided further evidence of a strong connection between wood and nematophagous fungi. Additionally, nematode-trapping fungi are readily isolated from decaying wood and leaves (see Swe *et al.*, 2009 for an example). Wood also proved to be a good food base for *Arthrobotrys superba*, as studies with radioactive tracers showed that this network-forming species could grow into soil from colonized birch wood discs (Persson *et al.*, 2000).

Since rotting wood is a nitrogen-poor substrate (with a C:N ratio usually >350:1), it has been suggested that the ability to capture nematodes and other soil organisms (e.g. other fungi, copepods, rotifers, springtails and amoebae) is an evolutionary response that enables cellulolytic and lignolytic fungi to overcome the nutrient stresses associated with their nitrogen-limited habitat (Thorn and Barron, 1984; Barron, 1992). Although this is a highly plausible hypothesis, there is little experimental evidence to support it. Jaffee (2004b) showed that *Arthrobotrys oligospora* could decompose wood chips on agar; that much more wood was degraded when nematodes were present than when they were absent; and that the fungus trapped virtually all the nematodes added to the agar/wood cultures. However, when similar pieces of wood were placed in soil, decomposition was unaffected by the addition of *A. oligospora* or nematodes, and few nematodes were trapped by the fungus. The reasons for the lack of predatory activity were never determined, but the author suggested that it may have been due the relatively high nitrate levels in the soil used for the experiment. Such a result highlights the need for much more work on the role of nitrogen in influencing the behaviour of nematode-trapping fungi while they are competing for carbon with other soil organisms. Nitrogen is an essential nutrient in agroecosystems, and it is important that we understand the impact of different concentrations and forms of nitrogen on predatory activity.

In considering how this issue might be tackled in future, an experimental system similar to that used by Quinn (1987) may be useful. He inoculated *Arthrobotrys oligospora* and *Dactylellina ellipsospora* onto leaf litter and then added nematodes, bacteria and competitor fungi to produce microcosms with various levels of complexity. Although trapping activity was not measured directly, the results showed that when competing saprophytes (in this case *Purpureocillium* and *Trichoderma*) were present, both of the nematode-trapping fungi reduced nematode populations to a greater extent than when competitors were absent. Imposing nitrogen treatments into such a system could provide insights into whether nematophagous fungi only resort to carnivory in nitrogen-limited habitats, or whether it occurs whenever these fungi face competition from other saprophytes.

Nematophagous Fungi as Agents for Suppressing Nematode Populations

Occurrence in agricultural soils

Numerous comprehensive surveys (cited by Gray, 1987) and more recent studies (e.g. Persmark *et al.*, 1995; Saxena, 2008) have shown that nematode-trapping fungi, together with endoparasitic fungi and oomycetes, are present in agricultural soils throughout the world. More targeted surveys of cyst and root-knot nematodes have also shown that saccate females and eggs are commonly parasitized by fungi (Morgan-Jones and Rodriquez-Kabana, 1988). Clearly, nematophagous fungi and oomycetes are widespread, and in many situations they are probably the most important group of nematode antagonists in soil. Nevertheless, it is likely that their occurrence has been grossly underestimated in many of the above studies, as the methods generally used to isolate these organisms are unsatisfactory.

Parasites and predators of vermiform nematodes are most commonly recovered from soil using a culture-based 'sprinkle-plate' method, in which a small quantity of substrate or soil is placed on agar and 'bait' nematodes are used to selectively enhance

the growth of nematophagous organisms. After incubation, fungal and oomycete parasites are identified under a microscope. Since each agar plate is really a microcosm containing numerous interacting soil organisms (e.g. fungi, bacteria, nematodes, enchytraeids and microarthropods), and the outcomes from those interactions are unpredictable, the antagonists isolated will vary with the number and type of bait nematode, incubation conditions, moisture content and many other factors. Also, sprinkle plates tend to be biased towards organisms adapted to growth on agar, and so they provide no more than an imperfect snapshot of the fungal and oomycete antagonists present in a soil. Additional information can be obtained by quantifying the organisms using a most-probable number procedure (Eren and Pramer, 1965; Dackman *et al.*, 1987; Jaffee *et al.*, 1998), but again the data obtained will be limited by the inadequacies of the sprinkle-plate method.

Recent studies with PCR primers that selectively amplify Orbiliales fungi not only show the limitations of the sprinkle-plate method, but also highlight the problems involved in monitoring nematophagous fungi in soil. Orbiliales-specific primers were used to selectively amplify, clone and sequence Orbiliales DNA extracted from soil, litter and wood, and results were compared to those obtained with the sprinkle-plate method (Smith and Jaffee, 2009). The two methods generally detected different species of Orbiliales, with only three species found in both assays. Five nematode-trapping species detected on sprinkle plates were never recovered as DNA sequences (including two of the most common taxa, *Arthrobotrys gephyropaga* and *Dactylellina parvicollis*), while seven additional species in the Orbilia clade were found only as DNA sequences. Three possible reasons were given for discrepancies between the two methods: (i) the molecular assay may only have detected species present at high population densities; (ii) some species may have been missed in the molecular study because the RFLP patterns used to screen clones for sequencing were quite similar; and (iii) many of the isolates obtained with the culture-based method were morphologically similar and may not have been assigned to the correct species.

In another study using similar techniques in sugarcane soils, the same molecular method detected many more nematode-trapping Orbiliales than sprinkle plates (Stirling *et al.*, 2011b). Given the limitations of culture-based methods, such a result is not surprising. However, finding additional complexity in what is only a small component of the nematophagous fungal and oomycete community is somewhat concerning from an ecological perspective, as it raises questions about how interactions between nematodes and some of their most important natural enemies in agricultural environments will ever be understood.

With regard to fungi that parasitize nematode eggs, our current knowledge is based on isolations made from eggs of sedentary endoparasitic nematodes, most commonly *Globodera*, *Heterodera* and *Meloidogyne*. Eggs of these nematodes are aggregated within cysts or in an egg mass, and so they can be readily retrieved from soil or roots and scanned for signs of parasitism. However, almost nothing is known about parasites of eggs that are laid singly in soil, as is the case with eggs of free-living nematodes and many plant parasites. Nematode eggs take at least several days to develop and hatch, and so there is sufficient time for some parasitism or predation to occur. Interestingly, the molecular methods used by both Smith and Jaffee (2009) and Stirling *et al.* (2011b) detected fungi in a clade that contains the well-known egg parasite *Brachyphoris oviparasitica*. Such observations raise numerous questions about the role of these fungi in the ecology of free-living and plant-parasitic nematodes: are they largely parasitic on the eggs of endoparasitic nematodes; do they parasitize eggs that are laid singly in soil or roots; can they survive saprophytically; and do they parasitize soil organisms other than nematodes? In the absence of answers to such questions, it is difficult to know whether the presence of these fungi in soil means that they are contributing in some way to the suppression of particular nematode populations.

Another problem with using the presence of nematophagous fungi as an indicator of predatory activity is that these fungi are nutritionally diverse, and a range of potential food sources other than nematodes are available to them in soil. As discussed earlier, many nematophagous fungi have saprophytic and parasitic modes of nutrition, and when parasitism does occur, it may not necessarily be directed at nematodes. *Fusarium oxysporum* is a good example: in addition to parasitizing nematode eggs, this species is an important plant pathogen and may also live in roots as a harmless symbiont. Several nematode-trapping fungi, including the widely distributed species *Arthrobotrys oligospora*, can coil around the hyphae of other fungi, kill the cells of their fungal hosts, and then derive nutrients from those cells (Tzean and Estey, 1978; Persson *et al.*, 1985; Persson and Bååth, 1992; Olsson and Persson, 1994). Some nematode-trapping species are also able to parasitize fungal sclerotia (Li *et al.*, 2003). Fungi that parasitize nematodes, or are closely related to nematophagous species, are also known to parasitize oomycetes (Drechsler, 1952, 1962, 1963), mites (Abiko *et al.*, 2005), collembolans (Drechsler, 1944), copepods (Barron, 1990), rotifers (Abiko *et al.*, 2005) and rhizopods (Drechsler, 1936; Liu *et al.*, 1996). Thus, whenever a 'nematophagous' fungus is found in soil, there are always doubts about its primary nutrient source.

Population density and predacious activity in soil

One issue that has consistently limited our ability to understand the role of nematophagous fungi as biocontrol agents is the difficulty of measuring the population density of these fungi in soil. Three quantification methods have been used, and they all have their limitations. Perhaps the most useful is a direct extraction technique, in which soil is wet-sieved and spores are separated by centrifugation using a solution with a high specific gravity such as magnesium sulfate. Such a method was used by Crump and Kerry (1981) to recover resting spores of *Nematophthora gynophila* and *Pochonia chlamydosporia* from soil.

Although only useful for fungi that produce readily recognizable spores, it proved valuable in this instance because it showed that spore numbers were higher in a soil that was suppressive to *Heterodera avenae* than in soils where populations of the nematode were increasing. Mean spore counts in the suppressive soil were 180 and 312 spores/g soil for *N. gynophila* and *P. chlamydosporia*, respectively.

Semi-selective media are commonly used to detect specific fungi in soil and study their population dynamics, and are available for *Purpureocillium lilacinum* (Mitchell *et al.*, 1987; Cabanillas and Barker, 1989; Gaspard *et al.*, 1990a) and *Pochonia chlamydosporia* (Gaspard *et al.*, 1990a; Kerry *et al.*, 1993). These media, or variants thereof, have played a vital role in efforts to commercialize these fungi for biological control purposes (see Chapter 12), but have rarely been used to quantify fungal populations in agricultural soils. Perhaps the best example of their use for such a purpose is an English study which showed that population densities of *P. chlamydosporia* were often greater than 2×10^3 cfu/g soil in soils suppressive to *Heterodera avenae* (Kerry *et al.*, 1993). Since methods for quantifying resting spores are also available (see previous paragraph), the tools are in place for studying both the inactive stage (chlamydospores) and the more active stages (mycelium and conidia) of this fungus. It is disappointing that such methods have not been used more widely, as studies quantifying indigenous populations of *P. chlamydosporia* could provide clues as to how soil factors or management practices influence fungal populations and suppressiveness to nematodes.

One of the only studies aimed at determining whether population densities of egg-parasitic fungi were related to soil factors or suppressiveness was briefly mentioned previously. Gaspard *et al.* (1990a) sampled 20 commercial tomato fields in California four times over 2 years, and measured populations of *Purpureocillium lilacinum* and *Pochonia chlamydosporia* using semi-selective media. The two fungi commonly occurred together (13 fields), but four fields had only one fungus or the other, and neither fungus was detected in three fields. Eighty soil samples were processed during the study, and populations of *P. lilacinum* and *P. chlamydosporia* greater than 10^4 cfu/g soil were only found in eight and three samples, respectively. Stepwise regression models showed that fungal population densities were sometimes correlated with soil factors such as percent sand, percent silt, percent clay, pH, cation exchange capacity, electrical conductivity and percent organic matter, but only a small proportion of the variation was explained by these models. Since fungal population densities increased markedly when soils were planted with tomato seedlings infected with *Meloidogyne incognita*, but barely changed when uninfected seedlings were used (Fig. 5.19), the presence of the nematode had a stimulatory effect on both *P. lilacinum* and *P. chlamydosporia*. As multiplication rates of first generation root-knot nematodes remained high in the presence of these high fungal populations (Fig. 5.19), and nematode reproduction did not increase when the field soils were sterilized (Gaspard *et al.*, 1990b), the authors concluded that these fungi were not effectively suppressing the nematode in California tomato fields. However, it is possible that a different result would have been obtained if nematode populations had been assessed after two or three generations, or if the soil had been left undisturbed so that the nematode and its parasites had an opportunity to interact during the following winter season, or until the next crop was due to be planted. Observations of this nature, which require monitoring of pest and parasite over more than one season, are rarely carried out.

Given that culture-based methods of quantifying nematode-trapping fungi are both time-consuming and tedious, and the data generated are often quite variable, it is not surprising that few attempts have been made to quantify these fungi in soil. The most comprehensive study was carried out by Persmark *et al.* (1996) in a Swedish field that grew spring rape in the first year, was fallowed mechanically the following year, and planted to potatoes in the third year. The field was sampled at monthly intervals for 29 months, and at each sampling time, soil was collected from four 10-cm layers to a depth of 40 cm. Nematodes were extracted and counted every month, and

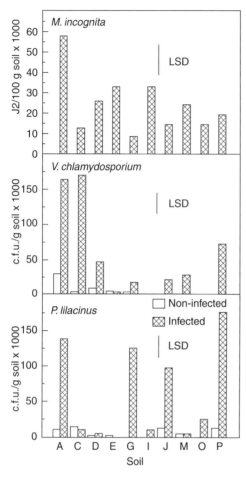

Fig. 5.19. Final population densities of *Meloidogyne incognita*, *Pochonia chlamydosporia* and *Purpureocillium lilacinum* in ten tomato field soils planted with *M. incognita*-infected and uninfected tomato seedlings under greenhouse conditions. Tukey's least significant difference (LSD) bar indicates differences among all means ($P = 0.05$). (From Gaspard et al., 1990a, with permission.)

nematophagous fungi were evaluated every second month. Five endoparasitic fungi and nine nematode-trapping species were recovered from the field, most commonly in the upper soil layer. Soil from this layer generally contained 2–4 nematode-trapping species/g soil, and their average population density in the 0–10 and 10–20 cm layers was 41 and 62 propagules/g soil, respectively. Populations fluctuated during the year, with the highest densities and greatest number of species recorded in autumn (Fig. 5.20).

Another intensive study using similar techniques was undertaken by Jaffee et al. (1998) to determine whether a change from conventional to organic management increased the population densities and diversity of nematode-trapping fungi. Two treatments in a long-term annual cropping trial in California were sampled three times per year for 2 years, and nematophagous fungi were quantified using a soil dilution and most-probable number technique. Sprinkle plates were also used in an attempt to detect fungi present at low population densities. Two endoparasitic species and seven nematode-trapping species were detected, but except for three species (*Arthrobotrys thaumasia*, *Drechslerella dactyloides* and *Dactylellina candidum*), population densities were less than 1 propagule/g soil. Populations of the most common species varied temporally and sometimes with management practice, but were always less than 20 propagules/g soil (Fig. 5.21). Similar low population densities were obtained in other Californian studies. *Drechslerella dactyloides* and *Dactylellina ellipsospora*, the nematode-trapping species most commonly found in a survey of vineyards and peach orchards in California, were usually present at levels of 5–50 propagules/g soil (Stirling et al., 1979), while population densities of the five species detected in untreated soils from two other California vineyards were less than 10 propagules/g soil (Jaffee, 2003). Since the Swedish and Californian studies were carried out in quite different environments and cropping systems, the results suggest that populations of fungal antagonists capable of attacking vermiform nematodes are relatively low in many agricultural soils. However, results of this nature may simply reflect the fact that these fungi do not sporulate prolifically in soil. The future challenge is to measure fungal biomass rather than numbers of propagules, with the critical questions from an ecological perspective being whether populations of nematode-trapping fungi are as low as they appear from studies with dilution plates; whether there is a relationship between trapping activity and some more appropriate estimate of fungal population density; and whether populations of these fungi can be enhanced through changes in management practice.

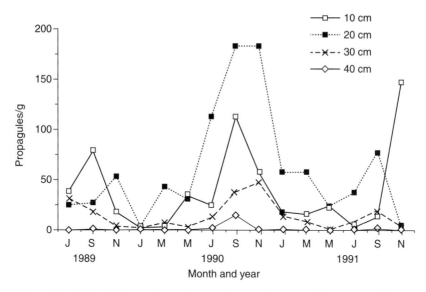

Fig. 5.20. Seasonal variation in the population density of nematode-trapping fungi (propagules/g soil) in a Swedish field at four depths in the soil profile. (From Persmark *et al.*, 1996, with permission.)

In the Swedish study discussed earlier (Persmark *et al.* (1996)), populations of some groups of nematophagous fungi declined during the 29-month sampling period, and this was associated with a decline in the total number of nematodes and populations of plant-parasitic trichodorids. Thus, in the upper layer of soil there was a positive correlation between nematode numbers and the frequency of endoparasitic fungi ($r = 0.48$), and also with the frequency of parasitic nematode-trapping species ($r = 0.59$). The 'parasitic' nematode-trappers included *Arthrobotrys gephyropaga* (an *Arthrobotrys/Gamsylella* species that produces adhesive hyphal branches and two-dimensional networks) and *Stylopage* sp. In contrast, there was no correlation between nematode numbers and the frequency of a saprophytic nematode-trapping group consisting of *Arthrobotrys musiformis*, *A. oligospora*, *A. superba* and *A. psychrophila*. Although these results suggested that the two most parasitic groups of nematophagous fungi may have been responsible for the decline in nematode populations, the authors believed that a much longer sampling period was required to determine whether the nematode population was being regulated by the fungi.

The issue of whether fungal population density (determined using a dilution plating and most-probable number technique) is a useful indicator of predacious activity in soil was addressed using alginate pellets containing assimilative hyphae of five different species of nematode-trapping fungi (Jaffee, 2003). Fungal populations were varied by adding different numbers of pellets to the soil, and trapping activity was assessed by adding assay nematodes. The results indicated that the two adhesive knob-forming fungi (*Dactylellina candidum* and *D. ellipsospora*) trapped many more nematodes as fungal population density increased. The relationship between the two parameters was weaker and inconsistent for the species that formed constricting rings (*Drechslerella dactyloides*) and was non-existent for the network-formers (*Arthrobotrys oligospora* and *A. eudermata*), as they trapped few or no nematodes, even when the fungus population density exceeded 10^3 propagules/g soil.

The results of Jaffee's (2003) study are important, because they confirm what had often been predicted by others: that nematode-trapping species at the saprophytic end of the nutritional spectrum (i.e. species that form adhesive networks) are poor predators,

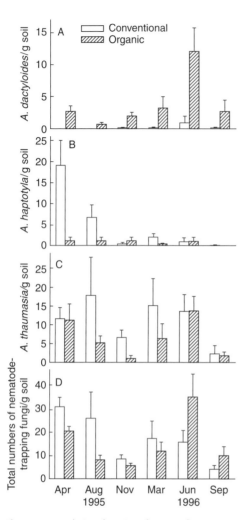

Fig. 5.21. Population densities of nematode-trapping fungi in conventional and organic plots in California. (A) *Drechslerella dactyloides*; (B) *Dactylellina candidum* (syn. *Arthrobotrys haptotyla*); (C) *Arthrobotrys thaumasia*; (D) Total numbers of nematode-trapping fungi/g soil. Values are means plus the standard error of four replicate plots. (From Jaffee *et al.*, 1998; with permission.)

densities measured in the studies cited, it is likely that the nematode-trapping fungi are capturing and killing a relatively small proportion of the nematodes in many agricultural soils.

The regulatory capacity of nematophagous fungi

Based on the results of the studies cited, it is clear that nematophagous fungi prey on nematodes in soil. However, that does not necessarily mean that they will have a major impact on nematode populations, as predation on nematodes may be coincidental to obtaining nutrients from other food sources. Nematophagous fungi may kill some nematodes, but they will only limit population growth if they interact with their hosts in a density-dependent manner. Thus, in situations where the host nematode is rare, encounters between hosts and parasites are unlikely, and so the parasite has little effect on host population density. When hosts are abundant, the parasite reproduces or aggregates near the host, host–parasite encounters become more frequent, and the parasite can limit host population growth. The regulatory forces associated with density-dependent suppression of a host by parasites are, therefore, the essence of biological control (Jaffee, 1992, 1993).

Given the taxonomic and nutritional diversity within the nematophagous fungi (discussed earlier in this chapter), it is not surprising that early studies (summarized by Cooke, 1968; Gray, 1987, 1988; and Stirling, 1991) indicated that these fungi differ markedly in their capacity to regulate nematode populations. Endoparasitic species and spontaneous trap-formers have little or no saprophytic ability and show a typical predator–prey relationship with their hosts, increasing and decreasing in response to changes in host density. In contrast, network-forming species show no such response. They tend to behave in much the same way as a saprophyte, increasing in population density as soon as the new nutrient source becomes available, and before the free-living nematodes, which are always associated with decomposing organic matter, begin to multiply. Thus, their predacious activity is not related to nematode population density.

whereas species that form constricting rings or knobs will capture nematodes, regardless of their population density. However, they also indicate that populations of predatory species must be greater than about 10^2 propagules/g soil to achieve high levels of predatory activity, and populations of this magnitude are rarely achieved by resident nematode-trapping fungi. Thus, based on the fungal population

These initial observations on the regulatory capacity of nematophagous fungi have generally been supported by the results of more recent studies. Some of those studies are discussed next.

Endoparasitic fungi

Most of the ecological studies on endoparasitic fungi have been carried out with *Hirsutella rhossiliensis*, as it is a parasite of several economically important plant-parasitic nematodes. Several lines of evidence have been obtained to show that levels of parasitism by this fungus are dependent on the population density of its nematode hosts:

- Spatial sampling in three California peach orchards showed that the percentage of the host nematode (*Mesocriconema xenoplax*) parasitized by *H. rhossiliensis* increased significantly as nematode population density increased (Jaffee *et al.*, 1989). In another study in one of these orchards, spatial sampling (but not temporal sampling) indicated a relationship between numbers of hosts and parasitism (Jaffee and McInnis, 1991).
- When different numbers of host nematodes (in this case, *Heterodera schachtii*) were inoculated at different times into heat-treated soil infested with the fungus, and spore acquisition was assessed using *Heterorhabditis bacteriophora* as the assay nematode, the percentage of assay nematodes with spores increased with host density (Jaffee *et al.*, 1992a; Fig. 5.22). A dynamic model developed by Jaffee *et al.* (1992a) to explain the interaction between *H. schachtii* and *H. rhossiliensis* predicted the dynamics observed in these microcosm experiments.
- When healthy juveniles of *Meloidogyne javanica* were added in different numbers to vials containing soil from a peach orchard, parasitism increased with host density (Jaffee *et al.*, 1993).

Although the above experiments provided evidence that high levels of parasitism can occur in nematode populations, *H. rhossiliensis* is a weak regulator of nematode population density, as it does not respond strongly to host density. Epidemics caused by the fungus will occur, but they are slow to develop and require substantial inputs of host nematodes (Jaffee *et al.*, 1989; Jaffee, 1992). Thus, *H. rhossiliensis* has its limitations as a biological control agent of plant-parasitic nematodes.

An experiment by Bouwman *et al.* (1994) provides evidence that other endoparasitic fungi also have the capacity to regulate nematode populations. Organic matter (lucerne and straw meal) was mixed into γ-irradiated soil, and microcosms were then established containing various combinations of microorganisms, bacterivorous

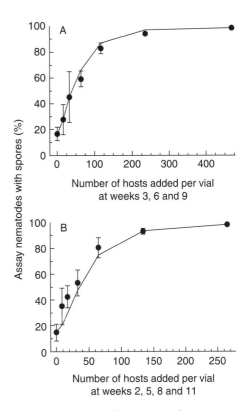

Fig. 5.22. Percentage of assay nematodes (*Heterorhabditis bacteriophora*) with *Hirsutella rhossiliensis* spores as affected by numbers of healthy host nematodes (*Heterodera schachtii*) added to soil microcosms in two experiments (A and B). Observed values (mean ± 1 SD) are represented by • and predicted values by a line. (From Jaffee *et al.*, 1992a, with permission.)

nematodes and *Drechmeria coniospora*. Observations over 6 months showed that in the absence of nematophagous fungi, populations of one of the nematodes (*Rhabditis* sp.) increased rapidly before declining to about 20 nematodes/g soil after 180 days. In the presence of *D. coniospora*, populations decreased to nearly zero on day 180. The other nematode (*Acrobeloides bütschlii*) increased to even higher population densities in the absence of nematophagous fungi (85 nematodes/g soil on day 180). *D. coniospora* had little impact on the nematode during the first 60 days, but after 180 days it had reduced the population by nearly 50%.

Van den Boogert *et al.* (1994) used a similar experimental system to confirm that *Rhabditis* sp. was more susceptible to *D. coniospora* than *A. bütschlii*, and also to show that the fungus had the capacity to reduce nematode populations. When the soil was amended with lucerne meal, the population density of *D. coniospora* increased in the presence of *Rhabditis* sp., and after 6 weeks, populations of the nematode were markedly reduced. In contrast, the fungus did not multiply as readily on *A. bütschlii* and had a lesser impact on its population density. Although these results indicated that *D. coniospora* was a highly effective parasite of certain nematodes, the authors noted that the fungus was rarely isolated from agricultural fields, even when nematode population densities were high, raising questions about whether its presence was being underestimated by the methods used to detect such fungi.

Nematode-trapping fungi

In a study of density-dependent parasitism by nematophagous fungi, Jaffee *et al.* (1993) added four species of nematode-trapping fungi to microcosms and found that their responses to host density were in accord with the observations and ecological classifications of Cooke (1963), Jansson and Nordbring-Hertz (1979, 1980) and Jansson (1982). Increases in the percentage parasitism of the assay nematode (*Heterodera schachtii*) with host density were greater for *Drechslerella dactyloides* (which forms constricting rings) and *Dactylellina ellipsospora* (an adhesive knob former) than for *Arthrobotrys oligospora* and *A. gephyropaga*, two

species that form adhesive networks. However, none of the trap-formers were as dependent on nematodes as the endoparasitic fungus *Hirsutella rhossiliensis*. The strong response of *Dactylellina ellipsospora* to host density was later confirmed using *Meloidogyne javanica* as the assay nematode (Jaffee and Muldoon, 1995b), leading the authors to conclude that it is an obligate parasite of nematodes and has little or no saprophytic activity in soil. A follow-on experiment in which *D. ellipsospora* was added to field soil in alginate pellets confirmed its predatory capacity, as the fungus reached high population densities (sometimes greater than 10^3 propagules/g soil) and suppressed the number of root-knot nematodes in roots by nearly 100% after 120 days.

Although it is clear that nematode-trapping species at the predacious end of the nutritional spectrum are important predators of nematodes, the significance of nematodes in the nutrition of saprophytic species such as *Arthrobotrys oligospora* remains a continuing source of contention. Jaffee *et al.* (1993) found that *A. oligospora*, the most widely studied network-forming fungus, did not respond to nematode density and concluded that it relied on substrates other than nematodes as a food source. That conclusion was supported by studies showing that, although relatively high populations of the fungus may have been present in soil, it failed to trap assay nematodes (*Heterodera schachtii* and *Steinernema glaseri*) in the presence of organic amendments (Jaffee, 2002, 2004a) or in their absence (Jaffee, 2003). In contrast, Koppenhöfer *et al.* (1996) added nematodes colonized by *A. oligospora* to soil, and found that the fungus trapped substantial numbers of the entomopathogenic nematode *Heterorhabditis hepialus*. *Arthrobotrys oligospora* also markedly reduced populations of two bacterivorous nematodes in soil amended with organic matter, with populations of *Rhabditis* sp. almost reduced to zero after 180 days (Bouwman *et al.*, 1994). When four rates of *A. oligospora* inoculum were included in a microcosm experiment using much the same methods, the number of nematodes recovered decreased as fungal inoculum density increased, with the highest level of inoculum reducing populations of *Rhabditis* sp. by 96% after 6 weeks (Bouwman *et al.* 1996).

Jaffee (2004a) argued that reconciling the discrepancies observed in these experiments requires a better understanding of the ecology of *A. oligospora* in soil. Although it is unlikely that an adhesive network-forming fungus would invest resources into trap production unless it stood to gain benefits from the traps, experimental evidence suggests that nematodes are not always captured. Thus, future research should focus on determining when trapping devices are produced in soil, how long they survive and whether they are effective in capturing nematodes. The ultimate aim must be to better understand the factors that cause the fungus to switch trap production on and off as it moves between carnivory and saprophytism. However, such studies will only be relevant from an ecological perspective if they are carried out in an environment where competing microorganisms are always present, fragments of decomposing plant material and the cadavers of soil organisms provide opportunities for saprophytism, and nematodes and other fungi are potential nutrient sources.

Cyst and egg parasites

Since most of the examples of natural suppression of plant-parasitic nematodes by nematophagous fungi involve parasites of nematode cysts and eggs (see Chapter 10), this group of fungi clearly has the capacity to regulate populations of its hosts. In situations where populations of the host nematode are high, they utilize females or their eggs as a food source, and levels of parasitism gradually increase to the point where the nematode is permanently suppressed. In studies of cereal soils in England that were suppressive to the cereal cyst nematode, this strong relationship between the fungus and its host was observed for both *Nematophthora gynophila* and *Pochonia chlamydosporia* (Kerry and Crump, 1998). Since the latter species is a facultative rather than obligate parasite, this result indicates that fungi with the capacity to grow saprophytically in soil or on roots will sometimes suppress nematode populations. *Brachyphoris oviparasitica* is another example of such a parasite. Although it has been isolated from roots

(Stirling, 1979a; Gaspard and Mankau, 1986), *B. oviparasitica* is most commonly found parasitizing the eggs of endoparasitic nematodes (Stirling and Mankau, 1978a; Westphal and Becker, 2001a) and as discussed in Chapter 10, is associated with the decline of *Meloidogyne* and *Heterodera* in some fields in California.

Although many other fungi are capable of parasitizing the females and eggs of endoparasitic nematodes (see Table 10.1 for examples of fungi commonly associated with *Heterodera glycines*), it is likely that in most cases, their association with nematodes is largely opportunistic. High levels of parasitism may sometimes occur, but parasitic activity is influenced by soil and environmental factors and is not always related to nematode population density.

Host specificity within the nematophagous fungi

One of the factors often not recognized as affecting the ecology of nematophagous fungi is their host specificity. It is frequently assumed that these fungi will capture any nematode in their vicinity, whereas there is considerable evidence to suggest that individual species have preferences for certain food sources. Some examples are:

- Root-knot and cyst nematodes differ in their susceptibility to some nematode-trapping fungi.
  - Observations on agar by Den Belder and Jansen (1994) showed that *Meloidogyne hapla* was more susceptible to *Arthrobotrys oligospora* than two species of cyst nematode (*Globodera pallida* and *G. rostochiensis*).
  - When three species of root-knot nematode were challenged with *Dactylellina ellipsospora* and *Arthrobotrys gephyropaga*, they were much more susceptible to predation than *Heterodera schachtii* (Jaffee and Muldoon, 1995a).
  - Jaffee (1998) observed that *Dactylellina candidum* and *Arthrobotrys thaumasia* were more likely to trap *Meloidogyne javanica* than *Heterodera schachtii*.

However, the two nematode species exhibited equal susceptibility to *Drechslerella dactyloides*.

○ *Meloidogyne javanica* was more susceptible to *Dactylellina lysipaga* than *Heterodera avenae* (Khan *et al.*, 2006b).

• *Rhabditis* sp. was more susceptible than *Acrobeloides bütschlii* to *Drechmeria coniospora*, *Arthrobotrys oligospora* and *Dactylaria* sp. (Van den Boogert *et al.*, 1994; Bouwman *et al.*, 1994, 1996).

• Third-stage infective juveniles of the entomopathogenic nematode *Heterorhabditis heliothidis* were less susceptible than *Steinernema feltiae* to the endoparasitic fungi *Hirsutella rhossiliensis* and *Drechmeria coniospora*, largely because they were more likely to retain their second-stage cuticles when moving through soil (Timper and Kaya, 1989).

• The bacterial-feeding nematode *Teratorhabditis dentifera* was a good host of *Hirsutella rhossiliensis*, but dauer juveniles were not infected by the fungus, despite the fact that they acquired conidia (Timper and Brodie, 1995).

• In experiments with isolates of the egg-parasitic fungus *Pochonia chlamydosporia* that were originally obtained from eggs of root-knot nematode and cysts of potato cyst nematode, the fungus was more abundant and parasitized more eggs in the nematode from which it was obtained, indicating that this parasite shows nematode host preference at the infraspecific level (Mauchline *et al.*, 2004; Siddiqui *et al.*, 2009).

• Three species of *Steinernema* and two *Heterorhabditis* species were tested in soil microcosms for susceptibility to four nematode-trapping fungi and two endoparasitic fungi. All nematodes were susceptible to *Gamsylella gephyropaga*, *Myzocytiopsis* sp. and *Catenaria* sp., except that *Heterorhabditis indica* was not affected by the two endoparasites. The response to *Arthrobotrys* varied with fungal and nematode species (El-Borai *et al.*, 2009).

Observations on the differential susceptibility of plant-parasitic nematodes to nematophagous fungi have important implications for biological control. Jaffee and Muldoon (1995a) noted that *Dactylellina ellipsospora* and *Arthrobotrys gephyropaga* adhered less readily to *H. schachtii* than to *M. javanica* juveniles, and suggested that differences in cuticular chemistry may explain why the two nematode species differed in susceptibility to the fungi. However, the basis of host specificity is poorly understood in this and other nematode–parasite interactions, and so it is difficult to predict whether a plant parasite of interest will be susceptible to a particular fungus. Differences in susceptibility of root-knot and cyst nematodes to some nematode-trapping fungi, together with the fact that different strains of *P. chlamydosporia* are required to control these nematodes, indicate that it is important to understand host specificity when studying interactions between nematodes and their natural enemies.

It is often assumed that maintaining high numbers of free-living nematodes is the key to maintaining populations of nematophagous fungi. However, the limited data currently available indicate that members of this group of nematodes vary in their susceptibility to nematophagous fungi. The reasons for these differences are not well understood, but traits such as the size of the nematode, motility, retention of the J2 cuticle and resistance to fungal adhesion are some of the factors that determine whether nematodes are susceptible (Timper *et al.*, 1991). The amount of inoculum of nematophagous fungi in soil will, therefore, depend on the composition of the free-living nematode community, and this, in turn, will have an indirect impact on biological control of plant-parasitic nematodes.

Interactions between nematophagous fungi and nematodes in the rhizosphere

Since all root-feeding nematodes are obligate parasites, they spend most of their lives in the rhizosphere or in soil that is closely associated with roots. Thus, nematophagous fungi are unlikely to have an impact on this group of nematodes unless they are active within the rhizosphere, a region where enormous numbers of microorganisms compete intensely for the soluble, carbon-rich compounds constantly being released from roots (see Chapter 2). Rhizosphere competence (a capacity to survive and compete in this zone) is, therefore, an important attribute of fungal biocontrol agents of plant-parasitic nematodes (Kerry, 2000).

Association of nematode-trapping and endoparasitic fungi with roots

Although nematode-trapping fungi are normally considered to be soil inhabitants rather than root associates, they have been isolated from the rhizosphere. The incidence of *Arthrobotrys oligospora* was much greater in the soybean rhizosphere than in adjacent soil (Peterson and Katznelson, 1965), and the population density of nematode-trapping fungi was 19 times higher in the pea rhizosphere than in root-free soil (Persmark and Jansson, 1997). Although more studies are obviously required, these results suggest that the rhizosphere of leguminous plants is a favourable habitat for nematode-trapping fungi. In contrast, the rhizosphere of wheat and barley appears to have little effect on populations of these fungi (Van den Boogert *et al.*, 1994; Persmark and Jansson, 1997), and may even have a negative impact (Peterson and Katznelson, 1965). However, it is important to note that the presence of nematode-trapping fungi on roots does not necessarily mean that they are actively trapping nematodes. In fact, Persmark and Jansson (1997) observed that even when nematodes were present and conidiophores of trapping fungi were abundant on roots, traps were rarely seen.

When the root-colonizing capacity of various nematode-trapping species and isolates was assessed in two greenhouse tests with tomatoes (Persson and Jansson, 1999), levels of colonization were poor. Only six of 38 isolates were retrieved from root segments in both tests, and 16 isolates did not colonize the rhizosphere in either test. Isolates with varying capacities to colonize the rhizosphere were then formulated as alginate pellets and introduced into a sandy agricultural soil, and their capacity to control root-knot nematode was assessed in pots. The population densities of both *Dactylellina ellipsospora* and *Drechslerella dactyloides* increased in the rhizosphere over a 12-week period, whereas *Arthrobotrys superba* and *A. gephyropaga* were barely detectable after 12 weeks. However, despite clear differences in their rhizosphere-colonizing ability, none of the fungi reduced nematode damage on tomato.

Based on the evidence available, it appears that most nematode-trapping fungi (except perhaps the constricting ring and knob formers at the predacious end of the nutritional spectrum) are not strongly associated with roots. Although some species can colonize roots endophytically under axenic conditions (Bordallo *et al.*, 2002), there is little evidence to suggest that this occurs in the competitive rhizosphere environment. This inability to flourish in the niche occupied by plant-parasitic nematodes is one of the limitations of these fungi as biocontrol agents.

There is limited information on the rhizosphere competence of the endoparasitic fungi, but the fact that they are obligate parasites of nematodes and relatively poor saprophytes suggests that the root surface is unlikely to be their preferred habitat. This contention is supported by observations that autoclaved peach roots were a poor substrate for *H. rhossiliensis*, and that despite being readily recoverable from seeds incubated in sterile soil, the fungus was not isolated from inoculated, autoclaved wheat seeds incubated in non-sterile soil (Jaffee and Zehr, 1985). Also, the population of another endoparasitic fungus (*Drechmeria coniospora*) was no higher in the rhizosphere of wheat and barley than in bulk soil (Van den Boogert *et al.*, 1994).

Rhizosphere competence of fungi and oomycetes capable of parasitizing nematode cysts and eggs

Since the eggs and cysts of endoparasitic nematodes are located in root tissue or on the root surface, the fungi that attack them could be expected to have a much closer relationship to the rhizosphere than other nematophagous fungi. Thus, it is not surprising that fungi such as *Brachyphoris oviparasitica*, *Purpureocillium lilacinum* and *Pochonia chlamydosporia* are sometimes seen on roots or have been isolated from soil closely associated with roots (e.g. Stirling, 1979a; Gaspard and Mankau, 1986; de Leij and Kerry, 1991;

Mertens and Stirling, 1993). However, the interactions that occur between plants, nematodes and egg parasites in the rhizosphere have only been seriously studied with the latter fungus.

De Leij and Kerry (1991) were the first to demonstrate that rhizosphere competence was important in determining whether an isolate of *P. chlamydosporia* was an effective parasite of root-knot nematode eggs. Working with three isolates that not only parasitized eggs *in vitro* but also survived well in soil, they showed that the isolate which extensively colonized the root surface was the only one to reduce nematode populations consistently in pot experiments. Plant species were also found to differ in their ability to support *P. chlamydosporia*, as fungal population densities in the rhizosphere were much higher in brassicas (e.g. kale and cabbage) than in solanaceous plants such as tomato and aubergine (Bourne *et al.*, 1996). Interestingly, the presence of root-knot nematode increased the abundance of *P. chlamydosporia*, either because the fungus benefited from the additional exudates leaking from nematode-infested roots, or because it was able to proliferate in developing egg masses. Thus, despite the fact that *P. chlamydosporia* is able to grow saprophytically in soil and can also colonize soil organic matter, the rhizosphere and the endoparasitic nematodes that inhabit the root–soil interface appear to be more important nutrient sources for at least some isolates of this fungus (Kerry and Bourne, 1996).

The fact that *P. chlamydosporia* can colonize roots endophytically (see references cited earlier in this chapter) also indicates that the fungus has a relatively intimate relationship with roots. Thus, it is now standard practice to test isolates of *P. chlamydosporia* for rhizosphere competence before using them in biocontrol experiments. However, it is important to recognize that efficacy is not necessarily related to fungal population densities in the rhizosphere. The susceptibility of the host and the initial inoculum density of the nematode determine whether egg masses remain embedded in gall tissue or are exposed to fungal attack on the root surface, and these factors probably affect the level of nematode control to a greater extent than the extent of fungal proliferation in the rhizosphere (Bourne *et al.* 1996; Kerry, 2000).

The oomycete *Nematophthora gynophila* is an important parasite of *Heterodera* cysts and eggs, but its association with the rhizosphere is coincidental to the fact that its host nematode is located on the root surface. *N. gynophila* attacks female nematodes as they emerge from roots, and since the disintegrating bodies of infected nematodes are eventually filled with oospores, fragile spore masses will occur in the rhizosphere until they are dispersed by other soil organisms (Crump and Kerry, 1977; Kerry and Crump, 1980). However, the parasite has an obligate relationship with its host, and so it is not sustained by the nutrients available in the rhizosphere.

Nematophagous fungi and entomopathogenic nematodes

Entomopathogenic nematodes are particularly important members of the soil nematode community because of their potential for development as biological control agents against insect pests. Free-living infective juveniles of *Steinernema* and *Heterorhabditis* spp. search for insects in soil and once a host is located, they enter through natural openings or by puncturing the cuticle. Symbiotic bacteria carried by the nematode multiply and kill the host insect, and then the nematodes and bacteria feed on the liquefied host and reproduce. When nutrients are exhausted, thousands of infective juveniles emerge from the cadaver, but in the process of searching for new hosts, they are vulnerable to any natural enemies that may be present in soil. Thus, from an insect management perspective, it is important that the impact of naturally occurring antagonists on entomopathogenic nematodes is understood. However, this knowledge is also vital from a broader perspective. Entomopathogenic nematodes occupy the same environment and are subject to the same natural enemies as plant-parasitic and free-living nematodes, and so any effects on one group of nematodes will directly or indirectly affect other members of the soil nematode community.

The following very limited summary of some of the work on fungal antagonists of entomopathogenic nematodes indicates that those interested in biological control of plant-parasitic nematodes can learn a considerable amount from research on this group of nematodes. The studies selected for inclusion examined interactions between an insect pest, entomopathogenic nematodes and nematophagous fungi in two quite different habitats. Further information on entomopathogenic nematodes as a model system for advancing our knowledge of soil ecology can be found in Campos-Herrera et al. (2012b).

Citrus root weevil, entomopathogenic nematodes and nematophagous fungi in citrus soil

Entomopathogenic nematodes are commonly used in Florida to control citrus root weevil and the results of two studies suggest that nematophagous fungi are one of the factors that influence their capacity to survive in soil. In the first of these studies, bioassays of nematodes that were regularly retrieved from the field and assessed for parasitism on agar indicated that the number of infective juveniles killed by nematode-trapping fungi (*Arthrobotrys oligospora, A. musiformis* and *A. gephyropaga*) and the endoparasitic fungus *Catenaria* sp. was sometimes higher in soil where the entomopathogenic nematode community had previously been augmented with *Steinernema riobrave*, than in untreated soil (Duncan et al., 2007). In a follow-up study in the laboratory using a similar assay system, the addition of entomopathogenic nematodes to soil initiated a trophic cascade that reduced the survival of a cohort of nematodes added 1 week later. Pretreatment of the soil with *S. riobrave* increased the number of infective juveniles of *S. diaprepsi* killed by nematophagous fungi, and this effect was more apparent as the number of pre-inoculated nematodes increased (El-Borai et al., 2007).

The above results were consistent with the hypothesis that fungal antagonists respond in a top-down manner when entomopathogenic nematodes are introduced into soil, temporarily reducing populations of the added nematode. However, it was also recognized that studies of this nature may have been compromised by a problem that has always plagued those working with nematophagous fungi: that the agar assay used to assess predation did not reflect predatory activity in soil, because it favoured the species best able to compete on an artificial substrate. Pathak et al. (2012) confirmed that this was a reasonable concern, as a bioassay of soil infested with *Arthrobotrys dactyloides* and *Catenaria* sp. indicated that the level of parasitism by each of these fungi depended on whether they were present alone or together in the sample being assessed. Molecular data obtained by quantifying fungal DNA in nematodes extracted from soil were considered more likely to reflect the relative importance of nematophagous fungi as parasites. However, when DNA methods were used to assess a field experiment in which two different soil types received different entomopathogenic nematodes or were left uninoculated, they failed to show a relationship between entomopathogenic nematodes and nematophagous fungi. In fact, two of the seven fungi assessed (*Purpureocillium lilacinum* and *Catenaria* sp.) were much more prevalent in one soil treatment than in the other, despite the fact that nematode abundance was similar in both treatments (Duncan et al., 2013). In contrast, samples of nematode DNA collected within a citrus orchard revealed positive correlations between total numbers of entomopathogenic nematodes and a complex of nematophagous fungi that included *Catenaria* sp. and *Arthrobotrys gephyropaga* (Campos-Herrera et al., 2012a). Thus, the hypothesis that nematophagous fungi were regulating populations of entomopathogenic nematodes was supported by the second study but not by the first.

Although the use of molecular tools in soil ecology is in its infancy, it is clear from the work of El-Borai et al. (2011), Campos-Herrera et al. (2012a) and Pathak et al. (2012) that such tools can be used to discriminate between the saprophytic and parasitic activities of the nematophagous fungi found in soil. Since additional primers and probes will eventually be developed and quantification methods will improve, a greater range of soil organisms will ultimately be characterized with greater resolution, a prerequisite to understanding the dynamics of interactions between fungi and nematodes in the soil food web.

Moth larvae, entomopathogenic nematodes and nematophagous fungi in natural shrub-land soil

At a natural shrub-land habitat in California (Bodega Marine Reserve), an endemic ento-mopathogenic nematode (*Heterorhabditis mar-elatus*) kills and consumes the soil-dwelling larvae of the ghost moth, a serious pest of bush lupine (*Lupinus arboreus*). Since the soil at the site also contains many species of nematode-trapping fungi, and lupine mortal-ity caused by ghost moth varies across the site, Jaffee *et al.* (1996b) set out to test the hypothesis that high plant mortality occurred when nematophagous fungi suppressed the entomopathogenic nematodes that normally kept the moth under control. *Arthrobotrys oligospora* was found to be the most abundant nematophagous fungus at the site, and 11 other species were also detected. Some soil sam-ples contained large numbers of nematode-trapping fungi (as many as 695 propagules/g soil), but as numbers did not differ signifi-cantly between sites with high and low plant mortality, it did not appear that these fungi were responsible for the patchy distribution of the entomopathogenic nematode.

Since a related study (Koppenhöfer *et al.*, 1996, 1997) showed that the common species of nematode-trapping fungi at the above site trapped substantial numbers of *H. marelatus* juveniles as they moved through soil, Jaffee and Strong (2005) focused on whether these fungi trapped nematodes as they emerged and dispersed from the moths. Wax moth lar-vae were used in place of ghost moths, and when larvae parasitized by entomopatho-genic nematodes were added to soil, the popu-lation densities of four nematode-trapping fungi increased substantially. The greatest increase occurred with *Arthrobotrys oligospora*, with its population density increasing about 10 and 100 times when the added moth larvae were parasitized by *H. marelatus* and *Steiner-nema glaseri*, respectively (Fig. 5. 23). The zoo-sporic fungus *Myzocytiopsis glutinospora*

was only seen when parasitized moth larvae were present, and was not detected in the driest soil (Fig. 5.23). However, despite the strong response of the fungi to the presence of parasitized moth larvae, they trapped fewer than 30% of the dispersing nematodes.

In an attempt to determine whether the response of *A. oligospora* to the presence of moth larvae depended on the presence of *H. marelatus*, Farrell *et al.* (2006) added differ-ent combinations of nematode and larval resources to soil from Bodega Marine Reserve. Population densities of the fungus consist-ently increased when moth larvae were added, but this occurred whether the larvae were infected with living nematodes, infected with nematodes that had been killed by freez-ing, or contained no nematodes but had been killed by freezing. In contrast, *A. oligospora* only responded to the addition of nematodes alone in one of three trials (Fig. 5.24). Although these results suggested that *H. marelatus* was relatively unimportant in the nutrition of *A. oligospora*, this could not be confirmed, because other potential hosts (especially bac-terivorous nematodes) increased to very large numbers when moth larvae without living *H. marelatus* were added to soil. Since bacter-ivorous nematodes also increased when *H. marelatus* was added, total nematode popu-lations in all treatments were consistently greater than 10^2/g soil (Fig 5.25).

In conclusion, these studies have shown that nematophagous fungi respond strongly to the presence of moth larvae, and kill at least some of the entomopathogenic nema-todes that disperse from the larvae. However, from a broader perspective, perhaps the most important contribution of these stud-ies is to demonstrate that interactions between fungi, insect cadavers and nema-todes occur at microsites in soil. The pres-ence of a single decomposing moth larva completely changes the composition of the soil biological community in its near vicinity, because bacteria; saprophytic and nemato-phagous fungi; entomopathogenic nema-todes; and other soil organisms compete intensely for the nutrients contained within the larva. Thus, at a microsite level, popula-tions of organisms may reach densities that

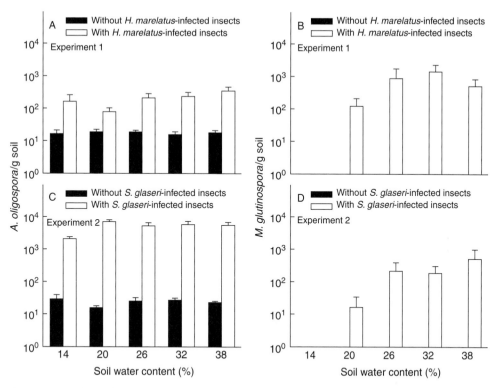

Fig. 5.23. Population densities of *Arthrobotrys oligospora* and *Myzocytiopsis glutinospora* in non-heat-treated soil as affected by the presence or absence of nematode-parasitized insects and soil water content. Experiment 1 (A and B) was done with *Heterorhabditis marelatus* and experiment 2 (C and D) with *Steinernema glaseri*. Solid and open bars indicate data from arenas without and with nematode-parasitized larvae, respectively. (From Jaffee and Strong, 2005, with permission.)

would never be seen in the bulk soil mass. Since similar intensive interactions will occur when other organic materials become available in soil (e.g. the body of a dead invertebrate, faeces from soil fauna, a root fragment, or a small piece of plant debris lying on the soil surface), soil is not a uniform entity but is made up of millions of unique microsites. Thus, our knowledge of nematophagous fungi would improve markedly if more effort was put into understanding how organisms interact within those microsites, and determining whether those interactions have an impact over larger spatial scales. For example, we need to know whether the natural enemies associated with decomposing litter on the soil surface have an effect further down the profile, or whether the nematophagous fungi that proliferate in an insect cadaver or on a piece of organic

debris will affect nematodes a few millimetres away on a nearby root.

Despite intensive studies of nematophagous fungi in soil from Bodega Marine Reserve over a period of more than 10 years, many questions about the ecology of the nematophagous fungi at that site remain to be answered (Jaffee and Strong, 2005). First, it is not clear why nematode-trapping fungi are more numerous and diverse at this site than in agricultural soils from a similar environment, or why *A. oligospora* in particular is especially abundant. Nguyen *et al.* (2007) suggested that it may have been due to the prevalence of isopods (an excellent food source for the fungus), but this conclusion was only tentative. Second, additional research is needed to determine whether *A. oligospora* and the other fungi respond to the presence of entomopathogenic

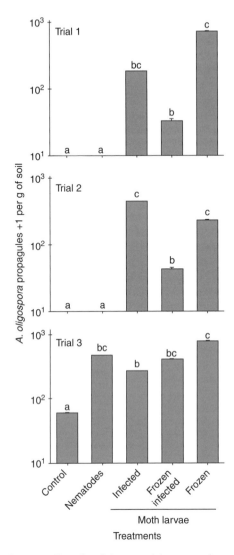

Fig. 5.24. Effect of moth larvae and the nematode *Heterorhabditis marelatus* on the population density of *Arthrobotrys oligospora* in three laboratory experiments, 35 days after one of five resources was placed in arenas containing soil from a nature reserve. The resources were: no resource (control); *H. marelatus* alone (nematodes); moth larvae containing *H. marelatus* (infected); moth larvae containing dead *H. marelatus* (frozen infected); or moth larvae containing no *H. marelatus* (frozen). Means within a trial followed by different letters differ at $P < 0.05$. (From Farrell *et al.*, 2006, with permission.)

nematodes, the dead insect, bacterivorous nematodes stimulated by the dead insect, or to a combination of these resources. Third, three fungi that formed network traps were

not enhanced by nematode-parasitized insects, but it is not clear whether this was due to competition between fungi that occupy similar ecological niches, as was shown to occur between *A. oligospora*, *A. eudermata* and *A. paucispora* (Koppenhöfer *et al.*, 1997). Fourth, the reasons that the dominant fungus at Bodega Marine Reserve (*A. oligospora*) responded much more to *Steinernema glaseri* than to *Heterorhabditis marelatus* are not known. Jaffee and Strong (2005) discussed several possibilities, but considered that the behaviour of the nematodes after emergence was probably involved. Unlike *H. marelatus*, *S. glaseri* aggregates on the cadaver after emerging and this may have increased exposure to adjacent fungi. The effect of the nematode on the cadaver may also be important, because cadavers parasitized by *S. glaseri* degraded much faster and seemed leakier than *H. marelatus*-parasitized cadavers, and soluble materials from the cadaver may have supported saprophytic growth of the fungi.

The impact of organic matter on predacious activity

In the years since the initial work of Linford and colleagues in Hawaii on nematophagous fungi and organic amendments (discussed in Chapter 9), many studies have shown that the population density or activity of various groups of nematophagous fungi is in some way related to either the organic matter status of the soil, or to the free-living nematode community that is always associated with decomposing organic matter. Some examples of that relationship are:

- In a study of 206 soil samples collected from a wide variety of sites and habitats in Ireland, Gray (1985) found that the relationship between the occurrence of nematophagous fungi and soil organic matter was almost statistically significant ($P = 0.051$). All groups of endoparasitic fungi were significantly associated with high nematode numbers, while endoparasites with adhesive and ingested conidia were both recorded more frequently in samples with high organic

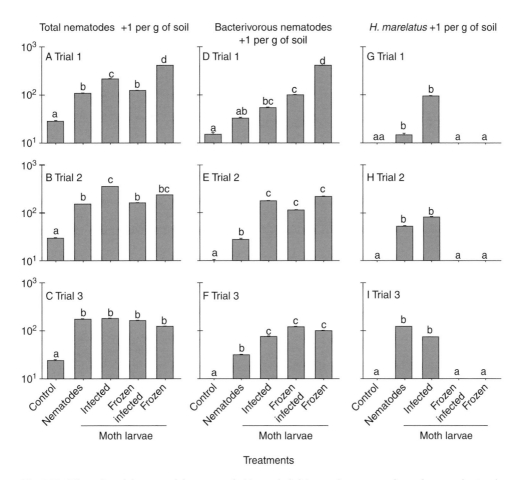

Fig. 5.25. Effect of moth larvae and the nematode *Heterorhabditis marelatus* on numbers of nematodes (total nematodes, bacterivorous nematodes or *H. marelatus*) per g of soil in three laboratory experiments, 35 days after one of five resources was placed in arenas containing soil from a nature reserve. See Fig 5.24 for an explanation of treatments. Means within a trial followed by different letters differ at $P < 0.05$. (From Farrell *et al.*, 2006, with permission.)

matter contents. The occurrence of predators with adhesive nets or constricting rings was related to the organic matter content of the soil, but not to nematode population density.

- Populations of predatory and endoparasitic fungi, bacteria and nematodes increased markedly when a Danish agricultural soil was fertilized with large amounts of farmyard manure for 6 years (Dackman *et al.*, 1987).
- *Pochonia chlamydosporia* proliferated significantly when chlamydospores were added to an organic soil, but populations remained static in two mineral soils (de Leij *et al.*, 1993).

- In a study of manure samples in Canada, nematodes became more numerous as manure piles decomposed. The number of species of nematophagous fungi and their prevalence was greater in decomposed than undecomposed manure (Mahoney and Strongman, 1994).
- Population densities of nematode-trapping fungi were ten times greater in a wheat field with a high organic matter content than in a barley field with a low organic matter status (Van den Boogert *et al.*, 1994).
- In a survey of agricultural soils in Central America, Persmark *et al.* (1995) found a positive correlation between the number and diversity of nematophagous fungi in

soil and nematode population densities. However, there was no correlation between organic matter content and the number of nematophagous fungi.

- A comprehensive study in a Swedish field showed that the population densities of nematodes, nematode-trapping fungi and endoparasitic fungi were consistently greatest in the upper layers of soil. These layers had a much higher organic matter content than soil further down the profile (Persmark *et al.*, 1996).

- Sunn hemp (*Crotalaria juncea*) is a useful green manure crop in tropical and subtropical environments because it grows rapidly and fixes considerable amounts of nitrogen. It is also a poor host of important nematodes such as *Rotylenchulus reniformis*, *Meloidogyne incognita*, *M. javanica* and *M. arenaria*, and its root and leaf exudates contain allelopathic compounds that are toxic to nematodes. *C. juncea* is also a useful organic amendment, because when fresh biomass is incorporated into soil, it enhances populations of some nematode-trapping fungi and increases fungal parasitism of eggs and vermiform stages of *R. reniformis* (Wang *et al.*, 2001). Enhancement of nematode-trapping fungi was greatest in a soil that had been growing pineapples for 4 years and did not occur in a soil that had recently been fumigated with 1,3 dichloropropene, indicating that responses to an amendment may not be obtained if soil microbial activities were not preserved by previous soil management practices (Wang *et al.*, 2003a).

- In a soil that had received an annual application of yard waste compost for 6 years, populations of nematode-trapping fungi were higher than the unamended counterpart. When *C. juncea* hay was added to both soils as an amendment, numbers of nematode-trapping fungi only increased in the soil that had previously received yard waste compost. Additionally, the *C. juncea* amendment did not increase populations of nematode-trapping fungi in a vegetable soil that had relatively low populations of these

fungi (Wang *et al.*, 2004b). These results suggest that populations of nematode-trapping fungi are more likely to increase in response to the addition of organic matter if initial populations are high due to the soil's previous history.

- In a survey of fungal pathogens of second-stage juveniles of *Heterodera glycines* in China, *Hirsutella* spp. were mainly detected in low pH soils with the highest levels of organic matter (Ma *et al.*, 2005).

Although these studies show general relationships between soil organic matter and the prevalence or activity of nematophagous fungi, those relationships are not always apparent when the responses of individual fungi to organic matter are considered. For example, in one of two long-term experiments comparing conventional and organic cropping systems in California, organically managed plots received much greater organic matter inputs than conventionally managed plots, and consequently had much greater microbial biomass and higher numbers of bacterivorous nematodes. They also supported more species of nematode-trapping fungi, although the difference between cropping systems was relatively small (Jaffee *et al.*, 1998). However, fungal species responded in different ways to the two cropping systems. *Drechslerella dactyloides* and *Nematoctonus leiosporus* were detected more frequently and in greater numbers in organic plots, whereas *Dactylellina candidum* and *Arthrobotrys thaumasia* tended to be more numerous in conventional plots.

In the second California trial at a different location, ten species of nematophagous fungi were detected, with *Arthrobotrys thaumasia* and *A. oligospora* the most abundant (10 and 1 propagules/g soil, respectively). However, except for *Nematoctonus leiosporus* and *Meristacrum* sp., which were detected more frequently in organic plots, detection frequencies and population densities of nematode-trapping fungi were similar in the organic and conventional farming systems (Timm *et al.*, 2001). Thus, when results from the two California studies are considered together, it is apparent that the additional inputs of organic matter in the organic farming systems did not substantially enhance nematode-trapping fungi. However, it was possible that the effects of organic amendments were transient or patchy, and so temporal or spatial peaks of activity may have been missed by the sampling methods employed (Timm *et al.*, 2001). Also, plots were cultivated regularly and samples were collected after treatments had been in place for only a few years. Longer times, or a greater period without disturbance, may have been needed to observe effects.

Endoparasitic fungi could be considered more likely than the nematode-trapping fungi to respond to organic amendments, as they tend to be dependent on nematodes as a food source. Results of a microcosm experiment with *Drechmeria coniospora* supported that hypothesis, as conidial density in soil amended with lucerne meal increased as populations of bacterivorous nematodes increased (Van den Boogert *et al.*, 1994). In contrast, experiments in the field and laboratory indicated that organic amendments did not enhance parasitism of nematodes by *Hirsutella rhossiliensis* (Jaffee *et al.*, 1994). In the field, parasitism of *Mesocriconema xenoplax*, one of its most common hosts, was slightly suppressed by the addition of composted chicken manure; while in the laboratory, numbers of bacterivorous and fungivorous nematodes increased in response to the addition of wheat straw or composted cow manure, but nematode parasitism usually decreased.

Since *H. rhossiliensis* is capable of parasitizing a wide range of plant-parasitic and free-living nematodes, the results of the experiments by Jaffee *et al.* (1994) were unexpected. Nevertheless, they highlight the fact that such results are always difficult to interpret, because organic matter influences many soil processes. Thus, the authors noted that in their experiments, organic matter may have stimulated some antagonists via density-dependent parasitism; may have been the source of ammonia and other nematicidal and fungicidal compounds; may have influenced several aspects of the biology of nematodes and *H. rhossiliensis* by altering soil porosity and soil water; and may have stimulated fungivorous nematodes and other antagonists of fungi. In addition, it was not

known whether all the free-living nematodes in the soil were hosts of *H. rhossiliensis*.

One of the difficulties involved in determining whether decomposing substrates stimulate nematophagous fungi is that the fungi are difficult to quantify, particularly in the field, where fungi are likely to occur in patches or hotspots associated with organic matter. Jaffee (2002) attempted to overcome this problem by applying amendment treatments to small segments of PVC pipe, sealing both ends with mesh, and then burying these 'soil cages' in the field. In addition to minimizing variability associated with the patchy distribution of organic matter, this system allowed the soil to be assessed in an undisturbed state. The latter point is important, because the physical act of mixing the soil during the collection process destroys trapping structures (McInnis and Jaffee, 1989; Jaffee *et al.*, 1992b).

To determine whether nematode-trapping fungi responded to organic inputs, Jaffee (2002) amended soil with dried grape leaves at a rate equivalent to 4500 kg/ha and then buried soil cages containing amended and unamended soil in the field. The cages were retrieved over a period of about 10 weeks and fungal population densities and levels of trapping activity were assessed using standard methods. Four experiments were carried out (the cages were buried in two vineyards and the experiment was repeated at each site the following year), and the results were inconsistent. At one site, where *Dactylellina candidum* had been added in alginate pellets, the grape leaf amendment stimulated its population density and trapping activity in one year but not in the other. At the second site, where *Arthrobotrys oligospora* was the dominant nematode-trapping fungus, the presence of grape leaves stimulated populations in one year but not in the other, but the fungus did not trap nematodes in either year. Although bacterivorous and fungivorous nematodes always increased when grape leaves were added to soil, these results typified the inconsistency often observed in experiments with organic amendments. Jaffee concluded that the biology of nematode-trapping fungi was poorly understood and suggested that future work should focus on fungivorous nematodes that may either enhance or suppress these fungi, depending on the relative rates of fungivory and trapping; the effects of environmental factors such as temperature and moisture on activity; and the impact of amendment quality and quantity on the behaviour of these fungi in soil.

In a follow-on study using similar methods, Jaffee (2004a) explored the latter issue. Grape and lucerne leaves were used as amendments, and they were mixed with soil at two rates (the equivalent of 4500 and 9000 kg/ha). Soil cages were filled with amended and unamended soil; one set was inoculated with alginate pellets containing *Dactylellina candidum*, while a second set received *Arthrobotrys oligospora*. The two batches were buried in separate vineyards and fungal population densities and trapping were assessed 25–39 days later. Results of experiments with *D. candidum* in successive years indicated that this fungus consistently trapped assay nematodes in field soil. The population density of *D. candidum* and its trapping activity were most enhanced with the smaller quantity of lucerne amendment, and a combined analysis of data from several experiments indicated that the mean percentage of assay nematodes with adherent knobs was correlated with the population density of the fungus (Fig. 5.26).

In contrast to the results with *D. candidum*, populations of *A. oligospora* generally increased in response to all amendments, although population density was most enhanced by the larger quantity of lucerne amendment. However, *A. oligospora* trapped few or no nematodes, regardless of amendment, a result that had also been obtained in previous studies (Jaffee, 2002, 2003). After discussing possible reasons for the different responses of the two fungi to organic amendments, Jaffee (2004a) concluded his paper with a question: why did *A. oligospora* trap so few nematodes in these experiments? The fact that someone who had studied this fungus for many years could ask such a question is testimony to the difficulties involved in understanding the behaviour of nematophagous fungi in soil.

Other factors influencing predacious activity

Analyses of data obtained from a survey of Irish soils showed that other than organic matter and nematode density (two factors considered in the previous section), the main soil factors that affected the distribution of nematophagous fungi were pH and soil

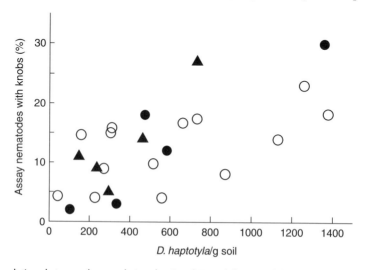

Fig. 5.26. Correlations between the population density of *Dactylellina candidum* (syn. *D. haptotyla*) and the percentage of assay nematodes with adherent knobs in vineyard soils. Correlation analysis for data from several experiments (n = 24) indicated that r = 0.66 and P < 0.01. The data were obtained from various experiments (represented by different symbols), with each value the mean of six replicates. (From Jaffee, 2004a, with permission.)

moisture (Gray, 1985). However, it was not clear why some nematophagous fungi were associated with acidic soils and others pre- ferred soils with a higher pH, while the asso- ciation with soil moisture was compromised by the use of moisture content rather water potential to assess the soil's moisture status. Nevertheless, it is quite clear that moisture content in particular can have a major impact on the occurrence and activity of nemato- phagous fungi. It is obviously important for parasites that produce zoospores, with the observations of Kerry *et al.* (1980) and Jaffee and Strong (2005) confirming this for *Nematophthora gynophila* and *Myzocytiopsis glutinospora*, respectively. With regard to non- zoosporic fungi, Jaffee *et al.* (1989) observed that *Hirstutella rhossiliensis* parasitized more nematodes in the drier soil between the irriga- tion furrows in peach orchards than in the wetter soil adjacent to the furrows. Later, it was found that soil water status substantially affected transmission of this fungus, appar- ently because sporulation was reduced at high water potentials (Tedford *et al.*, 1992). In con- trast, soil water content did not affect the detection of *Arthrobotrys oligospora* and five

other nematode-trapping fungi (Jaffee and Strong, 2005), although it has been suggested that these fungi may produce spores rather than traps when soil is dry (Jaffee *et al.*, 1992b). One important factor that influences the predacious activity of nematophagous fungi is soil disturbance. This issue was first recog- nized by McInnis and Jaffee (1989), who observed that spores produced by *Hirsutella rhossiliensis* will not adhere to nematodes unless they are attached to phialides. If soil containing this fungus is disturbed, the hyphae are disrupted and the spores are no longer able to infect nematodes. Jaffee *et al.* (1992b) confirmed that result and also showed that disturbance was detrimental to some nematode-trapping fungi. *Steinernema glaseri* parasitized by one of five fungi were used to inoculate small vials containing sandy soil. The soil was incubated for 21 days to allow the fungi to utilize the host and produce spores or traps outside the host, and then the soil was either shaken in a plastic bag and returned to the vial, or left undisturbed. Predacious activity was assessed by adding assay nematodes and recovering them 66 h later. The results, which are shown in Fig. 5.27,

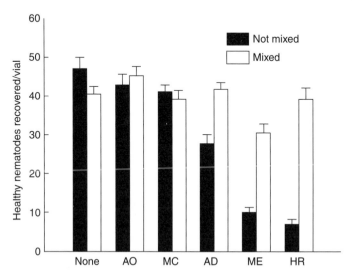

Fig. 5.27. Parasitism of assay nematodes (juveniles of *Heterodera schachtii*) in mixed or unmixed soil previously inoculated with no fungus (None), *Arthrobotrys oligospora* (AO), *A. gephyropaga* syn. *Monacrosporium cionopagum* (MC), *Drechslerella dactyloides* syn. *A. dactyloides* (AD), *Dactylellina ellipsospora* syn. *M. ellipsosporum* (ME) and *Hirsutella rhossiliensis* (HR). Fungi were added to soil in the form of parasitized nematodes, and after 21 days, soil was mixed or left undisturbed and then inoculated with assay nematodes. (From Jaffee *et al.*, 1992b, with permission.)

indicated that the number of healthy assay nematodes recovered from vials containing *Arthrobotrys oligospora* or *A. gephyropaga* was similar to the control, whether or not the soil was disturbed. *Drechslerella dactyloides* and *Hirsutella rhossiliensis* both reduced the number of healthy assay nematodes relative to the control, but only when the soil was disturbed. *Dactylellina ellipsospora* produced an intermediate response, reducing the recovery of healthy nematodes in disturbed soil, but having a greater effect in undisturbed soil.

Such results are important, because they indicate that disturbance events such as soil cultivation will at least temporarily reduce the predacious activity of some nematophagous fungi. Perhaps a more important effect is that the amount of fungal inoculum in soil may also be reduced. If a nematode-trapping fungus, for example, has converted all the resources available in a captured nematode into traps, the destruction of those traps may mean that reserves in spores and assimilative hyphae are no longer available for new growth. Thus, disturbing the soil could have a long-term impact on fungal inoculum density. It also means that estimates of fungal population densities may be low if they are based on soil samples that have been mixed or otherwise disturbed (Jaffee *et al.*, 1992b).

Maximizing the Predacious Activity of Nematophagous Fungi in Agricultural Soils

It is clear from the foregoing discussion that the ecology of nematophagous fungi is poorly understood. Populations of these fungi are difficult to measure in soil, while it is even more challenging to assess levels of parasitism and predation. Also, the soil biological community is so complex that it is almost impossible to predict how particular groups of nematophagous fungi will respond to management-induced changes in the soil environment. Nevertheless, our knowledge of the these fungi has increased markedly over the last 25 years, and the papers by Bruce Jaffee and colleagues cited in this chapter are an essential starting point for anyone

attempting to understand how nematophagous fungi interact with each other, and also with different sources of organic matter, plant roots, the abiotic environment, other soil organisms and different components of the nematode community (e.g. bacterivorous, fungivorous, entomopathogenic and plant-parasitic nematodes).

From a nematode management perspective, the critical question to be answered is whether we can use our ecological knowledge and our ever-improving understanding of nematophagous fungi to maximize their predacious activity in agricultural soils. Given the extreme complexity of the soil biological environment and the fact that plant-parasitic nematodes are only one of many potential food sources for nematophagous fungi, it would be easy to conclude that any attempt to enhance the predacious activity of these fungi will have unpredictable consequences. However, this chapter finishes on a positive note by considering how we might better utilize the suppressive services offered by the nematophagous fungi.

The first step in deciding what can be done to enhance the activity of nematophagous fungi is to reflect on what we have learnt from the studies discussed in this chapter:

- Nematophagous fungi are likely to occur in every soil, but their diversity and population density will vary with soil physical properties, environmental conditions and the way the soil has been managed in the past. These fungi may sometimes be difficult to detect, but they will be found if appropriate isolation methods are used. Results of some initial studies using molecular methods indicate that nematophagous fungi are likely to be much more prevalent and diverse than studies with traditional cultural methods would suggest.

- There is enormous diversity in the mechanisms used by fungi to parasitize or prey on nematodes. Some endoparasitic fungi produce zoospores that swim towards the host and encyst on the cuticle, while others produce adhesive spores that attach to passing nematodes. The egg-parasitic fungi infect their hosts by growing over the surface of the egg and then penetrating the egg shell, while the nematode-trapping fungi use a range of different trapping devices to capture nematodes.

- Some nematophagous fungi are obligate parasites of nematodes, but most are also able to utilize alternative food sources. Many are saprophytes, some are mycoparasites and some are parasitic on soil animals other than nematodes. Although it is often assumed that nematodes are the primary food source for nematophagous fungi, the nutritional preferences of individual species in the competitive soil environment are often not known.

- A few nematophagous fungi are relatively specialized parasites of specific nematodes, but most have a relatively wide host range. Nevertheless, there is invariably some specialization, and this may be manifested as a preference for particular bacterivorous or entomopathogenic nematodes, or as a capacity to parasitize or prey on some plant-parasitic nematodes but not on others.

- Nematophagous fungi parasitize nematodes for different reasons. Specialized parasites respond to increasing numbers of nematodes by using them as a food source. The response of facultative parasites is less predictable, but predacious activity seems to be associated in some way with intense competition for carbon and nitrogen within the soil biological community.

- Continuous inputs of organic matter are required to sustain the soil biological community. Nematophagous fungi are a component of that community, and they will respond to organic inputs through saprophytism, parasitism or a combination thereof. Organic matter is, therefore, the key resource determining whether nematophagous fungi are present and playing an active role within the soil food web.

- The effects of organic matter will depend on the quantity and quality of the organic inputs. Those inputs can come from various sources (e.g. root exudates, plant residues, carcases of soil organisms and organic amendments), but they must occur continuously if populations of nematophagous fungi are to be sustained over time.

- Since interactions between nematophagous fungi, organic matter and free-living nematodes occur at microsites within the soil profile, plant-parasitic nematodes are most likely to be affected by these fungi if those microsites are located near roots that are their food source.

- Inputs of organic matter are likely to have many indirect effects on nematophagous fungi. For example, adding organic matter to soil improves its moisture-holding capacity, and this will have subtle effects on nematode activity and the predatory behaviour of some fungi. In another example, organic matter inputs will improve a soil's nutrient status by adding key nutrients, improving its nutrient-holding capacity and enhancing nutrient cycling. Thus, plant growth will improve, and since this will result in higher carbon inputs to the soil, there will be flow-on effects to the soil biological community.

- Since any increase in fungal population density is inevitably followed by fungivory, populations of nematophagous fungi do not increase indefinitely when additional nutritional resources become available. The population density and level of activity of these fungi will eventually be determined by the bottom-up effects of organic matter, nematodes and other food sources, and the top-down effects of natural enemies.

- When soil is disturbed, the infective structures produced by nematophagous fungi are destroyed or displaced. Consequently, disturbing the soil reduces predacious activity and may also have a negative impact on fungal inoculum density. Continual soil disturbance has an even greater impact, as it also reduces the soil carbon reserves on which nematophagous fungi depend.

When the soil management practices used in agriculture are viewed from the perspective of whether they maintain an active and fully functioning community of nemato-phagous fungi capable of contributing to the suppression of nematode pests, it is clear that there are serious deficiencies in many current farming systems. Perhaps the biggest prob-lem is the continuing decline in levels of soil organic matter. The organic content of most agricultural soils has declined by more than 50% since they were first cultivated, and as carbon losses from erosion and tillage usually exceed the gains from inputs of crop residues, this decline will continue. In an effort to improve the situation, organic amendments are sometimes applied, particularly in high-value crops. However, amendments are usually incorporated into soil with tillage imple-ments, and high application rates are used in the expectation of obtaining an obvious and immediate response. Amendments applied in this way invariably cause rapid and sub-stantial increases in soil biological activity, but effects on nematophagous fungi and the nematode community are essentially unpredictable.

Another major problem with current soil management practices is the negative impact of tillage on nematophagous fungi. Although the detrimental effects of tillage must be con-firmed by monitoring nematodes, fungi and organic matter in long-term tillage experi-ments, the results of experiments in micro-cosms suggest that tillage is likely to suppress predacious activity temporarily and may also reduce population densities in the longer term. The disastrous effects of soil fumigation on beneficial soil organisms are well documented, but it is likely that that solarization and bare fallowing, for example, are also detrimental to organisms such as nematophagous fungi. Another issue deserving particular attention is the impact of nitrogen fertilizers on nematopha-gous fungi. Since some of these fungi may resort to carnivory in order to obtain nitrogen, it is important to know whether the type, rate and placement of nitrogen fertilizer influences their population density and predatory activ-ity. Because nitrogen addition and availability stimulate mineralization of available carbon (Fog, 1988; Recous et al., 1995; Henriksen and Breland, 1999), nitrogen may also be detrimen-tal to nematophagous fungi by increasing the rate of decomposition of organic matter.

In considering management prac-tices that are likely to enhance rather than deplete nematophagous fungi, the three key components of conservation agricul-ture (minimal soil disturbance, mainte-nance of plant residue cover and crop rotation) stand out as practices that are worthy of adoption. The soil health bene-fits obtainable from conservation agricul-ture were discussed in Chapter 4, but there is some evidence to suggest that another soil health benefit will be greater levels of predacious activity from nematophagous fungi.

The basis of conservation agriculture is that tillage is minimized and residues pro-duced by the principal crops and cover crops are left on the soil surface to decom-pose. When these practices are employed, fungi play an important role in the decom-position process for a number of reasons: (i) hyphal networks are no longer being disrupted by tillage; (ii) fungi are able to tolerate the dry conditions that often occur near the soil surface; (iii) the fungal energy channel is favoured in situations where decomposing organic materials have a rela-tively wide C:N ratio; and (iv) fungal hyphae can translocate nutrients across air gaps or between areas of high and low nutrient availability.

Although some studies have shown a shift towards fungal dominance within the microbial community under no-till agricul-ture, it is more common to find that both bac-terial and fungal biomass are higher under no-till than under intensive tillage (Helgason et al., 2009). However, the advantages gained by fungi when tillage is eliminated become more apparent when microbial communities within soil aggregates are assessed. Bacterial, fungal and total biomass were up to 32% greater in no-till than in conventional till aggregates, but a biomarker for arbuscular mycorrhizal fungi, one of the main con-tributors to micro-aggregate formation, was 40–60% higher in no-till aggregates (Helgason et al., 2010).

Data collected from a long-term experi-ment in China that included residue reten-tion, tillage and crop rotation treatments have also shown that the practices involved in conservation agriculture are beneficial to fungi. Species richness, fungal diversity and total number of fungal species were all much greater under no-till than under conventional till, and the resulting fungal community in the no-till treatment was also beneficial to micro-aggregate formation. Interestingly, these benefits were greater at depths of 5–10 and 10–15 cm than near the surface (Wang *et al.*, 2010).

Although nematophagous fungi were not considered in the studies cited, this group of fungi should be favoured by the lack of disturbance associated with no-till farming systems. The fact that nematophagous fungi are commonly associated with decomposing high-carbon substrates also suggests that they will be advantaged by the practices associated with conservation agriculture. Since a decrease in physical disturbance slows the rate of organic matter decomposition (Beare *et al.*, 1994), nematophagous fungi could also be expected to benefit from a more constant supply of nutrients in a no-till system. However, the presence of root channels in soils that are no longer tilled may have an even greater impact on nematophagous fungi, particularly with regard to their predatory activity against plant-parasitic nematodes. Plant roots are excellent tillage tools, and the term 'biological drilling' has been used to describe the process by which plant roots form channels during growth, before decaying to form open biopores (Cresswell and Kirkegaard, 1995). Since roots in no-till systems tend to follow channels produced by previous crops (Williams and Weil, 2004; Chen and Weil, 2011), root channels are microsites within the soil profile where new roots come into contact with decomposing roots from previous crops. Since interactions between organic matter, nematophagous fungi, free-living nematodes and plant-parasitic nematodes are most likely to occur at such microsites, the probability that nematophagous fungi will be exhibiting predatory activity at a microsite occupied by plant-parasitic nematodes should increase when soil is no longer tilled.

Another lesson we can learn from our knowledge of nematophagous fungi is that when organic amendments are added to soil in order to enhance biological mechanisms of nematode control, it may be better to place the amendments on the soil surface as mulch rather than incorporating them into soil with tillage implements. Although both application methods are likely to increase populations of nematophagous fungi, it seems logical to use the simplest application method and also retain the advantages likely to be gained from minimizing disturbance.

Crop rotation is an important component of conservation agriculture, and again, there is virtually no information on its effects on nematophagous fungi. However, one of the benefits of using a range of crops in a farming system is that crops vary in the architecture of their root systems and in the quality and quality of their biomass. Thus, the type of organic matter returned to the soil and its locations within the soil profile can be varied by including various crops in the rotation (e.g. perennial or annual crops, legumes or non-legumes, deep or shallow-rooted crops, and crops with taproot or fibrous-root systems). Although we cannot predict how various nematophagous fungi will respond to different sources of organic matter, a varied range of organic inputs from dissimilar crops is more likely to serve the nutritional needs of a diverse group of fungi than residues from a single crop.

In conclusion, the last section of this chapter is an attempt to consolidate what is known about farming systems and the ecology of nematophagous fungi into a body of knowledge that can be used to enhance the suppressive services provided by the soil food web. However, I recognize that many of the comments made with regard to the role of minimum tillage, residue retention and crop rotation in enhancing suppressiveness to nematodes are speculative. Nevertheless, they are made in the hope that scientists will be encouraged to consider the issues involved and begin working in this under-researched area.

Finally, the issues associated with modifying farming systems to enhance suppressiveness to plant-parasitic nematodes are discussed in Chapters 9, 10 and 11, and it is important that the points considered here are kept in mind when reading those chapters.

However, it is also important to have realistic expectations about what is achievable through changes to a farming system. The biology of agricultural soils can almost certainly be improved to the point where fungi contribute significantly to the suppression of nematode pests, but nematode populations will sometimes not be reduced to levels below the damage threshold. In those situations, the challenge is to find practices that can be added to the farming system to further reduce populations of the target pest without compromising the suppressiveness obtained from nematophagous fungi.

6

Nematodes, Mites and Collembola as Predators of Nematodes, and the Role of Generalist Predators

Nematodes have a diverse range of natural enemies, but the literature on biological control has been dominated by research on microbial antagonists for a number of reasons. First, most microorganisms are readily cultured on defined media, and so they are amenable to study in the laboratory. Second, commercial mass production techniques are available for bacteria and fungi, and so they tend to be the first organisms selected for research programmes on inoculative or inundative biological control. Third, organisms that exhibit a degree of specificity towards the target nematode have generally been considered more effective biocontrol agents than generalists, and this has meant that research has focused on bacteria such as *Pasteuria*, and relatively specialized fungi capable of parasitizing the females and eggs of sedentary endoparasitic nematodes.

This chapter focuses on another group of soil organisms that are best described under umbrella terms such as 'generalist', 'polyphagous' or 'omnivorous', and considers their role in suppressing nematode populations. It covers all the predatory micro-, meso- and macro-fauna discussed in a review by Small (1988), but focuses particularly on nematodes, Collembola and mites, the most widely studied predators of nematodes. Often neglected by those interested in finding 'silver bullet' solutions to nematode problems, these generalist predators dominate higher trophic levels

in the soil food web and provide a background level of nematode suppression in all soils. Consequently, they are vitally important when soils are being managed to conserve natural enemies of nematodes. The roles of nematodes, Collembola and mites as top-down regulators of nematodes and other members of the soil fauna are discussed here, together with the soil and crop management tactics that can be employed to enhance their activity within agroecosystems. The relative merits of specialists and generalists as biological control agents are also considered, and it is suggested that polyphagous predators are an important but underestimated component of the soil biological community, and must play a role in ecologically sound nematode management programmes. The chapter concludes with a discussion of best-practice farming systems and their capacity to provide the regulatory services needed to suppress nematode pests.

Predatory Nematodes

In the soil environment, nematodes that prey on other nematodes are found mainly in five taxonomic groups within the Nematoda: Mononchida, Dorylaimida, Diplogasteridae, Seinuridae and Tripylidae. Since these orders

and families are not only phylogenetically
different (van Megen *et al.*, 2009) but also ecologically diverse, predatory nematodes cannot be considered as a single entity. The main
characteristics of each group of nematodes
are, therefore, outlined next. Additional general information can be obtained from Khan
and Kim (2007).

Characteristics of the five major groups of predatory nematodes

The Mononchida, commonly referred to as
mononchs or mononchids, are unique among
the predatory nematodes in that all species are
predacious. Most species are relatively large
(usually greater than 1 mm in length) and are
considered K strategists by ecologists due to
their long life cycles and relatively low fecundity. In the successional or colonizer–persister
classification often used to denote the ecological role of nematodes, they occupy level 4
(see Table 3.1 in Chapter 3). Equipped with a
distinct and well-armed mouth cavity that is
instantly recognizable (Fig. 6.1), mononchids
either completely swallow their prey or use
one or more puncturing or grasping teeth to
tear the victim apart. A recent monograph by
Ahmad and Jairajpuri (2010) provides an excellent coverage of the morphology and phylogeny of this group of nematodes, and is also
an indispensable guide to the taxonomy of the
Order Mononchida, which contains eight families, 49 genera and hundreds of species.

The order Dorylaimida, which also contains large-bodied K strategists, is phylogenetically related to the mononchids, with both
orders residing in a single major clade within
the Nematoda (Holterman *et al.*, 2006). However, members of the Mononchida form a relatively tight taxonomic unit, whereas the order
Dorylaimida is much more diverse, with its
members displaying a mosaic of morphological
characters that make them notoriously difficult to identify (Holterman *et al.*, 2008). These
taxonomic and diversity issues mean that it is
almost impossible to make general statements
about the feeding habits of dorylaimids. Most
genera are considered omnivorous, but some
are predators and others are plant or fungal

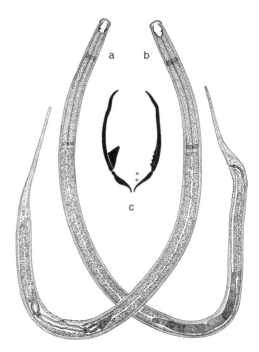

Fig. 6.1. A typical mononchid (*Parahadronchus
shakili*). (a) Male; (b) female; and (c) a magnified view
of the buccal cavity. (From Ahmad and Jairajpuri,
2010, with permission.)

feeders (Yeates *et al.*, 1993a). Predatory species
are often 1–4 mm in length, and they either
suck body fluids from their prey through an
opening in their narrow stylet, or use a mural
tooth and associated sheath to ingest liquid
food. A book by Jairajpuri and Ahmad (1992)
covers the morphology and systematics of
this group of nematodes, while a review by
McSorley (2012) provides a good summary of
the ecology of three of the most common
dorylaimid genera, *Aporcelaimellus*, *Eudorylaimus*
and *Mesodorylaimus*.

Since diplogasterids are close relatives
of the bacterial-feeding rhabditids, they share
with them characteristics such as a short life
cycle and a capacity to feed on bacteria.
However, the buccal cavity of most diplogasterids is armed with teeth, and although this
enables these species to prey on nematodes,
they are often selective in their feeding habits
(Grootaert *et al.*, 1977). Little is known about
their ecology in natural environments, but
the fact that they are readily cultured on
bacteria has meant that there has been some

interest in using them as inundative bio-logical control agents against plant-parasitic nematodes.

The genus *Seinura* is unusual in that it is one of very few predatory nematodes in the Aphelenchoididae, a family that predomin-antly feeds on fungi. Predation on nematodes is a three-stage process involving penetration of the cuticle by the stylet; injection of digest-ive enzymes to immobilize the victim and disintegrate its tissues; and bulb pulsation to withdraw the body contents of the prey and suck them into the intestine (Linford and Oliveira, 1937; Hechler, 1963; Wood, 1974; Esser, 1987). However, predatory activity has only been observed on agar, and little is known about the role of *Seinura* as a predator in natural environments.

The fifth group of predatory nematodes belong to the family Tripylidae, a group of pre-dominantly aquatic nematodes that also occurs in soil. There are seven genera in the Tripylidae, but a recent molecular study suggests that the family should be reorganized into two clades (Prado Vera *et al.*, 2010). Tripylids have a rela-tively narrow stoma armed with teeth, and their intestinal contents indicate that they ingest small microfauna, including nematodes. However, their impact as predators in agricul-tural soils has never been determined.

The prey of predatory soil nematodes

Nematodes likely to be predacious are nor-mally distinguished from other nematodes on the basis of their oesophageal structure and the type of armature used to capture and consume their prey (Fig. 6.2). However, as Small (1987) pointed out, the structure of the buccal cavity provides no more than an indi-cation that a nematode may consume other nematodes, as most predatory nematodes are to some extent omnivorous, and nematodes with similar mouth parts can have widely dif-ferent feeding habits. Thus, to fully understand the ecological role of a purported predator, its

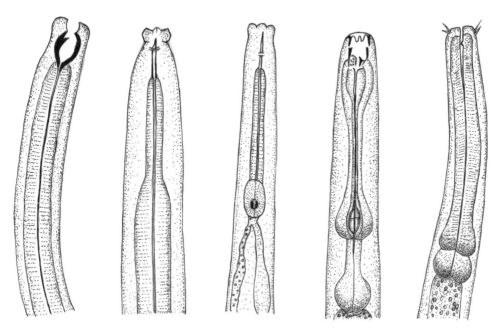

Fig. 6.2. The mouth parts and anterior region of common predatory nematodes. From left to right: mononchids have a wide buccal cavity armed with teeth and/or denticles; dorylaimids are equipped with an odontostyle; *Seinura* has a stylet similar to plant- and fungal-feeding aphelenchids; diplogasterids have a slender or spacious stoma that may be armed with moveable teeth; while the stoma of the Tripylidae is a simple tube with one or more teeth on the dorsal and ventral walls.

primary food sources must be determined in an environment where a full range of potential prey is available.

Much of our current knowledge of the biology, life history and feeding habits of predatory nematodes has been obtained by observing their behaviour on agar plates. However, from the perspective of feeding habits, prey availability on agar does not reflect the natural situation. Also, there is little evidence that the behaviour of the predator and prey in such a highly contrived environment bears any resemblance to activity in soil. Thus, only a limited number of studies in Petri dishes will be mentioned here, simply to indicate the type of information that has been obtained from this type of research. Further examples can be found in Khan and Kim (2007).

- Studies in New Zealand and India provide examples of the many different feeding habits that are possible within aporcelaimids. In the former study, an unidentified species of *Aporcelaimellus* took at least 12 weeks to complete its life cycle, but was only able to multiply on two of nine algal species (Wood, 1973a). The studies in India looked at the predatory activity of *A. nivalis* in the presence of nematodes, and showed that when this species was offered various plant-parasitic nematodes as prey, *Meloidogyne* and *Heterodera* juveniles were preferred, but predation was determined by factors such as prey number, temperature, agar concentration and starvation of the predator (Khan *et al.*, 1991). However, when encounters with 31 potential prey species from five feeding groups were assessed, the attack rate and strike rate was greatest, and the number of prey remaining was lowest, for bacterial-feeding nematodes (Bilgrami, 1993).
- Observations on the feeding behaviour of two dorylaimids on rice root nematode, *Hirschmanniella oryzae*, showed that *Discolaimus major* killed more prey, and fed and aggregated longer than *Laimydorus baldus*. Prey search and killing abilities were governed by temperature, prey density, starvation and prey incubation, and depended on feeding duration and

the number of predators aggregating at feeding sites (Bilgrami and Gaugler, 2005).

- *Mesodorylaimus* is a widely distributed genus that is usually considered omnivorous. In a laboratory study that is sometimes cited as evidence of omnivory, a species closely related to *M. lissus* was observed to feed predaciously on other nematodes and encysted amoebae; parasitically on fungal hyphae, algae, root hairs and epidermal cells of wheat; and microphagously in colonies of bacteria and actinomycetes (Russell, 1986). However, further work is required to confirm these observations, as in some of these cases, it is not clear whether the nematode was stylet-probing a potential food source, or feeding, growing and reproducing on that food source.
- Mononchids are usually categorized as relatively long-lived nematodes, but observations on agar have shown that *Mononchus aquaticus* can multiply rapidly on a diet of *Panagrellus redivivus*, with a generation time of only 14 days being recorded at 28°C. Prey were only detected when they touched the lips of the predator, and rhabditids were preferred as a food source over other nematodes (Grootaert and Mertens, 1976). However, in another study in which rhabditids were not included as prey, the least active nematodes were found to be the most vulnerable to attack (Bilgrami *et al.*, 1983).
- Prey preference studies with three diplogasterid predators (*Butlerius* sp., *Mononchoides longicaudatus* and *M. fortidens*) showed that all three species were capable of feeding on bacterial-feeding and plant-parasitic species (Grootaert *et al.*, 1977; Bilgrami and Jairajpuri, 1989). Thus, claims by Bilgrami *et al.* (2005) that juveniles of *Meloidogyne incognita*, *Heterodera avenae* and *Anguina tritici* were the 'highly preferred' prey of *Mononchoides gaugleri* must be treated with caution, as the only prey tested were plant parasites. Prey search duration was dependent on temperature and prey density, and was briefest (10–20 minutes) at a temperature of 20–30°C when 150–225 prey were present on a 5.5 cm-diameter Petri dish.

Another common method of determining the feeding habits of predatory nematodes is to measure the response of prey when predators are added to pots. However, such experiments have the same failings as observations on agar, as a full range of food sources will not be available to the predator when sterilized soil or potting mix is used as a substrate. When natural soil is added to pots, evidence that 'prey' numbers are reduced by the 'predator' is only circumstantial, as other predators are also likely to be contributing. Therefore, observations of gut contents provide the only direct evidence of predation in the soil environment. However, as indigestible or undigested remains of the prey must be seen (e.g. enchytraeid setae, nematode stylets), such analyses cannot be used for fluid feeders or for predators with difficult-to-identify prey in the gut.

One of the best examples of the systematic use of gut content analysis to characterize the feeding habits of a predatory nematode in a natural environment was a study in which the nematodes found in the gut of a mononchid (*Anatonchus tridentatus*) were compared with the community of nematodes present in the soil from which the predator was obtained. Twenty-three nematode species were found in soil from the sampling site, and 18 of those species were represented in the gut contents, indicating that this predator was polyphagous with respect to its nematode diet. However, *A. tridentatus* seemed to prefer some nematodes over others, as certain dorylaimids (e.g. *Pungentus*, *Aporcelaimellus*

and *Eudorylaimus*) were considerably more common in its gut than in the soil nematode community (see Small, 1987, for details of an unpublished study by P. Grootaert).

In the first serious attempt to define the prey of predatory nematodes, Small (1987) collated information derived from gut content analyses, direct observation, observations of nematodes in culture, feeding experiments and pot experiments, and used it to draw conclusions about the diet of various predatory nematodes. Although it was recognized that the methods used to assess feeding habits were not entirely satisfactory, the detailed tables included in that review provide a good summary of published information on the prey of various species of predatory nematodes. One of the summary tables (presented in modified form as Table 6.1) is particularly useful, because it indicates the prey organisms that have been reported for each of the five main groups of predatory nematodes.

When the information presented in Small (1987) is considered in its entirety, a number of conclusions can be drawn:

- Predatory nematodes may have preferred food sources, but most are polyphagous to some extent.
- *Seinura* is the only known monophagous predator.
- Nematodes are the most frequently recorded prey of predatory nematodes. However, it is not clear whether this represents a genuine preference by the predators, or merely reflects the greater

Table 6.1. An indication of the diet of five groups of predatory nematodes based on the percentage of records published prior to 1987 that referred to common soil organisms as prey.

Predator	No. of records	% of records referring to various organisms as prey					
		Nematodes	Oligochaetes	Rotifers	Tardigrades	Protozoa	Others[a]
Mononchids	143	75	7	7	1	5	6
Dorylaimids	91	47	24	4	1	4	19
Aphelenchids[b]	15	100	0	0	0	0	0
Diplogasterids	39	59	5	0	3	21	13
Others[c]	32	25	6	31	6	13	19

[a]Includes bacteria, fungal hyphae and spores, moss, algae, diatoms, mites, mite eggs and insect eggs
[b]Mainly *Seinura*
[c]Includes *Tripyla* and *Tobrilus* (Tripylidae); *Ironus* (Ironidae); *Synonchium* (Cyatholaimidae); and *Eurystomina* (Oncholaimidae)
(Modified from Small, 1987, with permission.)

interest of nematologists in studying nematodes as prey.

- Plant-parasitic nematodes are rarely, if ever, the primary prey of predatory nematodes. Most members of the nematode community are potential prey, and whether a particular species is successfully attacked depends on its proximity to the predator (in space and time), and factors such as body size, cuticle thickness, retention of cuticles from previous moults, and a capacity to elude attack by moving rapidly. Since nematode body size varies with life stage, juvenile and adult predators may have different food sources, and prey may be differentially susceptible to predation at different stages in their life cycles.
- Oligochaetes (mainly enchytraeids) often feature as prey, particularly for some of the larger nematodes.
- Dorylaimids that are often considered predatory (e.g. *Eudorylaimus*, *Labronema*, *Aporcelaimellus* and *Aporcelaimus*) are really polyphagous and may need to ingest algae or moss to reproduce (Wood, 1973b).
- Mononchids, diplogasterids and *Seinura* are commonly reported as cannibalistic, whereas this phenomenon is rarely seen in dorylaimids. However, it is not clear whether cannibalism is an artefact of culture conditions in which preferred prey are absent.

Small's (1987) review concludes with a comment that our knowledge of the natural prey of predatory nematodes is slight, and that much of the available information stems from chance observations. That conclusion is supported by Yeates *et al.* (1993a), who noted that the feeding habits of many nematodes have been inferred, and have not been confirmed by maintenance in biologically defined conditions for many generations.

The most common source of information on the feeding habits of nematodes is a paper by Yeates *et al.* (1993a) that lists the presumed feeding types of most nematode families and genera. Although this reference is a vital source of information for nematologists and is commonly cited in ecological studies, the detail in the paper is often overlooked. Thus,

the designation of mononchids as predators is usually interpreted to mean that their prey is nematodes, when the authors clearly state that they 'feed on invertebrates such as protozoa, nematodes, rotifers, and enchytraeids'. Also, the comments about the possible role of bacteria in the nutrition of mononchids are usually ignored. *Mononchus propapillatus* and *M. tunbridgensis* have been cultured on bacteria (Yeates, 1987a; Allen-Morley and Coleman, 1989), while first and second stages of *M. aquaticus* did not prey on nematodes but survived on bacteria (Bilgrami *et al.*, 1984), raising questions about the role of bacteria in the nutrition of mononchids. Yeates (1987b) argued that mononchids would have difficulty competing with rhabditids for bacteria as their large size would limit their scavenging range, while the energy requirements of later stages could probably only be met by predation. However, this does not preclude the possibility that bacteria are an important food source for small mononchids and the juvenile stages of large species.

Three other issues raised by Yeates *et al.* (1993a) must also be kept in mind when considering the feeding habits of predatory nematodes. First, in a habitat such as soil, predators with broad mouths are likely to coincidently ingest foods other than their invertebrate prey during the feeding process. Substrate ingestion, for example, has been reported for diplogasterids. Second, predatory nematodes may feed on diatoms and other algae, fungal spores, yeast cells and protozoa, but this is difficult to confirm because these organisms lack marker structures and cannot be seen in the opaque gut of a nematode. Third, feeding habits should ideally be determined at a species level, as food preferences are likely to vary within genera or families.

The difficulties involved in verifying the diet of soil organisms in an environment where many potential food sources are available raises questions about whether it is possible to better define the feeding habits of predatory nematodes. Small (1987) mentioned that faecal analysis, serological techniques and marking of potential prey were possibilities that had not yet been used, but with the technologies available today, the best option is molecular analysis of gut contents. PCR-based techniques are widely used to analyse predation by invertebrates in terrestrial ecosystems, and although it is not always easy to detect degraded, semi-digested DNA – and issues such as short-post-ingestion detection periods, cross amplification and the risk of false positives inevitably cause difficulties – these problems have been overcome to some extent for insects and mites (King *et al.*, 2008). A recent study in Hawaii indicates that such techniques are also useful for nematodes. When the gut contents of five omnivorous and predatory nematodes were examined using species-specific PCR primers, *Mononchoides*, *Mononchus*, *Neoactinolaimus*, *Mesodorylaimus* and *Aporcelaimellus* tested positive for DNA of *Rotylenchulus reniformis*, one of the most common plant-parasitic nematodes in Hawaiian soils (Cabos *et al.*, 2013). Since that study was done in soils where many potential prey were available, simple assays of this nature can be used to determine whether certain prey are utilized by predatory nematodes in natural environments. In future, next-generation sequencing technologies offer opportunities for much more powerful dietary analyses, as they provide precise taxonomic identification of the food items actually utilized when a diet is highly diverse (Pompanon *et al.*, 2012). Since DNA metabarcoding has the potential to reveal many prey species simultaneously, it may eventually provide information on the dietary range of predatory nematodes that is vital to understanding their function within an ecosystem.

In concluding this discussion, it is clear that most predatory nematodes are polyphagous to some extent. Individual species will vary in where they lie on a continuum from stenophagous specialist to ultra-generalist, but few species will be totally reliant on nematodes as a food source. Furthermore, it is unlikely that a predatory nematode will feed only on plant-parasitic nematodes, and this must be kept in mind when their potential as biocontrol agents is being considered. Predatory nematodes are generalists rather than specialists, but this does not mean that they cannot play a role in suppressing populations of economically important nematode pests.

Predatory nematodes as regulatory forces in the soil food web

Specialist and generalist predator nematodes occupy upper trophic levels within the soil food web, and one of their functions (along with many other organisms discussed elsewhere in this book) is to act as top-down regulators of herbivorous, bacterivorous and fungivorous nematodes occupying lower trophic levels. There is experimental evidence to show that mononchids can markedly reduce populations of bacterivores and fungivores in microcosms (see Chapter 3 for a discussion of experiments by Allen-Morley and Coleman, 1989, and Mikola and Setala, 1998), but given the biological complexity of soil and the polyphagous feeding habits of predatory nematodes, the critical question from a biological control perspective is whether they play a role in regulating or suppressing plant-parasitic nematodes. Yeates and Wardle (1996) argued that they were likely to play no more than a minor role for the following reasons: (i) mononchids and dorylaimids are K strategists with generation times of weeks or months, and so they do not have the reproductive capacity to respond to rapid increases in populations of plant parasites; (ii) predators may have multiple food sources; (iii) bacterial-feeding nematodes often make a greater contribution to the diet than plant-feeding nematodes; and (iv) large predators may be unable to access nematodes located in small soil pores. Nevertheless, the authors did conclude that plant growth could possibly be improved by increasing populations of predatory nematodes. However, this effect was most likely to be achieved through indirect mechanisms (i.e. predatory nematodes enhancing nutrient cycling and thereby improving the plant's capacity to cope with stresses associated with a high nematode burden), rather than through a reduction in populations of plant-feeding nematodes.

It would be unreasonable to expect polyphagous predators to regulate populations of plant-parasitic nematodes to the same extent as some of the more specialized parasites discussed elsewhere in this book, but a study by Sánchez-Moreno and Ferris (2007) indicates that they do have a suppressive function. Soil was collected from woodland and an adjacent vineyard, and analyses of the nematode assemblage indicated that taxa richness, the Structure Index and the abundance of omnivores and predators (e.g. *Discolaimus*, *Eudorylaimus*, Qudsianematidae and *Prionchulus*) were significantly higher in the woodland. Suppressiveness, as determined in a bioassay with *Meloidogyne incognita* juveniles, was greater in the woodland soil and was positively related to the Structure Index, relative abundance of omnivores and predators, and organic matter content, whereas it was negatively correlated with the abundance of bacterial and fungal-feeding nematodes. However, this does not mean that omnivorous and predatory nematodes were totally responsible for suppressing the plant parasite. The prevalence of omnivores and predators indicated that long and complex food webs were present in the natural soil, and that organisms in similar functional guilds would also have been contributing to a suppressive service that is ultimately provided by many different members of the soil biological community.

Analyses of soil nematode communities by many authors in many different environments have consistently shown that mononchids and dorylaimids are more common in natural and relatively undisturbed ecosystems than in agroecosystems, and are favoured by perennial rather than annual cropping (Freckman and Ettema, 1993; Wasilewska, 1994; Neher, 1999b, 2001). Given that these nematodes have a regulatory role, and that their presence is an indicator of a structured food web in which nematode predation and multitrophic interactions are occurring (Ferris *et al.*, 2001), it has been hypothesized that the relatively low abundance of these nematodes in agroecosystems is one of the reasons that their potential prey (bacterivorous, fungivorous and herbivorous nematodes) exhibit unregulated population increase in many agricultural fields. The following two sections briefly summarize the research that has been undertaken to understand the factors responsible for the decline in populations of omnivores and generalist predators under agriculture, and the continuing research effort to find management practices that can be used to restore the condition of soil food webs.

Impacts of agricultural management on omnivorous nematodes and generalist predators

Short- and long-term effects of soil fumigation

One way to start looking at the impact of agricultural management on the nematode community is to consider the short- and medium-term effects of perhaps the most biologically disruptive practice used in agriculture: fumigation of soil with a broad-spectrum biocide such as methyl bromide. A study by Yeates *et al.* (1991) is useful in this respect, as it followed the recovery in biological activity in undisturbed soil cores after they were fumigated with methyl bromide and then returned to their original pasture or forests sites. Microbial biomass and bacterial numbers recovered rapidly following fumigation, but after about 24 weeks, fungal hyphal lengths were 25% lower in the fumigated soils. Nematodes were eliminated by the fumigant, and although recolonization was detected by day 26, nematode abundance and the number of species present was still much lower in fumigated than untreated soils after 24 weeks (10 versus 62 nematodes/g soil; and 10 versus 31 nematode species, respectively). The cumulative number of nematode genera detected to 24 weeks (Table 6.2) is particularly interesting because it indicates the capacity of different nematode groups to respond to a major disturbance event. The predominant nematodes found in the fumigated soils were bacterial-feeding species (e.g. *Rhabditus*, *Cephalobus* and *Plectus*), as few aphelenchids and tylenchids had returned. With regard to the omnivorous and predatory component of the nematode community, some mononchids were observed in fumigated soil after 24 weeks whereas the dorylaimids were almost completely absent. However, samples taken more than 4 years after the soil was fumigated showed that both groups of nematodes had returned, as populations of predators at the high-rainfall pasture and forest sites were higher in fumigated soil than the control (Yeates and van der Meulen, 1996). However, there were differences in the recolonization capacity of genera, as *Iotonchus* failed to colonize either pasture or forest soil, whereas *Clarkus* recolonized well.

Table 6.2. Mean nematode population densities from day 0 to day 166 in untreated (U) and methyl bromide-fumigated (F) soil cores located at pasture and forest sites in moderate and high-rainfall environments; and the number of nematode genera (grouped by higher taxa) accumulated in soil collected 54,110, and 166 days after cores were fumigated and transferred to each site.

| | Moderate rainfall sites | | | | High-rainfall sites | | | |
| | Pasture | | Forest | | Pasture | | Forest | |
	U	F	U	F	U	F	U	F
Nematodes/g soil	42.3	2.4	8.2	1.2	304.5	5.0	10.8	0.3
No. of genera								
Enoplida	1	1	2	1	2	–	4	–
Mononchida	–	1	1	1	2	1	3	1
Dorylaimida	9	1	9	1	8	1	10	–
Chromadorida	4	2	7	2	6	3	7	1
Rhabditida	5	4	5	4	5	5	4	3
Diplogasterida	–	–	1	–	–	1	–	–
Aphelenchida	1	2	2	1	7	–	1	1
Tylenchida	9	2	6	–	1	–	4	–
TOTAL	29	13	33	10	31	11	33	6

(From Yeates *et al.*, 1991, with kind permission from Springer Science + Business Media B.V.)

A study in the Netherlands carried out in the field at about the same time as the previous study showed similar effects of soil fumigation. Nematodes were first detected 3 weeks after plots were fumigated with metham sodium, with bacterivorous and fungivorous nematodes in c-p 1 and c-p 2 guilds being observed. Nematode populations in fumigated plots then remained low for another 30 weeks, and even after 1 year, persisters (c-p 3–5) were still rare or below detection levels. The addition of cow manure (20 t/ha) to plots 4 weeks after the fumigation was completed did not markedly increase numbers of omnivorous or predatory nematodes (Ettema and Bongers, 1993). Another study in strawberry-growing soils with a long history of soil fumigation showed that fumigants being used as alternatives to methyl bromide reduced populations of omnivores and predators to almost undetectable levels, and that recovery did not occur during the time between annual fumigation events (Sánchez-Moreno *et al.*, 2010).

These observations are consistent with the notion that the soil biological community is radically changed by a major disturbance event such as fumigation. The community then undergoes a series of successional responses, with bacteria and their predators

(bacterivorous nematodes and protozoa) the first to recolonize, and organisms in upper trophic levels in the food web the last to recover. Interestingly, however, mononchids, which are usually considered K strategists, were observed more frequently in fumigated soil at 24 weeks than r-strategist diplogasterids (Table 6.2), raising questions about the factors that govern the colonizing ability of these generalist predators. The value of the results obtained more than 4 years after soil was fumigated (Yeates and van der Meulen, 1996) is that the data showed recolonization was both nematode-specific and site-specific. Some mononchid species recolonized well and others did not, while the structure of the re-formed nematode community was influenced as much by site (i.e. vegetation and rainfall) as by previous fumigation. This suggests that it is difficult to generalize with regard to how a nematode community will respond to a disturbance event.

Negative effects of other agricultural management practices

Although no other agricultural practice is likely to affect the nematode community to the same extent as soil fumigation, there is a large body of evidence to indicate that practices commonly used in agriculture have subtle and sometimes major effects on nematodes, particularly those at higher levels in the soil food web. Evidence for the negative impacts of some of those practices is summarized next. However, care is needed in interpreting the results of some studies, as nematodes are often only identified to family level and indices are used to characterize the nematode community. For example, a review of the literature by Neher (2001) showed that the Maturity Index was reduced by cultivation, nitrogen fertilization, liming, soil fumigants and herbicides, indicating that these practices are likely to have had negative effects on some cp-4 and c-p 5 nematodes. In such situations, the original publications must be consulted to determine whether particular nematodes were negatively affected.

SOLARIZATION. Heat at temperatures above 60°C is commonly used for experimental purposes to eliminate predatory nematodes and other antagonists of nematodes (e.g. McSorley et al., 2006). However, the much lower temperatures achieved when soil is solarized may also affect omnivores and predators. For example, Wang et al. (2006) found that Seinura was tolerant of the heat generated by solarization, whereas Mylodiscus, the dominant predator at the site where the work was conducted, did not survive in solarized plots. In another study, the number of omnivore + predator genera was reduced by solarization, with only one genus (Eudorylaimus) surviving the treatment (McSorley, 2011c).

FALLOWING. A long-term bare ground fallow (5 years of regular cultivation and occasional applications of glyphosate) markedly reduced numbers of omnivores + predators and the number of genera in that trophic group (McSorley, 2011c). However, the less severe fallowing practices commonly used in agriculture are likely to have far less impact. For example, a survey of 49 fields in Australia that were due to be replanted with sugarcane showed that the number of omnivores and predators (and the proportion of that group of nematodes within the nematode community) following a 5–12 month fallow maintained with herbicides or cultivation but supporting

some weeds, was similar to fields that had grown a legume crop or had been planted to sugarcane (Stirling et al., 2007).

TILLAGE. A review of 106 published studies indicated that tillage is more likely to be detrimental to large than small soil organisms (Wardle, 1995). Omnivore-predator nematodes, which are smaller than many other members of the soil fauna, showed a variable response to tillage, probably because a variety of tillage implements would have been used in the studies included in the analysis, and the frequency and depth of tillage would also have varied markedly. Results of some later studies (e.g. Yeates et al., 1999; Sánchez-Moreno et al., 2006) suggest there may be few adverse effects from tillage, possibly because the omnivores and predators now present in agricultural fields are taxa that are relatively tolerant of tillage. However, tillage more commonly has a negative effect on these nematodes, as shown by the following studies: (i) cultivating land that had been under shrub vegetation for many years reduced the abundance of predatory nematodes and quickly changed the nematode community structure to that of a continuously cultivated soil (Villenave et al., 2001); (ii) omnivores were more abundant in no-till than conventional till soils in North Carolina, but contrary to what might be anticipated based on their c-p value of 4, three mononchid genera (Anatonchus, Clarkus and Mylonchulus) were relatively tolerant of cultivation (Fiscus and Neher, 2002); (iii) in a field experiment in California, differences in the second and third trophic levels of the nematode community were achieved through differential management, but these differences disappeared after a single oat crop was planted and the soil was ploughed (Berkelmans et al., 2003); and (iv) after 6 years of no-till, omnivores were more abundant, and structure and maturity indices were higher than in soil that had been tilled conventionally (Okada and Harada, 2007).

NITROGEN. Experiments in which a range of different nematodes were exposed to assay solutions containing various nitrogen sources showed that nematodes in c-p groups 4 and 5 were more sensitive to nitrogen than lower c-p groups (Tenuta and Ferris, 2004). Since levels of nitrogen salts high enough to reduce the survival of c-p 4 and c-p 5 nematodes are likely to occur at microsites around fertilizer granules, and in some parts of the soil profile when nutrients are applied as a single annual dose in a band, the authors suggested that nitrogen fertilization may contribute to the low abundance of these nematodes in many agricultural soils. However, it is not clear whether the detrimental effects of nitrogen are confined to situations where it is used inappropriately or excessively. Results from a fertilizer trial in New Zealand indicated that nitrogen applied at 200 kg/ha/annum had little impact on populations of omnivorous nematodes, whereas the extremely high application rate of 400 kg/ha/annum reduced populations of *Aporcelaimus* (Sarathchandra *et al.*, 2001). Since enhancement of fertilizer use efficiency is currently a major focus of agricultural research, the critical question from an agroecological perspective is whether the omnivorous and predatory nematode community is detrimentally affected when fertilization practices are optimized to the point where soil nitrogen availability is relatively well synchronized with crop demand. Some of the critical questions to be answered with regard to the effects of nitrogen on these nematodes are:

- When nitrogen is concentrated in bands due to the application methods commonly used in row crops, what is the short- and long-term impact on the nematode community within and between the bands?
- Can the detrimental effects of nitrogen be minimized by reducing the volume of soil in direct contact with the fertilizer (i.e. by increasing the distance between fertilizer bands, with a consequent increase in the nitrogen concentration within the band)?
- Are any observed changes in nematode numbers and diversity due to the toxic effects of nitrogen, or to changes in the availability of the nematodes' food sources?
- Since toxicity to nematodes is most likely to occur as the soil dries and nitrogen concentrations increase within soil water films, what is the role of

anhydrobiosis in overcoming the toxic stress of nitrogenous compounds?

OTHER CHEMICAL STRESSORS. Numerous studies (reviewed by Nagy, 2009) have shown that omnivorous and predatory nematodes are especially vulnerable to heavy metal pollutants. For example, this group of nematodes was particularly sensitive to zinc, copper and nickel in soil (Korthals *et al.*, 1996), while experiments in water, dune sand and soil showed that sensitivity to $CuSO_4$ increased with c-p class (Bongers *et al.*, 2001). These detrimental effects on the nematode community can continue long after soil is contaminated with heavy metals, as omnivorous and predacious nematodes were either reduced or eliminated by zinc and copper treatments that had been applied 12 years previously (Georgieva *et al.*, 2002). However, when zinc and copper are used as microelements in agriculture, they can have positive rather than negative effects on c-p 4 and c-p 5 nematodes (Nagy, 1999), suggesting that it is the concentration rather than the presence of these elements that is important from a biological perspective.

CONCLUSION. It is clear that omnivorous and predatory nematodes are sensitive to a range of physical and chemical disturbances, and this is the reason they are often considered 'stress-sensitive' species. Since they have long generation times (months rather than days or weeks), and produce relatively low numbers of eggs, recolonization occurs slowly following a disturbance event. Thus, when populations of omnivorous and predatory nematodes are depleted by mismanagement, it will take time to restore the regulatory balance in the soil food web.

Management to maintain a well-structured soil food web

Based on the observations discussed, a food web with some capacity to regulate nematode populations through the activities of a biological community that includes omnivorous and predatory nematodes is only likely to be achievable in agricultural soils by enhancing the food sources that sustain these nematodes, and by minimizing the disturbance events that affect them. Thus, the first steps commonly taken to improve food web structure are to (i) increase organic inputs with cover crops, and (ii) reduce disturbance through conservation tillage practices such as no-till, minimum-till or strip till. Although it is too early to draw conclusions on the types of cover crops that should be included in the rotation sequence, the choice of crop appears to be important, as the crop grown in a long-term farming systems trial in California had significant effects on the nematode community in most years (Berkelmans et al., 2003). A capacity to produce biomass is likely to be another important attribute of the crop, but there is some evidence to suggest that legumes are beneficial. Bean, safflower, tomato and maize were included in the California trial mentioned above, and although differences in the Structure Index, which is primarily determined by omnivorous and predatory nematodes, were not consistent between crops, this index was generally higher following beans. Amending soil with sunn hemp, a leguminous amendment, increased the abundance of omnivorous and predatory nematodes in pots (Wang et al., 2003b), while the Structure Index was significantly greater following sunn hemp than bare ground in a strip-tilled living mulch system in Hawaii (Wang et al., 2011).

One advantage of including legumes in the rotation sequence is the additional benefit of building soil nitrogen pools. However, the biological problems associated with nitrogen usage will not be overcome by simply introducing a legume into the farming system. Given the detrimental effects of excessive nitrogen on c-p 4 and c-5 nematodes, overall nitrogen management must be improved, and to do this we need to know much more about the effects of application rate, placement, formulation method and nitrogen source, on these organisms. Applying nitrogen fertilizer in bands or as granular formulations creates high-nitrogen pockets that may impact negatively on omnivorous and predatory nematodes, so we need to know whether it is better from a biological perspective to spread the fertilizer evenly or to band it; to use powders

rather than granules; or to apply a controlled-release product. We also need to address the issue of low fertilizer use efficiency in modern industrialized cropping systems. Commonly, more than 50% of applied nitrogen is not used by the crop, and this excess nitrogen is not only detrimental to the off-farm environment, but may also have negative consequences for omnivorous and predatory nematodes. Improving nitrogen use efficiency by synchronizing soil nitrogen availability with plant nitrogen demand remains a major challenge in agriculture (Grandy et al., 2013), but improvements in this area could have major spin-offs in terms of enhancing the structure of soil food webs.

Producing biomass in situ by growing legumes and a diverse range of other crops in a minimum-till farming system is the most sustainable way of arresting the decline in levels of soil organic matter that has occurred under modern agriculture. However, an alternative that is economically feasible in some situations is to use locally available organic wastes as soil amendments. From the perspective of creating a more structured nematode community, there is some evidence that this approach can produce the desired effect. For example, when cow manure was applied at 20 t/ha to a coarse sandy soil in the Netherlands, predatory mononchids (genus Clarkus) increased from negligible levels to 3–5% of the nematode community after 44–60 weeks. Three dorylaimid genera also increased markedly from about 16 weeks, largely due to an increase of Microdorylaimus, which appeared to be feeding on algae (Ettema and Bongers, 1993). In another experiment with rice straw compost in Japan, predatory nematodes increased in abundance in comparison with chemical and no fertilizer treatments (Okada and Harada, 2007). However, excessive amounts of nitrogenous amendments can also have negative impacts. For example, mononchids were almost eliminated by the 10 t/ha poultry manure typically applied to ginger fields in Fiji (Turaganivalu et al., 2013).

Most organic amendments are usually incorporated into soil with tillage implements, but applying them to the soil surface as mulch may be a better option, as the soil is not physically disrupted. The main benefits from mulching are greater moisture retention and a more stable temperature and moisture environment, and this leads to increased soil microbial activity and biomass. Thus, when asparagus and maize crops were mulched with sawdust in New Zealand, populations of omnivorous nematodes and two groups of predatory nematodes (Mononchidae and *Nygolaimus*) were enhanced to the point where their predatory activities were thought to have been suppressing populations of bacterial and fungal-feeding nematodes (Yeates *et al.*, 1993b). Further observations after the mulch had been in place for 7 years indicated that populations of top predatory nematodes were consistently greater in mulched plots than other treatments in each of the last 6 years of the experiment (Fig. 6.3). Since an increase in microbial biomass was not matched by a corresponding increase in bacterial-feeding nematodes, it was presumed that the top predators were keeping these nematodes in check (Yeates *et al.*, 1999).

Although the impact of organic amendments and mulches on omnivorous and predatory nematodes has often been studied, many of the observations made are not as useful as they could be for a number of reasons: experiments are often done in soils where nematodes at higher trophic levels have been depleted by years of inappropriate management; the soil is cultivated aggressively when the experiment is established, or when the amendment or mulch is applied; and the soil nematode community is assessed before it has had time to respond fully to a treatment. The following are some of the many examples of such studies that could be mentioned: (i) in an experiment in which composted cotton gin trash, swine manure and a rye–vetch green manure were applied as amendments to a tomato field in North Carolina, the fact that the plots were tilled twice during each growing season probably explained why the already low populations of omnivorous nematodes did not recover sufficiently to affect maturity indices (Bulluck *et al.*, 2002); (ii) plots in a

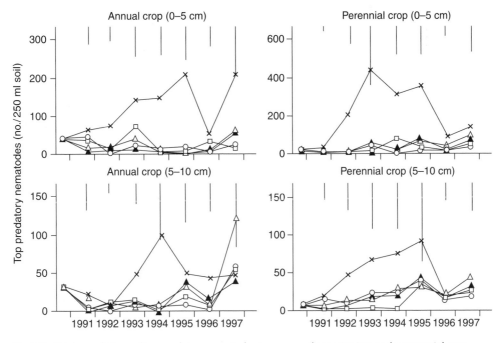

Fig. 6.3. Total populations of top predatory nematodes in an annual crop (maize) and a perennial crop (asparagus) when weeds were managed with sawdust mulch (×) or by four treatments that involved repeated cultivation (○), hand weeding (△) or herbicide application (▲ and □). Vertical bars represent least significant differences at $P = 0.005$. (From Yeates *et al.*, 1999, with permission.)

Florida experiment were rototilled before cover crop and mulch treatments were established, and this may have been one of the reasons that treatments had little impact on omnivores and predators (Wang *et al.*, 2008); and (iii) when maize litter was added to an arable soil in Germany, the micro-food web was so depleted that omnivores and predators at higher trophic levels did not respond (Scharroba *et al.*, 2012).

Since crops grown in organic farming systems are never exposed to pesticides, and plants obtain nutrients through mineralization of organic matter, it could be argued that food webs should be more structured on organic than conventional farms. However, this is not always the case, presumably because the soil is frequently disturbed by tillage. In a long-term experiment in California, predacious and omnivore nematodes were low in both organic and conventional farming systems (Ferris *et al.*, 1996), while differences in maturity indices were not observed in a survey of conventionally and organically managed farms in North Carolina (Neher *et al.*, 1999a).

The fact that soil food webs in perennial cropping systems are more structured than in annual crops (Neher and Campbell, 1994; Neher, 1999b) suggests that a combination of continuous organic inputs and an absence of tillage may be required to enhance omnivorous and predatory nematodes in annual cropping systems. However, it has proved quite difficult to consistently demonstrate this. A long-term experiment sampled by Okada and Harada (2007) in Japan clearly showed that K-strategy nematodes were enhanced by combining no-till with organic fertilizer, while similar benefits were obtained in Hawaii when mulch was combined with strip tillage, a tillage system in which only a small proportion of the seed row is cultivated and most of the field remains undisturbed (Wang *et al.*, 2011). In contrast, observations in a long-term farming systems experiment in California (Sánchez-Moreno *et al.*, 2006; Minoshima *et al.*, 2007), and at an adjacent organically managed site (DuPont *et al.*, 2009), indicated that omnivore and predator nematodes were scarce, and did not become more abundant under continuous cropping and no tillage. Thus, it is possible that in some situations, the desired response will never be obtained because nematodes at higher trophic levels in the soil food web are no longer present following decades of cultivation.

Maintaining the suppressive services provided by predatory nematodes and other generalist predators

What should be clear from the preceding discussion is that it will be quite difficult to restore food webs in agricultural soils to anything approaching the structure and diversity of natural soils. Some level of disruption is inevitable in agroecosystems, and since omnivorous and predatory nematodes are particularly vulnerable to disturbance, the regulatory services they provide are easily squandered. The key to maintaining the structured food webs that provide regulatory services is to develop a farming system where carbon inputs are continuous and the soil biological community is not constantly disrupted by tillage and harmful chemical inputs. Therefore, farming systems must embrace practices such as minimum tillage; optimal nitrogen management; legumes and biomass-producing cover crops in the rotation sequence; and a permanent cover of crop residue or organic mulch on the soil surface. Practices likely to be detrimental to soil organisms (e.g. solarization and bare fallowing; fumigants, nematicides and most soil-applied pesticides; some above-ground chemical inputs; and excessive amounts of nitrogen fertilizer), must be avoided. Since the benefits from modifying the farming system in this way are likely to take many years to eventuate, regular monitoring is required to confirm that a well-structured biological community eventually develops, and that full remediation occurs. In situations where omnivorous and predatory nematodes have been eliminated by decades of inappropriate management, the option of reintroducing them from undisturbed locations nearby may have to be explored (Tomich *et al.*, 2011). However, when this was attempted by transferring undisturbed cores from a natural forest to an agricultural soil, re-establishment was not immediately successful, as populations of the introduced nematodes had not increased after 1 year (DuPont *et al.*, 2009).

Ferris (2010) provides a good summary of the issues that must be considered when attempts are being made to maintain or enhance the structure and functions of the soil food web. What is clear from that review, and from any text in soil ecology, is that properly functioning soil food webs contain a multitude of organisms with apparently similar functions, and that these organisms collectively provide a range of services, including regulation of nematode populations. Enhancing soil biological diversity provides the system with functional resilience, as it ensures that a particular service will continue when conditions are unfavourable to one of the guilds that normally provide that service. Thus, a study by Griffiths *et al.* (2010) in the UK showed that when arable fields were subjected to various amendment and rotation treatments, there were times when faunal groups such as enchytraeids and predatory nematodes were reduced. However, when this occurred, there were compensatory increases in functionally equivalent groups such as earthworms and mesostigmatid mites. In the same way, populations of omnivorous nematodes were low and predatory nematodes were never observed in the soil used for a vegetable farming systems experiment in Australia, but suppressiveness to root-knot nematode was still enhanced by some treatments, presumably because a subset of the natural enemies normally found in a healthy soil were still effective (Stirling, 2013). Collectively, these observations suggest that when management practices are being considered to sustain regulatory services in agricultural soils, the focus should be on developing a diverse and active soil biological community rather than on increasing populations of specific predatory nematodes.

Predatory nematodes and inundative biocontrol

In the years since Cobb (1920) first suggested that plant-parasitic nematodes could be controlled by introducing predacious nematodes into infested fields, there have been many optimistic pronouncements about the potential of these nematodes as biocontrol agents. In recent years, attention has shifted from mononchids to the diplogasterids, and evidence that nematode populations can by reduced by species of *Mononchoides* has been presented (Khan and Kim, 2005; Bilgrami *et al.*, 2008). This work is discussed in Chapter 12, but as inundative biocontrol is likely to be limited by economics to small areas and valuable crops, future research should focus on understanding the feeding habits of predatory nematodes; the factors that influence their occurrence in the soils used for agriculture; and the practices required to maintain them at reasonable population densities in agricultural soils.

Microarthropods as Predators of Nematodes

The main members of the soil mesofauna: mites, Collembola and Symphyla

Mites, collembolans and symphylans (Fig. 6.4), together with a variety of small insects collectively known as microarthropods, are ubiquitous in soil. They are particularly abundant in natural habitats such as grasslands and forests, where soil organic matter has accumulated over many years, root biomass is high, roots are permanently present and the soil surface is always covered by a layer of decaying litter. In such situations, particular groups of microarthropods can reach population densities greater than $100,000/m^2$.

Mites are the most abundant arthropods in soil, and are found in four major taxonomic groups: the Oribatida (formerly Cryptostigmata), Prostigmata, Mesostigmata and Astigmata. More than 50,000 species have been described.

- Oribatids (suborder Oribatida) are by far the most numerous soil mites, but in contrast to other microarthropods, they reproduce relatively slowly and are, therefore, considered K strategists. Although mainly fungivores and detritivores, some are predacious.
- The Prostigmata are the most diverse group of soil mites, displaying a huge range of morphological and behavioural variation. Many above-ground species are serious pests of plants and vertebrates, but terrestrial species are also important,

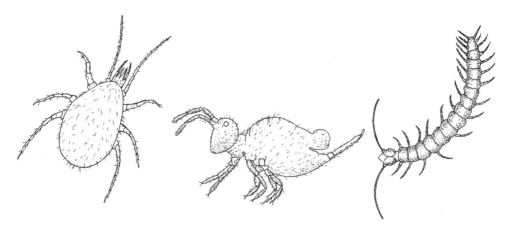

Fig. 6.4. Examples of common microarthropods in soil: a mite, collembolan and symphylan.

and sometimes outnumber all other soil mites. Soil-inhabiting prostigmatic mites generally feed on fungi or algae, or ingest particulate matter, but some are predators, feeding on other arthropods or their eggs, or on nematodes.

- Mesostigmatid mites are not as numerous as oribatids or prostigmatic mites, but are universally present in soil. A few species are fungal feeders, but most are predators. The larger species tend to feed on small arthropods or their eggs, whereas the smaller species are mainly nematophagous.
- The Astigmata are the least common of the soil mites. In agroecosystems, they are most abundant in situations where crop residues or manures have been incorporated into soil under moist conditions.

Collembolans (springtails) are primitive insects that sometimes rival mites in their abundance in soil. Many species are opportunistic r-strategists capable of rapid population growth when decomposing organic matter and its associated microbes are available as a food source. They primarily feed on fungi, but many species are omnivorous and some are known to consume nematodes.

Symphylans, relatively large translucent arthropods that resemble centipedes, are widespread in soil but nowhere near as common as collembolans. Since they are photophobic and can move rapidly, they hide in voids when soil is disturbed, and are, therefore, rarely seen. Although generally considered detritivores, they can also be crop pests, as they will consume plant roots. Some species are predators of arthropods and nematodes.

Evidence of nematophagy in various groups of microarthropods

Results from field observations, feeding studies and analyses of gut contents

In his comprehensive coverage of interactions that occur within the soil food web, Wardle (2002) pointed out that microarthropods commonly prey on other members of the soil fauna, and provided several examples where populations of springtails, enchytraeids and nematodes were regulated by top predators. However, determining who is eating what in soil is fraught with the same difficulties discussed earlier for predatory nematodes. Microarthropods are grouped into five feeding categories (bacterivores, fungivores, plant feeders, omnivores and predators), but omnivory is common, even among animals that may exist primarily on one food source. Thus, many arthropod predators are generalists, and may attack nematodes, lumbricid and enchytraeid oligochaetes, tardigrades, insects, myriopods and mites (Moore *et al.*, 1988). Determination of trophic behaviour requires a combination of laboratory observations, gut content analyses of field-collected specimens and experimentation, but as with all predators in soil, none of these methods is entirely satisfactory.

With regard to observations on arthropod–nematode interactions, studies by David Walter and colleagues in the late 1980s, but particularly the observations made by Walter and Ikonen (1989), are worthwhile summarizing here, as they provide a good summary of what was known about nematophagous microarthropods at that time. Although this research focused specifically on grassland soils in Colorado, the observations made over several years provide a good indication of the types of microarthropods that prey on nematodes, and their feeding habits and behaviour in soil.

- Feeding studies with 14 species of mesostigmatic mites showed that they ate 4–8 nematodes (*Acrobeloides* sp.) per day, which equates to between 22% and 109% of their body weight. The amount of nematode biomass consumed was strongly related to the body weight of the predator. When mites were given a choice of nematode prey, there was no significant indication of preference.

- The most important nematophagous taxa in Colorado grasslands were the Acari (mites) and Symphyla. The latter group are generally considered detritivores or root feeders, but observations of symphylans collected from grasslands showed that they primarily had nematodes and arthropods in their gut contents, and little fungal, detrital or plant material.

- All four main acarine sub-orders were found in grassland soils, and each order contained nematophagous and fungivorous species. This indicates that the feeding habits of mites are a species-level phenomenon, and cannot be inferred from morphology or relationships to higher-level taxa.

- Mesostigmatic mites appeared to be the most important group of nematophagous arthropods, as only six of 63 species tested did not readily feed on nematode prey. In the presence of only nematode prey, most of the nematophagous species produced eggs and developed into adults.

- Although it had previously been suggested that nematophagous mesostigmatids had developed specialized mouthparts (designated types 1, 2 or 3), and that generalist predators were equipped with type 4 mouthparts, observations on 30 mite species indicated that there was little relationship between the structure of their chelicerae and feeding habits. Most mesostigmatic mites consumed nematodes and arthropod prey, and some were also able to feed on fungi. Thus, three categories of nematophagous arthropods were recognized: omnivores feeding on microbes and nematodes; general predators of soil invertebrates; and nematode specialists. However, it was recognized that the specialized group could attack a variety of prey.

- Nematophagy could not be confirmed for tydeid mites, despite the fact that studies in other environments (Santos *et al.*, 1981) had previously shown that this group of mites were able to prey on free-living nematodes.

- Species in the families Rhodocaridae, Digamasellidae and Ascidae were considered to represent a 'small-pore nematophagous mite guild' characterized by: (i) an ability to maintain continuous cultures on nematode prey; (ii) the presence of chelicerae adapted to feeding on nematodes and arthropods; and (iii) a small body size and convergent body plan that allowed these animals to access prey residing in small pore spaces. Since species in this guild were able to colonize soils to a depth of 30 cm, they could access a much larger proportion of the soil profile than larger mites.

Other studies have also shown that nematodes are likely to be a component of the diet of many different mites. For example, mites from four different orders (Mesostigmata, Endeostigmata, Oribatida and Astigmata) were tested against entomopathogenic nematodes in the laboratory and the majority were able to consume *Steinernema feltiae* and *Heterorhabditis heliothidis*. Although most species also fed on arthropods, some required nematode prey for reproduction (Epsky *et al.*, 1988). Observations in a strawberry field in Turkey indicated that when white grubs were killed by entomopathogenic nematodes, a species of *Sancassania* (Acari: Acaridae) moulted to the adult stage and began feeding on the host tissues and/or microbes associated with the cadavers, and also consumed exiting infective juveniles. In the laboratory, this mite consumed about 40 nematodes per day;

was more likely to consume *S. feltiae* than *H. bacteriophora*; and was able to kill more nematodes in sand than in a loamy soil (Karagoz *et al.*, 2007). When the stylets of plant-parasitic nematodes were counted in the faecal pellets of an oribatid mite (*Pergalumna* sp.), the data obtained in a laboratory study showed that on average, about 18 *Meloidogyne javanica* and 42 *Pratylenchus coffeae* were consumed per day (Oliveira *et al.*, 2007). Collectively, these studies indicate that many mites are generalist feeders, but they are capable of killing significant numbers of nematodes in some situations.

Although the contribution of nematodes to the diet of Collembola has not been studied to the same extent as mites, there is certainly evidence that some species can consume large numbers of nematodes. For example, Gilmore (1970) showed that ten of the 12 Collembola species tested fed on nematodes in the laboratory, and that two species (*Entomobryoides similis* and *Sinella caeca*) consumed more than 25 nematodes/day in a mixture of soil and vermiculite. In another experiment in which two species of Collembola (*Tullbergia krausberi* and *Onychiurus armatus*) were introduced into a sandy soil, the number of *Tylenchorhynchus dubius* on ryegrass seedlings was reduced by about 50% after 77 days (Sharma, 1971). Although both these studies were carried out in highly artificial environments, the high numbers of prey consumed suggests that nematodes are likely to be part of the diet of many collembolans. However, it is important to recognize that consumption of nematodes in the laboratory does not necessarily mean that predation occurs in soil. *Proisotoma* nr. *sepulchris* reached high population densities in greenhouse cultures of root-knot nematode (more than 10,000 individuals/L soil) and was associated with decreased numbers of nematodes in pots, but observations of gut contents revealed only decaying roots and associated microflora (Walter *et al.*, 1993).

Detection of predation using stable isotope ratios and molecular techniques

Since animal tissues are more enriched in ^{15}N than their food source, one way of determining whether an animal occupies a distinct trophic niche within soil food webs is to study stable isotope ratios (i.e. ^{15}N/^{14}N). When this method was used to evaluate trophic niche differentiation in oribatid mites from four forest sites in Germany, the results suggested that these mites spanned three or four trophic levels, and that different species occupied different trophic niches (Schneider *et al.*, 2004). Several taxa had much higher ^{15}N signatures than their litter habitat, suggesting that they lived predominantly on an animal diet that possibly included nematodes. The value of this method is that it reflects nutrition over a long period of time and, therefore, allows an animal's predominant diet and trophic niche to be determined. However, stable isotope analysis is of limited value for identifying precise food resources, and so a more complete picture of an animal's trophic position can be obtained by combining it with molecular gut content analysis (Maraun *et al.*, 2011).

Numerous molecular approaches for detecting prey remains in the guts, faeces and regurgitates of predators are now available (see reviews by King *et al.*, 2008 and Pompanon *et al.*, 2012), and a study by Read *et al.* (2006) demonstrated for the first time that they can be used to measure predation by soil-dwelling arthropods on nematodes. A PCR-based approach was developed for detection of three species of entomopathogenic nematodes, and a collembolan and mesostigmatid mite were employed to calibrate post-ingestion prey detection times. Detection times for nematode DNA within the guts of the predator were found to be longer for Collembola than for mites, with half-lives (50% of the samples testing positive) of about 9 and 5 h, respectively. Predatory activity in the field was confirmed by retrieving microarthropods from a barley field where the nematode *Phasmarhabditis hermaphrodita* had been applied 12 h earlier. Prey DNA was detected in three of the four most abundant microarthropod taxa recovered from the field, indicating that some species had fed on the introduced nematodes (Fig. 6.5). Since a previous DNA-based study had shown that one of the collembolans was preyed upon by linyphiid spiders, there are clearly opportunities to use molecular technologies to better understand the trophic linkages that occur within highly complex soil food webs.

The primers designed by Read *et al.* (2006) have since been used to study predation

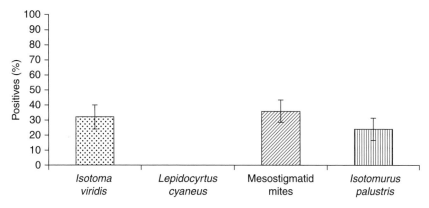

Fig. 6.5. Percentage of field-caught microarthropods (the Collembola *Isotoma viridis*, *Lepidocyrtus cyaneus* and *Isotomurus palustris*, and mesostigmatid mites) testing positive for predation on the nematode *Phasmarhabditis hermaphrodita* using species-specific primers targeting mitochondrial DNA. (From Read *et al.*, 2006, with permission.)

and scavenging by three mite taxa: Oribatida, Gamasina and Uropodina (Heidemann *et al.*, 2011). Several species of oribatid mites that are normally considered detritivores were found to consume nematodes when they were offered in a no-choice laboratory experiment. Living and freeze-killed *Phasmarhabditis hermaphrodita* and *Steinernema feltiae* were then added to forest soil and 48 h later, mites were retrieved and checked for the presence of the inoculated nematodes. Not surprisingly, the predatory gamasid and uropodid mites consumed nematodes, but interestingly, several of the oribatid mite species also ate living nematodes, even though they could have selected different food in the field (Fig. 6.6). DNA from dead nematodes was also found in their guts, indicating that these mites were also capable of scavenging for food.

One of the main conclusions that can be drawn from these studies is that current trophic groupings for Collembola and oribatid mites are inadequate. Many putative detritivores actually function as predators, fungivores or scavengers, and this has important implications for the structure and functioning of detritivore food webs. Although many of these omnivores will only feed opportunistically on nematodes, these findings suggest that a huge range of microarthropods is potentially predatory, particularly in situations where nematodes are readily available. From an agricultural perspective, this suggests that

if generalist predators are nurtured and soil structural characteristics allow them access to root systems, they could play an increasingly important role in suppressing plant-parasitic nematodes as populations rise. It also suggests that when entomopathogenic nematodes are applied in high numbers for insect pest control, microarthropods that normally survive on fungi, detritus and other food sources may limit their efficacy by switching to preying on recently added nematodes.

Studies of 'fungivorous' and 'predatory' arthropods in microcosms

A microcosm experiment by Hyvönen and Persson (1996) demonstrates some of the complexities that occur when microarthropods and nematodes interact in an environment where numerous species with a wide range of feeding habits are present. Microcosms containing a complete microbial, nematode and tardigrade community and a defaunated humus substrate were established and then three different arthropod treatments were imposed: a no-arthropod control, fungivores only and fungivores + predators. The arthropods used as inoculum were collected from the field, but in the fungivore treatment, prostigmatid, astigmatid and mesostigmatid mites were removed to avoid adding potential predators, and so the inoculum consisted of 14 collembolan species and 17 species of oribatid

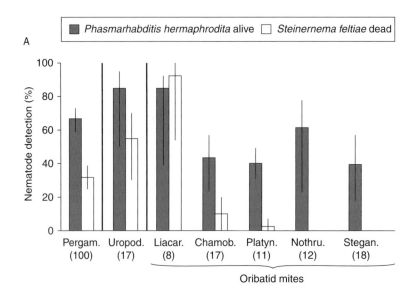

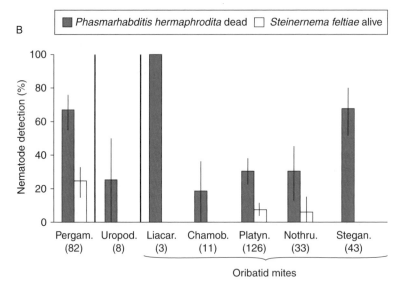

Fig. 6.6. Detection of (A) living *Phasmarhabditis hermaphrodita* and dead *Steinernema feltiae* and (B) dead *Phasmarhabditis hermaphrodita* and living *Steinernema feltiae* in the gut of seven mite species (Pergam. = *Pergamasus septentrionalis*; Uropod. = *Uropoda cassidea*; Liacar. = *Liacarus subterraneus*; Platyn. = *Platynothrus peltifer*; Nothru. = *Nothrus silvestris*; Stegan. = *Steganacarus magnus*; Chamob. = *Chamobates voigtsi*) in the field using part of the COI gene as a molecular marker. Data represent percentages of mite individuals in which nematodes were detected in the gut (the number of tested individuals is given in parentheses). The upper and lower confidence limits are indicated as error bars. (From Heidemann *et al.*, 2011, with permission.)

mite that were considered 'fungivorous'. Microcosms with fungivores + predators were supplied with the same fungivores, but also received a mixture of prostigmatid and mesostigmatid mites.

When nematode populations were assessed after 103 and 201 days, the results showed that the 'fungivorous' arthropod community reduced nematode abundance, and that this suppressive effect was further strengthened by the addition of predatory arthropods. The data did not support the hypothesis that the reduction in nematode populations was due to competition between fungivorous arthropods and fungivorous nematodes for food, as both bacterial- and fungal-feeding nematodes were affected (e.g. *Wilsonema, Metateratocephalus, Acrobeloides* and *Aphelenchoides* spp.), and estimates of fungal hyphae did not indicate clear differences between the arthropod treatments. Predation appeared to be the main reason that nematode populations declined, and predacious activity was probably due in part to microarthropods that would be better considered omnivores rather than fungivores.

Although the experiment was not set up to assess the impact of tardigrades, it became obvious during the study that their role in the system also had to be considered. Tardigrades seemed to be efficient predators of nematodes, but their numbers, in turn, were strongly reduced by predatory arthropods. Thus, trophic relationships between nematodes, tardigrades, arthropods and other soil organisms were more complicated than expected (Fig. 6.7), demonstrating the difficulties involved in working with multiple interactions in soil food webs.

Mesostigmata as predators of nematodes in agroecosystems

Although a wide range of mites are able to utilize nematodes as a food source, the Uropodina and Gamasina, two major taxonomic groups within the Order Mesostigmata (= Gamasida), are the most specialized predators of nematodes. However, it should be apparent from the previous discussion that most of the studies of nematode predation by mesostigmatids have been done in relatively undisturbed natural habitats, where these mites are always an important component of the microarthropod community.

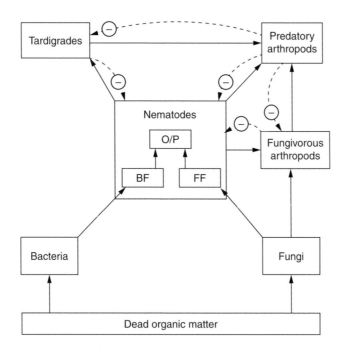

Fig. 6.7. Conceptual diagram of trophic relationships in a microcosm study with bacterial-feeding nematodes (BF), fungal-feeding nematodes (FF) and omnivore/predator nematodes (O/P). Unbroken arrows indicate material fluxes and broken arrows denote negative impacts. (From Hyvönen and Persson, 1996, with kind permission from Springer Science + Business Media B.V.)

Table 6.3. Approximate benchmark figures for mite abundance and species richness of mesostigmatid mites in grassland and arable field soil at a depth of 0–15 cm.

	Grassland	Arable field
Abundance (individuals/m²)		
Acari	100,000	50,000
Gamasina	10,000	5,000
Uropodina	5,000	2,000
Species number		
Gamasina	15	7
Uropodina	10	2

(From Koehler, 1999, with permission.)

Forests will support many more mesostigmatid species than any other habitat, and as can be seen from Table 6.3, the Gamasina and Uropodina occur in much greater numbers and are more speciose in grasslands than in agricultural fields. Thus, the nematophagous activity that has been observed in natural environments does not necessarily occur in the soils used for agriculture.

Of the two groups of mesostigmatids mentioned above, the Uropodina are the least important from an agricultural perspective, as they are only found in large numbers in organic materials such as compost, manure, older sewage sludge, decaying wood and deciduous forest litter. Thus, they are not abundant in most agricultural soils. Some genera of Uropodina are mycophagous, but most feed on nematodes and insect larvae. Their distribution is highly specific, as they are only found in rich organic habitats where populations of bacterivorous and fungivorous nematodes will always be high. Since such habitats are usually ephemeral, the mites must disperse to fresh sources of organic matter when decomposition is complete.

The Gamasina are much more important predators in agroecosystems, and a review by Koehler (1999) provides a good summary of their biology and ecology. The most important observations made in that review are:

- The majority of Gamasina are mobile predators. They utilize chemical or tactile stimuli to locate their prey, pierce or cut the body with their chelicerae, produce a digesting liquid to pre-orally digest their food, and then suck up the prey.

- Nematodes play a central role as prey for Gamasina in agroecosystems. Some species feed exclusively on nematodes, but most can also utilize other food sources (e.g. insect larvae and eggs, Enchytraeidae, Acari and Collembola).

- Size rather than motility may be critical for success when these mites prey on nematodes. The predator must be able to access soil pores and microhabitats where prey density is high.

- The low abundance and diversity of Gamasina in agricultural soils is the result of continual mechanical disturbance; the lack of a litter layer; and compaction of the soil by farm machinery.

- The Gamasina species that survive in the disturbed and compacted soils used for agriculture are relatively small in body size and have a high reproductive capacity.

- Pesticides toxic to arthropods or their prey (e.g. aldicarb) have much the same effect as soil compaction, reducing the Gamasina to an assemblage dominated by one or two species.

- The Gamasina are a good indicator of soil health. A healthy agricultural soil should contain a diverse range of Gamasina species that are adapted to different habitats. Such a community would contain surface dwellers; species that live in organic accumulations; species that survive readily in upper soil layers; and euedaphic species capable of living in deeper soil to depths of 20 cm.

Management to enhance microarthropod abundance and diversity in agricultural soils

Since soil microarthropods are unable to make their own burrows, they are entirely dependent on air-filled pores that are at least as large as their body width. Their abundance is, therefore, closely linked to soil physical conditions, particularly the volume of habitable pore space and the connectivity of soil pores. However, when soil is used for agriculture, soil structure is markedly altered due to tillage and compaction from farm machinery. Bulk density increases and total porosity declines, and these structural changes not only affect aeration, water movement, solute transport, nutrient availability and root growth, but also reduce the living space for microarthropods. Thus, changes in the physical structure of soils are almost certainly one of the main reasons that microarthropods are not abundant in many of the soils used for agriculture.

Two studies provide a good example of the impact of soil porosity on microarthropods. In the first, populations of various groups of microarthropods were monitored at sites in the Netherlands that varied in soil type and land use, and relationships between biomass, organic matter and pore size distribution were examined (Vreeken-Buijs *et al.*, 1998). In general, microarthropod biomass was higher in sandy soils than in loamy soils, and higher in grasslands than in wheat fields. Analyses of data from the grassland sites showed that biomass of the four principal functional groups of microarthropods was negatively correlated with the smallest pore size class (<1.2 µm), while biomass of three of the groups (predatory mites, non-cryptostigmatic mites and omnivorous Collembola) was positively correlated with the largest class (>90 µm). In a second study carried out in the laboratory, soil was compressed to six levels of bulk density (1.02–1.56 g/cm^3), and populations of three species of Collembola were found to be reduced by compaction due to the loss of coarse pores (>120 µm). The data obtained from bulk densities typically found in wheel tracks indicated that abundance was reduced by about 80% compared to loose soil (Larsen *et al.*, 2004).

Clearly, the only way to provide conditions suitable for microarthropods in an agroecosystem is to adopt a farming system that preserves and enhances the soil's physical structure. Based on our knowledge of the role of plants and organic matter in maintaining soil structure (see Chapter 2) and the management practices required to improve soil health (see Chapter 4), the key requirements for achieving structural characteristics suitable for microarthropods in agricultural soils are obvious: permanent plant residue cover;

continuous inputs of organic matter; no tillage; and controlled traffic. That knowledge also indicates that this package of requirements must be adopted in its entirety if soil structure is to be optimized, as all components are important for the following reasons.

- A cover of mulch or plant residues on the soil surface provides a habitat that is similar to the litter layer or 'forest floor' environment of natural ecosystems, which is dominated by microarthropods. It also increases the population density and activity of microarthropods by dampening moisture and temperature fluctuations (Badejo *et al.*, 1995).
- Organic inputs from crop residues and roots are the energy supply for the bacteria and fungi that enmesh soil particles into aggregates; and for the earthworms, enchytraeids and other ecosystem engineers that create the voids and macropores required by microarthropods. Plant roots also make a contribution by improving subsoil macroporosity, a process that has been termed 'biological drilling' (Cresswell and Kirkegaard, 1995).
- Eliminating tillage is essential because it accelerates soil carbon losses, eliminates earthworms and large invertebrates, destroys soil aggregates, and disrupts the fungal hyphae responsible for aggregation. Tillage is particularly harmful to microarthropods because its main negative impacts are on the upper 20 cm of the soil profile, their primary habitat.
- Farm machinery must be confined to traffic zones to minimize compaction problems. Many soils become uninhabitable to microarthropods following a single pass with a tractor or other pieces of farm equipment.

Although conservation tillage is now widely practised in some regions of the world, concern is sometimes expressed that soil microarthropods do not always increase in abundance under no-till, particularly in fine-textured soils. For example, a Japanese study in a humus-rich volcanic soil showed that no-till plots had more earthworms and enchytraeids than conventionally tilled plots, but population densities of mites and collembolans were similar in both treatments

(Miura *et al.*, 2008). A review of 162 German studies also showed that tillage adversely affected earthworm populations, whereas the response of microarthropods was dependent on soil texture (van Capelle *et al.*, 2012). Collembolans were not affected by tillage intensity in coarse-textured and silty soils, but were adversely affected in fine-structured loams. Although more work is required at a local level to understand the reasons for these effects, a series of Australian papers (Bridge and Bell, 1994; Bell *et al.*, 1995, 1997, 1998) are well worth consulting, as they address many of the issues involved in restoring the physical fertility of fine-textured soils that have been degraded by decades of continuous cropping.

The studies cited focused on the physical and chemical fertility of ferrosols (kraznozems) in a high-rainfall, subtropical environment. Ferrosols are some of the most fertile soils in Australia, but after years of continuous cropping, they showed reduced water infiltration rates, high bulk densities and increased penetration resistance. Organic carbon levels were often as low as 10 g C/kg soil, or about 25% of the level in untilled soil, and this was associated with low aggregate stability, crusting of the soil surface and low cation exchange capacity. Ley pastures were a useful tool in restoring the physical fertility of these soils, but the benefits were affected by ley management (e.g. the pasture species, ley duration and grazing management), as well as the tillage method used to return the ley area to cropping. The message to soil ecologists from these studies is that when soils have been degraded by many years of inappropriate management, it takes more than a change to no-till till and a few biomass-producing crops to correct the problem. In some soils, many years of careful nurturing, and perhaps even a pasture phase, may be required before there is any chance of re-creating a habitat suitable for soil microarthropods. In fact, it could be argued that the reappearance of large numbers of mites and Collembola is a sign that soil physical fertility has been restored.

The other important point that must be recognized when steps are taken to increase the number and diversity of microarthropods (or any other soil organism) is that the environment will always limit what is achievable. In a study at three sites in Western Australia, for example, Osler and Murphy (2005) recovered relatively few species of oribatid mites from natural vegetation, and found that population densities were extremely low (440–1720 individuals/ m^2). Only eight of the 22 species found in natural vegetation were recovered from adjacent fields cropped with grains, legumes and pastures, and oribatid population densities were generally reduced by more than 80%. These low populations are undoubtedly a response to the relatively harsh environment at the study sites (hot dry summers; annual rainfall of 355–420 mm; soils with >80% sand and low levels of soil organic matter), as oribatid populations are much higher and more speciose in more benign environments. Thus, in any particular environment, climate and soil type will limit what improved agricultural management might achieve. Nevertheless, data from natural vegetation provide a useful indication of what may be attainable.

One of the reasons for attempting to increase microarthropod abundance and diversity is to enhance the suppressive services provided by the soil food web. Fortunately, there is some evidence that this is a worthwhile approach, at least with regard to enhancing suppressiveness to plant-parasitic nematodes. For example, in a study of the nematode and arthropod communities in conventionally and organically managed vegetable plots in California, predatory, omnivorous and fungivorous mites were not only found to be more abundant in the organic treatment, but also exhibited very similar temporal patterns to predatory nematodes (Sánchez-Moreno *et al.*, 2009). Since previous work had shown that suppressiveness to root-knot nematode was related to the prevalence of omnivore and predator nematodes (Sánchez-Moreno and Ferris, 2007), it seems that the presence of organisms at a high trophic level in the soil food web is the critical issue with regard to achieving a suppressive service. It may not matter whether those organisms are mites, nematodes or some other generalist predator.

A study by McSorley and Wang (2009) provided further evidence that plant-parasitic nematodes are suppressed in soils with long and complex food webs. Final populations of root-knot nematode (*Meloidogyne incognita*) were 78–95% lower in natural soils from Florida than in adjacent agricultural soils, and suppressiveness was associated with the abundance of Collembola and mites. Given that nematodes are an important food source for some groups of microarthropods and are consumed opportunistically by many others, such a result is not unexpected. The fact that tardigrades were also more abundant in the suppressive soils is a reminder that the natural soil biological community is complex, and contributions from other members of that community cannot be ignored.

Miscellaneous Predators of Nematodes

Since nematodes are the most numerous multicellular organisms found in soil, it is hardly surprising that many groups of soil organisms, other than those discussed earlier, have been reported to prey on nematodes. Small (1988) reviewed the isolated reports in the literature on predation of nematodes by protozoans, turbellarians and tardigrades, and the contents of that review highlight how little we have learnt about some of these predators in the last 25 years. For example, a study of a turbellarian flatworm (*Adenoplea* sp.) by Sayre and Powers (1966) remains the only serious study of this group of generalist predators. Nevertheless, there has been a limited amount of work on tardigrades and protozoans, and it has extended our knowledge of these organisms and provided a clearer picture of their role in regulating nematode populations.

Based on the observations of Hyvönen and Persson (1996), in an experiment discussed previously, tardigrades can be both predators of nematodes and prey for microarthropods. Nematode–tardigrade interactions were later studied in a system where per capita feeding rates could be measured, and the results showed that consumption of nematodes by *Macrobiotus richtersi* depended on temperature, prey density, the individual body

mass of prey and predator, and the behavioural response of prey to attack (Hohberg and Traunspurger (2005)). More than 50 nematodes were consumed by *M. richtersi* per day on agar, but predation rates were reduced in fine sand because small pores provided a refuge for the prey. Follow-on work revealed that the functional response of the predator was best predicted by a model that accounts for increasing satiation of the predator as it feeds over time (Jeschke and Hohberg, 2008).

During a study by Sánchez-Moreno and Ferris (2007) on the suppressive services provided by omnivorous and predatory nematodes in California soils, it was observed that naturally occurring tardigrades also had an impact on the nematode community. This prompted further investigations which showed that two tardigrade species (*M. richtersi* and *M. harmsworthi*) had the capacity to significantly reduce nematode biomass (Sánchez-Moreno et al., 2008). Populations of tardigrades varied markedly between sampling times, probably because they have the capacity to feed on bacteria, fungi, protozoa, algae, yeasts, nematodes, rotifers and plant roots, and are, therefore, able to quickly respond to food sources when they become available. However, perhaps the most important result from this study was that when the regulatory function provided by predatory nematodes disappeared due to predation pressure and/or unfavourable environmental conditions, the suppressive service was maintained due to the high abundance of tardigrades. Thus, this observation reinforces the point made earlier that if diversity and functional resilience is built into the soil biological community, nematode populations will always be subjected to regulatory forces.

Small (1988) failed to mention copepods in his review, but Muschiol et al. (2008) cited four reports of copepod predation on free-living and plant-parasitic nematodes. These authors then presented results of laboratory experiments showing that *Diacyclops bicuspidatus*, a common copepod in northern regions of the world, was a voracious feeder on nematodes. Starved copepods were able to consume about 45 nematodes in 2 h, or the equivalent of 44% of their body mass. Although normally found in aquatic habitats, copepods also occur in soil, raising questions about their role as predators of nematodes in terrestrial habitats.

Most reports of protozoan predation on nematodes have involved *Theratromyxa weberi*, a naked amoeba with a feeding stage that can engulf nematodes in a few hours (Small, 1988). However, Yeates and Foissner (1995) observed that two testate amoebae were also able to prey on large filiform-tailed nematodes in forest litter. However, the broader ecological significance of this finding was uncertain, as the testacea and their nematode prey may have been restricted to moist open-textured habitats, and the food sources of the amoebae and their capacity to capture relatively motile nematodes were not assessed. Laboratory observations have also shown that when the common soil flagellate (*Cercomonas* sp.) and the nematode *Caenorhabditis elegans* compete for a bacterial food source, the nematode sometimes consumes the flagellate and becomes the dominant bacterial predator, while in other situations the flagellate attacks and kills its much larger competitor (Bjørnlund and Rønn, 2008). Thus, it is possible that interactions of this nature determine the relative importance of flagellates and nematodes in the soil environment, as both groups are always competing for the same food source.

Small (1988) considered whether oligochaetes were potential predators of nematodes, and cited several papers which showed that earthworms consumed nematodes or their eggs as they ingested their food, and that they sometimes markedly reduced populations of plant-parasitic nematodes. Most of those studies were also discussed in a later review by Brown (1995). However, it would be a gross misunderstanding to think of earthworms as predators of nematodes. Earthworms may inadvertently consume nematodes that are associated with plant litter, decomposing roots or humus (Hyvönen *et al.*, 1994), but their role as litter transformers and ecosystem engineers means that they also modify soil physical and chemical properties, soil water regimes and soil nutrient cycling processes to such an extent that they have many indirect effects on the nematode community (Brown, 1995). Thus, the effect of earthworms

on soil nematodes is likely to be quite variable, and will depend on the properties of the habitat, the feeding biology of the particular earthworm species, and the composition of the nematode community (Ilieva-Makulec and Makulec, 2002). Enchytraeids are also substrate ingesters, and although they may consume some nematodes during the feeding process, their main food source is fungal mycelium, decaying organic matter and the microbes associated with it (Bardgett, 2005).

Generalist Predators as Suppressive Agents

Based on the material presented in this chapter, it should be apparent that the nematodes and microarthropods capable of preying on nematodes are also able to utilize many other food sources. Although individual taxa vary in the range of organisms they consume and relative importance of nematodes in their diet, the feeding habits of predatory invertebrates are best described by terms such as omnivorous, polyphagous or generalist. With regard to food preferences within the nematode community, the prey chosen by invertebrates may be influenced by factors such as the size and motility of the victim, but most predators are relatively non-selective and are best considered as generalists. Certainly there is little evidence to suggest that they prefer plant-parasitic nematodes over free-living species. This generalist categorization also applies to the nematophagous fungi discussed in Chapter 5, as most have some capacity to live as saprophytes, and very few parasitize or prey specifically on particular nematodes.

Given the huge number of organisms that reside in soil, and the fact that they compete with and prey on each other, multi-species interactions are common. Diverse communities containing many different species share common food resources and interact through complex networks of competition, parasitism and predation. Plant-parasitic nematodes are one of the many parties involved, but the effects of these interactions on this group of herbivorous nematodes are difficult to predict because numerous trophic relationships may occur simultaneously (e.g. intra-specific preda-tion, intraguild predation and obligate secondary predation). Intra-specific preda-tion involves cannibalism and is, therefore, easy to comprehend, but the other types of interactions are more complex and are depicted in Fig. 6.8. If these trophic webs are viewed from the perspective of whether the herbivore will be suppressed by its associates within the soil food web, it is clear that obligate secondary predation will have an impact, as all biological control agents are subject to parasitism or preda-tion. However, intraguild predation may also influence the level of suppression, par-ticularly in soil where a diverse range of hosts and prey is present, because many predators will share a common prey (the herbivore) and will also prey on each other.

Rosenheim et al. (1995) used the trophic web in Fig. 6.9 to show that intraguild predation

may occur when several biological control agents interact with a plant-parasitic nema-tode. However, when soil organic matter and other soil organisms are also included in the picture, the interactions depicted in the figure are even more complex, as the effects of preda-tors on the target pest will be influenced by at least the following factors: (i) predators may show some preference for free-living nema-todes over the plant parasite; (ii) the predators and herbivores may be in different patches in the soil profile, particularly if the organic mat-ter is spatially separated from the root system; (iii) the prey selected by the mononchid may vary with its developmental stage; (iv) in some situations the mononchid predator may be can-nibalistic; (v) if a suitable organic substrate is available, the nematode-trapping fungus may survive saprophytically; (vi) one nematode-trapping fungus may impact on others thro-ugh mycoparasitism; and (vii) the omnivorous

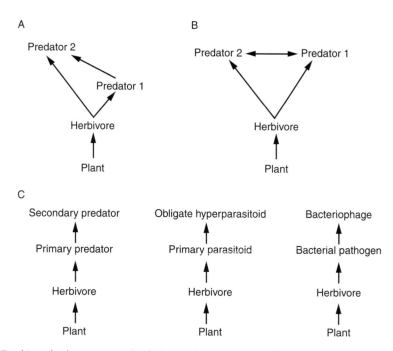

Fig. 6.8. Trophic webs demonstrating the distinction between intraguild predation and obligate secondary predation. (A) Unidirectional intraguild predation, in which predators 1 and 2 share a common prey (the herbivore), and predator 2 eats predator 1. (B) Bidirectional intraguild predation, which is the same as (A), except that predators 1 and 2 eat each other. (C) Obligate secondary predation, in which the secondary predator (which is shown as a secondary predator, an obligate hyperparasitoid or a bacteriophage) attacks the primary consumer but does not attack the herbivore. Thus, the primary and secondary consumers cannot compete for a common resource. (From Rosenheim et al., 1995), with permission.)

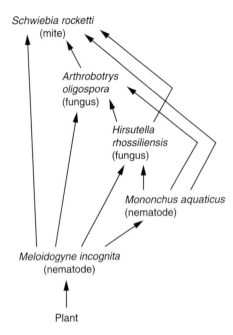

Schwiebia rocketti
(mite)

*Arthrobotrys
oligospora*
(fungus)

*Hirsutella
rhossiliensis*
(fungus)

Mononchus aquaticus
(nematode)

Meloidogyne incognita
(nematode)

Plant

Fig. 6.9. Intraguild predation among biological control agents of nematodes in soil ecosystems. The nematode-trapping fungus *Arthrobotrys oligospora* has a broad host range and may attack both plant-parasitic nematodes such as *Meloidogyne incognita* and predacious nematodes such as *Mononchus aquaticus*. *Arthrobotrys oligospora* may also be mycoparasitic, that is, it may parasitize other fungi such as *Hirsutella rhossiliensis*. *H. rhossiliensis* does not attack other fungi but may parasitize both plant-parasitic and predacious nematodes. Finally, the mite *Schwiebea rocketti* eats both nematodes and nematophagous fungi. (From Rosenheim *et al.*, 1995, with permission.)

that attack the females and eggs of major pests. The role of generalist predators as suppressive agents has received much less attention. In the same way, entomologists have concentrated on 'classical' biocontrol, where populations of pests are reduced by specialist natural enemies, especially parasitoids. However, there is now renewed interest in the long-recognized, but little exploited, background level of suppression provided by assemblages of mainly generalist predators. A review of this subject by Symondson *et al.* (2002) is, therefore, relevant. Although it mainly considers issues associated with general predation on aboveground insect pests, many of the issues that are likely to be critically important in the much more complex terrestrial ecosystem are addressed.

The main disadvantage of generalists as regulatory agents is that they do not respond as well as specialists to an increase in prey density. Their functional response levels off as they become satiated, and this response may be further modified when alternative prey become available. However, there are trade-offs between the respective merits of specialists and generalists, and so the disadvantages of generalists are offset to a great extent by the following features:

- The polyphagous food habits of generalist predators help sustain them when pest populations are low.
- Populations of soil organisms will fluctuate with temperature, moisture, season of the year and many other factors. Having a diverse range of generalist predators means that their function (providing suppressive services) will be maintained over time.
- Opportunistic feeding habits allow generalists to rapidly exploit a food resource that becomes available when pest populations increase.
- Generalists may drive pests to extinction at a local or microsite level, but because they have the ability to use alternative foods, predator numbers do not necessarily decline.

Symondson *et al.* (2002) also collated data from more than 40 years of manipulative

mite may consume the two nematophagous fungi or the predatory nematode, or may survive on alternative food sources within the soil community. In such circumstances, it is almost impossible to predict the outcome of these interactions, but what is clear is that the two-species interactions often studied by biological control researchers are a highly simplistic version of what actually happens in soil.

Because of the difficulties involved in studying numerous organisms in complex soil systems, research on biological control of nematodes has focused on relatively specific parasites (e.g. *Pasteuria penetrans*) and fungi field studies with insect pests, and used the results to show that single generalist predators, or assemblages containing several generalists, reduced pest numbers significantly in about 75% of cases. They also discussed how predator populations might be manipulated to enhance pest control, and interestingly, mulching and no-till, two of the practices promoted in this book as a means of enhancing suppressiveness to plant-parasitic nematodes, were found to be useful against insect pests.

Although generalist predators clearly have the potential to suppress populations of plant-parasitic nematodes, they will only do so if they are co-located with the target nematode. Soil biological communities are distributed in patches centred on resources (either roots or organic matter) and communities in different patches do not necessarily interact (Ferris, 2010). Thus, there may be high populations of microarthropods, predatory nematodes and nematophagous fungi in an organic litter layer or in the upper few centimetres of the soil profile, but management practices must ensure that they interact with plant-parasitic nematodes, which are mainly associated with roots further down the profile. Maintaining ecosystem engineers such as earthworms and enchytraeids must, therefore, be an important management objective, because these organisms not only create macropores but also move organic materials from the surface into deeper soil layers. Since the microfauna and mesofauna associated with decomposing roots of previous crops provide a food source for generalist predators, and the channels they occupy are large enough to be occupied by microarthropods, they must also be preserved. Thus, the management challenge in annual cropping systems is to grow the next crop in such a way that its roots, and the plant-parasitic nematodes that are inevitably associated with them, come into contact with these microsites of predatory activity (see also an earlier discussion with regard to nematophagous fungi in Chapter 5).

Concluding Remarks

Although this chapter has focused mainly on predatory nematodes and microarthropods,

these two groups should not be considered in isolation from perhaps the most important generalist predators in soil: the nematophagous fungi. Thus, these concluding remarks consider generalists collectively; reflect on their role in providing regulatory services; and argue that high population densities and diversity of these organisms provide an indication that an agroecosystem is self-regulatory, and has the capacity to prevent pests such as plant-parasitic nematodes from becoming dominant. Comments are also made on the crop and soil management practices required to develop self-regulating agroecosystems, and the need for soil ecologists, agricultural scientists and farmers to work more collaboratively to improve current best-practice systems.

Generalist predators as indicators of ecological complexity and a capacity to suppress pests

The nematode management component of this book is based on the premise that in a properly functioning ecosystem, any increase in the population density of a nematode pest will invariably be met with limitations that will be due to the actions of its natural enemies. If that premise is true, and there is a plethora of evidence to suggest it is, then the challenge of agricultural management is to create an active and diverse biological community capable of providing this regulatory service. One way of looking at that task is to consider the natural enemies that should be key components of that community. Although nematologists tend to focus on omnivorous and predatory nematodes, I suspect that nematophagous fungi and microarthropods are much more important, as they are by far the most numerous and diverse groups of nematophagous organisms in undisturbed soils. Since they have been reduced to low population densities in most agroecosystems, their collective abundance and diversity is likely to be the best indicator of whether regulatory services have been restored and are likely to be operating. Although there are dangers in singling out one or two groups of natural enemies in complex ecosystems, the advantage of doing so is that it focuses the mind on the specific practices required to enhance soil biodiversity. If nematophagous fungi and microarthropods are restored to agricultural soils, predatory nematodes, tardigrades and many other natural enemies will also have returned.

Since all agricultural soils were mechanically tilled until the 1980s, and most are still routinely cultivated, soil fungi are disadvantaged relative to bacteria, and decomposition channels in the soil are generally bacterially dominant. The widespread use of nitrogen fertilizers has exacerbated this trend. Microarthropods have also declined because the soil voids that harbour them are no longer habitable due to compaction, or have been destroyed by tillage. Additionally, the huge decline in levels of soil organic matter that has occurred since soils under natural vegetation were converted to agriculture has affected both groups of organisms: fungi no longer have access to the substrates required to support their saprophytic activity, while the loss of the energy source required by primary decomposers means that the food supply available to microarthropods is not only reduced in quantity, but is also less diverse. Since the practices that must be changed to return fungi and microarthropods to agricultural soils are well known, the creation of soils with biological activity and complexity resides in the hands of land managers. And the important point from a nematode management perspective is that if that job is done well, the end result will be a soil with an enhanced capacity to suppress plant-parasitic nematodes.

Conservation (or autonomous) biological control

In a review of the role of generalist predators in agroecosystems, Symondson et al. (2002) referred to the pest management approach discussed in the previous paragraph as conservation biological control. It is a process that uses a range of habitat management and diversification techniques to enhance natural enemy numbers and their ability to suppress

pests. Instead of using specialist biocontrol agents as magic bullets, assemblages of native natural enemies are exploited as one of several regulatory mechanisms for controlling pests. Other authors have used different terms, with Vandermeer (2011) referring to it as 'autonomous biological control'. The approach is the same, and is based on the premise that the internal functioning of the ecosystem will result in the regulation of all potential pests. The ecosystem is constructed by the farmer in a way that autonomously keeps the problems at bay. Note that regardless of the term used, ecological knowledge is being used to develop a self-regulating agroecosystem, and that the key to success is the aptitude of the land manager.

The basis of conservation biological control is that an increase in inputs of organic matter, together with a reduction in disturbance and various detrimental inputs, will result in an increase in populations of bacteria, fungi and numerous other soil organisms that were referred to by Ferris et al. (2012a) as amplifiable prey. These additional organisms provide resources for generalist and specialist predators in the system, and their greater abundance will increase top-down pressure on other potential prey: root-feeding organisms that were referred to as target prey. These authors then developed a simple model (Fig. 6.10) to explain how a cascading system of amplification of predacious nematodes could lead to the regulation of plant-feeding nematodes. Although analyses of data obtained from banana plantations in several countries suggested that an increase in levels of amplifiable prey resulted in greater predator pressure on the target prey, there were few obvious relationships among links in the above model. This led to a revised model being proposed (Fig. 6.11) to accommodate more fully the numerous interactions that occur within the soil food web. The value of such a model is that it encapsulates the key elements of conservation biological control: (i) systems level regulation of soil pest species occurs when resources to predators are provided through a trophic web of carbon and energy exchanges; (ii) the development and maintenance of an abundance of predators requires either a consistent supply of food to that trophic level or the presence of long-lived

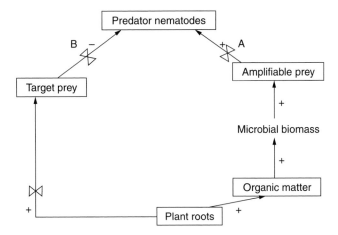

Fig. 6.10. Model of the cascading system of amplification of predacious nematodes leading to the regulation of target prey through apparent competition. Regulation of the target prey becomes suppression when the rate of flow from target prey to predators (a function of predator abundance and consumption rates) exceeds the rate of increase in target prey biomass (a function of target prey abundance, growth rates and resource availability). The triangles are regulator symbols indicating that flow down an arrow is regulated or controlled by some factor that is not explicitly included in the model diagram. Impact regulators marked A indicate the need for favourable conditions for sensitive predator species; those marked B indicate the need for co-location of predators and prey. (From Ferris *et al.*, 2012a, with permission.)

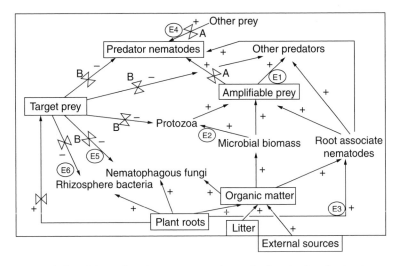

Fig. 6.11. An expanded model to reflect the importance of cascade diversion in a food web system that leads to regulation of target prey through subtle apparent competition and functional complementarity. See Fig. 6.10 for details of the regulator symbols. Expansions of the simple model depicted in Fig. 6.10 are: E1 – resource diversion to other predators of the amplifiable prey which may also consume target prey; E2 – resource diversion to other consumers of amplifiable prey, including protozoa of which some may consume target prey and some become resources for amplifiable prey; E3 – root associate nematodes are stimulated by roots and organic matter; they contribute to the amplifiable prey and function as resources for predators; E4 – access of generalist predators to other resources besides amplifiable prey; E5 – nematophagous fungi are stimulated by roots and organic matter and are generalist predators of both target prey and amplifiable prey; E6 – rhizosphere bacteria stimulated by roots may have antibiotic and other effects on target prey. (From Ferris *et al.*, 2012a, with permission.)

predators with slow turnover rates; (iii) regulation will not occur unless predators and the
target pest are co-located; and (iv) regulatory
services will decline if physical or chemical
constraints are imposed.

Practices associated with developing self-regulating agroecosystems

The management practices required to
enhance the regulatory services provided by
the soil food web will not be discussed in
detail here, as they are mentioned throughout
this book and are covered specifically in
Chapters 11, 13 and 14. Instead, they have
been categorized into two groups and are
listed below.

- *Key practices.* Permanent plant residue
  cover; a diverse rotation sequence; continuous inputs of organic matter; reduced
  tillage; and avoidance of compaction
  through traffic control.
- *Second-tier practices.* Biomass-producing
  cover crops; inclusion of legumes in the
  rotation; integrated crop and livestock
  production; organic mulches; improved
  nitrogen use efficiency; optimized water
  management; site-specific management of
  inputs; and integrated pest management.

Although all of the above practices are
important, they have been separated into two
groups because it is recognized that improving a farming system is an incremental process, and it is not feasible to implement all the
desired practices at once. Nevertheless, the
five practices listed first are essential, as without them it is impossible to produce a soil that
can harbour the fungi, microarthropods and
other organisms responsible for providing
suppressive services. Roger-Estrade *et al.*
(2010) provided a thorough overview of the
biological benefits associated with reduced
tillage, but also pointed out that soil management involves much more than minimizing
or eliminating tillage. Compaction must be
prevented, diverse organic inputs from rotation crops are required, and crop residues
must be managed effectively. Thus, the five
key practices must be adopted together. They

are the building blocks on which sustainable,
self-regulating agroecosystems can be built.

The key practices discussed here are not
really contentious, as they form the basis of conservation agriculture, a farming system that is
already widely practised in some countries
(see Chapter 4). Thus, the challenge of the future
is to integrate the second-tier practices into such
farming systems, and understand how they
influence the soil biological community. Critical
issues from an ecological perspective are
whether the move to conservation agriculture
has actually resulted in a more active and
diverse biological community; whether nutrients and other inputs can be managed in ways
that benefit, or at least minimize any detrimental effects on key indicators such as nematophagous fungi or microarthropods; and whether
the site-specific management practices associated with precision agriculture can be better
utilized to avoid the biological diminution associated with excessive inputs of nutrients. With
regard to the provision of suppressive services
against nematode pests, the key question is
what level of suppressiveness has been generated under current best practice, and is it sufficient to reduce populations of the target
nematode below the damage threshold? Once
the answer to that question is known, it will be
possible to decide whether other practices need
to be integrated into the farming system to
reduce losses from nematodes to acceptable levels. If that is necessary, practices that enhance
the activity of specific suppressive agents could
be utilized, thereby providing an even more
effective and resilient form of suppressiveness
(see Chapters 7 and 10).

The disconnect between agricultural scientists, soil ecologists and the farming community

Over the last three decades, there has been a
major increase in research in soil microbial
ecology, and the scientific literature is replete
with books, reviews and papers on soil biodiversity, decomposition processes, the soil
food web, rhizosphere effects, microorganism–
invertebrate interactions and many other related topics. We now know much more about

the soil biological community than we did 30 years ago, but this knowledge has not been translated into improvements in the way soil organisms are managed on-farm. For example, research has shown that nematode pests can be suppressed by naturally occurring parasites and predators, and ecologists often indicate that biological suppressiveness is an important ecosystem service, but most farmers have no knowledge of these issues. They assume that crop losses from nematodes are inevitable, and are usually not aware that nematodes have natural enemies that could possibly provide an alternative to nematicides and resistant cultivars.

In my opinion, the above situation is the result of the detachment that exists between agricultural scientists, soil ecologists and the farming community. The lack of interaction between the former groups is particularly concerning, and is manifested in some of the issues outlined as follows.

- Many agricultural scientists do not view agriculture from an ecological perspective. They focus on maximizing the provisioning services provided by cropping and livestock enterprises, and give little thought to the many other services that an agroecosystem should provide. Their understanding of the energy flows and nutrient cycling processes that occur within agroecosystems is generally relatively simplistic, and their knowledge of soil biological processes is limited. This lack of detailed ecological knowledge, particularly as it relates to soil biology, means that agricultural scientists are unlikely to read the ecological literature, and rarely talk to soil ecologists. Much the same criticism could be applied to discipline-based specialists who work with soil. They tend to work with a limited range of organisms, and their ecological knowledge is often limited to understanding how abiotic factors such as temperature or moisture influence particular soilborne pests or pathogens.
- Much ecological research is conducted in natural ecosystems and that is quite understandable. These systems cover a large part of the earth's surface and we need to understand the impact of soil organisms on processes such as plant productivity in forests and grasslands, plant species diversity, plant succession and plant invasiveness. Also, the knowledge generated in natural systems can be utilized in agroecosystems. However, disconnection occurs when ecologists fail to accept that agriculture is a managed system. Some level of disturbance is inevitable when land is cropped or animals are grazed, and so the biodiversity that exists in natural systems will never be recreated in an agroecosystem.
- Ecologists often compare agricultural systems to undisturbed grassland. Again, that is an understandable and acceptable practice, but such research sometimes does not address ecological issues of importance to modern agriculture. For example, Postma-Blaauw et al. (2010, 2012) showed that various biological parameters were detrimentally affected when land was converted from grassland to arable cropping. However, such a result is not surprising, given that the arable plots were deeply and regularly tilled. What is more relevant from an agricultural perspective is whether the same result would have been achieved if conversion had been done in a less aggressive manner (e.g. spraying the grassland with a herbicide such as glyphosate and then direct-drilling the first and following crops).
- In the introductory sections of soil ecology papers, or in reviews of the ecological literature, misleading statements are often made about agricultural practices. The most common fault is to use the general term 'pesticide' to cover the chemicals used against weeds, insect pests, plant pathogens and nematodes, and then imply that they are all detrimental to the soil biological community. Chemicals need to be considered individually, as some will have negative effects and others will have little impact. An insecticide/nematicide such as aldicarb, for example, will affect many components of the soil biological community, whereas the herbicide glyphosate is much more environmentally benign, binds tightly to

soil constituents, and has little or no effect on soil organisms at commercial rates (Haney *et al.*, 2000; Weaver *et al.*, 2007; Duke and Powles, 2008). Of course, we always need to be on the lookout for subtle impacts (Zabaloy *et al.*, 2012) or for long-term or cascading effects on soil food webs (Helander *et al.*, 2012), but such studies must be done with the knowledge that each chemical has different properties, and that effects will be influenced by rate of application and environmental conditions.

• In the same way, fertilizers need to be considered individually. Nutrients are exported when crops are harvested or farm animals are sold, and so they must be replaced if the agricultural system is to be sustainable. However, the ecological literature often argues that fertilizers are detrimental to the soil biological community, and usually ignores the beneficial effects that arise from nutrient inputs (e.g. increased biomass production and its flow-on effects to soil organic matter). The critical issue from an agroecological perspective is the type of biological community that can be sustained in systems where key crop management practices are optimized and fertilizer use efficiency is high. Knowledge of the effects of different formulations, application rates, timings and placement on various soil organisms is also important, because it may indicate how detrimental effects from fertilizers can be reduced.

• Many papers written by soil ecologists provide insufficient detail on the management practices used in the farming systems that were studied. Vague terms such as 'conventional', 'organic', 'integrated' or 'improved' are often used. To be useful to agriculturalists, authors must provide specific information on tillage practices, herbicide usage, fertilizer inputs and the many other practices likely to influence the soil biota.

• Although soil ecologists agree that it takes many years to restore biological communities in agricultural soils, most ecological studies in agroecosystems have been conducted in environments where disturbance has been minimized for only a few years. Observations made in such situations are simply not relevant to the many farmers who have been using some form of conservation agriculture for more than 20 years.

• Intensification of agriculture is almost invariably seen as a negative by soil ecologists. Thus, the detrimental effects of increasing fertilizer and pesticide inputs are highlighted and the positives are generally ignored. Intensification can take many forms, and may involve: (i) increased plant diversity through crop rotation, cover cropping or intercropping; (ii) nitrogen inputs from legumes; (iii) higher carbon inputs through greater biomass production, more intensive cropping, or the use of organic amendments or mulches; (iv) a reduction in the negative impacts of bare fallows by cropping more frequently; and (v) lower pesticide inputs due to the adoption of integrated pest management. Land managers require more than a list of negative impacts: they need to know what type of soil biological community is possible when the farming system is intensified under the best management practices currently available.

The disconnection between scientists with some understanding of soil ecology and the farming community is another concern. This issue was discussed in Chapter 4 and is a major impediment to progress. Closer collaboration between these groups would see farmers increasing their understanding of ecological processes, and soil ecologists becoming more aware of the many variables that interact in farming systems. Fortunately, this has already occurred in some regions of the world, with a sustainable farming systems project in California being one such example (Temple *et al.*, 1994). Additionally, there are excellent farmers in many countries who have long been committed to the principles of ecological agriculture. Since they are now well down the pathway towards enduring sustainability, soil ecologists with an interest in agriculture must make an effort to learn from their experiences.

Earlier in this chapter and elsewhere in this book, the basic requirements for improving the sustainability of farming systems, enhancing soil health and increasing suppressive services have been discussed. To reiterate, the key practices are permanent plant residue cover, a diverse rotation sequence, continuous inputs of organic matter, reduced tillage and avoidance of compaction through traffic control. In most regions of the world, farmers can be found who have been using most, if not all of these practices for many years, and are now dealing with a range of second-tier improvements to their farming system. Since few research sites with such histories are available, soil ecologists need to locate appropriate on-farm sites, study them intensively, and determine what is possible under best-practice management in specific environments. My own experience in the grain-growing soils of southern Australia suggests that it may be surprising what can be achieved. Based on the results of preliminary nematode community analyses, soils managed using most of the above practices for more than 15 years were better in some respects than adjacent soils under grassland or natural vegetation (Table 6.4). Free-living nematodes rather than plant parasites dominated both land uses, but soils in cropped fields tended to have more omnivorous and predatory nematodes. The lower nematode Channel Ratio in four of the five cropped fields indicated that fungi were playing a greater role in the detritus food web in these fields than in adjacent less-disturbed sites. Additionally, *Pratylenchus neglectus*, the main nematode pest of cereals in southern Australia (Vanstone *et al.*, 2008) was present, but population densities were relatively low in all of the cropped soils.

The data presented below are encouraging, as they suggest that the biology of cropped soils is not necessarily decimated by agricultural management. However, this conclusion must be confirmed by assessing more sites; by studying fungi, microarthropods and other soil organisms in more detail; and by monitoring the temporal population dynamics of some of the biota. From a nematode management perspective, the critical questions are whether the relatively low populations of *Pratylenchus* observed in these preliminary samples result from biological suppressiveness, and if so, whether the regulatory service that has been established is

Table 6.4. Nematode communities at five sites[a] in South Australia where grain-growing soils had been subjected to minimum-till, residue-retained management for more than 15 years[b], and in adjacent undisturbed areas maintained as grassland or natural vegetation[c].

	Nematodes/g soil									
	Pratylenchus neglectus		Dorylaimids		Mononchids		Total free-living nematodes		Nematode Channel Ratio[d]	
Site	Crop	Natural	Crop	Natural	Crop	Natural	Crop	Natural	Crop	Natural
1	1.2	0.3	0.26	0.16	0.020	0	12.7	13.2	0.41	0.78
2	0.2	0.2	0.12	0.25	0.028	0.010	14.2	14.5	0.50	0.73
3	0.1	0.1	0.39	0.19	0.000	0.008	12.7	14.2	0.66	0.79
4	2.6	6.4	0.48	0.42	0.031	0	14.6	13.0	0.73	0.58
5	0.4	0	0.23	0.03	0.016	0	9.8	9.3	0.70	0.95

[a]All sites were in a Mediterranean climate (predominantly winter rainfall with hot, dry summers). Sites 1–4 were sandy clay loam soils (chromosols or calcarosols) in areas with an annual rainfall of about 450 mm. Site 5 was a fine sand with an annual rainfall of 260 mm. Sites were sampled in early April 2012, about 1 month before the next crop was due to be planted. Data are means of three replicate samples from 0–15 cm.
[b]Crops in the rotation sequence included wheat, barley, oats, canola and peas. None of the soils had been tilled for at least 15 years; crop residues were retained on the soil surface; weeds were managed with glyphosate and other herbicides; and fertilizer inputs were determined on the basis of crop type, nutrients removed in the previous harvest and the results of soil tests. Farm traffic was not controlled, so some soil compaction would have occurred.
[c]Natural sites were adjacent each cropped field and consisted of undisturbed grassland at sites 1–4; and natural vegetation at site 5 consisting of mallee (*Eucalyptus*) with a *Psilocaulon* and *Atriplex* understory.
[d]Number of bacterivorous nematodes/(number of bacterivorous + fungivorous nematodes)
(G.R. Stirling and K. Linsell, unpublished data)

sufficient to maintain nematode populations at levels that do not cause economic damage. From the farmer's perspective, he or she needs to know whether subtle changes to the rotation sequence, fertilizer placement or some other practice can further improve what is currently considered a relatively productive and sustainable farming system.

One problem with the sites chosen for this study is that the cropped soils would have been compacted to some extent, as farm traffic had not been controlled. However, there are fields in the same region that have been under minimum-till for 25–30 years, and where controlled traffic using GPS guidance was introduced in the last 5 years. Such sites provide opportunities for soil ecologists to not only assess the biological impact of what has been considered in this book as best-practice management, but also add value to what progressive farmers are already achieving.

7

Obligate Parasites of Nematodes: Viruses and Bacteria in the Genus *Pasteuria*

Except perhaps for oomycetes such as *Nematophthora gynophila* and some of the more specialized predatory microarthropods, the organisms discussed in Chapters 5 and 6 do not have an intimate relationship with nematodes. They certainly parasitize or prey on them, but are also able to survive on other food sources. This chapter covers a group of organisms that are considered 'obligate' parasites because they are nutritionally dependent on nematodes and cannot lead an independent existence outside their host. Opinions differ as to whether viruses are living organisms, but they are considered obligate parasites here because the resources they use for reproduction are derived from living cells. However, the virus section is relatively brief, as viruses have largely been overlooked as agents capable of causing disease in nematodes. The main focus is on the genus *Pasteuria*, a widely distributed bacterium that is known to attack many different nematodes, including all economically important plant-parasites.

Viral Infectious Agents of Nematodes

Although virus-like particles have occasionally been observed in nematodes, and a virus-like pathogen was thought to have caused debilitating symptoms in root-knot nematode (Loewenberg *et al.*, 1959), it is only recently that viral infections in nematodes have been confirmed. In the first report, two novel RNA viruses distantly related to known nodaviruses were found infecting *Caenorhabditis* (Félix *et al.*, 2011). The viruses damaged the intestinal cells of the nematode, but infected hosts continued to survive, although reproduction was impaired. Four different RNA viruses were later found in soybean cyst nematode, *Heterodera glycines*, and since they are related to viruses that cause disease in other hosts, it is possible that they are detrimental to this economically important pathogen (Bekal *et al.*, 2011).

New high-throughput DNA-sequencing methods are facilitating viral genome discovery, and given that viruses infect many other invertebrates, a wide array of new and interesting viral pathogens of nematodes will almost certainly be found in the next few years. Since viruses have been deployed as a safe and environmentally friendly means of managing insect pests (Hunter-Fujita *et al.*, 1998), it is possible that they will also be useful against nematodes. Research on nematode viruses is in its infancy, but investigations into their infection mechanisms and pathogenicity should be encouraged.

Bacteria in the Genus *Pasteuria*

The bacterial genus *Pasteuria* contains a group of obligate parasites of invertebrates that are primarily found in nematodes. Hundreds of nematode species are known to host various species and strains of *Pasteuria*, but the bacterium is of particular interest because it can attack many widely distributed species of root-knot and cyst nematode, the world's most damaging plant-parasitic nematodes. Once these nematodes are infected, the parasite proliferates within their bodies and normally prevents reproduction. There is, therefore, considerable interest in the role of indigenous strains of *Pasteuria* in suppressing populations of these nematodes, and also in the use of the parasite as an inundative biocontrol agent. Since these topics are covered in Chapters 10 and 12, respectively, this chapter focuses on ecological issues that affect the way *Pasteuria* interacts with its nematode host. Reviews of early work on *Pasteuria* are available (Sayre and Starr, 1988; Stirling, 1991; Sayre, 1993), while more recent reviews include Chen and Dickson (1998), Trudgill *et al.* (2000), Chen and Dickson (2004b), Gowen *et al.* (2008), Davies (2009), Mateille *et al.* (2010) and Schaff *et al.* (2011).

Distribution, host range and diversity

Pasteuria-like organisms occur throughout the world and have been found in association with many different nematodes. The most comprehensive compilation of host records listed 323 nematodes in 116 genera (Chen and Dickson, 1998), but as the authors noted, lists of this nature are not entirely satisfactory because most reports are based on attachment of endospores to the nematode's cuticle rather than parasitism. The term 'host' should be reserved for situations where a nematode is actually infected by the parasite and mature endospores are produced within the body. Most hosts of *Pasteuria* belong to the Tylenchida and Dorylaimida, but many other nematode orders are represented, prompting Sturhan (1988) to comment that all terrestrial nematodes may be potential hosts.

The endospore is the most conspicuous morphological feature of *Pasteuria*, and when endospores are observed in nematodes or attached to the cuticle, it is apparent that there is enormous diversity within the genus. For example, Giblin-Davis *et al.* (1990) found six *Pasteuria* isolates associated with different nematode hosts in a survey of four bermuda-grass (*Cynodon* spp.) golf course fairways in southern Florida. Five of the isolates differed significantly in sporangium diameter and endospore width (Fig. 7.1). In other studies of endospores associated with a wide range of plant-parasitic and free-living nematodes from numerous locations, endospore diameter varied from 2.7 to 8.0 μm (Ciancio *et al.*, 1994), and the size of endospores was related to the body wall thickness of the corresponding host nematode (Ciancio, 1995a). This variability is typical of *Pasteuria* and suggests that the genus is an assemblage of numerous 'forms' or species, possibly each with a specialized and rather limited host range. It also explains why terms such as '*Pasteuria*-like' and the '*Pasteuria penetrans* group' are often seen in the scientific literature.

Taxonomy, systematics and phylogeny

Species in the genus *Pasteuria* are parasitic in water fleas and nematodes, but it took almost a century for the connection between the two groups of parasites to be made. The genus was initially erected by Metchnikoff (1888) to accommodate *Pasteuria ramosa*, a bacterial parasite of *Daphnia magna*, but other scientists failed to find the various life stages observed by Metchnikoff and concluded that his work was erroneous. A similar parasite was then observed in nematodes, but was thought to be a sporozoan and named *Duboscqia penetrans* (Thorne, 1940). Later studies with the electron microscope showed that the nematode parasite was prokaryotic and it was reassigned to the genus *Bacillus* (Mankau, 1975a, b; Mankau and Imbriani, 1975). Metchnikoff's organism was then rediscovered in *Moina rectirostris*, a member of the family Daphnidae (Sayre *et al.*, 1977, 1979). Once the similarities between the parasites of cladocerans and

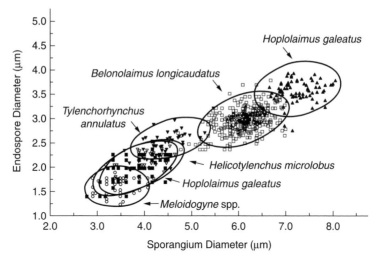

Fig. 7.1. Bivariate scattergram comparing the endospore and sporangium diameter of *Pasteuria* isolates associated with various plant-parasitic nematodes found on bermudagrass turf in southern Florida. (From Giblin-Davis *et al.*, 1990, with permission.)

nematodes were recognized (Sayre *et al.*, 1983), and various nomenclatural issues were attended to (see Sayre, 1993), the bacterial parasite of root-knot nematode was renamed *Pasteuria penetrans* (Sayre and Starr, 1985; Sayre *et al.*, 1988).

The systematics of *Pasteuria* and the phylogenetic relationships of several species were initially established through sequence comparisons of genes encoding 16S rRNA. Initial work showed that the type species for the genus (the cladoceran parasite *Pasteuria ramosa*) was related to other Gram-positive endospore-forming bacteria, and was closely related to members of the family Alicyclobacilliaceae (Ebert *et al.*, 1996). Further studies with two isolates of *P. penetrans* derived from root-knot nematodes, *Meloidogyne* spp. (Anderson *et al.*, 1999), an isolate from soybean cyst nematode, *Heterodera glycines* (Atibalentja *et al.*, 2000), and an isolate from sting nematode, *Belonolaimus longicaudatus* (Bekal *et al.*, 2001) then indicated that *Pasteuria* spp. belong to the *Clostridium–Bacillus–Streptococcus* branch of Gram-positive eubacteria (Preston *et al.*, 2003). This finding was supported through use of the *spo*0A gene, which placed *Pasteuria* with members of the supergenus *Bacillus* (Trotter and Bishop, 2003). Since phylogenetic analytical procedures that rely on one gene are

sometimes inaccurate, Charles *et al.* (2005) then used a multilocus approach with 40 housekeeping genes to show that *P. penetrans* clustered tightly in the *Bacillus* class of the Gram-positive eubacteria with low G+C content. These analyses also suggested that *P. penetrans* was ancestral to *Bacillus* spp. and may have evolved from an ancient symbiotic bacterial associate of nematodes.

Currently, *Pasteuria* is located in the Phylum Firmicutes, Order Bacillales, Family Pasteuriaceae (Vos *et al.*, 2009), but because there is considerable diversity within the genus, it is likely to contain many taxa, while also exhibiting diversity at an intra-species level (Nong *et al.*, 2007; Mauchline *et al.*, 2011). On the basis of morphological, morphometric, ultrastructural, developmental, host range and molecular data, Dickson *et al.* (2009) recognized five *Pasteuria* species: *Pasteuria ramosa*, *P. penetrans*, *P. thornei*, *P. nishizawae* and 'Candidatus P. usgae'. The provisional rank of the latter species was based on a decision in 1995 by bacterial taxonomists to establish 'Candidatus' as a new taxonomic category to accommodate putative taxa that were well-characterized but had not yet been cultured. 'P. hartismeri', a parasite of *Meloidogyne ardenensis*, had previously been described by Bishop *et al.* (2007), but was not included in the compilation by

Dickson *et al.* (2009) because it had not been validated according the *International Code of Nomenclature of Bacteria*. Since then, 'Candidatus P. aldrichii' has been discovered in Florida, parasitizing the bacterivorous nematode *Bursilla* (Giblin-Davis *et al.*, 2011).

The remainder of this chapter focuses on the ecology of the *Pasteuria* species listed earlier. However, as *P. penetrans* is the most widely studied species, it receives the most attention. Aspects of other species are only discussed when they differ markedly from *P. penetrans*.

Pasteuria penetrans: A Parasite of Root-knot Nematodes (*Meloidogyne* spp.)

Life cycle and development

The life cycle of *P. penetrans*, as originally described by Sayre and Wergin (1977) and Imbriani and Mankau (1977), is initiated when endospores adhere to the cuticle of second-stage juveniles of *Meloidogyne* spp. as they migrate through soil. Attachment is thought to be a 'Velcro®-like' process in which

collagen-like fibres on the surface of the endospore interact with fibrous structures on the cuticle of the nematode (Davies, 2009). Spore-encumbered nematodes then enter the roots of a host plant (Fig. 7.2), establish a feeding site, and become sedentary. The attached endospores germinate before the developing nematode moults to the third-stage juvenile, and infection occurs when a germ tube emerges through the basal ring of the endospore and penetrates the cuticle (Fig. 7.3). Cauliflower-shaped microcolonies of dichotomously branched, septate mycelium are then produced, and daughter colonies form by lysis of intercalary cells. As this process continues, the number of daughter colonies increases but they contain fewer, but larger vegetative cells. Eventually, quartets of developing sporangia predominate in the nematode's pseudocoelom and they give rise to doublets and finally to single sporangia containing a single endospore. This process is depicted in Fig. 7.4, and was confirmed in a study of four isolates of *P. penetrans* from Florida (Chen *et al.*, 1997a).

Davies *et al.* (2011) noted that the conventional description of the life cycle of *P. penetrans* implied that microcolonies were the vegetative growth stages of the

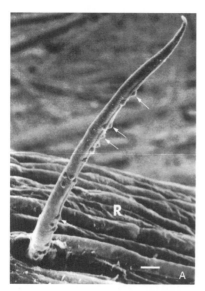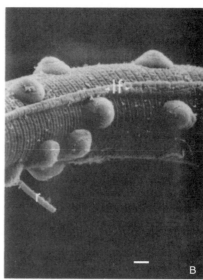

Fig. 7.2. Scanning electron micrographs showing (A) *Pasteuria penetrans* endospores (arrows) attached to a root-knot nematode juvenile that has partially penetrated a tomato root (R). Bar = 10 μm. (B) Numerous endospores on the cuticle, near the nematode's lateral field (lf). A rod-shaped bacterium (r) has attached to the surface of one endspore. Bar = 1 μm. (From Sayre and Starr, 1985, 1988, with permission.)

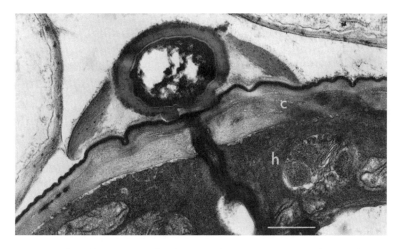

Fig. 7.3. Cross section through a germinated endospore of *Pasteuria penetrans*, showing the germ tube penetrating the cuticle (c) and hypodermis (h) of a root-knot nematode juvenile. Bar = 0.2 μm (From Sayre and Wergin, 1977, with permission.)

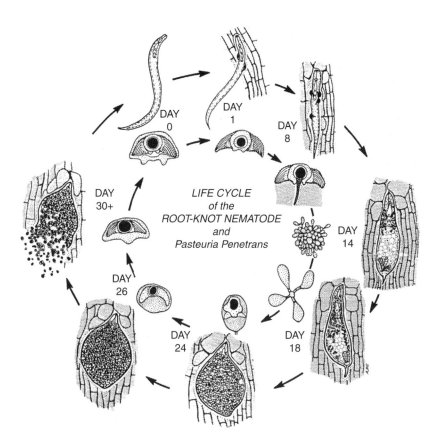

Fig. 7.4. A diagrammatic sketch of one version of the life cycle of *Pasteuria penetrans*, and the parasitized life stages of the root-knot nematode, *Meloidogyne incognita*. (From Sayre and Starr, 1988, with permission.)

parasite, with proliferation occurring when intercalary cells lysed to form daughter microcolonies. However, by removing *Pasteuria*-infected females from roots and compromising the nutrition of the nematode, these authors were able to observe other stages of the life cycle that had not previously been seen. These stages included a filamentous growth phase of the germ tube, and the production of *Bacillus*-like rods that eventually became committed to sporogenesis. Based on these observations, the life cycle was re-evaluated and the following three phases were recognized (see Fig. 7.5):

- *Phase I: Attachment and germination.* These processes are essentially the same as in previous descriptions, except that endospores may occasionally germinate on nematodes that have not yet located a plant root.
- *Phase II: Rhizoid formation and exponential growth.* In addition to forming microcolonies, the germ tube can also produce exploratory filamentous structures that are similar to the rhizoids present in some *Bacillus* spp. These structures radiate out into the developing nematode and then the rhizoid growth form terminates, with reproduction continuing in the form of rapidly reproducing rod-like structures typical of *Bacillus* spp. This exponential growth phase leads to hugely aggregated granular masses within the developing female and is a phase of growth that had not previously been observed in *Pasteuria*.
- *Phase III: Sporogenesis.* When sporulation is initiated by an environmental trigger, the granular masses form microcolonies that then go through five stages of developmental sporogenesis typical of *Bacillus* spp., to form mature endospores. Once this process is complete, endospores usually fill the entire body of the female nematode. When infected nematodes decompose, these endospores are released from the cadaver into the soil.

Although granular masses of rod-shaped bacilli can be seen in the bodies of nematodes

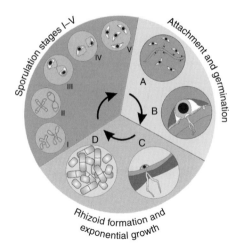

Fig. 7.5. A re-evaluation of the life cycle of *Pasteuria penetrans*. The main modifications to the life cycle depicted in Fig. 7.4 are the formation of rhizoids on the germination peg to help the bacterium proliferate throughout the pseudocoelom (C); and the subsequent formation of rods (D), which then proliferate exponentially to form granular masses. (From Davies *et al.*, 2011, with permission.)

infected with *P. penetrans* (Davies, 2009; Davies *et al.*, 2011) and similarly shaped cells have been observed in *in vitro* cultures (Hewlett *et al.*, 2004), the involvement of these cells in the life cycle remains open to question. It is presumed that they are the vegetative growth phase of the bacterium, but as helper bacteria also appear to be involved in the growth of *P. penetrans* (Duponnois *et al.*, 1999; Hewlett *et al.*, 2004), confirmatory evidence that these rod-shaped bacteria are *Pasteuria* is still required (Davies, 2009).

Pathogenicity, pathogenesis and the impact of temperature

Although the process of infection by *P. penetrans* commences when endospores attach to a nematode, this does not mean that the spore-encumbered nematode is infected. Only 27% of the females developing from juveniles encumbered with one spore were eventually parasitized, indicating that not all the spores that adhere to nematodes will germinate and initiate infection (Stirling, 1984). The number of spores required to ensure infection will obviously vary with the quality of spores and environmental conditions, but in this experiment, 74% and 90% of the females were eventually infected when four and five spores, respectively, were attached to juveniles.

Once the parasite breaches the cuticle and enters the vegetative growth phase, the infected nematode continues to develop normally and its feeding does not appear to be impaired. However, as resources are diverted from the nematode to the parasite, the nematode's reproductive system is replaced by more than two million *P. penetrans* spores. However, the impact of the parasite on fecundity is dependent on temperature. At 30°C, an isolate of *P. penetrans* proliferated extensively through females of *M. javanica* before they reached maturity, completing its life cycle in 20–30 days. In contrast, the life cycle took 80–100 days at 20°C, and females often developed ovaries containing eggs before infection prevented further egg production (Stirling, 1981).

Given the genetic diversity within isolates of *P. penetrans*, and the fact that various host nematodes will have different temperature optima, it is likely that temperature will have a range of effects on the host–parasite interaction. Nevertheless, *P. penetrans* is generally most pathogenic to nematodes at temperatures of 25–30°C. For example, an isolate from South Africa developed most rapidly in *M. javanica* at 28–31°C (Giannakou *et al.*, 1999), while the optimum temperature for spore attachment and development within the body was 28–30°C for a Florida isolate on *M. arenaria* (Hatz and Dickson, 1992; Serracin *et al.*, 1997). Since the minimum growth temperature for *P. penetrans* is around 18°C (Chen and Dickson, 1997b; Giannakou and Gowen, 2004), and this is higher than the minimum growth temperature for most warm-climate species of *Meloidogyne*, the parasite is unlikely to be an effective biological control agent at low temperatures.

All these observations were made with isolates of *Pasteuria* that proliferated within the sedentary stage of root-knot nematode and sporulated in the maturing female. However, exceptions to the normal developmental process have occasionally been reported. For example, 21% of the *Meloidogyne* second-stage juveniles recovered from one field site in Florida were filled with *Pasteuria* spores (Giblin-Davis *et al.*, 1990), raising the question of whether spore germination always occurs after the nematode establishes a feeding site within roots. Given the diversity that exists within *Pasteuria*, it is possible that this strain behaves like species that parasitize vermiform nematodes, and, therefore, differs from *P. penetrans*. Another possibility is that infection occurs in spore-encumbered juveniles that have moved out of roots after trying unsuccessfully to set up nurse cells in an incompatible plant host. Clearly, more work is required to understand the reasons for what appears to be a relatively rare phenomenon.

Host specificity

The specificity of the interaction between endospores of *P. penetrans* and its root-knot nematode hosts was recognized many years ago, when endospores were found to attach more readily to the *Meloidogyne* species from which they were obtained than to other species (Mankau and Prasad, 1977; Slana and Sayre, 1981). Those observations were extended by Stirling (1985), who examined the interaction of four isolates of *P. penetrans* with 15 *Meloidogyne* populations and found that endospores adhered to some nematode populations but not to others. Populations of *M. javanica*, *M. incognita* and *M. hapla* were included in the study, and the results showed that attachment specificity occurred at a sub-species level, with large differences in the number of endospores attached between different populations of the same nematode species.

Since these initial studies, many other authors (e.g. Davies *et al.*, 1988, 1994; Oostendorp *et al.*, 1990; Channer and Gowen, 1992; Tzortzakakis and Gowen, 1994; Tzortzakakis *et al.*, 1996; Espanol *et al.*, 1997; Trudgill *et al.*, 2000; Wishart *et al.*, 2004; Timper, 2009) have confirmed that endospores from individual isolates of *P. penetrans* do not adhere to all populations of *Meloidogyne*. More detailed studies with

M. incognita and *M. hapla* have also shown that progeny derived from a single individual can differ in spore adhesion, regardless of whether the nematode reproduces sexually or asexually (Davies *et al.*, 2008). The latter result suggests that root-knot nematodes may have evolved a specialized, epigenetic mechanism that allows them to modify their cuticle in some way and thereby circumvent infection by *P. penetrans*.

The results of surveys in southern Europe, Africa, South America and the Caribbean (Trudgill *et al.*, 2000) clearly show that the restricted and relatively unpredictable host range of *P. penetrans* and the innate variability of its host are a major impediment to using the parasite for biological control purposes. Root-knot nematode populations in the surveyed regions were highly variable, with two, three or four *Meloidogyne* species being detected in most regions. About 20% of the sites in some regions had detectable mixtures of two or three root-knot nematode species, and variants that may have been undescribed *Meloidogyne* species were also identified. *P. penetrans* was also variable, with several of the 23 isolates collected during the survey either not attaching to, or barely attaching to, two assay nematodes: *M. incognita* and *M. arenaria*. Introducing isolates from elsewhere and utilizing them as inundative control agents would not necessarily overcome the specificity problem, as six isolates from around the world did not attach to at least 25% of the 39 *Meloidogyne* populations included in the test.

Timper (2009) made an important contribution to our knowledge of the population dynamics of *P. penetrans* and its host specificity in a long-term study aimed at determining whether crop rotation could be used to sustain suppressiveness caused by the bacterium without compromising yield. Soil that had been subjected to various crop sequences was assessed every year for 8 years using a bioassay with second-stage juveniles of *Meloidogyne arenaria*. When soils were bioassayed with a nematode population from the greenhouse that was weakly receptive to the *P. penetrans* spores in soil, differences in spore densities between rotation treatments were detected, but differences were relatively small. However, when receptive nematode clones from the field were used to assay spore abundance,

the differences between rotations were much greater. Spore densities were highest when peanut, a host of *M. arenaria*, was grown continuously, but appeared to remain at levels sufficient to suppress the dominant nematode phenotype in a rotation that included maize, a poor host of the nematode. Therefore, this study demonstrates that *P. penetrans* populations can be influenced by rotation practice, and that populations of *M. arenaria* may be heterogeneous for endospore attachment.

Estimating endospore numbers in soil

Ecological studies of *P. penetrans* have long been hampered by the fact that populations of the parasite are not easily quantifiable in soil. Direct quantification methods such as the most probable number and soil dilution techniques commonly used to enumerate culturable soil microorganisms are not suitable for *P. penetrans*, as its endospores do not germinate on agar media. Indirect methods (e.g. the proportion of spore-encumbered *Meloidogyne* juveniles in a sample, or numbers of spores attached to assay nematodes) have been used as a surrogate, but generally they are not used in a manner that provides data on endospore population densities. Chen *et al.* (1996c) showed that this was possible by serially diluting a stock endospore suspension, adding *Meloidogyne* juveniles to each spore concentration, and then measuring the number of endospores attached in a bioassay under standardized conditions. Endospore attachment rates were subjected to regression analysis and the regression equation was used to determine endospore concentrations in unknown samples. Although this process was initially used to quantify endospores in root material, it can also be used to assess spore concentrations in soil (e.g. Schmidt *et al.*, 2003).

Although bioassays have often been utilized in work with *P. penetrans*, they are relatively time-consuming to carry out; are rarely used quantitatively; and will only detect populations that can adhere to the bioassay nematode. Simple methods that quantify spore populations in soil are preferable for epidemiological studies. The simplest option is to extract endospores directly from soil, and this can be done with small quantities of soil by a combination of sonication, sieving and centrifugation (e.g. Davies *et al.*, 1990). However, Dabiré *et al.* (2001a) showed that reasonable estimates of endospore population density could only be obtained by direct extraction if the aggregates that harbour the endospores were fully dispersed. When soil samples were shaken manually prior to the enumeration process, only 50% and 20% of the added endospores were recovered from a sandy clay and clay soil, respectively, whereas more vigorous disruption procedures (e.g. end-over-end shaking for 16 h in the presence of agate marbles or NaOH) increased endospore recovery to about 80%. The remaining endospores proved difficult to detect because they could not be distinguished from the mineral particles present in the spore suspension. Nevertheless, this method has been used to enumerate endospores in a range of field soils from Senegal (Dabiré and Mateille, 2004; Dabiré *et al.*, 2007).

Immunofluorescence techniques overcome the problem of finding endospores in environmental samples (Fould *et al.*, 2001; Schmidt *et al.*, 2003), and have been used to quantify endospores in natural sand dunes (Costa *et al.*, 2006b). Endospores were readily detected in suspensions containing organic matter and other debris, and the threshold of detection of 210 endospores/g soil was lower than for other methods available at the time. Although the technique proved particularly useful for quantifying the low numbers of endospores present in sand dune samples, its utility for other purposes is limited by the fact that only small quantities of soil can be processed: an important limitation when enumerating an organism that is likely to be highly aggregated in soil.

An enzyme-linked immunosorbent assay (ELISA) with polyclonal antibodies detected endospores in soil after they had been extracted by differential sieving (Fould *et al.*, 2001), but the effectiveness of this method was limited by the low recovery of spores from clay soil during the spore extraction phase. Nevertheless, it is useful in coarse-textured soils and was used to quantify *P. penetrans* in a field trial in Senegal (Mateille *et al.*, 2009). Schmidt *et al.* (2003) overcame the

problem of differential recovery in sands and clays by using a chemical extraction method that released bacterial endospore antigen from different soils with much the same efficiency, regardless of their physical and chemical properties. These extracts were quantified with a monoclonal antibody and a tertiary ELISA detection system, and the minimum detection limit was about 300 endospores/g soil. However, despite the authors' contention that the method was ideally suited to rapid screening of large numbers of environmental samples, it has not yet been used for epidemiological studies with *P. penetrans*.

Although molecular methods are now routinely used to detect and enumerate organisms in soil, the difficulties involved in obtaining DNA from endospores have hindered progress with *Pasteuria*. Most initial studies utilized bead-beating methods to disrupt endospores, and in one such study, as few as 100 endospores were detectable/g soil (Duan *et al.*, 2003). More recently, an alternative method was developed in which *P. penetrans* endospores were decontaminated with lysozyme and SDS before being lysed with a commercially available detergent mix (Mauchline *et al.*, 2010). Since the detergent-based system of disrupting endospores produced high-quality DNA suitable for enzymatic multiple strand amplification, it enabled whole genome amplification (WGA) techniques to be used with very small quantities of starting material. The chemical lysis and WGA-coupled method was sensitive enough to detect the equivalent DNA yield of a single endospore, making it a potentially powerful tool for monitoring *Pasteuria* populations *in situ*.

The interaction between *P. penetrans* and its nematode host in soil

As with all interactions between organisms in soil, the interaction between *P. penetrans* and its nematode host is influenced by environmental conditions. The impact of temperature on spore attachment and pathogenicity has already been mentioned, while reviews of the limited number of studies on other environmental factors indicate that the effects of soil moisture and pH are inconsistent and poorly understood (Chen and Dickson, 1998, 2004b). Since the concentration of endospores in soil is central to determining whether *P. penetrans* has a major impact on its nematode host, this discussion will focus on the factors that influence the production, location and dispersal of spores in soil, and the effects of spore concentration on infectivity and fecundity.

Endospore production and release into soil

The life cycle of *P. penetrans* concludes when endospore production ceases and the infected *Meloidogyne* female is little more than an opalescent bag containing more than two million endospores (Stirling, 1981; Darban *et al.*, 2004). However, despite the huge interest in *P. penetrans* as a biological control agent, we know very little about the factors that influence endospore production in the field. Studies aimed at optimizing *in vivo* mass production systems have shown that the number of spores produced within infected females depends on many factors, including the *Meloidogyne* species hosting the parasite; the susceptibility of the host plant to the nematode; the density of *Meloidogyne* females within the root system; soil moisture status; temperature; and nitrogen fertilization practices (Sharma and Stirling, 1991; Davies *et al.*, 1991; Hatz and Dickson, 1992; Chen and Dickson, 1997a; Giannakou *et al.*, 1999; Darban *et al.*, 2004). Presumably these factors also influence spore production in the natural environment. We also know very little about the time it normally takes for the cadaver of an infected female to break down and for spores to be released into the soil. Questions of this nature are important for modelling the population dynamics of the host and parasite, as we must be able to predict the size of the pool of spores available to infect the next generation of nematodes, and we also need to know whether these spores can affect later generations of nematodes. Nevertheless, it is clear that spores will not attach to nematodes as soon as they are released from infected females. The sporangial wall and exosporium must be disrupted before attachment will occur, and although this can be achieved in the laboratory by

sonication (Stirling *et al.*, 1986), microbial degradation is presumably involved in soil.

One point that is important to recognize when *P. penetrans*–*Meloidogyne* interactions are being studied is that these interactions occur at microsites rather than in the bulk soil mass. Since endospores are released into the soil from infected nematodes, there is likely to be extremely high spore concentrations in roots harbouring parasitized nematodes, or near a decomposing root system, and very few spores only a short distance away. In fact, if the spore concentration was measured at microsites around root systems harbouring large numbers of parasitized nematodes, the spore concentration would probably be much greater than the spore loads used in experiments on inundative biological control discussed later in this chapter and in Chapter 12. Thus, from a nematode management perspective, it is critically important to take advantage of this situation and to ensure that the next generation of nematodes is exposed to the high endospore populations that occur at microsites within the soil profile. As discussed previously for nematophagous fungi and microarthropods (see Chapters 5 and 6), one way of achieving this with annual crops may be to eliminate tillage, and to encourage roots of the next crop to grow into root channels formed by the previous crop.

The impact of the physical and chemical environment on endospores, and on the spore-attachment process

Although endospores of *P. penetrans* can be stored in the laboratory for many years (Giannoukou *et al.*, 1997), survival in soil is a different matter. Once endospores are released into soil, they enter a physical and chemical environment that influences their location within the soil profile, and their interaction with nematodes. Soil is a heterogeneous medium, and a whole range of interrelated factors including particle size, porosity, aggregation, clay content, water flow, organic matter status and composition of the soil solution influence the likelihood of an endospore coming into contact with a nematode and initiating an infection. These factors were discussed by Mateille *et al.* (2010) and are summarized as follows:

- Endospores of *P. penetrans* are only 3–4 μm in diameter, and numerous studies have shown that they are leached downwards by water flow in coarse-textured soils (Oostendorp *et al.*, 1990; Mateille *et al.*, 1996; Dabiré and Mateille, 2004; Dabiré *et al.* 2005a; Cetintas and Dickson, 2005). In the latter study, for example, some endospores moved more than 37.5 cm following three or four applications of water to a soil containing 93% sand, 3.5% silt and 3.5% clay. The positive effect of increasing amounts of irrigation water

on percolation of endospores through another sandy soil, together with the negative effects of the highest application rate on spore attachment and levels of parasitism, are shown in Fig. 7.6.

- The presence of clay markedly reduces the downward movement of endospores. When irrigation water was applied daily to a 20-cm column of sand, 53% of inoculated endospores were leached through the column in 1 month, whereas only 14% and 0.1% of the endospores were recovered in leachates from sandy clay and clay soils, respectively (Dabiré *et al.*, 2007).
- Retention of endospores in clay soils is not only due to their reduced porosity. Some endospores are also adsorbed on colloids

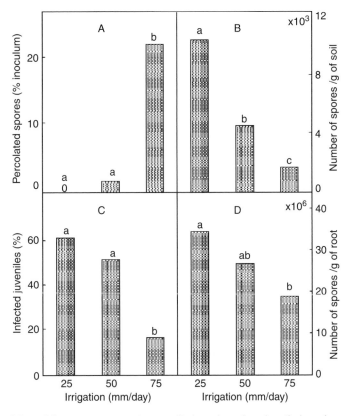

Fig. 7.6. Effect of three different irrigation regimes applied to tubes of sandy soil planted to tomato and containing both *Pasteuria penetrans* and *Meloidogyne javanica* on: (A) percolation of endospores; (B) the number of endospores remaining in soil after 35 days; (C) the proportion of *Meloidogyne* juveniles with endospores attached; and (D) the number of endospores produced in tomato roots. Data followed by the same letter are not significantly different, $P > 0.05$. (From Dabiré *et al.*, 2005a, with kind permission from Springer Science + Business Media B.V.)

or trapped in non-dispersible aggregates, and this means that they are unavailable for attachment to nematodes (Dabiré *et al.*, 2001a; Dabiré and Mateille, 2004). Bacteria commonly found in soil produce extracellular polysaccharides that enhance aggregate formation, and their presence results in endospores being retained in aggregates >200 μm (Dabiré *et al.*, 2001b).

- Since nematodes can only enter micropores greater than their own diameter, endospores will only attach to *Meloidogyne* juveniles when micropores are wider than about 10 μm. However, a water film that allows nematodes to move through the pore space and gain access to the endospores must also be present for spore attachment to occur (Dabiré *et al.*, 2005b).

- *P. penetrans* endospores are negatively charged, and their adhesion to the nematode cuticle, which is also negatively charged, is conditioned by the ionic composition of the soil solution. Both field and laboratory studies have shown that the spore-attachment process is favoured by the presence of high concentrations of cations (Ca^{2+}, Mg^{2+}, Fe^{3+}, etc.) and low concentrations of anions (Mateille *et al.*, 2009).

These inferences about the effects of physical and chemical factors on interactions between *P. penetrans* and *Meloidogyne* were made following studies in the laboratory, but have been confirmed by observations in the field. For example, survey data from Senegal showed that few *Meloidogyne* juveniles were encumbered with *P. penetrans* spores in the sandiest soils, and that more juveniles were infested as clay content increased (Mateille *et al.*, 2002). In a second study, numbers of free spores, leachable spores and adsorbed spores in different soil size fractions varied across a field used for vegetable production, and this affected the presence of both *P. penetrans* and its nematode host (Dabiré *et al.*, 2005b). A third study showed that several effects were apparent in a field with a variable soil texture after it was subjected to different irrigation regimes: (i) the population density of *P. penetrans* spores was lowest in intensively irrigated areas; (ii) the number of *Meloidogyne* juveniles encumbered with spores decreased as sand content increased, and spore

attachment was favoured by the presence of clay; and (iii) areas of the field with high *P. penetrans* population densities tended to have the highest cation concentrations in the soil solution (Mateille *et al.*, 2009).

To conclude, it is clear that soil physical and chemical properties have major effects on the interaction between *P. penetrans* and its nematode host. There is normally a density-dependent relationship between populations of host and parasite, but because of the factors depicted in Fig. 7.7, levels of infection by *P. penetrans* are not always related to endospore population density. Leaching and adsorption processes will always operate in soil, and the best spore percolation/retention balance occurs in soils with 10–30% clay (Dabiré and Mateille, 2004; Dabiré *et al.*, 2007).

The factors depicted in Fig. 7.7 also have implications for the way soils and crops should be managed to enhance infection by *P. penetrans*. Irrigation management is obviously critical in sandy soils, as applying water excessively, or in a manner that causes rapid drainage, may leach spores beyond the root zone. Tillage practices are also important, because excessive tillage will break up aggregates and mineralize carbon, with subsequent effects on the proportion of spores that are adsorbed on soil particles or trapped in aggregates.

Impact of spore concentration on nematode infectivity and fecundity

Since *Pasteuria* endospores must adhere to second-stage juveniles of root-knot nematode before infection can take place, an understanding of the spore acquisition process and its subsequent effects on the nematode is vital from a biological control perspective. Work in this area commenced in the early 1980s, when experiments with dried root powder containing millions of endospores showed that *P. penetrans* not only reduced the fecundity of infected females, but also affected the movement of second-stage juveniles into roots (Stirling, 1984). The number of juveniles penetrating roots decreased as both the spore concentration and the distance moved in soil increased, and the infectivity of juveniles with more than 40 spores attached was reduced. This effect was confirmed by Davies *et al.* (1988),

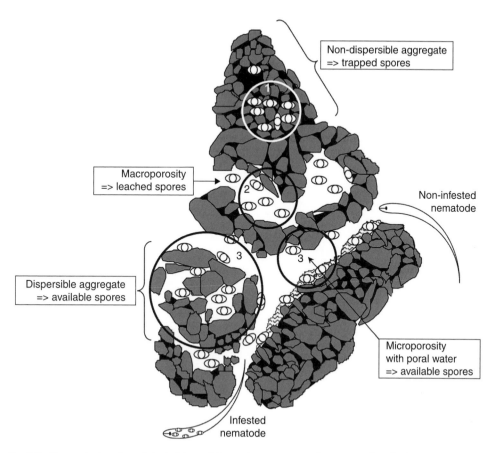

Non-dispersible aggregate => trapped spores

Macroporosity => leached spores

Non-infested nematode

Dispersible aggregate => available spores

Microporosity with poral water => available spores

Infested nematode

Fig. 7.7. Conceptual outline of the habitats of *Pasteuria penetrans* endospores and *Meloidogyne* juveniles showing the impact of various soil physical and chemical parameters on endospore attachment. (From Mateille *et al.*, 2010, with permission.)

who showed that invasion of tomato roots was reduced by up to 86% when nematodes were encumbered by more than about 15 spores. It was not clear why a large spore burden affected the infectivity of the nematode, but it was presumed that it impaired the capacity of the nematode to move to the root. The fact that adhering spores had a greater effect on invasion at high than low nematode inoculum densities in the latter study also suggested that spore-encumbered nematodes were less able to compete for feeding sites within the root.

More recent work in which nematode movement was tracked using an inverted microscope and a digital camera has clearly shown that the presence of endospores on the body disrupts the normal sinusoidal movement of the host nematode. Second-stage juveniles

of *M. javanica* did not show directional movement when 20–30 spores were attached, and directional movement was limited when nematodes were encumbered with 5–8 spores (Vagelas *et al.*, 2011, 2013). Thus, nematodes with no spores attached moved significantly longer distances than those encumbered with spores (Fig. 7.8).

Experiments in which second-stage juveniles of *M. javanica* were exposed to different concentrations of spores for various times provided an indication of the number of spores required in soil to achieve certain levels of spore encumbrance on nematodes (Stirling *et al.*, 1990). The number of spores adhering to juveniles increased linearly with spore concentration and time, and the rate of spore attachment at 27°C was approximately double the rate at 18°C (Fig. 7.9). Importantly, at a

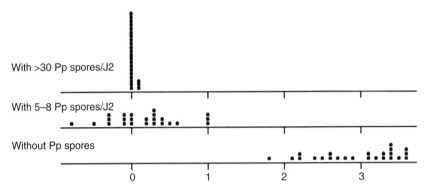

Fig. 7.8. Dotplots showing the travel distance of second-stage juveniles of *Meloidogyne javanica* without spores of *Pasteuria penetrans* (Pp), or with 5–8 or >30 adhering spores. The observation time was 60 seconds, while 1 on the x axis = the body length of a second-stage juvenile. (From Vagelas *et al.*, 2011, reprinted by permission of the publisher, Taylor & Francis Ltd.)

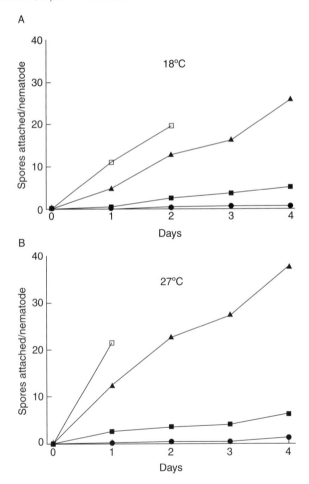

Fig. 7.9. Attachment of *Pasteuria penetrans* spores to *Meloidogyne javanica* juveniles after various times at 18°C or 27°C in soil containing one of four spore concentrations (10^3 •; 10^4 ■; $5×10^4$ ▲; or 10^5 □ spores/g soil). (From Stirling *et al.*, 1990, with permission.)

concentration of 10^5 spores/g soil, it took only 1 or 2 days (depending on temperature) for spores to attach in numbers that were likely to affect motility. A follow-up experiment, in which galling on tomato roots was measured after juveniles moved different distances through soil containing a range of spore concentrations, confirmed this result, as virtually no galls were formed when juveniles moved through soil containing 10^5 spores/g soil (Table 7.1). However, some of the nematodes that moved only 2 cm through that soil before entering the roots were not parasitized, indicating that even higher spore concentrations would be required to almost totally prevent reproduction.

Results of experiments with other populations of *P. penetrans* and other *Meloidogyne* species also suggested that spore concentrations greater than 10^5 spores/g soil may be needed to achieve a high level of control of root-knot nematode. For example, Chen *et al.* (1996d) introduced *P. penetrans* into field microplots infested with *M. arenaria* at application rates of 0, 10^3, 3×10^3, 10^4 and 10^5

Table 7.1. Number of galls caused by *Meloidogyne javanica*, and percentage of female nematodes infected with *Pasteuria penetrans*, after second-stage juveniles moved 2, 4 or 8 cm through soil infested with 0, 10^4, 10^5 or 10^6 spores/g soil.

Spore concentration (spores/g soil)	Distance (cm)	Galls/root system[a]		% infected females
0	2	315	(5.8)	0
	4	58	(4.1)	0
	8	127	(4.9)	0
10^4	2	164	(5.1)	15
	4	27	(3.3)	36
	8	14	(2.7)	75
10^5	2	50	(3.9)	76
	4	0.3	(0.3)	100
	8	0.2	(0.2)	100
10^6	2	0	(0.0)	–
	4	0.2	(0.2)	100
	8	0	(0.0)	–
LSD ($P = 0.05$)			(1.3)	

[a]Equivalent means, with transformed means [$\log_e (x+1)$] in parentheses.
(From Stirling *et al.*, 1990, with permission.)

endospores/g soil, and data collected from the first peanut crop showed that high numbers of attached spores and significant reductions in galling were only obtained with the highest inoculum level (Fig 7.10).

In this experiment, none of the treatments reduced final nematode population densities or significantly improved plant dry weights in the first year (Fig. 7.10B, G). However, improved control was observed in the second and third years, even though additional endospores were not added. In year 2, the level of galling on roots and pods was reduced and pod yield increased at the two highest inoculum densities. Data presented in Chen and Dickson (1998) indicated that in the third year, root and pod galls were absent from the 10^5 spores/g soil treatment, and that galling was reduced in all other treatments. Also, pod yields were 1.8–2.7 times greater in inoculated than control plots, and populations of second-stage juveniles at harvest were significantly reduced by both 10^4 and 10^5 endospores/g soil. These results indicate that amplification of *P. penetrans* in the microplots played a significant role in the increased suppression observed over a 3-year period. Similar cumulative increases in populations of *P. penetrans* have been observed in pots (Melki *et al.*, 1998), emphasizing the importance of continuing experiments for long enough to allow the potential of *P. penetrans* as a biological control agent to be fully expressed.

The fact that fewer *Meloidogyne* juveniles enter roots as they become encumbered with increasing numbers of spores is presumably a mechanism that has evolved to prevent *P. penetrans* from eliminating its host. Available evidence suggests that as spore populations increase, the motility-limiting effects of spore encumbrance will result in fewer infected females in roots and lower spore production. Since this will ultimately reduce the chances of later generations of nematodes being infected by the parasite, the impact of spore attachment on motility plays an important role in regulating populations of both the parasite and its nematode host.

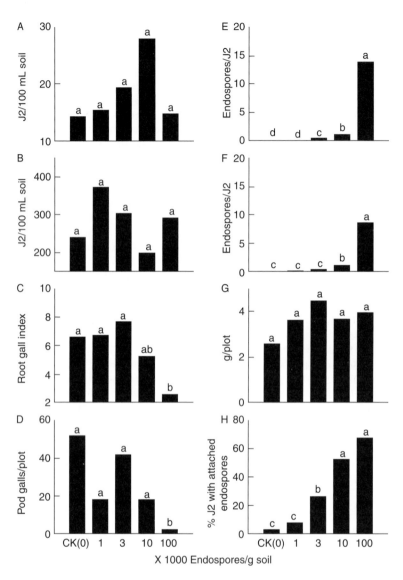

Fig. 7.10. Effect of *Pasteuria penetrans* inoculum levels on numbers of second-stage juveniles (J2) of *Meloidogyne arenaria*, gall ratings and peanut yield in the first year of a microplot experiment. Figures A and E contain data obtained at planting; all other data were collected at harvest. Mean values with different letters are significantly different (*P* > 0.05). (From Chen *et al.*, 1996d, with permission.)

Miscellaneous factors influencing the production and survival of endospores in soil

The endospores of *P. penetrans* have a remarkable capacity to survive adverse conditions, and have proved to be resilient to a wide range of physical and chemical treatments.

For example, soil fumigants such as 1, 3 dichloropropene (1,3-D), dibromochloropropene (DBCP) and ethylene dibromide (EDB) had little or no impact on endospores (Mankau and Prasad, 1972; Stirling, 1984), while non-volatile nematicides such as aldicarb, ethoprophos, fenamiphos and oxamyl had no more than a temporary effect (Mankau and Prasad, 1972; Brown and Nordmeyer, 1985; Walker and Wachtel, 1988). Also, some endospores remained viable after being heated to 100°C (Williams *et al.*, 1989), while a mild heat treatment (solarizing the soil with polyethylene sheeting for 15 weeks) had a positive rather than negative impact, increasing the proportion of *Meloidogyne* juveniles infested with *P. penetrans* spores (Walker and Wachtel, 1988). Such findings obviously have implications for how the parasite might be used in integrated nematode management programmes. If nematicides are used injudiciously they will eliminate the host nematodes on which spore production depends. However, low concentrations of most nematicides are likely to have little impact on endospore survival, and so it may be possible to use them for early-season nematode control without affecting endospore survival in soil, or endospore production in nematodes that enter roots after the chemical has dissipated. Integrated control strategies of this nature are discussed in Chapter 12.

Although all organisms are subject to predation in soil, no natural enemies of *P. penetrans* have been reported. Nevertheless, given the enormous biodiversity that exists in soil, biotic factors will almost certainly influence endospore production and survival. The most likely antagonists of *Pasteuria* are predators capable of severing the cuticle of infected root-knot nematode females before spore production is complete, and organisms that are able to consume endospores. *Paramecium* was seen to engulf endospores in a counting dish (Chen and Dickson, 2004b), and since all groups of protozoa are bacterivorous, it would be surprising if they did not consume *P. penetrans* endospores in soil. Talavera *et al.* (2002) observed amoebae with vacuoles containing structures similar to *Pasteuria* spores, and also suspected that filter-feeding rotifers may have consumed spores after they were applied to soil.

The potential of *P. penetrans* as a biological control agent

P. penetrans is the most extensively studied biological control agent of plant-parasitic nematodes for two reasons: (i) it is an effective and highly specialized parasite of root-knot nematode, the world's most damaging nematode pest; and (ii) its endospores are resilient enough to survive reasonably well in the hostile soil environment. Unfortunately, however, the *P. penetrans* research that has been undertaken over the last 40 years has to some extent been misdirected, as too many studies have focused on inundative biocontrol with exotic isolates (i.e. introducing endospores into soil in numbers sufficient to have an immediate effect on the host nematode). This has meant that issues associated with reducing nematode populations with indigenous isolates, or with inoculative approaches (i.e. introducing endospores in relatively low numbers and relying on natural increase in the parasite to suppress the pest), have been neglected.

Inundative biocontrol with *P. penetrans* clearly has promise, particularly now that commercial mass production methods are available (see Chapter 12). However, it is important to recognize that even if fermentation and formulation technologies are improved markedly, high production costs will limit the use of this approach to high-value crops. This means that if *P. penetrans* is to be utilized as a biological control agent in most of the world's root-knot susceptible crops, populations of indigenous isolates, or exotic isolates that are introduced at a relatively low but affordable application rates, will have to be increased through management. Thus, future research must focus on understanding host–parasite interactions in the soil environment, and using that knowledge to manipulate populations of the parasite. Soils that are suppressive to root-knot nematode due to the actions of *P. penetrans* are achievable (see Chapter 10), but they are most likely to be obtained by: (i) utilizing the molecular tools available to quantify populations of both host and parasite; (ii) manipulating populations of the two organisms with appropriate crop cultivars and rotational crops; (iii) optimizing irrigation and tillage practices so that the *P. penetrans* endospores remain in upper soil layers and are not leached down the profile; and (iv) taking into account the many environmental factors that influence interactions between the parasite and its host. Research on host specificity and *in vitro* culture should continue, but must be balanced with applied studies on how *P. penetrans* can be maintained and then utilized in sustainable farming systems.

Pasteuria as a Parasite of Cyst Nematodes (*Heterodera* and *Globodera* spp.)

Since cyst nematodes rival root-knot nematodes in their economic importance worldwide, it is not surprising that isolates of *Pasteuria* capable of parasitizing *Heterodera* and *Globodera* are of interest to nematologists. However, as was the case with *P. penetrans*, taxonomic problems, host specificity issues and the difficulties associated with quantifying populations in soil have hampered research on this group of parasites.

Taxonomy, phylogeny and host specificity

The results of a study in a Japanese field on the population dynamics of the upland rice cyst nematode, *Heterodera elachista*, indicated that populations of the nematode declined markedly after 5 years of continuous rice culture, and that this decline may have been caused by *Pasteuria* (Nishizawa, 1987). Spore-attachment tests with this isolate of *Pasteuria* showed that it differed from other members of the genus, as endospores attached to *Heterodera glycines*, *H. trifolii*, *Globodera rostochiensis* and several unidentified populations of *Heterodera*, but did not attach to root-knot nematodes or other plant-parasitic nematodes (Sayre *et al.*, 1991b). Since there were also differences in spore morphology and in some developmental characteristics within infected females (Sayre *et al.*, 1991a), the parasite was described as a new species, *Pasteuria nishizawae* (Sayre *et al.*, 1991b). In the taxonomic formalities associated with that description, the authors specifically limited the new epithet to members of the genus *Pasteuria* that were primarily parasitic in adult female nematodes belonging to the genera *Heterodera* and *Globodera*.

In 1994, *Pasteuria* was found parasitizing *Heterodera glycines* in Illinois (Noel and Stanger,

1994). This North American population was initially thought to differ from *P. nishizawae*, but later observations showed that both organisms had similar life cycles (Atibalentja *et al.*, 2004). However, light and transmission electron microscopy revealed subtle differences between the two parasites with regard to the size of the central body of the mature endospore, the appearance and apparent function of mesosomes during sporogenesis, and the surface features of endospores. Nevertheless, the two organisms were considered to be closely related, and this was confirmed when the nucleotide sequences of the 16S rRNA genes of the North American and Japanese strains were found to be almost identical. The original description of *P. nishizawae* was then modified to include new knowledge on the organism's life cycle, host specificity and endospore morphology (Noel *et al.*, 2005). However, that emended description narrowed the definition of *P. nishizawae* by restricting it to populations that completed their life cycle in the pseudocoelom of *Heterodera glycines*. Nevertheless, it was recognized that other hosts may eventually be confirmed, as endospores were known to attach to *Globodera rostochiensis*, *Heterodera elachista*, *H. lespedezae*, *H. schachtii* and *H. trifolii*.

Another population of *Pasteuria* that has been studied to some extent in cyst nematodes was collected from females of the pigeon pea cyst nematode, *Heterodera cajani*, in India. Endospores of this population attached to juveniles of *Globodera rostochiensis*, *G. pallida*, *Heterodera glycines*, *H. trifolii*, *H. schachtii* and *Rotylenchulus reniformis* (Sharma and Davies, 1996), but the capacity of these nematodes to host the parasite was not determined. However, a later study showed that this strain of *Pasteuria* completed its life cycle in *Globodera pallida*, with infected females remaining a creamy-white colour and failing to produce eggs (Mohan *et al.*, 2012). The 16S rRNA gene fragment of the *Pasteuria* from *H. cajani* was closely related to *P. nishizawae* (98.6% similarity), but until other genes are utilized in phylogenetic analyses, the identity of this parasite will remain uncertain.

Although most of the *Pasteuria* populations associated with cyst nematodes are similar to *P. nishizawae* in that they develop and sporulate in adult females, there are populations with a different life history. The first of these was found by Davies *et al.* (1990) in an English soil that was suppressive to the cereal cyst nematode, *Heterodera avenae*. Observations of nematodes collected from the field over a period of about 5 months showed that 38–56% of the second-stage juveniles had spores attached to their cuticles, but unlike *P. nishizawae*, infection and development was not observed in females or cysts. Instead, the life cycle was completed in second-stage juveniles, which were eventually filled with about 800 *Pasteuria* spores. In Germany, an isolate of *Pasteuria* parasitizing the pea cyst nematode, *Heterodera goettingiana*, was also found to have a similar life cycle, as parasitic stages and mature endospores were only found in second-stage juveniles (Sturhan *et al.*, 1994). Ultrastructural studies of this isolate showed that it was similar to *P. nishizawae*, but the fact that it developed exclusively in juveniles and also differed in endospore morphology and host range suggested that it was a distinct species.

Although there have been relatively few studies of the *Pasteuria* strains found associated with *Heterodera* and *Globodera*, it appears that the morphological and pathogenic diversity that exists within populations parasitizing root-knot nematodes is also present in populations associated with cyst nematodes. This diversity will continue to cause confusion until a number of taxonomic issues are resolved. Phylogenetic relationships between populations that are similar to *P. nishizawae* must be clarified, while populations that complete their life cycles within second-stage juveniles must be studied more intensively at a molecular level. One of the latter populations was included in a recent study by Mohan *et al.* (2012), with 16S rRNA sequence data being derived from a *Pasteuria* strain parasitizing pea cyst nematode in Italy (K.G. Davies, pers. comm.). However, the *Pasteuria* strain in the phylogenetic tree was referred to as *Pasteuria goettingianae*, despite the fact that a species with that name has never been validly published. Nevertheless, the observation that this strain was not in the same group as *P. nishizawae* and *P. penetrans* may be important from a taxonomic perspective, and should be confirmed.

Ecology and biological control potential

Since most of the resources available for research on *Pasteuria* parasites of cyst nematodes have been devoted to tackling taxonomic and host specificity issues, little is known about the ecology and biological control potential of these parasites. Nevertheless, two studies in the United States (Atibalentja *et al.*, 1998; Noel *et al.*, 2010) provide a clear indication that *P. nishizawae* can reduce populations of *Heterodera glycines* to levels that result in a yield response in the soybean crop. Both these studies are discussed in Chapter 10, but the results of the latter study are particularly important, because they show that *P. nishizawae* can induce suppressiveness to the host nematode when the parasite is transferred to a site where it does not occur. As *in vitro* production technologies are further developed (see Chapter 12), there may be opportunities to use a similar approach in commercial agriculture. One option would be to introduce *P. nishizawae* to a field on seed and then increase its population density to suppressive levels by managing populations of the host nematode.

In countries where soybean cyst nematode has been introduced and become a widespread problem, *P. nishizawae* is present on some farms (Noel and Stanger, 1994; Donald and Hewlett, 2009), and so there are opportunities to manage existing populations of the parasite. Such an approach is particularly relevant in Asian countries where soybean was first domesticated and *H. glycines* is widespread. For example, surveys have shown that *Pasteuria* is relatively common in Chinese soybean fields, and since a high proportion of the juveniles in some fields are heavily encumbered with spores (Ma *et al.*, 2005), some soils may already be reasonably suppressive to the nematode. Thus, it would be worthwhile assessing levels of suppressiveness in these fields, and determining whether suppressiveness is related to management practices.

One interesting observation from the above survey was that cysts containing *Pasteuria* spores were not found 2 months after spore-encumbered juveniles were inoculated onto soybean plants, raising the possibility that the isolates tested may have completed their life cycle within juveniles. Since this reproductive strategy has been observed with isolates parasitic in *Heterodera avenae* and *H. goettingiana*, the potential of such isolates as suppressive agents must be determined. Although their reproductive potential is likely to be much lower than isolates capable of multiplying in females, their capacity to reduce the number of juveniles invading roots is a valuable asset from a biological control perspective.

Candidatus Pasteuria usgae Parasitic on Sting Nematode (*Belonolaimus longicaudatus*)

Sting nematode (*Belonolaimus longicaudatus*) is widely distributed on the coastal plains of the south-eastern region of the United States, where it causes yield losses on crops such as carrot, maize, cotton, peanut, potato, soybean and strawberry. However, *B. longicaudatus* is best known for its destructive effects on turfgrasses (Crow and Brammer, 2008), and it was during a survey of golf course fairways at four locations in Florida that *Pasteuria* was first detected on this nematode (Giblin-Davis *et al.*, 1990). Initial studies suggested that this isolate of *Pasteuria*, together with one parasitizing *Hoplolaimus galeatus*, may have been new species, as their spores were relatively large and differed in size (Fig. 7.1), the two isolates were ultrastructurally different, and both isolates differed from known *Pasteuria* species. Later work showed that endospores from *B. longicaudatus* did not attach to *H. galeatus*, confirming that the two isolates were different, while ultrastructural observations continued to suggest that the isolate from *B. longicaudatus* was a new species (Giblin-Davis *et al.*, 1995).

Taxonomy and host specificity

The *Pasteuria* found on *B. longicaudatus* was variously designated as the 'large-spored' or 'S-1' isolate, and laboratory tests indicated that it was relatively host specific, as its spores

only attached to *B. longicaudatus* and *B. euthychilus* (Giblin-Davis *et al.*, 2003). Attachment was not observed on *Heterodera schachtii*, *Longidorus africanus*, *Xiphinema* sp. and several species of *Meloidogyne* and *Pratylenchus* (Giblin-Davis *et al.*, 1995, 2001). The fact that endospores did not adhere to the many other plant-parasitic nematodes found in association with *B. longicaudatus* on turfgrass in Florida also suggested that this isolate had a limited host range.

Endospores of the S-1 isolate of *Pasteuria* generally attached to the head or tail regions of *B. longicaudatus* (Fig 7.11), and ultrastructural studies showed that penetration, development and sporogenesis then proceeded in much the same manner as in other species of *Pasteuria*. However, S-1 endospores were distinctive with regard to the shape of the central body, the relatively thick outer spore coat and the conspicuous lateral thickening of the coat due to densely packed fibrous

Fig. 7.11. The spore-encumbered head region of sting nematode, *Benonolaimus longicaudatus*, with feeding stylet protracted. Arrows indicate spores that appear to have retained what looks like the exosporium. (From Giblin-Davis *et al.*, 2001, with permission.)

micro-projections (Fig 7.12). The sporangium and central bodies were also wider than other *Pasteuria* species, having mean diameters of 6.05 μm and 3.08 μm, respectively (Giblin-Davis *et al.*, 2001). Phylogenetic analyses of 16S rDNA sequences confirmed that the isolate from *B. longicaudatus* was related to strains of *Pasteuria* from nematodes and cladocerans, but also showed that it was unique (Bekal *et al.*, 2001). Therefore, this isolate was described as a new species and given the name *Candidatus* Pasteuria usgae (Giblin-Davis *et al.*, 2003).

Ecology and biological control potential

Initial studies of *Candidatus* Pasteuria usgae at a turfgrass site in Florida indicated that 40–80% of the *B. longicaudatus* at the site were encumbered with spores, and that the proportion of spore-infested nematodes remained reasonably constant over a period of 16 months.

However, poor bermudagrass performance and consistently high densities of the *B. longicaudatus* indicated that the parasite did not suppress nematode populations to acceptable levels (Giblin-Davis *et al.*, 1990). Also, populations of *B. longicaudatus* were not reduced 6 months after *Pasteuria*-infested soil was added to pots, even though 73% of the nematodes were encumbered with spores. The reasons why the parasite failed to suppress sting nematode populations were not known, but the authors noted that cadavers containing spores were slow to disintegrate and suggested that slow spore release may have impeded the cycling of spores to the next generation of nematodes in both the field and greenhouse experiment.

A later experiment in which the nematode and parasite were monitored in the field over a longer time period suggested that the parasite was having an impact on the nematode (Giblin-Davis *et al.*, 1998; Giblin-Davis, 2000). Soil infested with *B. longicaudatus* and

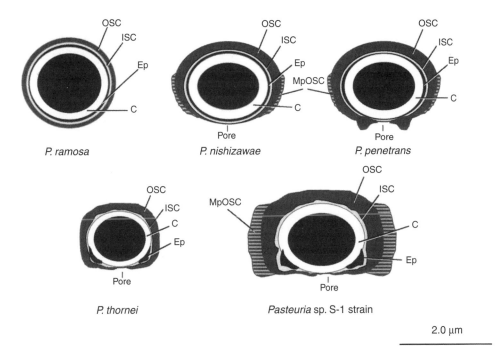

Fig. 7.12. Diagrammatic drawings depicting diagnostic characters of mature endospores of *Pasteuria* sp. S-1 strain (which was later named *Candidatus* Pasteuria usgae) and various *Pasteuria* species. (C = cortex; Ep = epicortical wall remnant; ISC = inner spore coat; MpOSC = micro-projections of the outer spore coat; OSC = outer spore coat). (From Giblin-Davis *et al.*, 2001, with permission.)

containing about 5000 *Pasteuria* endospores/ mL soil was autoclaved to kill both the nematode and *Pasteuria*, or heated at 46°C for 2 days to kill the nematode but retain *Pasteuria*; and then autoclaved or heated soil was added to a core in the centre of ten replicate plots on a bermudagrass green. Although initial sampling revealed that a low background level of *Candidatus* Pasteuria usgae was present at the trial site, the experiment was continued, and populations of *B. longicaudatus* and levels of spore encumbrance were monitored over the next 18 months by collecting samples at the inoculation point, and at 25 and 50 cm from that point. The results (Table 7.2) indicated that the percentage of *B. longicaudatus* encumbered with spores was significantly higher in the heated treatment than in the autoclaved treatment at all sampling positions at 6 and 13 months. Although populations of *B. longicaudatus* declined in plots treated with autoclaved soil due to contamination from *Pasteuria*, nematode populations in the heated treatment were significantly lower than the autoclaved treatment at the last two sampling times (Table 7.2).

In another study on a bermudagrass fairway where *Candidatus* Pasteuria usgae occurred naturally at different levels, regression analysis of data collected from monthly samplings at six locations over a period of 24 months showed that the parasite suppressed *B. longicaudatus* at all locations. Plots of the percentage of sting nematodes encumbered with *Pasteuria* endospores and sting nematode counts versus time showed that at survey locations with low initial levels of spore encumbrance, the proportion of spore-encumbered nematodes increased slowly over time and was accompanied by a corresponding decline in the number of sting nematodes. At the other locations, cyclic changes in spore encumbrance were observed that appeared to prevent the sting nematode population from resurging (Giblin-Davis, 2000).

Although these observations suggest that populations of *B. longicaudatus* are regulated in a density-dependent manner by *Candidatus* Pasteuria usgae, the factors that influence the interaction between the parasite and its host, and its capacity to suppress populations of sting nematode are poorly understood. Thus, future research must address the following issues:

- *Spore concentration in soil*. Unlike the *Pasteuria* species that parasitize females of root-knot and cyst nematode, *Candidatus* Pasteuria usgae produces relatively few endospores in infected nematodes. The parasite will develop to maturity in adults and third- and fourth-stage juveniles of *B. longicaudatus*, but spore-filled cadavers of juveniles, males and females contained an average of only 4700, 1483 and 3633 spores, respectively (Giblin-Davis *et al.*, 2001). Thus, even in a situation where the sting nematode population was relatively high (e.g. 500 *B. longicaudatus*/100 mL soil) and 4000 spores were produced per nematode, the

Table 7.2. Percentage of *Belonolaimus longicaudatus* encumbered with spores of *Candidatus* Pasteuria usgae (top two rows), and numbers of *B. longicaudatus*/100 mL soil (bottom two rows) 6, 13 and 18 months after autoclaved soil containing no *Pasteuria*, or heated soil infested with *Pasteuria*, was added to plots on a bermudagrass green. At each sampling time, cores were collected 0, 25 and 50 cm from where the spore-infested and non-infested soil had been placed.

Treatment	START	6 months			13 months			18 months		
		0 cm	25 cm	50 cm	0 cm	25 cm	50 cm	0 cm	25 cm	50 cm
Autoclave	0	8b	12b	10b	26b	51b	33b	72	61	70b
Heat treat	0.5	91a	75a	35a	85a	84a	77a	67	100	100a
Autoclave	296	100	183	139	143a	123a	109a	14a	9a	8a
Heat treat	294	88	103	129	7b	33b	87b	0.3b	0.4b	0.3b

For each parameter measured, means in columns with different letters are significantly different with Student's t-test at $P < 0.05$. (Adapted from Giblin-Davis, 2000, with permission from © 2000 American Chemical Society.)

Pasteuria population is unlikely to be more than about 20,000 spores/mL soil. Actual populations have rarely been reported, but soil from one bermudagrass fairway in Florida contained about 5000 endospores/mL soil (Giblin-Davis, 2000). Clearly, these spore concentrations are much lower than are achievable in soils infested with *P. penetrans*, raising questions about the spore loads present on golf courses and in natural environments, and the capacity of those populations to suppress sting nematode populations.

- *Release of endospores from cadavers.* Endospores of *Candidatus* Pasteuria usgae do not become available to infect the next generation of nematodes until they are released from the cadaver of a parasitized nematode. Since this only takes place slowly (Giblin-Davis *et al.*, 1990), factors that influence the disintegration of cadavers will affect the population dynamics of both the nematode host and the parasite. In highly managed environments such as golf course turf, we need to know whether fungicides and other commonly used pesticides influence the rate of release of endospores by affecting the soil microflora responsible for degrading the nematode cuticle. If such an effect is observed, the impact of cuticle-degrading microorganisms on the efficacy of *Pasteuria* must be determined.
- *Downward movement of endospores.* Research discussed earlier in this chapter has clearly shown that the retention and downward movement of *P. penetrans* endospores is influenced by many soil physical and chemical factors. However, leaching processes are particularly important in the ecology of *Candidatus* Pasteuria usgae, as the host nematode only increases to high population densities on sand-based putting greens or in fairways and recreational areas on coarse-textured soils of high porosity. All such sites are irrigated, and are also subject to the intense rainfall events that often occur in a subtropical environment. Since endospores applied to the surface of a simulated golf course putting green were leached down the profile to at least a depth of 30 cm

when 152 mm of irrigation water was applied, and downward movement was not hindered by the presence of a heavy thatch layer (Luc *et al.*, 2011), rainfall and irrigation practice will obviously play a major role in determining whether *Candidatus* Pasteuria usgae is an effective biological control agent.

- *Equilibrium population density and damage threshold.* Although there is evidence that *Candidatus* Pasteuria usgae can regulate populations of *B. longicaudatus*, the level of suppression is not always sufficient to reduce the nematode population density below the threshold for damage on turf grasses. Sting nematode is a serious pathogen of golf course turf, and there is a moderate to high risk of damage at population densities as low as 10–25 nematodes/100 mL soil (Crow, 2013). However, in one of the studies discussed earlier, equilibrium population densities in the presence of *Pasteuria* were about 120 nematodes/100 mL soil (Giblin-Davis *et al.*, 1990). Populations declined to acceptable levels in a second study (see Table 7.2), but as the decline occurred in both the heated and autoclaved treatment, it is not clear whether it was entirely due to *Pasteuria*. Therefore, long-term studies must be carried out to determine the range of equilibrium nematode population densities in *Pasteuria*-infested fields.
- *Host specificity.* Since *Belonolaimus longicaudatus* is endemic to the south-eastern region of the United States, the host-specific populations of *Candidatus* Pasteuria usgae obtained from that nematode are likely to be useful against sting nematodes from that region. However, it is not known whether they will also affect other belonolaimids. For example, the southern sting nematode *Ibipora lolii* is the most important pest of turfgrass in some areas of Australia (Siviour and McLeod, 1979; Stirling *et al.*, 2013), and it is not clear whether *Candidatus* Pasteuria usgae will be effective against this species. Also, specimens of *I. lolii* encumbered with *Pasteuria* spores have been observed in Australia, raising questions about the relatedness of this isolate to *Candidatus* Pasteuria usgae (Stirling *et al.*, 2013).

Commercial products created by *in vitro* culture

Now that *Candidatus* Pasteuria usgae has been cultured *in vitro* and a product containing the bacterium has been registered for use on turf in the United States (see Chapter 12), research attention appears to have shifted towards using *Pasteuria* as an inundative biocontrol agent. However, the failure of the first commercial formulations to control *B. longicaudatus* in an extensive series of field experiments in which *Pasteuria* was applied at rates as high as 3×10^5 endospores/mL soil (Crow *et al.*, 2011) raises serious concerns about the potential of that approach. The reasons for the lack of efficacy of Econem™ (Pasteuria Bioscience, Alachua, Florida) in the above experiments were not known, but one possibility is that endospores were moved below the root zone by irrigation water and rain, underlining the importance of thoroughly evaluating the efficacy of commercial bionematicides in the field. Since very high numbers of endospores are applied when Econem™ is used at the recommended rate (Crow *et al.*, 2011), the spore concentrations achieved in soil are many times higher than are ever likely to occur naturally. Therefore, the possibility that the presence of these spores stimulates unusually high levels of predation warrants investigation.

Although the factors discussed above may have contributed to the poor result obtained by Crow *et al.* (2011), data from experiments in a highly controlled growth-room environment suggest that the poor quality of commercially produced formulations of *Candidatus* Pasteuria usgae is the most likely reason for the failure to control sting nematode. When *in vitro*-produced bacteria were applied at rates up to 2.8×10^5 endospores/mL in two experiments, and at 1.38×10^6 endospores/mL in another two experiments, no more than about 20% of the sting nematodes were ever encumbered with spores (Luc *et al.*, 2010a, b). In contrast, naturally occurring populations of *Pasteuria* consistently attached to 40–80% of the nematodes in a field situation where the endospore population was likely to have been very much lower than in the above experiments (Giblin-Davis *et al.*, 1990). Endospores produced *in vitro* exhibit poor peripheral fibre development and are often much smaller than natural endospores (Luc *et al.*, 2010a), and since these fibres are involved in the attachment process (Giblin-Davis *et al.*, 2001), such endospores may not adhere readily to the host nematode. Therefore, it is important that mass production methods for *Candidatus* Pasteuria usgae are improved to the point where endospores produced *in vitro* have the full range of qualities exhibited by their naturally produced counterparts.

Pasteuria as a Parasite of Other Plant-parasitic and Free-living Nematodes

Although there are many reports of associations between nematodes and *Pasteuria*, in-depth information on these associations is limited to the three host–parasite relationships already discussed. Where *Pasteuria* has been studied on nematodes other than *Meloidogyne*, *Heterodera*, *Globodera* and *Belonolaimus*, most reports have been limited to morphological observations or to brief comments on levels of spore encumbrance, and have contained little information on the manner in which host and parasite interact in soil. The following is a brief summary of what can be gleaned from those reports from an ecological perspective.

Parasitism of root-lesion nematodes (*Pratylenchus* spp.) by *Pasteuria thornei*

In initial studies of bacteria belonging to the genus *Pasteuria* (as re-established by Sayre and Starr, 1985), it soon became apparent that there were morphological and developmental differences between the bacteria that parasitized *Meloidogyne* and *Pratylenchus* (Sayre *et al.*, 1988). Thus, the original description of *P. penetrans* was emended so that the name referred specifically to parasites of root-knot nematodes, with *Pasteuria thornei* being proposed to accommodate isolates that were parasitic in root-lesion nematodes (Starr and Sayre, 1988). Eleven *Pratylenchus* species were reported as hosts of *Pasteuria thornei*, including a number of widespread species such as

P. brachyurus, P. crenatus, P. neglectus, P. penetrans, P. thornei and *P. zeae* (Sayre *et al.*, 1988).

More recent studies have also shown that *Pasteuria* occurs on various *Pratylenchus* spp., but have added little to our knowledge of this parasite. Spore-encumbered nematodes were found during surveys in Turkey and Mexico (Elekcioglu, 1995; Franco-Navarro and Godinez-Vidal, 2008), but in the Turkish study, less than 2% of the *Pratylenchus thornei* in a wheat field were encumbered with endospores. *Pasteuria* was also found on *Pratylenchus thornei* in the northern grain-growing region of Australia, but again relatively few nematodes (<6%) had spores attached, and the parasite was detected in only 25% of fields (Li *et al.*, 2012). In contrast, Ornat *et al.* (1999) noted that 75% of the *Pratylenchus neglectus* in a commercial greenhouse in Spain were encumbered with endospores following a bean crop. Spores were also observed in the body cavities of fourth-stage juveniles and females. Interestingly, the level of spore encumbrance was higher in non-tilled than in tilled plots, reinforcing comments made earlier that *Pasteuria* spores may be more likely to attach to nematodes at undisturbed microsites where spore concentrations are high. In the case of *Pasteuria thornei*, these microsites are most likely to occur within the decomposing root system, as this is where infected nematodes would previously have been feeding.

Although the above studies suggest that *Pasteuria* does not generally infect large numbers of *Pratylenchus*, it is possible that the occurrence of the parasite is being underestimated by the methods generally used to extract vermiform nematodes from soil. Spore-encumbered nematodes, and nematodes that are infected by *Pasteuria*, are likely to be less motile than healthy nematodes, and are unlikely to be readily recovered with extraction methods that rely on motility (e.g. techniques based on Baermann funnels and trays). Centrifugation methods will overcome this problem to a certain extent, but as the specific gravity of *Pasteuria* is much higher than the specific gravity of the sucrose solutions mostly used in such methods, recovery of spore-filled nematodes is likely to be poor. This was elegantly demonstrated by Oostendorp *et al.* (1991a), who not only found that the specific

gravity of *Pasteuria* endospores was greater than 1.25, but also went on to show that greater numbers of *Pratylenchus scribneri* filled with *Pasteuria* spores were recovered when the specific gravity of the sucrose solution was increased from 1.14 to 1.26.

Another matter that has not been recognized in studies of *Pasteuria thornei* is that most of the nematodes infected by the parasite will be in roots rather than soil. Sporangia of the parasite do not overrun the anterior feeding apparatus of parasitized nematodes, and so infected nematodes will continue to feed (Sayre *et al.*, 1988). Extraction methods that recover *Pratylenchus* cadavers from roots may, therefore, provide a better indication of the levels of *Pasteuria thornei* that are actually present in soil and roots. Molecular methods of detecting other isolates of *Pasteuria* are available (Mauchline *et al.*, 2010), and could also be employed to detect and quantify *Pasteuria thornei* in *Pratylenchus*-infested root systems.

Parasitism of citrus nematode (*Tylenchulus semipenetrans*) by *Pasteuria*

Observations by Fattah *et al.* (1989) suggested that an isolate of *Pasteuria* from Iraq completed its life cycle within second-stage juveniles of *Tylenchulus semipenetrans* (Fig. 7.13). Later studies with populations from the United States and Spain confirmed that observation, and also showed that males were sometimes infected (Kaplan, 1994; Sorribas *et al.*, 2000). In a laboratory study with an isolate from Florida, Kaplan (1994) observed 320–400 endospores within the bodies of parasitized males and second-stage juveniles after 18 days at 25°C, but found that adult females were never infected. This isolate also appeared to be relatively host specific, as endospores did not adhere to the cuticles of several other nematodes tested.

The *Pasteuria* associated with *T. semipenetrans* in Florida was not widespread, as it was only found in one of 27 citrus orchards sampled (Walter and Kaplan, 1990). However, *Pasteuria* was relatively common in Spain, as it was present in the four orchards sampled in a nematode population dynamics study (Sorribas *et al.*, 2000) and occurred in 50% of

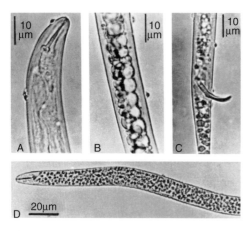

Fig. 7.13. *Pasteuria* associated with *Tylenchulus semipenetrans*, showing endospores attached to a second-stage juvenile (A, B) and male (C); and sporulation within the body of a second-stage juvenile (D). (From Fattah *et al.*, 1989, with permission.)

the 48 orchards sampled in a survey of microbial parasites associated with citrus nematode (Gené *et al.*, 2005). In one of the orchards in the former study, relatively high levels of spore encumbrance (20, 28, 34 and 47% of *T. semipenetrans* with spores attached) were observed on four of the 12 sampling occasions, raising questions about why the parasite was so common in that orchard. Since there was also a positive relationship between the number of nematodes in soil and the number with spores attached ($r^2 = 0.75$, $P = 0.00027$), this site would have been an ideal location for further studies on the role of *Pasteuria* in regulating populations of *T. semipenetrans*.

An isolate of *Pasteuria* parasitizing a reniform nematode (*Rotylenchulus reniformis*)

Rotylenchulus reniformis is one of the most widespread and damaging nematode pests of tropical and subtropical crops, but populations of *Pasteuria* capable of parasitizing this nematode appear to be relatively rare. Therefore, a report of *Pasteuria* endospores attached to the cuticle of *R. reniformis* in several cotton fields in the United States (Hewlett *et al.*, 2009) is of interest, as a follow-up study

(Schmidt *et al.*, 2010) showed that two isolates from Florida completed their life cycles in juvenile, male and female reniform nematodes. Germination occurred soon after attachment, and mature endospores were observed after 15 days at 28°C. Phylogenetic studies indicated that both isolates shared close homologies with *Candidatus* Pasteuria usgae and a *Pasteuria* isolate from *Hoplolaimus galeatus*. Populations of *R. reniformis* on cotton were reduced when extremely high numbers of *in vitro*-produced spores were applied to soil or seeds, but as efficacy in highly controlled environments is not necessarily related to field performance (see previous discussion on *Candidatus* Pasteuria usgae), it is not yet clear whether *Pasteuria* has the capacity to regulate populations of reniform nematode.

Density-dependent parasitism of *Xiphinema diversicaudatum* by *Pasteuria*

Pasteuria is known to parasitize many dorylaimid nematodes (Sayre and Starr, 1988), but interactions between *Pasteuria* and members of the Dorylaimida have mainly been studied in *Xiphinema*, largely because this nematode is a virus vector and also an important pest of many crops, especially perennials. One of those studies (Ciancio, 1995b) is particularly important, because it is one of the few comprehensive studies of the population dynamics of a plant-parasitic nematode and its *Pasteuria* hyperparasite in an undisturbed field environment.

Data obtained from monthly soil samples collected from a peach orchard in Italy showed cyclic fluctuations in the population density of *Xiphinema diversicaudatum* and also in the percentage of nematodes infected by *Pasteuria* (Fig 7.14). Statistical analyses using two different tests indicated that nematode abundance and rates of parasitism were density dependent. In a second component of the study, nematode populations and levels of parasitism were measured in spatial samples collected at one time from the same peach orchard, and these data were consistent with the cyclic nature of the *X. diversicaudatum*–*Pasteuria* interrelationship as described by the Lokta–Volterra predator–prey model of

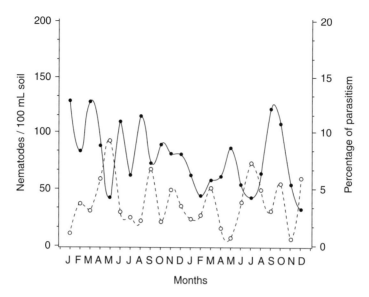

Fig. 7.14. Population densities of *Xiphinema diversicaudatum* (solid line) and the percentage of nematodes infected with *Pasteuria* (dotted line) in a 2-year study in an Italian peach orchard. (From Ciancio, 1995b, with permission.)

population dynamics. Such results indicate that although populations of *X. diversicaudatum* and *Pasteuria* will fluctuate over time, both host and parasite maintain a dynamic equilibrium in which neither of the interacting organisms is ever eliminated.

Associations between *Pasteuria* and other nematodes

This chapter has focused on species and isolates of *Pasteuria* that are parasitic in economically important nematode pests. However, *Pasteuria* is a diverse and ubiquitous organism and is known to be associated with many other plant-parasitic nematodes. In most cases, the association is little more than a record of endospores being attached to the nematode, but in other cases the interaction between *Pasteuria* and its nematode host has been studied in a little more detail. Reports of this work will not be discussed here, but examples include observations on a *Pasteuria* isolate parasitizing *Trophonema okamotoi* (Inserra *et al.*, 1992); a study on the dynamics of parasitism by *Pasteuria* on *Helicotylenchus lobus* (Ciancio *et al.*, 1992); a report on the morphology and ultrastructure

of *Pasteuria* in *Tylenchorhynchus cylindricus* (Galeano *et al.*, 2003); and the description of a *Pasteuria* isolate infecting *Mesocriconema ornatum* (Orajay, 2009).

The first report of *Pasteuria* parasitizing an entomopathogenic nematode came from India, where endospores were observed on the cuticle and in the pseudocoelom of infective juveniles of *Steinernema pakistanense* (Bajaj and Walia, 2005). Infestation levels were sometimes as high as 70%. *Pasteuria* was also found in *Pratylenchus* at the same site, but laboratory studies suggested that both isolates were host specific, as relatively few spores from *Steinernema* attached to *Pratylenchus*, and vice versa.

Pasteuria has also been found in association with many free-living nematodes, but it is only in the last decade that strains attacking these nematodes have been studied in detail. Sturhan *et al.* (2005) noted that parasitism of bacterial-feeding nematodes in the family Plectidae was uncommon, but infection was observed in several species of *Plectus* and *Anaplectus*. The sporangial and endospore morphology of these organisms and their developmental cycle in plectids was essentially the same as species and strains that parasitized tylenchid nematodes, but some morphological differences were observed. Sequence data from the 16S rRNA gene of a parasitized plectid indicated that this *Pasteuria* isolate differed from other species. Similar molecular analyses on a *Pasteuria* found in *Bursilla*, another bacterial-feeding nematode, showed that it was also a novel taxon, and in this case the name *Candidatus* Pasteuria aldrichii was proposed (Giblin-Davis *et al.*, 2011). Interestingly, a study that included the two isolates from bacterivorous nematodes suggested that they were more closely related to each other than to isolates from plant-parasitic nematodes (Schmidt *et al.*, 2010).

Concluding Remarks

It should be apparent from the material presented in this chapter that *Pasteuria* has many of the attributes required of a useful biocontrol agent: a multiplicity of strains capable of parasitizing all the world's key nematode pests; specificity towards particular nematodes; resilience to environmental stresses; and a capacity to reduce the fecundity of the host nematode. The host specificity exhibited by *Pasteuria* is perhaps its most important attribute, but could also be seen as a deficiency. On the one hand, the specificity of the host–parasite relationship means that the population density of *Pasteuria* is strongly linked to the population density of the host, and this means that the parasite can regulate populations of the host nematode in a density-dependent manner. On the other hand, the diversity that always exists within nematode populations means that some nematodes within a population will not be susceptible to a particular *Pasteuria* strain. Also, the capacity of the parasite to influence the host nematode is constantly changing because, as Davies (2009) pointed out, both host and parasite are involved in an 'arms race' where an increase in fitness of the parasite (e.g. new virulence against its host) is met by a reciprocal increase in the fitness of the host (i.e. genes for resistance against the parasite). Understanding and managing host specificity is, therefore, the greatest challenge in utilizing *Pasteuria* as a biological control agent.

To date, most research on *Pasteuria* has focused on strains that develop within females

of *Meloidogyne* and *Heterodera*, and since they are the world's most important nematode pests, this situation is unlikely to change. However, *Pasteuria* strains capable of completing their life cycles in second-stage juveniles of endoparasitic nematodes have also been reported, and their role in suppressing populations of the host nematode should be investigated, as they may be able to reduce the number of nematodes invading roots at lower spore concentrations than strains which sporulate in females. Studies with such strains should aim to understand the behaviour of juveniles after they are infected by *Pasteuria*; determine how many spores are produced in infected nematodes; establish when and where spores are released into the environment; and understand the relationship between spore concentration in soil and nematode mortality.

One area of nematode–*Pasteuria* ecology that is under-researched is the role of the parasite in regulating populations of nematodes that remain vermiform throughout their life cycle. Associations between *Pasteuria* and *Pratylenchus*, for example, have been recognized for more than 70 years, but little is known about the manner in which the host and parasite interact. Studies of this nature have been hampered in the past by an inability to find infected nematodes in soil or roots, or to quantify spore populations in soil. However, molecular detection techniques are now available that should overcome these problems, and it is vital that they are employed in ecological studies.

Another challenge facing anyone interested in using *Pasteuria* for biological control purposes is to understand the tactics required to enhance and then sustain indigenous populations of the parasite. *Pasteuria* is ubiquitous in soil, but levels of parasitism in agricultural fields are generally low, largely because of environmental factors and the way soils are managed (Mateille *et al.*, 2010). Future research must focus on understanding how populations of the host nematode can be kept at levels below the damage threshold at critical periods (e.g. at the time a nematode-susceptible crop is being planted), while maintaining populations of the parasite. This will only be achieved through a range of management tactics that may include implementing tillage practices that preserve microsites where *Pasteuria* populations are high; managing irrigation to prevent leaching losses; developing bed management practices that shed water from critical areas during major rainfall events; building up levels of soil organic matter to help retain endospores in the root zone; and managing populations of the host nematode with rotation crops and cultivars with varying levels of resistance and tolerance to the nematode. Such strategies form part of a farming systems approach to enhancing the suppressiveness of soils to plant-parasitic nematodes, and are discussed in Chapter 11.

Although *Pasteuria* has the capacity to regulate populations of its nematode host, levels of parasitism are rarely effective enough to prevent nematode damage, as equilibrium nematode population densities tend to be greater than the damage threshold for most crops. Thus, *Pasteuria* should not be expected to stand alone as a control agent: the suppressive service it provides must be integrated with the services provided by other natural enemies. However, truly integrated approaches to managing nematodes have not been possible in the past because data on the presence or population densities of various natural enemies have not been available. In such situations, management decisions were made on the basis of a nematode count provided by a diagnostic service, or by estimating the number of nematodes likely to be present, given a field's previous cropping history. Since molecular methods will form the basis of future nematode monitoring and diagnostic services, it is vital that such services also quantify natural enemies, including *Pasteuria*. Growers with access to such information can then start making management decisions that not only decrease populations of the pest nematode but also maintain or enhance populations of *Pasteuria* and other antagonists. Since some of the hosts of *Pasteuria* are particularly problematic in sandy soils (e.g. root-knot and sting nematodes), such services will also be useful for measuring the residual spore population in situations where major leaching losses are likely to have occurred (e.g. following a major rainfall event).

Although much of the recent research on *Pasteuria* has focused on improving *in vitro* culture techniques and developing methods of using the parasite as an inundative biological control agent (see Chapter 12), there is a danger that this approach to utilizing *Pasteuria* is being overemphasized. Applying endospores at the rates required for an inundative treatment to be effective (i.e. $>10^5$ endospores/g soil) is unlikely to be economic on most of the world's food and fibre crops. It may be feasible to introduce *Pasteuria* as an inoculative treatment (e.g. as a seed treatment, or by applying endospores to soil at planting using technologies associated with precision agriculture), but appropriate management practices will then have to be utilized to maintain and eventually increase endospore populations. Thus, the tactics discussed previously for managing indigenous populations of *Pasteuria* will also be needed when an exotic strain is introduced. Once the necessary tactics are implemented, long-term studies on the population dynamics of host and parasite will be required to demonstrate that benefits can be obtained from various inoculative treatments.

In conclusion, it is apparent from the literature cited in this chapter, and in Chapters 10 and 12, that the focus of *Pasteuria* research over the last two decades has been on improving methods of *in vitro* culture; developing molecular methods of quantifying *Pasteuria* in soil and roots; improving our understanding of host specificity; and exploiting genomics to understand host–parasite interactions. It is now time to use the knowledge that has been accumulated, and the molecular technologies now available, to address important ecological issues. The following is a list of possible research objectives:

- Integrate molecular quantification methods for *Pasteuria* into a commercial package that can be used to monitor and enumerate populations of the parasite and its nematode host *in situ*.
- Use the data generated from field studies with such quantification methods to model the population dynamics of the host and parasite.
- Study the distribution of *Pasteuria* in soils, particularly its occurrence at microsites associated with soil aggregates and plant roots.

- Determine how endospore distribution at a microsite level is influenced by tillage, irrigation, rainfall, clay content and the presence of organic matter.
- Enhance our understanding of the diversity of indigenous *Pasteuria* populations, and of strains used as inoculants.
- Study species-specific interactions between *Pasteuria* and its nematode host.
- Answer ecological and biological questions associated with the development of epidemics involving *Pasteuria*.

- Define the host ranges of various species and strains of *Pasteuria*.
- Develop a fingerprint of the organisms typically associated with a nematode-suppressive soil, and use it to ascertain whether *Pasteuria* plays a role in the development and maintenance of suppressiveness.
- Understand the co-evolutionary processes that shape various life-cycle strategies in different strains of *Pasteuria*.
- Identify the virulence genes that provide the genetic basis for host preference.

Section IV

Plant–Microbial Symbiont–Nematode Interactions

8

Arbuscular Mycorrhizal Fungi, Endophytic Fungi, Bacterial Endophytes and Plant Growth-promoting Rhizobacteria

Some plant-parasitic nematodes are above-ground parasites, but most obtain sustenance from roots and spend their entire lives below ground. Their root-feeding habit brings them into contact with a unique microbial environment that is quite different from the bulk soil mass. Between 25% and 40% of the energy captured by plants during photosynthesis is liberated as root exudates, and since the carbon compounds they contain are a rich source of nutrients for soil microorganisms, the root–soil interface (generally known as the rhizosphere) is by far the most microbially active zone in soil (see Chapter 2). It may seem surprising that plants would release a large proportion of their photosynthates into soil, but we now know that one of the reasons is to prime a defence system that provides protection from root-feeding herbivores. Large numbers of beneficial microorganisms utilize root exudates as a food source, and one of their functions is to interfere in some way with pests and pathogens of roots. Plants also have other ways of using soil microorganisms for their benefit. Some rhizosphere-inhabiting bacteria have plant growth-promoting properties; certain bacteria and fungi live endophytically in root cells and play a role in plant defence mechanisms; and symbiotic relationships are established with bacteria and mycorrhizal fungi to provide plants with nitrogen and to aid nutrient uptake. Thus,

nematode–plant interactions are not only influenced by antagonists that reside in soil or colonize the root surface, but also by a community of symbionts that live in roots or inhabit the rhizosphere. The most important symbionts: arbuscular mycorrhizal fungi, endophytic fungi, endophytic bacteria and plant growth-promoting bacteria, are discussed in this chapter. Although the various ways in which this group of organisms benefit plant growth are considered, the main focus is on their direct and indirect interactions with plant-parasitic nematodes, together with the management practices required to ensure that these symbionts play a role in reducing damage caused by nematode pests.

Arbuscular Mycorrhizal Fungi

In terms of ubiquity and partnerships throughout the plant kingdom, by far the most significant plant–microbe symbiosis is the relationship between plants and mycorrhizal fungi. Arbuscular mycorrhizal fungi are associated with more than 80% of plant species; ectomycorrhizae are abundant on tree roots in temperate forest ecosystems; ericoid mycorrhizae occur on heaths and shrubs; and orchid mycorrhizae form a relationship with plants in the family Orchidaceae. All of these groups have been the subject of a huge amount of research over many years, and a book by Smith and Read (2008) provides a good overview of what is known about each of them. This brief discussion focuses on arbuscular mycorrhizal fungi (hereafter referred to as AM fungi), as they form mutualistic associations with almost all crop plants and are, therefore, the most important mycorrhizal group from an agricultural perspective. A review by Willis *et al.* (2013) should be consulted for detailed information on their ecology.

Benefits to plants from a symbiotic relationship with arbuscular mycorrhizal fungi

The essence of the plant–mycorrhizal symbiosis is that plant-derived carbon is used by the fungus, and in return, the fungus acquires nutrients and water from the soil and supplies them to the plant. The key structural features of AM fungi are an extensive mycelial network that ramifies through the soil and absorbs nutrients at a distance from roots; and specialized, highly branched hyphae known as 'arbuscles' that form within root cortical tissue and facilitate the exchange of carbon, water and nutrients (Fig 8.1). Both parties benefit from this relationship: the fungus gains access to photosynthates, while the plant benefits directly through enhanced nutrient uptake, disease resistance and drought tolerance; and indirectly through improvements in soil structure.

Enhanced nutrient uptake

Phosphorus is an essential plant nutrient but is difficult to obtain from soil for a number of reasons. First, orthophosphate ($H_2PO_4^-$), the dominant form of phosphate absorbed by plants, occurs at low concentrations in the soil solution because it is either adsorbed to soil constituents or exists as poorly soluble precipitates. Second, uptake is an energy-dependent process because it takes place against a steep electrochemical gradient created by negative charges on cell membranes and the orthophosphate ion. Third, phosphate depletion zones form around roots because phosphate ions are taken up at a faster rate than they can be replaced by diffusion

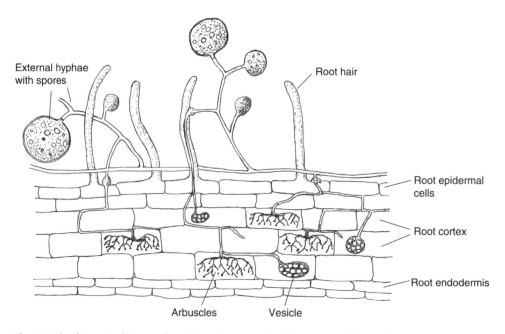

Fig. 8.1. The diagnostic features of an arbuscular mycorrhizal fungus colonizing a plant root.

from bulk soil (Bieleski, 1973; Sanyal and De Datta, 1991; Richardson, 2001). Thus, it is not surprising that plants have evolved a number of strategies to increase their capacity to obtain phosphorus from soil, and one of them is the establishment of a symbiotic relationship with AM fungi. Hyphae of these fungi act as an extension of the plant root, and studies with radioactive isotopes have shown that they can make a major contribution to plant phosphorus uptake (Smith and Smith, 2012).

Although the primary benefit to the plant from this mutualistic association is provision of phosphorus, there are also other nutritional benefits. Microcosm experiments have shown that AM fungi can capture large amounts of nitrogen from organic materials and transfer it to plants (Leigh et al., 2009), while studies cited by Willis et al. (2013) indicate that the hyphae of AM fungi are also involved in the uptake and transport of many important micronutrients, including zinc, copper, iron and selenium.

Drought tolerance

AM fungi can alleviate the unfavourable effects on plants of stresses such as heavy metals, soil compaction and salinity (Miransari, 2010), but the most commonly observed effect of AM fungi in stress situations is to improve drought tolerance. The impact of AM fungi on plant water relations was comprehensively reviewed by Augé (2001), and that review showed that the symbiosis altered rates of water movement into, through and out of host plants, with consequent effects on tissue hydration and leaf physiology. These effects were manifested in a number of ways, including: (i) higher transpiration rates and stomatal conductance to water vapour; (ii) increased photosynthetic rates; (iii) greater leaf and root hydration under drought stress; (iv) enhanced hydraulic conductivity and hyphal water transport; (v) greater soil water extraction and increased soil drying rates; and (vi) increased host growth rates and improved phosphorus acquisition during drought.

Although these effects are often subtle and transient, and vary with plant species,

fungal species and the circumstances in which the symbiosis occurs, they almost always improve plant fitness. In about 80% of studies reporting plant growth during drought, plants colonized by AM fungi were larger than non-mycorrhizal plants (Augé, 2001). Additionally, colonization by AM fungi enhances drought tolerance through its effects on soil structure (see the following section) and its flow-on effects to moisture retention (Augé, 2004).

Improved soil structure

The important role of AM fungi in improving soil structure and soil health is well documented (Andrade et al., 1998; Bethlenfalvay et al., 1999; Jeffries et al., 2003: Smith and Read, 2008). Hyphae of AM fungi grow into the soil matrix to create the skeletal structure that holds primary soil particles together through physical entanglement. These hyphae also produce copious amounts of glomalin, a recalcitrant glycoprotein that is deposited on the surface of the extraradical mycelium and adjacent soil particles. Glomalin acts as a long-term binding agent, and its concentration is strongly linked to aggregate stability (Wright and Upadhyaya, 1998, 1999; Nichols and Wright, 2004). Additionally, the carbon made available when mycorrhizal hyphae die and are degraded becomes a food source for microorganisms, and so they continue the process of developing and maintaining a water-stable and structurally sound soil.

Disease resistance

Studies with numerous plant pathogen–plant species combinations have shown that mycorrhizal symbioses can reduce the severity of symptoms caused by soilborne pathogens. Whipps (2004) reviewed these studies and found that Glomus intraradices and G. mosseae were the most commonly used AM fungi in research on plant disease control, and that Fusarium oxysporum, Rhizoctonia solani, Verticillium dahliae and Phytophthora spp. were the most common targets. AM fungi potentially influence plant pathogens through many different mechanisms (Table 8.1), but evidence that these modes of action are actually operative

Table 8.1. Potential modes of action of arbuscular mycorrhizal fungi involved in disease control.

Mode of action	Target of the action mechanism	Examples of the types of interactions involved
Direct competition or inhibition	Directly on the pathogen	Competition with the pathogen for carbon within or outside the root; for exudates external to the root; or for infection sites or space on roots
		Inhibition of pathogens through exudates from roots or from the mycorrhizal fungus (potentially including antibiotics and defence compounds)
		Competitive interactions with pathogens in soil
Enhanced or altered plant growth, nutrition and morphology	Indirectly through the plant	Alleviation of abiotic stress (e.g. increased nutrient or trace element uptake, drought tolerance, decreased toxicity to salt and heavy metals)
		Altered root branching and root morphology
		Hormonal changes
		Damage compensation
Biochemical mechanisms associated with plant defence mechanisms and induced resistance	Indirectly through the plant	Production of phenolics and phytoalexins
		Changes in amino acid levels
		Formation of internal structural barriers (e.g. lignins, callose)
		Production of defence-related proteins (e.g. β-1,3-glucanases, chitinases, peroxidases)
		Increased DNA methylation and respiration
		Induction of systemic resistance
Enhancement of antagonistic microbiota	Indirectly through the soil biota	Selective stimulation of the resident community of fungi and bacteria

Modified from Whipps (2004).

(either individually or collectively) is relatively limited. Also, disease control studies with AM fungi are generally undertaken in sterilized potting media in a laboratory or glasshouse environment, and it is not clear whether the results of such experiments are relevant to the much more complex field environment.

Interactions between plants, arbuscular mycorrhizal fungi and plant-parasitic nematodes

Since plant-parasitic nematodes and AM fungi are often found together in the same root system, nematologists have always been interested in whether they impact on each other. However, despite nearly 40 years of research and many reviews (e.g. Hussey and Roncadori, 1982; Smith, 1987; Ingham, 1988; Sikora, 1995; Pinochet *et al.*, 1996; Hol and

Cook, 2005; Hallmann and Sikora, 2011), few definitive conclusions can be drawn about the role of mycorrhizae in plant–nematode interactions. It is generally agreed that AM fungi enhance host tolerance to nematodes and slow down nematode development, but results of experiments with different fungal species and various hosts tend to be inconsistent. Hol and Cook (2005) suspected that some of this inconsistency was due to differences in the way various groups of nematodes interacted with plants, and so they used data from published studies to evaluate interactions between AM fungi and three key groups of root-feeding nematodes: sedentary endoparasites, migratory endoparasites and ectoparasites. Their main conclusions are:

• AM fungi reduced damage to shoots caused by both groups of endoparasitic nematodes by 5–14%, but had little effect on root damage. In contrast, AM fungi did not affect shoot damage caused by ecto-parasites, but significantly increased root damage caused by this group of nematodes.

- On average, AM fungi reduced nema-tode numbers by 21%, but the different nematode groups responded differently. Numbers of root-knot nematodes were reduced by an average of 33%, whereas cyst nematodes were not markedly affec-ted, and numbers of migratory endo-parasites increased by 9%.

Hol and Cook (2005) also prepared sev-eral detailed tables summarizing results of experiments in which plant biomass and pop-ulations of AM fungi and nematodes had been measured. When those tables are con-sulted, it is obvious that the effects of AM fungi are highly variable, presumably because the response is influenced by many factors, including nematode species, fungal isolate, plant genotype, nutritional status and envir-onment. This led the authors to question the value of such experimentation. They argued that measuring general effects provided no information on the mode of action of AM fungi, and suggested that it was time to start identifying and characterizing key aspects of the interaction process.

The range of potential mechanisms by which AM fungi influence plant-parasitic nematodes is somewhat similar to those out-lined in Table 8.1, but as for other soilborne pathogens, evidence that they actually oper-ate is somewhat limited. Nevertheless, there is increasing support for the hypothesis that induced resistance is involved, at least for some plant–nematode–mycorrhizal inter-actions. For example, split-root experiments by Elsen *et al.* (2008) showed that populations of *Radopholus similis* and *Pratylenchus coffeae* on banana were reduced when *Glomus intraradi-ces* and the two nematodes were spatially separated (Fig. 8.2). Similar methods were used to show that *Glomus mosseae* induced systemic resistance in tomato to both *Melo-idogyne incognita* and *Pratylenchus penetrans* (Vos *et al.*, 2012a). Regardless whether *G. mosseae* was applied locally (i.e. nematodes in the same compartment as the fungus) or systemi-cally (i.e. nematodes and the AM fungus physically separated), the number of nema-todes infecting the plant was significantly reduced for both nematodes (see Table 8.2 for results of the experiment with *M. incognita*). Further observations in a twin chamber set-up showed that juveniles of *M. incognita* accu-mulated in the soil of the mycorrhizal plant

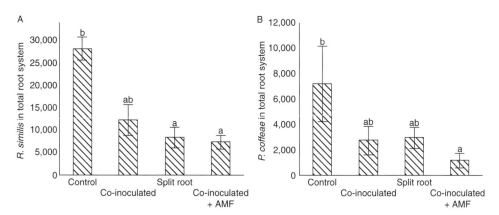

Fig. 8.2. Effect of *Glomus intraradices*, an arbuscular mycorrhizal fungus (AMF), on the total number of nematodes in the inoculated half of a split-root experiment with banana 8 weeks after inoculation with *Radopholus similis* (A) or *Pratylenchus brachyurus* (B): *control* = nematodes in right side of split root; *co-inoculated* = AMF and nematodes in right side of split root; *split root* = AMF in left side and nematodes in right side of split root; *co-inoculated + AMF* = AMF in left side and AMF and nematodes in right side. Error bars represent standard errors. In each graph, different letters indicate significant differences (*P* ≤ 0.05). (From Elsen *et al.*, 2008, with kind permission from Springer Science + Business Media B.V.)

Table 8.2. Effect of *Glomus mosseae*, an arbuscular mycorrhizal fungus (AMF) on growth of tomato, the number of nematodes associated with roots and the severity of galling in the right compartment of a split-root experiment, 8 weeks after inoculation with 200 second-stage juveniles of *Meloidogyne incognita*.

Treatment	Shoot weight (g)	Root weight (g)	Nematodes recovered from a 2 g sample of roots			Gall index (0–10 scale)
			Juveniles	Females	Eggs	
Control	47 a	11 a	2165 b	334 a	7625 b	5 b
Local AMF	52 a	8 a	626 a	133 a	3853 a	3 a
Systemic AMF	54 a	10 a	682 a	179 a	4120 a	3 a

Control = nematodes in right compartment of split root; Local AMF = AMF and nematodes in right compartment of split root; Systemic AMF = AMF in left compartment and nematodes in right compartment of split root.
Within each column, values followed by the same letter are not significantly different ($P \leq 0.05$).
(Modified from Vos *et al.*, 2012a, and used with permission.)

compartment, whereas they generally penetrated roots of the control plants (Vos *et al.*, 2012b). Nematode development was also slower in mycorrhizal roots (Vos *et al.*, 2012c). These results indicate that AM fungi are able to affect both sedentary and migratory endoparasitic nematodes, and that these effects are manifested through modifications to the host plant.

Although numbers of plant-parasitic nematodes are sometimes reduced when plants are inoculated with AM fungi, it is quite possible that no effect will be observed. In one such example, Westphal *et al.* (2008a) applied commercial mycorrhizal inoculants to watermelon transplants and did not reduce root-knot nematode infection in the glasshouse, or population densities of second-stage juveniles in a field infested with *Meloidogyne incognita*. However, one of the inoculants produced a slight but significant improvement in early growth in the field experiment, suggesting that mycorrhizal inoculation may be beneficial in situations where roots of seedlings are likely to be attacked by nematodes at a time when the plant is under transplantation stress.

In most studies aimed at minimizing nematode damage with AM fungi, soil or transplants are inoculated with commercially available inoculants that usually contain a limited number of mycorrhizal species, commonly isolates of *Glomus*. However, results of experiments by Affokpon *et al.* (2011a) in

Africa raise questions about whether that is an appropriate approach. When 20 strains of AM fungi native to West Africa and three commercial products containing different species of *Glomus* were evaluated for their protective effect against root-knot nematodes, the indigenous strains proved consistently superior to the commercial strains in both pot tests and in the field (see Chapter 12). Although it is possible that some of the local strains could eventually be cultured and made available commercially, it seems more logical to develop farming systems that nurture naturally occurring populations, as those strains are already well adapted to local environmental conditions. They will also be more diverse than any commercial product, and this appears to be particularly important, given that an analysis of more than 300 mycorrhizal studies showed that the response of plants to inoculation with AM fungi was greater with multiple species or whole-soil inoculum than with single species (Hoeksema *et al.*, 2010). Inoculation of plants with AM fungi almost certainly has a place in situations where native mycorrhizal populations have been depleted by years of soil fumigation, or when crops are established with transplants that have been grown in relatively sterile potting media. However, the possibility that additional benefits could be gained by planting into soil with a diverse range of indigenous mycorrhizal fungi deserves more attention.

Management to enhance arbuscular mycorrhizal fungi

The loss of biodiversity which occurs under agricultural management is manifested in many ways, but can be seen in soil as a decline in the number and multiplicity of a wide range of organisms, including the nematophagous fungi, predatory nematodes and soil microarthropods discussed in Chapters 5 and 6. AM fungi are no different: they are also affected by tillage practice, fertilizer use, pesticide inputs, the type of cropping system and crop rotation sequences (Plenchette *et al.*, 2005; Douds and Seidel, 2013). Since large-scale inoculation with AM fungi is impractical in most cropping systems, the only alternative is to utilize management practices that protect indigenous AM fungi and favour the development of this important, naturally occurring symbiosis. The various ways this can be achieved are discussed next.

Reduced tillage

The detrimental effects of tillage on AM fungi are well documented (Kabir, 2005; Douds and Seidel, 2013), and are mainly evidenced through effects on the extraradical hyphae. These hyphae are the main source of inoculum for AM fungi, especially when host plants are present. Providing they are not disrupted, the next crop is rapidly connected to the hyphal network, and a symbiotic relationship is quickly established. In contrast, spores are much less efficient in initiating colonization, as they take time to germinate and make contact with the roots. The extraradical mycelium is particularly important in cool climates, where the population of viable spores may be very low following winter; and also in Mediterranean climates, where the mycelia can survive the hot, dry summer. Deep ploughing (to more than 15 cm) hinders subsequent mycorrhiza formation by disrupting the extraradical mycelium developed in the previous season, and by reducing the density of propagules in the rooting zone. Other detrimental effects of tillage include a reduction in the diversity of AM fungi and a lower uptake of phosphorus and minor nutrients by plants.

A strong case for eliminating tillage from agriculture has been made several times in this book, as it is the first step that must be taken to improve soil health and enhance sustainability (Chapter 4); increase populations of various nematophagous organisms (Chapters 5 and 6); and avoid disrupting natural enemies in root-associated microsites (Chapter 7). Preservation of mycorrhizal fungi provides yet another reason to avoid tilling the soil. Although many growers will argue that some form of reduced tillage is not a practical option in their farming system, Morris *et al.* (2010) showed that most of the issues that were seen as obstacles to adopting non-inversion tillage could be overcome (see Chapter 11). Weed problems and the potential for herbicide resistance will always have to be dealt with in no-till systems, but cover crops provide one possible solution, as mulch capable of suppressing weeds and conserving moisture can be produced by rolling, undercutting or flail mowing (Douds and Seidel, 2013).

Fallow management, cropping intensity, crop sequence and cover cropping

Since AM fungi are strictly biotrophic, the survival of viable propagules (spores, colonized root fragments and hyphae) depends on the presence of host plants. Weed-free bare fallows, which are relatively common in agriculture, can, therefore, be expected to have a negative impact on mycorrhizal colonization. However, this effect does not necessarily occur in regions with hot, dry summers, as spores remain dormant during the crop-free period because germination is limited by moisture. In cold climates, where temperature limits spore germination, short fallows are mainly detrimental when uncharacteristically mild autumn or spring weather stimulates fungal activity before roots are available to colonize. Thus, bare fallowing is most likely to diminish AM fungi in warm, moist climates, as periodic rainfall stimulates growth of otherwise dormant spores and extraradical mycelium (Pattinson and McGee, 1997). In the subtropical grain-growing region of Australia, for example, where both summer and winter crops are grown, a problem known as 'long-fallow disorder' may occur when land is left fallow for 12–14 months to replenish subsoil moisture reserves after a previous crop (Thompson, 1987). Since this disorder is caused by a decline in viable propagules of mycorrhizal fungi, growers in that environment must give some consideration to maintaining indigenous AM fungi, particularly if they grow mycorrhizal-dependent crops such as cotton, sorghum, maize, sunflower, soybean, chickpea and mungbean (Thompson *et al.*, 2013). However, the yield benefits from a fallow with crop residues retained on the soil surface are so substantial (due to increases in soil water, mineralization of nitrogen and a reduction in root disease) that retention of AM fungi is only one of many issues they must consider. Nevertheless, it is important that growers eliminate practices that reduce the inoculum potential of AM fungi (e.g. tillage); utilize phosphorus and zinc fertilizers judiciously; and adopt rotations that not only reduce populations of key pathogens but also maximize the value from AM fungi (Thompson *et al.*, 2013). One option that also deserves consideration in such situations is to break the fallow with a short-term opportunity crop. This practice provides soil health and erosion control benefits (Bell *et al.*, 2006; Whish *et al.*, 2009), but is also likely to be beneficial to AM fungi. Experiments in Pennsylvania, an area that is subject to winter freezing, have shown that a similar practice (breaking the winter fallow with an oats cover crop) can improve mycorrhizal colonization, phosphorus uptake and yield of maize (Kabir and Koide, 2002).

Although most plant species host AM fungi, brassicaceous crops such as canola, cabbage, broccoli, mustard and radish, and some other crops (e.g. lupins, sugarbeet, beetroot and spinach) are exceptions, raising questions about whether they could impact negatively on a following crop by reducing mycorrhizal colonization. Ryan and Kirkegaard (2012) reviewed the situation in southern Australia, where canola is commonly grown in rotation with wheat and other cereals, and concluded that the advantages of including canola in the rotation far outweighed any benefits from AM fungi. They also pointed out that non-mycorrhizal break crops had been found to enhance yield of following wheat crops in many experiments

in North America and Europe. However, this does not mean that AM fungi should be neglected when crop-rotation sequences are planned, as these results may be due to the fact that the dominant crop (wheat) is not highly dependent on AM fungi. The response may be quite different on crops that are more mycorrhizal dependent, as shown by field trials in Japan, where sugarbeet, rape and mustard were found to depress populations of AM fungi and reduce the yield of the following maize crop (Arihara and Karasawa, 2000; Karasawa *et al.*, 2001).

To conclude, the key to increasing the population density and diversity of indigenous AM fungi is to crop intensively with host plants and to include different crop species in the rotation. AM fungi are obligate parasites of plants, and so hosts must be present to maintain their population density, while the inclusion of a range of crops enhances diversity. When crops are grown as a monoculture, AM fungi that are not necessarily effective symbionts tend to increase disproportionally, but they will be replaced by other members of the mycorrhizal community when a rotational cropping system is introduced (Douds and Seidel, 2013). In situations where temperature or water are not limiting factors, cover crops can also be used to achieve the same effect. Although such crops (commonly grasses or legumes) are usually maintained during the non-growing season to protect the soil from erosion, suppress weeds, provide nitrogen inputs, reduce nitrate leaching and enhance the organic matter status of the soil, they also increase the level of AM fungus inoculum and enhance colonization of succeeding crops (see Lehman *et al.*, 2012, and references therein).

Other crop and soil management practices

Many other agricultural practices influence AM fungi, and they must be kept in mind when attempts are made to enhance symbioses involving these fungi. Perhaps the most important is phosphorus fertilization, as numerous studies have shown that both mineral and organic phosphorus fertilizers reduce root colonization by AM fungi (Grant *et al.*, 2005). Pesticides are often deleterious, but effects will vary with active ingredient and application rate. For example, some fungicides actually preserve and stimulate AM fungi, while fungicides applied as a seed coating may be more damaging than fungicides applied to plants that are already mycorrhizal (Plenchette *et al.*, 2005).

Organic management is often seen as a way of maximizing colonization by AM fungi, and there is certainly evidence to support that view. Despite the fact that organically managed fields were routinely cultivated, they supported a greater diversity of AM fungi than conventionally managed fields (Verbruggen *et al.*, 2010), while results from several Australian studies showed that the percentage of root length colonized by AM fungi was often 2–3 times higher in organically managed annual crops than in neighbouring conventional crops (Ryan and Kirkegaard, 2012). However, it is questionable whether a move to organic farming is the best way of enhancing AM fungi. Yields of organic crops in Australia are strongly limited by phosphorus, despite high levels of colonization by AM fungi, and Ryan and Kirkegaard (2012) could find no evidence that organic farms developed an AM fungal community that was able to enhance phosphorus uptake to the point where it compensated for the reduced phosphorus inputs of organic agriculture.

Improving soil and plant health, and managing nematodes with arbuscular mycorrhizal fungi

Although results of experiments involving interactions between plants, nematodes and AM fungi tend to be highly variable, it is possible to conclude from numerous studies over many years that plant-parasitic nematodes cause less damage to plants when AM fungi are associated with roots than when they are absent. There is also evidence that AM fungi can reduce populations of some plant-parasitic nematodes, and that induced systemic resistance is one of the mechanisms involved. Such results have almost always been interpreted by nematologists as indicating that some level of nematode control can be achieved by inoculating plants with AM fungi. However, the problem with that approach is that comprehensive reviews have shown that the response of a plant to inoculation is almost impossible to predict, as the outcome is dependent on many factors including host plant species, nutrient availability, the identity and diversity of the fungi used as inoculum, and the composition of the soil biological community (Hol and Cook, 2005; Hoeksema *et al.*, 2010).

It is also obvious from the wider mycorrhizal literature that the effects of AM fungi on plant-parasitic nematodes are relatively insignificant when compared to the many other benefits of the symbiosis. I have, therefore, taken the view that AM fungi should not be seen as agents for nematode control, but rather as naturally occurring symbionts playing a key role in enhancing plant and soil health. Thus, the focus of agricultural management should be on managing crops and soil in ways that allow the many benefits of plant-mycorrhizal associations to be expressed. If that focus is maintained for a number of years, the outcome will be an agroecosystem with improved plant health, greater resilience to stress and a greater capacity to cope with the damage caused by soilborne pathogens, including plant-parasitic nematodes.

Managing indigenous mycorrhizal isolates is not straightforward, as AM fungi are negatively affected by many agricultural practices. Nevertheless, it is not an impossible task, as most of the management practices required to maintain or enhance AM fungi should form the basis of any sustainable agricultural system. Ideally, the practices involved (reduced tillage, shorter fallows, increased cropping intensity, appropriate crop sequences, cover cropping, and careful attention to phosphorus and pesticide inputs) should be adopted collectively, but there is good evidence to suggest that eliminating tillage is perhaps the most important change that should be implemented. Thus, in an experiment which examined the effects of tillage and crop sequence on AM fungi under Mediterranean conditions, colonization of wheat and triticale was affected more by tillage practice (conventional till versus no-till) than the combination of crop and previous crop (Brito *et al.*, 2012).

One advantage of taking a conservation or stewardship approach to managing AM fungi (rather than an inoculative approach) is that it provides opportunities to take advantage of mycorrhizal networks that connect plants with each other and allow resources to be distributed between them (Simard and Durall, 2004). Most AM fungi are not host specific, and can colonize a large number of plants from the same and from different plant species. Moreover, both small seedlings and large plants can be colonized by one mycorrhizal network. Provided this network is not disrupted by tillage (Jansa *et al.*, 2003), it could possibly be used to advantage in agroecosystems. Established plants are known to facilitate seedling establishment through such networks (van der Heijden and Horton, 2009), and when rice and watermelon are intercropped, colonization by AM fungi improves carbon transfer from rice to watermelon and increases phosphorus uptake by rice (Ren *et al.*, 2013).

Another advantage of enhancing indigenous AM fungi rather than relying on introduced isolates is that a more diverse mycorrhizal community can be developed. In an analysis of 306 studies in highly controlled greenhouse and growth-chamber environments, Hoeksema *et al.* (2010) found that the plant response was substantially greater when plants were inoculated with multiple AM fungi rather than a single species, suggesting that promoting a diverse range of AM fungi is advantageous. The above analysis also found that the response was greater when soil was used as inoculum rather than spores or other structures. Since soil inoculum presumably contains multiple fungal species as well as non-mycorrhizal microbes, protozoa and invertebrates, it appears that the effects of AM fungi on plants are more positive when the symbiosis occurs in a realistic biotic environment (i.e. multiple species of fungi and a diverse soil community) rather than in a highly contrived environment such as sterilized soil. Although we have a lot to learn about interactions between plants and AM fungi in biologically complex environments, this result is encouraging, because it suggests that introducing practices which promote mycorrhizal symbioses is likely to have a beneficial effect. The benefits will often be small and difficult to measure, but the extent of mycorrhizal colonization is one of many possible indicators of a well-functioning soil biological community.

Endophytic Fungi

In addition to the arbuscular mycorrhizal fungi, a wide range of other fungi can form symbiotic associations with plants. Generally known as endophytes, these fungi live in mycelial form within plant tissues without causing any obvious harm to their host (Maheshwari, 2006). They are often considered 'defensive mutualists', as their main role appears to be to defend their hosts from herbivorous insects, mammals, disease organisms and environmental stresses (White and Bacon, 2012). The most-studied fungal endophytes are a small number of ascomycete genera that are commonly associated with grasses and are often categorized as clavicipitaceous endophytes. However, there is a much greater number of non-clavicipitaceous endophytes, and although they have traditionally been treated as a single functional group, Rodriguez *et al.* (2009) provisionally reclassified them into three groups on the basis of their life history, ecological interactions and other traits. One of the non-clavicipitaceous groups includes *Fusarium* and the nematophagous fungi discussed later; a second group is mainly confined to above-ground tissues; while a third group contains dark pigmented fungi (now referred to as 'dark septate endophytes') that are commonly associated with plant roots.

Although a huge range of fungal endophytes has been found in association with plants, this discussion will focus on grass endophytes, *Fusarium* endophytes and endophytic nematophagous fungi, as they are known to have an impact on plant-parasitic nematodes. However, it is important to recognize that the symbiotic relationship between plants and fungal endophytes is a dynamic field of research, as it is being driven by the capacity of these fungi to produce an array of potentially useful bioactive metabolites and novel drugs (Suryanarayanan *et al.*, 2009); by the availability of molecular techniques that allow endophytes to be detected much more easily than in the past (Weiβ *et al.*, 2011); and by the need to find new disease-control strategies for major crops such as cereals (O'Hanlon *et al.*, 2012). Thus, new fungal endophytes are continually being found (e.g. Sánchez Márquez *et al.*, 2012), and some of them will almost certainly be active against nematodes. For example, there is now considerable interest in the order Sebacinales (Basidiomycota), a group of mycorrhiza-forming and non-mycorrhizal root colonizers that occurs worldwide (Franken, 2012). Recent evidence suggests that *Piriformospora indica*, the best-studied member of this order, has a negative impact on some plant-parasitic nematodes (Hallmann and Sikora, 2011).

Grass endophytes

The clavicipitaceous endophytes usually occur in above-ground tissues of plants, forming perennial systemic intercellular infections within their host plant and remaining endophytic throughout the life cycle of the host (Rodriguez *et al.*, 2009). Since most infected plants remain symptomless, this group of endophytes is typically seen as mutualist, protecting the host from insect herbivory, improving disease tolerance and enhancing the capacity of the host to cope with abiotic stresses such as drought (Schardl, 2010; Thom *et al.*, 2012). However, these endophytes also have negative effects, as the ergot alkaloids produced to deter insects are sometimes toxic to grazing animals. *Neotyphodium* (teleomorph *Epichloë*) is the most widespread and best-studied genus, largely because of its association with tall fescue (*Lolium arundinaceum*, syn. *Festuca arundinacea*; *Schedonorus arundinaceus*) and perennial ryegrass (*Lolium perenne*), two economically important grasses.

With regard to effects on plant-parasitic nematodes, experiments with root-knot nematodes and *Neotyphodium* endophytes have

produced conflicting results. On a positive note with respect to suppressive effects against nematodes, greenhouse tests by several researchers have shown that numbers of *Meloidogyne marylandi*, *M. graminis* and *M. incognita* on tall fescue and *M. naasi* on perennial ryegrass were reduced when *Neotyphodium* endophytes were present (Kimmons *et al.*, 1990; Stewart *et al.*, 1993; Elmi *et al.*, 2000; Jia *et al.*, 2013). In contrast, studies in the field and greenhouse by Cook *et al.* (1991) showed that invasion and development of *M. naasi* in perennial ryegrass was not affected by the presence of the endophyte. Nyczepir and Meyer (2010) obtained similar results with *M. incognita*, as nematode reproduction on tall fescue was not affected by endophyte status. Although reasons for these inconsistent responses are not known, differences between isolates of *Neotyphodium* spp. are probably involved, together with the fact that host genotype influences the growth of endophytes and their capacity to produce metabolites that may impact on nematodes.

Root-lesion nematodes (*Pratylenchus* spp.) are arguably the most economically important plant-parasitic nematodes on grasses, and the results of two studies suggest that *Neotyphodium* endophytes are highly effective against this group of nematodes. Observations in the field showed that infection of tall fescue by *N. coenophialum* reduced populations of *P. scribneri* from 224 to 2 nematodes/100 mL soil (West *et al.*, 1988), while a greenhouse study revealed that tall fescue infected by the same endophyte was essentially a nonhost of *P. scribneri* (Bacetty *et al.*, 2009). Although some ergot and loline alkaloids are toxic to *P. scribneri in vitro* (Bacetty *et al.*, 2009), experiments with genetically modified isolates of *Neotyphodium* that either lacked ergot alkaloids or had an altered alkaloid profile indicated that alkaloids did not contribute significantly to endophyte-associated suppression of *Pratylenchus* spp. (Panaccione *et al.*, 2006). Clearly, *Neotyphodium* endophytes have potential to be used as a tool for managing nematodes in grasses, but our lack of knowledge of their mode of action is hampering efforts to select useful isolates. If such isolates were identified, it should be relatively easy to incorporate them into commercial cultivars, as maternal plants pass the endophytes to offspring via seed infections.

Fusarium endophytes

Members of the *Fusarium oxysporum* complex are best known as devastating root rotting and vascular wilt pathogens of a wide range of crop plants, and as clinically important agents of human and animal mycoses. However, *Fusarium oxysporum* is primarily non-pathogenic, being ubiquitous in non-cultivated soils and commonly associated with the roots of non-symptomatic crop plants (Laurence *et al.*, 2012). The role of these non-pathogenic strains as biological control agents was first recognized in the 1970s when French scientists found they were present at high population densities in a soil that was suppressive to fusarium wilt. Reviews by Fravel *et al.* (2003) and Alabouvette *et al.* (2009) summarize what has since been learnt about interactions between plants and their fusarial associates, and also provide insights into the main mechanisms associated with *Fusarium*-mediated disease protection: competition for nutrients in soil; competition for infection sites on the root surface or within root tissue; and induction of systemic resistance.

One problem that has thwarted efforts to use *Fusarium oxysporum* as a biological control agent is the need to differentiate pathogenic from non-pathogenic strains. Although molecular tools can be used to identify known pathogens at *forma speciales* and race levels, determining pathogenicity still relies largely on bioassays. However, as more than 100 pathogenic forms of *F. oxysporum* are known, the large number of different plant species and cultivars that must be inoculated makes it almost impossible to establish that a particular strain is non-pathogenic.

Another complexity with regard to *F. oxysporum* is that the non-pathogenic nature and antagonistic potential of some strains is dependent on the presence of a multispecies consortium of rhizobacteria (predominantly γ Proteobacteria) that is associated with the mycelium. Evidence for this was provided by Minerdi *et al.* (2008) and Moretti *et al.* (2010),

who worked with strains of *F. oxysporum* that did or did not have bacteria attached to the hyphae (designated as wild-type and cured strains, respectively). When lettuce was inoculated with a known pathogen (*F. oxysporum* f. sp. *lactucae*), it produced typical disease symptoms. However, those symptoms were suppressed by the wild-type strain, whereas the cured strain did not affect disease development. In the absence of the pathogen, plants inoculated with the wild-type strain showed no symptoms, whereas the cured strain produced characteristic wilt symptoms 5 days after inoculation (Fig. 8.3). Typing experiments provided evidence that both the wild-type and cured strains were *Fusarium oxysporum* f. sp. *lactucae*. These results indicate that the antagonistic effect of the wild-type strain is not a fungal trait, but is due to interactions with ectosymbiotic bacteria. Such results have implications for biocontrol studies with *F. oxysporum*, as they suggest that the non-pathogenic nature and antagonistic potential of some isolates may be dependent on the presence and composition of a consortium of bacterial partners.

In another study that possibly has implications for *Fusarium*-mediated biological control, molecular analyses revealed lineage-specific (LS) genomic regions in *F. oxysporum* that include four entire chromosomes (Ma *et al.*, 2010). Two of these LS chromosomes were transferred between strains of *F. oxysporum*, converting a non-pathogenic strain into a pathogen. The relative ease by which new pathogenic genotypes were generated in controlled laboratory experiments supports the hypothesis that gene–chromosome transfer between strains of *F. oxysporum* may occur in nature. Although this process may be rare, the potential danger of horizontal gene transfer from a biocontrol perspective is that pathogenicity could be transferred to an introduced biological control agent. If the recipient was better adapted to the local environment than the donor, disease severity may increase.

Nematode control with endophytic strains of Fusarium oxysporum

Research on the use of endophytic strains of *Fusarium oxysporum* for control of plant-parasitic

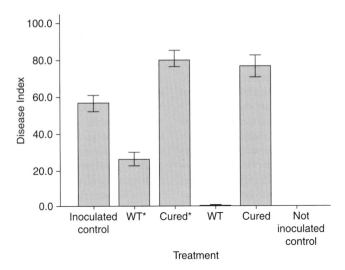

Fig. 8.3. Incidence of fusarium wilt on lettuce plants in pathogenicity assays. The inoculated control consists of plants transplanted into artificial soil inoculated with chlamydospores of *Fusarium oxysporum* f. sp. *lactucae* (Fuslat). The WT* and Cured* treatments consist of plants treated with wild-type and cured strains transplanted into soil inoculated with Fuslat. The WT and Cured treatments consist of plants inoculated with wild-type and cured strains and transplanted into artificial soil where Fuslat was not present. Error bars indicate standard deviations of the mean. (From Minerdi *et al.*, 2008, with permission.)

nematodes commenced in Kenya in 1989, when a lesion nematode (*Pratylenchus goodeyi*) and non-pathogenic strains of *F. oxysporum* were observed to produce unexpectedly low levels of root rotting when inoculated together on excised banana roots (see Sikora *et al.* (2008) for a brief account of those studies). Endophytic isolates of *F. oxysporum* from Kenyan tomato fields were then found to reduce populations of root knot nematode (*Meloidogyne incognita*) without adversely affecting plant health (Hallmann and Sikora, 1994), again suggesting that these mutualistic symbionts were potentially useful biological control agents. Since then, research on *Fusarium* endophytes has expanded exponentially and is detailed in reviews by Sikora *et al.* (2008) and Hallmann and Sikora (2011).

In most of the studies discussed in the above reviews, various strains of *F. oxysporum* were applied to seeds, seedlings or planting material with the aim of protecting the plant from nematode attack during the first few weeks of growth, a period when severe nematode damage is most likely to occur. Results from some of those experiments are presented in Chapter 12, and collectively they show that

fewer plant-parasitic nematodes are found in roots inoculated with non-pathogenic strains of *F. oxysporum* than in roots that are endophyte-free. Such results have prompted studies on the mode of action of *Fusarium* endophytes against nematodes, and although that work is ongoing, there is some evidence to suggest that the following mechanisms are operating:

- *Production of inhibitory or toxic metabolites.* When a strain of *F. oxysporum* was grown in a nutrient-rich artificial medium, secondary metabolites were produced that quickly inactivated all *Meloidogyne* and *Heterodera* juveniles, and about 65% of juvenile and adult *Pratylenchus zeae* and *Radopholus similis* (Hallmann and Sikora, 1996). However, it is not yet known whether these fungal metabolites are produced *in vivo* at concentrations sufficient to affect nematode development.

- *Changes in attractive/repellent properties of root exudates.* When *M. incognita* was inoculated at the centre of a substrate-filled tube linking two pots, one planted with a *F. oxysporum*-colonized tomato plant and the other with a non-colonized plant, the number of nematodes invading endophyte-infected roots was 36–56% lower than the non-fungal control. Additionally, the results of a choice test in which plant-free chambers were treated with tomato root exudates collected from either *Fusarium*-colonized or untreated plants, showed that nematodes moved away from the side containing root exudates obtained from endophyte-infected plants, indicating that these exudates were not as attractive to the nematode as exudates from untreated plants (Dababat and Sikora, 2007b).

• *Induced systemic resistance.* Split-root experiments with *Radopholus similis* on banana (Vu *et al.*, 2006) and *Meloidogyne incognita* on tomato (Dababat and Sikora, 2007c) showed that the capacity of the plants to defend themselves against nematode attack was enhanced by *F. oxysporum*, as the number of nematodes in non-treated roots was reduced by 25–45% when roots on the other side of the split-root chamber were pre-inoculated with the fungus. Such results indicate that *F. oxysporum* enhances resistance to plant-parasitic nematodes and that the plant's defensive response is expressed systemically throughout the plant.

Approaches to utilizing Fusarium-*mediated resistance to plant-parasitic nematodes*

One question that should always be considered when dealing with biological control is whether attempts should be made to mass-produce an antagonist and apply it as an inoculative or inundative treatment; or whether management strategies should be put in place to enhance naturally occurring populations of the antagonist. In the case of *F. oxysporum* and other similar endophytes, there is evidence that both approaches deserve consideration.

Tissue-cultured plantlets and nursery-produced transplants are often used to establish horticultural and vegetable crops, but because the planting material must be pathogen-free, it is generally grown in sterilized potting media. Consequently, young plants are introduced into the field without the symbionts that would normally provide some protection against nematodes and other soilborne pathogens. Inoculation of such material with protective strains of *F. oxysporum* and other useful symbionts is, therefore, a logical step, and should become standard practice in nurseries. Logistically, beneficial organisms are relatively easy to apply in such situations, and although extra costs will be involved, the final product should be more resilient to biotic and abiotic stresses than its non-inoculated counterpart.

Whether *F. oxysporum* should also be applied to seed or soil is a moot point. Mass-produced fungi are relatively easy to apply to seed, but to date there is no evidence to indicate that *F. oxysporum* would be useful against nematodes if applied in that manner. With regard to soil application, it will never be economically feasible to treat the bulk soil mass, but targeted applications to the soil surrounding seeds or transplants are worth considering. However, for this approach to be successful, researchers must work with typical agricultural soils and determine the spore concentration required, and the volume of soil that must be treated, to consistently reduce nematode numbers in roots. It will also be important to understand the impact of mycostatic factors on efficacy, and to establish whether soil treatment provides additional benefits that cannot be achieved by inoculating planting material.

Questions about whether *Fusarium*-mediated resistance to nematodes can be enhanced through management are much more difficult to answer, as we know almost nothing about the interactions between plants, nematodes and *Fusarium* endophytes in field situations. Species of *Fusarium* capable of forming endophytic relationships with plants are ubiquitous, and so it is possible that they are already having suppressive effects in the field. For example, research on banana-growing soils suppressive to *Radopholus similis* in greenhouse tests in Guatemala has shown that plants in those soils harbour isolates of *F. oxysporum* with high levels of biocontrol activity (Sikora *et al.*, 2008). However, what is needed in this situation are studies that relate nematode population densities and the severity of damage to the many biotic and abiotic factors that can influence nematode populations in the field. The degree to which roots are colonized by endophytes is one of the many biotic factors that could be measured. If a relationship between levels of *Fusarium* and suppressiveness was established, the management practices required to enhance endophytic colonization of roots could then be determined. However, from what is already known from the mycological literature, farming systems that provide regular carbon inputs and a year-round supply of roots are likely to be necessary. Non-pathogenic species of *Fusarium* are saprophytes as well as endophytes, and they will be present – often in high numbers – in any situation where healthy roots, decomposing roots and other sources of organic matter, are present.

Endophytic nematophagous fungi

Rhizosphere competence is an important attribute of all biological control agents of plant-parasitic nematodes, as it ensures that the antagonist is able to occupy the same microhabitat as the target pest. Since a capacity to colonize roots endophytically is really an extreme form of rhizosphere competence, such a trait could be expected to benefit any fungus that utilizes plant-parasitic nematodes as a food source. Endophytic root colonization has been reported for several nematophagous fungi, including *Pochonia chlamydosporia*, *Purpureocillium lilacinum* and *Arthrobotrys oligospora* (see Chapter 5), suggesting that these fungi may have a more intimate relationship with roots and perhaps nematodes than is possible through saprophytism. However, it is not yet clear whether a capacity to grow endophytically plays an important role in the ecology of these fungi, as endophytic growth has mainly been studied in well-managed greenhouse environments.

Concluding remarks on fungal endophytes: moving into uncharted waters

The fungal endophytes discussed in this chapter are no more than a small sample of the endophytes found in nature. DNA-sequencing technologies are making it easier to detect non-culturable or slow-growing endophytes in plant tissue, and this is resulting in a major increase in the number of endophytes found in association with plants. The situation with grasses provides a good example, as the diversity of the non-systemic, grass-associated endophytes is huge, and is growing rapidly (Sánchez Márquez *et al.*, 2012). Since many of these endophytes are found in roots and occupy the same ecological niche as nematodes, there is no doubt that in the field, plant–nematode–endophyte interactions are the rule rather than the exception. Thus, it is possible that population densities of plant-parasitic nematodes are already being influenced by endophytic fungi. The challenge of the future is to determine whether the incidence of these fungi, the identity of the endophyte, or the degree of colonization of plants or the root system influence nematode population dynamics or the level of nematode damage in agricultural environments.

Finally, in considering the role that fungal endophytes can play in nematode management, it should not be assumed that they will always have a positive impact on plant growth. Mayerhofer *et al.* (2013) analysed data from the literature and found a neutral to negative response in plant biomass to inoculation with root endophytes from the Ascomycetes (excluding the Clavicipitaceae). Plants sometimes responded positively to fungal endophytes, but the direction of the response and its magnitude depended on the identity of the host and endophyte, and experimental conditions, particularly the availability and sources of carbon and organic nitrogen.

Bacterial Endophytes and Rhizosphere-inhabiting Bacteria

The processes by which plants use root exudates and other root-derived organic materials to generate a soil microbial community capable of providing innumerable benefits was discussed in Chapter 2. The dominant members of that community are bacteria, and they play an important role in maintaining soil health, providing the plant with nutrients, and protecting roots from pathogens. The most widely studied group of beneficial bacteria lives on the root surface and in the rhizosphere, and its members are commonly referred to as plant growth-promoting rhizobacteria (PGPR). However, as some rhizobacteria can enter roots and become endophytic, root colonization should be viewed as a continuum (Fig. 8.4), with individual species occupying different ecological niches within the rhizosphere, rhizoplane and internal tissues of roots (Kloepper *et al.*, 1999; Compant *et al.*, 2005). Some bacteria are found in all these niches, and since it is not always clear whether they are acting as endophytes or rhizosphere inhabitants, occupants of both habitats are considered together here.

There is a large body of literature on plant-associated bacteria and it is impossible to cover it all in a book of this nature. However, numerous reviews provide a general overview of bacterial endophytes and rhizobacteria (e.g. Hallmann *et al.*, 1997; Sturz *et al.*, 2000; Vessey, 2003; Welbaum *et al.*, 2004; Rosenblueth and Martínez-Romero, 2006; Ryan *et al.*, 2008; Hartmann *et al.*, 2009; Hayat *et al.*, 2010; Compant *et al.*, 2010; Dennis *et al.*, 2010; Saharan and Nehra, 2011; Bhattacharyya and Jha, 2012), while others concentrate on

specific aspects such as plant disease control (e.g. Compant *et al.*, 2005; Pliego *et al.*, 2011; Maksimov *et al.*, 2011), or various groups of bacteria (e.g. Kloepper *et al.*, 2004; Hayward *et al.*, 2009; Qin *et al.*, 2011). This brief summary considers how root-associated bacteria influence plant growth, but mainly focuses on their role in protecting plants from damage caused by plant-parasitic nematodes and other soilborne plant pathogens.

Mechanisms associated with growth promotion by rhizobacteria

Huge numbers of bacteria in many different taxonomic groups are found in the rhizosphere (see Chapter 2 and particularly Table 2.3), and representatives from most of those groups are capable of endophytic growth in roots (Rosenblueth and Martínez-Romero, 2006). Given this enormous diversity, it is hardly surprising that root-associated bacteria have many different effects on plants. Some bacteria are important pathogens; others inhibit root-hair development or cause relatively superficial symptoms to root cortical tissue; and others are beneficial to the plant. The latter group acts in many different ways, and although its beneficial effects

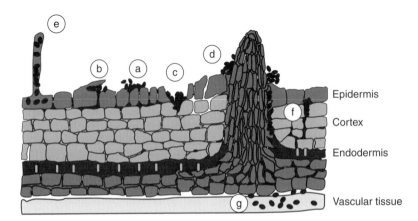

Fig. 8.4. External and internal colonization of plant roots by endophytic bacteria can be either: (a) random over the root surface, or (b) more specific below collapsed epidermal cells, (c) in association with wounds, and (d) at sites of lateral root formation. Internal bacterial colonization occurs (e) intracellularly in root epidermal cells including root hairs, (f) intercellularly within the root cortex, or (g) in association with the conducting elements. (From Hallmann, 2001, with permission.)

are discussed individually, multiple mechanisms are often involved.

Provision of nutrients

One of the most important roles of rhizobacteria is to provide the plant with nutrients, or facilitate their uptake from soil. The symbiotic relationship between nitrogen-fixing rhizobia and plants is widely recognized, as it accounts for more than 50% of the nitrogen currently used in agriculture. However, bacteria can affect plant nutrition in many other ways.

- More than 100 free-living, nitrogen-fixing bacteria are known (Sylvia *et al.*, 2005), and many live endophytically in roots, or reside in the rhizosphere. Rates of nitrogen fixation are relatively low (generally 5–25 kg/ha/year), but some crops, particularly rice and sugarcane, obtain substantial benefits from their diazotrophic associates (James, 2000).
- Phosphorus is an essential macronutrient for plant growth and development, but most soils are deficient in soluble forms that are readily available to plants. The role of arbuscular mycorrhizal fungi in enhancing phosphorus uptake has already been mentioned, but bacteria also mediate phosphorus availability by solubilizing insoluble mineral phosphate, and mineralizing and immobilizing the organic phosphorus pool (Rodríquez and Fraga, 1999; Richardson, 2001; Hayat *et al.*, 2010).
- In addition to their key role in nitrogen and phosphorus nutrition, rhizobacteria and other members of the soil biological community are constantly cycling all other macro- and micro-nutrients required by plants. Some bacteria specifically enhance iron uptake by sequestering Fe^{3+} from the soil and making it available to plants as a siderophore–iron complex.

Production of plant growth regulators

Higher plants produce hormones to regulate their growth and development, but most of those hormones are also synthesized by bacteria. For example, estimates by Patten and Glick (1996) suggest that 80% of the bacteria isolated from the rhizosphere are capable of producing indole acetic acid, an auxin that influences cell division, cell enlargement and root initiation. Cytokinins and gibberellins are also commonly produced, and collectively this group of hormones is considered partly responsible for the growth-promoting effects of many rhizobacteria. Ethylene, another potent plant growth regulator, is also produced by bacteria, and it may have a positive impact on plant growth through its effects on adventitious root and root-hair formation. There is also increasing evidence that one of the beneficial effects of PGPR is to reduce the high levels of ethylene associated with plants under stress. Excessively high concentrations of ethylene are inhibitory to plants, and many bacteria are able to lower plant ethylene levels through the production of the enzyme ACC deaminase (Glick *et al.*, 2007).

Suppression of soilborne pathogens

In addition to their direct effects on plants, bacterial endophytes and rhizobacteria can indirectly affect plant growth by suppressing soilborne pathogens. Since the mechanisms involved (competition for an ecological niche or substrate; production of inhibitory allelochemicals; and induction of systemic resistance) are much the same as outlined earlier for fungal symbionts, they will not be discussed here. Detailed information is available in reviews by Kloepper (1993), Hallmann *et al.* (1997), Raaijmakers *et al.* (2002), Weller *et al.* (2002), Kloepper *et al.* (2004) and Compant *et al.* (2005).

The impact of plant growth-promoting rhizobacteria on plant-parasitic nematodes

Interactions between rhizosphere-inhabiting bacteria and plant-parasitic nematodes were first studied seriously during the 1980s, when hundreds of bacterial isolates were screened for effects against root-knot and cyst nematodes (Becker *et al.*, 1988; Oostendorp and Sikora, 1989). Bacteria were tested for effects on motility *in vitro*, or were applied as drenches or seed treatments in greenhouse experiments, and a small percentage of isolates were found to exhibit some form of antagonistic activity (i.e. reduce the number of nematodes in roots or the level of galling).

After discussing those studies in my previous book (Stirling, 1991), I concluded that 'actinomycetes, some chitinolytic and proteolytic bacteria, and certain rhizosphere-inhabiting bacteria appeared to have potential as biological control agents, but bacteria were so diverse that other groups with activity against nematodes would undoubtedly be found'. I also argued that this biocontrol potential would only be harnessed if bacteria were tested more extensively in the field; formulation and application techniques were improved; and better information was available on their mode of action.

In the two decades since the above comments were made, bacteria continue to be screened for possible effects against nematodes. Isolates are often chosen by assessing the impact of culture filtrates on motility, and then candidates considered to have nematicidal activity are tested for efficacy in the greenhouse, usually in sterilized soil. Occasionally, a small number of isolates are included in field trials. Numerous such studies have been undertaken, and selected examples are summarized as follows, simply to indicate the range of bacteria that has been examined, the methods commonly used to identify useful isolates, and the results that can be expected from work of this nature:

- Culture filtrate of *Pseudomonas chitinolytica*, a bacterium isolated from chitin-amended soil, reduced the motility of second-stage juveniles of *Meloidogyne javanica*. Addition of bacteria to a natural soil reduced nematode infection and improved plant growth, but the mechanism did not appear to involve chitinolytic or proteolytic activity (Spiegel *et al.*, 1991).
- When *Bacillus cereus* was mixed with sandy loam soil at a rate of about 10^6 cfu/g soil, there was a small reduction in the number of second-stage juveniles of *Meloidogyne javanica* invading tomato roots. Nematicidal activity increased markedly in the presence of proteinaceous amendments, possibly because lethal levels of ammonia were produced during protein degradation (Oka *et al.*, 1993).
- An isolate of an unidentified species of *Streptomyces* applied as a drench, oatmeal

culture or seed coating reduced galling caused by *Meloidogyne incognita* and increased growth of tomato in sterile potting soil in greenhouse tests. Root galling was also reduced in the field (Dicklow *et al.*, 1993).
- *Pseudomonas chlororaphis* suppressed numbers of *Pratylenchus penetrans* in roots of strawberry plants grown in field soil in three experiments, and increased root and shoot growth in two of those experiments (Hackenberg *et al.*, 2000).
- Greenhouse tests in heat-treated soil amended with 0.6% (w/w) chitin indicated that five of 64 chitinolytic bacterial isolates consistently reduced numbers of *Heterodera glycines* compared to controls receiving neither chitin nor bacteria, or only chitin. However, there was no apparent relationship between chitinolytic activity on agar and suppression of the nematode in pots (Tian *et al.*, 2000).
- Culture filtrates of about one-third of a selected group of *Pseudomonas* strains were toxic to juveniles of *Meloidogyne javanica*. Some strains reduced nematode population densities and subsequent gall formation on mung bean when applied as a seed treatment or soil drench in the greenhouse. One strain of *P. aeruginosa* also reduced root-knot nematode development and nematode population density under field conditions (Ali *et al.*, 2002).
- In an initial screening test with field soil, 16 of 44 bacterial isolates with potentially useful biological control traits reduced numbers of *Paratrichodorus pachydermis* and *Trichodorus primitivus* on potato by 50–100%. When bacterized tubers were planted in soil naturally infested with the nematodes, four of those isolates (an unidentified bacterium and isolates tentatively identified as *Stenotrophomonas maltophilia*, *Bacillus mycoides* and *Pseudomonas* sp.) reduced nematode densities by 57–74% (Insunza *et al.*, 2002).
- An isolate of *Lysobacter enzymogenes* reduced the survival of eggs and juveniles of *Caenorhabditus elegans*, *Heterodera schachtii*, *Meloidogyne javanica*, *Pratylenchus penetrans* and *Aphelenchoides fragariae* when tested *in vitro*. The effects were

possibly due to the production of enzymes (chitinase, proteases and lipases) and other toxic metabolites (Chen *et al.*, 2006).

- Endophytic bacteria recovered from roots and seeds of rice were screened for effects against *Meloidogyne graminicola*, and an isolate of *Bacillus megaterium* was chosen for further study. Twenty days after this isolate was applied as a root dip and soil drench to rice seedlings grown in sterile soil, gall formation and the number of second-stage juveniles penetrating roots was reduced by about 40%. Culture filtrates of *B. megaterium* also reduced egg hatch (Padgham and Sikora, 2007).

- One isolate of *Bacillus thuringiensis*, a soil-dwelling bacterium commonly used as a biological insecticide, reduced galling caused by root-knot nematode in glasshouse and field tests when drenched onto soil or applied to seed (Zuckerman *et al.*, 1993). It was assumed that bacterially produced exotoxins were responsible for the effect, but this was never confirmed. In a later study with the same bacterium, Yu *et al.* (2008) found that some strains were toxic to *Meloidogyne hapla*, *Pratylenchus scribneri*, *Tylenchorhynchus* sp., *Ditylenchus destructor* and *Aphelenchoides* sp. in the laboratory. When root-knot nematode was treated with fluorescent-labelled crystal protein toxin, the fluorescent signal accumulated in intestinal tissue. However, the mode of action was unknown, as the toxic crystal was considered too large to have entered the nematode through its stylet.

- An antibiotic-producing strain of *Pseudomonas fluorescens* was tested as a seed treatment against *Meloidogyne incognita* on cotton, maize and soybean; *Meloidogyne arenaria* on peanut; *Heterodera glycines* on soybean; and *Paratrichodorus minor* on maize. In steam-treated soil, final numbers of *M. incognita* were consistently reduced by the bacterium, but there was no effect on the other nematodes. However, the bacterium did not suppress *M. incognita* in natural soil (Timper *et al.*, 2009).

- More than 300 endophytic bacteria were isolated from coffee roots in Ethiopia, and culture filtrates of 14 of 42 isolates were found to inactivate juveniles of *Meloidogyne incognita* in the laboratory. Tomato transplants were then dipped in bacterial suspensions, and in a series of greenhouse tests in sterilized soil, ten bacteria significantly reduced galling caused by root-knot nematode. *Bacillus pumulis* and *B. mycoidoes* were the most effective species, reducing the number of galls and egg masses by 33% and 39%, respectively (Mekete *et al.*, 2009).

- Culture filtrate of *Lysobacter antibioticus* reduced hatch of *Meloidogyne incognita* and increased mortality of second-stage juveniles in the laboratory, and when a culture of the bacterium was applied to tomato seedlings in pots, galling caused by the nematode and the number of nematodes in soil and on roots was reduced (Lee *et al.*, 2013).

These studies, together with many similar studies cited elsewhere (Chen and Dickson, 2004b; Tian *et al.*, 2007), demonstrate that a wide range of endophytic and rhizosphere-inhabiting bacteria are detrimental to plant-parasitic nematodes when introduced at high population densities into previously sterilized soils under highly controlled greenhouse conditions. What is disappointing is that in many cases, those studies have never been taken any further. Thus, possible mechanisms of action have never been seriously explored and it is not clear whether the bacteria would have been effective when applied in the field at realistic application rates. Given the biological complexity of soil, some level of simplification is needed when screening large numbers of isolates, but in terms of finding bacteria that could be incorporated into a biopesticide, we need to know much more than that bacteria A suppresses nematode B in a highly contrived environment (see Chapter 12). From an ecological perspective, information is needed on the manner in which bacteria interact with nematodes in the normal soil environment. Most of the bacteria considered detrimental to nematodes are ubiquitous in soil, the rhizosphere or roots; and so we also need to know whether indigenous populations impact on plant-parasitic nematodes in typical agricultural soils.

Interactions between rhizosphere- and root-inhabiting bacteria and plant-parasitic nematodes

Exudates produced by roots are used by plant-parasitic nematodes for host recognition, but they also play an important role in inducing the eggs of some nematodes to hatch. Since root exudates are also a nutrient source for rhizobacteria, the bacterial community at the root–soil interface will obviously interact with plant-parasitic nematodes by repelling, stimulating or inhibiting nematode movement, or by influencing egg hatch. In one such example, treatment of root exudates from sugar-beet seedlings with various Gram-negative bacteria reduced the capacity of the exudates to stimulate hatch of *Heterodera schachtii*, probably because the bacteria degraded compounds in the exudates that stimulated hatching (Oostendorp and Sikora, 1990).

Interactions between bacteria and nematodes may also occur within the root. Observations of root colonization by *Rhizobium etli* (syn. *Agrobacterium radiobacter*) using a green fluorescent protein marker showed that the bacterium preferentially colonized root tips, the lateral root base and root epidermal cells of *Arabidopsis*, and that bacteria also occurred in or near vascular tissue (Hallmann *et al.*, 2001). In the presence of *Meloidogyne incognita*, fluorescent signals emanating from the bacterium were most pronounced at juvenile penetration sites, suggesting that it was utilizing nutrients leaking from damaged root tissue. Bacteria were also observed within galled tissue, leading the authors to suggest that *R. elti* may have been interfering with nematode development and reproduction by intercepting and utilizing nutrients being transported to giant cells for use by the nematode.

Since root exudation patterns change quantitatively and qualitatively when plants are damaged by nematodes (Wang and Bergesen, 1974), the presence of plant-parasitic nematodes could be expected to change the biological community associated with roots. This was confirmed for endophytic bacteria by Hallmann *et al.* (1998) in studies with *Meloidogyne incognita* on cotton and cucumber. Internal communities of indigenous bacterial endophytes in cotton roots were significantly greater in the presence of the nematode and there was also a slight increase in species richness and diversity. Several hundred bacterial isolates were identified, but *Alcaligenes xylosoxydans*, *Burkholderia pickettii* and *Variovorax paradoxus* were either more common or only occurred in nematode-infested plants; *Brevundimonas vesicularis* was mainly isolated from nematode-free plants; and *Agrobacterium radiobacter* and *Pseudomonas* spp. were common in both treatments.

This study also investigated the impact of *M. incognita* on the root-colonizing capacity of a systemic endophyte (*Pseudomonas fluorescens* 89B-61) and a root cortical endophyte (*Enterobacter asburiae* JM22), applied as seed treatments. Recovery of *E. asburiae* from cucumber roots was positively, but not significantly, associated with the number of nematodes inoculated, except at 2 weeks after inoculation. The internal population of *P. fluorescens* on cotton roots also increased when the nematode was present. *E. asburiae* colonized cotton roots internally and was also found on the root surface around nematode penetration sites and on root galls, presumably because plant cell disruption resulted in leakage of nutrients favourable to the bacterium.

The study by Hallmann *et al.* (1998) is important, because it clearly shows that hundreds of indigenous endophytic bacteria are associated with roots growing in a typical agricultural soil. Since the soil used for the experiment with cotton had been fumigated 20 months previously with methyl bromide, it also shows that bacterial root colonization occurs in soils that have been subjected to this highly disruptive practice. However, no data were presented on whether the diversity of the bacterial community had been affected by the fumigant. Another important finding was that changes to the community of indigenous bacterial endophytes were observed within 7 days of inoculation with root-knot nematode, indicating that the bacterial community can respond quite quickly to the presence of the nematode. Such observations raise several important questions that must still be addressed by those interested in nematode–bacterial interactions:

• Do indigenous endophytic and rhizosphere-associated bacteria affect the capacity of plant-parasitic nematodes to establish feeding relationships with plants in agricultural soils, and does their presence influence the nematode's capacity to reproduce?

- Do management-induced changes in microbial population levels or microbial community structure affect plant–nematode associations?
- Do indigenous soil bacteria utilize the same mechanisms when they act against sedentary endoparasitic, migratory endoparasitic and ectoparasitic nematodes?
- What bacterial population densities and levels of diversity are necessary to significantly reduce the impact of plant-parasitic nematodes on their host?
- What factors influence root and rhizosphere colonization by indigenous bacteria in the first few days after roots come into contact with soil?
- Does agricultural management influence this colonization process?

To conclude, most of the nematological research on bacterial endophytes and rhizobacteria over the last 30 years has been directed towards finding bacteria that can be used as inoculants for nematode control. However, the work by Hallmann *et al.* (1998) indicates that a consortium of bacteria with far greater diversity than could ever be found in an inoculant begins colonizing the root system as soon as roots come into contact with soil. Surely our first priority should be to work with this consortium and better understand the way it interacts with nematodes and the plant?

Mechanisms by which root-associated bacteria influence plant-parasitic nematodes

Production of bioactive compounds

As mentioned earlier, bacterial endophytes and plant growth-promoting bacteria are commonly screened for possible effects against nematodes by adding a target nematode or its eggs to cell-free culture filtrate and measuring the response. A typical example is a study in which *Bacillus firmus* was grown in nutrient broth and then various dilutions

of cell-free culture filtrate were tested against several nematodes in laboratory assays (Mendoza *et al.*, 2008). Significant rates of mortality and paralysis were detected, and hatching of *Meloidogyne incognita* eggs was also reduced. However, the value of such tests as an initial screening procedure for identifying potentially useful antagonists, and the relevance of the results to activity in the soil environment is open to question for the following reasons:

- Various studies have shown that there is not necessarily a relationship between bacterial activity *in vitro* and *in vivo*. In one such example, bacterial strains that affected the vitality of *M. incognita* juveniles were selected in an *in vitro* test, but only about 20% of those strains reduced galling caused by the nematode in a subsequent test on cucumber seedlings (Becker *et al.*, 1988). In another study, Meyer *et al.* (2000) found that culture filtrates of *Burkholderia cepacia* strain Bc-2 contained extracellular factors that inhibited egg hatch and the mobility of second-stage juveniles of *M. incognita*. However, the bacterium did not reduce the number of juveniles or eggs on roots when it was applied as a seed coating or root drench to tomato in pots. Experiments with *Heterodera schachtii* have also shown that metabolites produced in artificial media are not a reliable indicator of antagonistic activity *in vivo*. In a laboratory assay, 139 bacterial strains reduced egg hatch, but only 7% of those strains reduced nematode infection in growth pouches (Neipp and Becker, 1999).
- In laboratory culture, bacteria may produce secondary metabolites that are toxic to nematodes, but the high concentrations and combinations of nutrients that are present in culture media are unlikely to occur in the nutrient-poor soil environment.
- The cell densities used to examine the effects of bacteria on nematodes in greenhouse tests are often unrealistically high. Bacterial populations of this magnitude are either unlikely to be sustained in soil for long periods, or are only likely to be achievable in specific microniches.

- Antibiotics bind rapidly and tightly to clay and humus colloids, and so bacterially produced antibiotics and toxins will often be inactivated in soil. However, the degree of inactivation will depend on pH, the clay mineralogy and other physicochemical factors (Williams, 1982; Stotzky, 2004).

Although bacteria are a well-recognized source of secondary metabolites with useful properties, the issue of whether these bioactive compounds are produced in soil has been debated for many decades (Williams, 1982). Plant pathologists have been involved in that debate, and it is only in the last 25 years that they have been able to confirm that bacterial antibiotics play a role in suppressing some soilborne pathogens (see reviews by Weller et al., 2002; Raaijmakers et al., 2002; Haas and Keel, 2003). The lines of evidence required to demonstrate this were discussed in Chapter 2, but data obtained using mutants with various levels of antibiotic production (e.g. antibiotic-negative mutants, complemented mutants with restored phenotypes, and over-producing strains) played a key role in those studies. The development of sensitive methods of detecting antibiotic production in natural environments (e.g. high-pressure liquid chromatography coupled with UV spectroscopy) has also contributed (see review by Thomashow et al., 2008).

To date, the techniques used to demonstrate the involvement of antibiotics in interactions between bacteria and soilborne plant pathogens have generally not been used in studies of bacteria–nematode interactions. The one exception is work by Cronin et al. (1997a) with Pseudomonas fluorescens, a bacterium known to inhibit the oomycete pathogen Pythium ultimum by producing the antibiotic 2,4-diacetylphloroglucinol (DAPG). When cysts of Globodera rostochiensis were exposed to a DAPG-producing strain of P. fluorescens, egg hatch increased but the percentage of mobile juveniles was reduced threefold. A transposon-induced DAPG-negative mutant did not affect juvenile mobility or hatch, but when this mutant was complemented by a plasmid-restoring DAPG synthesis, effects on hatch and mobility were observed. Although this result does not prove that DAPG influences the nematode under natural conditions, it does indicate that DAPG production by the bacterium affects egg hatch and juvenile mobility.

Given the enormous amount of work that has been required to show that antibiotics play a key role in the suppression of soilborne plant pathogens by bacterial antagonists, it will obviously be a major task to demonstrate that bacterially produced antibiotics influence nematode behaviour in the soil environment. Since the requisite studies have not yet been undertaken, the role of antibiosis as a mechanism for biological control of nematodes must, therefore, remain open to question.

Chitinolytic, proteolytic and lipolytic activity

The nematode cuticle is composed predominantly of proteins, and the egg shell contains chitin, collagen-like proteins and lipoproteins. Consequently, microorganisms with chitinolytic, proteolytic and lipolytic properties have long been considered to have potential as biocontrol agents against nematodes. Most of that work (reviewed by Stirling (1991) and Gortari and Hours (2008)) has focused on chitinolytic fungi and bacteria, as they are generally considered responsible for the decline in nematode populations that is often observed when chitin is added to soil. However, chitin contains large amounts of nitrogen and so it is likely that its action against nematodes, particularly at application rates greater than 1% w/w, is at least partly due to the release of ammoniacal nitrogen as the amendment decomposes (Mian et al., 1982). Nevertheless, a specialized microflora is known to develop during the late stages of the decomposition process, and there is some evidence to suggest that the organisms involved (fungi, bacteria and actinobacteria) contribute to the control obtained by producing metabolites or enzymes that act on the egg shell or nematode cuticle (Godoy et al., 1983a; Rodríguez-Kábana et al., 1987, 1989).

More recent observations of the microbial communities associated with nematode control in chitin-amended soil have also shown that populations of chitinolytic fungi and bacteria increase following the addition of chitin (Hallmann *et al.*, 1999). *Burkholderia cepacia* was recovered from both amended and non-amended soils and rhizospheres, but was most abundant with chitin amendment, while several other bacterial species, including *Aureobacterium testaceum*, *Corynebacterium aquaticum* and *Rathayibacter tritici*, were only recovered from amended soil. Amendment of soil with chitin also influenced the endophytic bacterial community in cotton roots. For example, *B. cepacia* rarely colonized roots obtained from non-amended soil but was the dominant endophyte in plants grown in chitin-amended soil. Such results indicate that chitin amendments capable of suppressing populations of plant-parasitic nematodes not only influence the microbial community in the soil and rhizosphere, but also change the bacterial community within plant tissue.

There are numerous bacteria with proteolytic, chitinolytic or lipolytic properties, but much of the more recent work on interactions between this group of bacteria and nematodes has focused on *Stenotrophomonas* and *Lysobacter*, two widely distributed genera considered to have potential for plant disease control (see review by Hayward *et al.*, 2009). Initial experiments showed that *S. maltophilia* was able to destroy root-knot nematode eggs sterilized by irradiation, but later work indicated that the bacterium did not prevent living eggs from hatching or reduce nematode multiplication on tomato in pots (Fegan, 1993). Nevertheless, there is evidence that *S. maltophilia* is detrimental to other nematodes, as it was one of the few chitinase-producing bacteria isolated from an Irish soil to almost completely inhibit *Globodera rostochiensis* in an *in vitro* test (Cronin *et al.*, 1997b). A follow-on study showed that when potato infested with *G. rostochiensis* was grown in soil that had previously been inoculated with *S. maltophilia*, eggs retrieved from cysts produced after one and two nematode generations did not hatch as readily as eggs from the non-inoculated control. However, the observed reduction in egg hatch, although significant, was considered insufficient to greatly influence nematode population dynamics in the soil environment. A Canadian study also showed that *S. maltophilia* was one of several

bacteria to reduce populations of trichodorid nematodes on potato (Insunza *et al.*, 2002).

With regard to interactions between *Lysobacter* and nematodes, studies by Fegan (1993) showed that *L. enzymogenes* could degrade the chitin layer of *Meloidogyne* egg shells *in vitro*, but was unable to reduce egg hatch or influence nematode multiplication on plants. Later studies with the C3 strain of *L. enzymogenes* indicated that the bacterium killed adults and juveniles of several plant-parasitic and free-living nematodes on agar plates (Chen *et al.*, 2006) and reduced the number of cysts produced by *Heterodera schachtii* in growth pouches (Yuen *et al.*, 2006). Since the bacterium was detrimental to vermiform stages of the nematodes as well as eggs, chitinase could not have been the only metabolite responsible for the effect. Proteases and lipases are also produced by the bacterium, while the antibiotic dihydromaltophilin may also have been involved (Yuen *et al.*, 2006).

Collectively, the above studies clearly show that chitinolytic, proteolytic or lipolytic bacteria are detrimental to nematodes in laboratory tests. However, they provide limited evidence of effectiveness in soil and even less evidence that the effects are due to enzymatic activity. Since scientists investigating the potential of chitinolytic bacteria for fungal control have had to deal with similar issues, a review by de Boer and van Veen (2001) is worthwhile reading, as it makes a number of points that are relevant to nematodes:

- Fungi and actinobacteria are the predominant chitin degraders in soil, largely because their filamentous habit allows them to bridge air-filled gaps in the soil matrix. Most chitinolytic bacteria are restricted in their motility, and cannot compete effectively with that group of microorganisms.
- Chitinolytic activity in soil is most likely to occur under conditions of carbon starvation, as chitinase production by many soil bacteria is repressed by small organic substrates such as sugars and amino acids. Rhizobacteria may not, therefore, produce chitinases in the presence of root exudates.
- Bacteria produce many different chitinases, and so the composition of the chitinase pool is likely to determine effects on particular target organisms.

• Chitinolytic bacteria produce both antibiotics and chitinases, and the susceptibility of an organism to lysis by a particular bacterial strain may be determined more by its sensitivity to the antibiotics than to the enzymes.

Given these observations, it is clear that chitinolytic, proteolytic and chitinolytic bacteria are not a panacea against nematodes. Bacteria with these properties almost certainly interact with nematodes in the natural environment, but their effects are likely to be relatively subtle and may not be manifested through enzyme production.

Induction of systemic resistance

As with the other groups of endophytes discussed earlier in this chapter, it is widely recognized that PGPR and bacterial endophytes elicit host defence mechanisms against bacterial, fungal and oomycete pathogens of plants through a process known as induced systemic resistance (see reviews by van Loon *et al.*, 1998; Kloepper *et al.*, 2004; Choudhary *et al.*, 2007). A study by Hasky-Günther *et al.* (1998) provided evidence that rhizobacteria can also induce resistance against nematodes. The bacteria chosen for this study (*Bacillus*

sphaericus strain B43 and *Agrobacterium radiobacter* strain G12) were isolated from the potato rhizosphere and shown to reduce penetration of *Globodera pallida* in potato roots. In split-root experiments, application of one of the bacteria to one side of the split-root system and *G. pallida* to both sides resulted in significant reductions of nematode penetration into both root systems. Effects were observed with both bacterial species, but the reduction tended to be greater in the bacteria-free side of the root system (Fig. 8.5A). In a second experiment, in which bacteria were applied to one side of the root system and nematodes were added to the bacteria-free side 24 h later, *B. sphaericus* and *A. radiobacter* caused a 58% and 55% decrease in nematode penetration, respectively (Fig. 8.5B).

Although the results of these experiments clearly show that rhizobacteria induce a form of resistance to nematodes that is systemic within the plant, the mechanisms involved are poorly understood. Numerous possibilities exist and they may act outside the plant or within root tissue. Mechanisms that act externally may cause the plant to produce secretions that are toxic to nematodes, or alter root exudation patterns, with flow-on effects to nematode hatching and host

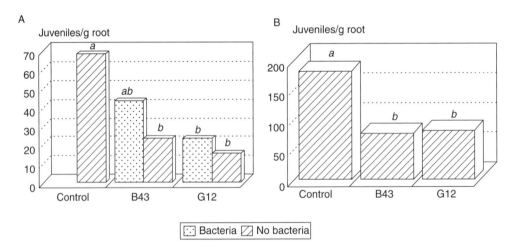

Fig. 8.5. Response of *Globodera pallida* to bacteria in two split-root experiments. (A) Impact of inoculating one side of a split root system with bacteria on penetration of the nematode into the bacteria-treated and untreated root sections. (B) The number of nematode juveniles invading potato roots on the untreated side when the other side was inoculated with bacteria. In both experiments, B43 indicates a strain of *Bacillus sphaericus* and G12 indicates a strain of *Agrobacterium radiobacter*. Mean values with different letters are significantly different, $P \leq 0.01$. (From Hasky-Günther *et al.*, 1998, with permission.)

recognition processes. Within the plant, phytoalexins may be produced that interfere with the feeding process or are toxic to an invading nematode, while physiological changes may also occur that strengthen cell walls or create barriers that limit nematode migration within roots. One particular advantage of induced systemic resistance as a defensive mechanism is that it often protects plants against a broad range of pathogens. Thus, when resistance is expressed against plant-parasitic nematodes, it may also provide wider benefits.

Manipulating populations of rhizobacteria for nematode management

As with arbuscular mycorrhizal fungi and fungal endophytes, there are differences of opinion with regard to whether rhizobacteria are best utilized for pest and disease management by developing inoculants or enhancing indigenous populations. However, as rhizobacteria and bacterial endophytes are readily cultured, the inoculation approach has generally been favoured, and some examples of its use against nematodes are discussed in Chapter 12. Bacterial inoculants are most likely to find a place in horticulture, as formulations can be developed that are readily applied to seed, transplants or tissue-cultured plantlets. Greenhouse experiments and field tests with vegetables, for example, have shown that whether bacteria are used alone or in combination with an appropriate organic amendment, they can reduce damage caused by plant-parasitic nematodes (Kokalis-Burelle *et al.*, 2002a, b) and induce systemic resistance to a range of plant pathogens (Jetiyanon and Kloepper, 2002; Jetiyanon *et al.*, 2003). However, given the ubiquitous distribution of rhizobacteria and the fact that inoculation will not be economically feasible for many crops, the possibility of enhancing indigenous bacterial populations through management should not be ignored.

Impact of crop rotation, organic amendments and other practices

When nematode-resistant rotation crops are used to manage plant-parasitic nematodes, it is usually assumed that their effectiveness is dependent on the extent to which they reduce populations of the target nematode. However, there is some evidence to suggest that rotation crops may also affect nematodes in the following crop through effects on rhizobacteria. Analyses of the bacterial community in the rhizosphere of plants that are susceptible (e.g. soybean) and resistant (e.g. velvet bean, castor bean, sword bean and Abruzzi rye) to plant-parasitic nematodes indicated that the resistant plants (referred to as antagonistic plants by the authors) had a distinct rhizosphere microflora (Kloepper *et al.*, 1991, 1992a). *Bacillus* species were predominant in soybean, whereas coryneform and Gram-negative genera tended to be associated with the antagonistic plants. When a selection of these bacteria were tested for their capacity to reduce symptoms caused by *Meloidogyne incognita* and *Heterodera glycines* on soybean in the greenhouse, about 20% of the bacteria from antagonistic plants exhibited activity against both nematodes compared with only 4% of the isolates from soybean. Kloepper *et al.* (1999) suggested that the activity of this antagonistic microflora may have explained why reductions of nematode populations following velvetbean (Rodríguez-Kábana *et al.*, 1992) sometimes persist for much longer than expected. Given that the rhizobacteria present on susceptible and antagonistic plants were not only markedly different, but also varied in their physiological profiles, this possibility deserves some attention.

Although the benefits of crop rotation are beyond reproach, modern farm businesses often have few rotation options, as marketing arrangements and machinery requirements tend to limit them to producing two or three specific crops. Nevertheless, bacterial endophytes and rhizobacteria can still play a role in such farming systems, because there is strong evidence to suggest that the best way of generating specific suppressiveness to root pathogens is to grow crops as a continuous or near-continuous monoculture (see Cook and Weller, 2004; Cook, 2006; and examples for nematodes discussed in Chapter 10). In this situation, the challenge is to understand the impact of crop sequence on the key organisms in the suppressive community, and then ensure that natural suppressiveness forms the foundation of a broader integrated disease management programme.

The use of chitin to promote suppressiveness to plant-parasitic nematodes was discussed earlier, and it provides yet another example of how the structure and function of a microbial community can be altered. There will be many other ways of achieving such changes, but with regard to enhancing bacteria that might be useful against nematodes, what we currently lack is an understanding of which bacteria are likely to be most useful, and their mechanisms of action. Once those issues are addressed, it will be possible to use molecular technologies to monitor the bacteria in soil and determine how they are influenced by various management practices. An example from plant pathology provides an example of what is possible.

Pseudomonas strains with the capacity to suppress a diverse range of soilborne pathogens have been studied for many years, and it is now known that they act by producing the antibiotic 2, 4-diacetylphloroglucinol (Weller *et al.*, 2002; Haas and Defago, 2005). This means that the *phlD* gene, which is essential for DAPG production, can be used to detect DAPG-producers in the rhizosphere. PCR-based assays for this gene have shown that the incidence and relative abundance of native populations of suppressive pseudomonads in the rhizosphere of maize and soybean is influenced in subtle but reproducible ways by crop rotation, tillage, organic amendments and chemical seed treatments (Rotenberg *et al.*, 2007b). Although practices that reduced root disease severity were not universally linked to increased root colonization by DAPG-producers, studies of this nature help to unravel the biological complexity that exists at the soil–root interface and will eventually provide an indication of how disease and nematode-suppressive bacterial communities can be managed.

Root-associated Symbionts: Only One Component of the Rhizosphere Microbiome

Evidence presented in this chapter clearly shows that fungal and bacterial symbionts benefit plants in many different ways, one of which is to reduce damage caused by nematodes and other soilborne pathogens. By necessity, this evidence has been obtained by studying interactions between plants, symbionts and pathogens in relatively simple experimental systems. However, the problem with such an approach is that in agricultural soils, these interactions take place in the presence of a vast array of competitive microbes. In the case of plant growth-promoting bacteria, for example, it is known that they can alter their gene expression when confronted with another bacterial species, and that the identity of the competitor is an important determinant of their behavioural and genetic response (Garbeva *et al.*, 2011b). Thus, the behaviour of a symbiont in a relatively simple microcosm may bear little relationship to how it behaves in the highly competitive environment at the root–soil interface.

Another problem in considering the role of individual symbionts is that they do not occur in isolation in natural environments. Mycorrhizal fungi, endophytic bacteria and fungal endophytes almost certainly occur together within root tissue, and will inevitably interact with each other. From the perspective of their effects on plant-parasitic nematodes, they also occupy similar ecological niches to many of the antagonists discussed in Chapters 5, 6 and 7, and so they are only one of many contributors to the suppressive services provided by the soil biological community. Therefore, it is time to take a more realistic view of interactions that occur within the soil food web. A huge number of soil organisms are constantly interacting with plant roots, with nematodes, and with each other, and we need to start thinking about the collective effects of these interactions on plant and soil health. Symbionts contribute by enhancing nutrient uptake, stimulating systemic resistance to pathogens and improving soil aggregate stability, but their capacity to perform those functions is influenced by the soil environment and by other members of the soil biological community.

Studies in which the community of soil organisms was considered as one entity could never be contemplated in the past. The difficulties associated with identifying various taxa, together with methodological differences in the way organisms were retrieved from soil, meant that soil biologists tended to be locked into working with certain groups of organisms. The advances in molecular detection technologies that have occurred over the last 20 years have removed that limitation. Molecular approaches are now being used to monitor the abundance and activity of soil organisms at a range of taxonomic levels; specific probes can be designed to identify the genes involved in specific functions; and high-resolution DNA-sequencing technologies are being used to explore the full range of biological diversity that exists in soil. When applied to issues associated with plant disease suppression, studies using these technologies have shown that the disease-suppressive ability of a soil is governed by microbial consortia and cannot simply be ascribed to a single microbial taxon or group (Mendes *et al.*, 2011). Thus, the community of organisms involved in suppressing root pathogens must be studied, rather than individual antagonists or symbionts.

Berendsen *et al.* (2012) provide a good overview of the current state of our knowledge of the rhizosphere microbiome and conclude that modern genomics techniques, which can now be used to study microbial communities in their natural environment, will deliver new insights into its activities and functions. Metagenomic studies (e.g. Mendes *et al.*, 2011) provide a comprehensive picture of the taxa comprising the microbiome and allow microbial communities in pathogen-suppressive and non-suppressive soils to be compared. Transcriptomics can contribute by helping to unravel the molecular mechanisms associated with interactions between plants and complex microbial communities (Schenk *et al.*, 2012). Ultimately, the use of these technologies should allow us to define disease-suppressive microbial 'signatures' and use them to determine how crops and soil should be managed to enhance suppressiveness to nematodes and other soilborne pathogens. Such a goal is realistic, but will only be achieved by focusing on indigenous biological communities, and by linking studies of the rhizosphere microbiome with cropping systems research.

Natural Suppression and Inundative Biological Control

9

Suppression of Nematodes and Other Soilborne Pathogens with Organic Amendments

Soilborne pathogens are an insidious problem in all agricultural crops. Numerous fungi, bacteria, oomycetes and nematodes debilitate root systems; cause wilt, root rot and damping-off diseases; and are associated with poor growth, yield decline and replant problems. In many crop production systems, losses from soilborne diseases are the norm, and preferred crops can only be grown successfully if specific management practices (e.g. crop rotations, disease-resistant varieties, soil fumigation) are implemented to limit damage. However, there are also situations where disease severity is not as high as expected, given the prevailing environment and the level of disease in surrounding areas. Sometimes this is due to subtle variations in soil physical or chemical factors (e.g. sand or clay content, pH or nutrient status), but in other cases it is a biological phenomenon. In these situations, naturally occurring soil organisms interact in some way with the plant or pathogen and either protect the plant from disease, or minimize disease severity.

The most common form of disease suppressiveness (often referred to as 'general' or 'non-specific' suppressiveness) is discussed in this chapter. It is found in all soils and provides varying degrees of biological buffering against most soilborne pests and pathogens. Since the level of suppressive activity is broadly related to total soil microbial biomass and is enhanced by practices that conserve or enhance soil organic matter, the term 'organic matter-mediated general suppression' is also commonly used (Hoitink and Boehm, 1999; Stone et al., 2004). This type of suppression can be removed by sterilizing the soil, and is due to the combined effects of numerous soil organisms acting collectively through mechanisms such as parasitism, predation, competition and antibiosis.

A second form of suppression (usually known as 'specific' suppressiveness) is discussed in Chapter 10. Although it can also be eliminated by sterilization and various biocidal treatments, it differs from general suppressiveness in that it results from the action of a limited number of organisms. This type of suppression relies on the activity of relatively specific antagonists and can be transferred by adding small amounts of the suppressive soil to a conducive soil (Weller et al., 2002; Westphal, 2005). Since specific suppression operates against a background of general suppressiveness (Cook and Baker, 1983), the actual level of suppressiveness in a soil will depend on the combined effects of both forms of suppression.

The organic matter required to sustain the activity of the organisms responsible for general disease suppressiveness can be imported from elsewhere and added to soil as an amendment, or it can be obtained from the residues of crops grown *in situ*. This chapter covers the use of organic amendments to enhance suppressiveness. Issues related to the way residues are managed within a farming system to improve physical, chemical and biological fertility, thereby reducing the impact of soilborne diseases, are covered in Chapter 11. Although general suppressiveness to plant-parasitic nematodes is the main focus of this chapter, there is also a brief discussion of broad-spectrum suppressiveness to other pathogens. Nematodes are only one of many soilborne constraints to crop production, and one advantage of organic matter-mediated suppressiveness is that it provides some protection against most, if not all, members of the soilborne pest/pathogen complex.

Organic Matter-mediated Suppressiveness for Managing Soilborne Diseases

Although organic inputs from roots, crop residues and animal manures are known to play a vital role in determining a soil's physical, chemical and biological properties (see Chapter 2), the practices employed in many modern farming systems have destroyed the link between organic matter and soil fertility. Nowhere is this more apparent than in the area of soilborne disease management. Levels of soil organic matter have declined, soils are now conducive rather than suppressive to soilborne diseases, and so natural forms of control have been replaced by soil fumigants, nematicides, fungicides and disease-resistant varieties. One possible exception is some sections of the container-based nursery industry, where compost-amended potting mixes are used to suppress root rots caused by *Pythium* and *Phytophthora*, and provide some suppression of *Rhizoctonia* damping-off (Hoitink and Boehm, 1999).

Organic matter-mediated suppressiveness has one main advantage over other forms of disease control: it is effective against a wide range of fungal, oomycete, bacterial and nematode pathogens that often act together to cause crop losses. Examples of its spectrum of activity are too numerous to

mention here, but include suppression of *Pythium* in Mexican fields following the application of large quantities of organic matter over many years (Lumsden *et al.*, 1987); the use of cover crops, organic amendments and mulches to suppress *Phytophthora* root rot of avocado in Australia (Broadbent and Baker, 1974; Malajczuk, 1983; You and Sivasithamparam, 1994, 1995); suppression of the same disease with *Eucalyptus* mulch in California (Downer *et al.*, 2001); the management of a fungal, bacterial and nematode-induced root disease complex of potato in Canada with chicken, swine and cattle manures (Conn and Lazarovits, 1999; Lazarovits *et al.*, 1999, 2001); the use of crop residues, animal manures and organic waste materials to reduce damage caused by plant-parasitic nematodes (reviewed by Muller and Gooch, 1982; Stirling, 1991; Akhtar and Malik, 2000; Oka, 2010; Thoden *et al.*, 2011; McSorley, 2011b; Renčo, 2013); changes in crop rotation and tillage practices to enhance suppressiveness to potato diseases caused by *Rhizoctonia, Fusarium, Helminthosporium* and *Phytophthora* (Peters *et al.*, 2003); and the use of fresh and composted paper waste residuals to suppress *Pythium* and *Aphanomyces*, the causes of common root rot of snap bean (Rotenberg *et al.*, 2007a). Tables in Litterick *et al.* (2004), Noble and Coventry (2005), and Janvier *et al.* (2007) provide other examples.

Sources of organic matter for use as amendments, and their beneficial effects

The only way to maintain levels of organic carbon in agricultural soils is to continually replace the organic matter that is lost through decomposition and removal of harvested product. Crop residues, biomass from cover crops and manures produced by on-farm livestock are the most sustainable sources of organic matter, as they can be produced on-farm, eliminating the transport costs involved in importing organic matter from elsewhere. However, agriculture is an export-oriented industry, and so the nutrient cycle loop can only be closed by ensuring that nutrients that leave the farm are ultimately returned to the farm (Gliessman, 2007). Thus, there is a place in agriculture for utilizing organic wastes associated with feedlots, dairies, sugar mills and other food production and processing operations, and the sewage produced in cities and towns, as a source of nutrients and organic matter. Composting or other means of stabilization or treatment may be needed to ensure that the organic material does not pose a threat to public health, and also to eliminate weed seeds and plant pathogens, but such treatments are an essential component of a process that produces useful amendments from wastes that would otherwise be incinerated or disposed of in landfills.

Many different organic materials are available for use as soil amendments in agriculture, and they can be subdivided into numerous categories (Table 9.1), or grouped more broadly into composts, uncomposted materials, manures, compost extracts and compost teas (Hue and Silva, 2000; Litterick *et al.* 2004). A list of the amendments commonly available

to landholders, together with data on their macro- and micro-nutrient content, can be found in Quilty and Cattle (2011). Many of these materials are commercial products, and are marketed on the basis that they will improve soil health and increase yields. However, the benefits from some of these products are often overstated, with claims made by suppliers and manufacturers frequently based on anecdotal rather than scientific evidence.

The role organic matter in promoting the physical, chemical and biological health of soil is widely recognized, as the practice of adding manures and crop residues to soil is as old as agriculture itself. However, this does not mean that amending soil with organic matter is always beneficial, or that it is useful in all environments. Organic amendments act in many different ways, and enhanced suppression of pests and diseases is only one of many possible effects. Some of those effects are briefly discussed next, and further detail can be found in Quilty and Cattle (2011).

Table 9.1. A selection of organic materials commonly available for use as soil amendments in agriculture.

Material	Examples	Common application rates	
		Solids	Liquids
Animal manures	Horse, poultry and cattle manures, and wastes from animal feedlots	10–100 t/ha	–
Plant residues	Grass and legume residues, wood chips, bark, green waste, sawdust and grass clippings	20–100 t/ha	–
Organic wastes	Sewage sludge, biosolids, spent mushroom compost, yard waste, winery wastes and paper-mill waste	20–100 t/ha	–
Biochar	Fine charcoal produced by slow pyrolysis of biomass from agricultural or forestry industries	1–140 t/ha	
Composts and compost teas	Composted materials from various sources, or teas produced by steeping compost in water	1–30 t/ha	50–1000 L/ha
Meat, blood, and bone meal	By-products of meat processing	0.1–1.2 t/ha	1–30 L/ha
Seafood by-products	By-products of fish processing	–	2–60 L/ha
Vermicompost	Vermicasts or worm castings	2–50 t/ha	10–100 L/ha
Humic substances	Humic acid, fulvic acid and humin	0.025–1 t/ha	1–30 L/ha
Seaweed extracts	Extracts from kelp and seaweed	–	0.5–20 L/ha
Bio-inoculants	Amendments containing microbial species considered beneficial	–	1–20 L/ha

Modified from Quilty and Cattle (2011).

- *A source of plant nutrients.* All organic materials contain plant nutrients, and when applied to soil as an amendment, they sometimes supply enough nutrients to maintain crop yields at the levels achievable with inorganic fertilizers. However, more commonly, the rate of mineralization is insufficient to meet plant demands, and the application rates required to obtain nutritional benefits are economically prohibitive. Other amendments such as biochar have no nutrient value, and most liquid organic amendments are applied at rates that are unlikely to produce more than minor nutritional effects.
- *Stimulation of plant growth.* Research has shown that some organic amendments are capable of eliciting hormonal growth responses in plants, or can lead to the production of plant growth-promoting substances in soil. Examples include extracts from seaweed, and low molecular weight humic substances, particularly those in vermicomposts.
- *Enhancement of soil organic carbon.* Most organic amendments, except those with relatively low carbon contents, will increase the organic content of soil. However, the longevity of this carbon depends on how the soil is managed and whether the amendment is continually reapplied. Soil carbon will be lost if the soil is repeatedly tilled, and reapplication is required to maintain microbial biomass carbon. Also, some amendments, especially materials with a relatively low C:N ratio, may stimulate microbial activity to such an extent that they increase carbon mineralization, potentially reducing the amount of organic carbon in soil.
- *Soil structural benefits.* The capacity of organic amendments to improve the physical condition of soil is well documented for materials such as composts and crop residues, which are applied at relatively high application rates. However, humic substances, which are usually applied at much lower rates, may produce similar effects, as they have the capacity to bind with clay minerals and promote aggregation.

Impact of organic source and application rate on disease suppression

In addition to these effects, the role of organic amendments in enhancing suppressiveness to a range of soilborne diseases is well documented. Bonanomi *et al.* (2007) assessed the results of 2423 experimental studies and found that amendments enhanced suppression in 45% of cases. Disease incidence increased in 20% of the experiments and there was no effect in the other experiments. Composts and organic wastes were the most suppressive materials and usually did not increase disease incidence. Crop residues were often suppressive, but could also be conducive to disease, while peat enhanced suppressiveness in only 4% of the experiments. The ability of amendments to suppress disease also varied with different pathogens. Suppression was commonly observed (>50% of cases) when the disease was caused by *Verticillium*, *Thielaviopsis*, *Fusarium* or *Phytophthora*, whereas it was rare for diseases caused by *Rhizoctonia solani*.

Application rate also has major effects on the capacity of an amendment to suppress disease. Thus, disease suppressiveness increased with application rate in about half the experiments assessed by Bonanomi *et al.* (2007), and was also noted by Noble and Coventry (2005) in their review of the suppressive effects of composts in field situations. This highlights the fact that in many published studies, high application rates were required to achieve immediate disease control. Work on organic amendments in Canada provides a good example. A variety of soilborne diseases of potato were controlled with solid materials (meat and bone meal, soy-meal and poultry manure) at rates of 15–66 t/ha, and liquid swine manure at 5500 L/ha (Lazarovits *et al.*, 2001). However, some of these amendments, particularly the meat and bone meal, and soy-meal, were not an option for most growers, as they were too expensive. Thus, despite good experimental evidence of effectiveness, the cost of purchasing a suitable amendment and transporting it to the required site is often prohibitive, even in high-value crops. Consequently, future research efforts must take a longer-term view of how such materials are used. We need to apply amendments at realistic application rates and recognize that the benefits of adding organic matter to soil are incremental, that the effects are cumulative, and that it takes time to develop disease-suppressive soils (Bailey and Lazarovits, 2003).

Effects on pathogen populations and disease

Organic amendments do not always reduce disease incidence or severity by reducing populations of the pathogen. Although this commonly occurs with *Thielaviopsis* and *Verticillium*, effects are variable with *Phytophthora* and *Fusarium*, while population densities of oomycetes and fungi that are aggressive saprophytes (e.g. *Pythium* and *Rhizoctonia*) often increase (Bonanomi *et al.*, 2007). Thus, there is not necessarily a correlation between disease suppression and pathogen population density, presumably because amendments sometimes act by inducing plant resistance, improving soil physical or chemical properties, or enhancing biostatic mechanisms that inhibit rather than kill the pathogen.

Variation in responses to organic inputs

It is obvious that many types and sources of organic matter can be used to enhance general suppressiveness and that they are effective in many different situations. However, results tend to be inconsistent, as plants and pathogens do not always respond in the same way to organic amendments, and efficacy is influenced by environmental conditions. Also, the level of suppressiveness varies with the nature of the inputs, the degree to which they have decomposed, and the rate at which they are applied (Scheuerell *et al.*, 2005; Bonanomi *et al.*, 2007, 2010). Pathogens that are good primary saprophytes but poor competitors (e.g. *Pythium* and *Fusarium*) will multiply on fresh organic matter and then be suppressed by organisms involved in the decomposition process, so the time taken for these competitive organisms to become dominant, and for

suppressiveness to develop, must be considered when organic matter is used to manage such pathogens. Enhancing suppressiveness to *Rhizoctonia*, a diverse genus that contains many plant pathogens, is also problematic. Aggressive isolates of *R. solani*, for example, have a number of intrinsic properties that make them difficult to control: a capacity to colonize fresh organic matter; mycelium that grows rapidly but does not degrade readily; and sclerotia that are insensitive to fungistasis and resistant to decomposition. Therefore, successful inhibition of this fungus relies on creating a biological environment where soil microorganisms compete with the pathogen for colonization sites on organic matter, and pathogen-specific antagonists are also active (Stone *et al.*, 2004).

The problems involved in achieving consistent control with organic amendments are illustrated by the results from a major study in Europe by Termorshuizen *et al.* (2006). Eighteen different composts originating from different countries and source materials were tested for effectiveness against seven different pathogens. A total of 120 assays were conducted and compost significantly enhanced disease suppression in 54% of cases. Disease levels were stimulated by composts in only 3% of cases. However, analyses of the data showed three important points:

- The disease suppressiveness due to composts in different pathosystems was generally not correlated (i.e. a compost may enhance suppressiveness to one pathogen, but that did not necessarily mean it would have the same effect on diseases caused by other pathogens).
- The level of suppressiveness of some of the composts was highly variable between pathosystems, with one compost simulating disease by 87% in the *R. solani*/cauliflower pathosystem, and inducing 84% disease suppression in the *R. solani*/pine pathosystem. Other composts showed relatively little variation in their effects on different pathosystems.
- Composts made from similar materials generally behaved in much the same way, particularly if they originated from the same composting facility. However, composts that

were removed from the same compost heap at different times behaved quite differently.

Mechanisms of action

One thing that is clear from the literature on organic matter-mediated suppression is that many organisms and multiple mechanisms are involved. Interactions between organisms occur on organic substrates, in the rhizosphere, and within roots (see Chapter 2), and they all contribute in some way to reducing the capacity of resident pathogens to destroy root systems and cause plant disease (see Chapter 3). The mechanisms involved are described in Box 9.1, and most, if not all, play a role in most examples of general suppressiveness.

Indicators of broad-spectrum disease suppressiveness

Since a capacity to suppress plant disease is one of the key attributes of a healthy soil, any move by farmers towards using organic amendments and other practices to improve soil health should result in enhanced disease suppression. However, suppressiveness will not become a mainstream disease management strategy until growers are confident that their farming system will enhance suppressiveness to levels capable of providing at least some control of the soilborne diseases they are likely to encounter. Consequently, it is important to be able to predict whether a soil is suppressive to key pathogens.

Numerous biotic and abiotic parameters have been proposed as indicators of suppressiveness, and although many of these parameters are associated with reduced disease incidence or severity, the strength of the relationships varies with soil type, environmental factors and the pathogen. Thus, despite considerable research effort in recent decades, there is a lack of identified, reliable and consistent indicators, and we are still far from being able to predict the disease-suppressive capabilities of a soil (Janvier *et al.*, 2007; Bonanomi *et al.*, 2010).

The physical properties most commonly associated with disease suppressiveness are the predominant clay type and the structure of the soil, while the most important soil chemical factors are pH, the amount of nitrogen and its form (either NO_3^- or NH_4^+), levels of calcium and the organic carbon content (Höper and Alabouvette, 1996; Scheuerell *et al.*, 2005; Janvier *et al.*, 2007). However, in most cases where an individual physico-chemical property has been correlated with suppressiveness, studies in other environments or with other pathogens have failed to show the same relationships. Parameters that collectively describe the overall abiotic environment are likely to be more useful, but their predictive value is limited by the complexity of interactions that occur within the soil environment.

From a biological perspective, the rate of hydrolysis of fluorescein diacetate (FDA) has proved to be one of the most useful indicators of suppression. General suppressiveness results from the activities of the whole microbial community, and since multiple enzymes present in both bacteria and fungi are responsible for FDA hydrolysis, this relatively simple method of estimating total microbial activity provides a relatively good indication of suppressiveness (Hoitink and Boehm, 1999). Other measures that provide information on the biomass, activity and diversity of the microbial community (e.g. soil microbial biomass, soil respiration, phospholipid fatty acid (PLFA) analysis and carbon degradation profiles) have also proved useful (Janvier *et al.*, 2007).

Since general disease suppression is mediated by the activities of many interacting microbial populations, it has been a challenging task to identify even a small proportion of the microorganisms involved. Biocontrol agents have been located using a 'recover and then identify' approach, but they tend to belong to a limited range of culturable genera (e.g. the fluorescent pseudomonads). Although there is evidence to suggest that such organisms are involved in some disease-suppressive situations (Weller *et al.*, 2002), the presence of a particular microbe is never strongly correlated with general suppressiveness. For example, farming practices can alter the relative incidence and abundance of DAPG-producing

Box 9.1. Mechanisms associated with organic matter-mediated disease suppression

The terms 'general', 'non-specific' and 'organic matter-mediated' suppression are used synonymously to refer to situations where the actions of the whole soil biological community reduce the severity of soil-borne diseases. Lack of specificity (i.e. activity against a broad range of pathogens including fungi, oomycetes, bacteria and nematodes), large and regular inputs of organic matter (from roots, crop residues and external sources), high levels of microbial activity and multiple mechanisms of action are features of this phenomenon. The mechanisms associated with this type of suppression were discussed earlier (see 'Plant–microbe–faunal interactions in the rhizosphere' in Chapter 2 and 'Interactions within the Soil Food Web' in Chapter 3), but are briefly outlined below:

Improved soil fertility and plant nutrition

Organic amendments improve soil physical properties, enhance a soil's capacity to hold water and nutrients, and are a source of nutrients that can be used by plants. Disease impacts are reduced because plants growing in amended soil are better able to cope with impaired root function due to soilborne pathogens.

Toxic decomposition products

Some plants contain pre-formed chemicals that are toxic to pathogens. By-products detrimental to pathogens may also be released when organic matter decomposes in soil. A wide range of chemical compounds are involved, including organic acids, phenolics, alkaloids, cyanogens, isothiocyanates, polythienyls, terpenoids and various nitrogenous materials (e.g. ammonia and nitrous acid).

Competition

Since root pathogens operate in a hostile soil environment, there are many competitive mechanisms that can influence the infection process. Infective propagules require nutrients for germination, but the soil microbial community competes with the pathogen for all available nutrient sources, including the root exudates that pathogens use to locate their infection sites. Thus, competition for nutrients is the norm in the rhizosphere, and it results in biostatic effects on spores, chlamydospores, sclerotia and other survival structures. Once pathogens reach the root surface, non-pathogenic and pathogenic strains may compete with each other (e.g. strains of *Fusarium oxysporum*), while endophytes and mycorrhizal fungi limit root colonization by other pathogens. Most plant pathogens also have some saprophytic capacity, but intense competition for organic substrates limits their ability to utilize non-living tissue as a food source.

Antibiosis and lytic activity

Many soluble and volatile antibiotics are produced by bacteria and fungi to help them survive in the competitive soil environment. Research with mutants obtained spontaneously or by genetic manipulation has shown that these antibiotics also act against plant pathogens. Extracellular enzymes produced by bacteria and fungi also act some distance from the site of production and may degrade hyphae, zoospores, conidia, sporangia, oospores and chlamydospores. Some soil fauna (e.g. amoebae, ciliate protozoans, nematodes and mites) feed on hyphae, or ingest or lyse spores and other resting structures.

Predation and parasitism

All soil organisms, including root pathogens, are subject to predation. Thus, bacteria are consumed by protozoa and bacterivorous nematodes; fungi are a food source for mites and fungal-feeding nematodes; and the soil fauna are attacked by higher-level predators. Parasitism is also common, as bacteriophages multiply within bacterial cells; fungal pathogens are subject to mycoparasitism; and plant-parasitic nematodes are parasitized by both bacteria and fungi, and consumed by predatory nematodes and microarthropods.

Induced systemic resistance

Disease-suppressive mechanisms do not necessarily have to act by killing the pathogen or reducing its fitness to attack the plant. In the case of induced systemic resistance, colonization of roots by certain endophytes and plant growth-promoting rhizobacteria induces plant defence responses that limit the capacity of the pathogen to cause disease.

pseudomonads in the rhizosphere, but practices that reduce root disease severity (e.g. chemical seed treatment, rotation and tillage) are not universally linked to increased root colonization by these bacteria (Rotenberg et al., 2007b).

To overcome the limitation of culture-based approaches and better categorize the microbial communities associated with suppressiveness, culture-independent molecular tools can be used to 'first identify and then recover' organisms with suppressive potential (Benítez and McSpadden Gardener, 2009). One culture-independent approach involves the use of genetic markers to locate functionally important microbial activities (e.g. antibiotic production) in field soils (McSpadden Gardener et al., 2005; Joshi and McSpadden-Gardner, 2006). A second approach involves defining a 'signature' or 'community profile' for a disease-suppressive microbial community. In this approach, ribosomal gene sequences are targeted, amplified from the rhizosphere environment and analysed. Terminal restriction fragment length polymorphism (T-RFLP) analyses then provide cost-effective methods of finding generalist populations that consistently contribute to suppression across environments. Such an approach has been used successfully to isolate microbes with pathogen-suppressing activities from soils suppressive to oomycete pathogens (Benítez and McSpadden Gardener, 2009). Another alternative is to combine PhyloChip-based metagenomics with culture-dependent functional analyses, an approach used by Mendes et al. (2011) to identify key bacterial taxa and genes involved in suppression of the fungal pathogen Rhizoctonia solani.

Given the complexity of the soil microbial community, the diversity of the pathogens responsible for soilborne diseases, the wide range of mechanisms involved in disease suppression, and the fact that antagonists at higher trophic levels in the soil food web (e.g. fungal-feeding arthropods and nematodes) are rarely assessed in studies of disease-suppressive soils, it is not surprising that universal indicators of suppressiveness have not been identified. However, this does not mean that efforts to fingerprint the microbial and faunal consortia responsible for general suppressiveness should be abandoned. Advances in high-throughput DNA sequencing and bioinformatics are providing opportunities to capture the heterogeneity and dynamics of complex biological communities in their natural environment. The challenge of the future is to find biomarkers that are consistently associated with disease-suppressive soils and use them to assess the impact of various management practices on the development of suppressiveness.

Organic Matter-mediated Suppressiveness to Plant-parasitic Nematodes

The use of organic matter for nematode control has a long history, as studies in the 1920s (cited by Linford et al., 1938) showed that organic mulches minimized root-knot nematode damage to orchard trees; that sugarbeet cyst nematode was killed by decomposing green manures; and that molasses, compost and a 'mud press' by-product of the sugar milling industry improved the growth of sugarcane and reduced the number of nematodes in roots. However, it was the seminal papers of Linford and his colleagues in Hawaii (Linford, 1937; Linford et al., 1938) that stimulated interest in this area by convincingly demonstrating that root-knot nematode populations declined when the nematode was exposed to decomposing organic matter.

In the Hawaiian experiments, pineapple leaves were the most common source of organic matter, and they were added at different rates to soil naturally infested with root-knot nematode. The amendments were then allowed to decompose for varying periods and their effects on the nematode were assessed by measuring the level of galling on bioassay plants. The main conclusions (based on results from several experiments) were as follows:

• The addition of large quantities of organic matter to soil (generally 1.2–3.6%

w/w, or the equivalent of 24–72 t dry matter/ha incorporated to a depth of 15 cm) consistently reduced the number of galls on indicator plants. The effects on galling were progressively greater as application rates increased.

- When indicator plants were planted after 1, 2, 3, 4 and 5 months of decomposition, galling was markedly reduced at all assessment times. However, the number of galls was lowest after 1 month and then increased. This result suggested that during the initial stages of the decomposition process, the amendment was not only killing juveniles, but also preventing juveniles from invading roots, or increasing the resistance of roots to invasion.

- Many common free-living nematodes (rhabditids, cephalobids and dorylaimids) were favoured by the abundant food supply in amended soil. These nematodes multiplied readily when organic matter was added to soil, and remained more numerous than in non-amended soil for at least 4 months.

- Numerous natural enemies of nematodes were observed in amended soils, including nematode-trapping fungi, predatory nematodes (*Discolaimus* and *Seinura*) and an unidentified predacious mite. Thus, Linford *et al.* (1938) hypothesized that organic amendments greatly increased populations of free-living nematodes and that those populations, in turn, supported an increase in a wide range of parasites and predators of nematodes. Since many of the natural enemies were non-specific, it was assumed that they were capable of killing plant-parasitic nematodes as well as free-living species.

In the years since these initial studies in Hawaii, there have been hundreds of reports on the use of organic amendments for nematode control (collated by Muller and Gooch, 1982; Stirling, 1991; Akhtar and Malik, 2000). More recent reviews of the topic (Oka, 2010; Thoden *et al.*, 2011; McSorley, 2011b; Renčo, 2013) are testimony

to the fact that organic materials are still seen as a potentially valuable tool for managing plant-parasitic nematodes, particularly in high-value annual crops. However, despite more than 70 years of research, organic amendments are still not used routinely in mainstream nematode management programmes; they are rarely discussed in extension articles on nematode management; and were barely mentioned in reviews by Porter *et al.* (1999) and Zasada *et al.* (2010) on the management practices that might be used as an alternative to fumigation with methyl bromide.

One of the reasons why organic amendments are not widely used by farmers is that large quantities must be applied to obtain immediate effects. Based on results from pot experiments, application rates of 1–5% w/w (the equivalent of approximately 20 and 100 t/ha) are commonly required to control nematodes (Stirling, 1991; McSorley, 2011b), and such high rates are a major limitation to the use of organic amendments (Whitehead, 1998). Cost constraints limit the amount of material that can be applied to a field at any one time, and when locally available organic materials are used at economically feasible rates, populations of plant-parasitic nematodes are usually not reduced, and will often increase.

Another reason why organic amendments have not been widely accepted as a nematode management tool is that results are inconsistent. Thus, in the studies assessed in the review by Thoden *et al.*, (2011), populations of plant-parasitic nematodes were reduced in only about half the experiments. However, problems of inconsistency became less of an issue when amendments were separated into groups based on their composition. Slurries prepared from animal manures and other waste products, or produced by anaerobic fermentation of organic materials, were most likely to be detrimental to plant-parasitic nematodes, whereas other materials (e.g. composts, green manures, cover crops and non-composted organic wastes) often had the opposite effect. Thus, in situations where a specific type of amendment is being evaluated in a particular soil type and environment, inconsistency of performance may be less of a problem than a cursory reading of the literature may suggest.

Given that the composition of specific organic amendments may vary from year to year, and that decomposition rates are affected by environmental conditions, the performance of amendments in field situations will always be difficult to predict. Nevertheless, it is important that nematologists come to grips with the inconsistent results often obtained with these materials. Organic matter plays a central role in soil physical, chemical and biological fertility, so as growers conserve organic matter or apply amendments to improve soil health, they need to know whether these practices will have an impact on the nematodes that are affecting their crops. Therefore, they need answers to questions such as:

- In what situations can amendments be reliably used as a primary nematode control tactic?
- Can amendments be integrated with other management practices to provide useful levels of control?
- How much time is required to reduce populations of nematode pests to acceptable levels?
- Are there other benefits from amendments that justify using them in situations where effects on plant-parasitic nematodes may be relatively limited?

The remainder of this chapter is devoted to a discussion of the issues raised by these questions. The manner in which amendments influence plant productivity, soil health and the nematode community is discussed initially. Our knowledge of the processes involved is then used to indicate the benefits likely to be obtained from various types of organic amendments, and the application strategies most likely to be successful against nematodes. Other ways of increasing levels of soil organic matter and enhancing suppressiveness to nematodes (e.g. by manipulating soil and crop management practices) are covered in Chapter 11.

Soil fertility and plant nutrition effects of organic amendments

The role of organic matter in contributing to soil health through its effects on soil physical, chemical and biological properties, and the flow-on effects to plant health and crop production, was discussed in Chapter 2, and is a constant theme that runs throughout this book. Thus, it is not surprising that when organic amendments are applied to soil in an attempt to control nematodes and other soil-borne pathogens, plant growth is improved and crop yields increase. Field studies in Florida and Canada (McSorley and Gallaher, 1995; Kimpinski et al., 2003) are two of many examples of this phenomenon. In the Florida study, a yard-waste compost applied at a very high rate (269 t/ha) increased the yield of transplanted squash and okra seedlings by 133% and 272%, respectively, but did not reduce final population densities of M. incognita, or the level of galling caused by the nematode. The Canadian study assessed the impact of compost applied annually for 7 years at a much more realistic rate (16 t/ha). Yields of potato and barley were consistently improved by the amendment (average yield increases of 27% and 12%, respectively), but populations of the three main plant-parasitic nematodes (Pratylenchus penetrans, Meloidogyne hapla and Heterodera trifolii) were either not affected or were significantly increased by the treatment.

As the authors of the Florida study pointed out in discussing their results, sophisticated experimentation and detailed measurement of numerous soil properties are required to determine whether yield improvements due to an amendment are the result of soil structural effects, improved water-holding capacity, changes in nutrient status or some other factor. Populations of plant-parasitic nematodes tend to increase as yields improve because larger or more highly branched root systems are produced by the plant, and so more food resources are available. In this situation, the improved capacity of plants to cope with relatively high nematode populations when growing in amended soil is testimony to the well-known fact that the environment has a major impact on crop losses due to soilborne pests and pathogens. Thus, rather than viewing organic amendments as a means of reducing nematode populations, it is perhaps better to think of them as conditioners that improve the soil environment (physical, chemical and biological), thereby enhancing crop growth and increasing the plant's capacity to tolerate nematode damage. Nematode populations may sometimes be reduced, but the situations in which this will occur are essentially unpredictable at present.

One conclusion that can be drawn from experiments with organic amendments is that there are invariably nutritional benefits from applying such materials to soil. All amendments contain nitrogen and the other macro- and micro-nutrients required for plant growth, and these nutrients are added to soil in relatively high amounts, given that amendments are usually applied at rates of more than 10 t/ha. The fate of these nutrients depends, in part, on the nature of the amendment. In the case of nitrogen, for example, the C:N ratio of the amendment is important. Nitrogen may become immobilized in microbial tissue when the organic material has a C:N ratio greater than about 20:1, whereas it is mineralized in the form of NH_4^+ or NO_3^- when the C:N ratio is less than that figure. Since the cumulative amount of nitrogen mineralized is directly related to the abundance of opportunistic bacterial- and fungal-feeding nematodes (Ferris and Matute, 2003), an active and enriched soil food web must be maintained to sustain mineralization processes. Nutritional effects are, therefore, most likely to be achieved by frequent inputs of labile organic sources.

The way organic inputs are managed is also important. Decomposition processes are mediated by microbes and other soil organisms (Laakso et al., 2000), and they can be influenced by management practices. Placing an amendment on the soil surface rather than incorporating it into soil causes energy channels within the food web to be more fungal dominant, thereby slowing mineralization rates (Hendrix et al., 1986; Holland and Coleman, 1987; Chen and Ferris, 1999, 2000). In winter-rainfall Mediterranean climates, conditions favourable for activity of the soil food web can be created by growing a cover crop in late summer and irrigating before

biological activity is limited by low winter temperatures (Ferris et al., 2004). Such practices increase the abundance of bacterivore and fungivore nematodes in the following spring, thereby increasing the concentration of mineral nitrogen available to the subsequent summer crop.

Nematicidal compounds from decomposing organic matter

In addition to their effects on soil health and plant nutrition, organic amendments may improve plant growth by releasing chemicals that are detrimental to soilborne pests and pathogens. The chemicals involved are plant metabolites and their products, and since they influence the growth, survival and reproduction of other organisms, they are often referred to as allelochemicals. Phytochemicals with nematicidal properties have received particular attention from nematologists, as they have been seen as a 'natural' alternative to the soil fumigants and nematicides that have formed the backbone of nematode management programmes in intensive agricultural production systems for many years. A huge range of plant-derived secondary metabolites has been identified, and these compounds may influence nematode behaviour by acting as repellents, attractants or toxicants, or by stimulating or inhibiting egg hatch (Chitwood, 2002).

Pre-formed chemicals in plant materials

There is a long history of experimentation with crude plant extracts for nematode control, and that work has resulted in the discovery of many natural chemicals with nematicidal properties, including alkaloids, diterpenes, triterpines, polyacetylenes, thienyl derivatives and sulfur-containing compounds (Akhtar and Mahmood, 1994). However, the complexity of plant by-products and crude preparations from plant tissues means that it is difficult to determine whether the chemicals within such preparations are acting individually or collectively as toxicants, or whether some other mechanism is also involved (e.g. induced plant resistance or biological control).

Nematicidal chemicals have been found in many plant species (Akhtar and Mahmood, 1994), and the leaves, shoots and seeds of these plants, and various by-products, are often used as soil amendments. The neem tree (*Azadirachta indica*) is perhaps the best-documented example. Seeds and other parts of this tree contain a group of chemicals known as limonoids, the most well known being the natural insecticide azadirachtin (Mordue and Blackwell, 1993). Azadirachtin is used to control insects in organic farming systems and is also registered as a nematicide in some countries, but few studies have convincingly demonstrated that it is effective against nematodes. Commercial formulations of azadirachtin did not kill juveniles of *M. javanica* or affect their motility, but they reduced the number of egg masses produced on tomato, and the number of eggs per egg mass (Javed *et al.*, 2007a; 2008). Nematode development was also affected, but effects were due to the induction of resistance mechanisms within roots, or transport of the chemical within the root system (Javed *et al.*, 2007b, 2008). Other studies with *M. incognita* showed that monthly applications of azadirachtin at the recommended dosage (6 L/ha) did not prevent the nematode from invading roots and developing to maturity (Ntalli *et al.*, 2009). Collectively, these results suggest that repeated applications and/or relatively high application rates of azadirachtin are likely to be required to maintain root-knot nematode populations at levels below the economic threshold. They also serve as a warning that pure compounds obtained from neem products are not necessarily nematicidal or nematostatic.

In addition to examining the effects of azadirachtin, Javed *et al.* (2007a, b, 2008) tested crude formulations prepared from dry neem leaves and neem cake and found that they often outperformed the refined product. Such results add to a large body of evidence (Akhtar, 2000) on the effectiveness of neem-based products (oils, oil-cakes, extracts, roots, leaves and root exudates derived from neem) against plant-parasitic nematodes. However, they also raise questions about the chemical basis of the anti-nematodal activity. Given the results of a recent study by Kosma *et al.* (2011),

it is likely that a number of secondary metabolites present in neem seeds (e.g. alkaloids, saponins, triterpenoids, steroids, tannins and phenolics) contribute to the nematicidal and nematostatic properties of such preparations. Some of these chemicals are lost when seed extracts are prepared, and this probably explains why neem seed powder often performs better than formulations produced through various extraction processes.

Polythienyls from marigolds (*Tagetes* spp.) and other Asteraceae are another well-studied example of a pre-formed phytochemical with nematicidal properties. Their effects on nematodes were first recognized more than 50 years ago (Uhlenbroek and Bijloo, 1958, 1959), and since then, many studies have shown that marigolds will reduce populations of plant-parasitic nematodes when they are used as a rotation crop, cover crop or green manure. They are mainly used against *Meloidogyne* and *Pratylenchus*, but since *Tagetes* species vary in their capacity to host various nematodes, efficacy is influenced by marigold species and cultivar (Ploeg, 1999). Also, marigold roots and shoots are relatively ineffective as an amendment, and best results are achieved by planting nematode-susceptible crops into soil that has recently grown marigolds (Ploeg, 2000).

*Chemicals released during
the decomposition process*

Glucosinolates are a class of sulfur-containing compounds that are found in a number of plant families, including the Brassicaceae. Normally they are non-toxic, but when *Brassica* tissue is disrupted, the glucosinolates come into contact with myrosinase (= thioglucosidase), an enzyme that hydrolyses these compounds to isothiocyanates and other biologically active products. Since many of these degradation products are broad-spectrum biocides, the process of suppressing soilborne pests and pathogens through the release of volatile breakdown products from *Brassica* residues has been termed 'biofumigation'.

Matthiessen and Kirkegaard (2006) provide a detailed review of the chemistry of glucosinolates and their derivatives; the biological activity of isothiocyanates; and the factors that affect the efficacy of the biofumigation process. With regard to effects on nematodes, Zasada and Ferris (2003) showed that isothiocyanates varied in their toxicity towards two widely distributed plant parasites (*Meloidogyne javanica* and *Tylenchulus semipenetrans*). When brassicaceous amendments were selected on the basis of their glucosinolate profiles and the sensitivity of the target nematodes, and application rates were based on the relative toxicities of the various isothiocyanates they contained, repeatable levels of suppression were obtained in the laboratory (Zasada and Ferris, 2004). However, results in the field tend to be much less consistent. While it is generally agreed that amending soil with *Brassica* tissue is beneficial (Ploeg, 2008; Oka, 2010), effective nematode control is only likely to be achieved when brassicaceous amendments are combined with other control measures (e.g. solarization or resistant cultivars). One major drawback is that brassicas are often hosts of the nematode being targeted (e.g. root-knot nematodes), and so there is always the danger that nematode populations will increase rather than decrease (McSorley and Frederick, 1995; McLeod and Steel, 1999; McLeod *et al*, 2001; Stirling and Stirling, 2003). Since nematicidal activity is largely localized to the soil layers into which the *Brassica* tissue is incorporated (Roubstova *et al.*, 2007), nematodes will also survive in the soil below the amended layer and later migrate back into the treated zone. The need to completely pulverize *Brassica* tissue to maximize isothiocyanate release, and the requirement for large amounts of water to move the isothiocyanates down the soil profile, are also important practical constraints (Matthiessen *et al.*, 2004; Matthiessen and Kirkegaard, 2006).

Enzymatic degradation processes similar to those in the Brassicaceae also occur in other plants. For example, a glycoside known as dhurrin is present in leaf tissue of Sudan grass, and in hybrids of sorghum and Sudan grass, and when that tissue is disrupted or begins to decompose, the enzyme β-glucosidase is released from plant cells and dhurrin is hydrolysed (Widmer and Abawi, 2000). One of the end-products of this reaction is HCN, which is toxic to many different organisms, including nematodes (Robinson and Carter, 1986). Studies by Widmer and Abawi (2000) showed that purified extracts containing cyanide were the only fractions from Sudan grass to suppress infection of lettuce roots by *Meloidogyne hapla*, and since egg hatch and infection were affected by cyanide concentrations as low as 0.1 ppm, the authors concluded that cyanide was the primary factor responsible for the suppressive effect of Sudan grass green manures.

The contribution of phytochemicals to the nematicidal effects of organic amendments

What should be clear from the previous discussion is that plants are a source of potentially useful chemical compounds with the capacity to kill nematodes or influence their behaviour in some way. However, only a small number of those chemicals have been mentioned here. Many other naturally occurring compounds with nematoxic properties have been identified, but as Chitwood (2002) pointed out in his review of the subject, evidence of activity against nematodes is often based on *in vitro* tests with relatively high concentrations of the chemical. What is more relevant from a practical perspective is whether the much lower concentrations that are present within plant cells have an impact on nematodes when green manures, rotation crops and plant-based amendments are incorporated into soil.

Halbrendt (1996) discussed this issue when reviewing the role of allelochemicals in nematode management. He pointed out that it is difficult to prove that compounds extracted from the leaves and roots of plants affect nematodes under natural conditions, as there is no satisfactory protocol that can be used to prove that allelopathy actually occurs. Nonetheless, it is possible to demonstrate a cause-and-effect relationship by showing that the following occur in sequence: (i) the allelochemical is produced; (ii) the chemical is transported through the environment to the target organism; and (iii) the target organism is exposed in sufficient quantity and for sufficient time to be affected.

268 Chapter 9

However, given that soil is a complex and dynamic medium and that the fate of any allelochemical will vary with soil type and environmental conditions, it is unlikely that a close relationship between the production of an allelochemical and an effect on a nematode pest will ever be consistently obtained. Allelochemicals probably play a major role in the nematode control that is obtained when soil is amended with large quantities of certain plant materials (e.g. neem, brassicas or forage sorghum), but they probably have much more limited and relatively subtle effects in most other situations, particularly those where amendments are applied at economically realistic rates.

Nematicidal properties of nitrogenous amendments

Of all the nematicidal chemicals released from organic matter during the decomposition process, ammonia is the most widely studied, and, therefore, deserves specific attention. Its nematicidal activity was first recognized in the 1950s, when anhydrous ammonia was being assessed as a nitrogenous fertilizer. High ammonia concentrations were recorded in localized areas near the injection point when anhydrous ammonia was injected into soil, and nematode numbers were reduced substantially when concentrations reached 300–700 mg/kg soil (Eno et al., 1955). This nematicidal effect has since been confirmed by others, including Rodriguez-Kábana et al. (1981, 1982), whose glasshouse experiments indicated that ammonia concentrations in excess of 300 mg/kg soil were required to produce good nematode control. Additional experiments with urea, which is readily converted to ammonia by the urease present in soil, showed that it is also a good nematicide if applied at levels in excess of 300 mg N/kg soil (Miller, 1976; Rodriguez-Kábana and King, 1980; Huebner et al., 1983; Chavarría-Carvajal and Rodriguez-Kábana, 1998).

Nitrogen is a constituent of almost all the organic materials used as soil amendments, and so ammonia is invariably produced when these materials decompose in soil. Since the amount of ammonia generated varies with

the level of nitrogen in a particular organic substrate, there is a relationship between the nitrogen content of an amendment and its effectiveness as a nematicide (Singh and Sitaramaiah, 1971; Mian et al., 1982; and Fig. 9.1). Animal manures, oil-cakes and green manures and hays from leguminous crops provide better nematode control than grassy hays and stubbles, and cellulosic materials such as sawdust. However, nitrogen content is not the only factor to be considered when organic materials are being used for nematode control. The carbon content of an amendment is also important, because there must be

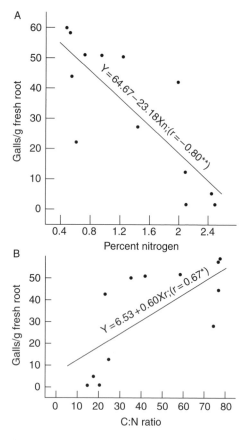

Fig. 9.1. Relationship between the nitrogen content of an organic amendment (A) or the C:N ratio of an amendment (B) and galling caused by *Meloidogyne arenaria* in squash plants grown in soil treated with 1% w/w of various amendments (From Rodriguez-Kábana et al., 1987, with kind permission from Springer Science + Business Media B.V.)

enough carbon available to permit soil micro-organisms to metabolize the nitrogen and convert it into proteins and other nitrogenous compounds. In the absence of a readily available carbon source, ammonia can accumulate and cause phytotoxicity. Materials such as chitin, urea and some oil-cakes and animal manures have a low C:N ratio and since they may be deleterious to plants at high application rates, these materials are almost always applied prior to planting. Amendments with a C:N ratio greater than 14 are not phytotoxic (Rodriguez-Kábana, 1986; Rodriguez-Kábana et al., 1987), and this means that the toxicity associated with the use of highly nitrogenous materials can be overcome by adding a carbon supplement (Rodriguez-Kábana and King, 1980; Huebner et al., 1983; Culbreath et al., 1985). The nematicidal activity of wastes with a high C:N ratio can also be improved by adding urea or other sources of nitrogen (Miller et al., 1968; Johnson, 1971; Stirling, 1989; Stirling and Nikulin, 1998), but as ammonia is released gradually from such amendments, the concentrations achieved in soil may be sub-lethal to nematodes.

Although the nematicidal effects of nitrogenous soil amendments are well recognized (Rodriguez-Kábana, 1986; Oka, 2010), it is questionable whether such amendments will play more than a minor role in future nematode management programmes. The major limitations are the costs involved in acquiring and applying large amounts of these materials, and the health and environmental problems likely to occur when the ammonia generated is oxidized to nitrate and finds its way into groundwater. Also, there are a number efficacy issues. First, ammonia has poor diffusion characteristics (Eno et al., 1955), and so the nematicidal effects of nitrogenous amendments tend to be limited to the zone of incorporation. Second, high concentrations of ammonia are difficult to achieve in some soils because the equilibrium between ammonia and ammonium is affected by pH. Thus, alkaline materials may have to be added to improve efficacy in soils of low to neutral pH (Oka et al., 2006a, b).

Third, ammonia released from nitrogenous amendments is rapidly oxidized by nitrification processes, and nitrification inhibitors may be needed, particularly in alkaline soils, to maintain ammonia concentrations at effective levels (Oka and Pivonia, 2002).

Although slaked lime and other alkaline materials have been used in Israel to enhance the efficacy of ammonia-releasing amendments (Oka et al., 2006a, b, 2007), it is not entirely clear whether ammonia is always responsible for the observed nematode mortality. In the United States, for example, research has shown that alkaline reagents such as cement kiln dust, fly ash or quicklime can be added to municipal biosolids to produce granular materials with desirable agronomic properties. Initial studies indicated that plant-parasitic nematodes were suppressed by such products and that chemically mediated mechanisms (including ammonia production) were involved (Zasada and Tenuta, 2004). However, later work showed that suppressiveness was primarily due to a rapid increase in soil pH to levels between 10 and 11.9 (Zasada, 2005).

Enhancing biological control mechanisms with organic amendments

Impact of amendments on natural enemies, particularly nematophagous fungi

Although the observations of Linford and his colleagues in the 1930s suggested that the nematode control achieved with amendments of pineapple residue was due to the activity of natural enemies, evidence to support that claim was largely circumstantial. Predacious animals (mononchid and dorylaimid nematodes and a mite capable of 'devouring nematodes freely') multiplied in soil amended with pineapple leaves, and mycelium and traps of several species of nematode-trapping fungi were observed on living and dead nematodes that were freshly washed from soil (Linford et al., 1938). However, an experiment in which nematode-trapping fungi were inoculated into soil failed to address the issue of whether they contributed to the nematode control obtained with organic amendments, as predacious activity in amended and non-amended soil was not compared. Also, all inoculated treatments and the control were contaminated with a number of nematophagous fungi and various predatory nematodes (Linford and Yap, 1939).

Today, enhancement of natural enemies is still considered one of the reasons that organic amendments often reduce populations of plant-parasitic nematodes. However, few studies have convincingly demonstrated that effects on nematodes are at least partly due to biological control mechanisms. It is certainly true that successional biotic changes take place within the soil food web when organic matter is added to soil, but proving that particular organisms are directly responsible for any observed nematode mortality remains a challenge.

Since Linford and his co-workers believed that the predatory activity of the nematode-trapping fungi played a major role in the nematode control associated with organic amendments, most early studies focused on this group of fungi, and that work (reviewed by Cooke, 1968; Gray, 1987; and Barron, 1992) resulted in a number of conclusions:

- Nematode-trapping fungi survive on organic substrates as saprophytes, but in the highly competitive environment associated with decomposing organic matter in soil, they are able to utilize nematodes as an alternative nutritional source.
- Fungi trap nematodes for only a relatively short period after the onset of decomposition.
- Nematode-trapping species differ in their growth rates, competitive saprophytic ability and predacious activity. Network-forming fungi grow rapidly and are relatively good saprophytes, but do not produce traps spontaneously and are inefficient predators. In contrast, branch-, knob- and constricting ring-forming species grow more slowly and are poorer saprophytes. They tend to produce traps spontaneously and are more efficient predators.
- Populations of free-living nematodes increase when organic matter decomposes in soil, but trapping activity and nematode populations are not always related. Predacious species tend to show a typical predator–prey relationship, whereas the

trapping activity of saprophytic species generally peaks before the peak in nematode numbers.

More recent studies (summarized in Chapter 5) continue to show that saprophytic and predacious species of nematode-trapping fungi respond quite differently to organic inputs. *Arthrobotrys oligospora*, a network-forming fungus and one of the most common saprophytic species, increased its population density when organic matter was added to field soil (Jaffee, 2002, 2004a), but only responded to substrates with a low C:N ratio (Nguyen *et al.*, 2007). Also, the relationship between fungal and nematode populations was relatively weak, as few nematodes were trapped by the fungus. Such results raise questions about why these fungi invest resources in specialized hyphae that are capable of capturing nematodes. Barron (1992) argued that the nematode-trapping habit is an evolutionary response by cellulolytic fungi to nutrient deficiencies in nitrogen-limiting habitats. The saprophytic nematode-trappers are efficient cellulase producers and survive well in carbon-rich habitats. They trap nematodes as a supplementary source of nitrogen and possibly other nutrients, and stop producing traps when their nutrient demands have been satisfied. Thus, the duration of trapping activity is usually relatively short, and is not related to nematode population density.

Dactylellina candidum, a predacious species that produces stalked, adhesive knobs, responds in a different way to inputs of organic matter. Populations of this fungus increase in amended soil and a period of trapping activity then follows (Jaffee, 2003, 2004a). However, the interactions between fungus and nematode do not exhibit classical predator–prey dynamics. After an initial increase, the fungal population declines while nematode populations continue to rise, possibly because the fungus is suppressed by the fungivorous nematodes that are usually abundant in organically amended soil (Jaffee, 2006). The possibility that fungivores or various mycostatic factors eventually limit the predatory capacity of these fungi is supported by the results of Jaffee (2004a), who found that the population density and trapping activity of *D. candidum* was most enhanced by small rather than large inputs of organic matter. Such results suggest that although an amendment may initially stimulate predatory activity in some nematode-trapping fungi, this positive effect will eventually be balanced by the negative effects of competing organisms.

Fungi that parasitize nematode eggs have not been studied to the same extent as the nematode-trapping fungi, but parasitism of root-knot nematode eggs by indigenous fungi increased when soil was amended with various organic materials (Rodriguez-Kábana et al., 1994). With regard to endoparasitic fungi, *Drechmeria coniospora* parasitized the bacterivorous nematodes that multiplied when a lucerne meal amendment was applied to soil (Van den Boogert et al., 1994), while the number of entomopathogenic nematodes parasitized by naturally occurring fungi increased markedly when soil in a citrus orchard was mulched with manure and straw (Duncan et al., 2007). In contrast, parasitism by *Hirsutella rhossiliensis* was not enhanced by applications of composted chicken manure, wheat straw or composted cow manure (Jaffee et al., 1994). Given the result with *H. rhossiliensis*, it should not be assumed that organic amendments will always increase the activity of endoparasitic fungi. Potential food sources (free-living nematodes), potential predators (fungal-feeding nematodes and microarthropods) and potential competitors (organisms with mycostatic capabilities) inevitably increase when organic matter is added to soil, and so the outcome in terms of parasitic activity is not easily predictable.

The capacity of different types of organic matter to enhance biological mechanisms of nematode suppression

Since most of the research on organic amendments for nematode control has focused on achieving nematicidal effects through chemical mechanisms, there is a dearth of information on the role of organic materials in stimulating a biological community capable of suppressing plant-parasitic nematodes. Nevertheless, there is a limited amount of evidence to suggest that amendments of quite different composition can act against nematodes through biological mechanisms.

CHITIN AMENDMENTS. Evidence presented earlier in this chapter indicated that chitin, one of the most abundant polysaccharides in nature and a universal waste product of the seafood-processing and fermentation industries, is effective against nematodes when applied to soil at rates greater than 1% w/w, largely because ammonia is produced during the decomposition process. However, chitinous materials may also act by stimulating a microflora that is antagonistic to nematodes, as detrimental effects on plant-parasitic nematodes are known to persist long after ammonia has dissipated, and are also observed at application rates that do not generate high concentrations of ammonia (Culbreath et al., 1986; Rodriguez-Kábana et al., 1987; Spiegel et al., 1987). Evidence to support the involvement of microbial antagonists in the suppressive effects of chitin is largely based on observations that a relatively specific community of chitinolytic fungi, bacteria and actinobacteria develops following the addition of chitin to soil (Rodriguez-Kábana et al., 1984; Spiegel et al., 1987); that some of these fungi are known parasites of nematode eggs (Godoy et al., 1983a; Rodriguez-Kábana et al., 1984); and that chitinolytic bacteria will kill nematodes in laboratory tests (see Chapter 8). However, if microorganisms were acting as suppressive agents in the studies cited above, their mode of action does not appear to have involved parasitism. Non-viable eggs of *Heterodera glycines* were not parasitized by fungi in chitin-treated soil, leading to the suggestion that metabolites produced during the decomposition process may have been responsible for the nematode control observed (Rodriguez-Kábana et al., 1984).

More recent studies with amendments containing chitin (e.g. Sarathchandra et al., 1996; Hallmann et al., 1999; Kokalis-Burelle et al., 2002a, b; Jin et al., 2005) have continued to show that they stimulate a chitinolytic microflora in the soil and rhizosphere, and that populations of plant-parasitic nematodes are reduced. However, the modes of action of chitinous amendments have not yet been clarified. Numerous *in vitro* studies have shown that culture filtrates and chitinases from chitinolytic fungi and bacteria affect egg morphology, egg hatch and the mobility and mortality of juvenile nematodes (see a review by Gortari and Hours (2008) on fungal chitinases, and the section on chitinolytic bacteria in Chapter 8), but there is little evidence to show that microbially produced enzymes and metabolites occur in soil at concentrations high enough to be detrimental to nematodes. Until transformants with varying capacities to produce particular enzymes or metabolites are compared for efficacy, both with each other and with wild-type strains, the role of microorganisms in the nematicidal effects of chitin amendments will remain unclear.

COMPOSTS. Crop residues, cover crops and manures are the most common organic inputs in many farming systems, and their role in enhancing general suppressiveness to nematode pests is discussed in Chapter 11. However, composts are often a better option for high-value horticultural and vegetable crops, as cover crops that might increase populations of plant-parasitic nematodes can be avoided, and valuable land can be used entirely for income-producing crops. Thus, in the nematological studies reviewed by Thoden *et al.* (2011), composts made from animal manures and various other wastes were the most common form of organic amendment investigated. Although it is difficult to draw firm conclusions from those studies due to the multiplicity of composts tested, the range of application rates used and the variation in assessment times following treatment, composts almost always increased numbers of microbivorous nematodes, and often had the same effect on populations of plant-parasitic nematodes. Since plant growth generally improved in compost-amended soil, the latter result was not unexpected, as increased root biomass would have provided more resources for herbivores. However, composts were generally incorporated into the soil by tillage, and since this practice is known to be detrimental to many natural enemies of nematodes (see Chapters 5, 6 and 7), it may have been the reason that suppressiveness was not enhanced. Thus, the critical question from a biological control perspective should be: is a soil's suppressiveness to nematodes more likely to be improved by avoiding tillage and continually applying compost amendments to the surface of an undisturbed soil? The long-term experiments required to answer

that question have, to my knowledge, never been established.

SAWDUST AND MOLASSES. The focus on using chitin and other nitrogenous amendments to achieve short-term nematode control through ammonia production, and the widespread interest in composts that are generally made from nitrogenous materials, has meant that options for enhancing suppressiveness through the use of high-carbon/low-nitrogen amendments have not been explored to the same extent. Nevertheless, studies on apples and asparagus cited later in this chapter showed that sawdust (C:N ratio of approximately 400:1) substantially reduced nematode populations on perennial crops, provided it was applied as mulch and kept in place for several years. Sawdust will also enhance suppressiveness to plant-parasitic nematodes in annual crops, as four successive applications of sawdust plus urea (22, 17, 7 and 1 month prior to planting) reduced populations of *Meloidogyne javanica* on tomato to the point where plants were almost free of galls (Vawdrey and Stirling, 1997). The mode of action of the amendment was not determined, but the presence of high numbers of fungivorous nematodes indicated that the biological community in the sawdust-amended soil was strongly fungal dominant, leading the authors to speculate that the cellulolytic and lignolytic nematophagous fungi commonly associated with rotting wood (Barron, 1992) may have been responsible for the high level of suppressiveness. In another study in a much colder environment, Brzeski and Szczech (1999) also observed high populations of fungivorous nematodes at a site where sawdust (8 t/ha) had been applied each autumn for six consecutive years. Improvements in the soil's physical properties (bulk density and water-holding capacity) and an increase in populations of *Trichoderma*, a well-known mycoparasite, were also observed, demonstrating that a range of soil health benefits can be obtained from continual inputs of sawdust.

Molasses, which has sometimes been used in the sugar industry as a fertilizer and soil improver, is another high-carbon amendment known to affect plant-parasitic nematodes.

Initial work in pots indicated that the number of eggs produced by root-knot nematode, and galling caused by the nematode, were reduced when molasses was applied at application rates of 1–10 g/L soil (Rodriguez-Kábana and King, 1980; Vawdrey and Stirling, 1997). These results prompted further work to determine whether molasses was effective in the field. In an experiment with tomatoes in Australia, molasses was injected into the trickle irrigation system every week for 12 weeks at an application rate of 375 L/ha/week, which was equivalent to applying 7 t molasses/ha over the life of the crop. Galling was significantly reduced and the level of control was similar to that obtained with the nematicide fenamiphos (Vawdrey and Stirling, 1997). In Hawaii, molasses applied through the irrigation system or by an overhead boom sprayer generally improved plant growth and also reduced populations of *Rotylenchulus reniformis* on papaya and *Heterodera schachtii* on Chinese cabbage (Schenck, 2001). Although the reasons for the detrimental effects of molasses on nematodes were not determined in either of these studies, it was presumed that at least some biological mechanisms were involved.

Amendments with a high C:N ratio: are they the key to more sustained suppressiveness?

Much of the work on organic amendments for nematode control (particularly the work with nitrogenous amendments discussed earlier) has sought to obtain immediate effects, and to achieve them through chemical rather than biological mechanisms. In an attempt to determine whether amendments could be used to induce longer-term biological suppressiveness, Stirling *et al.* (2003) added eight different organic materials to soil in a sugarcane field and assessed changes in various chemical and biological properties over the following 12 months. Changes in suppressiveness were monitored by periodically collecting soil from the field and conducting bioassays with root-knot nematode (*Meloidogyne javanica*) and lesion nematode (*Pratylenchus zeae*) in small containers and in pots. Results from both types of bioassay showed that

amendments with a high C:N ratio (e.g. sawdust, grass hay and sugarcane trash) induced suppressiveness to plant-parasitic nematodes 4 and 7 months after they were added to soil, whereas the soil amended with nitrogen-rich materials (e.g. lucerne hay, feedlot manure, poultry manure, chitin and mill mud) was not suppressive. Some of the sawdust and sugarcane-trash treatments were still suppressive after 12 months. Multiple regression analysis indicated that suppressiveness was associated with low levels of nitrate–nitrogen in soil, a fungal-dominant soil biology and high numbers of omnivorous nematodes.

Given that amendments with a high C:N ratio had a long-term impact on suppressiveness, and that crop residues remaining after harvest are a readily available source of such material in the sugar industry, the impact of this form of organic matter on suppressiveness to *Pratylenchus zeae* was assessed in a field experiment (Stirling *et al.*, 2005). Sugarcane residue (C:N ratio of 67:1) was incorporated into soil without nitrogen, or additional nitrogen was supplied as either soybean residue or ammonium nitrate. The carbon inputs from plant material (10 t C/ha) were the same in all treatments, while the plus-nitrogen treatments received 210 kg N/ha. Sugarcane was planted 23 weeks after amendments were incorporated, and 24 weeks later, population densities of *Pratylenchus zeae* in roots (averaged across all amendment treatments) were 95% lower than the non-amended control. All amendments increased readily oxidizable carbon, microbial biomass, microbial activity and numbers of free-living nematodes, but had no impact on dorylaimid or mononchid nematodes or three naturally occurring species of nematode-trapping fungi. However, an unidentified predatory fungus was found in soil amended with sugarcane trash, but was never seen in non-amended soil.

Stirling *et al.* (2005) suspected that fungal predation on nematodes was an important component of the suppressiveness observed in the above experiment, and hypothesized that nematophagous fungi became dominant when the C:N ratio of the amendment was high enough to immobilize most of the nitrogen in soil. However, the authors also indicated that amendments such as sugarcane trash probably act in many different ways, as carbon inputs have a cascading effect on all trophic levels in the soil food web. When organic matter is added to soil, bacteria and fungi (trophic level 1) are fed on by consumers such as protozoa and free-living nematodes in trophic level 2; they, in turn, are consumed by predators in trophic level 3 (e.g. nematophagous fungi, predatory nematodes and predatory arthropods). Thus, the food web in amended soils consists of many diverse and interacting elements, and its biological buffering capacity tends to prevent any component (e.g. a nematode pest) from becoming dominant. In this situation, it is never completely clear which organisms occupy the highest trophic level, as turnover is rapid and many unknown predators may be involved.

In another study comparing the capacity of different organic amendments to suppress plant-parasitic nematodes, Tabarant *et al.* (2011b) found that sugarcane bagasse (C:N ratio of 39) was much more suppressive than amendments with lower C:N ratios (e.g. plant residues, sugarcane refinery sludge and sewage sludge with C:N ratios of 25, 17 and 7, respectively). About 14 weeks after bagasse was incorporated into soil, the number of plant parasites (predominantly *Rotylenchulus reniformis*) in banana roots was reduced by 96%, while the abundance of carnivorous nematodes increased markedly. The plant residue amendment also had an impact on plant-parasitic nematodes, but the amendments with a low C:N ratio had no effect. Thus, suppressiveness was only enhanced by organic materials that decomposed slowly and favoured the development of a fungal decomposition pathway in soil.

Another high-carbon amendment that could possibly have a long-term impact on agricultural soils is biochar, a charcoal-like product that is created from organic wastes by pyrolysis (a process involving thermal degradation of biomass in the absence of air). Although initially developed as a means of sequestering carbon and mitigating climate change, it is now clear that biochar can improve soil health by influencing soil structure, changing soil pH, increasing cation exchange capacity and reducing ammonium

leaching (Atkinson *et al.*, 2010). Impacts on the soil biota are not as well researched, presumably because biochar is seen as a relatively inert material that is unlikely to influence soil organisms. Nevertheless, most studies have shown that biochar additions increase microbial biomass, change microbial community composition, influence soil enzyme activities and improve root growth (Lehmann *et al.*, 2011). Since the addition of biochar to soil induces resistance in plants to foliar pests and pathogens (Elad *et al.*, 2010) and increases the abundance of root-associated microbes with plant growth-promoting and/ or biocontrol properties (Graber *et al.*, 2010; Kolton *et al.*, 2011), it is possible that biochar will not only improve the capacity of crops to cope with nematode damage, but also enhance suppressiveness to nematodes. One advantage that biochar has over other organic amendments is that it remains relatively unchanged after application to soil. Thus, any soil fertility benefits obtained with biochar could be expected to remain in place for a long period.

To conclude, materials with a very high C:N ratio such as sawdust strongly immobilize nitrogen and this means that they are difficult to use as soil amendments. Nevertheless, the studies discussed suggest that cellulosic amendments with lower C:N ratios (e.g. residues from grass crops) could possibly be used to stimulate the soil fungal community and perhaps enhance suppressiveness to nematodes. However, the results of an experiment in which sorghum straw was used as an amendment and incorporated into the soil annually for 26 years raise questions about the utility of that approach. Although a fungal-dominant decomposition channel was generated and populations of plant-parasitic nematodes remained low, the latter effect was probably the result of low root biomass production by the sorghum crop due to inadequate nutrition (Villenave *et al.*, 2010). Thus, perhaps the best way of using crop residues and similar materials as amendments is to leave them on the soil surface as mulch (see discussion later in this chapter). A community of fungi and microarthropods will inevitably develop within and under the undisturbed mulch layer and act as suppressive agents, and the challenge is then to fertilize the crop in such a way that plant growth is optimized without compromising the soil's suppressiveness to nematodes.

Temporal effects of amending soil with organic matter

The studies cited highlight the difficulties involved in determining whether organic amendments impact on nematodes through biological mechanisms. However, the short-term nature of most experiments with amendments raises another important issue. Parasitic or predacious activity is usually assessed within a few weeks of applying an amendment, at a time when competitive forces induced by the added organic matter are at their peak, and before successional changes have taken place within a biological community that has been disturbed by organic inputs. This point is perhaps best illustrated by reference to work on succession within the nematode community. When residues of sunn hemp decomposed in a warm, subtropical environment, there was an initial proliferation of enrichment opportunists (bacterivorous nematodes with short life cycles). After about 14 days, these nematodes were replaced by general opportunists (fungivores and bacterivorous Cephalobidae), with omnivorous and predatory nematodes only becoming a relatively important component of the community from about 42 days (Wang *et al.*, 2004a). In a much cooler environment, persister nematodes (omnivores and predators) also responded to the addition of organic matter (in this case cow manure), but their populations did not peak until nearly 1 year after the manure was applied (Ettema and Bongers, 1993). In other studies, organic amendments applied prior to planting an annual crop had no measurable impact on the number of predatory nematodes present at harvest (Koenning and Barker, 2004; Ferris *et al.*, 2004). Such results demonstrate that it may take months or even years for the predatory component of the nematode community to respond to organic inputs. This time lag is likely to be particularly lengthy in situations where the soil biology has been compromised by agricultural practices such as

frequent tillage, infrequent use of cover crops and soil fumigation.

The importance of taking a long-term view of the effects of organic amendments on biological control agents is apparent from the results of a study by Minor and Norton (2004). Biosolids (lime-stabilized sewage sludge) and chicken manure compost were used as amendments, and 2, 3, 4 and 5 years after they were applied to soil, effects on soil mites were assessed. From the perspective of soil nematodes, the data on mesostigmatid mites are particularly relevant, as they are important predators of free-living and plant-parasitic nematodes (Koehler, 1997, 1999). Forty species of predacious Mesostigmata were identified at the experimental site; they were favoured by the application of biosolids and chicken manure compost; and the effects of the amendments persisted for several years after they were applied. Although predation on nematodes was not confirmed, the central role of nematodes in the nutrition of mesostigmatid mites (Fig. 9.2) suggests that significant numbers of nematodes must have been consumed. Such a result also highlights the difficulties involved in obtaining a true picture of the biological effects of organic amendments. Monitoring must continue for a considerable period after an amendment is applied, and the expertise required to monitor all the natural enemies of nematodes is rarely available.

Incorporation of amendments versus mulching

One problem with using organic amendments to enhance natural mechanisms of nematode suppression is that amendments are usually incorporated into the soil, presumably in an attempt to expand effects to greater depths in the soil profile. However, as has been mentioned many times in this book, tillage is detrimental to many soil organisms (Wardle, 1995; Lenz and Eisenbeis, 2000; Berkelmans *et al.*, 2003), with some natural enemies of nematodes (e.g. fungi, microarthropods and the larger predatory nematodes) being particularly sensitive to mechanical disturbance. Thus, the effects of organic amendments on biocontrol mechanisms are almost certainly compromised by tillage.

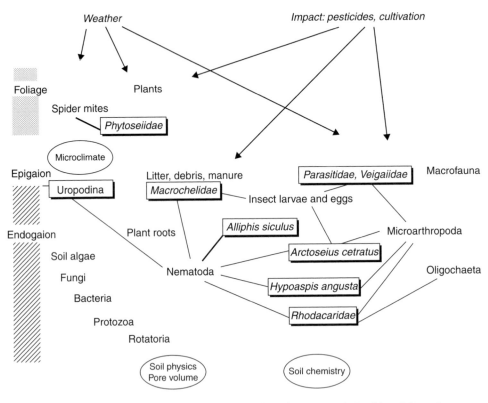

Fig. 9.2. Predatory mites (Gamasina) at an ecosystem level. Predator–prey relationships of Gamasina species (in boxes) are indicated by lines, with bold lines showing food preferences. (From Koehler (1999), with permission.)

Since the effects of tillage on organic amendments are rarely examined, the results of a microplot experiment in a tomato/soybean/maize cropping system (Treonis *et al.*, 2010a) provide some insights into the biological effects of annual applications of amendments in the presence or absence of tillage. A low C:N amendment of biosolid, compost or vetch was applied in spring and it was followed by a high C:N amendment of baled straw in autumn. Organic matter levels, respiration and numbers of protozoa, fungi and nematodes responded positively to the organic amendment, and although these effects were more pronounced at a depth of 0–5 cm, they were also observed at 5–25 cm. Incorporation of the amendments into soil rather than leaving them on the surface increased all these parameters at 5–25 cm. In amended soil, some PLFA biomarkers for bacteria and fungi were lower under no-till

than tillage at both depths, but nevertheless, the amendments improved most parameters, regardless of tillage. However, the proportion of omnivore and predator nematodes in amended soil at a depth of 5–25 cm was much higher in non-tilled than tilled soil, reinforcing the point that tillage may limit the response of this group of predators to the addition of organic matter.

Since the samples analysed by Treonis *et al.* (2010a) were collected a little over 2 years after amendment and tillage treatments were first imposed, and only months after the most recent additions of organic matter, the soil biological community would have been in transition to a new equilibrium when the data were obtained. Nevertheless, there is clear evidence from perennial cropping systems to show that suppressiveness to plant-parasitic nematodes is enhanced when undisturbed soil is mulched with organic matter to create a 'forest floor' environment at the soil surface. Mulches based on sawdust appear to be particularly effective, as populations of *Pratylenchus* on apple and *Helicotylenchus* on asparagus were markedly reduced after mulching for 5 and 9 years, respectively (Stirling *et al.*, 1995; Yeates *et al.*, 1999). Similarly, mulching citrus trees with composted manure reduced populations of citrus nematode (*Tylenchulus semipenetrans*) by 99% (Duncan *et al.*, 2007). In a microcosm experiment with sugarcane, mulching soil with sugarcane residue enhanced suppressiveness to *Meloidogyne javanica* and *Pratylenchus zeae* to a greater extent than incorporating the residue into soil (Stirling *et al.*, 2011a), while low populations of plant-parasitic nematodes and high levels of suppressiveness were observed in field soils that had been covered with sugarcane residue for several years (Stirling *et al.*, 2011b).

Microarthropods are one group of natural enemies that definitely benefit from the presence of a surface layer of organic matter, as evidenced by the following examples: (i) a study in which maize plots were mulched with woody residues of three plant species, or the stems and straw from maize and rice, clearly showed that populations of microarthropods increased, regardless of the type of mulch (Badejo *et al.*, 1995); (ii) populations of mites and collembolans increased markedly within a few months of establishing a living mulch of white clover between rows of maize. The communities of soil organisms under the mulch reached a more mature successional stage during the second season, and the microarthropod component was characterized by high population densities of higher trophic groups such as mesostigmatid mites (Nakamoto and Tsukamota, 2006); and (iii) observations in a field experiment in Japan showed that one of the main reasons the abundance of microarthropods is enhanced by living mulch is that the litter layer which develops under the mulch is an ideal habitat for Oribatida, Prostigmata and Collembola (Kaneda *et al.*, 2012). Thus, from a nematode-suppression perspective, it is clearly advantageous to apply organic amendments to the soil surface, as this should promote the activity of microarthropods, many of which are specialist or generalist predators of nematodes. Other benefits likely to be obtained from mulching include improved soil physical properties, better soil moisture retention and cooler soil temperatures during summer.

The way forward: combining multiple mechanisms of action

It is clear from the material presented in this chapter that growing a range of crops and preserving their residues, retaining cover crops as green manure or mulch, or adding an external source of organic matter to soil, initiates a series of transformations that directly or indirectly affect soilborne pathogens, and ultimately, plant growth. However, the complexity of the processes involved (see Fig. 9.3) tends to result in reductionist thinking. Thus, instead of considering the collective effects of multiple mechanisms of action, research on organic amendments for pest and disease control has usually focused on single mechanisms. The huge volume of literature on the nematicidal and nematoxic effects of plant-derived chemicals is testimony to that fact. A specific chemical is either identified in plant tissue or found to be released during the decomposition process, and the organic amendment is then simply seen as a means of delivering a 'natural' pesticide to a target.

This narrow view of organic amendments (as a repository of useful chemicals) is most apparent with the concept of 'biofumigation'. Initially considered to involve the release of biocidal compounds from brassica residues, and sometimes widened to cover degradation products from any organic material (Bello *et al.*, 2004), the term 'biofumigation' implies that the active ingredients are volatile and act in much the same way as broad-spectrum fumigants such as methyl bromide and 1, 3-dichloropropene. However, these criteria cannot be met for several reasons. First, methyl isothiocyanate (one of the chemicals released from decomposing *Brassica* residues) is not gaseous at ambient temperatures, and principally moves through soil by diffusing

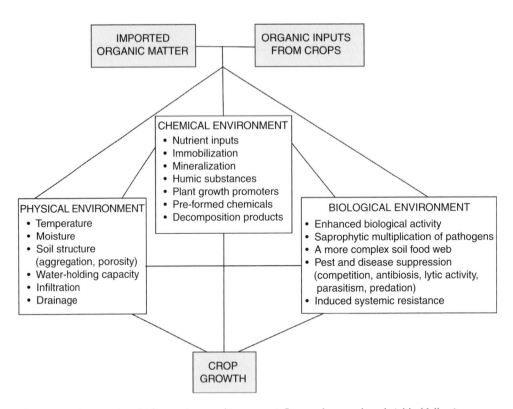

Fig. 9.3. Mechanisms by which organic amendments can influence the growth and yield of following crops.

in the water phase (Smelt and Leistra, 1974; Lembright, 1990). Many other chemicals that are released during the breakdown process are also non-volatile. Second, glucosinolates play no role in the suppression of some diseases (e.g. *Rhizoctonia* root rot of apple) by brassicaceous amendments. Instead, populations of resident *Streptomyces* spp. increase and they induce a defence response in the plant (Mazzola, 2007). Third, results from field trials over many years have shown that amendments with 'biofumigant' properties do not produce the consistent results achievable with commercial fumigants. Nevertheless, the term 'biofumigation' has been popularized to such an extent that it creates unrealistic expectations within the farming community, with growers using the tactic no longer thinking holistically about the range of soil health benefits that are achievable when organic matter is added to soil. It would, therefore, be better if the term 'biofumigation' was no longer used.

The use of organic amendments to produce ammonia is another example of a reductionist approach to organic matter management. Nitrogenous amendments are applied with specific and relatively short-term aims in mind: to produce concentrations of ammonia and other nitrogen-containing substances that are sufficient to control the target pathogen. The damage done to beneficial biota when the soil is mechanically disturbed to incorporate the amendment; the detrimental effects to non-target organisms of nitrogenous compounds that are present at high concentrations; and the potential for nitrate pollution of waterways and the environment are problems that are usually ignored.

The expectation that immediate improvements in plant growth or reductions in populations of plant-parasitic nematodes will be obtained when an organic amendment is added to soil is yet another example of a failure to think holistically about the long-term benefits of soil organic matter. Instead of encouraging farmers to change the practices that have resulted in huge losses of soil organic matter over many decades, researchers tend to focus on the short-term effects of organic amendments, and may not even use these materials in ways that contribute to long-term soil-building processes. Since most agricultural soils are physically, chemically and biologically degraded because carbon losses have exceeded carbon inputs for many years, achieving permanent rather than temporary improvements in levels of soil organic matter must always be the ultimate goal.

Given that there are many ways in which introductions of organic matter can influence soil health and productivity and also impact on nematodes and other soilborne pathogens, decisions on the manner in which organic amendments are used in nematode management programmes should not be made on the basis that a specific chemical will be released, or that a particular group of antagonists will be enhanced. Instead, a broad and relatively long-term view should be taken, with the ultimate goal being to instigate management practices that improve the health of the soil in a general sense. This means thinking holistically about the effects of organic matter on a whole range of soil properties, and then developing a management strategy that integrates organic amendments and other practices into a farming system that will ultimately minimize losses from plant-parasitic nematodes through effects on soil physical, chemical and biological fertility. Some of the practices that might be used in such an approach are considered in Chapter 11.

10

Specific Suppression of
Plant-parasitic Nematodes

Two major groups of top-down forces govern the abundance of plant-parasitic nematodes: those that operate independently of density, and those that are density dependent. In the soil environment, the most common density-independent forces are abiotic (e.g. extremes of moisture or temperature, and the disruptive effects of pesticides and practices such as tillage). However, many antagonists of nematodes (e.g. polyphagous predators and parasites that are able to survive saprophytically in soil) also act independently of density. They may parasitize or prey on plant-parasitic nematodes, but because they have alternative food sources, their level of activity is not affected by the pest's population density. Thus, they will affect nematode abundance (see Chapters 5 and 6), but are unlikely to individually regulate population levels of any of the nematodes that form part of their diet.

Antagonists that act in a density-dependent manner are the only organisms with regulatory capacity. Such organisms are relatively host-specific parasites that are capable of destroying an increasing proportion of the population as abundance increases. Jaffee (1992, 1993) explained the regulatory process in the following way:

- Numbers of the parasite increase as the population density of the host nematode increases, thereby increasing the probability that an average host will be parasitized.

- The host density declines because infection by the parasite causes mortality. Since the host is suppressed, the parasite then declines. Thus, negative feedback between the host and parasite affects the densities of both organisms.

- The parasite exerts little negative effect when host density is low but has a large negative effect when host density is high. Therefore, the host density tends to stabilize at some equilibrium level, and so the parasite 'regulates' nematode population density.

- Although the parasite suppresses its host, it does not cause it to become locally extinct (i.e. both host and parasite are able to coexist).

Since relatively host-specific parasites act through density-dependent mechanisms, they are more likely than other antagonists to have long-term effects on nematode populations. This chapter focuses on their role as suppressive agents and provides evidence that they can markedly reduce populations of plant-parasitic nematodes.

The Role of Fungi and Oomycetes in the Decline of *Heterodera avenae*

The failure of the cereal cyst nematode *Heterodera avenae* to increase in abundance when susceptible cereals are grown in intensive rotations is the best-documented example of naturally occurring control of a plant-parasitic nematode. The phenomenon was first reported from England by Gair *et al.* (1969), who followed populations of *H. avenae* in a field where oats and then barley were grown at high and low nitrogen levels for 13 years. Under both nitrogen regimes, nematode populations peaked 2 years after the start of the experiment and then fell rapidly, the decline continuing for the rest of the experiment (Fig. 10.1). This decline was not the result of a reduction in the capacity of injured root systems to support the nematode, because it was not arrested when barley, an efficient host of the nematode and a more tolerant crop, replaced oats after 8 years. Other studies showed that cereal cyst nematode also failed to increase under intensive cereal cultivation in other parts of Europe (Jakobsen, 1974; Ohnesorge *et al.*, 1974). Since formalin (a relatively ineffective nematicide) increased the multiplication of *H. avenae* when applied as a drench at 3000 L/ha (Williams, 1969; Kerry *et al.*, 1980, 1982b, c; Kerry and Crump, 1998; and Fig. 10.2), it was suspected that the chemical removed

a parasite or competitor that was limiting nematode reproduction.

Initial studies on the causes of this decline phenomenon showed that two nematode parasites (the oomycete *Nematophthora gynophila* and the fungus *Pochonia chlamydosporia* [syn. *Verticillium chlamydosporium*]) were widespread in fields infested with *H. avenae* (Kerry, 1975). *N. gynophila* invaded cyst-nematode females as they emerged from roots and completely destroyed them within 1 week. Effects of *P. chlamydosporia* depended on the developmental stage of the nematode when it was attacked. Cysts were small and devoid of eggs when immature females were infected, whereas egg production was reduced and many eggs were parasitized when nematodes were attacked after egg-laying commenced.

During surveys of cereal fields in which populations of *H. avenae* increased or declined, Kerry and Crump (1977) found that in both situations, similar numbers of new females were produced (Fig. 10.3a). In soils where nematode multiplication occurred, there was a positive linear relationship between the number of females per root system and the number of new cysts and eggs in

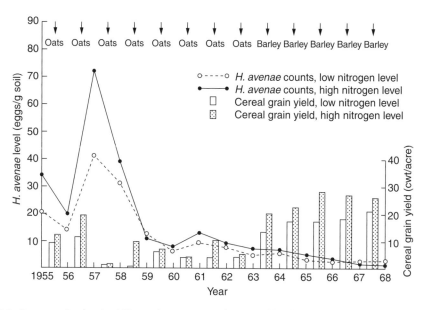

Fig. 10.1. Post-cropping levels of *Heterodera avenae* and grain yields when cereals were grown continuously in a field in England. (From Gair *et al.*, 1969, with permission.)

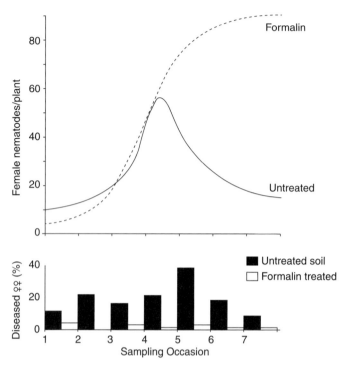

Fig. 10.2. Effect of formalin soil drenches (3000 L/ha) on the numbers of females of *Heterodera avenae* on barley roots and parasitism by *Nematophthora gynophila* in pots on seven weekly sampling occasions. (From Kerry, 1981, copyright Allanheld, Osmum and Co., now Rowman and Littlefield Publishers, Inc. Reprinted with the permission of the publishers.)

soil, whereas in soils where populations declined and females were heavily attacked by fungi, there was no such relationship (Figs 10.3b, c). Thus, the effect of parasitism by the fungus and oomycete was to reduce the number of females forming cysts, and the number of viable eggs that these cysts contained. Because of the difficulty of extracting fragile diseased females from soil and the disintegration of diseased cysts, rates of infection of nematodes by fungi and losses of females were generally underestimated. However, 50–60% of the mortality in females of *H. avenae* in some field plots was accounted for by *N. gynophila* and *P. chlamydosporia* (Kerry *et al.*, 1982a, and Fig. 10.4).

The activity of parasites of *H. avenae*, particularly *N. gynophila*, is influenced markedly by soil moisture. More than 90% of females and eggs were killed by the fungus and oomycete under moist conditions in pots, but levels of parasitism were considerably reduced

when soil was kept dry (Kerry *et al.*, 1980). Parasitism was, therefore, more severe when conditions were moist during May, June and July, the period when adult females of *H. avenae* were exposed on the roots. The two parasites were less active during dry summers or in well-drained soils and this may have accounted for the greater incidence of the nematode in free-draining light soils overlying chalk, and on coarse sands.

Kerry *et al.* (1982c) concluded that *P. chlamydosporia* and *N. gynophila* were the major, if not the only factor limiting multiplication of *H. avenae* on susceptible hosts in some English cereal fields. Their conclusions were based on the following evidence, collected from numerous surveys and experiments over a number of years:

• Numbers of spores of *P. chlamydosporia* and *N. gynophila* were much greater in soils where *H. avenae* failed to multiply

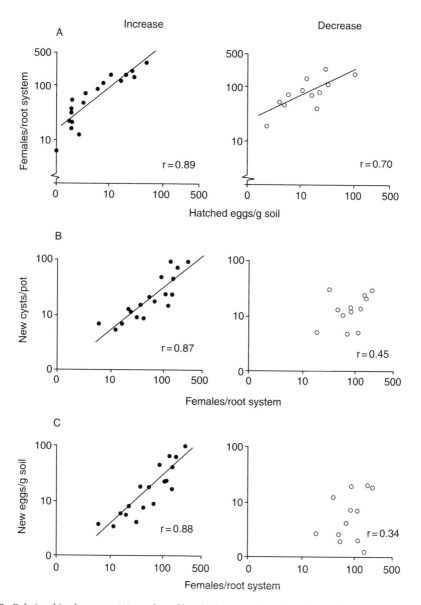

Fig 10.3. Relationships between: (A) number of hatched eggs and number of new females per root system; (B) females per root system and new cysts per pot; and (C) females per root system and new eggs/g of soil, in soils where populations of *Heterodera avenae* increased or decreased under spring barley in pots. (From Kerry and Crump, 1977, with permission.)

than in soils where the nematode was increasing (Crump and Kerry, 1981; Kerry *et al.*, 1982a).

- Populations of *H. avenae* declined in years when the level of parasitism was high and increased in years where there was less parasitism (Kerry *et al.*, 1982b).

- Application of formalin to soil reduced the numbers of spores of both parasites and increased nematode multiplication (Kerry *et al.*, 1980, 1982 a, b, c; Crump and Kerry, 1981).

- The 'formalin' effect was observed in soils where *P. chlamydosporia* and *N. gynophila*

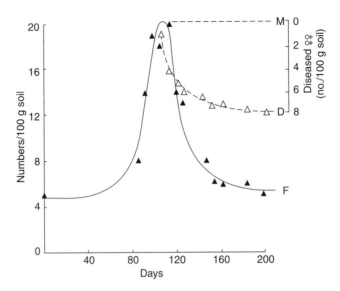

Fig 10.4. Changes in numbers of females and full cysts of *Heterodera avenae* and parasitism by
Nematophthora gynophila through a growing season on barley: ▲ numbers of females and cysts with
contents; △ accumulated numbers of females destroyed by *N. gynophila*. The number of females killed is
the difference between the maximum number of females and full cysts (M) and the number of full cysts
post-harvest (F). Females parasitized by fungi (D) account for approximately 50% of the lost females. (From
Kerry, 1980, with permission.)

were present but did not occur when
they were absent (Kerry *et al.*, 1980).

• Heavy watering increased levels of
parasitism and reduced the number of
females, cysts and eggs in pots. When
the two parasites were controlled with
formalin, watering did not influence nematode numbers (Kerry *et al.*, 1980).

Although parasitism of females and eggs
by an obligate parasite such as *N. gynophila*
could be expected to be density-dependent,
and despite evidence that the oomycete
was involved in regulating nematode populations, high rates of parasitism were not
always associated with high densities of
H. avenae females (Kerry *et al.*, 1982c). In fact,
in one particular season, a smaller proportion
of females were killed on oats than on wheat
or barley, even though oats had much higher
numbers of females on roots. In an attempt to
explain this result, the authors suggested that
the distribution of females on roots may have
been more important than their average
density in determining the spread of disease
through a population. Such a suggestion is

feasible because infections by *N. gynophila* are
initiated by overwintering resting spores and
are then spread by zoospores that have a limited capacity to move in soil.

After many years of study, it is now clear
that *P. chlamydosporia* and *N. gynophila* are
widely distributed in the cereal-growing soils
of England and northern Europe, and that in
many of those areas they have a significant
impact on populations of *H. avenae*. However,
as environmental factors influence the level of
parasitism, these parasites are not equally
effective in all situations. *N. gynophila* is more
active in wet seasons than in dry seasons,
because rainfall during the critical period
when white females are present on the root
surface is particularly important in providing
the soil moisture conditions needed for zoospore dispersal. Despite the fact that the activity of *N. gynophila* is limited by lack of moisture
in dry seasons and in some free-draining soils,
its effects, together with those of *P. chlamydosporia*, provide significant economic benefits to
the English and European cereal industries.

Although naturally occurring antagonists obviously have the capacity to control

cereal cyst nematode, the critical issue from a practical perspective is whether growers can be confident that suppressiveness will develop and that nematode populations will be maintained at levels that do not cause economic losses. Kerry and Crump (1998) addressed this issue by studying the population dynamics of both the nematode and its parasites over time in four quite different soils. Initial population densities were >100 eggs/g soil, but by the seventh year they had declined in all soils to <10 eggs/g soil, despite the continuous cropping of susceptible barley varieties (Fig. 10.5). Treatment of the soil with formalin markedly increased nematode population densities, but within 3 years of ceasing formalin applications, populations had declined to the same level as in untreated soil (<2 eggs/g soil). Thereafter, it was difficult to detect the nematode, despite the fact that susceptible cereals were grown continuously (Fig. 10.5). Such results indicate that natural suppressiveness to cereal cyst nematode can develop relatively quickly, and that once it is established, it has the capacity to permanently maintain nematode populations at levels below the economic threshold.

Given that microbial antagonists play an important role in suppressing populations of

H. avenae in some European countries (Kerry, 1982), it is surprising that natural enemies of this nematode have largely been ignored as potential control agents in other regions of the world. H. avenae is the most damaging nematode pest of wheat and barley worldwide, but control efforts have largely been directed towards the development of resistant or tolerant cultivars (Nicol and Rivoal, 2008). Although this strategy has been successful in Australia, where only one pathotype of the nematode occurs (Vanstone et al., 2008), diversity within H. avenae and the presence of other cyst nematode species (e.g. H. filipjevi and H. latipons) limits the usefulness of resistance in other countries. Consequently, natural biological control deserves more attention, with results from a recent study in China (Riley et al., 2010) supporting that assertion. H. avenae is common in cereal fields on the Tibetan plateau, but nematode population densities are variable and this variability cannot be explained by crop rotation effects. The presence of large numbers of parasitized cysts and small cysts, and disparities between total cyst and egg counts, suggest that nematode multiplication and survival are being determined by microbial antagonism. Although the authors did not

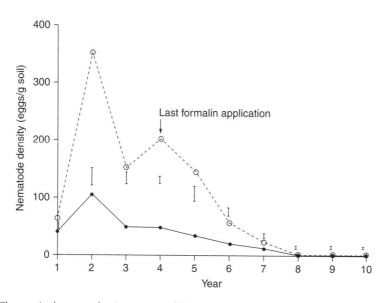

Fig. 10.5. Changes in the mean density (eggs/g soil) of *Heterodera avenae* on susceptible cereals in untreated soils (•) and in soils treated with formalin (o). (From Kerry and Crump, 1998, with permission.)

attempt to isolate or identify the antagonists, they concluded that future research should focus on developing an understanding of natural suppression and determining how it might be manipulated to produce favourable outcomes for growers.

Parasitism of *Meloidogyne* spp. on Peach by *Brachyphoris oviparasitica*

Lovell peach was once widely used as a peach rootstock in California, but because of its susceptibility to root-knot nematode, most growers changed to the *Meloidogyne*-resistant Nemaguard rootstock during the 1960s. However, when Ferris *et al.* (1976) studied nematode populations in a few of the remaining old orchards on Lovell rootstock, they had difficulty finding sites with high populations of root-knot nematodes. Since the soil, climate and host were suitable for the nematode, they speculated that these orchards may have been situated in areas biologically unsuited to *Meloidogyne*. The possibility that the nematode was under natural biological control prompted a search of these orchards for antagonists and led to the discovery of *Brachyphoris oviparasitica* (syn. *Dactylella oviparasitica*), a fungus parasitic in the eggs of root-knot nematode (Stirling and Mankau, 1978a). *B. oviparasitica* penetrated eggs from appressoria that formed on the surface of the egg, and then proliferated through the egg and eventually destroyed it. Because root-knot nematode eggs are aggregated in egg masses on the root surface, often all the eggs produced by a female were destroyed (Stirling and Mankau, 1978b, 1979).

Samples collected from the field throughout the year showed that between 20% and 60% of the eggs were parasitized by *B. oviparasitica* (Stirling *et al.*, 1979). However, the actual level of parasitism may have been much higher because the fungus destroyed eggs in less than 9 days at 27°C (Stirling, 1979b), and so some parasitized eggs probably disappeared before being counted. Addition of *B. oviparasitica* to sterilized soil in pots reduced root galling on peach and the number of nematodes in soil (Stirling *et al.*, 1979).

Although these observations suggested that *B. oviparasitica* was responsible for the low populations of root-knot nematode on peach trees grown on Lovell rootstock in California, they did not explain why the fungus had no apparent effect in adjacent vineyards where the nematode and parasite were also present. Greenhouse experiments then showed that *M. incognita* produced relatively small egg masses containing a maximum of 350 eggs on peach, whereas some egg masses on grape contained more than 1600 eggs. *B. oviparasitica* invariably parasitized most of the eggs in the small egg masses on peach, but on hosts such as grape only about 50% of the eggs were parasitized and many viable eggs remained (Stirling *et al.*, 1979). Thus, the authors suggested that successful suppression by *B. oviparasitica* depended as much on the inability of the nematode to produce large numbers of eggs on the host plant as it did on the activity of the parasite.

Unfortunately, there has only been one follow-up study on the role of *B. oviparasitica* in suppressing populations of root-knot nematode on peach. Trees on Lovell rootstock were planted on the experimental farm where Stirling *et al.* (1979) had observed high levels of parasitism, and growth, fruit production and populations of root-knot nematode (*M. incognita*, *M. javanica* and *M. arenaria*) were assessed for 6 years (McKenry and Kretsch, 1987). During the first year, *B. oviparasitica* was detected microscopically in egg masses from trees that had been inoculated with the fungus or planted with field soil known to have been infested with the fungus. However, the parasite was rarely observed in the second year and was not detected in the third year. Trees initially had relatively high populations of root-knot nematode, but populations had declined after 6 years, and this decline was not associated with *B. oviparasitica*. The authors indicated that the fungus may have been ineffective because a relatively saprophytic isolate had been used, but they also speculated that the rootstock had gradually become more resistant to the nematode as trees matured. Therefore, evidence for the involvement of *B. oviparasitica* in the suppression of root-knot nematode in California peach orchards remains largely circumstantial, and is based mainly on the results of green-house studies and on observations that high levels of parasitism occur naturally in some fields.

Suppression of *Heterodera schachtii* by *Brachyphoris oviparasitica* and Other Fungi

In 1975, soil infested with *Heterodera schachtii* was introduced to a research site in southern California. The field was then cropped inten-sively with hosts of the nematode in an attempt to develop a site suitable for evaluat-ing the efficacy of newly developed nemati-cides. Observations in the first few years indicated that this objective was achieved, as nematode populations increased to levels capable of causing significant yield loss. How-ever, population densities of *H. schachtii* and disease incidence then declined to the point where the field was no longer suitable for nematicide trials.

About 20 years after the nematode was introduced, the first steps were taken in a decade-long investigation aimed at identifying the causes of this decline phenomenon. The investigation (reviewed by Borneman *et al.*, 2004 and Borneman and Becker, 2007) com-menced with experiments demonstrating that the suppressiveness was biological in nature. Observations in the field showed that *H. schachtii* multiplied more readily in plots treated with a broad-spectrum biocide (metham-sodium) than in non-treated plots; while in the greenhouse, pre-plant treat-ments with heat and various fumigants reduced the suppressiveness of the soil (Westphal and Becker, 1999). Additional studies indicated that suppression could be transferred to a conducive soil using small portions of either suppressive soil or cysts (Westphal and Becker, 2000, 2001a; Yin *et al.*, 2003a, b; and Fig. 10.6). Also, *H. schachtii* was strongly suppressed at relatively high inoculum levels, suggesting that suppres-siveness was density-dependent (Westphal and Becker, 2001b). Culture and micro-scopic analyses of nematodes from the suppressive soil then revealed that eggs and cysts were colonized and parasitized by fungi. A range of nematode-destroying fungi, including *Brachyphoris oviparasitica, Fusarium oxysporum*, other *Fusarium* spp., *Purpureocillium*

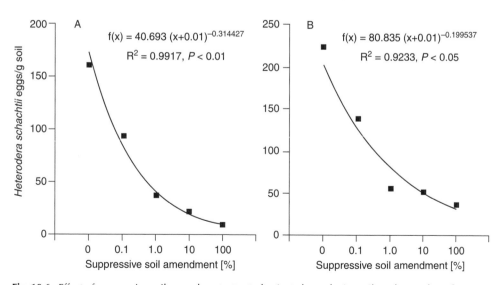

Fig. 10.6. Effect of suppressive soil amendment rates to fumigated, conducive soil on the number of *Heterodera schachtii* eggs/g of soil in greenhouse trials. Amended soils were infested with 5000 second-stage juveniles per pot and cropped to Swiss chard (A) followed by cabbage (B). (From Westphal and Becker, 2000, with permission.)

lilacinum and various unidentified fungi were isolated on agar media (Westphal and Becker, 2001a).

In follow-on studies, soils with different levels of suppressiveness were created with biocides, or by combining different amounts of suppressive and conducive soil, and oligonucleotide fingerprinting of rRNA genes (OFRG) was used to identify the main fungal phylotypes associated with different levels of suppression (Yin *et al.*, 2003a). Clones with high or moderate sequence identity to *Fusarium oxysporum*, *Brachyphoris oviparasitica* and *Lycoperdon* sp. were the predominant rDNA in cysts from suppressive soil treatments. However, when the amount of fungal DNA was quantified, levels of *B. oviparasitica*-like rDNA were about 100 times greater than levels of the other two fungi. This result was validated in a second phase of the study, as sequence-selective quantitative PCR assays showed that the largest amounts of *B. oviparasitica* PCR product came from soils possessing the highest levels of suppressiveness to *H. schachtii* (Yin *et al.*, 2003a). The key role of *B. oviparasitica* as a suppressive agent was then confirmed in greenhouse and microplot experiments (Olatinwo *et al.*, 2006a, b, c). The fungus reduced population densities of *H. schachtii* on sugarbeet when it was introduced into microplots, and populations were maintained at low levels when additional nematodes were added after the first harvest. Soil collected at the completion of the microplot experiments remained suppressive for a further 16 weeks in the greenhouse. In contrast, *F. oxysporum* did not affect *H. schachtii* in the greenhouse or in microplots at 11 weeks, but did have a significant effect in microplots at 19 weeks. Later work provided additional evidence to suggest that *B. oviparasitica* was the main suppressive agent, as high initial population densities of this fungus were associated with low final population densities of *H. schachtii* (Yang *et al.*, 2012b).

Although the above studies suggested that fungi, particularly *B. oviparasitica*, were largely responsible for the decline of *H. schachtii*, fingerprinting techniques were also used to obtain evidence that bacteria may have been involved (Yin *et al.*, 2003b). Five major taxonomic groups of bacteria were associated with *H. schachtii* cysts, and three assemblages

of clones (designated groups 1, 2 and 3) contained clones that were more prevalent in highly suppressive soil than less suppressive treatments. However, when these groups were examined with specific PCR analyses, only one group was consistently associated with high levels of suppression. This group had moderate to high sequence identity to rDNA from several *Rhizobium* species and to uncultured α-proteobacterial clones, but was later considered to be more closely related to several *Zoogloea* species (Borneman *et al.*, 2004).

Although *B. oviparasitica* appeared to be the primary cause of this decline phenomenon, the results of these studies also indicated that a relatively complex microbial community was associated with *H. schachtii* cysts in suppressive soil. Thus, suppressiveness to the nematode may not have been due totally to a single organism. For example, bacteria found in cysts may have produced enzymes or toxins that allowed associated fungi to more easily penetrate eggs, while initial colonization of cysts by fast-growing fungi such as *F. oxysporum* may have made it easier for secondary organisms to colonize cysts or eggs. Given the complexity of the soil biological environment, it is inevitable that organisms will interact, and so it is impossible to exclude the possibility that several organisms contributed to the suppressiveness observed.

Parasitism of *Mesocriconema xenoplax* and *Heterodera* spp. by *Hirsutella rhossiliensis*

Mesocriconema xenoplax is one of several factors contributing to the short life of peach trees in the south-eastern United States. Observations that nematode populations in some orchards had declined unexpectedly, even when customary weather patterns and farm practices prevailed, led to suggestions that the reduction in nematode numbers may have been caused by parasites and predators. Surveys of South Carolina peach soils showed that dead *M. xenoplax* with brown heads and distorted bodies filled with fungal hyphae were often present, and that occasionally a large proportion of the nematode population was parasitized. *Hirsutella rhossiliensis* was consistently isolated from these dead nematodes and was able to penetrate and kill juvenile and adult *M. xenoplax* under laboratory conditions (Jaffee and Zehr, 1982). The parasite also suppressed nematode multiplication in the greenhouse (Eayre *et al.*, 1983). The impact of *H. rhossiliensis* on population densities of *M. xenoplax* in the field was not determined, but Zehr (1985) suggested that the fungus might have been partially responsible for the rapid fluctuations of *M. xenoplax* often observed in peach orchards and might be involved in the occasional collapse of nematode populations observed in some orchards. However, even in the presence of the fungus, nematode populations generally remained large enough to cause significant injury to peach trees.

Studies of fungal and nematode population dynamics in California peach orchards (Jaffee *et al.*, 1989; Jaffee and Mclnnis, 1991) provided more data on the role of *H. rhossiliensis* in regulating populations of *M. xenoplax*. These studies focused on determining whether parasitism by *H. rhossiliensis* was affected by nematode density, as density-dependent parasitism is a key factor in all regulatory systems. Populations of interacting hosts and parasites were sampled repeatedly, and parasitism and host density were measured through time, but such studies showed that there was no relationship between the two parameters (Jaffee and Mclnnis, 1991). In contrast, spatial sampling in three orchards indicated that percent parasitism was dependent on nematode density. Thus, the authors questioned whether spatial sampling could be used to make inferences about temporal density-dependent parasitism, and concluded that *H. rhossiliensis* was no more than a relatively weak regulator of *M. xenoplax* populations. The mean nematode density was far above economic injury levels; high levels of parasitism tended to occur only at extremely high nematode densities; and signs of an impending decline in nematode numbers were never observed in plots with high levels of parasitism.

Another situation where a species of *Hirsutella* had been associated with low nematode populations was reported from oil radish fields in western Germany (Muller, 1982). In this case, as many as 90% of the second-stage juveniles of *Heterodera schachtii* were parasitized by a species later considered by Jaffee and Muldoon (1989) to be a synonym of *H. rhossiliensis*. Muller (1985) used fungicide treatments to obtain indirect evidence that *H. rhossiliensis* was suppressing cyst nematode populations in these fields, and Juhl (1985) provided supportive data showing that the fungus could substantially reduce the number of cysts formed. Jaffee and Muldoon (1989) also showed that natural populations of the fungus had the potential to suppress *H. schachtii*. Penetration of cabbage roots by the nematode was reduced by 50–77% in soil infested with *H. rhossiliensis*, and this reduction could be explained by mortality induced by the fungus.

Since Muller (1985) had shown that *H. rhossiliensis* was relatively widespread in German sugarbeet fields, similar surveys were undertaken in California. The results (Jaffee *et al.*, 1991) indicated that the parasite was present in 10 of 21 sugarbeet fields, but that the number of *H. schachtii* parasitized by the fungus was relatively low. Since damaging levels of cyst nematodes were found in most of the fields where *H. rhossiliensis* occurred, the authors concluded that the fungus was not providing adequate control of the nematode. This prompted investigations into the potential of inundative applications of the fungus for short-term control of *H. schachtii*. Although the results of those studies did not suggest that this was a worthwhile strategy (see Chapter 12), they did provide important information on the dynamics of the interaction between host and parasite. Microcosms were seeded with low levels of *H. rhossiliensis* and various numbers of *H. schachtii* were added at 3-week intervals. Density-dependent parasitism was observed, as levels of parasitism increased with the addition of increasing numbers of host nematodes, and declined in the absence of host recruitment (Jaffee *et al.*, 1992a). However, disease epidemics were slow to develop, presumably because the number of spores produced by the parasite was insufficient to ensure high transmissibility. Thus, natural infestations of *H. rhossiliensis* do not necessarily result in rapid (within-season) regulation and limitation of *H. schachtii*.

H. rhossiliensis is also known to parasitize second-stage juveniles of the soybean cyst nematode (*Heterodera glycines*), and initial observations in Minnesota suggested that levels of parasitism in one field (11–53% of the juveniles parasitized) may have been high enough to suppress nematode populations (Chen, 1997). A survey of 237 fields then showed that species of *Hirsutella* were relatively common, with *H. rhossiliensis* and *H. minnesotensis* detected in 43% and 14% of soybean fields, respectively (Liu and Chen, 2000). Since higher rates of fungal parasitism were observed in plots under soybean monoculture than in a maize/soybean annual rotation (Chen and Reese, 1999), a greenhouse study was conducted to compare the suppressiveness of soils under these two cropping sequences (Chen, 2007). Heat and formalin were used to remove suppressive factors, and both treatments increased the nematode egg population density relative to untreated soil. Mixing untreated and treated soil from two monocultured fields (10% untreated: 90% treated) reduced nematode population densities to the level of the untreated soil, indicating that suppressive biological factors were present, and that they were transferrable. *H. rhossiliensis* parasitized second-stage juveniles in soil from all fields, but levels of parasitism were particularly high in a field that had been in soybean monoculture for 35 years. However, the effect of the fungus on the nematode was much the same as had been observed by Jaffee *et al.* (1992a) for *H. schachtii*: levels of suppression were not high enough to reduce nematode population densities to non-damaging levels.

Decline of *Heterodera glycines* and the Possible Role of Egg-parasitic Fungi

Although soybean cyst nematode (*Heterodera glycines*) is a major pest of soybean in the United States, there have been several reports of situations where nematode populations on susceptible cultivars were much lower than expected:

- When resistant and susceptible soybean cultivars were grown continuously for 5 years at a nematode-infested site in Mississippi, cyst counts on the susceptible cultivar decreased after each year of production, with the count in the fifth year about 85% lower than in the first year (Hartwig, 1981).

- Carris *et al.* (1989) monitored cyst numbers every 3–4 weeks in two Illinois fields that had been under soybean monoculture for at least 4 years. Although the same susceptible cultivar was grown in both fields, the nematode population increased in one field but remained at low levels in the other.

- In 1987, populations of *H. glycines* were found to be unexpectedly low in a soybean field in Arkansas. Although the field had recently been cropped with a susceptible soybean cultivar, nematode populations had declined by 97% in 8 months (from 140 to 4 eggs/g soil). In a greenhouse test, application of formalin solution at a rate that was fungicidal but not nematicidal resulted in a 23- to 60-fold increase in the number of nematode eggs compared with untreated soil (Kim and Riggs, 1994).

- Observations that *H. glycines* did not increase to problematic levels in two Florida soils prompted Chen *et al.* (1996a) to use a greenhouse bioassay to compare the suppressiveness of these soils with three other soils where the nematode was undetectable. The results indicated that the nematode population densities (eggs/g soil) in four of the soils were 85–92% lower in untreated soil than microwaved soil, indicating a relatively high level of suppressiveness.

Hartwig (1981) suspected that a pathogen may have been responsible for the decline in *H. glycines* populations observed in his study, but did not determine the organisms involved. However, as fungi had previously been shown to be associated with the decline of cereal cyst nematode (see earlier sections of this chapter), and were seen growing within cysts and eggs, the other authors used standard mycological techniques to isolate and identify the fungi, and also considered their possible role as suppressive agents. The fungi and oomycetes most commonly found in two of the studies (Carris *et al.*, 1989; Chen *et al.*, 1996a) are listed in Table 10.1. For comparative purposes, data from a similar study at another suppressive site (Chen *et al.*, 1994a) and from three surveys in typical soybean fields (Morgan-Jones *et al.*, 1981; Ownley Gintis *et al.*, 1983; Chen and Chen, 2002) are also included. Since only one fungus (a sterile hyphomycete designated ARF18) was able to consistently parasitize eggs of *H. glycines* in laboratory tests, data on the fungi isolated in the Arkansas study were never published (Kim and Riggs, 1991).

Table 10.1. Fungi and oomycetes commonly isolated from females and cysts of *Heterodera glycines* in surveys conducted in various states of the United States.

	Morgan-Jones et al. (1981)[a]	Ownley Gintis et al. (1983)[b]		Carris et al. (1989)[c]	Chen et al. (1994a)[d]	Chen et al. (1996a)[e]	Chen and Chen (2002)[f]
		Young cysts	Old cysts				
Chaetomium cochloides		+					
Corynespora cassiicola							+
Cylindrocarpon tonkinense			+				
Cylindrocarpon destructans				+			+
Cylindrocarpon olidum							+
Dictyochaeta coffeae					+		
Dictyochaeta heteroderae					+		
Exophiala pisciphila		+		+	+		+
Fusarium oxysporum	+	+	+	+	+	+	+
Fusarium solani	+	+	+	+	+		+
Gliocladium roseum	+			+			+
Gliocladium catenulatum					+		
Humicola sp.						+	
Mortierella spp.							+
Neocosmospora vasinfecta	+		+		+		
Paecilomyces variotii			+				
Paraphoma radicina				+		+	
Phoma terrestris			+				
Phoma multirostrata	+						
Phoma macrostoma	+						
Phoma spp.							+
Phytophthora cinnamomi		+					
Pochonia chlamydosporia			+	+			+
Purpureocillium lilacinum			+				
Pyrenochaeta terrestris				+	+		+
Pythium sp.		+					
Scytalidium fulvum			+				
Stagonospora heteroderae	+			+	+		+
Trichosporon beigelii		+	+				
Black yeast	+				+	+	
Sterile mycelium		+				+	

[a]Soils from a number of soybean fields in Arkansas, Florida, Mississippi and Missouri
[b]One soybean field in Alabama
[c]A suppressive and non-suppressive soil from Illinois
[d]A suppressive field in Florida
[e]Five soils from Florida
[f]Forty-five fields in southern Minnesota

Based on the observations summarized in Table 10.1, it is possible to draw the following conclusions: (i) many different fungi are found in cysts and eggs of *H. glycines*; (ii) the degree of fungal colonization increases as nematodes become more mature; (iii) many of the fungi are known pathogens of nematodes; and (iv) there is no clear difference in the cyst mycofloras of suppressive and non-suppressive soils. Although Chen *et al.* (1996a) showed that the nematode population density was negatively correlated with the percentage of cysts colonized by fungi, the complexity of the mycoflora associated with cysts has meant that it is difficult to convincingly demonstrate that some of these fungi play a role in suppressing populations of the nematode.

Those who reviewed the early work on fungal parasitism of cyst nematodes (including *H. glycines*) concluded that the most frequently encountered genera were *Cylindrocarpon*, *Fusarium*, *Gliocladium*, *Paecilomyces*, *Pochonia* and black yeasts in the *Exophiala–Phialophora* complex. Kerry (1988) considered these fungi were facultative parasites, whereas Morgan-Jones and Rodriguez-Kábana (1988) argued that they had a relatively specialized relationship with nematodes and played a significant regulatory role in cyst nematode population dynamics. However, given our knowledge of the ecology of most of the fungi found in nematode females and eggs, it is difficult to sustain that argument. These fungi are primarily soil saprophytes and their association with nematodes is largely opportunistic. They can survive in the rhizosphere, the same ecological niche as the sedentary stages of cyst nematodes, and so are in an ideal position to invade developing nematodes. However, they are unlikely to regulate cyst nematode populations, as these fungi are not markedly affected by the presence or population density of the nematode (Nicolay and Sikora, 1989), and often parasitize relatively few females and eggs (e.g. Chen and Chen, 2002). Some genera (e.g. *Pochonia*) are more specialized than others, but isolates vary considerably in their capacity to colonize eggs, and so, as a group they should be considered relatively unspecialized parasites of

nematodes. In some situations, large numbers of cysts and eggs may be parasitized, but the level of parasitism is quite variable and tends to be influenced by soil and environmental factors rather than nematode population density.

Suppression of Root-knot Nematode by *Pochonia chlamydosporia* and Other Organisms

In a study by Bent *et al.* (2008), six Californian soils, selected on the basis that they appeared inhospitable to one or more plant-parasitic nematodes, were assessed for their capacity to suppress *Meloidogyne incognita*. Non-fumigated and methyl iodide-treated soils were inoculated with the nematode, and 2 months later, population densities on tomato and wheat in the non-fumigated soil from one of the sites were 5- to 16-times lower than in the fumigated soil. Distinct differences in nematode population densities were then observed when this soil was subjected to various fumigation, temperature and dilution treatments, confirming the biological nature of the suppressiveness.

A fungal rRNA gene analysis (OFRG) preformed on *M. incognita* egg masses collected at the end of the above experiments showed that a complex fungal community was associated with the reproducing nematode. Eleven phylotypes were identified, including *Pochonia chlamydosporia*, *Fusarium oxysporum*, *Saccharomyces* sp., *Tetracladium* sp., *Monacrosporium elliposporum*, *M. geophytum*, *Ceratobasidium* sp. and *Auricularia* sp., but *P. chlamydosporia* was the only phylotype that was negatively correlated ($r \leq 0.478$, $P \leq 0.01$) with at least one of the nematode population density measurements (egg masses, eggs or second-stage juveniles). A sequence-selective qPCR assay targeting this phylotype confirmed the negative association between *P. chlamydosporia* genes and nematode population densities. Given that *P. chlamydosporia* is a ubiquitous soil saprotroph and a well-known antagonist of *Meloidogyne* spp., it is likely to have been contributing to the suppressiveness observed in this soil. However, it was impossible to come to a more definitive conclusion, as it was not known whether the other fungi found in egg masses were pathogenic, and the possible role of bacteria and other antagonists was not investigated.

Suppression of *Heterodera glycines* and Sudden Death Syndrome of Soybean

Two important characteristics of soilborne diseases are that crop losses may be caused by more than one pathogen, and that pathogens may interact synergistically to increase disease severity. Thus, if biological suppressiveness is to be useful against soilborne pathogens, it must be effective against all components of a disease complex. This issue was addressed by Westphal and Xing (2011) in a study of *Fusarium virguliforme*, the cause of sudden death syndrome (SDS) of soybean in the United States, and *Heterodera glycines*, one of the major pests of soybean worldwide. Soybean was planted into fumigated and non-fumigated plots at two sites in Indiana, and population densities of *H. glycines* and the development of SDS were monitored for a period of 5 years. Results from one of the sites indicated that population densities of *H. glycines*

declined in non-fumigated plots and increased following fumigation (Fig. 10.7), while foliar symptoms of SDS were less severe in non-fumigated plots than in fumigated plots.

Although these observations suggested that one of the soils had become suppressive to two quite different pathogens under soybean monoculture, the authors were cautious when discussing their results for a number of reasons: (i) the decline phenomenon occurred at only one of the two study sites; (ii) the suppressive mechanisms were not determined, but are likely to be different for nematodes and fungi; (iii) it was not clear how interactions within the microbial community could favour beneficial fungi, the agents most commonly associated with suppressiveness to cyst nematodes, while simultaneously inhibiting a fungus that was a plant pathogen; (iv) complex interactions were likely to have been occurring in the suppressive soil, as the causal agent of SDS (*F. virguliforme*) can colonize cysts of *H. glycines* (Roy *et al.*, 2000); and (v) suppression of *H. glycines* may have indirectly affected SDS by weakening the nematode's role in exacerbating damage by the fungal pathogen. Further research to address these issues is obviously required.

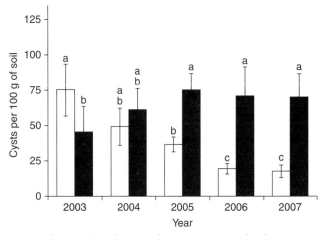

Fig. 10.7. Number of cysts of *Heterodera glycines* at harvest in 5 years of soybean monoculture in non-fumigated (white bars) and methyl bromide-fumigated (black bars) plots naturally infested with the nematode. Values are means with standard error bars. Columns with the same letter are not significantly different. (From Westphal and Xing, 2011, with permission.)

Suppression of Root-knot Nematodes
by *Pasteuria penetrans*

Most studies of naturally occurring populations of *Pasteuria* have been conducted with *P. penetrans*, a species that is a specific parasite of numerous *Meloidogyne* spp. (Starr and Sayre, 1988). From what is known of its biology, host–parasite relationships and ecology (see Chapter 7), *P. penetrans* is likely to have little impact on nematode populations at low spore densities, because only a few nematodes come into contact with spores and become infected. Nevertheless, more than two million spores may be produced in each infected female, and since these spores can survive for a long periods, spore population densities can be expected to increase over time, and eventually reach levels that possibly result in the local extinction of the host in micro-patches where the host nematode is aggregated. However, spore densities are unlikely to increase rapidly, as new spores are not released into soil until the body of the infected female nematode is ruptured by degradative soil microorganisms. Also, high spore concentrations will not persist indefinitely for a number of reasons: (i) spores can be leached downwards by water flow, particularly in coarse-textured soils; (ii) spores will be subject to predation; and (iii) once spore concentration increases to a level where second-stage juveniles are encumbered with large numbers of spores, the number of nematodes establishing feeding sites in roots will decline, and this will limit further spore production. Thus, spore population densities will eventually stabilize at an equilibrium level that will be determined by many factors, including soil texture, the intensity of cropping, the susceptibility of crops to the host nematode, the number of nematode generations possible each year, rainfall intensity, the capacity of the soil to infiltrate water and the biological status of the soil.

The relevant question with regard to biological control is whether there are situations where these equilibrium spore densities have ever been reached in cultivated fields, and whether at those densities, the parasite is able to reduce root-knot nematode populations to levels below economic thresholds. Initial studies in this area were conducted in situations likely to favour multiplication of *P. penetrans*, such as fields where a root-knot susceptible crop had been grown for many years, either as a stable monoculture or in a perennial cropping system. Williams (1960) noted that 34% of the *Meloidogyne* females in a sugarcane field in Mauritius were infected by *P. penetrans*, and Spaull (1984) later observed similar levels of infection in a number of South African fields. In a sugarcane field where there was a marked gradient in the number of nematodes across the field, Spaull (1984) was able to show that the proportion of females infected by *P. penetrans* increased as the number of nematodes increased (Fig. 10.8). However, *P. penetrans* appeared to have little impact on root-knot nematode populations because fields infested with the parasite generally had higher nematode populations than those where it was absent. Since *P. penetrans* tended to become more common as the number of years a field had been

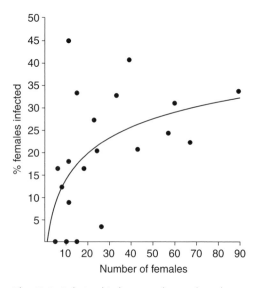

Fig. 10.8. Relationship between the number of female *Meloidogyne* per plot and the proportion infected with *Pasteuria penetrans* ($Y = -3.669 + 8.015 \log_n X$; $r = 0.4933$). (From Spaull, 1984, with permission.)

cropped to sugarcane increased, it was suggested that the parasite may still have been in an early phase of population build-up and that eventually it may have caused nematode populations to decline.

Cropping history was found to influence the presence of *P. penetrans* on grapevines in South Australia, as the parasite was widespread in vineyards more than 25 years old, but was difficult to detect in vineyards less than 10 years old (Stirling and White, 1982). However, in contrast to sugarcane in South Africa, there were fewer root-knot nematodes in old vineyards infested with *P. penetrans* than in young vineyards without the parasite. Although many factors could have been responsible for these differences in nematode numbers with vineyard age, Stirling (1984) went on to show that *P. penetrans* may have been responsible. He found that nematodes could be eliminated from old vineyard soil with nematicides (either DBCP or 1,3D) and by autoclaving, but that autoclaving was the only treatment detrimental to naturally occurring *P. penetrans*. When treated soils were potted, planted to grapes, and inoculated with *M. javanica*, more nematodes were found after 1 and 2 years in the autoclaved soil, where naturally occurring *P. penetrans* had been eliminated, than in the nematicide-treated soil where *P. penetrans* was still present (Table 10.2). In an experiment with soil from another site, root-knot nematode populations on grape reached a higher level in virgin soil containing little *P. penetrans* than in a similar textured soil that had been collected

from a 25-year-old vineyard naturally infested with *P. penetrans*. The suppressive effect of *P. penetrans* and other naturally occurring antagonists in vineyard soils from the same region of South Australia was confirmed in a pot experiment by Bird and Brisbane (1988). They planted tomato seedlings in several field soils, inoculated the pots with *M. javanica*, and found that the nematode's reproductive capacity was much lower in the soils from Cooltong (which were infested with *P. penetrans*) than in the non-infested soils from Loxton (Fig. 10.9A). Autoclaving removed the suppressiveness of two Cooltong soils but had no effect on one of the Loxton soils (Fig. 10.9B).

Two examples from Florida demonstrate that *P. penetrans* can also multiply on annual crops to levels capable of suppressing root-knot nematode. In the first example, a research site that had been planted to tobacco for 7 years was identified as likely to be suppressive to a mixed population of *Meloidogyne incognita* and *M. javanica* (Chen *et al.*, 1994b). Root-knot damage was initially severe, but yield reductions decreased over the years to a point where there was little difference in the growth of susceptible and resistant tobacco cultivars. The suppressive nature of this soil was confirmed by Weibelzahl-Fulton *et al.* (1996), who assessed the capacity of the two nematode species to reproduce and cause damage on a susceptible tobacco cultivar in autoclaved, microwaved, air-dried and untreated soils. The numbers of galls, egg masses and eggs increased

Table 10.2. Populations of *Meloidogyne javanica* and percent females infected in soil from a 60-year-old vineyard after naturally occurring levels of *Pasteuria penetrans* were reduced by autoclaving and retained following nematicide treatment.

	M. javanica juveniles/200 g soil[a]				M. javanica juveniles/plant[a]		% females infected
Treatment	Year 1		Year 2		Year 2		
DBCP	19	(2.87)	138	(4.61)	1561	(7.28)	59
1,3D	36	(3.27)	177	(4.78)	3261	(7.99)	62
Autoclaved	104	(4.47)	259	(5.30)	5973	(8.64)	1
LSD (*P* = 0.05)		(0.93)		(n.s.)		(0.48)	

[a]Numbers in parentheses are transformed [log$_e$ (no. nematodes + 1)]. (From Stirling, 1984, with permission.)

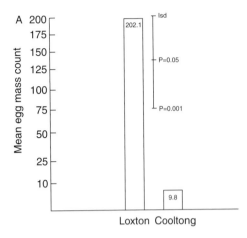

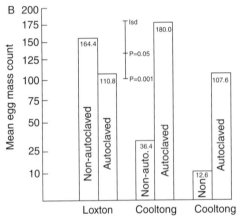

Fig. 10.9. (A) Means of egg mass counts of *Meloidogyne javanica* on tomatoes grown in soil from vineyards in the Riverland region of South Australia: three sites at Cooltong (where *Pasteuria penetrans* was present), and four sites at Loxton (where *P. penetrans* was not observed). (B) The effect on egg mass production of autoclaving two of the Cooltong soils and one of the Loxton soils. (From Bird and Brisbane, 1988, with permission.)

markedly when soil fungi and endospores were killed by autoclaving (Fig 10.10). Since numbers of egg masses and eggs were affected more than galling, the loss of suppressiveness was probably due to effects on *P. penetrans*. The microwave treatment markedly reduced fungal population densities, but did not affect endospores, and the fact that this soil remained suppressive (Fig. 10.10) suggested that suppressiveness was not caused by nematophagous fungi.

The second example of suppressiveness to root-knot nematode in an annual crop was observed in a Florida peanut field that had been used for nematode research for several years. The field was heavily infested with *M. arenaria* race 1, but in the late 1980s there were signs that the nematode population density was declining. Gradually, over a period of 4–5 years, peanut yields had increased and galling caused by the nematode had declined. Preliminary examination of second-stage juveniles extracted from the site indicated that 32% of the nematodes were infested with one or more endospores of *P. penetrans* (Dickson, 1998). The field was then planted to a winter crop of vetch (also a host of the nematode), and soil samples were later collected and either left untreated (i.e. stored at 10°C) or were autoclaved, air-dried or drenched with formalin. Suppressiveness was assessed using a bioassay in which tomato was inoculated with *M. arenaria*.

In the first bioassay (Dickson *et al.*, 1991; Dickson, 1998) there were fewer galls and egg masses on plants in untreated and air-dried soil than in autoclaved or formalin-treated soil. Soil from 0–15 cm was more suppressive than soil from 15–30 cm. In a second bioassay 8 months later, egg mass counts did not differ between treatments, but galling was less in untreated and air-dried soil from 0–15 cm than in autoclaved or formalin-treated soil. Since between 82% and 100% of the second-stage juveniles that had been added to air-dried soil were encumbered with *P. penetrans* endospores, these data suggested that *P. penetrans* was the major factor suppressing *M. arenaria*. However, parasitic fungi may also have also been playing a role, as 17% of the *M. arenaria* females and 50% of the egg masses collected from the field were infected by fungi. The two most common fungi were *Purpureocillium lilacinum* and *Penicillium* sp., but *Pochonia chlamydosporia*, *Neocosmospora* sp. and *Aspergillus* sp. were also isolated.

In a follow-on study at this site, the long-term persistence of suppressiveness induced by *P. penetrans* was assessed by planting the field to two non-hosts of *M. arenaria* race 1 (bahiagrass and rhizomal peanut) or leaving it as a weed fallow dominated by nematode-susceptible weeds such as hairy indigo and

Number/root system

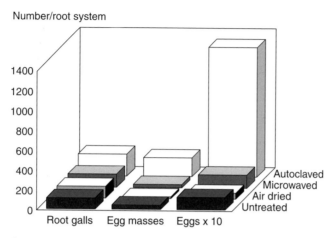

Fig. 10.10. Effect of treatments on a soil suppressive to *Meloidogyne* spp. and the expression of root galling on tobacco following inoculation with 2000 second-stage juveniles of *M. incognita* race 1. (From Weibelzahl-Fulton *et al.*, 1996, with permission.)

alyce clover. These treatments were maintained for 9 years and then the field was planted to four successive peanut crops. Bioassays at the end of the first peanut crop indicated that 63% of second-stage juveniles were encumbered with endospores in the weed fallow plots, and infestations of *P. penetrans* remained at high levels (77–87%) for the remainder of the experiment (Cetintas and Dickson, 2004). In the treatments where a nematode-resistant host had been grown, only 2% of bioassay nematodes had spores attached after the first peanut crop, and even after 3 of 4 years of peanuts, infestation levels were still lower than after weed fallow, ranging from 53–63%. These observations clearly demonstrate that populations of host-specific parasites such as *P. penetrans* will decline unless a population of the host nematode is maintained.

Another important observation from this study was that although populations of *P. penetrans* reached relatively high levels, particularly in plots that had previously been weed fallow, root-knot nematode was still causing economic damage, as yields were low and peanut roots, pods and pegs were heavily galled. The most likely reason for this is that 23–41% of the female nematodes at the experimental site were not *M. arenaria* but a race of *M. javanica* capable of attacking peanut (Cetintas *et al.*, 2003). Since endospores of

the *Pasteuria* strain present at the site did not attach readily to this race of *M. javanica* and were unable to infect the nematode (Cetintas and Dickson, 2004), a relatively large proportion of the *Meloidogyne* population was not being affected by the parasite. This highlights one of the difficulties involved in obtaining consistent control of root-knot nematode with *P. penetrans*. The genus *Meloidogyne* contains a diverse group of nematodes with a wide host range, and given the host specificity of *P. penetrans* (discussed in Chapter 7), there is always potential for nematode populations to appear that are not constrained by the parasite.

Suppression of *Heterodera glycines* by *Pasteuria nishizawae*

Infestations of soybean cyst nematode (*Heterodera glycines*) were established in 1982 in plots on a research farm in Illinois. Several years later, population densities of the nematode began to decline. Observations at this site in 1993 (Noel and Stanger, 1994) showed that more than 90% of the second-stage juveniles recovered from the soil were encumbered with endospores of a *Pasteuria* sp. that was later identified as *P. nishizawae* (Noel *et al.*, 2005). Microplots established at the site were then planted to a nematode-susceptible soybean cultivar and the population dynamics of both organisms was studied for 2 years by sampling every 2 weeks during spring, summer and autumn (Atibalentja *et al.*, 1998). The results suggested that the infestation of *P. nishizawae* may have developed more gradually in one of the microplots, as it had the highest population densities of *H. glycines* and the lowest levels of the parasite (15–31% of the second-stage juveniles encumbered with endospores). In contrast, an average of 74% and 81% of the nematodes in other microplots were infested with spores in 1994 and 1995, respectively. Numbers of second-stage juveniles per 250 mL soil (Y) decreased with the number of endospores per juvenile (X) according to the exponential decay model $Y = 67.4 + 220.1e^{-1.2X}$. Seasonal fluctuations in populations of both *H. glycines* and *P. nishizawae* were observed, but since peaks in nematode densities coincided with a decline in the number of endospores, the parasite appeared to be regulating populations of its host. On the basis of these results, the authors concluded that *P. nishizawae* was capable of maintaining populations of *H. glycines* at equilibrium densities that are below the damage threshold reported from field studies.

Confirmatory evidence that *P. nishizawae* suppressed populations of soybean cyst nematode was obtained in a study in which soil from these microplots was transferred to a trial site infested with *H. glycines* (Noel *et al.*, 2010). In 1999, a monoculture of soybean was initiated and a factorial experiment involving tillage (conventional or no till) × cultivar (susceptible or moderately resistant to *H. glycines*) × ± *P. nishizawae* was established. The population dynamics of the nematode and parasite were monitored for 7 years and the results showed that infestation of plots with *P. nishizawae* reduced initial nematode population densities (Pi) by about 43% overall. The reproductive factor (Pf/Pi) was also lower in infested than non-infested plots (8 ± 2.1 compared with 14 ± 1.9). With only one exception, yield of soybean in *P. nishizawae*-infested plots was similar to or greater than in plots without the parasite. Tillage method had no effect on population densities of *H. glycines*, while the absence of a cultivar × *P. nishizawae*

interaction suggested that the effects of the parasite on nematode populations and soybean yield were not affected by soybean cultivar.

Management Options to Enhance Specific Suppressiveness

The role of tolerance, resistance and crop rotation

The examples presented in this chapter provide clear evidence that when populations of plant-parasitic nematodes increase to high levels, host-specific fungal and bacterial parasites will utilize the nematodes as a food source and increase in population density. In some situations, the level of mortality caused by the parasite will increase to the point where nematode populations decline, sometimes to levels below damage thresholds. However, many of the suppressive situations discussed were observed at research sites, where susceptible crops were continually grown in a deliberate attempt to increase populations of the target nematode. Since several years of relatively high nematode populations are required before specific suppressiveness develops, and heavy crop losses can be experienced during that transition phase, such circumstances would not normally be allowed to develop on a commercial farm. Instead, the grower would control the pest with soil fumigants, nematicides, a bare fallow, resistant varieties or some other practice. However, the decision to use such control measures has ecological repercussions. Many beneficial organisms will be killed, populations of natural enemies will be depleted and nematode populations may be temporarily reduced to levels where host-specific parasites have little chance to operate. Thus, the soil loses its biological buffering capacity and becomes conducive to the pest. In situations where nematode-resistant varieties are used as the predominant means of control, resistance-breaking biotypes of the nematode will often develop, and so a continuing plant breeding effort is then required to stay ahead of the pest.

Given the shortcomings of these 'eradicative' approaches, it might be a better long-term strategy to try to cope with the pest nematode and introduce practices that encourage the development of both specific and general suppressiveness. However, for such a strategy to be effective, crop losses must be minimized in some way during the years when nematode populations are high and suppressiveness is developing. The most feasible way of doing this is to use cultivars or rootstocks that are able to tolerate nematode damage. Nematode-tolerant genotypes are available for many crops (see Thompson et al., 1999; McKenry and Anwar, 2008; Buzo et al., 2009; Pang et al., 2011; Blessit et al., 2012, for some examples), but are under-utilized in nematode management because nematologists have largely focused on using resistance as a control measure. If tolerance was combined with management practices that optimize nutrient and water availability to the crop (both well-recognized methods of alleviating plant stress and reducing the damaging effects of nematodes), it should be possible not only to produce high-yielding crops in the presence of relatively high populations of most plant-parasitic nematodes, but also to provide the conditions needed for specific suppressiveness to develop.

Since rotational cropping is a normal feature of most farming systems, and the crops included in the rotation sequence will vary in their capacity to host particular nematodes, their role in maintaining specific suppressive agents must be considered. The critical question with regard to rotational cropping systems is the impact of a resistant or non-host crop on the nematode populations required to maintain suppressiveness. This issue was addressed by Westphal and Becker (2001c) in their work with a soil that was suppressive to Heterodera schachtii. They found that cropping wheat (a non-host of the nematode) for 2 years reduced suppressiveness, whereas suppressiveness was maintained when soil was fallowed, or resistant cultivars of sugarbeet or oilseed radish were grown for the same period. Since similar nematode population declines were observed in all three rotations, the authors felt that the decline in suppressiveness following wheat

was probably not due to a reduction in the number of hosts available to the suppressive agents. However, the authors were unable to provide a plausible explanation for the result, although they surmised that some form of nematode activity during the period when resistance is being manifested in roots, or subliminal nematode reproduction, may have been necessary to maintain suppressiveness.

Limited studies with Heterodera glycines have also failed to clarify the role of crop rotation in maintaining suppressiveness to cyst nematodes, as results of American work have been inconsistent. Observations in Minnesota indicated that parasitism of second-stage juveniles by Hirsutella rhossiliensis was greater following soybean than maize (Chen and Reese, 1999), and that soils from two fields under soybean monoculture were more suppressive than soil from a maize/soybean rotation (Chen, 2007). In contrast, suppressiveness did not develop under soybean monoculture at one of the sites included in a study in Indiana, with final population densities of H. glycines only being reduced by a maize/soybean rotation (Xing and Westphal, 2009). At another Indiana site, soil became suppressive following continuous soybean, but nematode populations were only reduced to levels achievable under a maize/soybean rotation.

Clearly, much more work is required to understand the factors involved in maintaining specific suppressiveness in rotational cropping systems. Populations of relatively host-specific antagonists will decline if their food source is removed by growing resistant or non-host crops, so maintenance of biological suppressiveness is really a balancing act: nematode populations must be maximized to sustain suppressive activity, but should not be allowed to reach the point where unacceptable yield losses occur. Finding the appropriate balance will not be easy, as the acceptable nematode population must be viewed from two perspectives: the potential for nematode damage to the crop and the need for host nematodes to sustain the parasite. Perhaps the best way of addressing the issues involved is to establish field experiments with crops and cultivars that have varying levels of resistance, and monitor nematode populations and suppressive activity for several years. Ideally, such experiments should contain tolerant crops and cultivars that span the full range of the resistance–susceptibility continuum, but plant genotypes with intermediate levels of resistance and some tolerance are most likely to produce the desired result. Since the effects of such treatments will be crop-, site- and nematode-specific, experiments of this nature must be established at multiple sites.

The impact of tillage

In considering management practices that could be modified to enhance the development of specific suppressiveness, tillage stands out as the issue warranting attention. Obligate and relatively host-specific parasites of plant-parasitic nematodes have an intimate relationship with their host, and so they multiply and sporulate where these nematodes are found (i.e. in roots or the rhizosphere). Since roots tend to follow paths of least resistance in soil, nematodes and their parasites are not distributed uniformly through the soil profile. They exist in microniches, and this is where the infective propagules produced in a previously parasitized nematode are most likely to interact with the next generation of nematodes. Since tillage destroys this intimate relationship, it must hinder the development of suppressiveness. For example, endospores of *Pasteuria penetrans* are most likely to be present at effective concentrations in microniches where infected *Meloidogyne* females are decomposing, but host–parasite relationships will change markedly when endospores are dispersed by tillage, or physical conditions are created that cause the endospores to leach down the profile. In the case of a parasite such as *Hirsutella rhossiliensis*, tillage dislodges conidia from phialides, thereby preventing them from attaching to nematodes (McInnis and Jaffee, 1989). Since even relatively simple disturbances (such as mixing soil by hand) can markedly reduce suppressiveness (Stirling, 2008; Westphal and Xing, 2011; Stirling *et al.*, 2012), minimum tillage must be a component of farming systems that aim to maximize specific as well as general suppressiveness.

Integrated management to improve the efficacy of *Pasteuria*

The *Pasteuria* group of bacteria produces large numbers of resistant endospores that can survive periods when populations of the host nematode are low. However, that does not necessarily mean they will survive indefinitely, because, as mentioned previously, long-term use of rotation crops with nematode resistance markedly reduces populations of *P. penetrans* (Cetintas and Dickson, 2004). Nevertheless, the results of a study also discussed earlier (Noel *et al.*, 2010), provide cause for optimism that once high populations of *Pasteuria* have been established, they can be maintained. Even when cultivars resistant to *H. glycines* were grown continuously, *P. nishizawae* still had an impact on nematode population densities. This result is encouraging, because it suggests that provided the level of resistance is not excessive, nematode-resistant cultivars can be used to increase populations of *P. nishizawae*. In the above experiment, for example, the resistant cultivar Linford reduced nematode numbers by an average of 77% in the 6 years it was grown, but final population densities (450–975 eggs/ 100 mL soil) were still high enough to maintain the parasite.

Another factor that is likely to influence whether endospores are retained or lost in soil is its texture. The silty clay loam soil in which the above experiment was conducted may have been one of the reasons that populations of *Pasteuria* were maintained, as soils of this texture tend to retain most of the endospores produced by parasitized nematodes. Results may have been different in another soil type, as soil texture not only influences the number of endospores lost from the profile by leaching, but also the likelihood that a nematode will come into contact with an endospore (Dabiré and Mateille, 2004; Cetintas and Dickson, 2005; Dabiré *et al.*, 2007; Mateille *et al.*, 2009, 2010; Luc *et al.*, 2011).

One characteristic of the *Pasteuria* group of bacteria that gives it an advantage over other antagonists is the robust nature of its endospores. Thus, in situations where populations of *Pasteuria* are high, it should be possible to manage the host nematode using practices that would never be possible with other antagonists. For example, populations of *Pasteuria penetrans* can be increased by growing crops that are susceptible to the host nematode (Madulu *et al.*, 1994), but in such situations, it may be necessary to reduce the nematode population so that it is not above the damage threshold for the following crop. Since *Pasteuria* endospores are highly resistant to heat and most chemical treatments, it may be possible in those circumstances to reduce nematode populations with a short fallow, solarization or an appropriate nematicide, without affecting the parasite. Obviously, such treatments would have to be applied judiciously, as the aim is to reduce nematode populations to the point where the next crop will not be affected by the pest, but not to the point where populations of the parasite will decline. Unfortunately, integrated management strategies that involve natural populations of *Pasteuria* have rarely, if ever, been explored by nematologists. Instead, most research has focused on adding endospores to nematode-infested sites, a strategy that is discussed in Chapter 12.

Making Better Use of Natural Control: The Way Forward

It is now more than 30 years since the groundbreaking work of Kerry and colleagues showed that cereal cyst nematode was under natural control in many European countries. Examples of nematode populations being suppressed by *Pasteuria* began to appear in the scientific literature at about the same time. Thus, specific suppressiveness to plant-parasitic nematodes is a real phenomenon. It has been observed in many environments; it occurs in annual and perennial cropping systems; and it is manifested against all the key nematode pests of agricultural crops. In fact, natural mechanisms of nematode control are so widespread that it is reasonable to ask

why surveys for natural enemies, and assessments of suppressiveness, are not one of the first steps taken when management strategies to control a nematode pest are being developed. I would argue that documenting the parasites and predators that occur in a particular situation, and using bioassays to obtain estimates of their effects, is such a useful exercise that it should become standard practice in nematode pest management. The molecular tools required to assess the biological 'signatures' that define a suppressive soil are continually being improved (e.g. Smith and Jaffee, 2009; Xiang *et al.*, 2010b; Mendes *et al.*, 2011; Rosenzweig *et al.*, 2012), and the long-term research objective must be to use these tools to provide growers with more comprehensive data on the key groups of organisms (both pests and antagonists) that occur in their soils. There will be two main benefits from undertaking such a process:

- Creating awareness that nematode pests have natural enemies. Regardless of the situation, some parasites and predators will be present, and if the pest is to be managed in an ecologically sound manner, potential impacts of control measures on all components of the soil biological community must be considered.
- Providing information that can be used to enhance natural controls. Survey data showing that a particular parasite or predator appears to be more effective in a certain soil type or environment, or under particular soil or crop management systems, will provide clues as to how naturally occurring suppressive forces could be better exploited.

The advantage of recognizing that nematode populations are always suppressed to some extent by natural enemies is that it changes the mindset of those recommending management practices for a particular pest. Instead of just thinking about how nematode populations might be reduced with nematicides, resistant cultivars or some other practice, the impact of a proposed control measure on natural enemies must also be considered. Reducing crop losses from nematodes remains the objective, but consideration must be given to managing nematode populations in ways that allow natural enemies to play a role in accomplishing that goal. In other chapters in this book, such an approach has been termed 'integrated soil biology management', with the 'soil biology' component encompassing the pest nematode, its natural enemies, and the plants and other soil organisms that impact on either the pest or its antagonists.

If an integrated approach to managing the soil biology is taken in situations where nematode populations are high and crop losses are occurring, then the following checklist of questions needs to be answered before a management plan is developed:

- What natural enemies are present? Is the nematode being attacked by any relatively specific parasites, and if so, are they likely to have the capacity to regulate nematode populations? Are there already situations (within parts of a field or in comparable environments) where the pest is already being suppressed by natural enemies?

- Are density-dependent suppressive agents likely to increase to high levels and regulate nematode populations? How long will it take for suppressiveness to develop? Is the final equilibrium population density likely to be below the economic threshold?

- Can management practices be introduced that will have little impact on nematode populations but will improve the crop's capacity to cope with nematode damage?

- What additional practices are worth considering in the long term as a means of enhancing either general or specific suppressiveness? Is it worthwhile establishing field trials to confirm the effectiveness of these practices?

- If nematicides, resistant cultivars and other conventional control practices are used to reduce nematode populations, will they be economically effective? Is the soil likely to become conducive to nematode pests, and what are the long-term costs of losing biological suppressiveness? Will dependence on pesticides or nematode-resistance limit management options in future?

This list is almost certainly not comprehensive, as there will be situations where other issues should also be addressed. Nevertheless, the important points are that: (i) the positives and negatives of all possible nematode management options should be considered; and (ii) decisions are made with both a short- and long-term timeframe in mind.

With regard to practices likely to enhance specific suppressiveness to plant-parasitic nematodes, integrated management strategies that combine nematode-tolerant crops (or rootstocks) with tolerant rotation crops and optimized water and nutrient management are worth exploring. Two types of systems deserve consideration, even though they have rarely, if ever, been evaluated by nematologists: (i) rotational sequences in which all crops not only host the target nematode but are also tolerant of nematode damage; and (ii) rotations in which crops that are poor or non-hosts of the target nematode are included, but decisions on how often they are grown, or the length of the rotation period, are made on the basis of their effects on both the nematode and its specific parasites.

Given the detrimental effects of tillage on specific suppressiveness, minimum tillage should ideally be integrated into both these systems. There is also a strong argument for including practices that enhance soil organic carbon levels, as such practices will improve soil health and enhance general suppressiveness to nematodes and other soilborne pathogens.

It should be noted that the above strategies are not presented as a *fait accompli*, but as options that deserve consideration. Approaches of this nature are more likely to be effective in some situations than in others. For example, they may not be useful when the nematode pest is highly virulent and crops are damaged at very low population densities. It is also recognized that specific details will need to be determined for individual situations. For example, there are farming systems where income is derived from the principal crop, and a rotation crop is grown for its soil-health benefits. In such cases, the rotation crop may not need to be highly tolerant of nematode damage. Also, the survival qualities or carry-over capacity of the relevant parasite will have an impact on the management system most likely to be effective. *Pasteuria*, for example, has relatively long-lived endospores, and may be able to cope with the absence of its nematode host for longer periods than other host-specific parasites.

One issue that is likely to promote discussion is my contention that tolerance rather than resistance should be better promoted as a nematode management tool. Resistance is generally favoured by nematologists (Trudgill, 1991), but the case for greater use of tolerance is based on the premise that tolerant or partially resistant cultivars will maintain rather than deplete the nematode populations on which specialized parasites depend, and will, therefore, foster the development of naturally occurring biological control. The argument for tolerance also has a practical basis. As agriculture intensifies to meet the world's increasing demand for food, crops will be grown more frequently, minimum-till farming systems will become commonplace, and practices already widely used in precision agriculture will help ensure that crops suffer minimal moisture or nutrient stress. If tolerant (or partially resistant) cultivars were to be included in such farming systems, the required components would be in place to utilize specific suppressiveness in nematode management. Currently, though, data demonstrating the effectiveness of such farming systems are lacking. The critical question, which can only be answered with long-term field experiments, is whether well-managed farming systems based on nematode-tolerant cultivars and minimum tillage will produce satisfactory yields, and also sustain biological suppressiveness at levels capable of keeping populations of plant-parasitic nematodes at acceptable levels.

In concluding this chapter, one final point must be made. Perhaps the main impediment to exploiting specific suppressiveness against nematodes is the attitude of many nematologists towards biological control. As can be seen from the examples presented here, naturally occurring nematode control is a well-documented phenomenon. However, whenever suppressive agents are identified, they are invariably seen as the next 'silver bullet' in the war against nematodes. The possibility that naturally occurring parasites and predators could be manipulated to improve the level of control is rarely considered. Given the opportunities outlined in this chapter and the limitations of introducing organisms for biological control purposes (see Chapter 12), it is to be hoped that this discussion will stimulate interest in improving our understanding of the role of farming systems in developing and maintaining both specific and general suppressiveness. Currently, this is an under-resourced area of research.

11

Integrated Soil Biology Management: The Pathway to Enhanced Natural Suppression of Plant-parasitic Nematodes

---

Earlier in this book (Chapter 4), soil health and sustainable farming systems were discussed in terms of the role they will play in meeting the world's future demands for food. Perhaps the most important conclusion from that discussion was that long-term food security will depend on our capacity to manage soil biotic interactions in such a way that crop productivity is maintained or improved, soil health is enhanced, off-site environmental impacts are minimized, nutrient cycling processes are playing a role in crop nutrition and natural enemies are helping to minimize losses from pest and diseases. With regard to the latter point, examples presented in Chapters 9 and 10 demonstrate that populations of plant-parasitic nematodes can be suppressed by a range of general and relatively specific antagonists. The challenge of the future is to better utilize these suppressive forces in nematode management programmes.

In an essay on managing the antagonistic potential that exists in all agroecosystems, Sikora (1992) pointed out that biological control is achievable when integrated crop production techniques are used to increase the activity of specific antagonists, or groups of antagonists. Some of the points made in the introductory section of that review are well worth repeating today:

- The antagonistic potential of a soil is the capacity of a soil ecosystem, through biotic factors, to prevent or reduce the spread of a pathogen, parasite or other deleterious agent. Alternative terms with similar meanings include suppressiveness, fungistasis, anti-phytopathogenic potential and biological buffering.
- The biotic factors associated with suppressiveness are antagonists, an umbrella term for parasites, predators, pathogens, competitors and other organisms that repel, inhibit or kill plant-parasitic nematodes.
- In natural ecosystems, the buffering capacity of the soil is highly effective in suppressing most soilborne pathogen and nematode problems.
- Agroecosystems are managed systems, and are, therefore, distinctly different from natural ecosystems. Nevertheless, the soil still has some buffering capacity, and the level of suppressive activity is influenced by the way it is managed.

Sikora's concept of managing the antagonistic potential of the soil has been expanded in this book to encompass management practices that improve soil health and enhance the provision of important ecosystem services such as nutrient cycling. This chapter addresses some of the practical issues involved in adopting such an approach, which was discussed in Chapter 1 and termed 'integrated soil biology management'.

It considers how various soil and crop management practices can be integrated into a farming system so that it not only enhances the activity of naturally occurring antagonists but also improves the soil physical, chemical and biological environment to the point where crops are less likely to suffer nematode damage.

Assessing Soils for Suppressiveness to Plant-parasitic Nematodes

Historically, nematode management has had a limited time frame and has focused almost entirely on the pest: how many plant-feeding nematodes are present in a particular situation; what is the likelihood that they will cause economic damage; what control measures can be used to have an immediate impact on their numbers? Despite the fact that natural enemies occur in all agricultural soils, practitioners involved in nematode management never measure their impact on nematode population densities, and rarely consider how levels of suppressiveness could be enhanced. However, if nematode pests are to be managed in an ecologically sound manner, we must consider the impact of their natural enemies. This means that methods are required to assess the capacity of a soil to suppress populations of a target pest.

Survey methods to identify nematode-suppressive soils

Plant nematologists inevitably focus on situations where nematode pests are causing economic damage. Thus, it is easy to gain the impression from the nematological literature that plant-parasitic nematodes are a universal problem and that crop losses occur in every field. However, any comprehensive survey of a crop in a particular region will show that this is not the case, as populations of pest nematodes vary considerably from field to field. Such variability is often associated with differences in soil type or the abiotic

environment (e.g. temperature or moisture), but it may also be due to management, or to differences in the antagonistic potential of the soil.

The first step in finding soils that may be suppressive to a particular plant-parasitic nematode is to peruse survey data and look for situations where the environment is suitable for the nematode but the pest's population density is relatively low. In some cases, historical factors may be directly responsible (e.g. a previous fallow or a resistant rotation crop may have reduced nematode populations), but in other situations, management practices may be acting indirectly by enhancing the biological buffering capacity of the soil. Data on nematode distribution within a field should also be subjected to similar scrutiny, as subtle variations in distribution patterns may be due to biotic rather than abiotic factors.

Although it is an expensive and time-consuming process to obtain data on the variability of soil physical, chemical and biological parameters within a field, modern technologies are making this task easier. For example, the availability of high-throughput, DNA-based systems for quantifying populations of plant-parasitic nematodes and other root pathogens (e.g. Ophel-Keller et al., 2008) means that biological data are easier to obtain than in the past. Also, technologies now available in precision agriculture automatically generate data on biomass or yield variability within fields (Melakeberhan, 2002; Srinivasan, 2006), and such outputs often lead to efforts to determine why yield is variable. During this process, data on plant-parasitic nematodes and many other parameters with the potential to limit yield are collected. It is, therefore, possible to identify sites within fields where populations of plant-parasitic nematodes are low, and perhaps go on to determine whether that is due to poor plant growth; subtle changes in soil type; the presence of other root pathogens; or unknown suppressive agents. Since edaphic factors such as soil texture markedly influence crop yield and the potential of nematodes to cause damage (Monfort et al., 2007), field-mapping techniques in precision agriculture also provide opportunities to target areas within fields where efforts to enhance suppressiveness are likely to generate the greatest economic return.

Soils suppressive to plant-parasitic nematodes have been found in many different situations (see Chapter 10), but the recent work of Robinson *et al.* (2008) provides a good example of how survey data can be used to detect previously unrecognized suppressiveness to an important nematode pest. Surveys of more than 3000 cotton fields in the United States had previously shown that reniform nematode (*Rotylenchulus reniformis*) was present in most fields at relatively high population densities. However, some fields had inexplicably low nematode populations, despite the fact that soil characteristics and cropping history were favourable to the nematode. Other fields had much lower nematode population densities in surface soils than expected. Results of assays in pots showed that there was a biological reason for these differences in nematode distribution, as transferable agents from four fields in south Texas suppressed reniform nematode populations by 48%, 78%, 90% and 95%. A similar result was obtained with five soils from Louisiana, with nematode populations being suppressed by between 37% and 66%.

Bioassays for suppressiveness

Once a potentially suppressive soil has been identified, various techniques are available to verify suppressiveness and confirm its biological nature (Westphal, 2005). One of the most common methods is to sterilize or partially sterilize the soil with heat, gamma irradiation or a broad-spectrum fumigant; adjust nematode population densities in the treated and untreated soil to the same levels by inoculation; and then grow plants in the two soils and check for differences in nematode multiplication rates. Bioassays of this nature were used to identify Californian soils with biological activity against *Meloidogyne incognita* (Pyrowolakis *et al.*, 2002; Bent *et al.*, 2008) and *Heterodera schachtii* (Westphal and Becker, 2001a); and to show that the surface layer of some

cereal-growing soils in Australia is suppressive to *Pratylenchus thornei* (Stirling, 2011b). Another frequently used technique, which is most useful when the suppressive agent(s) has a relatively short life cycle and the capacity to multiply readily, is to add small quantities of the test soil to sterilized or partially sterilized soil and demonstrate that the transfer reduces nematode multiplication or results in high levels of parasitism or predation on nematodes (Westphal and Becker, 2000; Chen, 2007).

One weakness of these bioassay methods is that differences in plant growth between sterilized soil and the test soil may confound the detection of suppressiveness. For example, plant growth often improves when soil is partially or fully sterilized, and so it is not always clear whether lower numbers of nematodes in the test soil are due to suppressive organisms, or result from fewer feeding sites being available to the nematode.

An alternative approach to assessing suppressiveness is to eliminate plants from the test system. Sterilized and untreated soil is inoculated with an assay nematode and mortality is assessed following incubation in the laboratory (see Jaffee *et al.* (1998); Pyrowolakis *et al.* (2002); Sánchez-Moreno and Ferris, (2007); Stirling *et al.* (2011b); and Timper *et al.* (2012) for some examples). However, a disadvantage of this type of assay is that it focuses on suppressive forces that affect the migratory stages of a nematode's life cycle and takes no account of antagonistic interactions that may occur in the rhizosphere or within roots. Thus, the best way of confirming suppressiveness to a particular nematode is to demonstrate its occurrence in both plant and soil assays.

Indicators of suppressiveness

Since greenhouse and laboratory assays are time consuming and labour intensive, various biotic and abiotic parameters have been proposed as indicators of suppressiveness (see previous discussion in Chapter 9). A logical first step is to measure soil carbon, as it provides an estimate of the organic resources that are sustaining the soil food web. Measures of permanganate oxidizable carbon (an estimate of labile carbon) are particularly useful (e.g. Moody *et al.*, 1997; Weil *et al.*, 2003) as many soil parameters that are sensitive to soil management practices (e.g. microbial biomass C, microbial biomass N, soil respiration, mineralizable C, mineralizable N, particulate organic matter and light-fraction organic matter) are related to this carbon fraction (see references cited in Weil *et al.*, 2003, and Lucas and Weil, 2012). Other advantages of measuring labile carbon are that it is routinely measured when assessing soil health (Pattison *et al.*, 2008; Gugino *et al.*, 2009), and methods can be modified for use in the field (Weil *et al.*, 2003). Since suppressiveness to plant-parasitic nematodes is likely to increase as a soil's carbon status improves (Stirling *et al.*, 2011b), it should be possible to measure total or labile carbon and use the data to locate potentially suppressive soils or to identify management practices that may have enhanced suppressiveness.

From a biological perspective, microbial activity (measured as the rate of hydrolysis of fluorescein diactetate), is one of the most useful indicators of suppressiveness. Widely used in assessing suppressiveness to a range of plant pathogens (Hoitink and Boehm, 1999), this relatively simple technique is also associated with suppression of plant-parasitic nematodes (Stirling *et al.*, 2005, 2011a, b).

One feature of agricultural soils is that, because of the way they are managed, they tend to lack biological complexity. The nematode community, for example, is usually dominated by nematodes with short generation times and relatively high reproductive rates (i.e. plant-parasitic and microbivorous species). Such nematode assemblages are indicative of relatively simple, non-structured food webs, whereas suppressive mechanisms are most likely to operate in soils that have complex food webs with long food chains and many trophic links (Jaffee *et al.*, 1998; Berkelmans *et al.*, 2003). This hypothesis is supported by results obtained by McSorley *et al.* (2006) and Sánchez-Moreno and Ferris (2007), who showed that the prevalence of omnivorous and predatory nematodes was associated with suppressiveness. Thus, soils suppressive to plant-parasitic nematodes are most likely to be found in situations where there is a structured nematode community

containing a range of omnivorous and predatory species. Communities of this nature are most likely to occur under perennial crops, or in annual cropping systems where the land is cropped continually, residues are returned to the soil, soil disturbance is minimized, broad-spectrum biocides are never used and inputs of synthetic fertilizers are not excessive.

One problem with finding useful indicators of suppressiveness is that research teams often do not have the breadth of expertise required to monitor the range of organisms likely to be involved, and even when organisms known to be antagonistic to nematodes are assessed, the key suppressive agents may still not be identified. A study by McSorley *et al.* (2006) in a sandy soil that was suppressive to root-knot nematode provides an indication of the difficulties involved. Free-living nematodes (bacterivores, fungivores, omnivores and predators), enchytraeids, Collembola, mites, nematode-trapping fungi, egg-parasitic fungi, *Pasteuria* spp. and a variety of rhizosphere fungi and bacteria were monitored, and none of their population patterns were found to be consistently associated with suppressiveness. The authors concluded that the observed nematode suppression was due to the combined effects of many antagonists, and also indicated that other potential antagonists were likely to have been involved, even though their contribution had not been assessed.

The above situation, in which the suppressive services provided by the soil biological community are the result of the activities of many different organisms, is likely to be the rule rather than the exception in soils that are suppressive to soilborne pests and pathogens. The only way of dealing with this complexity is to use DNA-based approaches to identify and quantify the vast array of potential antagonists that occur in such soils. Fortunately, this is now possible, as a range of molecular methodologies have been developed in recent years to characterize microbial communities in the soil and rhizosphere (e.g. Daniel, 2005; Roesch *et al.*, 2007; Buée *et al.*, 2009b); identify antagonists of nematodes (e.g. Zhu *et al.*, 2006; Smith and Jaffee, 2009; Xiang *et al.*, 2010b; Mauchline *et al.*, 2011); assess soil nematode assemblages (e.g. Porazinska *et al.*, 2009; Sørensen *et al.*, 2009; Donn *et al.*, 2011;

Griffiths *et al.*, 2012); and identify microorganisms associated with suppressiveness to plant-parasitic nematodes and other soilborne pathogens (e.g. Bornemann and Becker, 2007; Sanguin *et al.*, 2009; Mendes *et al.*, 2011; Rosenzweig *et al.*, 2012). Given the rapid and continuing improvement in high-throughput DNA technologies, these tools will soon be used routinely to characterize biological communities in suppressive and non-suppressive soils. Ultimately, these molecular tools will also enable us to find specific suppressive agents in soil and better understand the complex nature of general suppressiveness. Once a 'signature' or 'fingerprint' of a suppressive biological community is available, it will be much easier to find soils likely to be suppressive to plant-parasitic nematodes, and to identify the key agronomic practices associated with the development of suppressiveness.

Modifying Farming Systems to Enhance Suppressiveness

Once a nematode-suppressive soil is identified and biological factors are found to be involved, a common reaction from nematologists is to see such situations as a potential source of biological control agents that could be mass produced and utilized at other locations. That reaction ignores the fact that suppressiveness is likely to have a complex biological basis (i.e. be due to a community of organisms acting together), and that suppressiveness will have developed under a particular set of circumstances and is, therefore, not easily transposed to another location. Thus, a more appropriate response would be to raise questions such as the following. Why has suppressiveness developed in this particular situation? Is the phenomenon associated with certain soil types or environments? Can differences in the way soils are managed explain why some soils are suppressive and others are conducive to the nematode pest? What management practices could be introduced to enhance levels of suppressiveness? Nematologists

who are largely dealing with nematode-conducive soils might ask a similar range of questions. Why are nematodes an unremitting pest in this particular soil or environment? Why are soils that are cropped in a certain manner invariably conducive to nematode problems?

During the process of considering these questions, one is forced to take a holistic view of the farming system and consider how it influences the development of nematode problems. In most cases, the answer is relatively clear cut: plant-parasitic nematodes multiply readily because susceptible crops are grown repeatedly in soils with little or no biological buffering capacity. Decisions on what should be done in such situations are sometimes difficult to make, as financial and human resources will limit the options available. Thus, it is usually easier to be reactive: to accept that the farming system is immutable and attempt to alleviate nematode problems with further inputs (chemical or biological), or by making minor changes to the way the crop is grown. However, I will argue in this chapter that modifying the farming system is a better long-term option. Crop and soil management practices influence soil physical and chemical fertility; they determine the quality of the food resources available to the nematode pest; and they also affect the regulatory forces provided by the soil biological community.

The task of modifying a farming system is not something that can be accomplished easily. All facets of the crop production process must be re-evaluated, an integrated set of alternative management practices must be formulated, and the new system must then be tested in the field. Specialists in nematology, plant pathology and soil biology can help redesign a farming system, but to ensure that the alternatives are likely to be feasible in practice, the final options should be endorsed by a multidisciplinary team that also includes innovative farmers, agronomists, soil scientists, crop nutritionists, agricultural engineers and economists.

Since a wide range of fungal, oomycete, bacterial and nematode pathogens can potentially damage the roots of most crops, developing a more active, diverse, resilient and pathogen-suppressive soil biology must be a major consideration when a farming system is being redesigned. The agents causing crop loss will vary from crop to crop and with soil type and environment, but it is important to recognize that usually there will be more than one key soilborne pest or pathogen. Thus, management practices that may be useful against nematodes, for example, must be considered in the light of their impact on other components of the pest–pathogen complex. One of the advantages of focusing on the suppressive services provided by the soil biota is that increasing a soil's biological buffering capacity usually enhances suppressiveness to a broad range of pathogens.

Numerous examples could be cited where changes in management practices have enhanced suppressiveness to nematodes and soilborne pathogens. However, the development of suppressiveness is dependent on soil type and the prevailing environment, and so it is impossible to provide recipes that are effective in particular situations. Thus, the remainder of this chapter concentrates on key practices that are known to influence suppressiveness, and how they might be integrated into different types of farming system.

Organic Matter Management: The Key to General Suppressiveness

Soil organic matter not only plays a central role in maintaining the sustainability of cropping systems (see Chapter 2 and also a review by Fageria, 2012), but also determines the biological status of cropped soils and their capacity to suppress soilborne pathogens (see Chapter 9 and reviews by Stone et al., 2004, and Bonanomi et al., 2011). Maximizing carbon inputs and minimizing carbon losses is, therefore, the key to enhancing the suppressiveness of agricultural soils. Unfortunately, however, agriculture has a long history of depleting rather than improving the amount of carbon sequestered in soil (Paustian et al., 1997). For example, the conversion of natural vegetation to agriculture results in a marked decline in soil organic matter levels, with

losses of more than 50% commonly recorded (Smith et al., 1993; Lal, 2004). These losses are partly due to the fact that carbon additions to soil are frequently less in agroecosystems than in natural systems. However, another reason for the decline is that carbon losses are accelerated by the way crops and soils are managed (Magdoff and Weil, 2004b). Management practices such as tillage, planting techniques, methods of handling crop residues, the level of cover cropping and the species used as rotation crops all influence soil organic matter, and the flow-on effects of changing the carbon status of a soil will eventually affect its biological status and its capacity to suppress soilborne pests and diseases.

Carbon levels in soil are determined by the balance of inputs from crop residues and organic amendments, and carbon losses through organic matter decomposition. However, organic matter affects so many ecological functions that there is a conflict between the need to use it and the need to conserve it. On the one hand, soil organic matter must decompose if it is to provide the nutrients required for plant growth and the organic compounds needed to fuel the soil food web and thereby promote microbial activity, biological diversity and disease suppression. On the other hand, soil organic matter must accumulate if these functions are to operate in the long term; if carbon is to be sequestered; and if the soil's water- and nutrient-holding capacity is to be improved. Thus, the management of soils to maximize the benefits from soil organic matter requires intelligent compromises, as two contradictory goals are being pursued (Weil and Magdoff, 2004). For this reason, the management practices discussed in the following sections cannot be considered in isolation. All available options must ultimately be integrated into a farming system so that it delivers the desired result.

Management impacts on soil carbon, and flow-on effects to the soil biota

The scientific literature on the effects of crop and soil management practices on soil carbon levels is voluminous, and is only summarized briefly here. From the perspective of carbon inputs, photosynthetic processes determine the amount of carbon entering a system, and so the crops that are grown and the perennial forages and cover crops included in the farming system are the only sources of carbon in most agroecosystems. Thus, the main options for increasing the amount of carbon that can be sequestered in soil are to reduce the frequency and duration of bare fallow periods and to include perennial forages, pasture, high-residue crops and cover crops in the rotation (Magdoff and Weil, 2004b). Practices that increase productivity (e.g. crop rotation, fertilizer inputs and irrigation) will also increase crop residue return, but the magnitude of the gains will depend on whether residues are retained or removed, and the proportion of the biomass partitioned into the harvested product (Paustian et al., 1997). Repeated application of animal manures, municipal biosolids, composts and various organic waste materials also have the potential to markedly increase soil carbon levels (Diacono and Montemurro, 2010), but amendments of this nature are not economically feasible in many farming systems.

Since carbon inputs from plants are constrained by environmental factors such as moisture and temperature, management practices that reduce carbon losses through decomposition play a major role in maintaining or increasing soil carbon levels in agroecosystems. Arguably, the most important of these practices is the frequency and intensity of tillage. Numerous studies have shown that soil organic carbon stocks increase under various forms of conservation tillage (see Franzluebbers, 2004; Chan et al., 2011; Caride et al., 2012; De Sanctis et al., 2012, and references therein), although the increase is sometimes limited to surface soils (Angers and Eriksen-Hamel, 2008; Powlson et al., 2011b). The magnitude of the difference in carbon levels between no-tillage and conventional tillage is variable, but is typically about 300 kg/ha per year in North America (Fig. 11.1), although it is sometimes even greater (Chatskikh et al., 2009). However, there are also situations where reducing tillage has minimal impact (Ogle et al., 2012), or where its adoption is constrained by problems caused by grass weeds, soil compaction and soil-borne diseases (Powlson et al., 2012).

While sequestering more carbon in soil is an admirable objective from the perspective of improving soil health and mitigating climate change, the difficulties involved in achieving that goal should not be underestimated. In many agricultural systems, the availability of water constrains plant productivity and sets

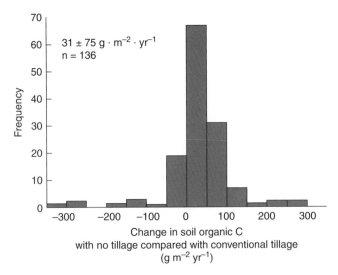

Fig.11.1. Frequency distribution showing the change in soil organic carbon with no-tillage compared with conventional tillage (based on 136 observations in North America). (From Franzluebbers, 2004, with permission.)

an upper limit on what is achievable. In such situations, carbon capture can be improved through appropriate inputs of fertilizers or other management practices that maximize water use efficiency (expressed in terms of dry matter production per mm of available water). With regard to reducing decomposition rates through changes in tillage practice, there are limitations to what is achievable. Changes in levels of soil organic carbon under conservation tillage are sometimes relatively small and are concentrated mainly in surface layers (Fig. 11.2); it can take 10–15 years to see small but significant effects (Angers and Eriksen-Hamel, 2008); and the sink capacity for soil carbon is finite and accumulation slows as soils approach their carbon saturation capacity (Chung *et al.*, 2010; Chan *et al.*, 2011).

One reason that many studies have shown surprisingly little response in soil organic matter levels to differences in residue input is that low nutrient availability may limit carbon sequestration. The more stable

fine fraction pool of soil organic matter (commonly referred to as soil humus) has more nitrogen, phosphorus and sulphur per unit of carbon than the plant material from which it originates, and the low levels of carbon sequestration often observed in long-term experiments in conservation agriculture may be due to the limited availability of these nutrients. Experiments with carbon-rich wheaten straw showed that addition of nutrients markedly increased net humification efficiency (Kirkby *et al.*, 2013), demonstrating that if carbon is to be sequestered into the more stabilized pool of soil organic matter, elements other than carbon are also required. Such results suggest that nutrient management practices which aim to optimise economic returns and limit the risk of environmental damage by matching nutrient supply to the crop's requirement may inadvertently limit the supply of nutrients required to synthesise the more stabilized pool of soil organic matter. Thus, from a soil health perspective, it may be advantageous to add supplementary nutrients when carbon-rich stubbles are the primary source of organic matter. However, the problem with this approach is that emissions of nitrous oxide, an important greenhouse gas, increase markedly when inorganic nitrogen and degradable organic materials are available, and soil moisture conditions are suitable for denitrification (Dalal *et al.*, 2003; Wang *et al.*, 2011).

Regardless of the practical difficulties involved in increasing soil carbon levels, there is overwhelming evidence to indicate that the effort is worthwhile from a biological perspective. The three most important soil improvement practices associated with conservation agriculture (minimal soil disturbance, permanent plant residue cover and crop rotation) not only increase total soil carbon levels, but also increase labile carbon and mineralizable nitrogen, with flow-on effects to microbial biomass and activity (Franzluebbers, 2004; Franchini *et al.*, 2007; Melero *et al.*, 2009; Roper *et al.*, 2010; Kirkby *et al.*, 2013), and to populations of meso- and macrofauna (Roger-Estrade *et al.*, 2010; Min and Toyota, 2013). What is not clear is whether the biodiversity of some conventionally managed agricultural soils has declined to such an

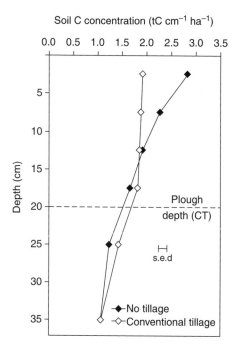

Fig. 11.2. Trends in soil carbon concentration with depth after 21 years of no-tillage or conventional tillage at a site in southern Brazil (mean values for two crop rotations). (From Machado *et al.*, 2003, with permission.)

extent that all ecosystem services provided by the soil biological community can be restored (Wall *et al.*, 2010).

Although it may never be possible to restore a soil's biodiversity to its original state in a reasonable time frame, most agricultural soils contain a wide range natural enemies of nematodes, so the critical question from a biological control perspective is whether the suppressive services they provide can be influenced by manipulating the distribution of organic matter within the soil profile; the quantity or quality of organic inputs; or decomposition processes. For example, levels of suppressiveness suitable for shallow-rooted crops could possibly be achieved by focusing on maintaining high levels of organic matter in upper layers of the soil profile. Suppressiveness might also be enhanced by manipulating the timing of key decomposition events. For example, residues from previous crops could be managed, or the C:N ratio of amendments varied, so that decomposition occurs at an appropriate time, thereby enhancing the biological buffering capacity at a critical stage in-crop development (e.g. in the first few weeks after planting, when nematode attack is likely to cause maximum damage). Irrigation has been used to manipulate the level of nutrient cycling from cover crop residues (Ferris *et al.*, 2004) and there is no reason why similar strategies could not be used to activate organisms associated with suppressiveness.

Tillage and its impact on suppressiveness

The effects of tillage on nematode communities and suppressiveness are often examined. However, it is difficult to draw firm conclusions from such studies because tilled systems vary in the frequency, intensity and depth of cultivation, and the amount of organic matter incorporated; while varying levels of disturbance occur in reduced-till treatments. Also, reduced-till treatments have often been in place for relatively short periods after many years of conventional cultivation, and so there are questions about whether the biological community has reached a steady state under the new tillage regime. Tillage effects are, therefore, best examined in long-term

trials. Westphal *et al.* (2009) cited experimental evidence to show that population densities of *Heterodera glycines* on soybean in the eastern maize-growing region of the United States were lower under no-till than conventional tillage, but noted that the crop and tillage management practices studied had been in place for less than 10 years. On the basis of this observation, the effects of tillage intensity on the nematode were examined by sampling an experiment under a maize–soybean rotation that had been established 28 years previously. The results of that study showed that the incidence of *H. glycines* was reduced by long-term no-till (Westphal *et al.*, 2008b: Seyb *et al.*, 2008). Final nematode population densities were proportional to tillage intensity, with numbers of cysts and eggs highest under mouldboard ploughing plus secondary tillage; intermediate under less aggressive tillage regimes; and lowest under minimum-till (Westphal *et al.*, 2009). No attempt was made to determine whether biological suppression was involved in the tillage effect, but data obtained with a penetrometer suggested that changes in soil physical properties may have contributed to differences in nematode population densities.

With regard to the more immediate effects of a change in tillage practice, Sánchez-Moreno and Ferris (2007) showed that suppression of *Meloidogyne incognita* juveniles was reduced when tillage was simulated by mixing soil for 20 minutes in a concrete mixer. In a field experiment on a ginger farm in Australia, bioassay results indicated that minimizing tillage enhanced suppressiveness to *M. javanica*. In the second and third years after tillage treatments were established, gall ratings and egg counts were usually significantly lower under minimum tillage than conventional tillage (Stirling *et al.*, 2012). Additionally, soil that was not disturbed during the process of setting up bioassays was more suppressive than soil gently mixed by hand. Stirling (2008) had previously shown that disturbed soil was not as suppressive to *Pratylenchus zeae* and *Meloidogyne javanica* as undisturbed soil, and Westphal and Xing (2011) obtained a similar result in a bioassay with *H. glycines*. Collectively, these observations raise questions about whether estimates of suppressiveness are compromised when soil is disturbed during the soil collection process.

In contrast to these results, Timper et al. (2012) used two different bioassays to compare suppressiveness to plant-parasitic nematodes in conventionally tilled and strip-tilled cotton, and found that tillage did not affect the survival of *Rotylenchulus reniformis* in soil or reproduction of *M. arenaria* on peanut. Although the omnivorous and predatory nematodes that are often associated with suppressiveness are usually favoured by a reduction in tillage (Lenz and Eisenbeis, 2000; Okada and Harada, 2007), data presented by Timper et al. (2012) indicated that tillage effects were inconsistent, with indices reflecting the proportions of these nematodes in the nematode community (maturity index and structure index) sometimes higher and sometimes lower in the conventional tillage treatment.

In summarizing the results of studies in which conservation tillage practices were used to reduce soil disturbance and increase the abundance of soil organisms in higher positions in the soil food web, Ferris et al. (2012b) noted that mechanical disturbance generally must be minimized for at least 2 years before omnivorous and predatory nematodes begin to increase in abundance. In a California study, for example, no-tillage plus continuous cropping increased microbial and fungal biomass in the surface layer (0–5 cm) and also increased concentrations of immobile nutrients such as phosphorus and potassium. However, this treatment did not increase the complexity of the nematode community, possibly because higher-trophic-level nematodes had been eliminated after decades of cultivation (Minoshima et al., 2007). A Hawaiian study produced similar results, as populations of omnivorous and predacious nematodes failed to increase, and root-knot nematode was not suppressed by non-tilled living mulch (Marahatta et al., 2010).

Although much more work is needed before we can conclude that suppressiveness to plant-parasitic nematodes is enhanced by long-term reductions in tillage, the general improvement in biologically active fractions of organic matter that occurs under minimum tillage, together with the fact that water is more readily available to plants and the soil biota, suggests

that key soil properties associated with biological activity are being moved in the right direction. The beneficial effects of reduced disturbance on nematophagous fungi and predators (McInnis and Jaffee, 1989; Jaffee et al., 1992b; Wardle, 1995), together with the protection that predatory microarthropods receive from residue cover (Koehler, 1997) are other reasons for eliminating tillage from the farming system, or markedly reducing its frequency and intensity. Since the soil biological community is likely to have been markedly affected by the tillage practices historically employed in agriculture, and biodiversity is slow to recover when tillage practices are modified (Adl et al., 2006), long-term tillage trials must be monitored to fully demonstrate the effects of minimum tillage on suppressiveness.

Using organic amendments, cover crops and mulches to enhance suppressiveness

Given the detrimental effects of tillage on detritus food webs and the slow recovery of organisms at higher trophic levels (Wardle, 1995; Ferris et al., 2012b), the work on organic amendments for nematode control (discussed in Chapter 9) deserves some comment. In most cases, amendments are mechanically mixed into the upper layers of the soil profile at high application rates. When used in this way, the amendment may stimulate the activity of some microbial antagonists, but the negative effects of the incorporation process and the products of organic matter decomposition will almost certainly be detrimental to many predators. Thus, instead of using organic amendments in much the same way as a nematicide (i.e. having an expectation that a response will be obtained within weeks or months), it may be better to consider how they might be used to generate more enduring changes to the soil biological community.

One way of using an organic amendment to deliver a soil food web with some biological buffering capacity is to apply the amendment to the soil surface as mulch. Numerous studies in perennial crops have shown that mulching is an effective nematode management technique (see discussion later in this chapter), while the benefits of mulching in terms of suppressiveness to plant-parasitic nematodes were recently confirmed in a microcosm experiment which showed that placing crop residue on the soil surface enhanced the soil's suppressiveness to *Meloidogyne javanica* and *Pratylenchus zeae* to a greater extent than incorporating the residue into the soil profile (Stirling *et al.*, 2011a). Data showing the mulch effect on root-knot nematode are presented in Table 11.1. An even better way of enhancing suppressiveness may be to apply the amendment more frequently but in smaller quantities, so that the mulch layer is continually replenished. Since the latter practice simulates to some extent the way organic matter is returned to soil in natural systems (i.e. the 'forest floor' effect), such a practice may eventually produce a diverse soil biological community that is able to suppress populations of plant-parasitic nematodes through multiple mechanisms. Evidence presented in Chapters 5, 6 and 9 suggests that generalist predators (nematophagous fungi, predacious nematodes and microarthropods) are likely to contribute to the suppressive effect of mulches.

Since imported organic amendments are relatively expensive, the most sustainable way of obtaining organic matter for use as mulch is to produce it at the site where it is required and then preserve it using a no-till management system. Although potential problems can arise with mulches of this nature (e.g. increased cutworm damage in soft-stemmed annuals; higher populations of trunk-gnawing voles and mice in vineyards), the benefits to be obtained from generating and preserving residues (increases in rainfall infiltration and water

storage; and reductions in surface sealing, water erosion and soil water evaporation) mean that residue retention is an indispensable part of sustainable crop production (Mitchell *et al.*, 2012a). Mulches produced on-farm could be expected to enhance suppressiveness to plant-parasitic nematodes, but to date, evidence to support that hypothesis has mainly been obtained from relatively short-term studies (e.g. Stirling and Eden, 2008; Wang *et al.*, 2008, 2011; Stirling, 2013). There is, therefore, an urgent need to assess levels of suppressiveness at sites where crop residues have been maintained on the soil surface for many years.

Impact of Management on Specific Suppressiveness

Although the first step towards enhancing a soil's suppressiveness to plant-parasitic nematodes is to introduce soil and crop management practices that increase levels of soil organic matter, the effect of these practices on specific suppressiveness also needs to be considered. Host-specific parasites can have a large impact on nematode population densities (see Chapter 10), and their effects add to the level of nematode control that is achievable through organic matter-mediated suppressiveness. Nematode management strategies must, therefore, be designed with both general and specific suppressiveness in mind.

Rotational cropping provides many benefits to a farming business, as more than one commodity is produced for sale; the workload is spread over more of the year; a wider variety of crop residues is returned to the soil; and losses from soilborne diseases are reduced.

Table 11.1. Effect of covering soil with sugarcane residue for 18 and 27 weeks, as assessed by planting a tomato seedling in the soil at the end of the mulching period, inoculating the bioassay plant with *Meloidogyne javanica*, and assessing root galling and egg production after 7 weeks.

Time in field	Root gall rating		Number of eggs/plant	
	No mulch	Mulch	No mulch	Mulch
18 weeks	3.83 a	2.75 b	102,329 a	23,933 b
27 weeks	4.67 a	3.47 b	249,460 a	97,273 b

Means in each row followed by the same letter are not significantly different ($P = 0.05$).
(From Stirling *et al.*, 2011a, with permission.)

Nevertheless, there are also advantages in concentrating on a single crop, as high returns can be obtained from cash crops that are ideally suited to local growing conditions; specialization provides economic benefits; and assured markets are available for staple crops (e.g. rice, wheat and potatoes). Thus, despite the benefits of growing several crops, there are justifiable reasons why agricultural production is often centred on a specific crop. In such situations, soilborne diseases are often a limiting factor. Consequently, host-specific organisms capable of suppressing the causal pathogens have an important role to play in reducing losses in farming systems where there is one dominant crop.

Pasteuria is the best-known host-specific parasite of plant-parasitic nematodes, and an example from Australia illustrates the dilemma faced by farm managers in attempting to maximize the activity of such a parasite. *Pratylenchus thornei* is an important pest of wheat in the northern grain-growing region, and nematode-resistant and tolerant cultivars are being developed to provide control (Thompson *et al.*, 2008). However, the recent finding that low populations of *Pasteuria thornei* occur in about 25% of fields in that region (Li *et al.*, 2012) raises questions about the best way of increasing its population densities and levels of activity. Will the use of resistant cultivars reduce populations of *Pratylenchus thornei* to the point where populations of its obligate parasite will decline? Would it be better to use nematode-tolerant cultivars, as they will maintain host nematode populations and presumably suffer minimal yield loss?

Questions of this nature are not commonly addressed by those interested in biological control of nematodes. However, available evidence suggests that the cultivars chosen and the rotation practices employed in a farming system will have an impact on levels of parasitism by *Pasteuria*. Host-specific parasites and their hosts maintain a dynamic relationship over time (Atibalentja *et al.*, 1998), and so any management practice that reduces nematode populations would be expected to be detrimental to the parasite. Thus, for root-knot nematodes, both the number of females parasitized by *P. penetrans* and the number of spore-encumbered juveniles increases on a good host

of the nematode (Madulu *et al.*, 1994; Cetintas and Dickson, 2004), but declines when a nematode-resistant crop or fallow is included in the rotation (Ciancio and Quénéhervé, 2000), indicating that these practices must be used judiciously if populations of the parasite are to be maintained. Nevertheless, results of work by Noel *et al.* (2010) indicate that nematode-resistance does have a place in integrated management programmes. Although populations of soybean cyst nematode (*Heterodera glycines*) were reduced by both resistant cultivars and *Pasteuria nishizawae*, the cultivar × *Pasteuria* interaction was not significant. This suggests that the parasite and the resistant cultivars could be used together to manage the nematode, presumably because the soybean cultivars were only moderately resistant to the nematode, and, therefore, did not reduce nematode populations to low levels. Non-host crops such as maize, sorghum, peanut and cotton also reduce populations of *H. glycines* (Kinloch, 1998), and their impact on *P. nishizawae* should be investigated, as they may have a greater impact on the parasite than soybean cultivars with some nematode resistance. Results of a crop sequence study at a site naturally infested with *Pasteuria penetrans* and *Meloidogyne arenaria* (Timper, 2009) also indicate that it is possible to sustain a nematode-suppressive soil with appropriate rotations (see Chapter 7).

Integrated Nematode Management or Integrated Soil Biology Management?

It is now more than 20 years since a previous version of this book was published (Stirling, 1991), and during that time there has been a change in the way nematode management is perceived by most nematologists and pest management consultants. Instead of aiming to decimate nematode populations with broad-spectrum biocides, more specific chemicals are used to reduce pest population densities to levels that do not cause economic damage. Also, we no longer rely on single control options such as crop rotation or nematicides. Instead, a number of cultural, chemical and biological controls are used together in management packages that are referred to under umbrella terms such as 'integrated nematode management', 'integrated crop management', 'bio-management', 'bio-system management' or 'natural pest management' (Sikora *et al.*, 2005).

Integrated nematode management is a variant of the widely used term 'integrated pest management' (IPM), and is commonly thought of as a strategy that uses all available tactics in a compatible and environmentally safe manner to maintain nematode populations at levels below the economic threshold (Ornat and Sorribas, 2008). However, the problem with the above definition is that it does not consider the available tactics from an ecological perspective. Some of the tactics used to manage nematodes are simply incompatible with the objectives considered important in this book: sequestering carbon in soil; improving soil health; enhancing the productivity and sustainability of farming systems; and increasing the capacity of agricultural soils to cycle nutrients and suppress soilborne pathogens. For example, it is possible to markedly reduce populations of plant-parasitic nematodes with soil fumigants, intensive tillage, solarization and long periods of fallow, but these practices also deplete soil carbon levels, impact negatively on many soil physical and chemical properties, and reduce the soil's biological buffering capacity. They are, therefore,

not generally compatible with ecologically based nematode management.

The concept of integrated soil biology management introduced earlier in this book (see Chapter 1) is a much more holistic way of looking at nematode problems and how they might be overcome through ecological processes. Instead of targeting the nematode, integrated soil biology management targets shortcomings in the farming system and the management-induced changes in soil properties that have allowed nematodes to become pests. It aims to improve soil health and enhance the activity, diversity and resilience of the soil biological community so that plant growth is not constrained by physical or chemical impediments, and nematodes and other soilborne pathogens are kept in check by natural regulatory forces. Management tactics such as those mentioned in the previous paragraph have the opposite effect and cannot, therefore, be accommodated within the concept of integrated soil biology management.

A wide range of management tactics are available for use against nematodes (Sikora *et al.*, 2005) and they can be subdivided into three groups, based on their compatibility with the objectives of integrated soil biology management (Table 11.2). However, there are caveats with regard to how each of these

Table 11.2. Tactics available for use in integrated nematode management programmes, and their compatibility[a] with the concept of integrated soil biology management.

Compatible	Possibly compatible	Not compatible
Exclusion, quarantine	Flooding	Intensive tillage
Diagnosis and prediction	Biofumigation	Long periods of clean fallow
Organic amendments	Removal of infested roots	Solarization
Crop rotation	Some nematicides	Soil fumigation
Mulching	Trap crops	Most nematicides
Multiple cropping	Fallow management	
Cover crops	Inundative biocontrol	
Antagonistic crops		
Nematode-free planting material		
Bio-enhancement		
Resistant cultivars and rootstocks		
Tolerant cultivars		
Improved crop husbandry		
Changes in planting and harvest dates		

[a]See text for caveats with regard to how each tactic has been categorized.

tactics has been categorized. For example, the impact of rotational cropping strategies depends on the level of carbon inputs (which will be affected by climate, soil type and management), the extent of carbon losses and the degree of biotic disruption due to tillage. Nematode-resistant cultivars and rootstocks can be used to reduce populations of the target pest and help maintain organic matter-mediated suppressiveness, but may be detrimental to the organisms responsible for specific suppressiveness. Uprooting and destroying the roots of a nematode-infested crop may sometimes be a useful post-harvest tactic, but unless it is done with minimal soil disturbance, perhaps by dragging a single tine along the planting row, it will impact negatively on the soil biota. Trap crops are likely to be beneficial to the system if the plants are killed by means other than cultivation, while fallow management strategies could have a range of effects, depending on the types of weeds present; the amount of biomass produced; the extent to which weeds remove water that could be used by the following crop; the potential of weeds to host damaging nematodes; and the tillage practices employed at the end of the fallow period. Given the level of pulverization and tillage required to achieve a biofumigation effect (see Chapter 9), and the fact that mineral nitrogen sometimes accumulates following crops such as canola (Ryan *et al.*, 2006), it is possible that the *Brassica* crops commonly used as biofumigant-producers would more beneficial to the farming system if they were simply used as rotation crops. Since fumigants and nematicides generally have broad-spectrum effects, they are incompatible with integrated soil biology management. However, the next generation of nematicides (e.g. chemicals that are applied in small quantities to seeds; products that are sprayed on foliage and move downwards systemically; and chemicals that target specific receptors unique to plant-parasitic nematodes) may be more compatible, as they are less likely to have major effects on non-target organisms. The possible role of inundative biocontrol in integrated soil biology management is also unclear, as biostatic forces are likely to increase in healthy soils and may prevent the

introduced organism reaching the population densities required for success (see Chapter 12).

The integrated soil biology management concept is somewhat similar to the eco-epidemiological approach to nematode management proposed by Mateille *et al.* (2008). These authors considered the nematode management strategies commonly used in tropical and subtropical Africa and noted that nematode communities in agroecosystems lacked diversity and were dominated by plant parasites. Control options largely focused on eradicating the target pest, and when they were employed, both the pest and its antagonists were killed, the biological community was rearranged, and biotic imbalances were induced. Additionally, resistance-breaking biotypes were selected when nematode-resistant cultivars were introduced, and nematode species that are normally not pathogenic also began to cause problems. Thus, this 'soil cleaning' approach was considered unsustainable. An alternative ecological approach was proposed in which the central focus was on enhancing biological activity and diversity. Under this scenario, pest populations were constrained by interspecific competition for space and resources, and by microbial antagonists. Both approaches are outlined in Fig. 11.3.

Integrated Soil Biology Management in Various Farming Systems: The Pathway to Enhanced Suppressiveness

Integrated soil biology management is a philosophy that is built around two key principles: (i) soil organic matter plays a central role in determining a soil's physical, chemical and biological fertility; and (ii) healthy soils are the key to sustainable crop production. Management practices that maintain or increase levels of soil organic carbon influence numerous soil properties that impact on soil health, but from a biological perspective, one of the most important is the capacity to suppress soilborne pests and pathogens (see Chapter 2). Thus, it is possible to argue that farming systems which increase the

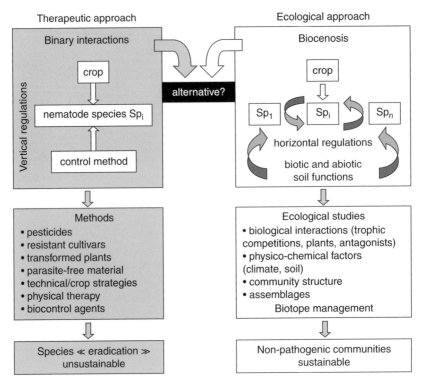

Fig. 11.3. Eradicative and ecological approaches to nematode management. Note that the term 'therapeutic' was used by the authors but the term 'eradicative' is used here. (From Mateille *et al.*, 2008, with kind permission from Springer Science + Business Media B.V.)

carbon content of soil will eventually result in the development of relatively robust systems of biological control.

Although integrated soil biology management has been proposed as an option for dealing with nematode problems in sustainable farming systems, it is little more than a concept, and is largely untested. Thus, we do not know whether such an approach is likely to be successful, or whether it is can be applied to different crop production systems. This issue is addressed in the remainder of this chapter. However, the applicability of the approach is considered for only selected farming systems, as it is impossible to cover the diversity of crops and the many different ways that crops are grown worldwide. Ultimately, however, the applicability of integrated soil biology management will be determined at a local level, as the management practices that can be used to influence soil health, crop growth and the biological status of soil are dependent on

environmental factors such as soil type and climate, and production-related issues such as economics, market requirements and the resources available to the grower.

Grains, oilseeds, pulses, fibre crops and pastures

The key role of conservation agriculture

Most of the world's grains, oilseeds, pulses and fibre crops (e.g. wheat, maize, sorghum, barley, oats, millet, sunflower, canola, soybean, lupin, chickpea, faba bean, field pea, mung bean, lentil and cotton) are grown in extensive cropping operations that specialize in two or three crops. Long-term studies of farming systems that include these crops have shown that the systems most likely to be agronomically productive and environmentally sustainable combine high cropping intensities, adequate fertilization, conservation tillage, crop residue retention and sound rotation practices (Reeves, 1997). Collectively, these practices improve soil physical properties such as aggregate stability, air-filled pore space and bulk density, and also sustain or increase soil organic carbon, although the extent of the latter effect depends on soil type and climate. Conservation tillage can only slow the loss of soil organic carbon, not halt or reverse it, and so continual cropping (within the limits of the environment) is important in increasing carbon inputs. Crop rotation, always an important component of any farming system, becomes even more critical when conservation tillage practices are adopted (Reeves, 1997). In humid regions, the ability of rotation to ameliorate limitations to productivity or to increase crop yield potential is usually associated with reductions in disease, whereas in semi-arid regions, the benefits of rotations are often the result of improved water use efficiency.

In some regions of the world, fear of reduced yields under conservation tillage is one of the main factors constraining the uptake of the sustainable farming systems discussed in the previous paragraph. This apprehension is understandable, as it is based on the results of studies in the UK, for example, which showed that direct drilling consistently produced high yields on light chalk soil, whereas on a clay soil, yields were 25–40% lower than conventional tillage in 2 of 3 years. Morris *et al.* (2010) commented on those studies and then reviewed the impact of non-inversion tillage systems in the UK, outlining their agronomic impact, the economic benefits that were possible and the advantages of those systems to the soil and the environment. The main issues that were limiting the adoption of conservation agriculture were also discussed in that paper. Those impediments, together with suggestions as to how they might be overcome, are:

- Difficulties associated with planting into large quantities of crop residue (e.g. blockages and handling concerns with direct drilling machinery, and poor crop emergence and establishment).
  - Equipment solutions are available (e.g. large-diameter disc coulters, residue-management wheels and star wheels) to cut through crop residues and clear crop residues away from the seed row. A wider spacing between rows will improve residue clearance between coulters.
- A reduction in soil temperature under the residue layer (due to its insulating and reflective properties, and its impact on evaporation rates), which slows germination and development in spring-sown crops.
  - Residue-free strips can be included to increase heat input into the soil.
- Crop emergence and growth problems associated with the allelopathic effects of water-soluble toxins produced by crop residues and microorganisms during the residue decomposition process.
  - Improved crop residue-management practices are available (e.g. shallow incorporation of residues; increasing the time between residue decomposition and sowing; changing the position of crop residues relative to the seed; removal of residue from the seedling row).
- Decreased rates of nitrogen mineralization due to slower residue decomposition and the immobilization of soil mineral nitrogen by residues with a high C:N ratio.
  - This problem, which is usually only temporary, can be overcome by changing residue-management and nitrogen fertilization practices. In some situations, it may be more important to retain nitrogen in soil microbial biomass than to risk losing it by leaching.

After collating the results from numerous studies in the UK, Morris *et al.* (2010) concluded that non-inversion tillage practices provided numerous benefits, including improved timeliness of operations; lower fuel usage; increased work rates; better erosion control; improved soil structure; and benefits to soil physical properties that improved water infiltration, increased water-holding capacity and reduced particle loss to streams and rivers. Most of the issues that were seen as obstacles to adoption could be overcome through appropriate management practices. This is an important conclusion, given that conservation agriculture is only widely practised in a few countries. It highlights the potential for improving many current farming systems, and demonstrates that it is possible to improve soil carbon levels on land where crops are grown on a large scale.

Integration of pastures into crop-based farming systems

Mixed farming systems involving both crops and livestock are an important component of large-scale cropping enterprises in many countries, and contribute in a number of ways to sustaining the productivity of the world's farmlands. Although the trend, particularly in developed countries, has been for increased specialization and separation of crop and livestock enterprises, pastures have an important role to play in crop-based farming systems, particularly if those systems incorporate biological nitrogen fixation and manure recycling

(Bellotti, 2001; Wilkins, 2008; Fisher *et al.*, 2012). The main reasons for including pastures in a farming system are listed in Table 11.3, but there are also potential negative impacts (e.g. removal of ground cover by grazing animals, soil compaction, drying of the soil profile, redistribution of nutrients to stock camps, redistribution of weed seeds). Since the effects of including pastures in a mixed farming system are influenced by many factors (e.g. pasture species, stock type, stocking rate and grazing practice), the skill and passion of the manager and the level of management inputs play a major role in determining whether the final outcome will be positive or negative (Fisher *et al.*, 2012).

In future, pastures will continue to play their traditional role in large-scale cropping systems (i.e. diversifying farm income and maintaining soil fertility), but increasingly will be utilized to maintain the sustainability of the production system as a whole. However, the move towards increased cropping intensity is placing pressure on pasture-crop

Table 11.3. The main benefits of integrating pastures into crop-based farming systems.

Parameters affected by pastures	Traits associated with pastures, and their benefits
Soil organic carbon	Carbon levels are maintained or improved due to the higher root: shoot ratio of pastures compared to crops; slower decomposition rates; absence of disturbance; and the high C:N ratio of roots.
Dryland salinity[a]	Salinity problems are reduced due to perennial plant growth; the deep-rooting habits of pasture species; growth during both summer and winter; and the capacity of roots to cope with potential subsoil constraints.
Weed control	Pastures have competitive effects on weeds and are an important component of integrated weed management. Herbicide-resistant weeds can be controlled by grazing following the crop phase.
Stubble and soilborne diseases	Pasture species are resistant and tolerant to many important crop diseases. Crop stubble harbours some pathogens, but stubble loads are reduced by grazing.
Nutrient supply	Legumes produce biologically fixed nitrogen. Pastures also enhance biological nutrient cycling and recover inorganic nitrogen from deep in the soil profile.
Biodiversity	Above-ground species abundance and diversity increases due to greater numbers of plant species in pastures. Similar effects occur below ground due to a build-up of organic carbon.
Farm diversification and risk management	Pastures provide an alternative income source, and less variability in farm income.

[a]Dryland salinity is the result of a hydrological imbalance that increases deep drainage and recharge to groundwater, rendering low-lying areas unproductive due to salinity and waterlogging.
(Modified from Bellotti, 2001 and Fisher *et al.*, 2012.)

systems that rely on self-regeneration of annual legumes following the cropping phase, and so alternatives are required. Possible options include phase-cropping systems, where a sequence of pasture years is followed by a sequence of cropping years; the use of dual-purpose crops that can be grazed in the early stages of growth; and intercropping, overcropping or companion cropping systems in which two or more plant species occur concurrently for at least part of their life cycle. With regard to grazing management, Fisher *et al.* (2012) suggested that key areas for future research included the development of decision-making tools that indicate when livestock should be moved to another area, thereby minimizing their negative impact; and the application of precision livestock technologies (e.g. virtual fencing, remote sensing and electronic tagging) to improve the efficiency and ease of livestock management.

One of the most important roles of well-managed pastures in agroecosystems is to increase levels of soil organic carbon. Conant *et al.* (2001) reviewed 115 studies on the effect of pasture management and conversion of land to

pasture on soil carbon stocks, and concluded that improved management increased soil carbon stocks in 74% of the studies (Fig. 11.4). Carbon sequestration rates were highest during the first 40 years after treatments began; they tended to be greatest in the top 10 cm of soil; and the rate of improvement was highly influenced by biome type and climate. Results from three long-term trials in temperate Australia (Chan *et al.*, 2011) provide an indication of some of the issues that affect soil organic carbon sequestration in mixed farming systems. First, the pasture phase is important in maintaining and increasing soil organic carbon. Second, when long-term pasture land with a relatively high soil organic carbon content is converted to cropping, the practices associated with conservation agriculture will tend to maintain rather than increase soil carbon stocks, largely because the rate of change in carbon concentration declines as it approaches steady-state equilibrium levels. Third, tillage is responsible for 80% of the total loss in organic carbon in upper soil layers, and so it has a much greater effect on carbon stocks than stubble management. Fourth, improved

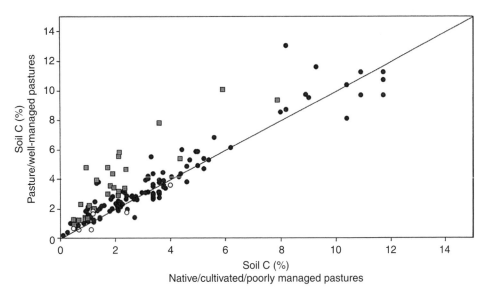

Fig. 11.4. Impact of management improvements (e.g. fertilization, grazing management, sowing of legumes or grasses, conversion of land from cultivation or native vegetation) on soil carbon sequestration in grasslands. The soil carbon concentration data were obtained from studies in 17 countries comparing improved versus unimproved pastures (•); grassland versus cultivated fields (■); or grassland versus native vegetation (○). Soil carbon concentration increased for points above the line and decreased for points below the line. (From Conant *et al.*, 2001, with permission.)

management practices (e.g. regular fertilization with phosphorus and potassium) play an important role in carbon sequestration, as they improve the growth of pastures based on subterranean clover and other legumes, and increase levels of nitrogen fixation.

Impact of management on soil biological parameters

The earlier discussion shows that when agricultural soils are managed appropriately, the amount of carbon sequestered will increase. However, does this mean that the soil's biological status (in terms of activity and diversity) will improve? This question is best answered by looking at the impact of soil and crop management practices on microbial biomass, as it is central to ecosystem function and is widely used as an indicator of soil quality (Gonzalez-Quiñones et al., 2011). One such study (Roper et al., 2010) showed that microbial biomass, microbial activity (estimated by soil respiration) and labile fractions of organic matter were more responsive to changes in management (in this case tillage practice) than total soil carbon and nitrogen, or crop yields. Franchini et al. (2007) came to a similar conclusion in a quite different environment, as parameters associated with microbiological activity (microbial biomass carbon and nitrogen, metabolic quotient, and soluble carbon) were affected to a greater extent by tillage and crop rotation than total stocks of carbon and nitrogen. However, one problem with microbial biomass measurements is that they treat the soil biological community as a single undifferentiated unit. Thus, they do not indicate whether there has been a change in specific components of that community (e.g. the fungi, nematodes and microarthropods capable of parasitizing or preying on nematodes).

Although plant-parasitic nematodes are economically important pests of many field crops (see relevant chapters in Evans et al., 1993; Barker et al., 1998; and Ciancio and Mukerji, 2008), there have been very few studies on the impact of management practices on suppressiveness to nematodes in large-scale farming systems. In one example, where maize was the dominant crop and wheat and clover were grown as cover crops during winter, omnivorous

and predatory nematodes were more abundant under no-till than conventional till (Fu et al., 2000). Also, predators responded relatively quickly (within <10 weeks) to the addition of leaf litter to the surface of no-till plots, whereas this response was not observed in cultivated plots. However, as with the situation in a maize–soybean rotation where populations of soybean cyst nematodes were found to decline with a reduction in tillage intensity (Westphal et al., 2009), data were not presented to show that suppressiveness was enhanced by changes to tillage or residue-management practices.

The grain-growing region of northern Australia is one area where there have been limited studies on suppressiveness to plant-parasitic nematodes. This region is characterized by cropping systems that are strongly dependent on stored moisture rather than in-crop rainfall, and tillage systems that are increasingly reliant on zero or minimum tillage. The dominant crops are wheat and sorghum (grown in winter and summer, respectively), with smaller areas of chickpea, cotton, barley, mung bean and maize. Since cropping intensity is limited by the need to replenish soil water reserves by fallowing, inputs of organic matter are often less than is required to maintain a neutral soil carbon budget (Dalal et al., 1994, 1995). It is, therefore, not surprising that measurements of various biological and biochemical parameters (microbial biomass, microbial activity, cellulase activity, free-living nematodes, total DNA and fatty acid profiles) show that the biological status of the soil is relatively poor (Bell et al., 2006). However, the same study did show that management practices such as stubble retention and zero tillage provided some benefits, although they tended to be relatively small and were mainly confined to the upper 5 cm of the soil profile. Also, there were positive effects of breaking long fallows with a short-term crop of field peas planted with minimal fertilizer inputs and then sprayed with herbicide at pod initiation to kill the crop and prevent weeds setting viable seed. The advantage of this practice, which is termed brown manuring, is that it can be implemented at relatively low cost, both in terms of crop establishment and, more importantly, use of stored soil moisture.

Despite the relatively low levels of biological activity in the soils discussed, assessments for suppressiveness to *Pratylenchus thornei* (Stirling, 2011b) showed that the nematode consistently multiplied to much higher population densities in sterilized cropped soils (typically 22–35 nematodes/g soil) than in untreated soils (typically 2–8 nematodes/g soil). Other bioassay results indicated that multiplication of *P. thornei* was sometimes reduced by transferring field soil to sterilized soil; and that the nematode failed to multiply in soil from some farms, particularly if it was collected from the upper 25 cm of the soil profile. Although increased root biomass and better root health in assay plants may have been partly responsible for the differences in nematode multiplication rates in bioassays of sterilized and non-sterilized soil, these results suggest that biological mechanisms of suppression are operating in many of these soils, at least near the soil surface.

The above hypothesis is supported by observations on the relationship between populations of *P. thornei* and root distribution. Measurements taken under a direct-drilled wheat crop showed that roots were concentrated in the upper 30 cm of the soil profile, whereas populations of *P. thornei* were more evenly distributed with depth. Relatively high nematode populations were maintained at 30–45 and 45–60 cm, despite a paucity of roots at those depths (Fig. 11.5). Since *P. thornei* is an obligate parasite of roots and its distribution did not mirror root distribution, it is possible that surface soil is more suppressive to the nematode than soil deeper in the profile. However, this hypothesis needs to be tested, as there are alternative explanations: (i) environmental conditions near the surface may be less suited to the nematode than further down the profile; and (ii) the nematode may not multiply readily in the structural roots responsible for most of the root biomass in upper layers of soil.

Although the results of studies mentioned suggest that the role of management in enhancing suppressiveness to nematode pests is a fruitful area for further research, it is important that the term 'management' is not interpreted too narrowly. Most previous studies have concentrated on the impact of stubble retention and zero or minimum tillage, and have not taken a holistic view of the farming system. Changes to one component of the system (e.g. reducing tillage) are likely to produce no more than incremental improvements in soil carbon levels, biological activity, soil health and suppressiveness, and major changes will only occur when multiple

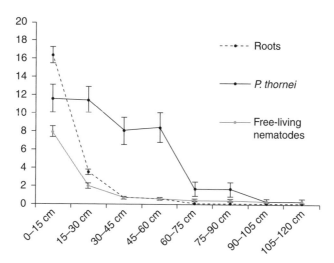

Fig. 11.5. Populations of free-living nematodes and *Pratylenchus thornei* (nematodes/g soil) and root biomass (g/0.25 m²) at various depths in the soil profile following a wheat crop (bars indicate standard errors). (From Stirling, 2011b, with permission.)

improvements are made. Thus, from a biological control perspective, nematologists need to encourage farmers to reassess their soil and crop management practices, and help them fine-tune their farming system to the point where it is not only consistently productive and economically viable, but also sequesters soil carbon (i.e. carbon gains are greater than losses). Farmers would be more likely to move in this direction if we could provide answers to the following questions:

- Are soils collected from situations where the soil carbon budget has been positive for many years more suppressive to nematodes than soils where carbon levels are static or declining?
- What level of suppressiveness is achievable in situations where conservation tillage, sound rotation practices, high cropping intensities, crop residue retention and adequate fertilizer inputs have been integrated into the farming system?
- In situations where soil carbon levels have been maximized for the soil type, environment and farming system, is the level of suppressiveness sufficient to reduce populations of the key nematode pest(s)?
- Is suppressiveness largely confined to upper soil layers, and if so, is the level of suppression sufficient to impact on nematode population densities, and what practices could be introduced to improve suppressiveness with depth?
- Do the soil-health benefits that accrue from the farming system (e.g. improved soil physical properties, reduced compaction, better water capture and storage, increased storage and cycling of nutrients, enhanced biological buffering) improve growing conditions for crops and thereby increase damage thresholds for key nematode pests?
- Is it necessary to include resistant cultivars, tolerant cultivars or non-host crops in the rotation, and if so, do the collective benefits from the whole farming system reduce the impact of plant-parasitic nematodes to economically acceptable levels?

In concluding this discussion on integrated soil biology management as it relates to large-scale farming systems, it is important to recognize that the management practices mentioned are designed to promote soil health and enhance organic matter-mediated suppressiveness. However, in circumstances where specific suppressiveness must also be managed, the crop rotation component of the farming system may need to be modified. Thus, in situations where a relatively host-specific parasite of a key nematode must be maintained (e.g. *Pasteuria nishizawae* on *Heterodera glycines*; *Pasteuria thornei* on *Pratylenchus thornei*; *Nematophthora gynophila* and *Pochonia chlamydosporia* on *Heterodera avenae*), it may be necessary to choose crop cultivars and rotation crops that maintain populations of the pest. In such situations, farm managers should aim to enhance suppressiveness through a combination of general and specific mechanisms.

Vegetable crops

Vegetables are a vitally important component of world food production, with about 900 million tonnes being produced globally from a cultivated area of more than 50 million ha. However, the vegetable industry is almost impossible to characterize in a meaningful way, as more than 20 crops are widely grown; there is a diverse range of growing environments; and management practices vary from country to country, and between climatic regions. Nevertheless, modern vegetable production has a number of key features:

- Limited land resources, particularly in areas near major population centres. This means that land is valuable and pressures to intensify production are always present. Intensification reaches a maximum in the greenhouse vegetable industry, which now occupies more than 400,000 ha of land worldwide.
- High cost structures due to the large array of inputs (e.g. irrigation, plastic mulch, machinery, labour requirements, nutrients, pesticides, refrigeration, packaging and transport).

- Demanding supply chains that require products to be consistent and of high quality, produced in large quantities, delivered to markets daily and capable of meeting stringent food-quality standards.
- Industry fragmentation and the relatively small size of many individual vegetable-producing operations, a factor that limits communication with the industry and also between growers.

Soilborne diseases are always one of the major constraints to vegetable production, and growers have increasingly dealt with such problems by resorting to nematicides, fungicides and broad-spectrum fumigants. However, the recent phase-out of methyl bromide and the increasing level of scrutiny of all agricultural chemicals by regulatory bodies are prompting moves towards biologically based pest management tactics (Zasada *et al.*, 2010). The main options (as they pertain to root-knot nematodes, the most important nematode pest of vegetables worldwide) are discussed below.

Organic amendments

In most agricultural production systems, economic constraints prevent organic amendments being used for soilborne disease management. However, they are an option in high-value crops, with most of the work on organic amendments for nematode control (discussed in Chapter 9) having been done with vegetable crops. Nevertheless, the role of amendments in managing nematode pests of vegetables remains unclear, largely because numerous sources of organic matter can be used; multiple mechanisms of action are involved; results tend to be inconsistent; application rate influences efficacy; long-term impacts on soil health and the environment are largely unknown; and efficacy varies with the crop, the key nematode pest and the environment. Available evidence suggests that organic amendments have a role to play in future nematode management strategies, but the sources of organic matter and the way it is used will need to be determined at a local level.

Much of the work on organic amendments for nematode control has been directed towards selecting materials that generate toxic metabolites and decomposition products in concentrations sufficient to kill nematodes. Strategies of this nature offer the possibility of relatively immediate and reliable nematode control, and should, therefore, be pursued. However, much more work is required on how such amendments are best integrated into the vegetable farming system to achieve long-term effects. It is possible, for example, that continued application of organic materials with a chemical mode of action will eventually increase soil carbon levels to the point where biological mechanisms of control begin to operate. It is also possible that the nutritional and soil-health benefits from amendments will over-ride the need for nematode control, as presumably occurred in a 7-year Canadian study with beef manure and compost amendments (Kimpinski *et al.*, 2003). These amendments increased populations of *Pratylenchus penetrans* and had no effect on *Meloidogyne hapla*, but both treatments increased potato yield by 27%. This example demonstrates the need for long-term studies, and also shows that when an amendment improves growing conditions for the plant, it may not be necessary to reduce nematode populations.

With regard to amendments that are unlikely to act by releasing nematicidal compounds during the decomposition process, McSorley's (2011b) summary of results from numerous experiments in Florida provides an indication of the efficacy of these materials. Root-knot nematode populations were reduced in only 5 of 11 studies with composts and crop residues, and were never reduced in 26 tests with composted municipal yard wastes. However, the amendments released nutrients when they decomposed, and also improved the soil's water-holding capacity. Thus, crop yields increased in more than half the above tests, again confirming the soil-health benefits that accrue from inputs of organic matter.

Crop rotation, cover cropping and other practices

Crop rotation and cover cropping are important components of sustainable vegetable production systems, and Lu *et al.* (2000) provide an excellent review of their benefits, particularly as they relate to fresh-market tomato production. Experiments in Maryland showed that when hairy vetch was combined with crimson clover and rye, the cover crops produced about 9 t/ha biomass containing 150–200 kg N/ha. When these crops were sprayed with herbicide or killed by mowing, the residues produced mulch that suppressed weeds during the early period after tomatoes were transplanted. In comparison to plants grown on polyethylene, the hairy vetch cropping system was much more profitable because yields were 22% higher over a 5-year period; nitrogen fertilizer requirements were reduced; crops matured when fruit prices were higher; and the costs involved in purchasing and laying polyethylene were avoided. Similar benefits have been obtained in other regions. For example, conservation tillage systems are also becoming increasingly attractive to producers of processing tomatoes in California's Central Valley, as costs are lower than standard tillage methods and yields in the two systems are comparable (Mitchell et al., 2012b).

Vegetable production systems such as those described above are feasible, but in situations where root-knot nematode is a major pest, cover crops with resistance to the common species of Meloidogyne must be included, as nematode populations have to be reduced to low levels prior to planting the vegetable crop. Fortunately, numerous resistant rotation crops are available, including castor (Ricinus communis), crotalaria (Crotalaria spectabilis), hairy indigo (Indigofera hirsuta), maize (Zea mays), marigold (Tagetes spp.), oat (Avena sativa), pangola grass (Digitaria decumbens), sesame (Sesamum indicum), sorghum (Sorghum bicolor), sorghum–sudangrass (S. bicolor × S. sudanense), sunn hemp (Crotalaria juncea) and velvetbean (Mucuna pruriens), although the level of resistance varies with crop cultivar and nematode species (McSorley, 2001). The most useful rotation crops in warm climates, where root-knot nematode problems generally occur, are selected cultivars of sorghum–sudangrass hybrids. In addition to producing enough biomass to improve soil carbon levels, they grow quickly enough to outcompete weeds, which are often alternate hosts of root-knot nematode in vegetable cropping

systems. Since legumes contribute to sustainability by producing biologically fixed nitrogen, they should also be included in the rotation sequence. However, this option is limited to some extent by a paucity of Meloidogyne-resistant legumes.

Weed-free fallows are another excellent tactic for managing root-knot nematode (Johnson et al., 1997), as second-stage juveniles have limited food reserves and soon die of starvation in the absence of a host (van Gundy et al., 1967). Fallow periods as short as 8 weeks are, therefore, an effective control measure, provided the soil is warm, moist and weed-free (Stirling et al., 2006). The negative effects of a tilled fallow on soil structure and fertility and its role in promoting wind and water erosion are disadvantages of this practice, but these problems can be overcome by retaining surface residues using a farming system in which crops, cover crops and weeds are killed with herbicides or managed with non-chemical methods other than tillage.

Meloidogyne-resistant cultivars have been available in vegetable crops such as tomato, cowpea, lima bean and sweet potato for more than 20 years (Roberts, 1993), but because the choice of resistant crops and cultivars is relatively limited, they play no more than a minor role in many current nematode management programmes. In situations where resistant cultivars are not available, an alternative approach, which is only relevant to transplanted vegetable crops, is to induce nematode-resistance mechanisms in transplants. Initial work in this area has shown that combining an organic amendment (chitin) with a phyotchemical (benzaldehyde) improves transplant growth and reduces galling caused by root-knot nematode in the subsequent crop. Some plant growth-promoting rhizobacteria have similar effects when applied alone or in combination with these treatments (Kokalis-Burelle et al., 2002a, b).

Integrated management

Since the individual management practices discussed do not consistently reduce losses from nematodes in vegetable crops, a nematode management programme based on multiple tactics is likely to be more effective. Various integrated management packages have been suggested (see Roberts, 1993; Chellemi, 2002; Litterick *et al.*, 2004; Ornat and Sorribas, 2008; Wang *et al.*, 2008; Zasada *et al.*, 2010; and Collange *et al.*, 2011 for examples), but in some cases, the proposed approaches are little more than a list of control practices. What is really needed are truly integrated management packages that address the practical issues involved in integrating a number of nematode management tactics into a vegetable farming system that is not only productive and sustainable, but also profitable. Since the combination of tactics required will be site-, market- and environment-specific, there is no simple recipe for success, and the final integrated packages that are adopted will depend on the needs of individual enterprises.

Australia's subtropical vegetable industry provides an example of the challenges involved in developing effective and reliable nematode management programmes at a local level. This industry supplies most of Australia's winter and spring production of tomatoes, capsicums, zucchinis, rockmelons, watermelons, snow peas, potatoes and sweet potatoes, and all of its ginger. Each commodity is produced by specialist growers who crop their

land every year and rely on irrigation water, soil fumigation and nutrient inputs to sustain productivity (Stirling and Pattison, 2008). Not all components of an alternative and more sustainable farming system are currently in place, but progress is being made. For example, we know that: (i) some cultivars of forage sorghum, Rhodes grass (*Chloris gayana*) and soybean are useful rotation crops; (ii) these crops are resistant to the dominant root-knot nematode species in vegetable-growing soils (*M. javanica* and *M. incognita*); (iii) amendments of poultry manure, sawdust, compost and sugarcane residue enhance suppressiveness to these nematodes; and (iv) tillage is detrimental to any organic matter-mediated suppressiveness that is induced by amendments or biomass from a rotation crop (Stirling *et al.*, 2006; Stirling, 2008; Stirling and Eden, 2008; Stirling *et al.*, 2012). What has been lacking are attempts to fully integrate these practices into a vegetable farming system.

A recent study (Stirling, 2013) has shown that when minimum tillage, compost amendments and nematode-resistant rotation crops were used together in what was considered a sustainable vegetable farming system (Table 11.4), capsicum yields increased root-knot nematode populations declined

Table 11.4. Amendment, rotation and tillage management practices used in a 2-year experiment in subtropical Australia comparing a conventional and a more sustainable capsicum farming system on a clay loam soil naturally infested with *Meloidogyne incognita*[a].

Treatments	Amendment[b]	Rotation crop[c] Year 1	Year 2	Tillage[d]	Mulch for vegetable crop
Conventional	– –	Forage sorghum	Forage sorghum	T	Plastic
Sustainable	+ +	Rhodes grass	Forage sorghum + soybean	M	Plastic over crop residue[e]

[a]At the commencement of the experiment, the site was found to be infested with a low population of *M. incognita*. In both years, plots were either inoculated (+RKN) or not inoculated (–RKN) with additional root-knot nematodes prior to planting capsicum.

[b]The conventional treatment received no amendment (– –). In the sustainable treatment, compost (10 t/ha) and sugarcane residue (20 t/ha) was applied in year 1, and an amendment made from mill mud, sawdust and sugarcane residue was applied at 16 t/ha in year 2 (++).

[c]In the conventional treatment, forage sorghum was grown for 11 weeks and then incorporated by tillage. In the sustainable treatment, Rhodes grass (year 1) and forage sorghum followed by soybean (year 2) were grown for 23, 9 and 9 weeks, respectively, before being sprayed with glyphosate.

[d]T = tillage prior to planting the rotation and vegetable crops; M = minimum tillage (one cultivation to prepare beds in the second year).

[e]The rotation crops were flail mown and plastic was placed on the undisturbed bed.

(Modified from Stirling, 2013, and presented with permission.)

and the level of nematode damage was reduced, while suppressiveness to the nematode was enhanced in the second year in comparison to the conventional system (Table 11.5). However, there are still several problems to overcome before such farming systems can be utilized by growers. For example, organic amendments and rotation crops change the soil's physical structure and may exacerbate symphylan problems (Stirling *et al.*, 2012); root rotting caused by *Pythium* and *Rhizoctonia* may kill plants in soils where root-knot nematodes are suppressed (Stirling, 2013); cutworm damage increases when crop residues are retained (Stirling and Eden, 2008; Smith *et al.*, 2011); and equipment is required to lay plastic mulch onto undisturbed beds covered with crop residue (Stirling, 2013). Also, some physically degraded soils do not self-repair easily when pastures, rotation crops and amendments are used to improve their organic matter status, and so soil compaction may limit root growth and yield when minimum tillage is first introduced (Smith *et al.*, 2011).

These studies demonstrate that many agronomic, soil, nutritional, economic, farm equipment, pest management and disease issues must be addressed when attempts are made to improve the sustainability of well-established farming systems. A number of nematode management tactics could possibly be included in such a system (Table 11.6), but because a greater level of planning and skill is required to make them successful, adoption is poor and most growers continue to use conventional practices. Biologically sound nematode management programmes suitable for particular soils, climates and crops can be developed, but in addition to continuing research, nematologists and other professionals need to work with growers and help them address the many issues involved.

With regard to the management practices listed in Table 11.6, the possibility of incorporating biological products into future integrated nematode-management programmes for vegetable crops warrants some comment. Products of this nature have not been considered in this chapter, as the focus has been on developing farming systems that enhance biological diversity and increase the level of control provided by naturally occurring antagonists. Also, as farming systems improve and soil carbon levels increase, competitive effects from a more active and diverse soil biological community will make it increasingly difficult to establish an introduced organism. Nevertheless, specific biological products applied to seeds or seedlings have a role to play in some situations, particularly during transition to more sustainable vegetable production systems. The options available are discussed in Chapter 12.

Perennial crops

Perennial crops differ from annual crops in many respects, but the main difference from a nematode management perspective is that

Table 11.5. Effects of a conventional and sustainable capsicum farming system on crop performance, populations of *Meloidogyne incognita*, and the level of damage caused by the nematode in successive years, and the impact of inoculating capsicum with additional root-knot nematodes prior to planting.

	Treatment	Fruit yield (kg/plant)		Root gall index		Number of root-knot nematodes/200 mL soil	
		– RKN	+ RKN	– RKN	+ RKN	– RKN	+ RKN
Year 1	Conventional	1.03 a	0.82 a	3.1 a	4.5 a	97 a	446 a
	Sustainable	1.09 a	1.03 a	1.1 a	2.9 a	22 a	41 a
Year 2	Conventional	0.81 b	0.90 b	4.0 ab	5.3 a	9120 a	5508 ab
	Sustainable	1.20 a	1.14 a	1.9 c	3.2 b	1174 bc	311c

For each parameter in each year, means followed by the same letter are not significantly different (*P* = 0.05).
(Prepared from Stirling, 2013, but modified to include data not previously published.)

Table 11.6. Practices likely to be suitable for inclusion in integrated management programmes for root-knot nematode in sustainable vegetable production systems.

Management practice	Reasons for inclusion	Possible issues requiring consideration
Biomass-producing rotation and cover crops with nematode resistance and tolerance	Reduce nematode populations and soilborne diseases; tolerate nematode damage; compete with weeds; increase soil carbon; add biologically fixed nitrogen (legumes)	Producing sufficient biomass quickly, and at an appropriate time of the year; input costs (seed, water, fertilizer); residue management
Residue retention and minimum tillage	Limit soil erosion losses; maintain soil moisture; dampen temperature fluctuations; provide opportunities to replace plastic mulch with plant material; benefit top predators; provide a refuge for microarthropods	Exacerbation of some pest and disease problems; compaction-related issues in physically degraded soils; methods for handling residues; equipment for planting into or laying plastic onto undisturbed beds; weed management
Short fallow periods at an appropriate time	Reduce root-knot nematode populations at critical times (i.e. in the few weeks before planting)	Ensuring that enough ground cover is retained to prevent erosion problems
Organic amendments	Increase soil carbon; provide nutrients; enhance suppressiveness	Costs and availability; application method (incorporate or apply as mulch?); application rate; timing
Nematode-resistant or tolerant cultivars	Increase threshold levels for damage; reduce population increase during the crop (resistant cultivars)	Lack of availability for many crops; their impact on host-specific parasites of the target nematode; costs of producing grafted seedlings on resistant rootstocks
Chemical seed treatments	Provide protection during the critical period within a few weeks of planting	Availability of safe and effective chemicals with minimal impact on non-target species
Biological seed treatments, and biologically based mixes for transplants	Populate roots with beneficial organisms; induce disease-resistance mechanisms.	Development of relevant transplant production systems; competition with indigenous organisms

nematode populations must be maintained at acceptable levels for a much longer period: several years for crops such as sugarcane, banana and pineapple, and for a much longer period with trees and vines. Also, perennial tree and vine crops have a much deeper root distribution than annuals, which means that biological control mechanisms must operate deep in the soil profile and are not necessarily enhanced at depth by manipulating the biology of surface soils. Nevertheless, enhancing biological suppressiveness should be easier in perennials than in annual crops, as organisms at higher trophic levels in the soil food web are more easily maintained because the soil is not tilled to the same extent, and disruption due to planting occurs less frequently. Since it is

impossible to cover all perennial crops in a book of this nature, this discussion will focus mainly on apple and banana, two important crops that have been chosen as examples because they are grown in quite different climates and cropping systems.

In general terms, the tactic most likely to generate general suppressiveness in perennial crops is to grow a companion crop or maintain a permanent pasture in the inter-row space between trees or vines, and manage it by regular mowing. However, companion crops and inter-row plantings need to be appropriate for the situation, as they must be tolerant of shading and competition from the perennial crop, and be able to tolerate trafficking by farm machinery. Also, the sward must be well managed, as companion plants may compete with the crop for water and nutrients, and the additional costs (e.g. seed requirements, mowing and the nets required in some crops to collect products harvested with tree-shakers) impact on the economics of the farming system. Nevertheless, the benefits of such a practice in terms of reducing soil losses due to erosion and improving the physical, chemical and biological properties of soil are well established. The effects of cover or companion crops and permanent pastures on plant-parasitic nematodes capable of attacking the perennial crop are not as clear, as they may not only host important nematodes (McLeod and Steel, 1999), but also enhance suppressiveness. The latter effect has rarely been studied, but suppressiveness is most likely to increase in upper soil layers. However, there will probably be situations where deep-rooted species could be used to achieve impacts further down the profile.

Enhancement of general suppressiveness

APPLE.   Research on apple (reviewed by Mazzola, 2007) demonstrates that it is possible to enhance suppressiveness to soilborne pathogens by manipulating the rhizosphere bacterial community through modifications to the cropping system. Greenhouse trials with soils conducive to apple replant disease showed that successive, short-term wheat cropping enhanced the growth of apple seedlings and reduced infestations of *Rhizoctonia solani* and *Pratylenchus penetrans*, two of the many pathogens commonly associated with replant problems. Significantly, the suppressive effect of wheat was quite specific, as some cultivars induced disease suppression by enhancing populations of specific fluorescent pseudomonad genotypes with antagonistic activity towards *R. solani*, whereas others did not. Rhizoctonia root rot was also suppressed by amendments of *Brassica* seed meal, but in this case, it was not a biofumigation effect. Instead, suppressiveness was associated with an active resident microbial community and elevated populations of *Streptomyces* spp. in the apple rhizosphere. Collectively, these results indicate that there is potential to enhance the suppressiveness of orchard soils by modifying their biological status with companion crops and organic amendments. However, this is not an easy task due to the complex aetiology of apple replant disease (Mazzola, 1998). A 1-year wheat crop and a 3-year *Brassica* green manure both reduced *Rhizoctonia* infection and improved tree growth and yield, but suppression of *Pratylenchus* was not maintained beyond the first growing season (Mazzola and Mullinix, 2005).

With regard to studies directed specifically at *Pratylenchus*, Stirling *et al.* (1995) showed that it was possible to reduce losses from the nematode in apple replants without resorting to fumigant nematicides. Thus, when apple trees were removed 12 months prior to replanting, an amendment of feedlot manure and a cover crop of *Lablab purpureus* (both incorporated into soil with additional nitrogen as urea) reduced populations of *P. jordanensis* on 2-year-old apple trees. A longer-term experiment with two mulch treatments (sawdust alone or sawdust + feedlot manure) produced even better results. Although these treatments were outperformed by methyl bromide in the first 2 years after planting, mulched trees not only had the highest growth rates after 5 years, but also produced yields that were as good as, or better than, methyl bromide (Table 11.7). Nematode populations were also reduced by the mulch treatments, but other benefits (reduced soil temperatures during summer; improved water infiltration; increased root density near the soil surface and improved soil water-holding capacity) also contributed to the growth and yield response. Thus, in contrast to the short-term effects of methyl bromide, mulching provided multiple benefits that are likely to have continued long after the experiment was terminated. Thus, the authors concluded that an integrated strategy was effective for managing *Pratylenchus* in an apple replant situation. The key tactics were: (i) early removal of the previous crop; (ii) growth of a green manure crop or the addition of an animal manure amendment; (iii) choice of a vigorous, nematode-tolerant rootstock; and (iv) maintenance of a layer of organic mulch around trees.

BANANA.   Gowen *et al.* (2005) listed a number of nematode control practices that were available for use on banana, and Stirling and Pattison (2008) described how some of those

Table 11.7. Effect of various soil treatments and the impact of rootstock on the growth and yield of apple trees and the number of root lesion nematodes (*Pratylenchus* spp.) in soil during the fourth and fifth years after trees were replanted.

Rootstock	Treatment	Yield (kg/tree)		Number of *Pratylenchus*/200 mL soil	
		Year 4	Year 5	Year 4	Year 5
MM106	Methyl bromide (1000 kg/ha)	7.5 c	8.1 de	111 abc	218 a
	1,3D (340 L/ha)	4.6 de	5.8 ef	178 a	88 abc
	Sawdust –manure mulch[a]	7.2 cd	9.3 cd	72 bc	28 bc
779	Untreated	4.1 e	4.7 f	180 a	397 a
	1,3D (340 L/ha)	9.0 bc	13.1 b	162 ab	111 abc
	Sawdust-manure mulch[a]	13.6 a	16.1 a	56 c	24 c
	Untreated	10.2 b	11.3 bc	114 abc	192 ab

[a]Mulch maintained regularly to a depth of 10 cm.
In each column, numbers followed by the same letter are not significantly different ($P \leq 0.05$).
(From Stirling *et al.*, 1995, with permission.)

practices (e.g. clean planting material; crop destruction; fallowing; crop rotation; retention of crop residues; improved nutrient management) are being used in an integrated manner by Australian banana growers. However, the banana industry as a whole is still a long way from integrating soil biology management into its crop production system. Potentially useful organic amendments have been identified in pot experiments (Tabarant *et al.*, 2011a, b; Pattison *et al.*, 2011), but their long-term effects on productivity, soil properties, root health and nematode populations have not been determined. One of the few long-term experiments with banana (a 6-year field trial in Uganda) showed that mulching with a mixture of chopped maize and grass markedly increased plant biomass and yield but did not affect root necrosis caused by *Radopholus similis* and *Helicotylenchus multicinctus* (McIntyre *et al.*, 2000). The mulch effect was thought to be due to an improvement in soil nutrient status and lower soil bulk density, faster water recharge and greater water uptake, a result that underscores the need to undertake long-term field trials when evaluating the effects of soil treatments on a perennial crop. It also highlights the need to take a broad view of treatment effects and consider their impact on parameters other than plant-parasitic nematodes.

Enhancement of specific suppressiveness

One difference between perennial crops and annual crops is that the nematodes that attack perennials interact with the plant over many years in a relatively undisturbed environment. Thus, there is considerable opportunity for specific relationships to develop between a nematode pest and its naturally occurring parasites. Situations where such relationships result in high levels of parasitism are discussed in Chapter 10, but some examples from perennial crops are:

- In peach orchards in California, the fungal parasite *Brachyphoris oviparasitica* can invade root-knot nematode egg masses and destroy most of the eggs. Since the egg masses produced by the nematode on peach contain relatively few eggs, fungal parasitism is partly responsible for the low nematode populations observed in some peach orchards (Stirling *et al.*, 1979; and also Chapters 5 and 10).
- A large proportion (about one-third) of the *Meloidogyne* females in Mauritian and

South African sugarcane fields are infected by the bacterial parasite *Pasteuria penetrans* (Williams, 1960; Spaull, 1984).

- In South Australia, *Pasteuria penetrans* is widespread in vineyards more than 25 years old, possibly explaining why root-knot nematode populations are lower than in young vineyards (Stirling and White, 1982; Stirling, 1984; and also Chapter 7).

- A large proportion (23–87%) of the *Meloidogyne* eggs developing in grape and kiwifruit fields in Queensland, Australia, were parasitized by fungal parasites, predominantly *Pochonia chlamydosporia* and *Paecilomyces lilacinus* (Mertens and Stirling, 1993).

Although high levels of parasitism do not necessarily mean that nematode populations will be reduced to levels below the damage threshold, these examples demonstrate that nematode pests of perennial crops are often subject to attack from specific parasites. Thus, the pathway to developing biologically based nematode management practices in perennial crops is to: (i) recognize that specific parasites of the target nematode are likely to be present on well-established crops; (ii) measure parasitic activity due to host-specific parasites in various situations (e.g. different fields or regions) and identify the factors (e.g. climate, soil type or management) responsible for variation in the level of activity; and (iii) develop an understanding of the management practices most likely to enhance parasitic activity.

Unfortunately, the above pathway is rarely taken. Host-specific parasites are often viewed from a relatively narrow perspective (i.e. as a source of microbial agents for commercial production and release). However, questions related to the management practices likely to enhance specific suppressiveness are vitally important, and will only be answered by understanding why these parasites are effective in some situations and not in others. As with annual crops, we also need to know whether the use of nematode-resistant cultivars or rootstocks (with the concomitant danger of developing resistance-breaking biotypes of the nematode) is a better long-term strategy than using nematode-tolerant material,

which will maintain both the pest and its specific parasites but suffer little yield loss. Finally, there is the question of how established nematode–parasite relationships can be maintained when a perennial crop is replanted. In the case of *Pasteuria*, for example, spore distribution will always be linked to nematode distribution (i.e. both spores and nematodes will be located in microsites near roots), and it will have taken many years for spore concentrations to reach their current levels. In such situations, it may be best to minimize disruption during the replanting process (i.e. by eliminating aggressive tillage or by sawing-off trees and vines and treating the remaining stumps with herbicide), as such practices may help to maintain the intimate relationship that is likely to have developed between host nematodes and their parasites. However, as soilborne pathogens and the allelopathic effects of old roots have the potential to damage replants, research will be required to confirm the practicality of such replanting systems.

An example of progress: nematode-suppressive soils in sugarcane

The Australian sugar industry is relatively small by world standards (32 million tonnes of cane produced from about 400,000 ha of land), but the changes it has made since the turn of the century are a good example of the benefits of taking a holistic approach to addressing soil-related constraints to productivity. In the early 1990s, relatively large yield responses (15–40%) were obtained in experiments with soil fumigants, rotation crops and nematicides, indicating that nematodes and other soilborne pathogens were associated with an industry-wide problem known as yield decline. However, it was also recognized that the problem had multiple causes. Soils were physically and chemically degraded after more than 100 years of sugarcane monoculture; the soil was being compacted by harvest and haul-out machinery; and soil organic matter levels were declining due to excessive tillage, and because valuable crop residues were being burnt to facilitate the harvest operation. These issues were addressed by a multidisciplinary team of researchers (see Chapter 4) and the end result was an improved sugarcane farming system based on four key principles: minimum tillage; residue retention; a leguminous rotation crop; and controlled traffic using global positioning system guidance. The new system was quickly adopted by growers because it increased sugar yields, reduced costs, improved soil health and produced additional income from rotation crops such as soybean and peanut (see Garside *et al.*, 2005 for a summary of the research). From a nematological perspective, losses from *Pratylenchus zeae* and *Meloidogyne javanica* were reduced because the rotation crops reduced nematode population densities at planting. However, damage thresholds were also likely to have increased due to improvements in soil health (Stirling, 2008).

Prototypes of the improved sugarcane farming system were first tested in 2001 (Bell *et al.*, 2003), and since that time it has become increasingly apparent that one of the long-term benefits of reducing tillage and retaining crop residues is enhanced suppressiveness to plant-parasitic nematodes. Evidence to support that contention is summarized as follows:

- When sugarcane is harvested green and crop residues are retained (a process known as green cane trash blanketing), soil organic carbon levels gradually increase during a cycle of one plant crop and four ratoons. However, when the soil is tilled prior to planting the next crop, the benefits of retaining crop residues are negated and soil carbon returns to levels

observed 5 years previously (Table 11.8). A direct-drill planting system does not impact negatively on soil carbon levels, and so when tillage is discontinued, soil carbon should increase and eventually stabilize at a higher equilibrium level.

- Organic amendments with a high C:N ratio (e.g. sawdust, sugarcane residue and grass hay) enhance suppressiveness to plant-parasitic nematodes, and are more effective than nitrogenous amendments (e.g. feedlot manure, poultry manure, chitin and organic wastes from sugar mills) in enhancing long-term suppression (Stirling *et al.*, 2003).
- An amendment of sugarcane residue has a major impact on the nematode community, increasing populations of free-living nematodes and reducing populations of plant parasites. After 47 weeks, roots growing in soil amended with sugarcane residue contained 95% fewer *Pratylenchus zeae* than roots in the unamended control (Stirling *et al.*, 2005).
- Mulching soil with sugarcane residue enhances suppressiveness to a greater extent than incorporating the residue in soil, possibly because the soil environment under mulch is more amenable to biological activity because it is cooler and moister (Stirling *et al.*, 2011a).
- Most sugarcane roots suffer damage from nematode pathogens. However, roots immediately under the trash blanket are unusually healthy, while populations of *Pratylenchus zeae* in those roots are 5–16 times

Table 11.8. Effect of tillage and several years of green cane trash blanketing (GCTB) on concentrations of total and labile carbon (means ± SE) in the surface 25 mm of three different soil types on sugarcane farms in Queensland, Australia.

Soil	Trash management	Number of samples	Total C (g/kg)[a]	Labile C (g/kg)[a]
Red Ferrosol	Burnt sugarcane	8	16.74 ± 0.68	1.23 ± 0.09
	GCTB	6	22.53 ± 0.61	2.07 ± 0.03
	GCTB (after tillage)	4	16.95 ± 1.19	1.19 ± 0.05
Red Podzolic	Burnt sugarcane	7	14.20 ± 1.36	1.06 ± 0.09
	GCTB	5	16.06 ± 3.33	1.64 ± 0.38
Yellow Podzolic	Burnt sugarcane	3	15.33 ± 0.52	1.61 ± 0.05
	GCTB	6	18.92 ± 1.63	2.05 ± 0.16
	GCTB (after tillage)	3	13.10 ± 1.15	1.56 ± 0.12

[a]Values are means ± standard error. (From Bell *et al.*, 2001, with permission.)

lower than in roots a few cm further down the profile (Stirling *et al.*, 2011a, b).

- Bioassays with *Radopholus similis* indicate that the surface soil is highly suppressive to plant-parasitic nematodes. At one site, inoculated nematodes were readily recovered from heated soil after 8 days, but 84% fewer nematodes were recovered from untreated soil. The mean carbon level in this highly suppressive soil was 38.2 g C/kg (Stirling *et al.*, 2011b).

- In comparisons of sugarcane planted using standard tillage methods or by direct drilling, populations of *Pratylenchus zeae* were initially reduced by tillage, but due to strong resurgence, tended to be higher in tilled than non-tilled plots at the end of the plant crop. Even greater tillage effects were observed with *Meloidogyne javanica* and *Xiphinema elongatum*, as populations of both nematodes at the end of the season were significantly higher following conventional tillage. Conventional tillage was also detrimental to omnivorous and predatory nematodes (Stirling *et al.*, 2011c).

- The organisms responsible for suppressing nematode populations have not been determined, but suppressiveness is almost certainly due to multiple mechanisms. The amount of organic matter in soil (total carbon, total nitrogen, labile carbon) and the size of the free-living community, particularly the number of predatory nematodes, were both correlated with suppressiveness. Nematode-trapping fungi may also be involved, because when terminal restriction fragment length polymorphism (TRFLP) profiles were generated using Orbiliales-specific primers and cloned PCR products were sequenced, the level of suppression was correlated with the number of Orbiliales clone groups, the number of Orbiliales species, and the number of terminal restriction fragments (TRFs) (Stirling *et al.*, 2011b).

Although these studies suggest that soils will become more suppressive to nematodes under the new sugarcane farming system, it is also clear that improvements will not occur immediately. Most Australian

sugarcane soils have been cultivated aggressively for more than 100 years, and soil carbon levels are much lower than they could be. In a study that used perennial grass pasture as a standard reference, the average carbon level in 11 conventionally cropped sugarcane soils was 14 g C/kg compared with 27 g C/kg in soils from paired sites under pasture (Stirling *et al.*, 2010). Soil carbon levels had not increased after 5–7 years of the new sugarcane farming system (although there had been an improvement in some biological properties), indicating that it will take time to rectify the damage caused by years of conventional farming. Nevertheless, the results of the above survey are encouraging, as unpublished data from that study showed that carbon levels under perennial grass pasture at five of the eleven survey sites were greater than 30 g C/kg soil (range 30.5–43.8 g C/kg soil), levels that are similar to the highly suppressive soil identified by Stirling *et al.*, 2011b). Since sugarcane is somewhat similar to an undisturbed perennial grass pasture when it is grown using the improved farming system, it is not unreasonable to expect that when soils are managed under this system, organic matter levels will eventually increase and soils will become more suppressive to plant-parasitic nematodes, particularly in the upper 10 cm of the profile. This assertion is supported by data from South Africa showing that the organic matter content of sugarcane soil is negatively correlated with populations of *Pratylenchus zeae* (Rimé *et al.*, 2003).

With regard to specific suppressiveness, *Pasteuria penetrans* and *P. thornei* are generally present at low levels in Australian sugarcane fields, raising questions about why levels of parasitism have not increased, despite the widespread presence of their host nematodes and more than 100 years of sugarcane monoculture (Stirling, 2009). The move of the sugar industry towards minimum tillage provides opportunities to determine whether these parasites will be more effective suppressive agents when the soil is no longer tilled. Future work is also needed to determine whether endospores are being leached below the root zone by irrigation water or rainfall, or whether they are being consumed by predators. However, the ultimate research objective should be to understand the management practices required to develop sugarcane soils capable of suppressing populations of plant-parasitic nematodes to acceptable levels through a combination of general and specific mechanisms.

Organic farming systems

A brief discussion of organic farming is relevant here, as it is often promoted as a means of achieving healthy, disease-suppressive soils; the same goals discussed in this chapter under the banner of integrated soil biology management. It is, therefore, appropriate to consider whether organic farming provides similar benefits to those that could be achieved through the practices promoted throughout this book.

Proponents of organic farming argue that it is friendlier to the environment, more sustainable, equally productive, and more profitable than conventional farming systems. There is also a widespread belief among consumers that organic food is healthier and safer than conventional food. However, such claims are rarely based on objective scientific data (Trewavas, 2001, 2004). From an environmental perspective, nitrate leaching losses from organic farms are often lower than from conventional farms (Gomiero et al., 2011a), although some studies have shown that there is little difference between the two farming systems (Stockdale et al., 2002; Stopes et al., 2002; Kirchmann et al., 2007). Various microbial parameters (e.g. microbial biomass, microbial activity, and biodiversity) are usually higher in organically managed soils (Gomiero et al., 2011b), but since soils on organic farms are regularly tilled, organisms at higher levels in the soil food web are likely to be negatively affected (see Chapter 6). Some pesticides used in conventional agriculture clearly have undesirable side effects (Sánchez-Bayo, 2011), but the biological gains that are made from not using them are difficult to ascertain, as the negative effects of pesticides will vary with active ingredient, application rate and the attitudes and management skills of the user. Also, the benefits from pesticides and synthetic fertilizers in conventional agriculture are sometimes not recognized by advocates of organic farming (e.g. the environmental gains from herbicide-dependent no-till farming systems, or the additional organic matter inputs generated by optimally fertilized crops). With regard to food safety, most studies do not support anecdotal claims that food obtained from organic farms is safer, or of superior quality, than food produced using other farming systems (Magkos et al., 2006; Herencia et al., 2011).

When productivity is considered, it is generally recognized that organic farming is less productive than conventional farming, but the extent of the reduction is arguable. Badgley et al. (2007) reported that yields for organic plant foods in developed countries were about 8% lower than for non-organic systems, but Avery (2007) argued that this report was seriously flawed, as some results were misreported, data were used selectively and some non-organic yield studies were claimed as organic. Comparisons of organic and conventional production systems in Wisconsin and other northern states of the United States, for example, showed that organic maize, soybean and winter wheat crops produced about 10% less grain than their conventionally managed counterparts. However, the difference was much larger when wet weather reduced the effectiveness of mechanical weed control. A no-till maize-soybean rotation proved to be the most profitable grain system, although profitability was influenced by the size of organic price premiums and the level of government subsidies (Posner et al., 2008; Chavas et al., 2009).

Although numerous studies have been published in recent years comparing soil physical, chemical and biological properties on conventional and organic farms, they will not be reviewed here for several reasons. First, farming systems vary enormously, and when general terms such as 'conventional', 'organic', 'sustainable' and 'integrated' are used, they provide no indication of the soil and crop management practices that were actually used in each experiment or comparison. Second, those promoting alternative farming systems tend to make comparisons with worst-case scenarios, and often characterize conventional agriculture in an ultra-negative way. For example, van Bruggen and Termorshuizen (2003) indicated that 'conventional farming systems are biologically impoverished ecosystems', despite the fact that many practices common to organic farming are widely used in other forms of agriculture. They also claimed that 'in conventional, intensive agriculture, eradication of pests and diseases is frequently attempted' and suggested that only organic farmers were 'ecologically minded', when any non-biased survey would show that most farmers care about their local environment and use a range of pest control tactics other than pesticides. Third, the outcomes from any farming system are largely dependent on the skills of the land manager, so a good manager is likely to produce better outcomes, regardless of the system. Thus, rather than using dichotomous and potentially polarizing terms such as 'organic' and 'conventional' to characterize farming systems, it would be better to focus on 'best management practice' systems of all types, and consider how they might be improved.

Although the word 'organic' has been a barrier to the wider acceptance of this form of farming by those involved in other forms of agriculture, many of the practices used in organic farming are ecologically sound and fit comfortably under the banner of sustainable agriculture (Youngberg and DeMuth, 2013). For example, growers in many forms of agriculture use crop rotations, crop residues, animal manures, legumes, green manures, integrated crop and animal production, off-farm organic wastes, and biological pest control to maintain productivity, practices that contribute to achieving healthy soils and an active and diverse soil biological community. Thus, the main differences between organic farming and the most sustainable forms of the conventional agriculture are its reliance on mechanical tillage for seedbed preparation and weed control, and its avoidance of synthetic fertilizers and pesticides. However, with continuing developments in areas such as non-chemical weed management, nutrient management and biological pest control, there may eventually be little difference between organic farms and the best conventional farms with regard to effects on the soil biota.

In terms of effects on nematode pests, the limited work that has been done suggests that plant-parasitic nematodes cause similar problems in organic agriculture as they do in other farming systems. Thus, in one study in the south-eastern region of the United States,

there were more plant parasites in soils managed organically than in those managed conventionally (Neher, 1999a). In another study with vegetables on certified organic land, root-knot nematode increased in most of the cropping systems examined (including systems that included intercrops and living mulches), and there was little evidence of nematode suppression (Bhan *et al.*, 2010). In contrast, there were more plant-parasitic nematodes in the conventional system than the organic alternative in a Californian experiment (Ferris *et al.*, 1996). However, it is possible that this result was simply a reflection of the different crops grown in the two farming systems, as follow-up studies showed that the level of suppressiveness in the two soils was similar (Jaffee *et al.*, 1998; Berkelmans *et al.*, 2003). These studies also showed that the proportion of omnivorous and predatory nematode in the nematode community did not differ between the two treatments. Additionally, there was no consistent effect on nematode-trapping fungi, as population densities of some species increased in organic plots and other species decreased. Other reports also suggest that populations of omnivorous and predatory nematodes are similar under organic and conventional farming (Nair and Ngouajio, 2012; Coll *et al.*, 2012).

The only comprehensive study of nematode problems on organic farms was conducted in Germany by Hallmann *et al.* (2007). Soil samples taken from 55 vegetable and cereal farms indicated that several potentially damaging nematodes were present. *Pratylenchus* and *Meloidogyne* were common in 90% and 51% of the samples, respectively, while very high populations of *Heterodera avenae* were found in some cereal fields. A follow-up survey of growers indicated that yield losses could exceed 50% on carrots, onions and cereals, and were most pronounced on sandy soils. Interestingly, many growers considered that nematode problems worsened as the length of time under organic farming increased, possibly because: (i) legumes were a necessary component of the rotation and most are good hosts of root-knot nematode; (ii) high levels of weed infestation provided alternative hosts for the nematodes; and (iii) vegetation-free periods were avoided because of the need to avoid nutrient leaching. Although new non-chemical options for managing weeds in minimum-till farming systems (discussed later in this chapter) may allow organic growers to overcome the negative impact of tillage on some natural enemies of nematodes, it is possible that reduced-till organic systems will ultimately be no more sustainable than variants of some of the farming systems discussed in this chapter.

Impediments to the Development and Adoption of Farming Systems that Improve Soil Health and Enhance Suppressiveness

Although suggestions for improving current farming systems have been made throughout this book, it has been made clear that changing the way crops and soils are managed is not an easy task. Some of the problems likely to be encountered have been discussed elsewhere (see Chapters 4 and 13, and other parts of this chapter), but the main impediments to progress are listed next. They are summarized here in the hope that researchers, extension specialists, land managers and government policy makers will find ways of overcoming some of the many obstacles that are limiting the development and adoption of ecologically based systems of managing crops and soil.

- In some regions of the world, many farmers grow crops on land that is leased for relatively short periods, and this limits their capacity to undertake long-term stewardship initiatives. The continual expansion of corporate agriculture also raises questions about whether the owners (often city-based shareholders with little agricultural knowledge) will judge the performance of an enterprise on its capacity to generate immediate profits, rather than on its long-term sustainability.
- Farmers are naturally conservative, and are usually reluctant to change their current farming system. However, there are many other reasons why they may fail to adopt new technologies. Economic constraints are a major factor, particularly if a major investment (e.g. new machinery)

is required. Also, the free-market system into which agricultural products are sold does not necessarily reward those who enhance the sustainability of their farming operations.

- Sustainable farming systems suitable for particular crops, soil types and environments must be developed at a local level. Since this process is not prescriptive, a degree of risk and uncertainty is involved that is unacceptable to many land managers.
- A range of new problems will almost certainly occur during the transition to a new farming system, and this limits the number of farmers prepared to evaluate alternative practices. Examples of issues that may have to be addressed include compaction problems associated with poor soil structure; replanting diseases caused by damping-off pathogens; autotoxicity problems associated with undecomposed plant residues (Bonanomi *et al.*, 2011); and decreased availability of plant-available nitrogen due to immobilization (see Doane *et al.*, 2009, and references cited therein).
- Issues associated with weed management in minimum-till agriculture limit the adoption of farming systems known to improve soil health. The main concerns are the lack of alternatives to chemical herbicides and the ever-increasing number of reports of herbicide-resistant weeds (topics that are discussed in the next section of this chapter).
- Although research in many different areas is required to improve the sustainability of current farming systems, it is difficult to assemble the multidisciplinary research teams required to tackle this issue in a holistic manner. Also, the 2–4-year timeframe of most research projects limits opportunities to undertake the long-term investigations that are required.
- The complexity of ecologically based agriculture means that most farmers are ill-equipped to make the required management decisions, while scientists are often unaware of the practical issues that limit the adoption of potentially useful practices. Agricultural scientists, soil

ecologists and farmers must work together to improve best-practice farming systems, but there is a dearth of structured arrangements that encourage this to happen.

- Adoption of integrated, non-chemical management practices for nematodes and other soilborne pathogens is hindered by the reluctance of growers to commit resources to the collection and management of biological information; a lack of commercial services capable of quantifying soilborne pests and their natural enemies; and a dearth of crop consultants with a capacity to interpret such data and provide advice in areas where they might not have specialist knowledge.
- Current extension models are not always effective. Even when technologies are well researched, information does not necessarily filter through to the farming community. For example, conservation agriculture was first adopted in the United States in the 1960s (Uri, 1999), but a survey of California growers in 2002 indicated that only 35% were 'quite familiar' with the concept (Mitchell *et al.*, 2007).
- A huge agro-industrial complex services agriculture, and its marketing power, means that farmers are continually receiving information on products that they perceive as providing immediate solutions to their soil-related problems. This limits the likelihood that they will consider practices that provide long-term and more sustainable solutions.

Sustainable Weed Management Systems for Minimum-till Agriculture: A Priority for Research

Minimal soil disturbance, permanent plant residue cover and crop rotation are key components of sustainable crop production systems because, when used together, they improve soil health through effects on the soil's physical, chemical and biological properties. However, farming systems that utilize these practices are generally herbicide-dependent, and since more than 200 herbicide-resistant weed species have been reported worldwide (Heap, 2013),

there are serious questions about the long-term sustainability of such systems. Conservation agriculture has provided innumerable benefits to both agriculture and the environment over the last few decades, but maintaining those benefits and increasing the adoption of residue-management practices that preserve rather than deplete the soil biological community will ultimately depend on whether weeds can be managed in a sustainable manner.

From an integrated soil biology management perspective, the first decision to be made with regard to weeds is whether their removal is justifiable. Weeds have negative impacts in cropping systems because they remove water reserves that may be required by the crop, a major concern in water-limited environments. They also compete with the crop for light and nutrients, and can potentially host plant-parasitic nematodes and other soilborne pathogens. However, weeds also have positive effects. Carbon inputs from roots and root exudates are an energy source for the soil biological community, and this contribution to the long-term sustainability of the system is lost when weeds are removed. Thus, there are situations where it will be better to retain weeds and manage them by mowing or some other practice, than to remove them with herbicides. The decision to apply a herbicide should never be made on short-term or aesthetic grounds: it should always be a reasoned judgment that aims to maximize the benefits from weeds and minimize their negative effects.

Since herbicide-resistant weeds are now one of agriculture's most pressing global issues, non-chemical weed management is an active field of research and many alternatives to herbicides are being investigated. Some of those options include mechanical roller-crimping (Ashford and Reeves, 2003; Mirsky *et al.*, 2011); flame weeding (Datta and Knezevic, 2013); bioherbicides (Ash, 2010); mulching (Rowley *et al.*, 2011); microwave technologies (Brodie *et al.*, 2012); weed detection systems for site-specific management (López-Granados, 2010); automatic intra-row mechanical weed-knife machines (Pérez-Ruiz *et al.*, 2012); and breeding crops that can outcompete weeds (Gressel, 2011;

Worthington and Reberg-Horton, 2013). Ultimately, however, single-tactic approaches are unlikely to be successful, and integrated weed management systems that are underpinned by sound ecological principles (Liebman *et al.*, 2001) will be required. Thus, multiple tactics involving crop rotation, cover crops, competitive crop cultivars, appropriate planting patterns, the judicious use of tillage, targeted herbicide application and some of the practices currently being researched, will eventually be used together not only to reduce weed populations but also to minimize the selection pressures that drive the evolution of herbicide-resistant weeds (Gressel, 2011; Mortensen *et al.*, 2012; Harker and O'Donovan, 2013). Given the urgent need for weed management systems that minimize the need for tillage in organic agriculture, new developments will almost certainly occur, particularly with regard to the specialized equipment required to manage crop residues and target weeds in a site-specific manner. In the meantime, zonal tillage offers opportunities to overcome many of the agronomic challenges facing organic no-till systems (Luna *et al.*, 2012).

The widespread and rapid development of herbicide resistance in minimum-till cropping systems is a salutary lesson that emphasizes once again the dangers of relying on a single chemical tactic to control a pest. Agriculture has already faced the problems that developed when insecticides and fungicides were used excessively, and the same scenario is now being been repeated with herbicides. The reduced weed competition achieved with herbicides is one of the cornerstones of modern agriculture, but herbicide usage is almost always motivated by the short-term goal of reducing weed impact on the current crop. Historical evidence suggests that land managers need to lengthen their planning horizons and consider integrating a range of non-chemical weed management practices into their farming systems. Cover cropping, mulching, crop diversification and many other potentially useful practices will not only help overcome the ever-increasing problem of herbicide resistance in weed populations, but will ultimately prove to be vital components of integrated soil biology management.

The Way Forward: A Farming Systems Approach to Managing Nematodes

The most important message from this chapter is that plant-parasitic nematodes only become pests when susceptible crops are grown too frequently, and the management practices used to grow those crops weaken regulatory mechanisms to the point where they no longer maintain nematode populations below damaging levels. Thus, the only long-term solution to nematode problems is to improve the farming system. Fortunately, it is becoming easier to deal with the cropping sequence component of a farming system, as we now have good information on the capacity of various crops and cultivars to host most key nematode pests. For example, there are many highly resistant plants that can be used to reduce the common species of root-knot nematodes in vegetable cropping systems (Stirling *et al.*, 1996; McSorley, 2001), and so it should be possible to find appropriate rotation crops for a particular environment and then integrate them into the farming system (Cherr *et al.*, 2001). In the same way, resistance and tolerance ratings to *Pratylenchus thornei* and *P. neglectus*, for example, are available for crops that are a component of grain farming systems (Thompson *et al.*, 2008). However, it is important that such information is continually updated in extension literature as new cultivars are released, more crops are tested and resistant cultivars are developed.

Although appropriate rotation practices form the foundation of sustainable farming systems, there are limitations to their effectiveness, particularly in situations where there is more than one potentially damaging nematode pest. Thus, it is important that an effective level of suppressiveness to nematodes and soilborne pathogens is also maintained. This can only be achieved through better management of organic matter (Peters *et al.*, 2003; Bailey and Lazarovits, 2003; Stone *et al.*, 2004; Ghorbani *et al.*, 2008), as it plays a central role in the chemical and biological processes that improve crop growth and enhance competitive effects within the soil biological community. Since sequestering carbon in soil contributes to climate change mitigation and influences numerous soil properties relevant to ecosystem functioning (Powlson *et al.*, 2011a), there are many other compelling reasons for farmers to value soil organic matter and manage it better.

Fortunately, the main pillars of farming systems that will enhance soil carbon (crop rotation, minimum tillage and residue retention) are in place in many parts of the world, as the principles of conservation agriculture have already been widely adopted. The challenge in such situations is to build on the improvements already made and integrate other relevant practices into current farming systems. However, there are much bigger challenges in farming systems that are still reliant on cultivation, as constant tillage not only depletes soil carbon, but also disturbs the microbial community and kills predators. In such situations, modification of tillage practices is an essential first step to improving soil carbon levels, as importation of organic matter from elsewhere is not always an option.

It is easy to argue from an ecological perspective that modifying farming systems to enhance soil carbon is a pathway towards nematode- and disease-suppressive soils. However, achieving change and demonstrating effects is much more difficult. Science tends to be fragmented and discipline-based, and so nematologists, for example, rarely work in the multidisciplinary teams required to tackle the complex issues involved in improving farming systems. Also, specialists in areas such as nematology tend to take a reductionist approach to experimentation, and use the results of relatively simple comparisons (e.g. amendment versus no amendment; till versus no-till; rotation crop versus fallow) to explain the effects of individual management practices. The effects of a fully integrated set of management tactics are rarely assessed.

One possible way forward is to establish multidisciplinary research groups (led by agronomists, soil scientists and successful farmers) and commission them to develop productive, profitable and sustainable farming systems suitable for local conditions. One of the key measures of sustainability would be maintenance or enhancement of soil carbon over time. Nematologists, microbiologists and other specialists would contribute to the research programme by monitoring the soil biological community and assessing temporal changes in suppressiveness to nematodes and other soilborne pathogens.

One issue that is often overlooked by scientists is the contribution that farmers can make to the development of improved farming systems. Farmers' initiatives have played an important role in the rapid expansion of conservation agriculture over the last 20 years, but if farmers are to make an even greater contribution in this area, they need to have some understanding of the issues involved in managing systems that are inherently complex, not only at the crop and business management level, but also at an ecological level. This issue was discussed by Shennan (2008), who argued that current research and extension institutions are not well suited to helping farmers deal with this complexity. Ecologically based management systems will only be successful if farmers understand the ecological processes involved in plant–soil–organic matter–organism interactions, as they need to be able to adapt management tactics to suit their own situation. Thus, future research models will need to put more emphasis on field-based adaptive research. Such research will be conducted through researcher–farmer partnerships, with resources supplied to monitor system performance as adaptations are made to the farming system. From an educational perspective, participatory research and interactive learning processes should be encouraged, as they help to integrate the knowledge held by both farmers and researchers.

Once appropriate farmer–researcher collaborations are established at a local level, attempts can be made to improve current farming systems. The key question from a nematological perspective is whether crop and soil management practices that provide a physical and chemical environment amenable to plant growth, and generate a biological community capable of suppressing populations of plant-parasitic nematodes, result in a soil that is able to prevent nematodes from causing crop losses. Some of the key issues that should be addressed in answering that question are:

- How much biomass must be returned to the soil annually to increase the soil's carbon content, and is that achievable in the target environment, given that plant productivity will be constrained by factors such as temperature, the availability of water and the amount of sunshine?

- What soil carbon levels and forms of carbon are required to sustain a biological community capable of suppressing plant-parasitic nematodes, and are those levels realistically achievable?
- When rotation sequences are being determined, is a crop's capacity to minimize the multiplication of a particular nematode a more important attribute than its capacity to return organic matter to the soil through biomass production?
- Given that biological benefits from an improved farming system will accrue incrementally over many years, what practices can be used in the short term to limit losses from nematodes without forfeiting the long-term benefits from the new system?
- What indicators can be used to demonstrate that the biological status of the soil is improving, and is there evidence that biological activity and diversity will eventually reach a point where natural suppressiveness provides useful levels of nematode control?

The last bullet point about indicators deserves some comment, as there is a tendency for nematologists to use predatory nematodes as indicators of general suppressiveness. Although this is understandable, given that predatory nematodes are readily quantifiable in studies of nematode communities, it ignores the fact that this group of nematodes may be relatively unimportant antagonists in many soils. For example, in studies of soils that were suppressive and conducive to root-knot nematode, the population

patterns of omnivorous and predatory nematodes suggested that they were not associated with suppressiveness (McSorley *et al.*, 2008; McSorley and Wang, 2009). However, in the latter study, other invertebrate predators, especially mites and Collembola, were recovered more commonly in suppressive soils and increased in abundance over time. Clearly, the full range of antagonists in nematode-suppressive soils should be quantified in studies of this nature, and fortunately, this can now be done using a range of molecular techniques, including high-throughput genetic sequencing (Mendes *et al.*, 2011; Rosenzweig *et al.*, 2012), and possibly through gut content analyses (Heidemann *et al.*, 2011).

One final point that should be made in concluding this discussion is that holistic, biologically based nematode management systems minimize crop losses through multiple mechanisms: (i) by reducing nematode populations through crop rotation, resistant cultivars and the regulatory forces associated with suppressive soils; and (ii) by increasing damage thresholds through improvements in soil physical and chemical fertility, and the use of nematode-tolerant cultivars. Thus, none of the components of these integrated systems has to stand alone as a control measure. It is important to recognize this, because there will be situations where multiple control measures are required (e.g. sandy soils and water-limited environments, where adequate levels of suppressiveness may be difficult to achieve). Similarly, in situations where a particularly virulent nematode pest is present (e.g. some root-knot and cyst nematode species on annual crops), nematode-resistant cultivars and rotation crops are likely to be the mainstay of the farming system, particularly in the initial years when soil health is being improved and suppressiveness enhanced.

12

Biological Products for Nematode Management

As was pointed out in the introduction to this book (Chapter 1), biological control is a relatively broad concept that encompasses a range of strategies that ultimately result in a reduction in pest populations, or the capacity of the pest to cause damage, through the actions of parasites, predators and other antagonistic organisms. Biological control is achievable in many different ways, and strategies that conserve or enhance the activity of naturally occurring biological control agents were discussed in Chapters 9, 10 and 11. However, those forms of biological control are generally not recognized by growers or the general public, and have never captured their imagination. Instead, biological control is simply seen as an alternative to chemical pesticides. Thus, there is a widely held expectation that safe, efficacious and environmentally friendly biological products will be available to reduce populations of any pest or pathogen. Given the successes with some biological products (e.g. bio-insecticides based on *Bacillus thuringiensis*), this attitude is not surprising. However, it does not necessarily mean that such attitudes are realistic, or that effective biological products will be developed for all pests, or that they will be useful in all environments.

The question to be addressed in this chapter is whether the perceived optimism about biological products is justified when dealing with soilborne pests such as nematodes. What is clear is that there is a plethora of opinions on that subject. On the one hand, small companies around the world are selling products with purported effects against nematodes, and some of the world's major agro-industrial companies are in the process of developing new microbial products. On the other hand, ecologists with an understanding of the complexity of the soil environment, and most of the scientists who have worked extensively with nematophagous organisms, are much more cautious about the potential of inundative methods of biological control.

Since many of the issues associated with the development of biological products for nematode control (e.g. research approaches; selection of useful agents; experimental methodologies; assessment methods; establishment problems; and issues associated with mass-production and registration) were covered in detail by Stirling (1991), this discussion concentrates on progress made in the last 20–25 years. However, as the development of biological products is still being constrained by deficiencies in the experimental procedures used to evaluate potentially useful agents, and by a poor understanding of how biological control agents behave once they are introduced into the competitive soil environment, the chapter commences by focusing on these problems. It then goes on to discuss some of the commercial issues associated with the development of biological products, and finishes with a brief overview of the organisms most commonly used in biocontrol studies, and some of the recent work that has been done with them.

One point that must be clarified at the commencement of this chapter is that it focuses on living organisms and how they might be used to control nematodes. Thus, it is important to recognize that this discussion does not cover the full spectrum of products that are often referred to as 'biopesticides'. Preparations containing bioactive compounds produced by microorganisms have been deliberately excluded, as they are more correctly termed 'biochemical pesticides' (Box 12.1). Thus, products based on abamectin, azadirachtin, extracts from the soap tree (*Quillaja saponaria*) and fermentation by-products of *Myrothecium verracaria* are not discussed here, despite being referred to as 'biological nematicides' by López-Pérez *et al.* (2011). Such compounds are often considered 'natural' because they can be extracted from plants or are produced by microorganisms

when grown in large fermentation chambers, but they are essentially similar to synthetic chemicals. While it is true that they are found in the natural world, these compounds are not necessarily less toxic than synthetic compounds, and their non-target effects and modes of action are similar to synthetics (Chitwood, 2002). Also, when these compounds are used for pest control, they are applied at much higher concentrations than occur in the natural environment, where microorganisms produce antibiotics and metabolites in almost undetectable amounts as they interact with pests and various competitors.

Experimental Methods

Although there were many early attempts to introduce fungi, bacteria and other antagonists of nematodes into soil for biological control purposes, interest in this area began to increase during the 1980s, when the health and environmental problems associated with the use of nematicides became apparent.

Box 12.1. Biopesticides and the biopesticide market

The term 'biopesticide' has historically been associated with biological control, and by implication, involves the interactions of living organisms and their activities. However, the term is now used much more broadly to cover microbial pest control agents and bioactive compounds that are produced by microbes. Thus, the following terms, which are used by the United States Environmental Protection Agency (EPA) to categorize biopesticides, are more meaningful, as they provide a better indication of the type of product under consideration:

- *Microbial pesticides* contain microorganisms that are introduced to control pests.
- *Biochemical pesticides* contain naturally occurring substances that are used to control pests. This category includes compounds such as pyrethrins, neem oil, avermectin, phenolics and many other plant- or microbe-derived chemicals. Semiochemicals such as the insect sex pheromones, which are used to monitor, confuse or control pests, are also included.
- *Plant-incorporated protectants* consist of pesticidal substances produced by plants, and the genetic material required by the plant to produce that substance.

Recent reviews (e.g. Bailey, 2010; Chandler *et al.*, 2011; Glare *et al.*, 2012) provide data to show that the global biopesticide market is growing at a much greater rate than the market for synthetic pesticides. However, that market still represents a small proportion (about 3.5%) of the overall pesticide market. Also, biochemical pesticides, particularly products based on *Bacillus thuringiensis*, dominate the market, accounting for more than half of total sales. The microbial products component is increasing, but it is mainly based on a limited number of organisms (commonly *Agrobacterium*, *Bacillus*, *Pseudomonas*, *Streptomyces* and *Trichoderma*), and is sustained by small companies that often do not have the resources to scale-up production and successfully market the products. However, this situation may change in future, as several large agro-technology companies are now undertaking biopesticide research, or have recently acquired small biopesticide companies.

However, there were early concerns about the inadequacies of some of the work that purported to demonstrate biological control, with Kerry (1989, 1990) noting that few experiments met the basic requirements considered essential to properly estimate the potential of a biological control agent. At about the same time, Stirling (1988, 1991) provided guidelines on how biological control experiments should be conducted, and discussed the issues that must be considered when evaluating biological control agents. Those guidelines and issues remain relevant today, and are outlined in Box 12.2.

It is disappointing that 25 years after these issues were first discussed, the development of inundative biological control is still constrained by deficiencies in experimental methods. Studies are most commonly carried out in sterilized soil under controlled conditions in the greenhouse; unrealistically high rates of inoculum are often used; the temporal population dynamics of organisms following their introduction are rarely monitored; and the ecological issues associated with establishing an introduced organism in a competitive soil environment are usually ignored. The situation with biological products is even worse, as it is almost impossible to find peer-reviewed publications on the efficacy of products that are registered in some countries for nematode control. If inundative biological control is to have a future, scientists working in this area must assess potentially useful agents/products in an appropriate manner, and be prepared to undertake the field experiments required to demonstrate efficacy.

General Soil Biostasis and the Fate of Introduced Organisms

The term 'suppressiveness' has been used throughout this book to describe a process by which a nematode's capacity to proliferate is limited to some extent by the actions of the community of soil organisms that occupy the same ecological niche. 'Soil biostasis' (and its more specific variants, 'fungistasis' and 'microbiostasis') is a similar term, but is more commonly used when discussing the competitive interactions that occur between bacteria and fungi in the soil and rhizosphere. Regardless of the name given to the process, it is important to recognize that suppressive or biostatic factors are present in all soils and they act against all organisms, including beneficial organisms that are introduced into soil for biocontrol purposes. Since the interactions that occur between soil organisms and the mechanisms that impact on their growth and reproduction (primarily antibiosis, nutrient depletion, parasitism and predation) were discussed in Chapter 2, they will not be considered here. Instead, competitive forces associated with the indigenous microflora and fauna are seen as a normal part of the soil environment, and their role in influencing the effectiveness of inundative biological control is examined.

Biostasis is not a newly recognized phenomenon. Biostatic effects on fungi, for example, were first studied in the 1950s (Dobbs and Hinson, 1953), and early work soon showed that they were a major impediment to establishing fungal biological control agents in the soil environment. In fact, Cooke and Satchuthananthavale (1968) and Giuma and Cooke (1974) considered that fungistatic effects in natural soils were so severe that they were likely to prevent the successful use of nematode-trapping and endoparasitic fungi as biological control agents. Nevertheless, interest in using nematophagous fungi as biological nematicides did not diminish, prompting Jaffee and colleagues to commence investigations into the factors that influence the efficacy of fungi when introduced into soil for biocontrol purposes. The formulated product used in these studies was produced by growing fungi in shake culture, removing the nutrient medium, and then pelletizing macerated mycelium in 1% alginate (Lackey *et al.*, 1993).

Initial work with pelletized products containing *Hirsutella rhossiliensis* and several species of nematode-trapping fungi showed that nematophagous fungi performed well against some nematodes in laboratory microcosms (Jaffee and Muldoon, 1995a; Tedford *et al.*, 1995b), but were not as effective in field

Box 12.2. Evaluation of biological control agents and biological products against nematodes

The methods used to evaluate the effectiveness of chemical nematicides are relatively straightforward, as the response of both the nematode pest and its host crop is simply measured weeks or months after the chemical is applied. Evaluating biological alternatives is much more difficult, as the activity of the introduced agent will be influenced by: (i) soil physical and chemical factors; (ii) environmental conditions such as moisture and temperature; and (iii) the community of organisms that reside in the soil and rhizosphere. Anyone wishing to demonstrate the potential of a biological control agent must, therefore, show that it can survive and compete in a range of soil types and environmental conditions, and that it will remain active in a relatively hostile biological environment for long enough to affect the target nematode. Thus, experiments with biological agents for nematode control must address the following issues:

- *Assay system*. Much of the work on biological control agents is undertaken in the laboratory, or in sterilized soil or potting mix in the greenhouse. Although relatively simple assessment systems are essential for initial screening and development studies, they provide little indication of potential in the much more complex soil environment. Experiments in microplots or field plots are essential.
- *Soil type*. Interactions between an introduced organism and the resident soil biota will be influenced by the direct effects of factors such as clay content and levels of soil organic matter, and the indirect effects of such parameters on available moisture and other soil properties. Biological control agents must, therefore, be tested at sites that cover the range of situations where the target nematode is likely to be found.
- *Experimental design*. Microbial agents are usually mass-produced in a culture medium, and this substrate is sometimes added to soil with the organism, or in a formulated product. To determine whether observed effects are due to the organism or the substrate, the following treatments should ideally be included in the experiment: test organism alone; test organism and substrate; test organism and substrate (sterilized by heat or irradiation); substrate alone; and untreated control. An experiment discussed in this chapter provides a good example of why the substrate should be included as a treatment, as the corn grits substrate used to grow the test organism consistently produced a response (see Table 12.1).
- *Application rate*. Biological products are often tested in pots at application rates greater than 1% w/w (i.e. 10 g product/kg soil or the equivalent of about 20 t/ha). Such application rates are never likely to be acceptable, even for high-value crops. Application rates must be reduced to more realistic levels, either by markedly reducing the amount of product applied or concentrating the product in an appropriate location (e.g. a narrow band near the plant).
- *Survival and activity*. When a biological control agent is introduced into soil, it has to reach the site where the target nematode is located. It must also survive in an active state for long enough to impact on nematode populations. Consequently, temporal and spatial data on the presence and population density of the introduced organism should be collected as evidence that it is present in the appropriate niche at the right time.
- *Evidence of mortality*. An effective biological control agent will either kill the target nematode or prevent it from developing in the plant and causing damage. To confirm that these effects have occurred, evidence of mortality to egg, juvenile or adult stages of the nematode must be obtained, or data must be collected to show that the number of nematodes developing or feeding in roots has been reduced.
- *Isolation from soil or nematode carcasses*. Many of the organisms introduced into soil for biological control purposes are common soil inhabitants. Thus, to confirm that hyphae, spores or other structures found in soil belong to the introduced organism, or that the biocontrol agent has caused nematode mortality, the organism must be re-isolated from soil or the carcasses of nematodes, and specific markers used to differentiate it from naturally occurring strains.

microplots (Jaffee *et al.*, 1996a; Jaffee and Muldoon, 1997). Since pellets incubated in the field were damaged, presumably by soil organisms, and *H. rhossiliensis* formed smaller colonies in untreated field soil than in heated soil, Jaffee *et al.* (1996a) suspected that poor performance in the field was due to biotic inhibition. Enchytraeid worms were found to be partly responsible for this effect, as they increased in numbers and destroyed fungal colonies growing from pelletized hyphae (Jaffee *et al.*, 1997a). Also, experiments in the field showed that both *Hirsutella rhossiliensis* and *Arthrobotrys gephyropaga* did not perform as well when enchytraeids were present as when they were excluded from the intro-duced fungi with 20-μm mesh (Jaffee *et al.*, 1997b). It was also suspected that other soil organisms were involved in the fungistatic effect, as nematophagous fungi did better in heat-treated soil than in untreated soil, and fungivorous collembolans and dorylaimid nematodes often occurred in the same loca-tions as enchytraeids. Follow-on work then showed that, in contrast to the previous study, populations of the two fungi did not increase when protected by fine mesh (Jaffee, 1999). However, heat treatment increased the diame-ter of colonies growing from pellets (Fig. 12.1) and consistently increased fungal populations and levels of parasitism (Jaffee, 1999, 2000). Therefore, nematophagous fungi are clearly sensitive to biotic inhibition when formulated as pelletized hyphae. Worms that digest organic matter, and fungivorous microfauna, are sometimes responsible, but the main inhib-itory factors are likely to be microbial.

With regard to biostatic effects on other nematophagous fungi, a soil-membrane tech-nique was used to assess the capacity of sev-eral isolates of *Pochonia chlamydosporia* and *Purpureocillium lilacinum* to grow and estab-lish in sterile and non-sterilized sandy soils. Again, there was clear evidence of biotic inhi-bition (Monfort *et al.*, 2006). Both fungi showed much higher growth rates in sterilized soil (Fig. 12.2), with the colony area of most iso-lates reduced by 83–98% in non-sterilized soil from two locations. Some isolates were less sensitive to the soil microbiota than others, but growth in non-sterilized soil was always reduced by more than 50%. Unpublished

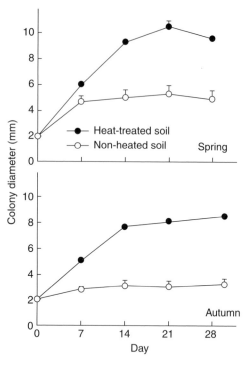

Fig. 12.1. Colony diameter of *Hirsutella rhossiliensis* growing from alginate pellets in observation chambers containing heat-treated (60°C for 2 h) or non-heated soil. Soil was collected twice during the year (in spring and autumn). Each bar is the mean of six replicate chambers. Vertical bars equal one standard error. Bars that do not appear are smaller than the symbol. (From Jaffee *et al.*, 1996a, with permission.)

data from the same authors also showed that *Pochonia chlamydosporia* and *Purpureocillium lilacinum* did not grow at all when the inocu-lum was in direct contact with soil. Given the results of the membrane assay, the authors speculated that toxic metabolites diffused through the membrane and arrested fungal growth. That hypothesis is supported by the results of Chinese studies with volatile organic compounds of microbial origin. Tests in Petri dishes showed that antifungal com-pounds are produced in natural soils, and also by numerous bacteria isolated from those soils; and that the volatiles are produced in concentrations that markedly reduce the spore germination and mycelial growth of both *Pochonia chlamydosporia* and *Purpureocillium lilacinum* (Xu *et al.*, 2004; Zou *et al.*, 2007).

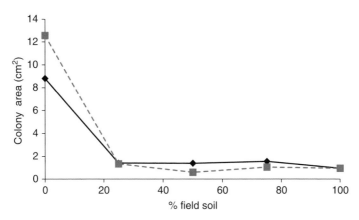

Fig. 12.2. Colony area (cm²) 4 weeks after isolates of *Pochonia chlamydosporia* (solid line) and *Purpureocillium lilacinum* (dashed line) were introduced on a colonized ryegrass seed into Petri dishes containing sterilized soil or increasing amounts of a field soil from Spain. (Adapted from Monfort *et al.*, 2006, with kind permission from Springer Science + Business Media B.V.)

Although these results indicate that facultative parasites such as *Pochonia chlamydosporia* and *Purpureocillium lilacinum* are subject to biostasis, this does not mean that they cannot be established in field soil. Populations of *Pochonia chlamydosporia* reached the presumed carrying capacity in gamma-irradiated soil, regardless of the number of chlamydospores inoculated, and this was also achieved in an untreated field soil, provided the initial inoculum density was high (Mauchline *et al.*, 2002). Thus, from a biocontrol perspective, the important point was that the fungus could remain viable in field soil for at least 8 weeks, indicating that it was a competent saprophyte capable of surviving in situations where plants and nematode hosts were absent. Studies discussed in this chapter have also shown that *Purpureocillium lilacinum* can be established for a limited period in some soils at relatively high population densities (see Figs 12.6 and 12.7 later in this chapter).

One major change that has taken place in agriculture over the last decade is a general recognition that soil health is an important component of long-term sustainability. Thus, steps are being taken to increase biological activity and diversity by improving a soil's carbon status, as this has flow-on effects to soil physical and chemical fertility.

However, biostatic forces also increase as the microbial community becomes more diverse (Wu *et al.*, 2008; Scala *et al.*, 2011), and so it is likely to become increasingly difficult to establish biological control agents in a new environment as soil health improves. This should be kept in mind by those interested in developing biological products for nematode control, as inundative biological control and sustainable agriculture are not necessarily compatible.

Data on biotic inhibition of the organisms present in commercial products are rarely published, but a study in California shows that it is a real issue in the field. In an experiment aimed at finding possible replacements for methyl bromide, various fumigants and compost treatments were compared in a replanted peach orchard (Drenovsky *et al.*, 2005). The compost was either applied alone, or with commercial inoculants that were injected through the drip irrigation system five times during the growing season. The inoculants typically contained 10–40 microbial taxa. Data from the compost and compost + inoculants treatments are of interest because assessments using phospholipid fatty acid analysis showed that the microbial community in the latter treatment was similar to its non-inoculated counterpart. Clearly, this result did not support manufacturers' claims that products of this nature can be used to enrich soil microbial communities. The reasons that a unique microbial fingerprint was not observed in inoculated soil were not determined, but the poor result with commercial microbial inoculants was probably due to the introduced organisms being unable to compete with the native microbial community in compost-amended soil.

It is clear from this summary that problems associated with establishing biological agents in soil must be addressed by anyone attempting to develop biological products for nematode control. In the early 1990s, there were various suggestions as to how such problems might be overcome (see reviews by Baker, 1990; Deacon, 1991; Stirling, 1991; and Cook, 1993), but biotic inhibition remains a constraint to inundative biological control. Screening methods for selecting biological control agents have been modified so that the process is carried out in field soil; rhizosphere-competent microbes have been specifically selected; research on biostatic mechanisms has been undertaken; organisms have been applied repeatedly rather than as a single dose; several strains have been applied together; and the soil environment has been modified prior to application; but these changes to selection processes and application procedures have not always improved consistency. Thus, the important message from this work is that biological control agents can be successfully introduced into field soil, but establishment will be always be limited by biotic factors. One of the keys to improving consistency may be to focus on using biological control agents in situations where their activity is not likely to be markedly constrained by biostasis. Such situations could possibly be identified by comparing biostatic effects in different soils, measuring parameters such as soil carbon or microbial activity, and determining the situations where biological activity and diversity is so low that the introduced organism will be able to grow and proliferate to some extent. Another option is to select biocontrol agents that either produce resistant spores as survival structures, or are good competitive saprophytes. Endophytes are also an alternative, as they occupy an ecological niche that is unavailable to many soil organisms.

Monitoring Introduced Biological Control Agents

Given the difficulties involved in establishing biological control agents in soil, it is obviously important to monitor survival and proliferation following their introduction. If an organism fails in field trials, the key question is whether it has survived but is ineffective, or whether it did not establish. A number of methods are commonly used to detect and quantify biocontrol fungi in soil, and they include direct extraction of spores (Crump and Kerry, 1981); isolation onto selective media (Mitchell et al., 1987; Kerry et al., 1993; Hirsch et al., 2001); and dilution plating followed by quantification with most-probable number techniques (Eren and Pramer, 1965; Stirling et al., 1979; Jaffee and Muldoon, 1995b). However, none of these methods are entirely satisfactory. Not all biocontrol fungi have large, readily identifiable spores, and isolation media are usually only partially selective; while dilution plating and most-probable number techniques are time-consuming and results are often variable. Estimates of levels of cyst and egg mortality (Mertens and Stirling, 1993; Kerry and Bourne, 1996; Westphal and Becker, 1999), and bioassays in which nematodes are added to soil and checked for attached spores or evidence of infection (McInnis and Jaffee, 1989; Rodriguez-Kabana et al., 1994) can also be used to show that an introduced fungus is present and active in soil, although it is not always clear whether the infection and mortality observed is due to the introduced fungus, or to closely related fungi that occur naturally in soil.

Given the limitations of traditional methods for monitoring biological control agents in soil, researchers have attempted to develop methods of visualizing biocontrol fungi during the process of root colonization and nematode infection. Monoclonal antibodies specific for hyphae and chlamydospores of *Pochonia chlamydosporia* were developed, and the hyphal-specific antibody was used to show that the fungus was able to grow in the rhizosphere (Fig. 12.3). Since cross-reactivity with other fungi and lack of specificity at the isolate level meant that the antibodies could not be used to monitor individual isolates of *P. chlamydosporia* in field soil, attempts were

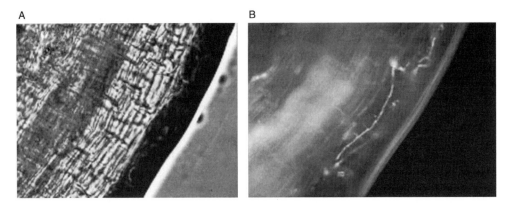

Fig. 12.3. Bright-field (A) and fluorescent (B) images of *Pochonia chlamydosporia* probed with monoclonal antibodies specific to hyphae of the fungus, showing growth on barley root. Magnification = 110×. (From Hirsch *et al.*, 2001, with permission.)

then made to develop transformants containing a visible marker gene. Protoplast-mediated transformation was unsuccessful (Atkins *et al.*, 2004), but later work with an *Agrobacterium tumefaciens*-mediated protocol produced a transformant containing the green fluorescent protein (GFP) gene that was visualized using laser confocal microscopy (Maciá-Vicente *et al.*, 2009a). Although such transformants are useful for studying root colonization and egg-infection processes (see Chapter 5), their status as genetically modified organisms limits their use in the field.

Molecular technologies for detecting and quantifying *P. chlamydosporia* have also been developed. Initial work showed that specific primers designed from an amplified and cloned fragment of the β-tubulin gene were specific to the fungus and could be used to identify it on tomato roots infected with root-knot nematode (Hirsch *et al.*, 2000, 2001). In follow-on studies, these primers were used in a competitive PCR assay that proved more reliable than a selective medium for quantifying the fungus in field soil (Mauchline *et al.*, 2002). Since differences in colonization of tomato plants infested with root-knot and cyst nematodes were observed, this study also provided evidence that the isolate used was specific for some nematodes, an important attribute of a biological control agent.

Following the taxonomic studies of Zare *et al.* (2001a, b), two sub-species of *Pochonia*

chlamydosporia were recognized (varieties *chlamydosporia* and *catenulata*). Atkins *et al.* (2003a) then developed a PCR assay based on ITS sequences that was able to detect *P. chlamydosporia* var. *chlamydosporia* in soil to which it had previously been added. However, in situations where densities of the fungus were low, information on population size could be obtained only by combining PCR diagnostics with other methods (e.g. baiting and selective media).

One issue that was recognized throughout these studies was the innate variability of *P. chlamydosporia*. Various strains can be identified on the basis of cultural characteristics (Kerry *et al.*, 1986) and DNA fingerprinting (Arora *et al.*, 1996), and since host-related genetic variation and nematode host preferences occur at the infraspecific level (Morton *et al.*, 2003; Mauchline *et al.*, 2004), it is important to be able to distinguish different isolates of the same fungus. Atkins *et al.* (2009) distinguished *P. chlamydosporia* var. *chlamydosporia* from *P. chlamydosporia* var. *catenulata* using real-time quantitative PCR (qPCR) with variety-specific primers targeted to β-tubulin and the ribosomal ITS region, respectively, and were able to measure the relative abundance of isolates of the two fungal varieties when inoculated singly or together on tomato plants. Sequence-characterized amplified polymorphic regions (SCARs) were then used to develop a primer set diagnostic for one isolate of *P. chlamydosporia* var. *chlamydosporia*.

However, primers specific to another isolate were not identified, highlighting the difficulties involved in developing methods for monitoring closely related fungi.

The arbitrary primers used in the randomly amplified polymorphic DNA (RAPD) technique are perhaps more useful in ecological studies, as they provide more detailed genetic information on the species of interest, and are useful for molecular fingerprinting at both interspecific and intraspecific levels. Zhu *et al.* (2006) used this approach to develop SCAR primers that could differentiate specific strains of *P. chlamydosporia* var. *chlamydosporia* and *Paecilomyces lilacinus* from many other isolates of these fungi. This relatively targeted and cost-effective method has potential for use in biological control field trials, as strains introduced as inoculants must be distinguished from indigenous isolates.

Amplified fragment length polymorphism (AFLP) analysis can similarly be used to detect genetic differences between isolates, and offers improved resolution and reproducibility. The method involves digesting DNA with restriction enzymes, ligation of adapters and PCR amplification of a subset of genomic DNA fragments. Isolate-specific DNA fragments are then cloned and sequenced, enabling the development of strain-specific primers based on the unique fragments (similarly to SCAR). Although not yet widely used for nematophagous fungi, such methods have been used to develop strain-specific primers capable of differentiating a mycoherbicide strain of *Fusarium oxysporum* from the various plant-pathogenic and non-pathogenic strains found in soil (Cipriani *et al.*, 2009).

A further example of the use of molecular tools to study the fate of biocontrol fungi in soil is described by Cordier and Alabouvette (2009), who used terminal restriction fragment length polymorphism (T-RFLP) fingerprinting to identify a specific biocontrol strain of *Trichoderma atroviride* and assess its impact on native microbial soil communities. Collectively, these examples show that molecular technologies provide powerful tools for tracking microorganisms in the environment, and

with further development, they will be used routinely in ecological studies with biological control agents. We know that an introduced organism will ultimately succumb to competition from the resident soil microflora, but if populations were monitored, and the factors influencing establishment were better understood, it may be possible to increase the length of time an organism remains active in soil by manipulating the environment, thereby improving efficacy.

Commercial Implementation of Biological Control

Bringing a biocontrol product to market and demonstrating its effectiveness is not a simple process. In fact, Bailey (2010) considered that the innovation chain for developing a biopesticide involved nine steps: (i) discovery and selection; (ii) proof of concept; (iii) technology development; (iv) market identification; (v) technology transfer; (vi) application development; (vii) commercial scale-up; (viii) registration; and (ix) technology adoption. He also noted that this multi-tiered process requires contributions from both science and business. However, the need for a business partner has largely been ignored by those interested in developing biological controls for nematodes. Instead of forming the collaborative links needed to commercialize their research, nematologists have remained trapped in the early stages of the innovation chain: locating potentially useful organisms; selecting virulent isolates; and confirming efficacy under relatively controlled conditions. This reluctance to form partnerships with commercial interests is one of many reasons why there are relatively few biological products available for nematode control. Whipps and Davies (2000) listed four such products, of which only one was registered in the United States (Fravel, 2005). Since then, other products have been developed (Hallmann *et al.*, 2009), but the number remains relatively small. Also, lists of registered products are often inflated by the inclusion of preparations that are biochemical in nature rather than microbial (e.g. avermectin, the fermentation product of *Streptomyces avermitilis*; and DiTera™ (Valent Biosciences, Illinois), a preparation made from killed fermentation solids of *Myrothecium verracaria*).

Box 12.3 outlines the issues that must be addressed in developing a biocontrol product for nematodes. Clearly, there are many challenges, as the commercialization process requires time, money and inputs from many parties. Since the challenges are much the same for nematodes as they are for other pests, additional information on the route to commercialization can be obtained from other texts (e.g. Deacon, 1991; Stirling, 1991; Powell and Jutsum, 1993; Cook, 1993; Cook *et al.*, 1996; Cross and Polonenko, 1996; Avis *et al.*, 2001; Fravel, 2005; Mensink and Scheepmaker, 2007; Kiewnick, 2007; Bailey, 2010; Ash, 2010; Glare *et al.*, 2012).

One particular point that is often not mentioned in discussions on biopesticides is the need to continue research after a product is registered. In many countries, registration authorities simply aim to ensure that the product is safe to use and is not likely to be detrimental to non-target organisms. They presume that issues related to efficacy will be determined in the marketplace. Consequently, it is a concern that many biocontrol researchers feel that their job is complete once a product is registered. Biological products are likely to be relatively specific, and will never be effective against every nematode, in all soil types, on every crop and in every environment. It is, therefore, important to define the situations where they are likely to be useful. Since the company marketing a product has a vested interest in selling it into all markets, it is up to nematologists to undertake this painstaking and often unrewarded task. Evidence-based information is needed to advance the cause of biological control, and when it is not available, growers will simply purchase products on the basis of the many glowing testimonials that are available on the internet. Unjustified claims are often made in internet-based advertising material, so it is important to balance these unfounded statements with credible information on effective products. Thus, once a biological product is registered, research must continue, and it

should aim to monitor the introduced organism in soil; demonstrate establishment; confirm efficacy; and define the situations where the product is most likely to be effective.

Inundative Biological Control of Nematodes: An Assessment of Progress with a Diverse Range of Potentially Useful Organisms

Given the huge number of antagonists of nematodes found in soil (see Chapters 5–8) it is not surprising that many different organisms have been considered to have potential as biological control agents. However, few of these organisms have been taken down the pathway to commercialization outlined in Box 12.3, with claims of biocontrol potential often made on the basis of a few simple tests in the laboratory or greenhouse. Nevertheless, some nematophagous fungi and bacteria have been relatively well researched, and the following is an attempt to assess their potential when applied in an inundative manner for control of plant-parasitic nematodes.

Nematode-trapping fungi

Linford and Yap (1939) made the first serious attempt to examine the potential of nematode-trapping fungi as agents for nematode control when they grew five species of *Arthrobotrys* and *Dactylellina* on sterilized sugarcane bagasse, added them to soil, and assessed their capacity to reduce damage caused by root-knot nematode on pineapple. One of these fungi (the adhesive knob-producer *D. ellipsospora*) significantly improved root and top growth, and increased the proportion of the root system that was free of nematode galls, while some of the other species seemed to reduce nematode injury. Although this was the first experiment to demonstrate that nematode-trapping fungi might be useful biological control agents, the following comments made by the authors are perhaps more

Box 12.3. Steps involved in commercializing a microbial product and demonstrating its efficacy against nematodes

Bringing a biocontrol product to market and demonstrating that it is effective is a complex process involving innumerable steps, and it does not finish with registration. Once a product is registered, further work is required to demonstrate efficacy in the field. Thus, it takes many years to commercialize a biological control agent, and during that process, the following issues must be addressed:

1. *Identifying potentially useful biocontrol agents.* Numerous isolates are collected and screened against a target nematode under controlled conditions in the laboratory or greenhouse. Efficacy is confirmed by measuring the impact of the biocontrol agent on nematode populations, initially in sterilized or pasteurized soil in pots and then in microplots and field soil.

2. *Technical and commercial issues associated with registration.*

- *Target market.* Commercial decisions on whether to develop a particular product are made on the basis of the size of the potential market and the likelihood of a return on an investment. Market size considerations are particularly important for biological products that might be useful against nematodes, because the worldwide market is relatively small, particularly in comparison to products used to control weeds, insect pests or fungal pathogens. Also, the specificity of biological control will limit its applicability to specific nematode pests, certain environments and niche markets; and restrictions on the movement of live organisms across geopolitical borders will limit market access. Once the target market has been identified, economic projections can be made by estimating production costs and future sales. This economic analysis is a crucial part of the development process, because the proposed product must generate enough income to recover investment costs and return a profit.

- *Mass-production and formulation.* This part of the product development process is a blend of science and engineering, although many believe it is also an art. The aim is to grow the biological control agent in submerged or solid-state fermentation systems and then formulate the biomass into a dust, granule, pellet, wettable powder, encapsulated product, suspension or emulsion suitable for commercial use. Key requirements are that the final product must (i) deliver an appropriate dose of the biocontrol agent to its target in a viable and virulent form; (ii) be relatively easy to apply; (iii) have an adequate shelf-life; and (iv) be cost effective to produce. Technology development in this area is often species-specific, and the methods used are often proprietary in nature and, therefore, unavailable. Nevertheless, some of the issues involved are addressed by Auld *et al.* (2003) and Hölker *et al.* (2004).

- *Technology transfer and protection of intellectual property.* At some point in the product development process, licensing arrangements may have to be drawn up to protect the interests of collaborating parties. The advantages and disadvantages of patenting the biocontrol agent, or part of the technology, must also be considered. In many cases, the costs involved will not be warranted, as the intellectual property resides in the organisms themselves. Also, fermentation and formulation methods are often held internally as a trade secret.

- *Registration.* Although the requirements of registration authorities vary from country to country, the data set required for microbial products generally includes information on (i) the origin, derivation and identity of the microorganism; (ii) its biological properties; (iii) manufacturing processes; (iv) stability in storage; (v) quality assurance and guarantees of potency; (vi) presence of unintentional organisms or ingredients; (vii) toxicology, safety issues and possible impacts on human health; (viii) environmental fate; and (ix) efficacy. Given the amount of data required and the prolonged nature of the registration process, the costs involved must always be kept in mind, particularly if the product is only likely to be used in a niche market.

3. *Efficacy of the registered product.* Registration authorities are mainly concerned about whether a microbial product poses a risk to human health or the environment. Efficacy is a lesser concern, as it is usually determined by performance in the marketplace. Thus, registration is not the end of the development process: further research is required to ensure that the microbial product is effective in the hands of the end user. This research is perhaps the most important part of the product development process, as extension personnel and pest management consultants are unlikely to recommend a product unless there is evidence that is effective. Therefore, research addressing the following issues is obligatory:

- *Establishment and proliferation in soil.* Data are required to show that the microorganism can be established in situations where it is likely to be used, and that it will survive in soil for long enough to affect the target nematode.
- *Efficacy demonstrated.* The product must be evaluated in field trials in different soil types and environments, with the primary aim being to define the situations where nematode populations are likely to be reduced and yield benefits obtained. Evidence of performance relative to established products must also be obtained.
- *Mechanisms understood.* The mechanisms of action of the biological control agent must be understood, and evidence obtained that they actually operate in soil and are responsible for nematode mortality when the product is applied.
- *Use guidelines determined and recommendations available.* All biological control agents will be relatively specific with regard to: (i) the nematode they will control; (ii) the plant they will protect; and (iii) the environment in which they will work. Therefore, it is important to provide growers with specific guidelines on where to use the product; how to apply it; what results to expect; and how the product can be integrated into nematode management programmes for specific crops and nematode pests. The product development process is not complete until information of this nature is included in extension material prepared by agencies that are independent of the company selling the product.

significant and are worth repeating, as they remain relevant today:

- Observations on agar indicated that fungi which captured nematodes in three-dimensional nets were more aggressively predacious than those with adhesive knobs. However, the reverse was true in soil, indicating that activity in simple laboratory microcosms is not necessarily related to performance in the field.
- Nematode-trapping fungi will capture some nematodes, but their degree of effectiveness falls far short of eradication.
- The potential rate of multiplication of root-knot nematodes under favourable conditions is so great that even when a large percentage of the juveniles of every generation is destroyed by its natural enemies, severe injury to the host plant may still occur.
- Nematode-trapping fungi are unlikely to be effective on plants that are highly susceptible to root-knot nematode, particularly if the susceptible host produces large galls as a result of repeated reinvasion by the juvenile progeny of adult nematodes. Efficacy is likely to be greater on hosts where galls are rarely infested by juveniles of succeeding generations, and on nematode-tolerant plants.
- Nematode-capturing fungi are beneficial members of the soil flora, but direct

application is not the only way of utilizing them. Agricultural practices also need to be shaped in ways that maximize their effectiveness.

In the 50 years following Linford and Yap's publication, the nematode-trapping fungi dominated the biological control literature. Ecological studies (reviewed by Gray 1987, 1988) focused on the nutritional requirements of these fungi, particularly the factors that determined whether they produced traps and began to capture nematodes. There were also numerous biocontrol studies, culminating in the development of commercial formulations containing *Arthrobotrys robusta* and *A. superba* (Cayrol *et al.*, 1978; Cayrol and Frankowski, 1979; Cayrol, 1983). Stirling (1991) discussed this work, included tables that summarized the results of a representative sample of experiments with various fungal species and different target nematodes in a range of environments, and concluded that the consistently high level of nematode control required in modern agriculture had not been achieved. The main reasons for the poor results were believed to be: (i) many species of nematode-trapping fungi did not interact with the target nematode in a density-dependent manner; (ii) biostatic forces inhibited establishment; and (iii) sedentary endoparasitic nematodes were often targeted, and the level of predacious activity required to control this group of nematodes was difficult to obtain. Spatial factors are also likely to have played a role, as the nematode and the introduced biocontrol agent can only interact if they are in the same place at the same time.

In studies aimed at improving our understanding of the factors perceived to be limiting the activity of nematode-trapping fungi in soil, Jaffee *et al.* (1993) added species with different trapping mechanisms to untreated field soil, varied the number of host nematodes by adding different numbers of *Meloidogyne javanica* juveniles every 3 weeks, and assessed the capacity of the fungi to capture an assay nematode (juveniles of *Heterodera schachtii*) after 15 weeks. Although background levels of *Hirsutella rhossiliensis* confounded the results of some experiments, the number of assay nematodes with adhesive knobs of *Dactylellina ellipsospora* or constricting rings of *Drechslerella dactyloides* increased with host density in inoculated soil, indicating a density-dependent relationship with these fungi. In contrast, the network-forming species *Arthrobotrys oligospora* and the branch/network-former *A. gephyropaga* failed to respond to nematode density. Since later work (Jaffee and Muldoon, 1995a) showed that the assay nematode used in the above study (*Heterodera schachtii*) was a poor host of *D. ellipsospora* and *A. gephyropaga*, additional experiments were then established using another soil and different host and assay nematodes (*Steinernema glaseri* and *Meloidogyne javanica*, respectively). Again, *D. ellipsospora* clearly responded to nematode density, indicating that it is opportunistically dependent on nematodes for nutrients, whereas regressions of fungal population density on host density were not significant for *A. gephyropaga* (Jaffee and Muldoon, 1995b). However, when suppressiveness was measured over time in a follow-up experiment, the latter fungus strongly suppressed nematode penetration of roots for 30 days, after which its predacious activity declined. In contrast, suppression by *D. ellipsospora* increased gradually throughout the study and was nearly 100% at day 120.

Following observations by Jaffee and Muldoon (1995a), which showed that the cyst nematode *Heterodera schachtii* was much less susceptible to *D. ellipsospora* and *A. gephyropaga* than three species of root-knot nematode (*Meloidogyne javanica*, *M. incognita* and *M. chitwoodi*), further studies indicated that these nematodes responded differently to nematode-trapping fungi (Jaffee, 1998). Although it was not clear why the two nematode genera differed in their sensitivity to some fungi, these studies were important because they demonstrated that the target nematode must be considered when selecting candidates for inclusion in biological control programmes.

The question of whether biostatic forces were inhibiting the establishment of nematode-trapping fungi was also addressed in these experiments. Assessments using dilution plates and nematode baiting indicated that the four fungi added to loamy sand and loam soils by Jaffee *et al.* (1993) were still present after 15 weeks. Also, estimates of fungal population density by Jaffee and Muldoon (1995b) showed that both *Dactylellina ellipsospora* and *A. gephyropaga* established rapidly in soil and persisted for at least 120 days. In a later study, two network-forming fungi (*Arthrobotrys oligospora* and *A. eudermata*); two fungi that produce adhesive knobs (*Dactylellina ellipsospora* and *D. candidum*); and a constricting ring-producer (*Drechslerella dactyloides*) were formulated in alginate pellets and added to soils from two California vineyards. The five fungi established in soil, their population densities were correlated with the number of pellets added, and all fungi except *Drechslerella dactyloides* were detected at populations greater than 10^2 propagules/g soil (Jaffee, 2003). However, the network-forming fungi trapped few nematodes in soil, regardless of their population density, whereas the two knob-producers produced large numbers of propagules and trapped nearly all of the assay nematodes (Fig. 12.4).

One of the main conclusions from this work is that nematode-trapping fungi differ in their dependency on nematodes, a result that confirmed observations made by others (e.g. Cooke, 1968; Jansson and Nordbring-Hertz, 1980; Jansson, 1982; Gray, 1987). Species that

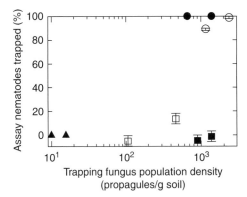

Fig. 12.4. Relationship between the population density of four species of nematode-trapping fungi and the percentage of assay nematodes (*Steinernema glaseri*) trapped in two experiments where the fungi were formulated in alginate pellets and added to field soil. Values are means ± standard error. The fungi included and their trapping mechanisms were: *Dactylellina candidum* ∘ and *Dactylellina ellipsospora* • (adhesive knobs); *Drechslerella dactyloides* ▲ (constricting rings); and *Arthrobotrys oligospora* ■ and *A. eudermata* □ (adhesive networks). (From Jaffee, 2003, with permission.)

produce constricting rings or adhesive knobs rely on nematodes as a food source and respond to an increase in nematode population density by parasitizing increasing numbers of nematodes. Species that produce adhesive nets are less dependent on nematodes: they tend to survive on other substrates and often do not trap nematodes when added to soil. These broad ecological groupings were also supported by results from Australian studies that assessed the predacious activity of 15 isolates in microcosms containing field soil. Species that produced detachable rings and knobs, or constricting rings (e.g. *Dactylellina candidum* and *Drechslerella dactyloides*) consistently produced traps and reduced the number of root-knot nematode juveniles recovered from soil, whereas network-forming species produced few traps and had little impact on nematode numbers (Galper *et al.*, 1995).

The second important finding of Jaffee and colleagues was that lack of predatory activity was not the result of a failure to establish. All the nematode-trapping fungi tested

were detected in soil weeks and sometimes months after they were introduced, and in some cases, relatively high population densities were achieved (Jaffee, 1998). Nevertheless, the persistence of introduced fungi was affected by the natural microflora and fauna. For example, *Dactylellina ellipsospora* and *Arthrobotrys gephyropaga* suppressed the number of root-knot nematodes in roots by 80% and 98%, respectively, in a laboratory experiment (Jaffee and Muldoon, 1995a), whereas they were less effective in field microplots (Jaffee and Muldoon, 1997). *A. gephyropaga* increased to nearly 3000 propagules/g soil after 8 days in those microplots, but a ten-fold reduction occurred in the next 14 days, possibly because the fungus was consumed by enchytraeid worms. Later work (discussed earlier in the section on general soil biostasis) showed that enchytraeids and smaller biotic agents reduced the persistence of all species of nematode-trapping fungi, although the severity of the biostatic effects varied from experiment to experiment.

When these studies are viewed from a biological control perspective, they indicate that high populations of nematode-trapping fungi can be established in soil, and that it is possible to achieve high levels of predacious activity, provided the target nematode is susceptible to the chosen fungus, and biotic constraints are not excessive. The results also showed that adhesive knob-producers were the most consistent performers and were also effective in a range of soils, suggesting that species in the genus *Dactylellina* warrant more extensive testing as biological control agents. However, Jaffee (2003) also warned that inundative biological control with nematode-trapping fungi has its limitations. Although high levels of activity were achieved with both *Dactylellina ellipsospora* and *D. candidum*, substantial trapping only occurred when fungal population densities exceeded those normally attained by resident nematode-trapping fungi. Thus, it may be possible to establish these fungi in soil, but abnormally high population levels have to be sustained for a considerable period to obtain useful levels of control. Given the biostatic factors present in all soils, this is likely to be difficult to achieve.

The alginate pellets used by Jaffee and colleagues in their experiments with nematode-trapping fungi (Jaffee and Muldoon, 1995a, b, 1997) and by others in their work with other nematophagous fungi (Kerry *et al.*, 1993; Lackey *et al.*, 1993; Meyer, 1994; Kim and Riggs, 1995) were a convenient way of introducing fungi into soil for ecological and efficacy studies. However, they also proved useful for addressing issues associated with formulating nematode-trapping fungi into products suitable for commercial use. Two fungi selected by Galper *et al.* (1995) as having biocontrol potential (*Dactylellina candidum* and *Drechslerella dactyloides*) were formulated as alginate/kaolin granules and then applied to soil, and observations in Petri dishes and microcosms showed that the presence of nutrients and the quantity of biomass in the granule determined the level of activity against nematodes (Stirling and Mani, 1995). The best formulation, which contained at least 10^7 colony-forming units/g of product, was obtained by re-fermenting the fungi after they were encapsulated in alginate, as this process increased fungal population densities by a factor of 10. Both fungi grew strongly from this formulation, producing large numbers of traps that were maintained for at least 10 days and extended at least 5 mm from the granule into soil. Since the number of fungal propagules was not markedly reduced by the process used to dry the granules, and the dried formulation could be stored for at least 8 months without loss of viability, these results suggested that it was feasible to produce a commercial product containing a nematode-trapping fungus.

Since it is difficult to scale-up the alginate formulation system to a level suitable for commercial use, Stirling *et al.* (1998b) investigated other options using the isolate of *Drechslerella dactyloides* found to be efficacious in the previous study. Initial work showed that the fungus grew readily in shake culture, but it proved difficult to convert this biomass into a viable, granulated product. Formulations prepared using kaolin and vermiculite as carriers and gum arabic as a binder showed poor viability when biomass was harvested from liquid culture, mixed with formulation ingredients, granulated

and then dried to a moisture content of less than 5%. However, inclusion of a solid-phase incubation step following granulation (incubation of moist granules for 3 days in a sterile plastic bag aerated with sterile air) markedly improved biological activity, presumably because it allowed the fungus to regenerate after being damaged during the formulation process. Observations in the laboratory then showed that granules produced in this manner were effective enough to warrant further testing, as hyphae always grew from granules and produced traps, and that the number of juveniles of *Meloidogyne javanica*, and the number of nematode-induced galls on tomato, were generally reduced by more than 90% in microcosm and greenhouse tests. Since these initial tests were carried out at a commercially unrealistic application rate (10 g granules/L soil), evidence of efficacy at lower application rates was then required. Thus, different rates of the formulated product were tested for effects on galling caused by root-knot nematode on tomato in the greenhouse. The experiment consisted of six application rates of granules (0, 0.4, 0.8, 1.6, 3.2 and 10 g/L soil) × 2 soil types (sand and sandy clay loam soils from the field) × 2 planting times (immediately after granules and root-knot nematodes were inoculated, or 2 weeks later) × 5 replicates, with the number of galls on roots being assessed 4 weeks after seedlings were transplanted. The results were encouraging because, although the level of control increased as the application rate increased, some control was obtained at rates as low as 0.8 g granules/L soil (Fig. 12.5).

Given the promising results obtained in this study, the next step was to evaluate the efficacy of these formulations in the field. Consequently, granules containing *D. dactyloides* were applied at rates ranging from 55 to 880 kg/ha in five field experiments in different soil types and environments, and their effectiveness was assessed by measuring yield of tomato and the level of galling caused by root-knot nematodes (Stirling and Smith, 1998). Various application rates and methods were used, and although the formulated product was not as effective as the nematicide fenamiphos, root galling 4–8 weeks after planting was reduced by at least one of the

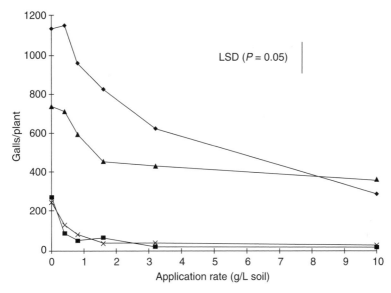

Fig. 12.5. Effect of a formulation of *Drechslerella dactyloides* applied at various application rates on galling caused by *Meloidogyne javanica* on tomato in two soils and at two planting times. Seedlings transplanted immediately after the formulation was applied to sand ♦ and clay loam soil ▲; or 2 weeks later (sand ■ and clay loam ×). Bar indicates the least significant difference between treatments and times. (From Stirling *et al.*, 1998b, with permission.)

formulations in four of five experiments. However, as is common with many control measures that are used against root-knot nematode, these early effects had often disappeared by the time the crop was harvested, with a significant reduction in galling only observed at harvest in one experiment.

Although the commercial partner funding the work did not consider the clay-based product was effective enough to warrant continuing this research, the authors believed the results were promising for the following reasons:

• The number of nematodes in roots early in the growing season was reduced by 75–80%. Although this is less than is achievable with some chemical nematicides, the results suggested that substantial numbers of nematodes had been captured by the fungus.

• Reductions in galling were observed at application rates of 40 and 80 g/m row (the equivalent of 220 and 440 kg granules/ha). This is a commercially realistic application rate, and certainly an improvement on the much higher application

rates (>2 t/ha) that had been required in previous experiments with nematode-trapping fungi.

• Since only about 1% of the weight of the granules consisted of fungal biomass and the remainder was an inert carrier (kaolin), reductions in nematode populations were achieved with the equivalent of about 2.2 kg of dried fungal material/ ha. This amount of biomass is easily produced in a bioreactor.

Although the laboratory-scale culture system used in the Australian studies cited provided formulations with good viability and vigour, it is doubtful whether such a diphasic system (initial growth in liquid culture followed by transfer to and incubation in a solid substrate) would be a commercially viable option for mass-producing *Drechslerella dactyloides* or other nematode-trapping fungi. Solid-state fermentation may be an alternative, as major advances have been made with this technology (Mitchell *et al.*, 2006), but it remains to be seen whether it is useful for these fungi. Thus, to conclude this discussion, it should be possible to develop useful products containing nematode-trapping fungi, but success is most likely to be achieved by focusing future research in the following areas: (i) selecting species and isolates from the predacious end of the nutritional spectrum (i.e. fungi that produce detachable rings and knobs, or constricting rings, rather than two- or three-dimensional networks); (ii) improving mass-production techniques, and evaluating formulation technologies that could possibly increase shelf-life; and (iii) defining the nematode species most likely to be amenable to control by an agent that targets motile stages of the nematode in soil; identifying situations where the introduced agent is not likely to be limited by natural biostatic forces; and defining the crops, soil types and environmental conditions where such an agent is most likely to be successful.

Although progress towards controlling plant-parasitic nematodes with nematode-trapping fungi has been relatively slow, the advances that have occurred in their use against animal-parasitic nematodes (see reviews by Soder and Holden, 2005; Larsen, 2006; Waller, 2006) suggest that success is achievable. Several nematode-trapping species have been tested against animal parasites, but most of the work has been done with *Duddingtonia flagrans*, a species that produces an abundance of thick-walled chlamydospores. When these resting spores are given to animals as a daily supplement, they pass through the gastrointestinal tract and germinate in faeces. Three-dimensional sticky networks then trap larvae of a range of nematodes pathogenic to animals, including species of *Haemonchus*, *Teladorsagia*, *Trichostrongylus* and *Nematodirus*. Since *D. flagrans* can be deployed in mineral or nutrient blocks, or mixed with food (Waller *et al.*, 2001; Sagüés *et al.*, 2011; Santurio *et al.*, 2011; de Almeida *et al.*, 2012), current evidence suggests that when combined with improved grazing management and the selective use of existing drugs, the fungus can increase the sustainability of animal production systems. If products containing nematode-trapping fungi are ever developed for use against plant-parasitic nematodes, it is likely that they will be used in much the same way: as a component of an integrated nematode management programme.

Endoparasitic fungi

Many endoparasitic fungi are known to attack nematodes, but this discussion will concentrate on the genus *Hirsutella*, primarily on *H. rhossiliensis*, as it is the only member of the genus to have been used extensively in biocontrol studies. Spores of *H. rhossiliensis* adhere to the cuticle of passing nematodes, and spore-encumbered nematodes die soon after infection. After the carcase has been occupied by the fungus, hyphae grow out from the dead nematode and adhesive spores are borne on bottle-shaped phialides that are spaced at regular intervals along the emergent hyphae (Jaffee and Zehr, 1982).

Much of our knowledge of the potential of *H. rhossiliensis* as a biological control agent has come from studies by Bruce Jaffee and colleagues in California. Although this research was not primarily aimed at developing biological products for nematode control, it showed that there are limitations to what is likely to be achievable with this fungus. For example, *H. rhossiliensis* was found to be no more than a relatively weak regulator of *Mesocriconema xenoplax* and *Heterodera schachtii* in naturally infested soils (see Chapter 10), suggesting that high levels of nematode parasitism were only likely to be achieved with high population densities of the fungus. Similarly, studies of mycostatic factors (discussed earlier in this chapter) indicated that *Hirsutella rhossiliensis* was quite sensitive to biotic inhibition when formulated as pelletized hyphae. Interestingly, the fungus was not inhibited by the soil biota when added to soil as parasitized nematodes (Jaffee, 2000), leading the author to speculate that formulations might perform better if the pellets more closely resembled natural forms of inoculum. Alginate pellets containing *H. rhossiliensis* differed from naturally parasitized nematodes in several respects (pellets were larger and drier and did not have a protective barrier such as the cuticle), and these factors may have made the fungus more sensitive to biostasis.

However, the condition of the hyphae in the pellet is likely to have been more important. The thin-walled assimilative hyphae are fragmented during the formulation process and are likely to be particularly vulnerable to biostatic forces. Thus, attempts to develop products based on *H. rhossiliensis* should probably focus on producing formulations containing intact rather than damaged hyphae.

Since all biological control agents are innately variable, the first step in developing a useful microbial product is to identify useful isolates (Box 12.3). In the case of *H. rhossiliensis*, Tedford *et al.* (1994) approached this task by collecting 25 isolates from soil mites and many different nematodes. They then assessed the genetic variability of the isolates using RAPD markers, measured their growth rates on agar, and compared their pathogenicity on agar and in soil. The results indicated that, although most isolates were morphologically and pathogenically similar, some were weakly parasitic or non-parasitic to nematodes. There were also substantial genetic differences between isolates, but pathogenicity assays in soil indicated that isolate A3a, which was used by Jaffee and his research group in many subsequent studies, performed as well as, or better than, all other isolates tested.

In the same way, Chen and colleagues began their studies on biological control of soybean cyst nematode by screening 93 isolates of *H. rhossiliensis* and 25 isolates of *H. minnesotensis* in laboratory assays on agar and in microwaved soil. The best isolates were then compared in a greenhouse assay with field soil. Most isolates of *H. rhossiliensis*, except those obtained from bacterial-feeding nematodes, parasitized a high percentage of second-stage juveniles of *Heterodera glycines* on agar and in pasteurized soil. In contrast, *Hirsutella minnesotensis* was less effective, particularly in the soil assay (Liu and Chen, 2001a). One isolate of *H. rhossiliensis* was highly effective in suppressing soybean cyst nematode in the greenhouse. Interestingly, and perhaps importantly, it was obtained from a site where the nematode had been suppressed by naturally occurring organisms after 27 years of continuous soybean culture.

Although the California group obtained promising results with *H. rhossiliensis* in laboratory microcosms, it proved to be ineffective in field microplots. When introduced into a loamy soil in the form of colonized nematodes (*Steinernema glaseri*), the fungus initially parasitized about 40% of the juveniles of *Heterodera schachtii*. However, levels of parasitism soon declined and the fungus did not reduce the number of nematodes detected in sugarbeet roots. In a concurrent experiment on tomato, juveniles of *Meloidogyne javanica* were rarely parasitized by *H. rhossiliensis* and nematode populations remained high during two growing seasons (Tedford *et al.*, 1993). Later experiments with alginate formulations of the fungus also produced disappointing results, as the fungus failed to suppress *Heterodera schachtii* on cabbage in loamy sand (Jaffee *et al.*, 1996a) and *Meloidogyne javanica* on tomato in a sandier soil (Jaffee and Muldoon, 1997).

The isolate of *Hirsutella rhossiliensis* selected in the Minnesota study was not tested in the field, but the performance of cultures grown on corn grits was assessed in field soil in the glasshouse (Chen and Liu, 2005). At an application rate of 1% w/w, corn grits alone reduced populations of soybean cyst nematode in some experiments, but corn-grit cultures of *H. rhossiliensis* were even better, sometimes almost eliminating the nematode (Table 12.1). However, when applied at lower, and possibly commercially realistic rates (0.2% and 0.04% w/w), the fungus was ineffective. Since some of the control obtained in these experiments was almost certainly due to the direct or indirect effects of the corn grits on which the fungus was grown, liquid suspensions containing mycelial fragments and spores were also evaluated. These formulations also reduced nematode populations, but the application rates required to obtain control were unlikely to have been cost effective in soybean production (Liu and Chen, 2005).

Velvis and Kamp (1996), working with isolates of *H. rhossiliensis* from the Netherlands, found that very high spore densities (about 10^5 spores/g soil) were required to achieve a 50% reduction in the number of *Globodera pallida* invading potato roots, while Liu and Chen (2002) noted that the slow growth of the fungus limited mass-production for field application. These observations, together with the results from California and Minnesota

Table 12.1. Population densities of *Heterodera glycines* on soybean in four pot experiments with a sandy loam soil from a field in Minnesota. The soil was either left untreated, treated on the day of planting, or treated 2 weeks before planting with a corn-grit amendment or a solid culture of *Hirsutella rhossiliensis* (isolate OWVT-1) grown on corn grits (both at an application rate of 1% w/w).

Experiment no. and application time	Nematode parameter	Untreated	Corn grits	Corn grits + fungus
1 (at planting)	Eggs/mL	783 ± 156	533 ± 93	63 ± 46
	100 eggs/g plant	1086 ± 565	277 ± 74	26 ± 18
	J2/10 mL soil	755 ± 56	321 ± 123	36 ± 29
	J2/g of plant	2659 ± 1652	465 ± 212	43 ± 33
2 (at planting)	Eggs/mL	1698 ± 596	785 ± 171	39 ± 38
	100 eggs/g plant	211 ± 63	120 ± 37	4 ± 3
	J2/10 mL soil	298 ± 157	237 ± 134	11 ± 8
	J2/g of plant	365 ± 166	359 ± 234	12 ± 8
3 (2 weeks pre-plant)	Eggs/mL	443 ± 241	115 ± 16	17 ± 10
	100 eggs/g plant	341 ± 210	21 ± 4	4 ± 5
	J2/10 mL soil	220 ± 186	52 ± 12	16 ± 6
	J2/g of plant	1360 ± 521	99 ± 48	36 ± 21
4 (2 weeks pre-plant)	Eggs/mL	491 ± 86	259 ± 78	0 ± 0
	100 eggs/g plant	117 ± 31	45 ± 11	0 ± 0
	J2/10 mL soil	136 ± 99	40 ± 14	5 ± 3
	J2/g of plant	296 ± 521	71 ± 48	7 ± 21

Values are means ± standard deviation of four replicates. The cited publication should be consulted for details of statistical analyses. (Data are from Tables 3 and 5 of Chen and Liu, 2005, and are used with permission.)

discussed previously, raise questions about the commercial potential of products based on *H. rhossiliensis*. It may be possible to improve methods of mass-producing and formulating the fungus, but as the regulatory capacity of *H. rhossiliensis* is limited (Jaffee and McInnis, 1991; Jaffee *et al.*, 1992a), it will be difficult to obtain the spore densities likely to be required to consistently suppress nematode populations. *H. minnesotensis* does not appear to be a better option, as it was generally inferior to *H. rhossiliensis* when the two species were compared in greenhouse tests (Chen and Liu, 2005; Liu and Chen, 2005). However, the former species did reduce numbers of *Meloidogyne hapla* on tomato when tested in steam-sterilized soil in a greenhouse test (Mennan *et al.*, 2006).

Cyst and egg parasites

Pochonia

Ever since *Pochonia chlamydosporia* was found to be associated with the decline of cereal cyst

nematode in England (see Chapter 10), there has been considerable interest in using it for inundative biological control of plant-parasitic nematodes. Some of the features that make it potentially useful for that purpose are its capacity to: (i) grow readily in culture; (ii) survive as a saprophyte when populations of a host nematode are low; (iii) produce thick-walled chlamydospores as survival structures; (iv) colonize the rhizosphere, the ecological niche occupied by endoparasitic nematodes as they develop and reproduce; and (v) parasitize the cysts of many cyst nematode species and the eggs of root-knot nematode.

As with other biological control agents, careful strain selection is important, as isolates of *P. chlamydosporia* vary in their optimum temperature for growth; production of chlamydospores; saprophytic and rhizosphere competence; host preference; and pathogenicity (Kerry *et al.*, 1986; Irving and Kerry, 1986; Bourne *et al.*, 1996; Kerry and Bourne, 1996; Mauchline *et al.*, 2004; Dallemole-Giaretta *et al.*, 2012). Since selected isolates could be grown relatively easily on grain and other solid substrates, simple mass-production methods were initially used to demonstrate that nematode populations were reduced when *P. chlamydosporia* was added to soil (Godoy *et al.*, 1983b; Kerry *et al.*, 1984). However, to overcome the problem of determining whether control was due to the fungus or to its nutrient base, later studies were generally done with chlamydospores, as large numbers (10^7 chlamydospores/g) were readily produced on a grain–sand medium (Crump and Irving, 1992).

Heterodera avenae was targeted in initial studies with *P. chlamydosporia*, but it was soon recognized that inundative biological control was unlikely to be economically viable for nematode pests of relatively low-value field crops. Numbers of cereal cyst nematode were reduced by between 26% and 80% when *P. chlamydosporia* was grown on oat grains and applied to soil in pots. However, unrealistically high application rates (the equivalent of 62.5 t/ha) were required to achieve this result, and the number of propagules of the fungus rarely reached the densities observed in soils that were naturally suppressive to the nematode (Kerry *et al.*, 1984). Much higher numbers of *P. chlamydosporia* were obtained in later studies with alginate granules containing wheat bran (Kerry *et al.*, 1993), but again the application rate was very high (1% w/w). Therefore, attention shifted to cyst nematode species that attack crops of higher value. Tests in small pots showed that populations of the beet cyst nematode (*H. schachtii*) and potato cyst nematode (*Globodera pallida*) could be reduced by about 75% (Crump and Irving, 1992), but again, the practicality of achieving good control on a field scale was questioned. Although the application rate used in the above experiment (10,000 chlamydospores/g soil) appeared realistic (equivalent to 100–200 kg dried spores/ha), the difficulty of producing chlamydospores in liquid culture was a major impediment to developing a biological product. Attempts to commercialize *P. chlamydosporia* have since been made (e.g. Crump, 2004), but products containing the fungus are not yet available for use against cyst nematodes.

Experiments on the impact of chlamydospore inoculum on root-knot nematode commenced in the early 1990s and an isolate that was originally obtained from eggs of *Meloidogyne incognita* was found to be effective against all widely distributed species of *Meloidogyne* (de Leij and Kerry, 1991). When this isolate was tested in pots, it readily established in the rhizosphere and on galled tissue, but was only effective when eggs were exposed on the root surface rather than embedded in galls. Since the proportion of eggs embedded in gall tissue increases as nematode population density increases, and is also influenced by nematode species, temperature and the host plant (de Leij, 1992; de Leij *et al.*, 1992b, 1993), the efficacy of this fungus depends on the conditions under which it is tested. Establishment and proliferation in soil and the rhizosphere is also influenced by soil type (de Leij *et al.*, 1993), and collectively, these factors make it impossible to predict with certainty whether *P. chlamydosporia* will be effective as a control agent. Nevertheless, the first field test of the fungus gave encouraging results. Inoculum was incorporated into a sandy loam soil to achieve a concentration of 5000 chlamydospores/g of soil to a depth of 30 cm, and estimates of fungal population density using a selective medium showed that the population increased to about 50,000 cfu/g of soil by day 50, before stabilizing at that level for the following 70 days. Populations of *M. hapla* were reduced by more than 90% (de Leij *et al.*, 1993).

At about the same time as de Leij and colleagues began their studies in England, isolates of *Pochonia chlamydosporia* were selected in Australia (Stirling and West, 1991) and used in a study aimed at developing commercially acceptable formulations of the fungus. Since chlamydospores could not be produced in submerged culture, hyphae and conidia were grown in a bioreactor, the fungal biomass was mixed with a carrier (kaolin) and a binder (gum arabic), and then the ingredients were granulated and dried to a moisture content of less than 2% (Stirling *et al.*, 1998a). The formulated product retained its viability when stored in vacuum-sealed plastic bags for 12 months, and when it was added to soil at a rate of 10 g/L, population densities of *P. chlamydosporia* increased to about 10^4 cfu/g of soil. Although 37–82% of the egg masses produced by *M. javanica* contained parasitized eggs in pot tests, the level of parasitism was insufficient to reduce populations of the nematode on tomato. Follow-up studies with the same formulations in field experiments showed that the fungus did not increase egg parasitism or reduce galling or nematode numbers at harvest (Stirling and Smith, 1998).

More recent studies have confirmed that *P. chlamydosporia* can be established in soil from inoculum, and that some isolates will parasitize substantial numbers of root-knot nematode eggs (e.g. van Damme *et al.*, 2005; Moosavi *et al.*, 2010; Carneiro *et al.*, 2011; Yang *et al.*, 2012a). On the basis of such results, authors of papers on *P. chlamydosporia* commonly conclude with optimistic pronouncements about the potential of the fungus as a bionematicide. However, they tend to downplay the issues that must be addressed to achieve that goal. A review by Kerry (2000) should, therefore, be read more widely, as it lists the key factors affecting the performance of *P. chlamydosporia* as a biological control agent against root-knot nematode. When these factors are considered (Box 12.4), it is clear that *P. chlamydosporia* is a potentially useful organism, but is only ever likely to be consistently effective in certain situations, generally as a component of an integrated nematode management programme. Commercial formulations of *P. chlamydosporia* are available in Cuba and Brazil (Dallemole *et al.*, 2011) and so the challenge facing nematologists is to define the market niche for such products in other countries, and then provide the field trial data and other evidence required to demonstrate that commercial investment in this area is warranted.

Purpureocillium

Purpureocillium lilacinum (syn. *Paecilomyces lilacinus*) was first recognized as an effective parasite of cyst and root-knot nematodes by Jatala *et al.* (1979). The fungus was then studied in many countries (Jatala, 1986), with work during the 1980s culminating in an isolate of *P. lilacinum* being commercially marketed in the Philippines for biocontrol purposes (Jatala, 1986; Timm, 1987). A table showing details of those early experiments can be found in Stirling (1991), together with a comment that the results were often difficult to interpret because either the data were

not published in full, or the experiments lacked treatments that enabled effects of *P. lilacinum* to be separated from those of the substrate on which it was grown.

There was little progress in the next decade, as formulations suitable for applying the fungus on a commercial scale were unavailable. Solid-substrate fermentation using low-cost substrates, and liquid culture followed by encapsulation in alginate, were then used to produce formulations containing more than 10^{10} and 10^9 propagules/g, respectively (Brand *et al.*, 2004; Duan *et al.*, 2008). However, major progress was not made until water-dispersible granules and wettable powders were manufactured. These formulations were not only readily applied using standard farm equipment, but also contained high concentrations of *P. lilacinum* spores (>10^{10} and >10^{11} spores/g for the water-dispersible granules and wettable powder, respectively). Products based on *P. lilacinum* strain 251 were then registered under trade names such as BioAct™ (Prophyta, Malchow/Peol, Germany) and MeloCon™ (Certis, Maryland), and are now available in many countries (Kiewnick, 2010).

The availability of commercial formulations of *P. lilacinum* has been a major step forward, as it has meant that the efficacy of the fungus could be evaluated in a wide range of situations (e.g. different soil types, environmental conditions and target nematodes). That work has shown that numerous factors affect the biocontrol efficacy of *P. lilacinum* strain 251 (PL251):

- *Application rate.* Experiments with different dosages of PL251 showed that populations of *M. incognita* and levels of galling on tomato were reduced by 58–74%, depending on the parameter measured (Kiewnick and Sikora, 2006a). The EC_{50} (the concentration of the fungus 6 days after planting required to produce a 50% reduction in the gall index, the number of egg masses and the final nematode population) was about 10^6 cfu/g soil, confirming results obtained by Gomes *et al.* (1991) with a Peruvian isolate of the fungus. However, in a later study, a lower concentration of PL251 (2×10^5 cfu/g soil) reduced root galling

Box 12.4. Key factors likely to affect the efficacy of a formulated product containing *Pochonia chlamydosporia*, and their role in determining the situations where the fungus is most likely to be useful in controlling root-knot nematode.

Factor	Impact on efficacy	Relevance to biological control
Isolate	Isolates vary in saprophytic capacity, chlamydospore production, rhizosphere competence and pathogenicity to eggs	Appropriate isolates must be selected
Nematode population density	When nematode populations are high, egg masses are embedded within large galls, and so the fungus does not come into contact with eggs	Success is most likely when nematode populations are low to moderate
Nematode species	All *Meloidogyne* species are susceptible to the fungus. However, species that produce small galls with an external egg mass are most susceptible, as eggs are available for colonization by the fungus	*M. hapla* is a better target than *M. javanica*, *M. incognita* and *M. arenaria*, as egg masses are more likely to be on the surface of a gall rather than embedded in galled tissue
Rhizosphere colonization of the host	The rhizosphere of some plants (e.g. cabbage, maize, bean, potato and tomato) supports higher fungal populations than others	Better results are likely to be obtained with good hosts of the fungus
Susceptibility of the host to the nematode	Egg parasitism may have little impact on nematode populations when multiplication rates are high	The fungus is unlikely to be effective on crops that are excellent hosts of the nematode
Tolerance of the host to the nematode	Since the fungus does not reduce initial nematode invasion of roots, it does not reduce damage on the crop to which it is applied	A tolerant crop is more likely to cope with the damage that occurs before egg parasitism commences
Carry-over effects	Since eggs become increasingly available as a food source as the growing season proceeds, levels of parasitism in annual crops tend to increase with time	The role of the fungus in reducing nematode carry-over to the next crop should be given more attention
Egg maturity	Immature eggs are more susceptible than eggs containing second-stage juveniles	The fungus is most likely to be successful when present on the gall surface at the time egg masses are first produced
Temperature	The fungus is less effective at temperatures >30°C because eggs will develop and hatch before the egg mass is colonized	Success is most likely at temperatures at or below the optimum for the nematode
Soil type	The fungus proliferates in most soils but may multiply more readily in some soils than others	Further work on the effect of soil type is needed, as it is likely to influence efficacy
Type of formulation	Formulations containing chlamydospores, or preparations on solid substrates, are more likely to establish and proliferate than formulations made from biomass produced in liquid media	Some formulation methods are better than others, but the more active formulations may be more expensive to produce

This information was derived from Kerry (2000), and from other references cited in the text

and the number of egg masses by 45% and 69%, respectively (Kiewnick *et al.*, 2011).

- *Persistence.* Once *P. lilacinum* is applied to soil, populations begin to decline.

Previous work had shown that very high concentrations of the fungus cannot be maintained in soil for long periods (Hewlett *et al.*, 1988; Gomes *et al.*, 1991), and a study with PL251 indicated that populations of this strain declined by about 55% in 60–70 days (Kiewnick and Sikora, 2006a). Other studies cited by the latter authors had shown that rapid decline began to occur 14–21 days after application, and that populations typically declined by at least 90% in 60–100 days. These observations were later confirmed by Rumbos *et al.* (2008), whose results are presented in Fig. 12.6. This lack of persistence of the fungus is desirable from an environmental perspective, because it minimizes the risk of detrimental effects on non-target organisms. However, it also means that multiple applications may be required to achieve satisfactory control.

- *Soil characteristics.* Gaspard *et al.* (1990a) reported that population densities of *P. lilacinum* in 20 California tomato fields were positively correlated with the silt content of the soils and negatively correlated with the amount of sand. Studies with PL251 also indicated that it preferred fine-textured soils, as the fungus persisted longer in silty loam and clay soil than in sand (Fig. 12.7). Since the addition of an organic seedling substrate had a positive effect on persistence, it was presumed that the higher rate of decline in sand was associated with low levels of saprophytic activity in soils with low nutrient status (Rumbos *et al.*, 2008).

- *Time of application.* Kiewnick and Sikora (2003, 2006a) obtained good control of *M. incognita* on tomato in pots by applying PL251 6 or 7 days before transplanting, whereas a 16-day delay between application and transplanting was most effective in another study (Anastasiadis *et al.*, 2008). A different isolate gave the best result when soil was treated 10 days before planting and again at planting (Cabanillas and Barker, 1989). Collectively, these observations indicate that is important to apply *P. lilacinum* before rather than after a crop is planted.

- *Nematode inoculum density.* When the efficacy of PL251 was tested against *M. incognita*, control on tomato was best when the initial nematode population density was in the range 100–400 nematodes/100 mL soil. At higher inoculum densities, levels of nematode damage on roots increased and biocontrol efficacy decreased (Kiewnick *et al.*, 2006).

- *Temperature.* When PL251 is added to soil, it declines at a greater rate at temperatures of 28°C than at lower temperatures (10–15°C), indicating that soil temperature is an important factor affecting the persistence of the fungus (Kiewnick, 2010).

- *Root-knot nematode species.* *P. lilacinum* is endemic to most tropical and subtropical soils and has mainly been used to control *Meloidogyne* species that are

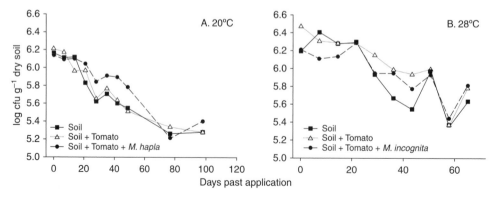

Fig. 12.6. Decline in populations of *Purpureocillium lilacinum* strain 251 in the presence or absence of a tomato plant and the nematode host in a mix of silty loam and sand at temperatures of 20°C and 28°C. (From Rumbos *et al.*, 2008, reprinted by permission of the publisher, Taylor and Francis Ltd.)

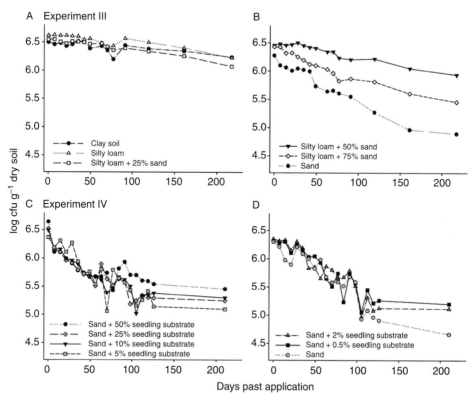

Fig. 12.7. Persistence of *Purpureocillium lilacinum* strain 251 in different soil substrates over 31 weeks (Experiment III, A and B) and 30 weeks (Experiment IV, C and D). (From Rumbos *et al.*, 2008, reprinted by permission of the publisher, Taylor and Francis Ltd.)

common in warm climates (e.g. *M. incognita*, *M. javanica* and *M. arenaria*). Since the fungus has a relatively high optimum temperature (24–30°C), a study was conducted with PL251 to determine whether it could control *M. hapla*, a species that can penetrate host roots at 10°C and has an optimum temperature for root invasion and growth of 20–25°C. The results indicated that *M. hapla* was an unsuitable target for PL251, as even at temperatures just below the optimum for the fungus (i.e. 19–21°C), it did not control the nematode on tomato (Kiewnick and Sikora, 2006b).

In addition to these observations, other studies have added to our knowledge of how PL251 acts against root-knot nematode. In contrast to results obtained by Cabanillas *et al.* (1988) with another isolate of *P. lilacinum*,

hyphae of PL251 were never found within roots, indicating that the fungus lives in the rhizosphere and infests nematodes directly from soil (Holland *et al.*, 2003). This lack of a close relationship with plant roots was later confirmed by Rumbos and Kiewnick (2006), who found that when PL251 was added to soil, the presence of plants did not influence the rate of decline in fungal population densities. The presence of root-knot nematode did not influence this parameter, either (Fig. 12.6), indicating that *P. lilacinum* differs from other egg-parasitic fungi (e.g. *Pochonia chlamydosporia*) in that it does not need to grow in the rhizosphere or on gall tissue to infect nematode eggs (Kiewnick *et al.*, 2011). These observations may explain why PL251 is more effective than *P. chlamydosporia* in reducing nematode damage on the crop to which it is applied. Provided it is applied before planting, PL251 can parasitize eggs that may be present in soil, thereby reducing the number of juveniles invading the plant. In contrast, *P. chlamydosporia* proliferates in the rhizosphere and only affects eggs produced by the first and later generations of the nematode.

The findings discussed are valuable because they indicate that formulations containing PL251 are potentially useful, provided that: (i) they are used in situations where populations of root-knot nematode are not excessively high; (ii) population densities of PL251 are in the root zone of the crop are maintained above 10^6 cfu/g soil; and (iii) the target is not *M. hapla*. Results to date also suggest that multiple treatments must be applied when control is required for more than 6–8 weeks, or in situations where the fungus declines rapidly (e.g. in sandy soils or when temperatures are high). However, it is important to remember that all the work discussed was done in short-term experiments (generally only one generation of the nematode) under controlled conditions in a growth chambers or greenhouse. Thus, there are still several steps to be taken before the full product development process outlined in Box 12.3 is completed. Some of the important questions to be answered in the field are:

- What is the rate of decline of the fungus when exposed to the fluctuations in moisture and temperature that occur in a field environment?
- How effective is the fungus in hot climates and in sandy soils with low levels of organic matter? This is an important question, because root-knot nematodes are most likely to cause severe damage in these situations.
- Will the fungus be effective on all crops, particularly those that are highly susceptible to root-knot nematode?
- Given that the fungus does not survive for long periods in soil, what application strategies should be used in situations where two or three generations of the nematode are expected to occur on a crop, and will that approach be effective?

The need to demonstrate efficacy of PL251 in the field is highlighted by the results of the first field test of the product published in a peer-reviewed journal. In an experiment in a commercial plastic house in Greece, PL251 did not provide satisfactory control of root-knot nematode on cucumber, as yield (number of fruits/plant) and the number of nematodes/g root did not differ significantly from the untreated control at either 60 or 100 days after planting (Anastasiadis *et al.*, 2008). Of even more concern is that the product was ineffective when used in combination with solarization. Since PL251 is promoted as a component of an integrated nematode management programme (Kiewnick, 2010), there is clearly a need for field studies that not only demonstrate its effectiveness in such programmes, but also reveal the management practices with which it is compatible. The availability of commercial products containing *P. lilacinum* provides opportunities to field test such products in a range of environments, and the successes and failures of those tests will help to better define the environmental conditions in which they are effective.

Although most of the work with *P. lilacinum* strain 251 has focused on root-knot nematode, experiments in pots have shown that the fungus also has the capacity to control burrowing nematode (*Radopholus similis*) on banana. Results were somewhat similar to those obtained by Kiewnick and Sikora (2006a) with *M. incognita*, as high levels of suppression were obtained when the fungus was applied three times (6 days before planting, at planting and as a plantlet drench) at a dosage of 6×10^6 cfu/g dry soil (Mendoza *et al.*, 2007). However, as the authors indicated when discussing their results, further research is needed to determine the long-term efficacy of the fungus in banana production.

In concluding this discussion, one issue that has been the subject of debate for many years is the possibility that the strains of *P. lilacinum* used for biological control of nematodes could also be opportunistic human pathogens. *P. lilacinum* is most commonly associated with ocular, cutaneous and sub-cutaneous infections, but upper respiratory tract and pulmonary infections have also been reported (Khan *et al.*, 2012).

Although most cases of disease occur in patients with compromised immune systems, evidence is accumulating that infections in immunocompetent patients are becoming more frequent (Carey *et al.*, 2003). It is, therefore, a concern that a recent molecular study found that *P. lilacinum* isolates used for biocontrol of nematodes are identical to those causing infections in humans (Luangsa-ard *et al.*, 2011). Although those working with the fungus as a biological control agent have argued that the fungus does not have the capacity to grow at human body temperature, and *P. lilacinum* has also been considered safe to use by registration authorities in several countries, the latest findings suggest that commercial products containing the fungus should be handled carefully.

Trichoderma

The capacity of *Trichoderma* spp. to attack and control plant-pathogenic fungi was first recognized in the 1930s (Weindling, 1932), and in a book published 40 years later, *Trichoderma* was considered to be the fungus most commonly associated with antagonistic effects against soilborne plant pathogens (Baker and Cook, 1974). Since that time, studies of its taxonomy, genetic diversity, antagonistic mode of action, interaction with plants and biocontrol potential have generated a huge number of review articles and books (e.g. Papavizas, 1985; Harman and Kubicek, 1998; Harman, 2000; Howell, 2003; Harman *et al.*, 2004; Lorito *et al.*, 2010; Shoresh *et al.*, 2010), and various strains of the fungus have been formulated into products that are used to promote plant growth and control a range of soilborne pathogens, including *Phytophthora*, *Pythium*, *Rhizoctonia*, *Thielaviopsis* and *Cylindrocladium*.

Although *Trichoderma* is normally not one of the many opportunistic fungi known to colonize cyst and root-knot nematodes (Morgan-Jones and Rodriguez-Kabana, 1988), its capacity to enhance plant growth and reduce losses from root diseases prompted studies of its effects on nematodes. Initial greenhouse experiments with different species and strains of the fungus produced varying results: improved growth of maize and a reduction of about 50% in the number of eggs produced by *Meloidogyne arenaria* (Windham *et al.*, 1989); no effect on reproduction of *M. incognita* on cotton (Zhang *et al.*, 1996); improved plant growth in both short- and long-term experiments and reduced galling in the short-term experiments (Spiegel and Chet, 1998); and no impact on the number of eggs and second-stage juveniles produced by *M. incognita* on tomato (Meyer *et al.*, 2000). Sharon *et al.* (2001) then provided convincing evidence that some Israeli strains of *T. asperellum* and *T. atroviride* were effective against root-knot nematodes. Peat-bran preparations of the fungus were applied at 1% w/w in a series of experiments targeting *M. javanica* on tomato, and the results showed that: (i) plant growth improved and root galling was drastically reduced when the preparation was applied to small pots, whereas the carrier alone had no effect; (ii) similar effects were observed when the fungus was applied to 50-L containers filled with field soil naturally infested with root-knot nematode; (iii) the fungal population remained high throughout the experiment in 50-L containers, declining from an initial population of 10^6 cfu/g soil to 7×10^5 cfu/g soil after 3 months; (iv) 28 days after treatment, the number of female nematodes in *Trichoderma*-treated roots was markedly reduced and early developmental stages of the nematode were absent, indicating that the fungus had mostly affected the capacity of root-knot nematode to penetrate roots; (v) extracts from *Trichoderma*-treated soils inhibited the mobility of second-stage juveniles; (vi) the proteolytic activity of *Trichoderma*-derived soil extracts was much higher than the untreated control; and (vii) when tested *in vitro*, two wild-type strains of *Trichoderma* coiled around and penetrated second-stage juveniles and also colonized separated eggs, but were unable to penetrate egg masses or the eggs inside them. In a later study, the same isolates, together with other isolates of *T. asperellum* and *T. harzianum*, exhibited biocontrol activity in pot experiments (Sharon *et al.*, 2007).

In the years since the Israeli studies cited above, *Trichoderma* has been considered by others to be potentially useful against plant-parasitic nematodes. For example, an isolate of *T. harzianum* from Iran, when applied as a spore suspension, reduced galling and the number of egg masses produced on tomato, with the effect increasing as the spore concentration increased (Sahebani and Hadavi, 2008). Twelve days after *Meloidogyne graminicola* was inoculated onto rice seedlings that had previously been dipped and drenched with unidentified isolates of *Trichoderma* from Vietnamese rice-growing soils, three of nine isolates were found to have reduced galling relative to the untreated control (Le *et al.*, 2009). In a more comprehensive study in Belgium and Benin, 17 African isolates of five different species of *Trichoderma* were screened for antagonistic potential in pots, and the most promising isolates were then assessed in a double cropping system of tomato and carrot. Most isolates were able to colonize the rhizosphere and reduce nematode populations in pots, while some isolates (particularly *T. harzianum* T-16) reduced galling and increased the yield of tomato. However, in contrast to the results of Sharon *et al.* (2001, 2007), the level of galling was not markedly reduced (Affokpon *et al.*, 2011b). Endophytic isolates of *Trichoderma* have also been widely tested on tissue-cultured banana plantlets, and the results of that work indicate that they not only reduce the number of *Radopholus similis* in roots, but also improve plant growth (Chaves *et al.*, 2009).

In summarizing the current situation with regard to *Trichoderma* and nematode control, there is certainly evidence from short-term experiments under relatively controlled conditions that some isolates reduce nematode populations and the level of damage. However, in comparison to some other biological control agents, the fungus has not been widely studied and little is known about its efficacy in the field. Nevertheless, *Trichoderma* is a prime candidate for further development as a nematode-control agent for a number of reasons:

- *Trichoderma* products are widely available, so the commercial issues associated with production, formulation and registration have already been addressed.

- Large culture collections have been established and isolates in those collections have been characterized for a range of attributes (Stewart *et al.*, 2010), so it should be relatively easy to obtain isolates that may be useful in particular situations.
- Many *Trichoderma* species promote plant growth, so benefits will possibly be obtained in the absence of nematode control.
- *Trichoderma* is already used against other soilborne pathogens, and since they are often involved in disease complexes with plant-parasitic nematodes, it makes sense to target these pathogens collectively.
- *Trichoderma* is an opportunistic symbiont and many strains have the capacity to colonize plant roots, the same ecological niche that is occupied by plant-parasitic nematodes.
- Most of the mechanisms of action suggested for biocontrol activity against phytopathogenic fungi (e.g. parasitism, enzymatic lysis, antibiosis, stimulation of plant defence mechanisms) are also relevant to nematodes (Sharon *et al.*, 2011).

In view of these reasons, future research should not focus on developing *Trichoderma* products specifically for nematode control. Instead, it should aim to add nematode suppression to the plant growth promotion and disease-control properties already available.

Other fungi

Many fungal genera other than *Pochonia* and *Purpureocillium* have been isolated from nematode females, cysts and eggs, and although some of these fungi simply grow saprophytically on dead and injured nematodes, others are capable of parasitizing healthy nematodes. However, few have been seriously considered for inclusion in biological products for nematode control, and when that has occurred, little progress has been made down the product development pathway outlined in Box 12.3. For this reason, only two of the many other fungi found parasitizing nematodes or their eggs are discussed next.

BRACHYPHORIS AND THE ARF FUNGUS. *Brachyphoris oviparasitica* (syn. *Dactylella oviparasitica*) is an efficient parasite of root-knot nematode eggs and was found to be associated with suppressiveness to the nematode in a California peach orchard (see Chapter 10). Although the fungus parasitized eggs of *Meloidogyne incognita* when introduced into sterilized soil in pots, it was more effective as a biological control agent on peach, a relatively poor host of the nematode, than on more susceptible hosts such as tomato and grape (Stirling *et al.*, 1979). However, when an isolate was grown in liquid culture and applied to newly planted peach trees in the field, it failed to control the nematode (McKenry and Kretsch, 1987). Attention then turned to assessing its potential against *Heterodera schachtii*, as the fungus was also known to suppress this nematode in the field (see Chapter 10). Initial studies in the greenhouse showed that *B. oviparasitica* reduced populations of *H. schachtii* when introduced into fumigated and non-fumigated soils (Olatinwo *et al.*, 2006a, b), while follow-up studies in field microplots and the greenhouse indicated that a single application prior to planting kept population densities of *H. schachtii* at a low level for the entire cropping season, which comprised at least four or five nematode generations (Olatinwo *et al.*, 2006b).

The Arkansas fungus (often referred to as ARF or ARF18) is an unnamed sterile hyphomycete that was isolated in the late 1980s from a soil that was suppressive to *Heterodera glycines* (Kim and Riggs, 1991). Initial tests in the laboratory showed that it had the capacity to parasitize eggs of several *Heterodera* species, and also *Meloidogyne incognita*. Later surveys indicated that the fungus is widely distributed in southern soybean-growing states of the United States (Kim *et al.*, 1998), and that it parasitized 64% of the eggs of *H. glycines* in a bioassay of soil from a field where nematode populations were low following 14 years of nematode-susceptible soybeans (Wang *et al.*, 2004c). Isolates varied in pathogenicity to *H. glycines* (Kim *et al.*, 1998; Timper and Riggs, 1998; Timper *et al.*, 1999), and when alginate–clay pellets containing mycelium of ARF18 were applied to steam-sterilized soil in the greenhouse, nematode populations on soybean were reduced by 86–99% (Kim and Riggs, 1995). Experiments

with homogenized mycelium showed that in addition to parasitizing eggs, ARF isolates (particularly those designated ARF-L) also parasitized young females, and third and fourth stage sedentary juveniles (Timper *et al.*, 1999). Follow-up work in microplots confirmed that phenotypes of ARF differ in their capacity to suppress *H. glycines*, as the ARF-L isolate parasitized more eggs than the ARF-C isolate, particularly in the first year of the experiment (Wang *et al.*, 2004d).

Surveys of Arkansas cotton fields have also shown that ARF parasitizes eggs of reniform nematode (*Rotylenchulus reniformis*), and that some of these isolates differ from those previously obtained from *H. glycines*. One of those isolates suppressed the number of eggs and vermiform nematodes for two generations when homogenized mycelium was applied to heat-treated soil at application rates as low as 0.05% w/w (Wang *et al.*, 2004e).

Although the above studies were not set up to evaluate the potential of *B. oviparasitica* and ARF as inundative biological control agents, they showed that both fungi can be grown in culture and that some isolates will parasitize large numbers of root-knot, cyst and reniform nematode eggs in pots and microplots. Since the formulations used were relatively crude (biomass from liquid cultures, hyphae and spores from agar cultures, homogenized mycelium, or alginate pellets), product development work with better formulations is clearly warranted. In fact, there is a compelling case for in-depth studies on these fungi and their relatives because: (i) *B. oviparasitica* and the ARF fungus belong to the same clade of nematophagous fungi (Yang *et al.*, 2012b); (ii) isolates of both fungi vary in pathogenicity; (iii) some isolates are not only parasitic but are also good competitive saprophytes (Wang *et al.*, 2004c); (iv) both fungi have the capacity to suppress more than one economically important plant-parasitic nematode; and (v) suppressiveness associated with these fungi has been observed in diverse situations.

FUSARIUM. Species of *Fusarium*, particularly *F. oxysporum* and *F. solani*, are frequently recovered from eggs of root-knot and cyst nematodes (see Bursnall and Tribe, 1974; Nigh *et al.*, 1980a; Godoy *et al.*, 1983b; Morgan-Jones *et al.*, 1984b; Morgan-Jones and Rodrigeuz-Kabana, 1988; Chen *et al.*, 1994a; Chen and Chen, 2002; and Kokalis-Burelle *et al.*, 2005 for some examples). Isolates capable of parasitizing nematode eggs have been found (e.g. Nigh *et al.*, 1980a, b; Chen *et al.*, 1996b; Gao *et al.*, 2008), but there has never been much interest in commercializing them, largely because they have not proven to be highly virulent in laboratory and greenhouse tests. Some of the above isolates may have been capable of endophytic growth, but most recent studies with *Fusarium* have focused specifically on endophytes (see Chapter 8), as they colonize roots when applied to seed, transplants and other types of planting material, and are considered to have more potential for development as biocontrol agents. Research on *Fusarium* endophytes and nematodes is reviewed later in this chapter.

Pasteuria

Bacteria in the genus *Pasteuria* have many attributes that make them potentially useful as inundative biological control agents (see Chapter 7), but the most important are: (i) species and strains of *Pasteuria* are host specific and so they can be used to target particular nematodes; (ii) specialized systems of storage and handling are not required, as *Pasteuria* endospores are relatively resilient; (iii) high spore loads can prevent endoparasitic nematodes from invading roots; and (iv) once nematodes are infected by *Pasteuria*, they generally do not reproduce. Results of field experiments in Australia stimulated interest in this group of bacteria, as tomatoes grown in soil treated with dried-root preparations of *P. penetrans* were almost free of galls caused by *Meloidogyne javanica* (Stirling, 1984). In the following years, research summarized by Stirling (1991) indicated that root-knot nematodes could be controlled with spore concentrations of about 10^5 spores/g of soil. The parasite was effective against many species of root-knot nematode, with a review by Chen and Dickson (1998) showing that results were overwhelmingly positive in experiments with

five species of *Meloidogyne* (*M. acronea*, *M. arenaria*, *M. graminicola*, *M. incognita* and *M. javanica*). Numerous parameters were measured in those experiments (numbers of second-stage juveniles invading roots; levels of root galling; nematode reproduction; numbers of eggs per root system, root biomass and yield), and *P. penetrans* was consistently beneficial, sometimes almost eliminating the nematode from pots in the greenhouse.

Many host-specific populations of *Pasteuria* have now been identified (see Chapter 7), and there has been some interest in using these species and strains to control nematodes other than root-knot nematode. The role of *Pasteuria nishizawae* in suppressing populations of *Heterodera glycines* was discussed in Chapter 10, while the use of *Candidatus* Pasteuria usgae for control of *Belonolaimus longicaudatus* on turfgrass will be covered later. However, because of its economic importance, root-knot nematode remains the main target of biological control studies with *Pasteuria*.

IN VIVO CULTURE. In the absence of *in vitro* production methods for *P. penetrans*, most of the work on biological control of root-knot nematode has been based on a mass-production system in which a host plant is inoculated with spore-encumbered juveniles, and when female nematodes have developed to maturity and are filled with endospores, the root systems are dried, ground into a fine powder, and stored for future use (Stirling and Wachtel, 1980). Spores in root systems can be quantified in many different ways (Chen *et al.*, 1996c), but concentrations $>1 \times 10^8$/g root are consistently obtained with this method (Sharma and Stirling, 1991). Spore yields may be improved by ensuring that appropriate numbers of spores are attached to inoculated nematodes (Hewlett and Dickson, 1993; Bu *et al.*, 2011) and through attention to issues such as nematode inoculum density, plant host, size of inoculated plants, harvest date and greenhouse temperature (Darban *et al.*, 2004; Cho *et al.*, 2005; Gowen *et al.*, 2008). Also, clean endospore suspensions can be obtained for experimental work by either sieving the powdered preparation to remove root material (Ciancio and Bourijate, 1995; Netscher and

Duponnois, 1998), or by retrieving spore-filled nematodes from roots using enzymatic maceration techniques (Dabiré et al., 2001a; Cetintas and Dickson, 2005).

Over the years, glasshouse production methods such as those described have been widely used to produce inoculum for small-scale experiments, but have generally been considered too expensive for commercial use. However, an in vivo production system was commercialized in 1995 by Nematech Ltd, a Japanese company. P. penetrans was mass-produced on root-knot nematode in heated greenhouses, but rather than applying dried-root powder to soil, the company developed a method of retrieving spores from roots, suspending them in water and applying the spores to soil. The product was sold in liquid form, had no special storage or handling requirements, and was marketed as a soil-improvement agent for horticultural crops. Information supplied by the company indicated that 2 years after P. penetrans was applied to crops such as tomato and cucumber, root-knot nematode populations decreased, and the level of galling caused by the nematode also declined. By the third year, yield was significantly better than the control and in the fourth year, yields in P. penetrans treatments were equal to or superior to a nematicide treatment. However, data to support these claims have never been published.

Field experiments in Japan with commercial inoculum obtained from Nematech indicated that high application rates of P. penetrans (5×10^{10} spores/m^2) resulted in M. incognita juveniles being heavily encumbered with spores for the duration of a tomato crop and after a 6 month fallow (Talavera et al., 2002). Damage caused by the nematode was reduced, and yield increased by 45% in the first year after spores were applied. Estimates of spore numbers at planting supported previous observations that spore concentrations of about 10^5 spores/g soil were required to achieve satisfactory control of root-knot nematode. At lower application rates, spore concentrations in the upper 10 cm of the soil profile increased after planting, and were greater than 2.5×10^4 spores/g soil at harvest. Importantly, spore concentrations were still at that level following a 6-month winter fallow (i.e. nearly 11 months

after the inoculum was applied). This result indicates that when P. penetrans is introduced at application rates that are not sufficient to achieve control, spore numbers will increase when a host of the nematode is planted. The data also showed that in a climate where vegetable cropping is restricted to the spring–summer–autumn period, spore concentrations can still be relatively high when the next crop is due to be planted. In this situation, only relatively low levels of inoculum would have to be applied to achieve control in the second crop.

Many other examples of the use of in vivo-produced spores of P. penetrans for biological control of root-knot nematodes could also be discussed, but a few results from experiments carried out in the field, or in field microplots, will be briefly mentioned here. The first involves a series of studies in the United States, which showed that P. penetrans is a useful agent for suppressing Meloidogyne arenaria race 1 on peanut. In an experiment in which peanut was grown in summer, and the soil was bare fallowed in winter (or cover crops of rye or vetch were grown), populations of P. penetrans increased over time from the relatively low levels added initially. Although peanut yield was not affected in the first 2 years, yield increased by 64% in the third year, indicating that it may be feasible to use in vivo-produced endospores in integrated nematode management schemes (Oostendorp et al., 1991b). Later studies showed that when P. penetrans was introduced into microplots at 10^5 endospores/g soil, root and pod gall indices were reduced by 60% and 95% in the first year, and by 81% and 90% in the second year, respectively. A lower inoculum level (10^4 endospores/g soil) also gave good control, particularly in the second year, when pod yield was significantly increased by both treatments (Chen et al., 1996d). In another experiment, suppression was proportional to infestation levels of P. penetrans, and was not affected by nematode population densities (Chen et al., 1997b).

Experiments in Ecuador and Tanzania (Gowen et al., 1998) demonstrated that root powder formulations of P. penetrans provided satisfactory control of root-knot nematode when applied at a relatively low application rate (about 10^3 spores/g soil). In Ecuador, six successive crops (two cycles of *Phaseolus*, *Phaseolus* and tomato) were cultivated over a 30-month period and the addition of *P. penetrans* reduced levels of galling and the numbers of nematodes in soil, particularly on the third to sixth plantings. In the Tanzania experiment, where five successive tomato crops were grown, *P. penetrans* reduced root galling and nematode numbers, and increased the yield of every crop. In another example from Lebanon, a root powder formulation of *P. penetrans* was applied at application rates up to 4×10^4 spores/g soil to a commercial tunnel house, and three successive crops of cucumber were grown. Roots were macerated and reincorporated into the soil at the end of each crop, and the result showed that the number of spores attached to second-stage juveniles increased with application rate and the number of crop cycles (Fig. 12.8). After three crop cycles, nematodes extracted from soil inoculated with the highest application rate had an average of eight spores attached (Melki *et al.*, 1998). In a similar experiment in a Moroccan greenhouse naturally infested with *M. javanica*, tomato, melon and tomato were grown in succession but roots were left to decompose in the soil following each crop. A single application of a dried-root preparation of *P. penetrans* resulted in high levels of parasitism in the second tomato crop, and lower

gall ratings and increased yield relative to the untreated control (Eddaoudi and Bourijate, 1998). Therefore, this group of experiments demonstrates that in situations where nematicides and biopesticides may be prohibitively expensive, *P. penetrans* can be introduced into a field using simple *in vivo* production systems, and populations of the parasite can be increased and then maintained by growing crops that are susceptible to root-knot nematode. Since *P. penetrans* has also been introduced to a new site through the transfer of infested root material (Kariuki and Dickson, 2007), there is clearly potential to build on these observations and include *P. penetrans* in integrated nematode management systems. Approximately 10^5 endospores of *P. penetrans*/g soil are required to achieve control (Stirling *et al.*, 1990; Chen *et al.*, 1996d), and future research should focus on maintaining these levels of the parasite through rotation practices; finding treatments that minimize nematode damage during the first few critical weeks after a vegetable crop is planted but do not markedly affect multiplication of the parasite; and then evaluating the effects of such treatments when used together in an integrated manner. Since *P. penetrans* will survive when soil is solarized or fallowed for short periods, and is not affected by some of the nematicides and other pesticides commonly

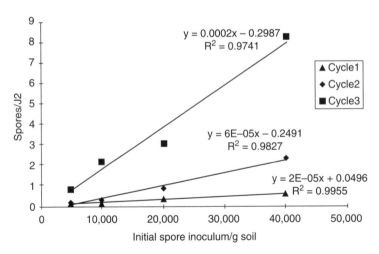

Fig. 12.8. Effect of the initial rate of application of *Pasteuria penetrans* and root reincorporation on the number of spores attached to second-stage juveniles (J2) of *Meloidogyne* spp. over three cucumber crop cycles. (From Melki *et al.*, 1998, with permission.)

used in vegetable production, numerous options are available for inclusion in such systems.

IN VITRO CULTURE. Given the limitations of mass-production methods that require the presence of a nematode host, *in vitro* production of *Pasteuria* was always seen as the most likely route to commercialization. In one of the first attempts to culture the organism, spores or vegetative mycelial bodies of *P. penetrans* were inoculated onto a diverse range of simple and complex media and incubated in a range of different environments, but spore germination or growth was never observed (Williams *et al.*, 1989). A later study resulted in the development of endospores, but exponential growth was not obtained (Bishop and Ellar, 1991). Rhizosphere bacteria such as *Enterobacter cloacae* were then found to promote the development of *P. penetrans* in plants infected with *M. incognita* (Duponnois *et al.*, 1999). Interestingly, this bacterium, and recognizable structures of *P. penetrans*, were found together in media that had been inoculated with *P. penetrans* (Hewlett *et al.*, 2004). Culture filtrate of *E. cloacae* was then found to support growth of *P. penetrans*. The critical chemical components and characteristics of the culture medium supplied by the 'helper'-bacterium were identified, and a defined growth medium suitable for commercial production of *P. penetrans* was developed (Hewlett *et al.*, 2004; Gerber *et al.*, 2006). Pasteuria Bioscience, a company located in Florida, then developed products based on this technology until it was acquired in 2012 by Syngenta, one of the world's largest crop-protection companies.

Results of some of the first tests with *in vitro*-produced *P. penetrans* indicated that at an endospore density of 10^5 spores/mL soil, galling caused by *M. incognita* was reduced significantly compared with the untreated control (Hewlett *et al.*, 2006). Encouragingly, the effective spore concentration was similar to that observed previously with endospores produced *in vivo*. However, attention then turned to evaluating the efficacy of *Candidatus* Pasteuria usgae against sting nematode (*Belonolaimus longicaudatus*), as it was a major

problem on turfgrass and other major crops in the sandy coastal plains of the south-eastern United States. *In vitro* methods of producing endospores of this species of *Pasteuria* were developed, and these endospores were able to attach to and infect *B. longicaudatus* (Hewlett *et al.*, 2008a, b). However, *in vitro*-produced endospores were more variable in size than endospores produced *in vivo*, presumably because endospores were at different stages of sporogenesis and peripheral fibre development when the *in vitro* fermentation process was halted (Luc *et al.*, 2010a).

When *Candidatus* P. usgae was tested in the field at application rates as high as 33×10^9 endospores/m², some formulations reduced sting nematode populations 15 days after application, whereas others did not (Hewlett *et al.*, 2008c). However, the biological control agent was effective against sting nematode on creeping bentgrass (*Agrostis palustris*) in a growth-chamber study in which the nematode was inoculated 13 days after endospores were incorporated into sand at concentrations ranging from 0.28×10^5 to 2.8×10^5 endospores/mL of soil. Populations of *B. longicaudatus* declined as inoculum levels increased, with the highest inoculum level reducing populations by 47%, 60% and 74% at 28, 56, and 84 days after the nematode was inoculated, respectively. Sporangium size did not affect the level of nematode suppression (Luc *et al.*, 2010a). A later test comparing liquid with granular formulations at a higher application rate (1.38×10^6 endospores/mL of soil) showed that the liquid formulation was more effective, reducing populations of *B. longicaudatus* by 59% and 63% in two experiments, and increasing endospore attachment to the nematode (Luc *et al.*, 2010b). However, because the experiment was terminated after 84 days, it was not possible to determine whether the parasite had any long-term effects on the nematode population.

In June 2009, a product with *Candidatus* P. usgae as its active ingredient was registered in the United States, and Econem™ (Pasteuria Bioscience, Alachua, Florida) biological nematicide was launched the following year. Econem™ contained *Pasteuria* endospores in a clay carrier (0.002%), and according to the manufacturer, 1×10^5 and 3×10^5 endospores/mL were added to the top 5 cm of soil when the product was applied at label rates of 98 and 293 kg/ha, respectively (Crow *et al.*, 2011). Although application rates of this magnitude had reduced populations of sting nematode in growth-chamber studies discussed previously (Luc *et al.*, 2010a), the product proved ineffective in the field. When applied at recommended rates in experiments on golf-course putting greens; fairways and tee boxes; and in pots, Econem™ reduced population densities of *B. longicaudatus* on only a single sampling date in one of eight field trials (Crow *et al.*, 2011). Turf health did not improve in any of the trials.

In attempting to explain the discrepancy between the negative results obtained in the field and the positive results previously obtained with *in vitro*-produced endospores in pots, Crow *et al.* (2011) made several important observations. First, Econem™ was based on an isolate of *Pasteuria* that was morphologically similar to *Candidatus* P. usgae, but its identity had not been confirmed with molecular sequencing data. Second, endospores used previously for pot experiments (Luc *et al.*, 2010 a, b) were produced using laboratory-scale fermenters, whereas Econem™ was produced in commercial fermenters and formulated into a commercial granular product. Thus, it is possible that during the scale-up in production, or during the formulation process, endospores had lost virulence. Third, visual improvements in the turfgrass were observed in one of the greenhouse experiments, but evidence that *Pasteuria* endospores attached to or infected *B. longicaudatus* was not obtained, suggesting that the high rates of clay included in the formulation may have improved plant growth. Fourth, the endospores in Econem™ were incorporated into the clay carrier, and the rate at which they were released into the environment from the carrier under field conditions was not known. Finally, previous studies had shown that *Pasteuria* endospores are readily leached down the profile in the sand specified for use on golf courses in the United States (Luc *et al.*, 2011), and so it is possible that irrigation and rainfall moved the endospores beyond the upper layers of the soil profile, where sting nematodes are mainly found. Therefore, these observations support the comments made earlier about the need to confirm the efficacy of registered biopesticide products by thoroughly testing them in the target cropping system (Box 12.3).

Although the development of *in vitro* methods of producing *Pasteuria* endospores is an exciting advance, there are clearly many issues that will need to be addressed before *Pasteuria*-based products find a place in nematode management programmes. Some of the more important issues are discussed next.

COMMERCIAL PRODUCTION AND FORMULATION.   In the ten years since *Pasteuria* was first cultured *in vitro*, there have been major improvements in the way the biological control agent is mass-produced, and these improvements will continue as more is learnt about its nutritional requirements. However, if *Pasteuria*-based products are to be consistent and efficacious, there is also a need for research at a commercial level, particularly with regards to fermentation and formulation methods.

TARGET CROPS AND NEMATODES.   Given that most crops experience nematode problems; that those problems are caused by a wide range of plant-parasitic nematodes; and that strains of *Pasteuria* are known to parasitize most nematode pests; commercial decisions will have to be made on which nematodes and crops are targeted by *Pasteuria*-based products. The decision to initially target sting nematode on turfgrass is interesting, because it raises questions about the situations in which a product such as Econem™ would be economically viable. On an area basis, golf-course putting greens in Florida are perhaps the most valuable pieces of agricultural real estate in the world, and so it may be feasible to apply *Pasteuria* at rates of 3×10^{10} endospores/m². However, it remains to be seen whether application rates of this magnitude are an option for agricultural crops.

ENDOSPORE QUALITY.   Observations by Luc *et al.* (2010a) indicated that *Pasteuria* endospores produced in commercial fermenters exhibited poor peripheral fibre development, were often smaller than natural endospores, and did not always attach readily to the host nematode. Thus, research on improving the quality of *in vitro*-produced endospores must continue. Steps must also be taken to ensure that *Pasteuria* maintains its virulence through many generations of *in vitro* culture.

CONTROL APPROACHES. Given that the high cost of *Pasteuria*-based products is likely to limit their use for inundative biological control, alternative approaches will have to be considered. An inoculative or classical approach is likely to be more feasible from an economic perspective, but if such an application strategy was used, the question to be answered is whether, under the cropping practices currently in place, *Pasteuria* will increase to levels capable of suppressing the target pest. There are many situations where *Pasteuria* already occurs naturally but does not suppress nematode populations or the amount of damage to levels acceptable to land managers, and there is no reason to believe that endospores produced *in vitro* will be more effective.

APPLICATION METHODS. One way of reducing application rates of *Pasteuria* to economically acceptable levels would be to apply endospores to seeds, transplants or other planting material. However, roots grow from the planting material and are then attacked by nematodes, and so such application methods are unlikely to deliver enough endospores at the appropriate site to control nematodes on the crop to which *Pasteuria* is applied. Nevertheless, they may be a suitable way of introducing *Pasteuria* into a field. However, such inoculative methods are unlikely to be effective in the long term unless endospore populations are increased by modifying the farming system to include more crops capable of hosting the nematode.

HOST SPECIFICITY. It has been recognized for many years that endospores of *Pasteuria* do not attach to all populations of a given nematode species, and that nematodes encumbered with spores do not always become infected (see Chapter 7). Also, endospores of some populations of *Pasteuria* only attach to and infect particular nematodes, while other

populations have a much broader host range, attaching not only to the nematode from which they were isolated but also to nematodes from a different genus. Since major nematode pests such as root-knot nematode frequently occur as mixed populations of two or three species, and crops are often attacked by nematodes in different genera, the specificity of *Pasteuria* will always be a stumbling block to using it as a biological control agent. It remains to be seen how this issue is best handled. One option is to select strains with a broad host range, while another is to include multiple strains within one product. Perhaps the ultimate solution is to develop a simple test to 'fingerprint' the target nematode population, use the data to determine the range of *Pasteuria* isolates likely to be efficacious in a given situation, and then prepare a 'designer' bionematicide based on that information.

FATE OF ENDOSPORES APPLIED TO SOIL. Although *Pasteuria* endospores must remain in the root zone to be effective, they are readily leached to lower depths by percolating water, particularly in soils with a high sand content (Oostendorp *et al.*, 1990; Mateille *et al.*, 1996; Talavera *et al.*, 2002; Dabiré and Mateille, 2004; Dabiré *et al.*, 2005a; Cetintas and Dickson, 2005: Luc *et al.*, 2011). Also, endospores are almost certainly subject to predation. Little is known about the biotic factors that influence endospore survival (Chen and Dickson, 1998), but Talavera *et al.* (2002) suggested that amoebae and filter-feeding rotifers may have been responsible for reducing endospore densities when *P. penetrans* was applied at high rates. Although little can be done to prevent endospores from being leached downwards in heavy rainfall events, there is a need for research on the fate of endospores applied to soil, as it may lead to the identification of management practices that can be used to minimize endospore losses.

Predatory and entomopathogenic nematodes, and microarthropods

Small (1987) prepared a thorough review on the feeding habits of predatory nematodes and concluded that because of our lack of knowledge of predation rates and their pre-ferred prey, it was difficult to estimate the potential of predatory nematodes as biological control agents. However, he commented in a later review (Small, 1988) that a 'magic bullet' control agent was unlikely to be found. Since then, it has generally been agreed that of the four main groups of predatory nematodes (mononchids, dorylaimids, diplogasterids and tylenchids), diplogasterids are most suited for inundative biocontrol of nematodes (Khan and Kim, 2007). Low fecundity, canni-balism, susceptibility to changing environ-mental conditions and culturing difficulties restrict the use of mononchids and stylet-bearing predators, whereas diplogasterids are easy to culture, have a relatively short life cycle, reproduce rapidly, have high rates of predation and rarely cannibalize each other (Siddiqi *et al.*, 2004; Bilgrami *et al.*, 2005, 2008). Assertions that diplogasterids may be effective biocontrol agents are supported by test results showing that *Mononchoides fortidens* and *Koerneria sudhausi* reduced galling caused by root-knot nematode in pots (Khan and Kim, 2005; Bar-Eyal *et al.*, 2008).

In the first report of a predatory nema-tode being field-released for control of plant-parasitic nematodes, Bilgrami *et al.* (2008) set up 10 cm-diameter microplots in a turfgrass field and applied 2000 laboratory-cultured *Mononchoides gaugleri* to each microplot (the equivalent of 250,000 nematodes/m^2). After 30 days, populations of plant-parasitic and non-parasitic nematodes were reduced by an average of 22% and 10%, respectively. On the basis of these results, claims were made that *M. gaugleri* possesses significant biocontrol potential. However, given the relatively low level of control obtained and the long list of issues that must be addressed in commercial-izing a biological control agent (Box 12.3), such a claim is premature and overly optimis-tic. Nevertheless, further research is war-ranted, as other closely related nematodes (e.g. *Steinernema* and *Heterorhabditis*) are pro-duced commercially and are widely used for insect control in high-value crops.

With regard to the entomopathogenic nematodes *Steinernema* and *Heterorhabditis*, there have also been reports in the literature of their potential to control plant-parasitic nematodes. For example, *H. bacteriophora* reduced populations of *Tylenchorhynchus* spp. and *Pratylenchus pratensis* on turfgrass under irrigated conditions in the field (Smitley *et al.*, 1992); *S. riobrave* was detrimental to plant-parasitic nematodes in another turf-grass study (Grewal *et al.*, 1997); and two species of *Heterorhabditis* reduced the total number of plant-parasitic nematodes in an experiment on golf-course turf (Somasekhar *et al.*, 2002). There have also been reports of entomopathogenic nematodes having a negative impact on root-knot nematodes in greenhouse studies (e.g. Perez and Lewis, 2004; Molina *et al.*, 2007). Various hypotheses have been advanced to explain these effects, including competition with plant parasites for space in the rhizosphere; stimulation of naturally occurring parasites and predators; and the release of toxic metabolites by either the nematodes or their symbiotic bacteria. However, the mechanisms involved are still not understood.

Despite promising results with ento-mopathogenic nematodes in these studies, there have also been many situations where results have been poor or inconsistent. Thus, these nematodes had little effect on a range of plant-parasitic nematodes in ten experi-ments on golf courses in Florida (Crow *et al.*, 2006); they did not suppress ring nematode (*Mesocriconema xenoplax*) on peach in pots or pecan in field microplots (Nyczepir *et al.*, 2004); and they were only marginally effective against the pecan root-knot nematode (*Meloi-dogyne partityla*) in the greenhouse (Shapiro-Ilan *et al.*, 2006). Nevertheless, further research is warranted because effects on plant-parasitic nematodes have been observed at an applica-tion rate of 25 infective juveniles/cm^2 or the equivalent of 2.5×10^9 infective juveniles/ha, a rate considered effective for insect control (Perez and Lewis, 2004). If consistent levels of efficacy could be demonstrated, it would be relatively simple to expand the use of entomopathogenic nematodes to include plant-parasitic nematodes, as product devel-opment costs have already been incurred.

Although mites are widely commer-cialized for above-ground release against pest mites of foliage, there have been no attempts to mass-produce mites or other microarthropods and apply them to soil for nematode control. There are almost certainly species that are voracious predators of nematodes, and they could probably be multiplied in one of the many systems that are available to culture bacterial, fungal or plant-parasitic nematodes. However, the critical questions are whether effective mass-production methods can be developed, and whether microarthropods can be established in agricultural soils that often lack the pore spaces and surface cover of organic matter required by microarthropods as a habitat (see Chapter 6).

Plant growth-promoting rhizobacteria and endophytes

Several endophytes and symbionts that act against plant-parasitic nematodes through a variety of mechanisms, or enable the plant to better tolerate attack by nematodes, were discussed in Chapter 8. Most of these bacteria and fungi live in the rhizosphere or grow endophytically in root tissue and are prime candidates for use in integrated nematode management programmes for a number of reasons: (i) they are naturally associated with plant roots, having coevolved with plants to form mutualistic associations that benefit both the organism and the plant; (ii) they colonize the same ecological niche as plant-parasitic nematodes; (iii) many have growth-promoting properties; (iv) endophytes are often absent from the roots of crop plants, either because agricultural practices such as soil fumigation, intensive tillage and pesticide treatment of seed are detrimental, or because transplants and tissue-cultured plantlets used in vegetables and other horticultural crops are produced under relatively sterile conditions; (v) most can be readily cultured *in vitro*; and (vi) they can easily be applied to seeds, transplants and other planting material. The full range of organisms tested against nematodes is too great to cover here, so this part of the chapter focuses on bacteria and fungi that have been widely researched, or have shown

biological control potential in extensive field or greenhouse tests.

Rhizobacteria and bacterial endophytes

There is a large body of literature on plant-associated bacteria that are capable of stimulating plant growth and reducing losses from soilborne diseases (reviewed by Hallmann *et al.*, 1997; Sturz *et al.*, 2000; Welbaum *et al.*, 2004; Fravel, 2005; Compant *et al.*, 2005, 2010; Rosenblueth and Martinez-Romero, 2006; Ryan *et al.*, 2008; Saharan and Nehra, 2011; Pliego *et al.*, 2011; Maksimov *et al.*, 2011; Bhattacharyya and Jha, 2012). Often referred to separately as plant growth-promoting rhizobacteria (PGPR) and bacterial endophytes, they are discussed together here because (i) there is a continuum of root-associated organisms from the rhizosphere to the rhizoplane to the epidermis to the cortex (Kloepper *et al.*, 1992b); and (ii) endophytic bacteria in roots are mostly derived from the rhizosphere (Compant *et al.*, 2010).

Large-scale screening programmes are almost certain to detect bacterial isolates that are antagonistic to plant pathogens (see Becker *et al.*, 1988, and Neipp and Becker, 1999, for examples with nematodes), and studies summarized in Chapter 8 give an indication of the range of bacteria that has been investigated. Most of the focus in the plant pathology literature has been on strains of *Pseudomonas* and *Bacillus*, with both of these genera having been tested against many different soilborne pathogens (Kloepper *et al.*, 1991, 1992a, 2004; Weller *et al.*, 2002; Jacobsen *et al.*, 2004). However, nematodes have mainly been targeted with *Bacillus* species (Meyer, 2003; Tian *et al.*, 2007), presumably because they are readily isolated from soil and are easily cultured. Also, their spores are durable, and so it is relatively easy to formulate them into biocontrol products (Schisler *et al.*, 2004).

The *Bacillus* species screened for activity against nematodes include *B. subtilis*, *B. firmus*, *B. amyloliquefaciens*, *B. sphaericus*, *B. thuringiensis*, *B. megaterium* and various strains of unnamed species. However, *B. firmus* is the most widely studied species, as it is registered as a biopesticide in some countries, with one isolate being marketed under the trade-name Nortica™

by Bayer CropScience (Monheim, Germany), a major agro-industrial company. Laboratory studies have shown that bacteria-free extracts of *B. firmus* reduce mobility and cause paralysis of *Radopholus similis, Meloidogyne incognita* and *Heterodera glycines* (Mendoza *et al.*, 2008; Schrimsher *et al.*, 2011), while secondary metabolites produced during fermentation, and aqueous suspensions of a commercial formulation of *B. firmus*, both reduce hatch of *M. incognita* (Mendoza *et al.*, 2008; Terefe *et al.*, 2009). Such results suggest that bioactive compounds produced by *B. firmus* may play a role in the control of plant-parasitic nematodes, but this is yet to be confirmed.

Initial studies with one commercial formulation of *B. firmus* originally developed in Israel and later marketed as BioNem™ by Bayer CropScience have produced promising results. In Greece, for example, the product reduced populations of root-knot nematode and levels of galling on cucumber in commercial greenhouses when it was applied in a 20-cm band over the planting row at the recommended rate of 70 g/m row (Giannakou *et al.*, 2004). In a later trial in the same cucumber production system, BioNem™ also provided some root-knot nematode control but was not as effective as solarization or the nematicide dazomet (Giannakou *et al.*, 2007). The product was also effective in Ethiopia, as it reduced galling by more than 75% and increased plant biomass by about 20% when applied at 200 and 400 kg/ha (Terefe *et al.*, 2009). Unfortunately, impacts on yield were not reported in any of these studies. Also, BioNem™ consists predominantly of animal and plant extracts (97% of the product weight), and it is not clear how much of the response was due to the direct or indirect effects of these organic materials.

Given the results already obtained with BioNem™, formulations containing *B. firmus* will almost certainly be tested on more crops and in different soils and environments, and its mechanisms of action will be further evaluated. Since a strain of *B. firmus* applied to cotton seed at 1.4 mg spores per seed protected cotton roots from reniform nematode for 30 days in field soil under greenhouse conditions (Castillo *et al.*, 2011), its potential as a seed treatment is also likely to be examined. The results of such

research will provide a much better indication of whether products based on *B. firmus* have a role to play in future integrated nematode management programmes.

One area where PGPR and bacterial endophytes clearly have potential is in the treatment of transplants used for vegetable production. The level of galling caused by root-knot nematode was not always reduced when tomato and pepper seedlings were grown in a potting mix containing chitin and different combinations of *Bacillus* species and then transplanted into a nematode-infested field, but most formulations improved plant growth and some also improved root health (Kokalis-Burelle *et al.*, 2002b). One of the formulations tested (*Bacillus subtilis* strain GBO3 + *B. amyloliquefaciens* strain IN973a) was commercialized as BioYield™ (Gustafson LLC, Plano, Texas), and given time and an appropriate amount of research, it is likely that other products, perhaps with greater effects on nematodes, will be developed.

Although other studies with PGPR for nematode control (summarized by Timper, 2011) have produced both positive and negative results, there are so many potentially useful bacteria in the rhizosphere that research in this area is certain to continue. Nine endophytic bacteria were listed by Sikora *et al.* (2007) as having been investigated for biological control of nematodes, and given that most plants are known to host at least one bacterial endophyte, and almost nothing is known about their effects on nematodes, this will become an increasingly important area of research. One bacterial group that deserves particular consideration is the actinobacteria, as they are known to produce bioactive compounds, trigger induced systemic resistance, and also have the capacity to colonize roots (Franco *et al.*, 2007; Qin *et al.*, 2011). Ultimately, a plethora of commercial products based on bacteria is likely to be developed, and they will be marketed on the basis that they promote plant growth or control nematodes and other soilborne pathogens. Given the range of situations in which they could be used, it will take time and a considerable amount of research to evaluate these products and determine whether they are effective in diverse environments. Introduced bacteria will be infected by bacteriophages and preyed on by protozoa and bacterial-feeding nematodes, and so their efficacy will almost certainly be affected by the presence of these indigenous competitors.

Arbuscular mycorrhizal fungi

There have been many studies on interactions between arbuscular mycorrhizal fungi (AM fungi) and plant-parasitic nematodes (Hallmann and Sikora, 2011), and although the effects of these fungi vary with environment, plant genotype, nematode species and fungal isolate, they are generally considered to slow nematode development and enhance host tolerance to nematodes. Analyses of data in 65 papers published after 1996 (Hol and Cook, 2005) confirmed these beneficial effects, as AM fungi offset the effects of nematode damage on shoot biomass by an average of 7%, and reduced nematode numbers by an average of 21%. However, ectoparasites tended to cause more damage to plants infected by AM fungi than to non-mycorrhizal plants, and numbers of migratory endoparasitic nematodes increased on plants colonized by AM fungi. Nevertheless, mycorrhizal infection generally reduced numbers of sedentary endoparasites, with the effect more apparent for root-knot nematodes than cyst nematodes. Since plants affected by diseases caused by fungal and oomycete pathogens also respond to mycorrhizae (see review by Whipps, 2004), AM fungi clearly have a role to play in improving plant health and reducing damage caused by nematodes and other soilborne pathogens.

Although many studies have shown that AM fungi either reduce nematode populations or improve plant growth when inoculated onto nematode-infested plants in sterilized soil in a greenhouse, evidence of efficacy in the field is much harder to find. Thus, a paper by Affokpon et al. (2011a) is of interest, as it addresses some of the issues that must be considered when AM fungi are used to reduce damage caused by plant-parasitic nematodes. Twenty strains of AM fungi native to West Africa and three commercially available species of Glomus were multiplied on sorghum plants, and their capacity to control root-knot nematodes was evaluated in separate pot

experiments, one with sterilized soil and one with non-sterilized soil. Results from the experiment in sterilized soil showed that many of the native strains suppressed nematode reproduction, and when the level of galling was averaged across all native strains, the gall rating (on a 0–10 scale) was reduced from 7.5 to 6.2. Nematode populations were much higher in non-sterile soil, but again, most of the native strains reduced nematode reproduction and the level of galling.

Based on the results of these pot tests, four native strains were selected and tested in a field experiment in which a tomato crop was followed by carrot. Some of those strains reduced nematode population densities in roots and soil, with one of them (a strain of Kuklospora kentinensis) suppressing egg production on tomato, and numbers of juveniles in soil and roots by 89%, 84% and 41%, respectively. In the follow-on carrot crop, three of the four AM fungi significantly suppressed numbers of nematodes in soil compared with the non-mycorrhizal control. Crop performance was also enhanced, as inoculation with AM fungi increased tomato yields by 26% and carrot yield by over 300%.

Although these results are encouraging, there are important lessons to learn from these experiments:

- Native strains from West Africa were consistently superior to the imported commercial strains, suggesting that adaption to local environmental conditions may be important for success.
- Differences in efficacy were observed among native strains, and also between strains of the same species, indicating that strain selection is likely to be important.
- Strains that were effective in pots were not necessarily effective in the field, raising questions about the most appropriate method for selecting useful strains.
- The biocontrol effect of AM fungi is not dependent on the level of mycorrhizal colonization, although the results of this and other work suggest that root mycorrhization must be above a threshold level of about 30%.
- The benefits observed over two crop cycles indicated that the positive effects of AM fungi can carry over to the next crop,

raising questions about whether this effect could be further enhanced by minimizing the disturbance factors known to be detrimental to these fungi.

This study is significant, because it demonstrates that AM fungi could possibly play a role in future management programmes for root-knot nematode on vegetable crops. However, it is also important to be realistic about what is achievable. In all three experiments, the better performing strains only reduced galling by about one rating unit, suggesting that AM fungi are unlikely to provide more than a limited level of nematode suppression. In fact, it is possible that the other functions of AM fungi (e.g. improved nutrient uptake) may have been responsible for some of the benefits, as the observed yield responses were much higher than expected, given the relatively small impact of mycorrhizal colonization on root galling.

One limitation of AM fungi is that they are obligate parasites, and so they must be grown on plants. Nevertheless, products containing AM fungi are available commercially. Inoculum is usually produced in pot culture, but soilless culture systems such as aeroponics, hydroponics and nutrient film techniques are also used (Hung and Sylvia, 1988; Dalpé and Monreal, 2004; Lee and George, 2005). However, high production costs limit the sale of AM fungi to small niche markets, while the presence of colonized root segments that are potentially contaminated with pathogens is a major limitation to the more widespread commercialization of these fungi. Although there is potential to expand the use of mycorrhizae to the biopesticide market, registration hurdles would have to be overcome if claims of disease-control activity were made. Mycorrhizal products are, therefore, deliberately sold as growth promoters, biofertilizers, soil conditioners and biological activators. Thus, mycorrhizae may very well be acting to some extent as biocontrol agents, but promoting them in that way could limit rather than enhance their commercial availability (Whipps, 2004).

Another issue that must be considered by anyone arguing the case for using AM fungi to control nematodes and other soilborne pathogens is that the effects of AM fungi on plants span the continuum from mutualism to parasitism, while the outcome of the symbiosis is affected by many factors, including host plant characteristics, fungal characteristics, soil biotic and abiotic conditions, and experimental procedures. Hoeksema *et al.* (2010) used meta-analyses to better understand this variability in outcomes and made several important observations: (i) the identity of the host plant was one of the most important predictors of plant responses to mycorrhizal inoculation, with non-nitrogen-fixing forbs, woody plants and C_4 grasses responding more positively to mycorrhizal inoculation than C_3 grasses and plants with nitrogen-fixing bacterial symbionts; (ii) plant response was influenced by nitrogen and phosphorus availability; and (iii) AM fungi had greater effects on plants when the inoculum contained multiple species and the symbiosis took place in a diverse soil biological community. These points are important because they indicate that the impact of inoculants containing AM fungi is likely to vary with crop species and a soil's nutritional status, and that results of experiments with single species in sterilized potting media may not reflect what is likely to occur under field conditions.

In conclusion, it is clear that AM fungi benefit plants in many different ways, and that these fungi should, therefore, be thought of holistically as general promoters of plant health (see Chapter 8). Since crops are usually colonized by mycelial growth from existing spores and hyphal networks, the focus of future research should be on ensuring that plants are colonized by the appropriate mycorrhizal species and isolates rather than on mass-production and inoculation. Mycorrhizal inoculants are likely to be most useful in crops that are transplanted (e.g. vegetables, bananas and fruit trees), but even when AM fungi are applied in these situations, the primary objective should be to improve plant growth rather than to control nematodes and other soilborne pathogens. Nematode populations may sometimes be reduced and mycorrhizal plants may be better able to tolerate the effects of nematode attack, but more efficient nutrient use and better plant health must always be the primary reasons for inoculating crops with AM fungi.

Fusarium *endophytes*

Since infection by nematodes during the early stages of plant growth can have a major impact on subsequent growth and yield, one of the most important principles of nematode management is that the roots of a crop must be protected from attack for at least the first few weeks after planting. Thus, one vital feature of any nematode control agent, whether it is chemical or biological, is that it must limit damage to roots as they grow into the hemisphere of soil within 25–50 cm of the base of the plant (Sikora *et al.*, 2007). From a biological control perspective, one of the best ways of achieving this is to use fungal endophytes to enhance the plant's resistance to nematodes, an approach that has been termed 'biological enhancement' (Sikora *et al.*, 2008).

Numerous endophytic fungi are associated with plant roots, but one endophyte considered to have biological control potential against nematodes is *Fusarium oxysporum*, a cosmopolitan species that has been isolated from the roots of many plants. Although *F. oxysporum* is perhaps best known as a cause of vascular wilt disease in crops such as tomato, tobacco and potato, most strains of the fungus are non-pathogenic, and some are antagonistic to nematodes (see Chapter 8 for

further information). On the basis that some strains were known to protect plants from fungal pathogens, Hallmann and Sikora (1994) isolated *F. oxysporum* from tomato roots in Kenya and found that four isolates reduced galling caused by *Meloidogyne incognita* by 52–75%, and had similar effects on the number of egg masses produced by the nematode. Later studies with one of these isolates (*F. oxysporum* strain 162) showed that these same parameters were reduced relative to non-treated controls when the fungus was applied to seed and the seedlings were later transplanted into soil into which *M. incognita* was inoculated (Dababat and Sikora, 2007a; and Fig. 12.9). Dual inoculation at sowing and transplanting did not improve the level of control in the above experiment, but was beneficial in other experiments, with inoculum levels of 10^5 cfu/g seedling substrate at sowing and an additional 10^4 cfu/g at transplanting producing the best result (Fig. 12.10). These effects were thought to be due to the production of compounds that had a repellent effect on the nematode (Dababat and Sikora, 2007b), or the induction of systemic induced resistance (Dababat and Sikora, 2007c). The reaction of a tomato cultivar to fusarium wilt did not affect the efficacy of this strain of *F. oxysporum* as a biological control

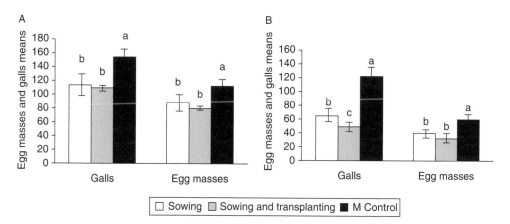

Fig. 12.9. Effect of *Fusarium oxysporum* strain 162 when inoculated at sowing, or at sowing and transplanting, on the number of galls and egg masses formed by *Meloidogyne incognita* on tomato. (A) and (B) contain data from separate experiments. 'M control' refers to *M. incognita*-infested pots without the fungus. Means followed by the same letter are not significantly different ($P \leq 0.05$; n = 10). (From Dababat and Sikora, 2007a, with permission.)

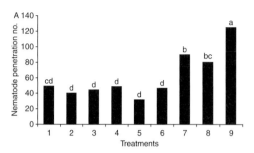
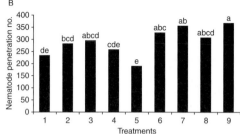

Fig. 12.10. Effect of *Fusarium oxysporum* strain 162 inoculation time and inoculum density on the number of *Meloidogyne incognita* per g tomato roots 2 weeks after nematode inoculation. Data are from two experiments. All treatments received *M. incognita* but treatment 9 was not inoculated with the fungus. Treatments 1–6 were inoculated with one of two concentrations of *F. oxysporum*, either only at sowing, or at both sowing and transplanting. Treatment 5 received a dose of 10^5 cfu/g substrate at sowing and a further 10^4 cfu/g soil at transplanting, while treatments 7 and 8 were only inoculated at transplanting. Means followed by the same letter are not significantly different ($P \leq 0.05$; n = 10). (From Dababat and Sikora, 2007a, with permission.)

agent, as the fungus was equally effective on resistant and susceptible cultivars (Dababat *et al.*, 2008).

Non-pathogenic strains of *F. oxysporum* have also shown potential against other plant-parasitic nematodes, with most of the work having been done with burrowing and lesion nematodes on banana. Pocasangre *et al.* (2000) isolated 132 endophytic fungi from banana roots and corms in Central America and found that when plantlets were treated with conidial suspensions of these fungi and later inoculated with *Radopholus similis*, five *Fusarium* isolates reduced the number of *R. similis*/g roots by more than 80%. Later studies with one of these isolates (H-20, a strain of a *Fusarium* species of uncertain identity), and a range of *F. oxysporum* strains from banana and tomato, showed that pre-inoculation with the fungi reduced the number of *R. similis* invading roots of tissue culture-derived plants (Vu *et al.*, 2006; Athman *et al.*, 2007). Some strains were more effective than others; and interestingly, *F. oxysporum* 162, which had previously been effective against root-knot nematode on tomato, was one of the better isolates in one of these experiments (Fig. 12.11). Similar effects have been observed with lesion nematode, as strains of *F. oxysporum* originating from banana in Kenya reduced population densities of *Pratylenchus goodeyi* by 47–60%, and lessened the damage caused by the nematode when tested in steam-sterilized soil in pots (Waweru *et al.*, 2013).

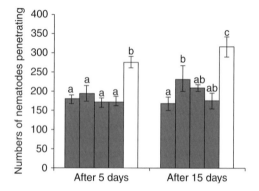

Fig. 12.11. Penetration of *Radopholus similis* into roots of untreated banana (white bars) and roots of plants inoculated with one of three isolates of *Fusarium oxysporum* or an isolate of an unidentified *Fusarium* species (dark bars). Plants growing in sterile soil were inoculated with 5×10^6 spores/plant, *R. similis* was inoculated 2 weeks later, and nematodes in roots were assessed after another 5 and 15 days. Means followed by the same letter are not significantly different ($P \leq 0.05$; n = 6). (From Vu *et al.*, 2006, with permission.)

Although there is concern that the non-pathogenic strains of *Fusarium oxysporum* used in biocontrol studies could possibly change to plant-pathogenic forms (see Chapter 8), the results discussed clearly indicate that selected strains of this and other *Fusarium* species have potential to control plant-parasitic nematodes in crops grown from transplants, or from plantlets derived from tissue culture. Conidia are readily produced in liquid or solid substrates and can easily be applied in commercial plant production systems. However, the fungus must be applied several weeks before planting, to allow time for the endophyte to establish before roots are subject to nematode attack. Seed application has also been suggested (Dababat and Sikora, 2007a), but endophytic fungi will probably be less effective on direct-seeded crops, as nematodes are likely to enter roots before the endophyte has established.

As is the situation with many promising biocontrol agents, what is currently missing are data showing the efficacy of *F. oxysporum* in long-term experiments. In most of these studies, effects on nematodes have been measured within a few weeks of planting, whereas in the field, nematodes continue to invade plants for many weeks, and several nematode generations may occur within 3–4 months of planting. Since the levels of control obtained in short-term pot experiments (see Figs. 12.9, 12.10 and 12.11) are unlikely to have major impact in the field, the application of *F. oxysporum* to transplants will almost certainly have to be supported by other nematode management practices. One practice worth investigating is to plant pre-inoculated material into well-managed soils with good biological diversity. In addition to containing numerous natural enemies, such soils will almost certainly have a good complement of bacterial and fungal endophytes. Since these naturally occurring endophytes can continue to colonize roots as they grow into deeper layers of the soil profile, endophytic activity against nematodes is likely to be sustained for a much longer period.

Combinations of Biocontrol Agents

Since biological control agents rarely provide high levels of nematode control when applied on their own, there has been considerable interest in using combinations of organisms in biocontrol studies. In some cases, agents with different modes of action have been used together, while in other cases, multiple strains of a single organism have been tried. Many different combinations have been utilized and no attempt will be made to cover them all here. However, some examples are:

- Spores of *Pochonia chlamydosporia* and *Pasteuria penetrans* were applied alone and in combination for control of *Meloidogyne incognita* on tomato in pots (de Leij *et al.*, 1992a).
- *Pochonia chlamydosporia* and *Dreschslerella dactyloides* were applied as formulated products and tested in the field for control of root-knot nematode on tomato (Stirling and Smith, 1998).
- Suspensions containing *Pasteuria penetrans* and two rhizosphere bacteria (*Enterobacter cloacae* and *Pseudomonas mendocina*) were applied to tomato in pots, and effects on *Meloidogyne incognita* were assessed (Duponnois *et al.*, 1999).
- Formulations containing two different plant growth-promoting *Bacillus* strains were applied to tomato and pepper transplants, and their effects on *Meloidogyne incognita* were assessed in the field (Kokalis-Burelle *et al.*, 2002b).
- The arbuscular mycorrhizal fungus *Glomus coronatum* and the endophytic fungus *Fusarium oxysporum* were tested for control of *Meloidogyne incognita* on tomato under greenhouse conditions (Diedhiou *et al.*, 2003).
- The biocontrol capacity of *Purpureocillium lilacinum* and the nematode-trapping fungus *Dactylellina lysipaga* were assessed in three nematode/plant systems: *Meloidogyne javanica* on tomato, *Heterodera avenae* on barley and *Radopholus similis* on banana (Khan *et al.*, 2006a).
- *Meloidogyne incognita* was inoculated into fumigated soil in pots, and the effects of commercial products containing *Purpureocillium lilacinum* and *Bacillus firmus* were assessed on tomato planted 0, 7 and 14 days after the products were applied (Anastasiadis *et al.*, 2008).
- The plant health-promoting rhizobacterium *Rhizobium etli* and the mycorrhizal fungus *Glomus intraradices* were assessed singly and in combination against *Meloidogyne incognita* on tomato (Reimann *et al.*, 2008).
- The mutualistic endophyte *Fusarium oxysporum*, the egg pathogen *Purpureocillium lilacinum*, and the antagonistic bacterium *Bacillus firmus* were evaluated

against *Radopholus similis* on banana under glasshouse conditions (Mendoza and Sikora, 2009).

- Dual strains of *Trichoderma atroviride* and *Fusarium oxysporum* were tested against *Radopholus similis* on banana in pots (zum Felde *et al.*, 2006; Paparu *et al.*, 2009).
- In greenhouse tests with banana plantlets, combinations of endophytic fungi (*Fusarium oxysporum* and *Trichoderma viride*) and bacteria (*Bacillus* or *Pseudomonas*) were evaluated against *Radopholus similis* (Chaves *et al.*, 2009).
- Four different groups of biocontrol organisms (the bacterium *Pseudomonas fluorescens*, the fungus *Purpureocillium lilacinum*, a yeast and a strain of cyanobacteria) were tested alone and in various combinations for control of *Meloidogyne incognita* on tomato (Hashem and Abo-Elyousr, 2011).

Unfortunately, most of these studies have never been taken further than simple screening tests under highly controlled conditions, and so there is no conclusive evidence that using more than one biocontrol agent results in better nematode control than using a single agent. Also, it seems that in many cases, the organisms included in these tests have been chosen arbitrarily, whereas it may have been better if they were selected on the basis that the biocontrol agents complemented each other in some way. There are many possibilities, but the following are some examples of multiple biocontrol systems that may be worth studying:

- An agent that acts quickly enough to reduce the number of nematodes invading the first roots produced by the crop (e.g. *Purpureocillium lilacinum*, *Trichoderma* spp. or an appropriate nematode-trapping fungus) and a parasite capable of reducing nematode population increase by attacking eggs later in the growing season (e.g. *Pochonia chlamydosporia*).
- One of the many antagonists available that parasitize or prey on nematodes, together with a bacterial or fungal endophyte that induces systemic induced resistance.

- Agents with a limited capacity to survive in soil due to the impact of biostatic forces (e.g. some nematophagous fungi), together with endospore-producing bacteria that are capable of surviving and remaining potentially infective for much longer periods (e.g. species of *Pasteuria* and *Bacillus*).
- Organisms with growth-promoting properties (e.g. PGPR), together with parasites and predators that act against nematodes rather than impacting on the plant.

Once a decision has been made to evaluate the combined effects of two or more biocontrol agents or products, there are some important considerations that should be addressed in such studies. First, if the organisms that are being applied together normally occupy the same ecological niche, they may compete with each other during the establishment phase. Thus, data on establishment should be obtained so that the timing or method of application can be changed if one of the organisms is found to be interfering with the other. Second, it is useful to know whether the end result of using a combination of organisms is additive or synergistic. Third, the introduced organisms will interact with each other, with the established soil microflora and fauna, and with the plant, and so these interactions must be understood if biocontrol efficacy is to be maximized.

In conclusion, it is apparent from even a cursory reading of the literature on biological control of nematodes that there are limitations to what can be achieved with a single biocontrol agent. Using multiple organisms is one possible pathway to improved reliability and efficacy. Kiewnick (2010) put this succinctly: 'the combination of antagonists with multiple modes of actions attacking different development stages of the target nematode and applied at different growth phases of the host plant seems a potential approach to overcome the limitations of biological control of nematodes'. We now need experimental results to show that such a strategy is consistently effective in the field.

The Role of Organic Amendments in Enhancing the Performance of Biological Products for Nematode Control

The literature on inundative biological control of nematodes is replete with examples of organisms being introduced to soil with the organic substrate on which they were grown. Stirling (1991) listed many such examples, and hundreds more could be added today. The most common substrates are wheat, oat, maize or rice grains, bran, straw, sugarcane bagasse, molasses, sawdust, chitin, animal manure and compost, but many other agricultural and industrial wastes and by-products have also been used. There are also examples of organic materials being included in, or applied with a formulated product in an attempt to enhance performance. Since organic matter becomes a substrate for resident organisms once it is introduced into soil, the critical question is whether including it in a formulated product or adding it with a biological control agent aids or hinders establishment. On the one hand, it could be argued that the biological control agent is favoured, as either it is already growing on the substrate, or it can be placed in a position where it can quickly utilize any organic materials that are applied with it. On the other hand, there is an argument that resident soil organisms will compete intensely for any added substrate, and in the process the biological control agent is likely to be inhibited.

In addressing the above question, it is clear from an earlier discussion of organic amendments and biological control (see Chapter 9) that the effects of amendments are essentially unpredictable. A complex series of events takes place at all levels of the soil food web when organic matter is added to soil, and although there is evidence that some natural enemies increase in population density, various components of the biological community respond in different ways to both the quantity and quality of the amendment, and the ensuing changes may continue for months after the amendment is applied. Thus, there is little more than circumstantial evidence to support the claim that the effectiveness of a biological control agent can be enhanced by adding an organic amendment. There may very well be benefits to be gained from such an approach, but proving that those benefits are due to the introduced organism rather than the direct or indirect effects of the amendment requires rigorous and well-monitored experiments, and better methods of determining the causes of nematode mortality than are currently available.

Although methodological problems and the complexity of the soil environment mean that is difficult to interpret the results of experiments with biological control agents and organic amendments, this does not mean that such work should not continue. Nematophagous fungi, for example, can easily be grown on grains, plant residues and other readily available organic materials, and so simple solid-substrate fermentation systems are perhaps the best way of mass-producing locally selected biological control agents and delivering them to farmers at a reasonable cost. Although not suitable for fungi such as *Hirsutella rhossiliensis* that are poor competitive saprophytes (Jaffee and Zehr, 1985), a paper by Khan *et al.* (2011) provides an example of what is possible with organisms with greater competitive ability. *Trichoderma harzianum*, *Pochonia chlamydosporia*, *Bacillus subtilis* and *Pseudomonas fluorescens* were grown on an autoclaved mixture of sawdust, soil and molasses, and then a powder formulation was produced by mixing one part of the stock culture with 20 parts of carrier (a mixture of flyash, soil and molasses). The resulting biopesticides contained very high concentrations of each organism ($>10^8$, $>10^{10}$ and $>10^{12}$ colony-forming units/g product for *P. chlamydosporia*, *T. harzianum* and the two bacteria, respectively). The formulations also remained viable with minimal microbial contamination when they were stored at room temperature for 12 weeks. In another example, Atkins *et al.* (2003b, c) grew *P. chlamydosporia* on cracked rice grain and then used it immediately as an unformulated product containing 6×10^6 chlamydospores/g growth medium. Clearly, simple methods such as this are ideally suited to small-scale production of biological control agents, and are already used in some countries. Cuba is one such example (Atkins *et al.*, 2003b, c), as it has a network of more than 200 small operators producing several microbial control agents, including *Bacillus thuringiensis*, *Lecanicillium lecanii*, *Metarhizium anisopliae* and *Trichoderma harzianum*.

The key question from a nematode management perspective is whether formulations of this nature are effective. Unfortunately, there is not enough information from well-monitored field trials to provide an answer with any degree of certainty. The four biopesticides prepared by Khan *et al.* (2011), for example, were only tested in autoclaved soil in pots, while the *P. chlamydosporia* grown on cracked rice by Atkins *et al.* (2003c) was only applied to small plots at one site. Products of this nature can probably be used in some countries with minimal attention to the commercial, intellectual property and registration issues outlined in Box 12.3, but results demonstrating that they are consistently effective in the field are still needed. Since organic substrates are being added to soil with the biological control agent, it is also important that data are obtained to show that the product consistently performs better than the substrate alone.

Inundative Biological Control as a Component of Integrated Nematode Management

Based on the results of the studies discussed in this chapter, it is clear that inundative biological control will rarely be effective enough to keep nematode pests under control for long periods. Thus, it will generally have to be combined with other management strategies. Nematode monitoring, various cultural controls, organic amendments, crop rotation, environmental modification and genetically resistant hosts are already a component of many integrated nematode management programmes (Bird, 1987; Duncan, 1991; Bridge, 1996; Barker and Koenning, 1998; Duncan

and Noling, 1998; Sikora *et al.*, 2005; Ornat and Sorribas, 2008; Westphal, 2011), and so once biological products are developed, they will have to be used in combination with at least some of these practices. Thus, we need to consider whether these practices, particularly those associated with the concept of integrated soil biology management, are compatible with inundative biological control.

It has previously been pointed out that many nematode management practices are an essential component of integrated soil biology management, but that inundative biological control may be one of the exceptions (see Chapter 11, but particularly Table 11.2). Integrated soil biology management aims to improve soil health, increase soil biological activity and diversity, and enhance the natural mechanisms that suppress pests and pathogens, but in the process, it is likely to increase the biostatic forces that limit the establishment of introduced organisms. Thus, biological products may not be effective (or even necessary) in situations where soil biological fertility has been improved by implementing some of the practices discussed in Chapters 9, 10 and 11. In fact, inundative biocontrol is most likely to find a place in biologically depleted environments: coarse-textured soils that are naturally low in organic matter; soils where carbon levels have declined due to excessive tillage or lack of organic inputs; situations where fallowing, solarization, soil fumigation, soil-applied pesticides and other such practices have diminished the soil biological community; and soils that are being rehabilitated after many years of mismanagement.

Given that nematicides are the primary nematode control tactic in many high-value crops, it is surprising that more effort has not been put into determining whether it is possible to integrate biological and chemical controls. It is clearly an option worth exploring for *Pasteuria*, which can withstand fumigant and non-volatile nematicides (Mankau and Prasad, 1972; Stirling, 1984; Brown and Nordmeyer, 1985; Walker and Wachtel, 1988), and for other bacteria that produce endospores as survival structures (e.g. *Bacillus* spp.). However, it is also an option for fungal biocontrol agents, as most non-volatile nematicides have limited anti-fungal activity. For example, *Pochonia chlamydosporia* is compatible with fosthiozate (Tobin *et al.*, 2008) and when it was applied with a low rate of aldicarb, the chemical did not affect the activity of the fungus, while control of *Meloidogyne hapla* was better than either the fungus or nematicide alone (de Leij *et al.*, 1993). Unfortunately, integrated systems of this nature have rarely been seriously investigated, as most biocontrol researchers seem intent on developing stand-alone systems rather than trying to use biological products in combination with chemicals. In vegetable crops, for example, it is quite likely that some non-volatile nematicides could be applied at much lower rates than are currently used (e.g. by seed treatment; by targeting planting material or the zone where roots first develop; or by reducing overall application rates) and still provide some nematode control during the first few weeks after planting. Combining such a treatment with a biological product may not only provide the level of reliability and efficacy that is required by growers, but is also likely to markedly minimize or completely remove any negative environmental impacts of the chemical. The challenge is to determine the chemical dosage required to inhibit nematodes in some way without affecting the biological control agent, and then use the data to find ways of utilizing chemical and biological products together. Another integrated option is to combine fungi capable of growing saprophytically on plant residues (e.g. *P. chlamydosporia*) with cover cropping (Dallemolle-Giaretta *et al.*, 2011). Since integrated management strategies of this nature are likely to enhance rather than hinder the uptake of biological products for nematode control, research in this area should be encouraged.

One final point that should be made with regard to integrating biological products with other nematode management practices is that the final package must be relatively simple. Farmers want to use the minimum number of tactics for the maximum benefit (Chandler *et al.*, 2011), and so a newly developed biopesticide will need to be effective when used in combination with no more than two or three other control practices. Thus, from the options listed in Table 11.2, a farmer might choose to reduce nematode populations with an appropriate rotation crop, apply a biopesticide, and then plant a nematode-tolerant crop. Another possible option would be to cultivate the soil and impose a short bare fallow to reduce nematode populations and minimize the impact of biostatic factors likely to prevent the introduced organism from establishing, and then apply the biopesticide. Thus, the ultimate test of any biological product for nematode control is whether it contributes to the effectiveness of simple integrated management systems such as those outlined.

Summary and Conclusions

From the examples presented in this chapter, it is clear that over the last two decades, considerable progress has been made in the area of inundative biological control of nematodes. Products based on *Purpureocillium lilacinum*, *Trichoderma* spp., *Bacillus firmus*, various *Bacillus* spp. and *in vitro*-produced *Candidatus* Pasteuria usgae, have been registered in the United States, Europe and elsewhere; *Pasteuria penetrans* has been marketed in Japan using an *in vivo* production system; and biological control agents produced in simple solid-substrate systems are being used by growers in some countries. Also, *Pochonia chlamydosporia*, various fungal endophytes and PGPR have shown promise in greenhouse studies. However, it is questionable whether the biological products currently on the market have 'proven efficacy', as claimed by Hallmann *et al.* (2009). It is certainly true that these products improve plant growth, reduce symptom severity or reduce nematode populations under controlled experimental conditions, but evidence of efficacy in the field is generally lacking. Thus, in the same way that extensive field testing was required in the 1960s and 1970s to identify the situations where chemical nematicides were efficacious and economically viable, the final steps in the evaluation process for biological products (Box 12.3) must now be completed. There are hundreds of examples of biological control agents that show promise in greenhouse tests, but the ultimate criteria for success are consistent results and widespread acceptance in the target market (Stirling, 2011a).

Since the current suite of biological products for nematode control have not been extensively tested in the field, it is difficult to predict where they are most likely to be useful. However, given that a microbial pesticide is unlikely to be produced, registered and distributed more cheaply than a chemical product, the market will mainly be limited to organic agriculture and to situations where nematicides are currently used (e.g. vegetable production; the nursery industry; horticultural replants; crops grown in glasshouses and other protective structures; and golf courses and other high-value components of the turf industry). It is difficult to see such products being used in the broader horticulture industry (e.g. perennial tree and vine crops), or in most low-value field crops. However, there may be possibilities in situations where seed treatment is a viable option (e.g. cotton and some cereal, grain and oilseed crops), or where technologies in precision agriculture provide opportunities to apply a product to a specific target (e.g. the soil in contact with a newly planted seed). The critical issue here will be to demonstrate that treating the seed or planting row with a biological product will enhance plant growth or improve establishment, and also provide roots with some protection from nematode attack during the first few weeks after planting. Such treatments may also find a place in agriculture as an inoculative rather than as an inundative biocontrol strategy. For example, endospores of a *Pasteuria* species capable of targeting a crop's key nematode pest could be applied at planting in the expectation that over time, the parasite will increase and eventually provide some nematode control.

Regardless of the situation in which biological products are used against nematodes, they usually provide only short-term or relatively moderate levels of control. Thus, inundative methods of biological control are only likely to be successful as a component of an integrated nematode management programme. However, as mentioned previously, this form of biological control is not necessarily compatible with the concept of integrated soil biology management as promoted in this book. Practices that improve a soil's biological status (e.g. minimum tillage, organic amendments, crop rotation and cover cropping) are likely to enhance biostasis, thereby reducing the efficacy of any biocontrol agent that is introduced into soil.

Although biostatic forces exist in all soils and will always limit the effectiveness of inundative biocontrol, the research discussed in this chapter demonstrates that introduced organisms can be established in soil, provided they are fast-growing species capable of competing saprophytically in the soil environment (e.g. *Purpureocillium lilacinum* and *Trichoderma* spp.), or have resistant spores that enable them to survive (e.g. *Pasteuria* and *Bacillus* species). Although population densities of the biocontrol agent will always decline following introduction, there are situations where populations remain at high levels for long enough to affect the target nematode, or where the organism can be reapplied if unacceptable decline does occur. Research with organisms that form mutualistic associations with plants has also shown that problems due to biostasis can be avoided to some extent through the use of endophytes. Unlike many other biocontrol agents, endophytes have developed intimate associations with plants and do not need to survive for long periods in the soil or rhizosphere.

Since this book has focused to a large extent on how natural biocontrol mechanisms can be enhanced with appropriate management practices and modifications to the farming system (see Chapters 1–4 and 9–11), it is reasonable to consider where inundative release of antagonists fits among the options available for managing nematodes through biological mechanisms. Unlike many contemporary nematologists, who see the development of biological products as the only feasible way of achieving biological control, I would argue that inundative biological control is only a stopgap measure until more sustainable farming systems are developed. Many options are available to improve current farming systems and enhance soil health, while integrated soil biology management is the pathway to long-term suppression of plant-parasitic nematodes through natural mechanisms (see Chapter 11). The advantage of taking such an approach is that crop losses due to nematodes are minimized through multiple mechanisms (i.e. nematode populations are reduced through crop rotation, resistant cultivars and the regulatory forces associated with suppressive soils; while nematode damage thresholds are increased through improvements in soil physical and chemical fertility, or the use of nematode-tolerant cultivars). Seedling development and early plant growth may be improved by some of the practices discussed in this chapter, but our ultimate objective must be to develop farming systems in which nematodes can be managed without 'silver bullets', whether they be chemical or biological.

Section VI

Summary, Conclusions, Practical Guidelines and Future Research

13

Biological Control as a Component of Integrated Nematode Management: The Way Forward

Plant-parasitic nematodes are important pests of agricultural crops and are generally managed by crop rotation, resistant cultivars or chemical nematicides, although a range of other tactics may be employed in some situations (see Table 11.2). Biological control is generally perceived as playing little or no role in current nematode-management programmes, largely because natural enemies are seen as constituents of products that are deployed in much the same way as chemical nematicides. This narrow view of biological control ignores the fact that: (i) the soils used for agriculture contain a vast array of natural enemies of nematodes; (ii) predacious or parasitic activity will be found in agricultural soils if attempts are made to detect it; (iii) nematode populations are always affected to some extent by the soil microfauna and fauna, as evidenced by the way nematode populations respond when their natural enemies are eliminated; and (iv) naturally occurring antagonists sometimes suppress populations of plant-parasitic nematodes to levels where the pest no longer causes economic damage. Unfortunately, the latter situation is not as common as it could be, because the resident organisms capable of suppressing plant-parasitic nematodes are continually being disrupted or destroyed by the management practices used in modern agriculture. Thus, biological control is not just about developing biopesticides targeted at specific plant-parasitic nematodes; it also involves improving soil and crop management practices to the point where the suppressive services provided by the soil biological community begin to operate effectively, as they do in natural ecosystems.

This chapter provides a brief summary of what we have learnt in this book about both forms of biological control. It begins with the premise that after years of inappropriate soil and crop management, the organic inputs required to sustain a community of organisms capable of regulating nematode populations have declined to the point where many agricultural soils are now conducive to nematode pests. Alternative farming systems that increase carbon inputs from roots and crop residues, and reduce carbon losses due to tillage, offer the opportunity to sequester rather than to deplete the food source on which natural enemies ultimately depend, and, therefore, form the basis of a process that is sometimes referred to as conservation biological control. The role of biopesticides that target nematodes is also discussed, as such products are seen as a useful stopgap measure during the transition to more sustainable farming systems. The chapter also considers the institutional, educational and logistical issues that prevent us from achieving sustained biological control, and outlines some of the research that should be undertaken to improve our understanding of interactions between plants, soil organic matter, nematodes and their natural enemies.

Ecosystem Services Provided by the Soil Biological Community, and the Key Role of Organic Matter

The primary role of agriculture is to supply human needs for food, fibre, biofuel and pharmaceuticals. However, its capacity to supply this provisioning service is highly dependent on a suite of supporting and regulating services that maintain the long-term stability and productivity of the non-renewable soil resource that ultimately sustains agricultural production (Power, 2010; Powlson *et al.*, 2011a). These additional services are provided by the soil biological community, with three of its most important functions being to maintain soil structure and fertility, store and cycle nutrients, and regulate populations of soilborne pests and pathogens.

Although the services provided by the soil biota determine the long-term stability and productivity of agricultural soils and hence the sustainability of agroecosystems, the organisms in the soil biological community cannot function without organic inputs from plants. Thus, the quantity and quality of inputs from plant residues and root exudates determine whether a soil will provide a full range of ecosystem services, a truism that is evident from the following sentence in Chapter 2.

> Organic matter plays a critical role in creating and stabilizing soil structure, which in turn produces good tilth, adequate drainage and the capacity to resist erosion; it increases soil water availability by enhancing water infiltration and increasing a soil's water-holding capacity; it is the principal long-term nutrient store and the primary short-term source of nearly all nutrients required by plants and soil organisms; and it also provides a food source for the organisms that cycle carbon and protect plants from disease.

Loss of organic matter is one of the many reasons for the continuing decline in the physical and chemical fertility of agricultural soils, and is certainly the main reason that soilborne diseases are now a chronic problem in most forms of agriculture. With regard to the suppressive services that limit nematode populations, organic matter is critically important because the population density and levels of activity of most of the parasites and predators that suppress nematode populations are linked directly or indirectly to levels of soil organic matter (see Chapters 5 and 6).

Once a soil loses its capacity to suppress nematode pests, poor growth, yield decline and other symptoms of nematode damage will begin to appear. However, losses in crop production are not just the result of high populations of plant-parasitic nematodes. Physical and chemical constraints associated with poor soil health place additional stress on the plant, and so fewer nematodes are required to cause crop losses. Thus, as soil health declines, the grower must cope with ever-increasing populations of plant-parasitic nematodes and ever-decreasing nematode-damage thresholds. In this situation, increasingly efficacious control measures are required. However, simply focusing on reducing populations of the pest is clearly the wrong response. The only long-term solution is to identify the management practices responsible for poor soil health and then take steps to address the problem.

Farming Systems to Improve Soil Health and Sustainability

Since the practices that can be used to influence the level of organic inputs into an agroecosystem are dependent on soil type, climate, the principal crop, various production-related issues and economics, it is impossible to be prescriptive about farming systems that will improve soil health at a particular location. Nevertheless, there is evidence to suggest that three soil improvement practices (minimal soil disturbance; permanent plant residue cover; and crop rotation, preferably including legumes) play a key role in improving the long-term sustainability of crop production systems. These practices were discussed in Chapter 4 and form the basis of a management system known as conservation agriculture. The economic and soil-health benefits from those practices are so great (see Table 4.1) that they should be an integral component of any farming system. Once they are adopted, a range of other practices (cultivars with improved yield potential; optimized nutrient and water management; controlled traffic; site-specific management of nutrient and pesticide inputs; integrated pest management; integrated crop and livestock production; amendments of organic matter; and mulching) can be incorporated into the system to further improve soil health and also enhance its productivity and sustainability (see Chapters 4 and 11).

Integrating these practices into a sustainable farming system is not a simple process, but farmers in many countries have shown that it is achievable. A major change in mindset is required to make the change to no-till agriculture, but once that hurdle is overcome, incremental improvements to the farming system can be made in ensuing years. In regions where plant growth is not constrained by rainfall or temperature, cover crops, perennial forages and high-residue crops have a role to play in improving the soil's organic matter status, while organic amendments can contribute in some situations. Perceived problems associated with managing large quantities of crop residues and controlling weeds in no-till systems can be overcome by modifying farm equipment, adjusting residue management practices and utilizing crop residues and other tactics in an integrated weed-management programme (see Chapter 11).

In concluding this brief summary of soil health and sustainable farming, it is important to be realistic about the amount of carbon that can be sequestered in soil through modifications to a farming system, and to recognize that it may take many years to reap the benefits that accrue from making the required changes. In situations where irrigation water and sunshine are plentiful and it is economically feasible to utilize organic amendments acquired from elsewhere, it may be possible to repair degraded soils relatively quickly. However, when biomass production is limited by rainfall, for example, only small amounts

of carbon will be sequestered and it may take many years for soil-health benefits to be observed. Nevertheless, it is still worthwhile making the change, because, in addition to the biological benefits that accrue from incremental improvements in soil organic carbon levels, there will be many other benefits (e.g. improved timeliness of operations, lower fuel usage, better erosion control and reduced fertilizer costs due to nitrogen inputs from legumes).

Will Suppressiveness be Enhanced by Modifying the Farming System?

Although there is considerable evidence to indicate that adopting the practices associated with conservation agriculture will improve soil health, evidence that this occurs is generally obtained by measuring physical parameters such as aggregate stability, soil water-holding capacity and penetration resistance; and chemical parameters such as pH, total carbon, labile carbon and potentially mineralizable nitrogen. Evidence of biological improvements is usually based on changes in microbial biomass carbon, microbial activity, nematode assemblages and a range of other parameters (see Chapters 3 and 4). Since none of these tests indicate whether suppressiveness to plant-parasitic nematodes has been enhanced, the reasons why a switch to a more sustainable farming system will produce such a result are outlined next.

The impact of plant residues, root exudates and other sources of organic matter on natural enemies of nematodes

Above-ground plant materials are often considered the main source of carbon inputs into soil, but about 40% of plant photosynthates are partitioned below-ground (Jones et al., 2009). Much of this plant biomass occurs as fine roots, and their high turnover rate means that they are the principal food source for heterotrophs below the soil surface. Since about 27% of the carbon allocated to roots finds its way into rhizodeposits, roots are also the main source of labile carbon compounds in soil, and their impact on the soil biological community is evidenced by the rhizosphere effect (see Chapter 2). The continual presence of plant roots is, therefore, vitally important in maintaining an active and diverse soil food web. Plant nematologists would rightfully argue that one of the consequences of continual cropping is an increase in populations of root-feeding nematodes, but what is often forgotten is that roots also support an active microbial community, and organisms that are antagonistic to nematodes are a component of that community. Thus, egg parasites such as *Pochonia chlamydosporia* will multiply in the rhizosphere; symbionts with the capacity to induce nematode and disease-resistance mechanisms within the plant (e.g. vesicular arbuscular mycorrhizae and endophytic microorganisms) will occupy root cells; obligate parasites such as *Pasteuria* will multiply; and generalist predators will increase because more herbivorous and free-living nematodes are available as prey. The role of plants in contributing to the latter effect is shown by a recent study in a soil under organic vegetable production. Cover crops had a greater effect on the soil food web than compost additions, and changes at the entry level of the web were transferred to higher trophic links, thereby increasing top-down pressure on plant-parasitic nematodes (Ferris *et al.*, 2012c).

Since plants have evolved with the soil biological community, it is not surprising that they have developed mechanisms of defending themselves against root feeders, or that they produce signals that guide beneficial microbes and soil fauna to the rhizosphere (see Chapter 2). Thus, roots can emit signal compounds to attract entomopathogenic nematodes capable of attacking root-feeding insects, and will recruit bacterivorous nematodes to spread plant growth-promoting bacteria or help ensure that legume roots are inoculated with nitrogen-fixing bacteria (Bonkowski *et al.*, 2009). Although we are only just beginning to understand the signalling processes involved, the clear message from these observations is that intimate root–microbe–faunal relationships occur to some degree in all soils, and play an important role in protecting the plant from root pathogens.

Another important feature of sustainable farming systems is the presence of a permanent layer of plant residues on the soil surface. In addition to protecting the soil from erosion, this organic material plays a vital role in sustaining antagonists of nematodes. Nematode-trapping fungi are commonly found on substrates with a high C:N ratio (see Chapter 5), and the presence of a litter layer should provide them with a much better habitat than is available in agroecosystems that are subjected to tillage. The litter layer, with its ameliorating effects on soil moisture and temperature, also provides an ideal habitat for mites and Collembola, some of which are predacious on nematodes. Microarthropods occur at very high densities in situations where surface soils are covered with plant residues (e.g. grasslands and forests), but populations are generally much lower in agroecosystems. Thus, retaining plant residues on the soil surface is beneficial to many different natural enemies of nematodes, and will enhance suppressiveness to plant-parasitic nematodes (see Stirling *et al.*, 2011a, b; and also Chapter 11).

The role of continual cropping and increased cropping intensities

Situations where host-specific antagonists increase to high population densities and suppress a nematode pest commonly occur on perennial crops, and on annual crops that are grown intensively or as a monoculture. Examples include the suppression of root-knot nematode on grapevines and sugarcane by *Pasteuria penetrans*; the decline of cereal cyst nematode in intensively cropped cereal fields due to parasitism by *Nematophthora gynophila* and *Pochonia chlamydosporia*; and the suppression of *Heterodera schachtii* by *Brachyphoris oviparasitica* and other fungi in soil under sugarbeet monoculture (see Chapter 10). Thus, continual cropping may increase populations of plant-parasitic nematodes in the short term, but is also the key to increasing populations of host-specific parasites capable of suppressing these pests.

It could be argued that many annual crops are cropped intensively but still suffer losses from plant-parasitic nematodes. However, this is primarily due to the way these crops are grown, rather than to a shortage of natural enemies. In the case of root-knot nematode on vegetable crops, for example, high populations of *Pochonia chlamydosporia* and *Purpureocillium lilacinum*, and high levels of egg parasitism, are often observed in field soils, particularly late in the growing season after the nematode has multiplied for several months (see Gaspard *et al.*, 1990b; Mertens and Stirling, 1993; and Bent *et al.*, 2008 for some examples, and also Chapter 5). Thus, when the crop is harvested, a community of organisms with considerable suppressive potential is likely to be present in the rhizosphere. Since these organisms are then disturbed or destroyed by the management practices commonly used in vegetable production (e.g. bare fallowing, aggressive tillage and soil fumigation), the challenge of the future is to redesign annual cropping systems so that the suppressiveness that has developed on the first crop continues to act during the non-crop period and is still present when the next crop is planted.

As agriculture intensifies to meet the world's increasing demands for food and fibre, crops will have to be grown more frequently. Losses from nematodes are likely to increase if current production practices are maintained, but the process of sustainable agricultural intensification (discussed in Chapter 4) provides opportunities to prevent this from occurring. Evidence presented in Chapter 10 indicates that specific suppressiveness is likely to be enhanced when land is cropped more intensively, while the organic matter produced by the additional crops, if managed appropriately, should enhance general suppressiveness. Thus, it should become easier to develop a soil that is suppressive to plant-parasitic nematodes as cropping intensity increases. The challenge is to ensure that the appropriate management practices are adopted.

One factor that may limit the soil carbon and biotic improvements that will occur when cropping systems are intensified is the crop cultivars currently available to farmers. Unlike their wild relatives, high-yielding cultivars do not need an active soil biota when grown in the heavily fertilized soils in which they are selected. These cultivars have been bred for high-input agriculture, and photosynthates that would normally move to roots and stimulate the soil food web are redirected into harvestable product. Thus, plant breeders need to consider developing cultivars that are better adapted to sustainable agriculture. Characteristics that deserve attention include plants with higher root–shoot ratios, more feeder roots and a surface rooting habit, as these features would enable crops to more readily obtain nutrients from the biological activity that exists in and immediately under the surface organic layer in sustainable production systems.

Reducing tillage results in multiple benefits that will improve soil health and enhance suppressiveness

It is often assumed that the detrimental effect of tillage on the soil biological community is mainly due to its role in hastening the mineralization of organic matter, the energy source that fuels the soil food web. Although that is important, it is now clear that the abrasive and physically disruptive effects of tillage, and its role in modifying habitat, are also a contributing factor. Tillage is most likely to affect large soil organisms such as earthworms, spiders, mites and Collembola (Wardle, 1995), but studies discussed in Chapter 5 indicate that it is also detrimental to many nematophagous fungi. The predacious activity of *Hirsutella rhossiliensis* and some species of nematode-trapping fungi is markedly reduced when soil is physically disturbed, and the destruction of trapping devices may reduce the population densities of these fungi (McInnis and Jaffee, 1989; Jaffee *et al.*, 1992b). Since the organisms detrimentally affected by tillage occupy high trophic levels within the soil food web, elimination of tillage from the farming system should enhance the suppressive services provided by the soil biological community.

It is also important to recognize that a permanent rather than a temporary change to no-till is required, as it may take months or years for the larger soil fauna to respond to a lack of physical disturbance. However, if that long-term change is made, the benefits obtained from eliminating tillage will be manifested in a number of ways: (i) increased faunal biomass due to higher levels of organic matter; (ii) elimination of mortality caused by mechanical abrasion; (iii) a more diverse biological community with a greater range of food sources for generalist predators; (iv) greater pore space for soil organisms due to a reduction in compaction and the preservation of root channels and macropores; (v) an improved habitat for soil organisms due to a permanent cover of plant residue; (vi) lower fluctuations in temperature as a result of greater surface cover; and (vii) a more favourable soil moisture status due to the soil's improved water-holding capacity and a reduction in evaporative losses.

The Role of Inundative and Inoculative Biological Control

Populations of plant-parasitic nematodes are often inordinately high in agricultural soils, but the advantage of the conservation approach to biological control (discussed earlier) is that it targets the cause of the problem (i.e. the loss of the biological buffering capacity that should be keeping the pest in check) rather than the effect (i.e. high populations of the pest). Targeting the pest provides no more than a temporary solution, but given the challenges involved in changing many farming systems, and the length of time required to observe benefits from such an approach, products that have an immediate effect will always find a place in the agricultural marketplace. Thus, there is a role for biological products that are applied in an inundative manner to either provide short-term nematode control or to improve the capacity of the plant to cope with nematode damage. Since there are many situations where the soil food web has been severely depleted by practices such as soil fumigation and excessive tillage, and there are sometimes fields where host-specific parasites such as *Pasteuria* are not present, biological products can also be used in an inoculative manner to introduce an antagonist that can then be increased

by manipulating populations of prey nematodes.

The studies cited in Chapter 12 provide a clear indication that inundative biological control has been the dominant component of research programmes on biological control of nematodes for the last 30 years. Fortunately, there is evidence that progress is being made, as products based on *Purpureocillium lilacinum*, *Trichoderma* spp. *Bacillus firmus*, various *Bacillus* spp. and *in vitro*-produced *Candidatus* Pasteuria usgae are now registered as biopesticides in some countries. However, one limitation of much of the work to date is that evidence of efficacy has generally been obtained under relatively controlled experimental conditions. Thus, as pointed out in Chapter 12, there is an urgent need to confirm that the registered products are efficacious and cost-effective in the target markets.

Since biostatic effects will always limit the effectiveness of introduced biological control agents, there has been a major focus in recent years on using organisms that live naturally in roots, as they do not have to compete to the same extent with the multitude of organisms that live in the soil or rhizosphere. Given the progress that has been made in understanding the interactions between some of these endophytes and plant-parasitic nematodes, it is likely that in the near future, selected endophytes will be commercialized for application to seed, seedlings and other forms of planting material (see Chapters 8 and 12).

Moving from Theory to Practice: Issues Requiring Attention

It is clear from the material presented in Chapter 4 and observations made by numerous authors (e.g. Gliessman, 2007; Pretty, 2008; Royal Society, 2009; Godfray *et al.*, 2010; Power, 2010; Vandermeer, 2011; Lal *et al.*, 2011; Powlson *et al.*, 2011a, b) that agriculture faces huge challenges in the 21st century. Global food production must be dramatically increased, while at the same time, soil and crop management practices must be modified to prevent soil health from deteriorating, mitigate and adapt to climate change, minimize off-site environmental impacts, and enhance the regulatory services provided by the soil biota. Farming systems will have to become more sustainable, and some of the options available to achieve this are discussed in Chapter 11. However, as agricultural production systems are modified to meet these diverse goals, one of the greatest challenges will be managing below-ground biodiversity to achieve greater levels of suppressiveness to nematodes and other soilborne pathogens. Some of the issues that must be addressed in achieving that goal are discussed next.

Assessment of suppressive services in long-term trials

Many of the assertions made in this book about the way suppressiveness to plant-parasitic nematodes can be enhanced are based on observations made in microplots, or by retrieving soil from the field and assessing suppressiveness in a greenhouse test. In cases where management effects on suppressiveness have been examined in field experiments, assessments usually involve a limited range of practices (e.g. till versus no-till; compost amendment versus no amendment), and since the practices examined have generally been in place for a limited period, there are always doubts about whether the soil's suppressive potential has been fully expressed. Thus, there is clearly a need for long-term field trials to demonstrate that particular management practices enhance suppressiveness. The problem with that approach is that long-term field trials containing appropriate treatments are rarely available, while 2- or 3-year research funding cycles mean that it difficult to obtain the funds required to establish and maintain such experiments.

With regard to field experimentation, perhaps the most important need is to demonstrate that a range of practices must be integrated to achieve maximum levels of suppressiveness. I have argued that a sustainable farming system would accommodate many of the management practices listed earlier in this chapter, and so assessing the contribution of individual components is less important than determining how the whole system performs.

Whatever systems are compared, those systems must represent best management practice, with the input levels and timing of all crop and soil management operations being optimized for the specific soil and climatic conditions at the research site.

A survey approach in which productivity, carbon inputs and various soil properties (including suppressiveness) are assessed on farms using different management practices is also useful for making comparisons between farming systems. However, variability in soil type, climate, historical management and many other factors are likely to make it difficult to interpret the results of such studies. Perhaps a better option is to work with progressive farmers; document what is achievable under best practice in a given soil type, environment and production system; and then extend that information to others. However, when the results of such studies are presented, it is important to be specific about the management practices that are actually utilized in the farming system. For example, it is unhelpful to use vague terms like 'organic' or 'conventional', as they provide no information on important issues such as the quantity, quality and timing of various inputs; the nutrient status of the crops; the manner in which surface residues were managed; the severity of tillage; and the weed and pest management practices used in each of the systems being assessed.

One issue that must always be recognized when attempting to manage nematode pests within a sustainable farming system is that the ultimate goal is to minimize losses from nematodes. Thus, it does not really matter whether this is achieved by reducing nematode populations through crop rotation; by minimizing plant stress through improvements in soil health; by developing a nematode-suppressive soil; by incorporating a specific control measure into the system; or through the combined effects of a number of practices and mechanisms. Nevertheless, once a farming system is established and optimized, it is worthwhile assaying the soil for suppressiveness (see Chapter 11), as it will indicate whether suppressiveness is contributing to the nematode control that is being obtained. However, given the response of the biological community to disturbance, and the fact that relatively simple disturbances (such as mixing soil by hand) can markedly reduce suppressiveness (Stirling, 2008; Westphal and Xing, 2011; Stirling *et al.*, 2012), suppression assays should be carried out using undisturbed soil cores.

Relationships between soil carbon status, biological activity, biodiversity and general suppressiveness

Nematode populations are sometimes suppressed by specific antagonists, but suppressiveness is more commonly the result of the combined actions of numerous parasites and predators. Although it is worthwhile knowing the identity of the major players in the suppressive community, it is more important to understand the general characteristics of that community and the management practices required to achieve it. Thus, rather than concentrating on the role of individual antagonists, attention should focus on finding useful indicators of suppressiveness. In the initial stages of such studies, the soil biological community is usually treated as a black box and the focus is on finding parameters that are broadly related to suppressiveness (e.g. labile carbon, microbial biomass carbon or microbial activity). Once that is achieved, it may be possible to improve the predictive value of such indicators by enumerating predatory nematodes, microarthropods or some other easily measured component of the antagonistic community; by incorporating one of the diversity indices commonly applied to nematode assemblages (Neher and Darby, 2009; Ferris and Bongers, 2009); or by analysing the rhizosphere microbiome and identifying a suite of organisms that are consistently associated with suppressiveness (Mendes *et al.*, 2011; Chaparro *et al.*, 2012). However, because of the biological complexity that is normally associated with suppressiveness, it is possible that, at a practical level, a soil's organic matter status, or general indicators of soil biological activity or diversity, may prove to be more useful indicators than the presence or population density of certain groups of antagonists. For example, Stirling *et al.* (2011b) argued that soil carbon is the simplest and most realistic

indicator currently available, and provided a limited set of data to show that Australian sugarcane soils with carbon contents of 30–43 g/kg were likely to have reasonable levels of suppressiveness to plant-parasitic nematodes.

Once relationships between soil carbon status, biological activity, biodiversity and suppressiveness are understood, the next step is to determine the levels of suppressiveness that are attainable in a given crop production system, soil type and environment, and then optimize management to achieve a soil that is somewhat suppressive to nematodes. Since attainable levels of suppressiveness will be influenced by net primary productivity, highly suppressive soils are most likely to be developed in situations where rainfall is plentiful or irrigation water is available, and plant growth is not constrained by sunlight or temperature. In low-input farming systems, where primary productivity is often limited by rainfall, much less soil carbon will be sequestered and this will increase the time it takes to develop a suppressive soil and limit the level of suppressiveness that is achievable. Nevertheless, it is still vitally important to move down that pathway, as few other nematode-management options are available in such situations, and other soil-health benefits come hand-in-hand with increasing levels of suppressiveness.

Management of specific suppressiveness

Suppressiveness is not only linked to the biological activity and diversity associated with soil organic matter: it is also due to the actions of host-specific parasites and predators. Thus, in addition to ensuring that a farming system enhances general suppressiveness, it is important that it also sustains organisms capable of specifically targeting key nematode pests. Populations of host-specific antagonists will decline in the absence of a host nematode, and so populations of the host must be managed so that they provide a continuing food supply for the antagonist without causing unacceptable levels of damage to the primary crop. Such a process requires a good understanding of interactions between specific suppressive agents and their hosts; knowledge of whether these agents can utilize food sources other than nematodes; and an awareness of nematode-damage thresholds.

One of the best examples of managing specific suppressiveness involves a study of the suppression of *Heterodera glycines* by *Pasteuria nishizawae* in Illinois (see Chapter 10). Noel *et al.* (2010) worked with field plots infested with the nematode and used data collected from a soybean monoculture over a 7-year period to show that *P. nishizawae* reduced population densities of *H. glycines* and generally increased yields, regardless of whether resistant or susceptible cultivars were used. This observation on the effect of crop cultivars is important, because it shows that provided the level of resistance is not excessive, populations of *P. nishizawae* will increase on nematode-resistant cultivars. It is also a reminder that the introduction of cultivars with high levels of resistance will almost certainly deplete populations of the nematodes on which host-specific parasites depend. Tolerant or partially resistant cultivars may, therefore, be a better long-term management option. Since rotation with crops that are not hosts of the target nematode is standard practice in many farming systems, the challenge of the future is to understand how the population dynamics of host-specific parasites is influenced by the presence of susceptible host crops, non-hosts and cultivars with varying levels of resistance. It should then be possible to develop crop-rotation practices that enhance general suppressiveness, maintain specific suppressive agents and keep nematode populations at levels below the damage threshold.

With regard to parasites that show some specificity to particular nematodes but are not obligate parasites (e.g. *Pochonia chlamydosporia*), other management options are available. Dallemole-Giaretta *et al.* (2011) provided the basis for one such system by showing that a potentially useful rotation crop (Surinam grass, *Brachiaria decumbens*) reduced the severity of galling caused by root-knot nematode, and that the level of galling was further reduced by the presence of *P. chlamydosporia*. Since the grass allowed only poor to moderate development of the fungus in the rhizosphere, and other crops appeared to support better rhizosphere colonization, opportunities exist to select rotation crops not only on the basis of their resistance to a target nematode, but also on their capacity to support a useful facultative parasite such as *P. chlamydosporia*.

Understanding interactions between the nematode community, natural enemies and organic matter

One of the hypotheses commonly advanced to explain why populations of plant-parasitic nematodes are sometimes reduced when soil is amended with organic matter is that free-living nematodes multiply on the microorganisms involved in the decomposition process, and their presence stimulates parasites and predators that then suppress all members of the nematode community, including plant parasites (see Chapters 5 and 9). However, the results of a microcosm study by Venette *et al.* (1997) show how little we know of interactions between organic matter, the nematode community and organisms at higher trophic levels in the soil food web. In this study, organic inputs (in this case exudates from plant roots) had quite specific effects on nematodes, and flow-on effects to natural enemies did not necessarily occur. Five plant species were included in the experiments and only two of them (sunn hemp and vetch) increased populations of bacterial-feeding nematodes. However, the rhizosphere effect was nematode-specific, as *Acrobeloides bodenheimeri* was selectively enhanced while populations of two rhabditids were not affected. Although *A. bodenheimeri* was susceptible to parasitism by *Hirsutella rhossiliensis*, an increase in its population density did not result in greater parasitic activity against *Heterodera schachtii*.

Several reasons were advanced as to why the bacterial-feeding nematodes used in this study may have escaped parasitism, and they demonstrate the complexity of nematode–parasite interactions in soil and the need for more work in this area. Differences in the locomotive capacity of the nematodes may have influenced transmission of spores to nematodes, while differences in developmental rates, cuticular chemistry or the presence or absence of a cuticular sheath, may have influenced their susceptibility to parasitism. Regardless of the reasons for the result, these observations indicate that enhancing populations of free-living nematodes does not necessarily increase parasitism of plant-parasitic nematodes by *Hirsutella rhossiliensis*. Thus, predicting the success of such a strategy requires knowledge of the ecology of the nematode species present in soil, and an understanding of how their antagonists react to various organic inputs.

Another reason more work is needed on interactions between organic matter, the nematode community and various natural enemies is that these interactions occur at a microsite level, and so parasitism or predation at one site in the soil profile may not necessarily affect nematodes occupying another niche that may be only be a short distance away. This point was elegantly demonstrated by Jaffee and Strong (2005) in their studies of moth larvae, entomopathogenic nematodes and nematophagous fungi (see Chapter 5). The presence of a single nematode-parasitized insect completely changed the soil biological community in its near vicinity, with one nematode-trapping fungus (*Arthrobotrys oligospora*) increasing to population densities that sometimes exceeded 10^4 propagules/g soil. Such population densities are never seen in the bulk soil mass, raising questions about whether nematophagous fungi proliferating in an insect cadaver or on a piece of organic debris will affect plant-parasitic nematodes that may be located only a few millimetres away on a nearby root.

The fact that nematode–parasite interactions occur at microsites is a particularly important issue, because it has implications for the way organic matter should be managed in agricultural soils. For example, based on observations discussed earlier in this chapter, it is clear that a community of organisms with considerable suppressive potential will be present in the rhizosphere when an annual crop reaches maturity. I would argue that it does not make sense to then cultivate the soil and disturb those organisms for the following reasons: (i) large numbers of parasites and predators will have already aggregated in the rhizosphere because of the presence of plant-parasitic nematodes, and will, therefore, be situated at a location where they can act on those nematodes; (ii) during the period before the next crop is planted, parasitic and predacious activities will continue, and those activities will add to the decline in nematode populations that will occur through natural attrition; (iii) many of the natural enemies will be nematophagous fungi, and since they are facultative parasites, they will also be able to survive by competing saprophytically for the resources present in the decomposing root system; and (iv) if the soil is left undisturbed until the next crop is planted, the root system of the new crop will tend to grow down root channels left behind by the previous crop (see Chapter 5), and so the suppressive community developed on the previous crop is ideally positioned to affect nematodes attacking the following crop. Thus, in no-till farming systems, plant roots are obviously an important microsite for biological activity, and there is clearly a need to understand the way nematophagous organisms, nematodes and roots interact at such microsites, and the implications that this has for the development of suppressiveness. Ideally, studies of this nature should be carried out over several cropping cycles.

Food preferences of parasites and predators in the soil environment

One issue limiting our capacity to manage parasites and predators of nematodes is an inadequate understanding of their food preferences in soil. Parasitism and predation can easily be observed in relatively simple systems such as agar plates, but it is not clear whether such observations are relevant to the soil environment, where many potential food sources will exist. The often-debated issue of the relative importance of saprophytism and parasitism in the nutrition of many nematophagous fungi is one such example, but there are many others. *Pochonia chlamydosporia* is a parasite of nematode eggs, but can also live as a saprophyte, grow endophytically in roots, and parasitize mollusc eggs and the oospores of oomycetes (see Chapter 5). However, we do not know whether the fungus utilizes all these nutrient sources in the soil environment.

Mononchid nematodes are considered obligate predators of nematodes, but the preferred prey of individual species in soil is poorly understood, and there is a possibility that some species also use bacteria as a food source (see Chapter 6). The situation is even worse with dorylaimids, with hundreds of species categorized as omnivores on the basis of their morphological features, feeding habits on agar plates and a few observations of gut contents. Questions about food preferences in soil will always be difficult to answer, but PCR-based techniques have commonly been used to identify organisms in the gut contents of arthropods (Symondson, 2002; King *et al.*, 2008), and are now being applied to predators of nematodes (Cabos *et al.*, 2013). PCR primers can be used to examine predation on specific prey or prey groups, but next-generation sequencing technologies will play an increasingly important role in understanding the dietary ranges of predators and their role within soil food webs, as they provide precise taxonomic identification of food items that are actually consumed when a diverse range of potential prey is available (Pompanon *et al.*, 2012).

Improved monitoring and diagnostic services for nematode pests and their natural enemies

In the past, nematode-management decisions have commonly been made by estimating the number of nematodes likely to be present in a field given its soil type, environment and previous cropping history; and then using information on damage thresholds to predict the level of damage on the next crop. This process can be improved by using a nematode diagnostic service to assess nematode populations prior to planting. In some countries, such services are now provided using molecular technologies, and have been expanded to include a range of soilborne pathogens (e.g. Ophel-Keller *et al.*, 2008).

One of the problems with the above approach is that it takes no account of the natural enemies that regulate and suppress nematode populations. Ideally, both the nematode and its main natural enemies would be assessed so that the suppressive effects of those organisms could be taken into account when making management decisions. This was recognized by Yang *et al.* (2012b), who argued that planting decision models for sugarbeet infested with *Heterodera schachtii* would be enhanced by incorporating population densities of *Brachyphoris oviparasitica* and other fungal parasites. These authors then went on to suggest that this could be achieved by developing biologically meaningful measurements of *B. oviparasitica* population densities, and by using field-based experiments to determine the impact of cropping regime and time on *B. oviparasitica*–*H. schachtii* associations.

Molecular technologies are already being used in ecological studies of nematodes and their parasites (e.g. Zhang *et al.*, 2006, 2008; Xiang *et al.*, 2010a, b), and with their continuing development, it will eventually be possible to monitor populations of large numbers of soil organisms at a price that is not prohibitively expensive. Although one of the benefits of such an approach will be a better understanding of the interactions between nematodes and their natural enemies, perhaps the greatest benefit from a practical perspective will be the provision of diagnostic services that give a more holistic view of the soil biological community. Rather than seeing soil as a haven for pests, growers and their advisers will become aware that natural enemies are present in their soils, and this may encourage them to think more seriously about management practices that reduce pest populations without harming beneficial organisms.

Coping with biological complexity

Two of the issues that are readily apparent from the literature on biological control of nematodes are: (i) most of the research has been done with single organisms; and (ii) experiments have generally been carried out with potted plants in sterilized or highly modified soils. Given the difficulties involved in working with soil organisms and the limited techniques that have been available, these approaches are understandable. Nevertheless, it is now time to recognize that there are huge numbers of organisms in soil; that they interact with each other; and that they act collectively against nematodes. Until now, it has been impossible to study this diverse biological community, but continually improving molecular technologies are providing the tools required to unravel the complexity of disease- and pest-suppressive soils (Mendes *et al.*, 2011). What is becoming apparent is that suppressiveness is associated with a biological consortium rather than a single organism. Once the taxa consistently associated with suppressiveness are identified, it should be possible to define a suppressive 'fingerprint' or 'signature' that will indicate whether a soil is likely to be suppressive. When such signatures are available, we should be able to use them to identify the management practices that need to be put in place to maintain a suppressive biological community.

One issue that is not clear with regard to suppressiveness is the importance of biodiversity. Most research has focused on achieving suppressiveness through high levels of biological activity, and it is not known whether a certain amount of diversity is also required. Agricultural soils may already be diversity-deficient (Wall *et al.*, 2010), and it is possible that predators at higher trophic levels in the soil food web have disappeared or declined to very low levels. Since recovery from many years of perturbation will take a long time (Adl *et al.*, 2006; Minoshima *et al.*, 2007; Sánchez-Moreno *et al.*, 2009), and in some cases may never occur, one option that may have to be considered when attempts are made to restore a full range of ecosystem services is to add soil from undisturbed locations to the fields under remediation (Tomich *et al.*, 2011). This option has already been tried and did not appear to be successful (DuPont *et al.*, 2009), but we really need to know whether biodiversity must always be fully restored. For example, when rotation, tillage and amendment practices were used to induce suppressiveness to root-knot nematode in a soil used for vegetable production, populations of omnivorous dorylaimids remained low and predatory mononchids were never observed, suggesting that suppressiveness can result from the activities of a small subset of the natural enemies normally found in a healthy soil (Stirling, 2013).

Multidisciplinary research, innovation networks, research/extension models and the role of farmers

Not all the impediments limiting the development of more robust and durable systems of nematode control are technical. The fragmentation of science into disciplines means that at an organizational level, it is difficult to form research groups with the expertise required to tackle the range of problems involved in developing more sustainable farming systems. Scientists with expertise in soil biology and ecology must be included in such groups, but the complexity of the soil biological community means that participants with different skills will have to be found to deal with the biodiversity that exists below-ground. The long-term trend toward specialization in science and the continuing transfer of resources into molecular biology and biotechnology is another potential problem. Although these changes are providing the analytical expertise, molecular tools and new-generation sequencing technologies required to solve hitherto intractable problems in soil biology, it is important that new disciplines are integrated with traditional areas of science. Ecological and agronomic input will always be required to ensure that key questions are being addressed and that research outcomes are relevant to agriculture.

One of the main objectives of this book has been to consider how suppressiveness to plant-parasitic nematodes can be enhanced within a sustainable farming system. Based on what has been written in this summary chapter and elsewhere, there are obviously many technical issues that are hindering progress. However, it is also apparent that the ultimate goal will not be achieved without a strong contribution from farmers. Most farmers have an abiding interest in ensuring that their cropping system is sustainable; they are adept at managing systems that are inherently complex; and they have access to information on soil health and sustainable agriculture. The most innovative already have farming systems in place that incorporate many of the practices considered vital to long-term sustainability. Since those farmers have experience and knowledge that is valuable to researchers, and the soils on their farms provide an indication of the biology that is achievable in a particular production system and environment, it makes sense to learn from that experience and also use such farms as a resource for research.

Although some segments of the farming community will always be willing to consider alternatives to their current practices, the challenges involved in making widespread changes to farming systems should not be underestimated. Integrated approaches to managing insects, weeds and other pests have been promoted for more than 50 years, but the majority of growers still use single tactics (usually chemicals) as a management tool. Another impediment to change is that farmers who lease land, and corporate landholders whose task is to generate profits for shareholders, are unlikely to be interested in initiatives that will lead to long-term sustainability.

Possible models that might increase the adoption of ecologically based crop management systems and also help farmers improve their understanding of interactions between plants, organic matter and soil organisms were discussed in Chapter 11, but they must involve a partnership between researchers and farmers. Once such partnerships are established, it becomes possible to monitor system performance as adaptations are made to the farming system. Many different parameters could be monitored, but from a nematode-management perspective, measurements of carbon inputs together with data obtained from regular suppression assays and nematode community analyses would provide useful information. Another necessary part of any new research/extension model would be the establishment of interactive learning initiatives with local farm organizations. Such initiatives would be designed in such a way that farmers and researchers could learn from each other and also integrate information held by both parties.

The efficacy of inundative biological control in complex and dynamic soil environments

Although there are strong arguments to indicate that future research should focus on conservation rather than inundative approaches to biological control of soilborne pests, the commercial incentives associated with developing effective biological products will ensure that there is continuing interest in the development of biopesticides suitable for use against nematodes. However, as pointed out previously (see Boxes 12.2 and 12.3), assessing the commercial potential of biological control agents requires more than a few simple tests in the laboratory or greenhouse. In-depth studies are required, with the basic requirements being to demonstrate that: (i) efficacy is observed at realistic application rates; (ii) the effects are due to the organism being introduced rather than to organic materials in the formulation or the substrate on which the organism was grown; (iii) the organism is not only effective in short-term greenhouse experiments with sterilized soil or potting media, but is also efficacious in microplot or field experiments where the host crop is maintained for a full growing season; (iv) the introduced agent can survive in soil for long enough to have an impact on nematode populations and is actually responsible for the nematode mortality observed; and (v) there is some understanding of how the organism being introduced reacts to biostatic forces that will vary with soil moisture, clay content, organic matter status and many other factors. Since it is a long and relatively complex process to bring a biological product to market, it is also important that collaborative links are developed with companies capable of commercializing the research, so that the target market can be defined, issues associated with mass production and formulation can be addressed, and products with adequate shelf-life can be produced.

Although the natural enemies of nematodes discussed in Chapters 5–8 provide a huge arsenal of organisms that could possibly be incorporated into biological products, the range of potentially useful organisms will be even greater in future. New bacterial and fungal endophytes are continually being discovered, while improved culture methods will eventually enable mass production of organisms that are currently difficult to grow *in vitro*.

The soil is an enormous reservoir of microorganisms and the possibilities for finding new candidates for biological control studies are endless. However, focusing on individual organisms may not necessarily be the best strategy. It may be better to obtain a whole suite of fungi and rhizobacteria from a healthy soil, mass-culture them, and transfer that community to a site with a similar soil type, climate or environment.

One question that must be answered with regard to biological products for nematode control is whether they have a role to play in reducing nematode populations in situations where farmers are making every effort to improve soil health and increase the sustainability of their farming system. There will obviously be situations where such farmers will be unable to enhance natural suppressiveness to levels that will permanently maintain populations of a target pest below the damage threshold, and so we need to know whether a biological nematicide is an appropriate control measure in such circumstances. The process of integrated soil biology management promoted in this book aims to increase soil biological activity and diversity, and enhance suppressiveness, and in the process is likely to increase the biostatic forces that limit the establishment of introduced organisms in soil. Thus, it is likely that biological products will be most useful during the transition from biologically impoverished systems, with inundative biological control becoming less effective as the farming system becomes more sustainable.

Concluding Remarks

Well-functioning ecosystems provide a wide range of 'provisioning', 'regulating' and 'supporting' services, and biological control is one of those services. Thus, whenever plant-parasitic nematodes utilize roots as a food source and increase in population density in a natural ecosystem, there is a response from resident antagonists and the nematode population is regulated by predation, parasitism or some other process (Costa *et al.*, 2012).

Biological control is, therefore, a natural ecological phenomenon that should occur in all agroecosystems. However, it will only take place if the soil biological community responsible for providing that service is nurtured rather than depleted by the practices used to grow crops, a process that has been referred to in this book as 'conservation biological control'.

The introduction of mass-produced organisms into soil for the purpose of reducing nematode populations is also a form of biological control. However, it is not a natural process. It is nothing more than a 'silver bullet' that aims to provide an ecosystem service that is no longer being supplied. Biological products with activity against plant-parasitic nematodes may eventually find a place in nematode-management programmes, but their role is likely to be limited to: (i) protecting transplanted crops from nematode attack while a natural microflora establishes on roots; (ii) providing some biological antagonism to nematodes when relatively high-value crops are planted into soils in which natural enemies have been severely depleted by years of mismanagement; and (iii) introducing a beneficial organism into situations where it does not occur or natural populations are low, as a prelude to increasing its population density through management. Thus, it is important that biological control does not become the province of the microbial products industry. Instead, the primary focus of biological control research must be on restoring the regulatory services that have been lost from most agroecosystems. This means focusing on the soil and crop management practices that deplete soil carbon, decrease soil biological activity, diminish biodiversity and impair soil health.

It is possible to eliminate the major deficiencies in many modern farming systems by adopting the principles of conservation agriculture. Farmers in many countries have made that change over the last 20–30 years. The next step is to gradually improve the new farming system, continue to restore soil health, and adopt practices that enhance biological activity and diversity. Many options are available, and they are summarized in the next chapter. However, given the constraints faced by farmers and the time it will take to restore soils that have been damaged by decades of inappropriate management, there will be situations where the level of suppressiveness is insufficient to limit crop losses. This does not mean that the process should be abandoned. Natural suppressiveness should always be the first form of defence against nematode pests, but it will sometimes have to be supported by other control measures. In those situations, the practices included in an integrated nematode-management package must be compatible with the overall objective of improving soil health and enhancing sustainability.

14

A Practical Guide to Improving Soil Health and Enhancing Suppressiveness to Nematode Pests

This concluding chapter is written for anyone with an interest in agriculture, but is targeted at land managers who wish to farm in a more sustainable manner. It aims to encourage those who grow crops or manage agricultural land to view their enterprise from an ecological perspective and consider how current management practices could be modified to achieve the following:

- improved productivity, better soil health and greater sustainability;
- fewer losses from soilborne pests and diseases;
- lower labour costs and reduced dependence on external inputs (e.g. fuel, pesticides and fertilizers); and
- fewer negative impacts on the surrounding environment.

In addition to summarizing the main messages from the book, this chapter provides a pathway to achieving healthier soils and more sustainable farming systems. However, given the range of crops that are grown across the world and the variety of soil types and environments in which agriculture is practised, it is impossible to provide guidelines for every situation. Therefore, the material included focuses on general principles. It is up to each land manager to use them as a framework for developing management practices that are

ecologically sound and relevant to their own crop production system.

Sustainable Agriculture and its Ecological Basis

Biological communities and ecosystem services

Plants are the lifeblood of all ecosystems. They use energy from the sun to convert atmospheric CO_2 into organic matter, which then supports all other organisms in the ecosystem. This community of organisms (plants, animals and microbes) interact with each other, and with non-living components of the environment such as air, water and soil, to provide many services, all of which are important in sustaining life on the planet (Fig. 14.1).

Although the primary role of agriculture is to provide human needs for food and fibre, it is important to remember that an agroecosystem is a community of interacting organisms, and if it is managed appropriately, it will provide all the services depicted in Fig. 14.1. Thus, managers of agricultural land must aim to ensure that in addition to being productive, their land also stores and releases nutrients, suppresses pests and pathogens,

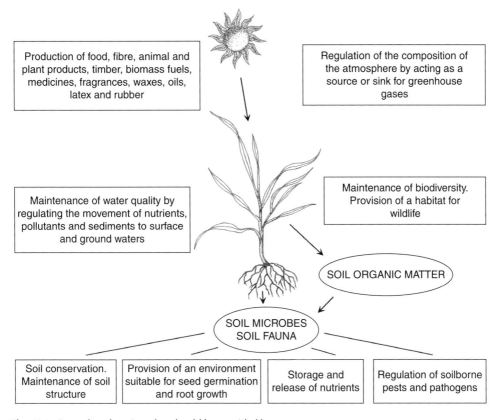

Fig. 14.1. Examples of services that should be provided by an agroecosystem.

maintains water quality, degrades pollutants and sequesters carbon.

Soil biological communities

Plants may be the driving force in an agroecosystem, but it is the unseen organisms that live in soil which make the system operate effectively. Organic matter produced by plants is decomposed by soil organisms, and during that process, the soil's physical structure is maintained and improved; vital nutrients are released for use by plants; roots are colonized by organisms that help the plant defend itself against pathogens; and natural biological control systems begin to operate against soil pests. Thus, plants and the soil biological community are mutually dependent on each other: plants provide the energy source required by soil organisms and, in turn, those organisms provide services that are essential to the plant.

Enormous numbers of organisms are found in soil, and some of the most important groups are listed in Table 14.1 and depicted in Fig. 14.2. However, from a farm-management perspective, it is probably better to think of these organisms as an interacting community. If soil and crops are managed appropriately, members of that community will perform their individual functions, but it is their collective contribution that is important in maintaining plant and soil health.

The soil food web

All the organisms that live in soil interact in what is known as the soil food web. Given the complexity of the soil biological community, it is almost impossible to fully understand the way this food web operates, but a simplified example showing interactions

Table 14.1. Some of the important microbial and faunal groups in soil.

Group	Size and number	Role in soil
Microbes (bacteria, actino-bacteria, fungi)	Microscopic organisms (< about 5 μm in diameter). Thousands of species. Occur in very high numbers (10^{11} bacteria and 10^5 fungi/g soil)	Decompose organic matter. Improve soil structure by binding soil particles together to form aggregates. Form associations with roots that fix nitrogen from air, or help plants retrieve nutrients from soil
Microfauna (algae, protozoa, nematodes)	Protozoa are small (10–100 μm in diameter) and live in water films around soil particles. There are commonly >10^4 protozoa/g soil near the soil surface. Nematodes live in the same environment but are larger (usually 0.5–2 mm long). Populations range from 5 to 50 nematodes/g soil, and more than 50 species are commonly present at one site	Algae convert sunlight into organic compounds. Protozoa and free-living nematodes feed on microbial cells and release nutrients from microbial biomass. Other nematodes are plant-parasites, and some are predators
Mesofauna (mites, springtails, symphyla, enchytraeids)	Most of these organisms are too small to be seen with the naked eye, but the larger animals (up to 2 mm long) are readily visible. Populations under natural vegetation are often >100,000/ m^2, but are much lower in agricultural soils	Feeding habits are diverse. Many groups feed on fungi and others live on decomposing organic matter. Some mites are important predators of nematodes
Macrofauna (ants, millipedes, centipedes, termites, beetles, earthworms)	Although these animals are present in lower numbers than other soil organisms, they are an important component of the soil biological community due to their large size	A variety of roles. Some are predators. The burrowing/channelling activities of others improve drainage and aeration, and change the physical architecture of soil. Litter transformers fragment plant debris, making it more accessible to microbes

between nematodes and other organisms is given in Fig. 14.3. Note that plants have been placed at the top of the diagram, as the whole food web is dependent on organic matter derived from plants. Microorganisms that live on the root surface utilize sugars and other organic substances that are exuded by roots, while other bacteria and fungi decompose dead leaves, shoots and roots. These microbes then become a food source for another group of organisms: bacterial and fungal-feeding nematodes, protozoa, mites and springtails. Since some nematodes are also capable of feeding on living roots, the nematode community at this level of the food web consists of three groups: bacterial feeders, fungal feeders and plant feeders. Those nematodes, in turn, become a food source for a whole range of predators, including fungi, nematodes and small arthropods. Thus, the number of predators

in soil and their diversity depends on the availability of prey, and this means that a constant supply of bacterial and fungal-feeding nematodes is the key to maintaining natural enemies that in many cases may also feed on plant-parasites.

Although actual food webs are much more complex than the structure depicted in Fig. 14.3, they have a number of important features:

- The level of activity and diversity within the soil food web is dependent on carbon inputs from plants.
- Food resources are generally limited in soil, and this means that soil organisms will compete for those resources.
- Organisms interact by taking nutrients from competitors; producing antibiotics or enzymes; or preying on each other.

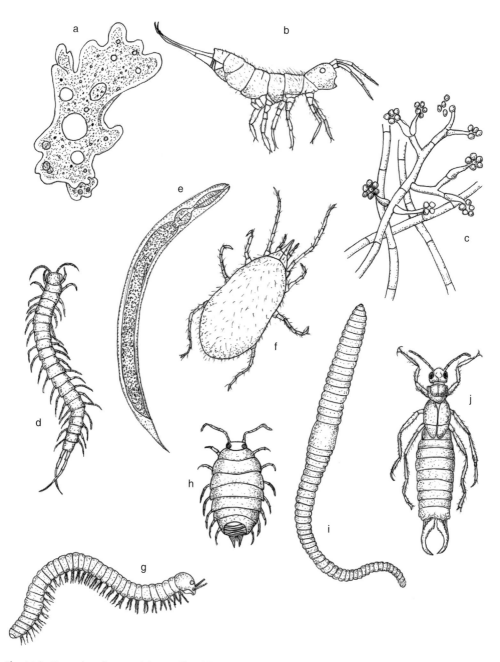

Fig. 14.2. Examples of some of the small and large organisms found in soil (not drawn to scale). From top, left to right: (a) protozoa; (b) springtail; (c) fungus; (d) centipede; (e) nematode; (f) mite; (g) millipede; (h) isopod; (i) earthworm; and (j) earwig.

- During these interactions, which will inevitably occur within the soil food web, nutrients such as nitrogen and phosphorus are continually recycled. These nutrients are contained within the bodies of soil organisms, but when those organisms are consumed by predators, the nutrients are

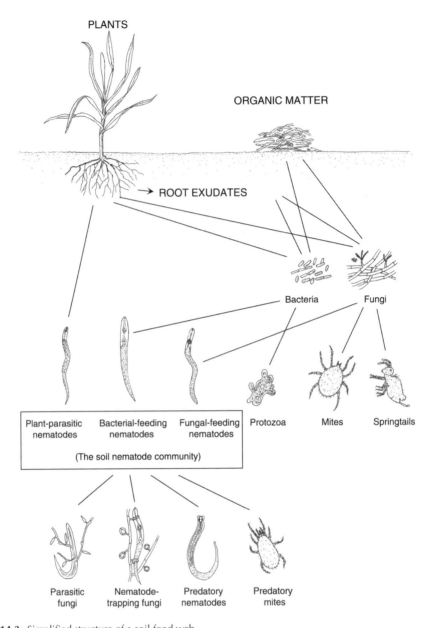

Fig. 14.3. Simplified structure of a soil food web.

released in mineral form and can be used by plants.

- Populations of individual organisms are subject to a constant series of checks and balances, and this prevents any particular organism (including pests such as plant-parasitic nematodes) from becoming dominant.

Soil physical and chemical fertility, and the role of organic matter

Interactions between organic matter and organisms in the soil food web are of interest from a biological perspective, but it is important to note that these interactions also have many positive effects on a soil's physical and chemical properties (see Box 14.1). Thus, better management of organic matter is the key to improving the overall health of agricultural soils so that they ultimately provide all the important functions required for crop production and environmental protection.

and that its activity and diversity is dependent on inputs of organic matter. However, when we look at the soils currently used for agriculture, we find that their biological status has declined markedly over the last 100 years or so. There are numerous reasons for this, but the following are some of the most important.

Soil fertility decline and the impact of management

It is clear from the material presented that the soil biological community is important,

Excessive tillage

The tillage practices commonly used to grow crops are the single most important reason for low biological activity and diversity in

---

Box 14.1. Impact of organic matter on soil physical and chemical properties

Organic matter is present in relatively small amounts in most agricultural soils, but it has profound effects on soil physical and chemical fertility. In most cases, these effects are manifested through interactions with organisms in the soil food web.

Physical fertility

Good soil structure is a key attribute of a fertile soil, and organic matter contributes in several ways:

- *Structural stability.* Roots and soil microbes bind soil particles and organic compounds together into stable aggregates, so that they are not dispersed when exposed to rainfall impact or disturbed by cultivation. In the absence of organic matter, slaking and dispersion occurs, and this leads to surface crusting, hard setting and other structural problems.
- *Water infiltration.* Earthworms and other soil fauna consume organic matter, and in the process create macropores and channels that allow water to rapidly infiltrate and drain. Any impairment of this process results in loss of topsoil and nutrients.
- *Storage of water.* Soils with high carbon contents have a high water-holding capacity because humus particles, which are formed as a result of microbial breakdown of organic matter, can hold several times their own weight in water.
- *Penetration resistance.* Plant roots will only grow readily in soils where resistance to penetration is minimal. The action of soil organisms creates pathways that allow roots to grow through soil without being impeded.

Chemical fertility

Although external inputs are necessary in agriculture to replace nutrients removed in the harvested product, soil organic matter plays an important role in crop nutrition:

- *Source of nutrients.* Soil organic matter is a source of all the nutrients required by plants and is the principal long-term storage medium for important elements such as nitrogen, phosphorus and sulfur. Most of the nitrogen taken up by the crop comes from organic pools rather than current-year fertilizer (except when the system is overwhelmed by excessively high nutrient inputs, as often occurs in intensive vegetable production).
- *Storage and release of nutrients.* During the decomposition process, nutrients in organic matter may be immobilized within microbial biomass or released in mineral form. Thus, the supply of nutrients to plants, and the potential for leaching losses to groundwater, are both governed by the interaction between organic matter and the soil biological community.
- *Buffering effects.* Humic substances have the capacity to resist changes in soil pH, thereby limiting the acidifying effects of fertilizers.

agricultural soils. When the soil is mechanically disturbed, much of the organic matter required to support the soil biological community is lost as carbon dioxide. Tillage also kills some soil organisms, particularly the larger ones, and many of them are predators. Since large organisms have long life cycles, they may take months or years to return. Although the impact of tillage depends on its frequency and intensity, any disturbance event will have a negative impact on many organisms within the soil food web.

Inadequate residue management

Many farmers do not value the residues produced by crops, and fail to include biomass-producing crops in their rotations. Burning crop residues, moving residues off-site or incorporating them into soil so that they decompose quickly depletes soil organic matter, the food source that supports the soil biological community.

Excessive fertilizer and pesticide inputs

Some practices used in modern agriculture (e.g. soil fumigation) are disastrous to the soil biological community, as beneficial organisms are killed as well as pests. Since the nutrients removed in harvested products must be replaced, nutrient inputs are essential in agriculture. However, fertilizers may have a negative impact on the soil biology, particularly if all the nutrients required by the crop are applied at the time of planting. From a biological perspective, it is best to apply small quantities of nutrients over time (e.g. by using slow-release formulations or adjusting application procedures), as this reduces the chance that toxic concentrations will occur in the water films inhabited by soil organisms. Pesticides may also be detrimental, but effects will vary with application rate, frequency and the spectrum of activity.

Soil compaction

Small arthropods occupy soil pores and crevices because they need space to move. When soils are compacted by farm machinery,

their homes are destroyed. Since root growth is impeded as soil strength increases, soil compaction problems must be addressed. However, tilling or deep ripping to break up hardpans and compacted layers is only a temporary solution, as the soil will be recompacted if traffic is not controlled. Also, such tillage operations further exacerbate organic matter decline and kill more predators.

Given the management practices commonly used in modern agriculture, it is not surprising that many farmers have to deal with soil-related problems such as compaction, soil loss due to erosion, poor infiltration of rainfall and irrigation water, excessive leaching of nutrients, dryland salinity, low water- and nutrient-holding capacities, and reduced yields due to soilborne pests and diseases. When farmers are faced with such problems, there are two choices: (i) continue with current practices and obtain temporary solutions by continually increasing fuel, nutrient and pesticide inputs; or (ii) consider an alternative farming system that provides a long-term solution.

Sustainable farming systems

Responsible and enduring stewardship of the land is the essence of sustainable agriculture. When a farming system is sustainable, agricultural land is managed in a manner that maintains its long-term productivity, resilience and vitality while minimizing adverse environmental impacts.

Sustainable agriculture is not a prescribed set of practices. It is a concept that challenges land managers to:

- View agriculture from an ecological perspective and try to understand how management practices not only influence plants, but also soil properties, the soil's organic matter status and the soil biological community.
- Recognize that biological processes play a key role in maintaining the soil resource.
- Balance the requirements for productivity and profitability with the need to sustain the soil and the wider environment.

- Use the ingenuity and practical skills of farmers, the basics of agricultural science, our understanding of soil ecology and the knowledge gained from on-farm research, to continually modify management practices and ultimately develop a farming system that is appropriate for a particular soil type, climate and production goal.

A Guide to Improving Soil Health and Minimizing Losses from Soilborne Diseases

Poor soil health and ongoing losses from soilborne diseases are a product of inappropriate farming systems and their harmful effects on soil organic matter and the soil biological community. Thus, land managers facing such problems need to assess their soil's carbon levels; reflect on the probable impacts of past and current management practices on the soil's biological status; decide on the factors most likely to be responsible for reducing biological activity and diversity; and then consider possible solutions. Modifications to the farming system will almost certainly be needed, and this is likely to be a daunting task, particularly if major adjustments are required. Nevertheless, change is possible. Many farmers have demonstrated this in recent years by adopting practices such as minimum tillage, residue retention, controlled traffic and site-specific management. Useful rotation crops (e.g. legumes and biomass-producing cover crops) have also been introduced into many farming systems.

The following section is a step-by-step guide to the issues that should be considered in determining whether an agricultural soil is being degraded, whether unnecessary losses from soilborne diseases are being incurred, and whether improvements to the current farming system are possible.

Step 1. Assess soil health

Healthy soils are the basis of sustainable agriculture. Soil physical, chemical or biological

constraints prevent crops reaching their yield potential, but also result in soil particles and nutrients moving off-site, and a loss of productive capacity in the long term. Assessing soil health is, therefore, the first step in determining whether current farming systems can be improved.

Detailed manuals and test kits for assessing soil health are available in many countries. Some examples are given in the reference section at the end of this chapter, and they can be consulted for detailed information on assessment methods. Soil samples can also be forwarded to local diagnostic laboratories. The following are examples of some of the properties that can be assessed:

- *Soil physical properties*: Soil structure, aggregate stability, water infiltration rate, bulk density and penetration resistance.
- *Soil chemical properties*: pH, electrical conductivity and plant macro- and micronutrients.
- *Soil organic matter*: Total carbon and labile carbon.
- *Soil biological properties*: Microbial biomass, soil respiration and the composition of the nematode community.
- *Root health*: Root biomass, root length, colour of roots, the degree of branching and the number of fine roots.
- *Soilborne diseases*: Presence of root lesions, the extent of root rotting and the severity of symptoms caused by specific pathogens (e.g. galling caused by root-knot nematode).

Step 2. Assess impacts of the farming system on soil health and consider options for improvement

The above process is likely to show that the soil being assessed is suboptimal in at least some respects. The next step is to identify the main soil health problems and the factors likely to be contributing to those problems (Table 14.2).

Having identified the factors contributing to poor soil health, it is possible to put together a list of management practices that

Table 14.2. Common soil problems that limit crop productivity, farm sustainability and environmental quality, and the main contributing factors.

Constraint	Contributing factors
Compacted soil	Uncontrolled traffic, particularly when soil is wet; use of heavy farm equipment
Surface crusting	Intensive tillage; low soil organic matter; insufficient use of biomass-producing crops; removal of crop residues
Acidification	Removal of harvested products from the field; leaching of nitrate below the root zone; use of acidifying nitrogenous fertilizers
Poor water infiltration rates	Lack of surface cover; low soil organic matter; absence of earthworms and root channels; compaction; excessive tillage
Low water-holding capacity	Low soil organic matter; high sand content
Poor nutrient retention	Low soil organic matter; excessive tillage
Soilborne pest and disease problems	Excessive use of susceptible crops; inadequate crop rotation; low soil organic matter

will help to alleviate the problem. Although these practices are listed individually in Table 14.3, they should be thought of holistically, as several practices will almost certainly have to be used together to achieve the desired outcomes.

The most important factor determining the health of an agricultural soil is the level of soil organic matter, as it affects a wide range of soil physical, chemical and biological properties. Thus, many of the practices listed in Table 14.3 are designed to increase inputs of plant and animal residues, and decrease losses of organic matter due to decomposition and erosion.

Step 3. Modify soil and crop management practices and assess the outcomes

There are many pathways to agricultural sustainability, and no single configuration of technologies, inputs and management practices is likely to be applicable in all situations. Nevertheless, the practices listed in Table 14.3 are consistently associated with healthy soils and sustainable crop production systems, and when applied together, or in various combinations, they will work synergistically to provide all the ecosystem services required of a soil: production of food and fibre; adsorption and infiltration of water; retention and cycling of nutrients; sequestration of carbon; detoxification of harmful chemicals; and suppression of soilborne pests and diseases.

When changes to a farming system are being considered, it is important to recognize that sustainable agriculture is not prescriptive. Many potentially useful technologies and practices are available, and it is up to the land manager to adapt them to local conditions and constraints. Nevertheless, the task does not have to be tackled alone. Extension personnel and farm-management consultants can provide information on farming systems that are likely to be successful, and will be able to list the benefits and limitations of various practices. Local research trials are also a useful resource, if they exist. Finally, it should be possible to find growers who are familiar with all the practices listed in Table 14.3, as they are commonly used in many forms of agriculture around the world.

Once decisions have been made to modify the farming system, the process should be prioritized, as it may not be possible to make all the desired changes immediately. Also, it is wise to test any new system on a small scale to sort out initial problems that inevitably occur.

Most agricultural soils are physically, chemically and biologically degraded, and it will take many years to repair them. Soil improvement is, therefore, a long-term process, and the benefits will not be seen immediately. For this reason, it is worthwhile monitoring key soil properties, as the results will indicate whether the system is moving in the right direction. Total carbon

Table 14.3. Benefits and limitations of soil and crop-management practices likely to improve soil physical, chemical and biological properties.

Management practice	Benefits and limitations
Reduced tillage	Various options are available to reduce the depth, intensity and frequency of tillage operations, with no-till being associated with the least physical disturbance. The main benefit of reducing tillage is increased levels of soil organic matter, and this has flow-on effects to many important soil properties. Also, organisms in the soil food web are no longer continually disrupted. Reduced fuel consumption provides economic and environmental gains, while reduced erosion and improved water and soil conservation are also important benefits
Controlled traffic with GPS guidance	Reduces soil compaction problems, particularly in farming systems with heavy machinery. Controlled traffic systems are normally an integral component of no-till agriculture
Maintenance of a permanent cover of plant residues	Moderates soil temperature fluctuations; conserves water and nutrients; protects soil from erosion; minimizes weed problems; and promotes soil biological activity. In cool climates, a disadvantage of retaining residues from cover crops on the soil surface is that soil temperatures may fail to warm quickly in spring. One requirement is that satisfactory methods of managing weeds must be available
Crop rotation	Rotational cropping is the simplest way of reducing populations of crop-specific pests and pathogens. Also, all rotation crops provide carbon inputs, while legumes will fix nitrogen
Increased plant diversity	Intercropping, planting of hedgerows and retention of windbreaks are examples of practices that increase biodiversity and ecosystem resilience. These practices also help sustain predators of above-ground pests
Organic amendments	Composts and other amendments add nutrients, stimulate nutrient cycling, provide carbon inputs, and enhance biological suppression of soilborne pests and pathogens. Maintaining organic mulch on the soil surface provides the same benefits, but also conserves soil moisture and minimizes soil temperature fluctuations
Improved crop yield potential through plant breeding and genetic modification	Provides resistance to major pests and pathogens. Increased biomass production also results in greater carbon inputs, although larger external inputs are often required to obtain higher yields. Narrowing of the genetic base decreases ecosystem resilience
Effective water and nutrient management	Produces healthy, productive plants that are better able to cope with the stress associated with loss of root function caused by soilborne pests and pathogens
Site-specific management	Enables crop requirements to be better matched to variable soil properties, thereby saving money and minimizing unintended impacts on the environment
Integrated crop and livestock production	Improves levels of soil organic matter; provides nitrogen inputs if legumes are included in the pasture phase; and reduces populations of crop-specific pests and pathogens
Integrated pest management	Reduces the need for pesticides that may have a negative impact on non-target species, or become environmental pollutants. A disadvantage is that more intensive management is required

and labile carbon are the logical parameters to assess, as they are relatively easy to measure and have direct and indirect effects on many soil properties. However, some other useful indicators are presented later.

Biological Control of Nematodes: One of Many Important Ecosystem Services

This chapter commenced by pointing out that a large and diverse group of organisms live in soil (Table 14.1) and that they co-exist within the soil food web. In their fight for survival, they prey on each other and compete for resources. Since plant-parasitic nematodes are one component of the soil food web, they are a potential food source for other members of the soil biological community. Thus, if their populations increase, they should be kept under control by the wide range of parasites and predators that are capable of attacking them (Fig. 14.4).

What should be obvious from Fig. 14.4 is that many different soil organisms are capable of attacking nematodes. Bacterial and fungal parasites produce spores that can adhere to nematodes as they move through soil. After those spores germinate, the parasite proliferates through the body, killing the nematode. A unique group of fungi produce ring traps, mycelial networks and other structures that are used to capture nematodes. Some predatory nematodes have teeth that enable them to eat other nematodes, while many large nematodes have feeding spears that are used to suck out the contents of nematode eggs. Some small mites and springtails have mouth parts that enable them to consume nematodes. Since there are many different species in each of these major groups of parasites and predators, the soil food web should be seen as a repository for hundreds of beneficial organisms. The challenge is to ensure that they survive in agricultural soils and play a role in limiting populations of nematode pests.

Another important message from Fig. 14.4 is that few natural enemies attack only the nematodes that feed on plants. Most are generalist predators, which means that they also consume bacterial and fungal-feeding nematodes. Thus, the number of predators within the soil food web is dependent on the presence of these nematodes (often referred to as free-living nematodes). Since free-living nematodes consume the bacteria and fungi that decompose organic matter, their numbers are determined by the quality and quantity of carbon inputs. Continual inputs of sugars and other readily utilizable forms of organic matter are, therefore, the key to maintaining the food supply for these generalist predators.

Although the natural enemies depicted in Fig. 14.4 play an important role in limiting populations of plant-parasitic nematodes, plants have also evolved protective mechanisms that help prevent their roots from being destroyed by pests. They allow a group of bacteria and fungi known as endophytes to colonize their root system, and the presence of these microbes induces the plant to turn on resistance mechanisms that will reduce nematode multiplication rates.

The important message from the preceding paragraphs is that in a properly functioning soil food web, the nematodes that feed on plants will be suppressed by the activities of parasites and predators, and also by plant resistance mechanisms that are switched on when appropriate microorganisms colonize roots. The fact that plant-parasitic nematodes dominate the nematode community in most agricultural soils is an indication that the soil food web is not functioning as it should.

Nematode-suppressive soils

Most of the organisms depicted in Fig. 14.4 can be found in the soils used for agriculture, but levels of predatory activity are usually not high enough to reduce nematode populations substantially. Nevertheless, there are situations where populations of a nematode pest are strongly suppressed by naturally occurring organisms. In some cases, relatively specialized parasites capable of attacking only certain nematode species will build up to high levels and permanently suppress the pest. Since control is achieved through the action of host-specific parasites, this phenomenon is termed *specific suppression*. In other cases, inputs of organic matter increase soil biological activity and diversity to the point where nematode pests are kept under control by the collective activities of a wide variety of natural enemies. Since the development of this type of suppressiveness is dependent on organic inputs, it is referred to as *organic matter-mediated suppression*. Land managers wishing to reduce populations of plant-parasitic nematodes through natural mechanisms must aim to ensure that both forms of suppression are operating in their soil.

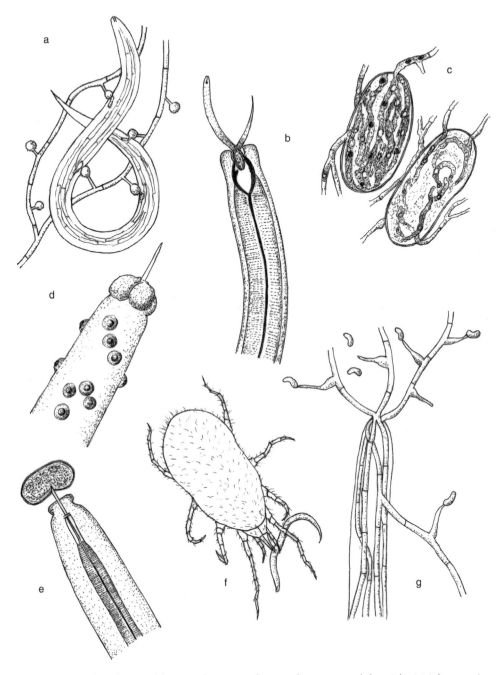

Fig. 14.4. Examples of some of the natural enemies of nematodes. From top, left to right: (a) A fungus using adhesive knobs to capture a nematode; (b) a predatory nematode eating another nematode; (c) fungi parasitizing nematode eggs; (d) spores of a bacterial parasite on the surface of a nematode; (e) a predatory nematode using its feeding spear to prey on a nematode egg; (f) a mite consuming a nematode; (g) a parasitic fungus growing from a nematode cadaver.

Nematode Management Within Sustainable Farming Systems

The first reaction of farmers faced with nematode problems is to find some way of eliminating the pest, or substantially reducing pest populations. What is often forgotten is that many of the control measures that are available (e.g. soil fumigants, nematicides, solarization and long fallow periods) are non-selective, and also kill the nematode's natural enemies. Thus, the end result is that the soil loses its natural capacity to suppress the pest, and populations will return to high levels when the next susceptible crop is grown.

Practices such as those listed above provide only temporary control and must, therefore, be reapplied before every crop is planted. From an ecological perspective, it makes more sense to manage the system as a whole so that the pest is suppressed by naturally occurring parasites and predators. This means thinking holistically about how plants, crop residues, nematode pests, their natural enemies and other members of the soil biological community can be manipulated to achieve the desired result. If that can be achieved, soil health will be enhanced and losses from nematodes will be minimized.

Examples of potentially sustainable farming systems

As mentioned previously, sustainable agriculture is not something that can be prescribed, as practices that are relevant in one locality, soil type or environment may not be suitable in another situation. Nevertheless, when the ingenuity and practical experience of farmers is combined with ecological knowledge, it is possible to improve the sustainability of most farming systems. The examples given next provide an indication of the types of changes that are possible in three quite different forms of agriculture. Issues related to nematode management in sustainable farming systems are also discussed.

Large-scale production of grains, oilseeds, fibre crops and pastures

Most of the world's food and fibre crops (wheat, maize, sorghum, barley, oats, millet, sunflower, canola, soybean, lupin, chickpea, faba bean, field pea, mung bean, lentil and cotton) are produced in large-scale cropping operations that specialize in two or three crops. The most productive and environmentally sustainable systems for producing these crops combine high cropping intensities, minimum tillage, controlled traffic, biomass-producing cover crops, crop residue retention, sound rotation practices and optimal nutrient inputs through technologies associated with precision agriculture. Collectively, these practices sustain or increase soil organic carbon levels, improve soil physical properties, reduce soil losses due to erosion, provide biologically fixed nitrogen, improve water infiltration, increase water-holding capacity and reduce losses from soil-borne diseases. Although there has been a trend in recent years towards separating cropping and livestock enterprises, mixed farming systems also have an important role to play in sustainable agriculture, particularly if biological nitrogen fixation and manure recycling are included in the pasture phase.

The widespread adoption of the above practices in many countries over the last few decades is testimony to the fact that regardless of soil type or climate, farming systems can be improved. There are certainly difficulties that will have to be overcome, but they are not insurmountable. For example, problems associated with planting into large quantities of crop residues can be overcome with appropriate equipment, while reduced soil temperatures under the residue layer and crop emergence problems associated with residue decomposition can be solved by changing the location of crop residues, or the way they are managed.

Plant-parasitic nematodes cause serious losses in many of the world's major field crops, with cyst nematodes (*Heterodera*), root lesion nematodes (*Pratylenchus*) and reniform nematode (*Rotylenchulus*) some of the most important pests. Thus, when a new farming system is being developed, appropriate nematode management practices must be integrated into the system. Crop rotation and the use of nematode-resistant cultivars are standard methods of reducing populations of particular nematode pests, and information on crops and cultivars that are suitable to the local area can be obtained from nematologists and extension specialists. Nematode-tolerant crops and cultivars should also be considered, as they suffer minimal losses from nematodes and help maintain pest-specific natural enemies (see comment at the end of this section). One advantage of using tolerance rather than resistance is that resistance-breaking biotypes of the nematode are never likely to develop.

Once a farming system is in place that incorporates all these measures, the critical question is whether plant-parasitic nematodes are still causing unacceptable levels of damage. Nematode pests will certainly be present, but all available evidence suggests that because crops are no longer growing in physically degraded soils, water-use efficiency has improved, and crop nutrition is managed more professionally, then losses from nematodes will be markedly reduced. In fact, when soil health is improved and plants suffer less moisture and nutrient stress, it may not be necessary to reduce nematode populations to achieve productivity gains.

Although levels of organic matter will gradually increase as the farming system is improved, it may be many years before organic matter-mediated mechanisms of suppression begin to operate effectively. Nevertheless, nematode populations should eventually decline, particularly in surface soils where organic inputs are highest. Nematode-specific biological control agents should also start to increase once the soil is no longer cultivated, as the intimate interactions between the pest and its highly specialized natural enemies are no longer being continually disturbed by rotary hoes, ploughs and other tillage implements. Increased cropping frequencies should also be beneficial, provided cultivars with high levels of nematode resistance do not dominate the rotation. Tolerant or partially resistant cultivars are a better option from a biocontrol perspective, as they maintain low to moderate populations of the pest, thereby ensuring that the food source required by host-specific parasites is always available.

Vegetable crops

In contrast to field crop industries, the vegetable industry has been slow to adopt improved and sustainable soil management practices. Most vegetable-growing soils are cultivated aggressively, and so soil biological activity has declined to the point where the soil is little more than an inert medium to physically support the crop. The principal vegetable crops are also grown repeatedly, with the inevitable nematode and soilborne disease problems being dealt with using soil fumigants and soil-applied pesticides. Nevertheless, there are examples of relatively intensive vegetable production systems where cover crops (including legumes to supply nitrogen) and other sources of organic matter are used as mulches in no-till or reduced-tillage systems to replace plastic mulch, suppress weeds, maintain better soil moisture, reduce soil erosion and nurture beneficial soil organisms.

In vegetable production systems that are not highly intensive, there is no reason why practices considered suitable for more extensive cropping systems could not be applied (i.e. minimum tillage, controlled traffic, crop residue retention, sound rotation practices and site-specific management). In fact, vegetable growers are likely to be able to implement the required changes and receive the benefits more quickly than farmers involved in dryland agriculture. Their crop is more valuable and irrigation water is usually available, and so they can afford higher inputs and exert greater management control.

Once a more appropriate vegetable farming system is in place, perhaps the most challenging issue in making it more sustainable is improving the way crop nutrition is managed. Vegetable growers traditionally apply large amounts of fertilizer, whereas a sustainable system would be more reliant on nitrogen inputs from legumes and mineralization from soil organic matter. Additional nutrient inputs will still be required, but application rates and timing must not only be determined by the crop's requirements, but also by the need to preserve beneficial organisms in the soil food web.

The strategy most likely to improve the biological status of soils used for highly intensive vegetable production is to utilize crop residues, composts, biosolids and other suitable waste products as organic amendments. By varying the quality, quantity and timing of organic inputs, and the degree of disturbance (i.e. whether organic materials are incorporated into soil or applied to the soil surface as mulch), it should be possible to produce a soil that is not only suitable for plant growth but also suppresses soilborne pathogens and provides the nutrients required by the crop.

Since root-knot nematodes (*Meloidogyne*), cyst nematodes (*Heterodera* and *Globodera*) and several other plant-parasitic nematodes cause problems on many vegetable crops, effective nematode management must be a component of a sustainable vegetable farming system. Fortunately, nematode-resistant and tolerant rotation crops are obtainable; fast-growing biomass-producing crops can be used to increase levels of soil organic matter; no-till and strategic tillage systems that preserve soil carbon are being used successfully by vegetable growers in some countries; organic amendments can be applied to further increase soil carbon levels and enhance suppressiveness to nematodes; nutrients in soil and plant tissue are easily monitored using commercial services; irrigation can be managed effectively using computerized probes that measure soil moisture at various depths in the profile; and analytical services are available to monitor populations of key soilborne pests and pathogens. Thus, many of the practices required to manage nematodes, and minimize the nutrient and moisture stresses that exacerbate nematode problems, are now available. Thus, there is no reason why innovative growers should not accept the challenge of trying to integrate these practices into their farming system. The first step is to establish some on-farm experiments and determine which practices are most likely to be useful.

Perennial trees and vines

Since some perennials have cropping cycles of about 5 years (e.g. sugarcane, banana and pineapple), and most tree and vine crops are in place for a much longer period, long-term sustainability is easier to achieve with perennials than with annual crops. Nevertheless, it will not happen automatically. Planning is required from the outset to ensure that the rows or beds used to grow the crop are established in such

a way that drainage is satisfactory and traffic is controlled, as this will minimize the need for tillage when the site is eventually replanted. Once the crop is established, a permanent layer of organic matter must be maintained on the soil surface, either by retaining residues produced by the crop; mulching with organic materials introduced from elsewhere; or growing companion crops or pastures in the inter-row space. In the latter case, shade-tolerant plants may be required and the sward must be well managed so the inter-row plantings do not compete with the perennial crop for water and nutrients.

Provided steps are taken to ensure that surface cover is well maintained and compaction is limited to traffic zones, soil health should be sustained and may even improve during the life of the crop. However, all crops must eventually be replanted, so the one remaining challenge is to find ways of doing this without losing the soil-health benefits gained during the long cropping phase. Since soil disturbance must be minimized, one option worth exploring is to kill the perennial crop with herbicide, remove above-ground biomass, grow a rotation crop that will fix nitrogen or contribute organic matter, and then replant in the spaces between the original plants. If deep ripping or cultivation is necessary to remove old stems and roots, it should be limited to the narrowest possible zone. An organic amendment would then be applied to replace the soil carbon that is inevitably lost during the tillage process. Once the crop is replanted, it would be mulched with a thick layer of organic matter, as this helps retain moisture, encourages root development near the soil surface, and provides food resources for the soil biological community.

Numerous plant-parasitic nematodes attack perennial crops, but problems are most likely to occur during the replanting process, as the crop is establishing a new root system in a soil that is already infested from the previous crop. Thus, it may be necessary to reduce populations of key nematode pests by growing a resistant rotation crop for at least 6–12 months. Once the perennial crop is re-established, the aim is to utilize organic inputs to maintain a food web in upper layers of the soil profile that is capable of maintaining nematode populations at relatively low levels. However, enhancing organic matter-mediated suppressiveness at lower depths is likely to be difficult, and this is where specific suppressiveness may be useful. When a nematode pest is associated with a crop for a long period and the soil is not disturbed, host-specific parasites often increase to levels that provide some nematode control. The best-known example is *Pasteuria*, a bacterium that can increase to high levels on root-knot nematode, one of the most important soilborne pests of perennials. Thus, biological management of nematode pests in perennial crops involves managing the soil food web with organic inputs to achieve suppression in surface soils, and encouraging the build-up of nematode-specific parasites at depth.

Other crops

In a book of this nature, it is obviously not possible to cover all of the crops grown around the world. Nevertheless, the examples presented provide an indication of the types of practices that should find a place in a sustainable farming system. Although farmers in developing countries will not have access to some of the technologies that have been mentioned, the principles involved in producing acceptable crops, maintaining healthy soils and managing nematode pests are still applicable. Thus, soil disturbance must be minimized; crops that are poor hosts of the target nematode must be included in the rotation; crop residues must be maintained on the soil surface; legumes must be used where possible to provide nitrogen inputs; and nutrients removed in harvested product must be replaced.

Indicators of improvement

Making improvements to a farming system is a challenging process. Since continual fine-tuning is likely to be required and results will not be seen immediately, it is worthwhile considering some indicators that could be used to assess progress. Crop yields will always be measured, as maintaining or improving yield

is always an important objective. However, this should not be the overriding goal, as the benefits of alternative farming systems are often manifested in other ways. Examples include increased profitability due to reductions in fuel and labour costs when tillage is minimized; quicker access to fields after rain because a controlled traffic system has been installed; or a more stable financial situation because a rotation crop provides an additional income stream.

From a soil-health perspective, numerous soil-health kits are available (see references at the end of this chapter), and they can be used to measure improvements in soil physical properties such as aggregate stability, water-holding capacity and penetration resistance. Roots can also be assessed to check whether they are showing symptoms of damage caused by soilborne pathogens.

One of the most useful indicators of progress is total soil carbon, and it should always be one of the parameters measured when soil is sent to a commercial laboratory for nutrient analysis. Regularly measuring labile carbon is perhaps even more useful, as this organic fraction sustains the soil microbial community.

From a biological perspective, the parameters that can be measured will depend on the diagnostic services that are locally available. In Australia, for example, a commercial service that quantifies soil organisms using DNA technologies has been operating since 1998. More than 100 tests are now offered and they cover most of the important fungal pathogens of roots, many plant-parasitic nematodes, several groups of free-living nematodes and other beneficial organisms such as mycorrhizae. If such services can be accessed, they can be used to monitor changes in the soil biological community and provide an indication of whether improvements are occurring. If a nematode diagnostic service is available, it should not only be used to monitor populations of plant-parasitic nematodes, but also to assess the omnivorous and predatory component of the nematode community. The latter group of nematodes is particularly important, as the presence of high numbers of predatory nematodes indicates that the soil food web has some capacity to suppress plant-parasitic nematodes.

Potential problems and possible solutions

Maintaining productivity, improving sustain-ability and managing pests in an ecologically sound manner is not a simple process, par-ticularly if the soil has been degraded by many years of inappropriate management. Growers intending to proceed down that path need to read widely, talk to professionals with experience in sustainable agriculture, and if possible, discuss their options with growers who have already begun the process of improving their farming system.

As could be expected when changes are being made to a farming system, some teeth-ing problems will occur during the transition to the new system. Some examples, and pos-sible solutions, are:

- Compaction and soil structural problems are common in soils that have been physic-ally degraded by many years of inappro-priate management. Thus, it may not be possible to switch immediately to a no-till system. Some form of tillage, additional inputs of organic matter and perhaps even a few years of a well-managed pas-ture, may be needed to restore the soil's physical fertility to an acceptable level.
- In situations where organic amendments with a high C:N ratio are incorporated into soil, or crop residues are used as mulch, fertilizer inputs may have to be carefully managed to avoid nitrogen immobiliza-tion problems.
- One problem that sometimes occurs when crop residues are retained as a source of organic matter is that root rot-ting caused by pathogens such as *Pythium* and *Rhizoctonia* may increase. Both these pathogens are capable of multiplying on fresh plant tissue, and so the solution usually involves ensuring that residues from the previous crop have decomposed before the next crop is planted.
- Weeds are always a potential problem, and in no-till farming systems they can be suppressed through the competitive effects of cover cropping and mulching, targeted herbicide application and the judicious use of tillage. However, herbicide-resistant weeds can develop when products

such as glyphosate are used excessively. Non-chemical weed management options such mechanical roller-crimping, flame-weeding and mechanical weed-knives may, therefore, be worth considering.
- Soft-stemmed annuals often suffer cut-worm damage when soil is mulched, and so control measures may be required in the first few weeks after planting.

Despite the potential problems, imple-menting the practices discussed in the last few pages will almost certainly improve the over-all health of the soil. However, it is important to recognize that we still have much to learn about how crops, soil and organic matter should be managed to optimize the services provided by the soil biological community. For example, estimating rates of nutrient min-eralization from organic matter remains prob-lematic, so it is difficult to optimize fertilizer rates and timing so that the nutrient inputs from all sources match the crop's require-ments. We also know that injudicious use of nitrogen fertilizer, for example, is likely to have negative effects on some components of the biological community (e.g. some preda-tors of nematodes), but the best way of over-coming such problems is not yet clear.

As soil health improves and the soil bio-logical community becomes more active and diverse, the soil will become more suppressive to plant-parasitic nematodes. However, as there have been relatively few studies of nematode-suppressive soils, it is difficult to predict whether levels of suppression will be sufficient to keep nematode pests under control. Current evidence suggests that a reasonably high level of suppressiveness is achievable, and that it will be effective in many circumstances. How-ever, in situations where highly pathogenic nematodes cause damage to crops at very low population densities, additional control meas-ures (e.g. cultivars with resistance or tolerance to the target nematode) may be an essential component of the new farming system.

Conclusions

Improving the sustainability of a farming sys-tem is not a simple task. It requires time and effort; a long-term commitment; good planning; ecologically sound objectives; a capacity to solve individual problems in a way that does not jeopardize the whole system; and a willingness to continually update management practices as new information become available. Nevertheless, the journey is worth taking, as the process is challenging, informative and satisfying. There will almost certainly be stumbling blocks along the way, but there is overwhelming evidence to suggest that the end result will be healthier soils, a more profitable farming enterprise and fewer impacts on the surrounding environment.

Questions Related to Soil Health, Soil Organic Matter and Nematode Management

This discussion concludes with some questions that could possibly be asked by farmers; I hope the answers will help to clarify some of the points made in this chapter.

It is all very well to focus on practices that supposedly improve the biology of soils. However, I can't see the organisms that live in soil, so how do I know that a soil's biological status is improving?
The soil biological community uses decomposing organic matter as a food source. Soil organisms are also reliant on sugars and other carbon compounds exuded from roots. Thus, it may not really be necessary to measure the biological status of soil. The biology will improve if plants are continually present and their residues are returned to the soil. However, if the soil is cultivated or subjected to long fallow periods, organic matter is dissipated and the soil biology will decline.

What indictors can I use to show that my soil is biologically healthy?
If there is a permanent cover of leaves, stems and other organic materials on the soil surface, or if the soil is covered with a mulch layer, a diverse range of soil animals will be present. Look carefully through the litter and surface layer of your soil and check for ants, beetles, spiders, slaters, millipedes and mites. Then use a spade to collect some soil. It should be a dark 'organic' colour and contain plenty of fine roots. Check that it is friable, breaking into crumbs of varying size when subjected to finger pressure. Place some pea-sized aggregates in water and check that they remain intact, a sign of aggregate stability. If the soil is friable and aggregates do not disperse in water, it is an indication that the soil is structurally sound and that larger organisms will be able to move about between soil particles. The presence of earthworms is also a good sign.

Soil carbon, particularly labile carbon, is perhaps the most useful indicator, as all important soil physical, chemical and biological properties are strongly influenced by carbon content. Soil nutrient-testing laboratories can measure total carbon; while labile carbon, the active fraction that is the food source for soil microorganisms, can be measured in the field using relatively simple methods.

The material in this chapter indicated that soil carbon levels vary with soil type and environment. Why?
Inputs of organic matter into soil are determined by plant biomass production, and are, therefore, much greater in high than low rainfall environments. Organic matter decomposes more rapidly in warm than cool soils, and in moist rather than dry soils. Clay particles protect organic matter from decomposition, and so decomposition rates are lower in clay soils than sandy soils. Thus, many different factors will influence soil carbon levels.

How can I determine what soil carbon levels are achievable in my particular soil type and environment?
One useful method is to find some undisturbed grassland or native vegetation in the neighbourhood and measure the soil carbon content. Suitable sites can often be found along fencelines, or next to a creek or roadway. Provided these areas have had continual organic inputs from plants; have not been disturbed for many years; and are exactly the same soil type as the soil being cropped, their carbon status will provide an indication of what carbon levels are possible in that particular environment.

Is it possible to increase soil carbon in cropped soils to the levels achievable in undisturbed grassland or native vegetation?

Yes, it is possible. However, carbon inputs must be greater than in the natural system, because when land is cropped, some biomass is removed as harvested product. Also, those inputs must not be dissipated through practices such as tillage. If irrigation is available, soil carbon levels should be much higher than a comparative non-irrigated soil, because organic inputs will not be limited by rainfall. Unfortunately, data collected from many countries shows that carbon levels in cropped soils are often 50–90% lower than achievable levels.

I understand that a soil cannot be considered healthy unless it contains more than 2% carbon. Is that true?

This figure is sometimes used as a threshold value, but because soil carbon levels are influenced by soil type and environment, it is not appropriate for all situations. For example, in a sandy soil cropped annually in a warm climate with 300 mm of rainfall, it may only be possible to achieve soil carbon levels of 1% using best-practice management. However, carbon levels of 3–4% may be achievable if the soil was a more fertile clay loam, or if the sandy soil received more rainfall, or was in a cooler climate. Also, soil health may be more closely related to labile carbon levels than to total carbon.

Can organic materials be introduced from elsewhere to improve soil health and biological activity?

Yes, but the degree of decomposition of the amendment (i.e. whether it is undecomposed or has been composted); its application rate; its C:N ratio; its composition (i.e. the proportion of recalcitrant compounds such as lignin and tannin); and the manner in which it is applied (i.e. incorporated with a tillage implement or left on the soil surface) will markedly influence the result obtained.

My soil was sent to an analytical laboratory and five species of plant-parasitic nematodes were found. Do I need to worry about them?

Plant-parasitic nematodes will always be present in agricultural soils, but most species cause little or no damage because they do not reduce root biomass, limit root elongation or produce major symptoms. Others will remain at relatively low population densities. A few species known to transmit plant viruses are particularly problematic, but in most cases, plant-parasitic nematodes should be seen as a normal component of the soil food web, and not something that must be eradicated. The important issue from a production perspective is to know which species have the capacity to cause economic losses in your soil type and environment, and then manage them so that their population densities are kept at reasonable levels.

How can I determine which nematodes may cause problems?

Use local extension services to find out which nematode pests are capable of damaging the crops of interest. Also, try to determine whether they are likely to cause problems on your farm. Many nematodes only damage crops in particular soils (e.g. sands, clay loams or clays) or in certain environments (e.g. cool, warm or hot climates).

How do I decide whether nematode damage is severe enough to justify action?

Look for symptoms that could be caused by nematodes (e.g. poor growth, wilting, chlorosis, stunting and yield decline). Also, check roots for signs of nematode damage (e.g. galling, presence of lesions, distortion of root tips, excessive root branching and low root biomass). If symptoms could possibly be caused by nematodes, confirm that they are one of the main causal factors by checking for other root pathogens and other possible exacerbating factors (e.g. inadequate nutrition or moisture stress). Forward soil and root samples to a nematode diagnostic service, and then seek advice from a professional about whether the results of the test suggest that nematodes are likely to be causing economic damage.

How can I check whether my management practices are enhancing suppressiveness to nematodes?
Scientists are able to test a soil for suppressiveness in the laboratory. Since farmers generally cannot access this type of test, one way of assessing whether a soil is suppressive is to set up an on-farm test in some small plots or a few rows. Use a diagnostic service to check the population density of the target nematode, and if it is not excessively high, plant a susceptible crop and monitor the response. The nematode will multiply rapidly to high levels and cause damage if the soil is conducive, whereas populations will remain low and the crop will not suffer damage if it is suppressive

Another option is to find a competent nematode diagnostic service and ask for a complete nematode community analysis (i.e. a count of plant-parasitic nematodes and the major groups of free-living nematodes). If plant-parasites predominate, it is a clear sign that the soil is not suppressive. The presence of a nematode community dominated by free-living species and containing many omnivores and predators indicates that the soil is supplying some suppressive services. In the future, DNA tests are likely to be developed that provide a signature or fingerprint of the types of biological communities normally present in a suppressive soil. Once tests of this nature are commercialized, it will be possible to determine whether a soil has the appropriate signature.

I understand that biological products can now be used to control nematodes. Are they effective?
It is true that some parasites and predators of nematodes have been mass-produced commercially, formulated as a granule or wettable powder, and are now being marketed for biological control purposes. However, because the agent being applied to soil must compete with organisms that are already present, populations of the introduced organisms soon decline to relatively low levels. Nematode populations are usually reduced for no more than a few weeks, but in some situations, this may be sufficient to protect the crop from damage. However, high levels of control are rarely achieved. If you consider such products are worth trying, treat a few small, randomly selected areas of a field and leave the adjacent areas untreated. Measure the yield response and use the results to determine whether treatment is economically feasible.

Does that advice also apply to soil improvers, organic fertilizers, microbial inoculants and other products sold by my retailer?
Yes. It is always worthwhile testing a new product on-farm before using it on a large scale. The effectiveness of any product that acts through biological processes will vary with the crop, soil type and environmental conditions.

Useful Information on Soil Health

Gugino, B.K., Idowu, O.J., Schindelbeck, R.R., van Es, H.M., Wolfe, D.W., Moebius-Clune, B.N., Thies, J.E. and Abawi, G.S. (2009) *Cornell Soil Health Assessment Training Manual.* 2nd edn. http://soilhealth.cals.cornell.edu/extension/manual.htm, accessed 8 November 2013. Cornell University, Geneva, NY.

Pattison, T. and Lindsay, S. (2006) *Banana Root and Soil Health User's Manual: FR02023 Soil and Root Health for Sustainable Banana Production.* Department of Primary Industries and Fisheries, Queensland, Australia.

Soil Quality Website (2013) Soil Quality Website. Coordinated by Dr Daniel Murphy, University of Western Australia. http://soilquality.org.au, accessed 8 November 2013.

USDA Soil Quality Test Kit Guide (2010) The Soil Quality Test Kit Guide: Measuring soil quality in the field. http://soils.usda.gov/sqi/assessment/files/test_kit_fact_sheet.pdf, accessed 8 November 2013.

References

Abbott, L.K. and Murphy, D.V. (eds) (2007) *Soil Biological Fertility. A Key to Sustainable Land Use in Agriculture.* Springer, Dordrecht, the Netherlands.

Abiko, S., Saikawa, M. and Ratnawati, M. (2005) Capture of mites and rotifers by four strains of *Dactylella gephyropaga* known as a nematophagous hyphomycete. *Mycoscience* 46, 22–26.

Acea, M.J., Moore, C.R. and Alexander, M. (1988) Survival and growth of bacteria introduced into soil. *Soil Biology & Biochemistry* 20, 509–515.

Adamchuk, V.I., Hummel, J.W., Morgan, M.T. and Upadhyaya, S.K. (2004) On-the-go soil sensors for precision agriculture. *Computers & Electronics in Agriculture* 44, 71–91.

Adl, S.M., Coleman, D.C. and Read, F. (2006) Slow recovery of soil biodiversity in sandy loam soils of Georgia after 25 years of no-tillage management. *Agriculture, Ecosystems & Environment* 114, 323–334.

Affokpon, A., Coyne, D.L., Lawouin, L., Tossou, C., Agbèdè, R.D. and Coosemans, J. (2011a) Effectiveness of native West African arbuscular mycorrhizal fungi in protecting vegetable crops against root-knot nematodes. *Biology & Fertility of Soils* 47, 207–217.

Affokpon, A., Coyne, D.L., Htay, C.C., Agbèdè, R.D., Lawouin, L. and Coosemans, J. (2011b) Biocontrol potential of native *Trichoderma* isolates against root-knot nematodes in West African vegetable production systems. *Soil Biology & Biochemistry* 43, 600–608.

Ahmad, J.S. and Baker, R. (1987) Rhizosphere competence of *Trichoderma harzianum. Phytopathology* 77, 182–189.

Ahmad, W. and Jairajpuri, M.S. (2010) *Mononchida. The Predatory Soil Nematodes.* Nematology Monographs & Perspectives No. 7. Brill, Leiden, the Netherlands.

Ahren, D. and Tunlid, A. (2003) Evolution of parasitism in nematode-trapping fungi. *Journal of Nematology* 35, 194–197.

Ahrén, D., Ursing, B.M. and Tunlid, A. (1998) Phylogeny of nematode-trapping fungi based on 18S rDNA sequences. *FEMS Microbiology Letters* 158, 179–184.

Akhtar, M. (2000) Nematicidal potential of the neem tree *Azadirachta indica* (A. Juss). *Integrated Pest Management Reviews* 5, 57–66.

Akhtar, M. and Mahmood, I. (1994) Potentiality of phytochemicals in nematode control: a review. *Bioresource Technology* 48, 189–201.

Akhtar, M. and Malik, A. (2000) Roles of organic soil amendments and soil organisms in the biological control of plant-parasitic nematodes: a review. *Bioresource Technology* 74, 35–47.

Alabouvette, C. (1999) Fusarium wilt suppressive soils: an example of disease-suppressive soils. *Australasian Plant Pathology* 28, 57–64.

Alabouvette, C., Olivain, C., Migheli, Q. and Steinberg, C. (2009) Microbiological control of soil-borne phyopathogenic fungi with special emphasis on wilt-inducing *Fusarium oxysporum. New Phytologist* 184, 529–544.

Alexander, M. (1977) *Introduction to Soil Microbiology*. 2nd edn. Wiley, New York.

Ali, N.I., Siddiqui, I.A., Shaukat, S.S. and Zaki, M.J. (2002) Nematicidal activity of some strains of *Pseudomonas* spp. *Soil Biology & Biochemistry* 34, 1051–1058.

Allen-Morley, C.R. and Coleman, D.C. (1989) Resilience of soil biota in various food webs to freezing perturbations. *Ecology* 70, 1127–1141.

Altieri, M.A. (1999) The ecological role of biodiversity in agroecosystems. *Agriculture, Ecosystems & Environment* 74, 19–31.

Anastasiadis, I.A., Giannakou, I.O., Prophetou-Athanasiadou, D.A. and Gowen, S.R. (2008) The combined effect of the application of a biocontrol agent *Paecilomyces lilacinus*, with various practices for the control of root-knot nematodes. *Crop Protection* 27, 352–361.

Anderson, J.M., Preston, J.F., Dickson, D.W., Hewlett, T.E., Williams, N.H. and Maruniak, J.E. (1999) Phylogenetic analysis of *Pasteuria penetrans* by 16S rRNA gene cloning and sequencing. *Journal of Nematology* 31, 319–325.

Andrade, G., Mihara, K.L., Linderman, R.G. and Bethlenfalvay, G.J. (1998) Soil aggregation status and rhizobacteria in the mycorrhizosphere. *Plant & Soil* 202, 89–96.

Angers, D.A. and Eriksen-Hamel, N.S. (2008) Full inversion tillage and organic carbon distribution in soil profiles: a meta-analysis. *Soil Science Society of America Journal* 72, 1370–1374.

Arihara, J. and Karasawa, T. (2000) Effect of previous crops on arbuscular mycorrhizal formation and growth of succeeding maize. *Soil Science & Plant Nutrition* 46, 43–51.

Arora, D.K., Hirsch, P.R. and Kerry, B.R. (1996) PCR-based molecular discrimination of *Verticillium chlamydosporium* isolates. *Mycological Research* 100, 801–809.

Ash, G.J. (2010) The science, art and business of successful bioherbicides. *Biological Control* 52, 230–240.

Ashelford, K.E., Norris, S.J., Fry, J.C., Bailey, M.J. and Day, M.J. (2000) Seasonal population dynamics and interactions of competing bacteriophages and their host in the rhizosphere. *Applied & Environmental Microbiology* 66, 4193–4199.

Ashford, D.L. and Reeves, D.W. (2003) Use of a mechanical roller-crimper as an alternative kill method for cover crops. *American Journal of Alternative Agriculture* 18, 37–45.

Athman, S.Y., Dubois, T., Coyne, D., Gold, C.S., Labuschagne, N. and Viljoen, A. (2007) Effect of endophytic *Fusarium oxysporum* on root penetration and reproduction of *Radopholus similis* in tissue culture-derived banana (*Musa* sp.) plants. *Nematology* 9, 599–607.

Atibalentja, N., Noel, G.R., Liao, T.F. and Gertner, G.Z. (1998) Population changes in *Heterodera glycines* and its bacterial parasite *Pasteuria* sp. in naturally infested soil. *Journal of Nematology* 30, 81–92.

Atibalentja, N., Noel, G.R. and Domier, L.L. (2000) Phylogentic position of the North American isolate of *Pasteuria* that parasitizes the soybean cyst nematode, *Heterodera glycines*, as inferred from 16S rDNA sequence analysis. *International Journal of Systematic & Evolutionary Microbiology* 50, 605–613.

Atibalentja, N., Jakstys, B.P. and Noel, G.R. (2004) Life cycle, ultrastructure, and host specificity of the North American isolate of *Pasteuria* that parasitizes the soybean cyst nematode, *Heterodera glycines*. *Journal of Nematology* 36, 171–180.

Atkins, S.D., Clark, I.M., Sosnowska, D., Hirsch, P.R. and Kerry, B.R. (2003a) Detection and quantification of *Plectosphaerella cucumerina*, a potential biological control agent of potato cyst nematodes, by using conventional PCR, real-time PCR, selective media, and baiting. *Applied & Environmental Microbiology* 69, 4788–4793.

Atkins, S.D., Hidalgo-Diaz, L., Clark, I.M., Morton, C.O., Montes De Oca, N., Gray, P.A. and Kerry, B.R. (2003b) Approaches for monitoring the release of *Pochonia chlamydosporia* var. *catenulata*, a biocontrol agent of root-knot nematodes. *Mycological Research* 107, 206–212.

Atkins, S.D., Hidalgo-Diaz, L., Kalisz, H., Mauchline, T.H., Hirsch, P.R. and Kerry, B.R. (2003c) Development of a new management strategy for the control of root-knot nematodes (*Meloidogyne* spp.) in organic vegetable production. *Pest Management Science* 59, 183–189.

Atkins, S.D., Mauchline, T.H., Kerry, B.R. and Hirsch, P.R. (2004) Development of a transformation system for the nematophagous fungus *Pochonia chlamydosporia*. *Mycological Research* 108, 654–661.

Atkins, S.D., Clark, I.M., Pande, S., Hirsch, P.R. and Kerry, B.R. (2005) The use of real-time PCR and species-specific primers for the identification and monitoring of *Paecilomyces lilacinus*. *FEMS Microbiology Ecology* 51, 257–264.

Atkins, S.D., Peteira, B., Clark, I.M., Kerry, B.R. and Hirsch, P.R. (2009) Use of real-time quantitative PCR to investigate root and gall colonisation by co-inoculated isolates of the nematophagous fungus *Pochonia chlamydosporia*. *Annals of Applied Biology* 155, 143–152.

Atkinson, C.J., Fitzgerald, J.D. and Hipps, N.A. (2010) Potential mechanisms for achieving agricultural bene-fits from biochar application to temperate soils: a review. *Plant & Soil* 337, 1–18.

Atkinson, H.J. and Dürschner-Pelz, U. (1995) Spore transmission and epidemiology of *Verticillium balanoides*, an endozoic fungal parasite of nematodes in soil. *Journal of Invertebrate Pathology* 65, 237–242.

Augé, R.M. (2001) Water relations, drought and vesicular-arbuscular mycorrhizal symbiosis. *Mycorrhiza* 11, 3–42.

Augé, R.M. (2004) Arbuscular mycorrhizae and soil/plant water relations. *Canadian Journal of Soil Science* 84, 373–381.

Auld, B.A., Hetherington, S.D. and Smith, H.E. (2003) Advances in bioherbicide formulation. *Weed Biology & Management* 3, 61–67.

Avery, A. (2007) 'Organic abundance' report: fatally flawed. *Renewable Agriculture & Food Systems* 22, 321–329.

Avis, T.J., Hamelin, R.C. and Bélanger, R.R. (2001) Approaches to molecular characterization of fungal bio-control agents: some case studies. *Canadian Journal of Plant Pathology* 23, 8–12.

Bacetty, A.A., Snook, M.E., Glenn, A.E., Noe, J.P., Hill, N., Culbreath, A., Timper, P., Nagabhyru, P. and Bacon, C.W. (2009) Toxicity of endophyte-infected tall fescue alkaloids and grass metabolites on *Pratylenchus scribneri*. *Phytopathology* 99, 1336–1345.

Badejo, M.A., Tian, G. and Brussaard, L. (1995) Effect of various mulches on soil microarthropods under a maize crop. *Biology & Fertility of Soils* 20, 294–298.

Badgely, C. and Perfecto, I. (2007) Can organic agriculture feed the world? *Renewable Agriculture & Food Systems* 22, 80–82.

Badgley, C., Moghtader, J., Quintero, E., Zakem, E., Chappell, J.M., Avilés-Vázquez, K., Samulon, A. and Perfecto, I. (2007) Organic agriculture and the global food supply. *Renewable Agriculture & Food Systems* 22, 86–108.

Badri, D.V. and Vivanco, J.M. (2009) Regulation and function of root exudates. *Plant, Cell & Environment* 32, 666–681.

Bae, Y.-S. and Knudsen, G.R. (2000) Cotransformation of *Trichoderma harzianum* with β-glucuronidase and green fluorescent protein genes provides a useful tool for monitoring fungal growth and activity in natural soils. *Applied & Environmental Microbiology* 66, 810–815.

Bae, Y.-S. and Knudsen, G.R. (2001) Influence of a fungal-feeding nematode on growth and biocontrol effi-cacy of *Trichoderma harzianum*. *Phytopathology* 91, 301–306.

Bahme, J.B., Schroth, M.N., Van Gundy, S.D., Weinhold, A.R. and Tolentino, D.M. (1988) Effect of inocula delivery systems on rhizobacterial colonization of underground organs of potato. *Phytopathology* 78, 534–542.

Bailey, K.L. (2010) Canadian innovations in microbial pesticides. *Canadian Journal of Plant Pathology* 32, 113–121.

Bailey, K.L. and Lazarovits, G. (2003) Suppressing soil-borne diseases with residue management and organic amendments. *Soil & Tillage Research* 72, 169–180.

Bais, H.P., Park, S.-W., Weir, T.L., Callaway, R.M. and Vivanco, J.M. (2004) How plants communicate using the underground information superhighway. *Trends in Plant Science* 9, 26–32.

Bais, H.P., Weir, T.L., Perry, L.G., Gilroy, S. and Vivanco, J.M. (2006) The role of root exudates in rhizosphere interactions with plants and other organisms. *Annual Review of Plant Biology* 57, 233–266.

Bajaj, H.K. and Walia, K.K. (2005) Studies on a *Pasteuria* isolate from an entomopathogenic nematode, *Steinernema pakistanense* (Nematoda: Steinernematidae). *Nematology* 7, 637–640.

Baker, C.J., Saxton, K.E., Ritchie, W.R., Chamen, W.C.T., Reicosky, D.C., Ribeiro, M.F.S., Justice, S.E. and Hobbs, P.R. (2006) *No-tillage Seeding in Conservation Agriculture*, 2nd edn. CAB International/FAO, Oxford, UK.

Baker, K.F. and Cook, R.J. (1974) *Biological Control of Plant Pathogens*. W.H. Freeman & Co., San Francisco, California.

Baker, R. (1990) An overview of current and future strategies and models for biological control. In: Hornby, D. (ed.) *Biological Control of Soil-Borne Plant Pathogens*. CAB International, Wallingford, UK, pp. 375–388.

Baldock, J.A. and Skjemstad, J.O. (1999) Soil organic carbon/soil organic matter. In: Peverill, K.I., Sparrow, L.A. and Reuter, D.J. (eds) *Soil Analysis and Interpretation Manual*. CSIRO Publishing, Collingwood, VIC, Australia, pp. 159–170.

Balis, C. and Kouyeas, V. (1979) Contribution of chemical inhibitors to soil mycostasis. In: Schippers, B. and Gams, W. (eds) *Soil-borne Plant Pathogens*. Academic Press, London, pp. 97–106.

Barber, D.A. and Martin, J.K. (1976) The release of organic substances by cereal roots into soil. *New Phytologist* 76, 69–80.

Bardgett, R.D. (2005) *The Biology of Soil: A Community and Ecosystem Approach*. Oxford University Press, Oxford, UK.

Bardgett, R.D. and Griffiths, B.S. (1997) Ecology and biology of soil protozoa, nematodes and microarthro-
    pods. In: van Elsas, J.D., Wellington, E.M.H. and Trevors, J.T. (eds) *Modern Soil Microbiology*. Marcel
    Dekker, New York, pp. 129–163.

Bardgett, R.D. and McAlister, E. (1999) The measurement of soil fungal:bacterial biomass ratios as an indicator
    of ecosystem self-regulation in temperate meadow grasslands. *Biology & Fertility of Soils* 19, 282–290.

Bar-Eyal, M., Sharon, E., Spiegel, Y. and Oka, Y. (2008) Laboratory studies on the biocontrol potential of the
    predatory nematode *Koerneria sudhausi* (Nematoda: Diplogastridae). *Nematology* 10, 633–637.

Barker, K.R. and Koenning, S.R. (1998) Developing sustainable systems for nematode management. *Annual
    Review of Phytopathology* 36, 165–205.

Barker, K.R. and Olthof, T.H.A. (1976) Relationships between nematode population densities and crop
    responses. *Annual Review of Phytopathology* 14, 327–353.

Barker, K.R., Pederson, G.A. and Windham, G.L. (1998) *Plant and Nematode Interactions*. American Society
    of Agronomy Inc., Madison, Wisconsin.

Barnes, G.L., Russell, C.C. and Foster, W.D. (1981) *Aphelenchus avenae*, a potential biological control agent
    for root rot fungi. *Plant Disease* 65, 423–424.

Barron, G.L. (1973) Nematophagous fungi: *Rhopalomyces elegans. Canadian Journal of Botany* 51, 2505–2507.

Barron, G.L. (1977) *The Nematode-Destroying Fungi*. Lancaster Press Inc., Lancaster, Pennsylvania.

Barron, G.L. (1990) A new predatory hyphomycete capturing copepods. *Canadian Journal of Botany* 68, 691–696.

Barron, G.L. (1992) Lignolytic and cellulolytic fungi as predators and parasites. In: Carroll, G.C. and Wicklow, D.T.
    (eds) *The Fungal Community, Its Organization and Role in the Ecosystem*. Mycology Series 9, Marcel-
    Decker, New York, pp. 311–326.

Barron, G.L. and Dierkes, Y. (1977) Nematophagous fungi: *Hohenbuehelia*, the perfect state of *Nematoctonus.
    Canadian Journal of Botany* 55, 3054–3062.

Barron, G.L. and Thorn, R.G. (1987) Destruction of nematodes by species of *Pleurotus. Canadian Journal of
    Botany* 65, 774–778.

Baxter, R.I. and Blake, C.D. (1969a) Some effects of suction on the hatching eggs of *Meloidogyne javanica.
    Annals of Applied Biology* 63, 183–190.

Baxter, R.I. and Blake, C.D. (1969b) Oxygen and hatch of eggs and migration of larvae of *Meloidogyne
    javanica. Annals of Applied Biology* 63, 191–203.

Beakes, G.W. and Glockling, S.L. (1998) Injection tube differentiation in gun cells of a *Haptoglossa* species
    which infects nematodes. *Fungal Genetics & Biology* 24, 45–68.

Beakes, G.W., Glockling, S.L. and Sekimoto, S. (2012) The evolutionary phylogeny of the oomycete "fungi".
    Protoplasma 249, 3–19.

Beare, M.H. (1997) Fungal and bacterial pathways of organic matter decomposition and nitrogen mineraliza-
    tion in arable soils. In: Brussaard, L. and Ferrera-Cerrato, R. (eds) *Soil Ecology in Sustainable Agricultural
    Systems*. CRC Press/Lewis Publ., Boca Raton, Florida, pp. 37–70.

Beare, M.H., Hendrix, P.F. and Coleman, D.C. (1994) Water-stable aggregates and organic matter fractions in
    conventional and no-tillage soils. *Soil Science Society of America Journal* 58, 777–786.

Beare, M.H., Hu, S., Coleman, D.C. and Hendrix, P.P. (1997) Influences of mycelial fungi on soil aggre-
    gation and organic matter storage in conventional and no-tillage soils. *Applied Soil Ecology* 5,
    211–219.

Becker, J.O., Zavaleta-Mejia, E., Colbert, S.F., Schroth, M.N., Weinhold, A.R., Hancock, J.G. and van
    Gundy, S.D. (1988) Effects of rhizobacteria on root-knot nematodes and gall formation. *Phytopathology*
    78, 1466–1469.

Bekal, S., Borneman, J., Springer, M.S., Giblin-Davis, R.M. and Becker, J.O. (2001) Phenotypic and molecular
    analysis of a *Pasteuria* strain parasitic to the sting nematode. *Journal of Nematology* 33, 110–115.

Bekal, S., Domier, L.L., Niblack, T.L. and Lambert, K.N. (2011) Discovery and initial analysis of novel viral
    genomes in the soybean cyst nematode. *Journal of General Virology* 92, 1870–1879.

Bell, M.J., Harch, G.R. and Bridge, B.J. (1995) Effects of continuous cultivation on ferrosols in subtropical
    southeast Queensland. 1. Site characterization, crop yields and soil chemical status. *Australian Journal
    of Agricultural Research* 46, 237–253.

Bell, M.J., Bridge, B.J., Harch, G.R. and Orange, D.N. (1997) Physical rehabilitation of degraded kraznozems
    using ley pastures. *Australian Journal of Soil Research* 35, 1093–1113.

Bell, M.J., Moody, P.W., Connolly, R.D. and Bridge, B.J. (1998) The role of active fractions of soil organic matter
    in physical and chemical fertility of ferrosols. *Australian Journal of Soil Research* 36, 809–819.

Bell, M.J., Halpin, N.V., Orange, D.N. and Haines, M. (2001) Effect of compaction and trash blanketing on rainfall
    infiltration in sugarcane soils. *Proceedings Australian Society of Sugarcane Technologists* 23, 161–167.

Bell, M.J., Halpin, N.V., Garside, A.L., Moody, P.J., Stirling, G.R. and Robotham, B.G. (2003) Evaluating combinations of fallow management, controlled traffic and tillage options in prototype sugarcane farming systems at Bundaberg. *Proceedings Australian Society of Sugarcane Technologists* 25: CD ROM.

Bell, M., Seymour, N., Stirling, G.R., Stirling, A.M., van Zwieten L., Vancov, T., Sutton, G. and Moody, P. (2006) Impacts of management on soil biota in Vertosols supporting the broadacre grains industry in northern Australia. *Australian Journal of Soil Research* 44, 433–451.

Bell, M.J., Stirling, G.R. and Pankhurst, C.E. (2007) Management impacts on health of soils supporting Australian grain and sugarcane industries. *Soil & Tillage Research* 97, 256–271.

Bello, A., López-Pérez, J., García-Álvarez, A., Sanz, R. and Lacasa, A. (2004) Biofumigation and nematode control in the Mediterranean region. *Nematology Monographs & Perspectives* 2, 133–149.

Bellotti, W.D. (2001) The role of forages in sustainable cropping systems of southern Australia. *Proceedings, XX International Grassland Congress, São Pedro, Brazil*, pp. 729–734.

Benítez, M.-S. and McSpadden Gardener, B.B. (2009) Linking sequence to function in soil bacteria: sequence-directed isolation of novel bacteria contributing to soilborne disease suppression. *Applied & Environmental Microbiology* 75, 915–924.

Benítez, T., Rincon, A.M., Limón, M.C. and Codón, A.C. (2004) Biocontrol mechanisms of *Trichoderma* strains. *International Microbiology* 7, 249–260.

Bent, E., Loffredo, A., McKenry, M.V., Becker, J.O. and Borneman, J. (2008) Detection and investigation of soil biological activity against *Meloidogyne incognita*. *Journal of Nematology* 40, 109–118.

Berendsen, R.L., Pieterse, C.M.J. and Bakker, P.A.H.M. (2012) The rhizosphere microbiome and plant health. *Trends in Plant Science* 17, 478–486.

Berkelmans, R., Ferris, H., Tenuta, M. and van Bruggen, A.H.C. (2003) Effects of long-term crop management on nematode trophic levels other than plant feeders disappear after 1 year of disruptive soil management. *Applied Soil Ecology* 23, 223–235.

Bernard, E.C. (1992) Soil nematode biodiversity. *Biology & Fertility of Soils* 14, 99–103.

Bernard, E.C. and Arroyo, T.L. (1990) Development, distribution, and host studies of the fungus *Macrobiotophthora vermicola* (Entomophthorales). *Journal of Nematology* 22, 39–44.

Bethlenfalvay, G.J., Cantrell, I.C., Mihara, K.L. and Schreiner, R.P. (1999) Relationships between soil aggregation and mycorrhizae as influenced by soil biota and nitrogen nutrition. *Biology & Fertility of Soils* 28, 356–363.

Bhan, M., McSorley, R. and Chase, C.A. (2010) Effect of cropping system complexity on plant-parasitic nematodes associated with organically grown vegetables in Florida. *Nematropica* 40, 53–70.

Bhattacharyya, P.N. and Jha, D.K. (2012) Plant growth-promoting rhizobacteria (PGPR): emergence in agriculture. *World Journal of Microbiology & Biotechnology* 28, 1327–1350.

Bieleski, R.L. (1973) Phosphate pools, phosphate transport, and phosphate availability. *Annual Review of Plant Physiology* 24, 225–252.

Bilgrami, A.L. (1993) Analysis of the predation by *Aporcelaimus nivalis* on prey nematodes from different prey trophic categories. *Nematologica* 39, 356–365.

Bilgrami, A.L. and Gaugler, R. (2005) Feeding behaviour of the predatory nematodes *Laimydorus baldus* and *Discolaimus major* (Nematoda: Dorylaimida). *Nematology* 7, 11–20.

Bilgrami, A.L. and Jairajpuri, M.S. (1989) Predatory abilities of *Mononchoides longicaudatus* and *M. fortidens* (Nematoda: Diplogasterida) and factors influencing predation. *Nematologica* 35, 475–488.

Bilgrami, A.L., Ahmad, I. and Jairajpuri, M.S. (1983) Some factors influencing predation by *Mononchus aquaticus*. *Revue de Nématologie* 6, 325–326.

Bilgrami, A.L., Ahmad, I. and Jairajpuri, M.S. (1984) Observations on the predatory behaviour of *Mononchus aquaticus*. *Nematologia Mediterranea* 12, 41–45.

Bilgrami, A.L., Gaugler, R. and Brey, C. (2005) Prey preference and feeding behaviour of the diplogasterid predator *Mononchoides gaugleri* (Nematoda: Diplogastrida). *Nematology* 7, 333–342.

Bilgrami, A.L., Brey, C. and Gaugler, R. (2008) First field release of a predatory nematode, *Mononchoides gaugleri* (Nematoda: Diplogastrida), to control plant-parasitic nematodes. *Nematology* 10, 143–146.

Bird, A.F. and Brisbane, P.G. (1988) The influence of *Pasteuria penetrans* in field soils on the reproduction of root-knot nematodes. *Revue de Nématologie* 11, 75–81.

Bird, G.W. (1987) Role of nematology in integrated pest management programs. In: Veech, J.A. and Dickson, D.W. (eds) *Vistas on Nematology*. Society of Nematologists, Hyattsville, Maryland, pp. 114–121.

Bishop, A.H. and Ellar, D.J. (1991) Attempts to culture *Pasteuria penetrans* in vitro. *Biocontrol Science & Technology* 1, 101–114.

Bishop, A.H., Gowen, S.R., Pembroke, B. and Trotter, J.R. (2007) Morphological and molecular characteristics of a new species of *Pasteuria* parasitic on *Meloidogyne ardenensis. Journal of Invertebrate Pathology* 96, 28–33.

Bjørnlund, L. and Rønn, R. (2008) 'David' and 'Goliath' of the soil food web – flagellates that kill nematodes. *Soil Biology & Biochemistry* 40, 2032–2039.

Blessit, J.A., Stetina, S.R., Wallace, T.P., Smith, P.T. and Sciumbato, G.L. (2012) Cotton (*Gossypium hirsutum*) cultivars exhibiting tolerance to the reniform nematode (*Rotylenchulus reniformis*). *International Journal of Agronomy* Article ID 893178. DOI:10.1155/2012/893178.

Blevins, R.L and Frye, W.W. (1993) Conservation tillage: an ecological approach to soil management. *Advances in Agronomy* 51, 33–78.

Bonanomi, G., Antignani, V., Pane, C. and Scala, F. (2007) Suppression of soilborne fungal diseases with organic amendments. *Journal of Plant Pathology* 89, 311–324.

Bonanomi, G., Antignani, V., Capodilupo, M. and Scala, F. (2010) Identifying the characteristics of organic soil amendments that suppress soilborne diseases. *Soil Biology & Biochemistry* 42, 136–144.

Bonanomi, G., Antignani, V., Barile, E., Lanzotti, V. and Scala, F. (2011) Decomposition of *Medicago sativa* residues affects phytotoxicity, fungal growth and soil-borne pathogen diseases. *Journal of Plant Pathology* 93, 57–69.

Bongers, T. (1988) *De nematoden van Nederland.* KNNV-bibliotheekuitgave 46, Pirola, Schoorl, the Netherlands.

Bongers, T. and Bongers, M. (1998) Functional diversity of nematodes. *Applied Soil Ecology* 10, 239–251.

Bongers, T., Ilieva-Makulec, K. and Ekschmitt, K. (2001) Acute sensitivity of nematode taxa to $CuSO_4$ and relationships with feeding-type and life-history classifications. *Environmental Toxicology & Chemistry* 20, 1511–1516.

Bonkowski, M., Villenave, C. and Griffiths, B. (2009) Rhizosphere fauna: the functional and structural diversity of intimate interactions of soil fauna with plant roots. *Plant & Soil* 321, 213–233.

Boosalis, M.G. and Mankau, R. (1965) Parasitism and predation of soil microorganisms. In: Baker, K.F. and Snyder, W.C. (eds) *Ecology of Soil-borne Plant Pathogens.* University of California Press, Los Angeles, California, pp. 374–391.

Bordallo, J.J., Lopez-Llorca, L.V., Jansson, H.-B., Salinas, J., Persmark, L. and Asensio, L. (2002) Colonization of plant roots by egg-parasitic and nematode-trapping fungi. *New Phytologist* 154, 491–499.

Borneman, J. and Becker, J.O. (2007) Identifying microorganisms involved in specific pathogen suppression in soil. *Annual Review of Phytopathology* 45, 153–172.

Borneman, J., Olatinwo, R., Yin, B. and Becker, O. (2004) An experimental approach for identifying microorganisms involved in specified functions: utilisation for understanding a nematode suppressive soil. *Australasian Plant Pathology* 33, 151–155.

Bourne, J.M. and Kerry, B.R. (1999) Effect of the host plant on the efficacy of *Verticillium chlamydosporium* as a biological control agent of root-knot nematodes at different nematode densities and fungal application rates. *Soil Biology & Biochemistry* 31, 75–84.

Bourne, J.M., Kerry, B.R. and de Leij, F.A.A.M. (1994) Methods for study of *Verticillium chlamydosporium* in the rhizosphere. *Journal of Nematology* 26, 587–591.

Bourne, J.M., Kerry, B.R. and de Leij, F.A.A.M. (1996) The importance of the host plant on the interaction between root-knot nematodes (*Meloidogyne* spp.) and the nematophagous fungus *Verticillium chlamydosporium* Goddard. *Biocontrol Science & Technology* 6, 539–548.

Bouwman, L.A., Bloem, J., van den Boogert, P.H.J.F., Bremer, F., Hoenderbloom, G.H.J. and de Ruiter, P.C. (1994) Short-term and long-term effects of bacterivorous nematodes and nematophagous fungi on carbon and nitrogen mineralization in microcosms. *Biology & Fertility of Soils* 17, 249–256.

Bouwman, L.A., Hoenderboom, G.H.J., van der Maas, K.J. and de Ruiter, P.C. (1996) Effects of nematophagous fungi on numbers and death rates of bacterivorous nematodes in arable soil. *Journal of Nematology* 28, 26–35.

Bowen, G.D. (1979) Integrated and experimental approaches to the study of growth of organisms around roots. In: Schippers, B. and Gams, W. (eds) *Soil-borne Plant Pathogens.* Academic Press, London, pp. 209–227.

Bowen, G.D. and Rovira, A.D. (1976) Microbial colonization of plant roots. *Annual Review of Phytopathology* 14, 121–144.

Bowen, G.D. and Rovira, A.D. (1999) The rhizosphere and its management to improve root growth. *Advances in Agronomy* 66, 1–102.

Bowen, G.D. and Theodorou, C. (1973) Growth of ectomycorrhizal fungi around seeds and roots. In: Marks, G.C. and Kozlowski, T.T. (eds) *Ectomycorrhizae: Their Ecology and Physiology.* Academic Press, New York, pp. 107–150.

Brady, N.C. and Weil, R.R. (2008) *The Nature and Properties of Soils*. 14th edn. Pearson-Prentice Hall, Upper Saddle River, New Jersey.

Brand, D., Roussos, S., Pandy, A., Zilioli, P.C., Pohl, J. and Soccol, R. (2004) Development of a bionematicide with *Paecilomyces lilacinus* to control *Meloidogyne incognita*. *Applied Biochemistry & Biotechnology* 118, 81–88.

Bridge, B.J. and Bell, M.J. (1994) Effect of cropping on the physical fertility of kraznozems. *Australian Journal of Soil Research* 32, 1253–1273.

Bridge, J. (1996) Nematode management in sustainable and subsistence agriculture. *Annual Review of Phytopathology* 34, 201–225.

Brito, I., Goss, M.J. and de Carvalho, M. (2012) Effect of tillage and crop on arbuscular mycorrhiza colonization of winter wheat and triticale under Mediterranean conditions. *Soil Use & Management* 28, 202–208.

Broadbent, P. and Baker, K.F. (1974) Behaviour of *Phytophthora cinnamomi* in soils suppressive and conducive to root rot. *Australian Journal of Agricultural Research* 25, 121–137.

Brodie, G., Ryan, C. and Lancaster, C. (2012) Microwave technologies as part of an integrated weed management strategy: A review. *International Journal of Agronomy* 2012, Article ID 636905.

Broeckling, C.D., Broz, A.K., Bergelson, J., Manter, D.K. and Vivanco, J.M. (2008) Root exudates regulate soil fungal community composition and diversity. *Applied and Environmental Microbiology* 74, 738–744.

Brown, G.G. (1995) How do earthworms affect microfloral and faunal community diversity? *Plant & Soil* 170, 209–231.

Brown, S.M. and Nordmeyer, D. (1985) Synergistic reduction in root galling by *Meloidogyne javanica* with *Pasteuria penetrans* and nematicides. *Revue de Nématologie* 8, 285–286.

Bruinsma, J. (ed.) (2003) *World Agriculture: Towards 2015/2030: An FAO Perspective*. Earthscan Publications, London.

Brzeski, M.W. and Szczech, M. (1999) Effect of continuous soil amendment with coniferous sawdust on nematodes and microoorganisms. *Nematologia Mediterranea* 27, 159–166.

Bu, X., Jian, H., Chen, Z., He, Q. and Liu, Q. (2011) System optimization for attachment of *Pasteuria penetrans* to *Meloidogyne incognita*. *Biocontrol Science & Technology* 21, 509–521.

Buée, M., Rossignol, M., Jauneau, A., Ranjeva, R. and Bécard, G. (2000) The pre-symbiotic growth of arbuscular mycorrhizal fungi is induced by a branching factor partially purified from plant root exudates. *Molecular Plant-Microbe Interactions* 13, 693–698.

Buée, M., de Boer, W., Martin, F., van Overbeek, L. and Jurkevitch, E. (2009a) The rhizosphere zoo: An overview of plant-associated communities of microorganisms, including phages, bacteria, archaea, and fungi, and some of their structural factors. *Plant & Soil* 321, 189–212.

Buée, M., Reich, M., Murat, C., Morin, E., Nilsson, R.H., Uroz, S. and Martin, F. (2009b) 454 Pyrosequencing analyses of forest soils reveal an unexpectedly high fungal diversity. *New Phytologist* 184, 449–456.

Bulluck III, L.R., Barker, K.R. and Ristaino, J.B. (2002) Influences of organic and synthetic soil fertility amendments on nematode trophic groups and community dynamics under tomatoes. *Applied Soil Ecology* 21, 233–250.

Burg, R.W., Miller, B.M., Baker, E.E., Birnbaum, J., Currie, S.A., Hartman, R., Kong, Y., Monaghan, R.L., Olson, G., Putter, I., Tunac, J.B., Wallick, H., Stanley, E.O., Oiwa, R. and Omura, S. (1979) Avermectins, new family of potent anthelmintic agents: Producing organism and fermentation. *Antimicrobial Agents & Chemotherapy* 15, 361–367.

Burges, A. and Raw, F. (eds) (1967) *Soil Biology*. Academic Press, London.

Burghouts, T.H. and Gams, W. (1989) *Vermispora fusarina*, a new hyphomycete parasitizing cyst nematodes. *Memoirs of the New York Botanical Garden* 49, 57–61.

Bürgmann, H., Meier, S., Bunge, M., Widmer, F. and Zeyer, J. (2005) Effects of model root exudates on structure and activity of a soil diazotroph community. *Environmental Microbiology* 7, 1711–1724.

Burns, R.G. (2010) Albert Rovira and a half-century of rhizosphere research. In: Gupta, V.V.S.R., Ryder, M. and Radcliffe, J. (eds) *The Rovira Rhizosphere Symposium*. The Crawford Fund, Deacon, ACT, Australia, pp. 1–10.

Bursnall, L.A. and Tribe, H.T. (1974) Fungal parasitism in cysts of *Heterodera*. II. Egg parasites of *H. schachtii*. *Transactions of the British Mycological Society* 62, 595–601.

Buzo, T., McKenna, J., Kaku, S., Anwar, S.A. and McKenry, M.V. (2009) VX211, a vigorous new walnut hybrid clone with nematode tolerance and a useful resistance mechanism. *Journal of Nematology* 41, 211–216.

Cabanillas, E. and Barker, K.R. (1989) Impact of *Paecilomyces lilacinus* inoculum level and application time on control of *Meloidogyne incognita* on tomato. *Journal of Nematology* 21, 115–120.

Cabanillas, E., Barker, K.R. and Daykin, M.E. (1988) Histology of the interactions of *Paecilomyces lilacinus* and *Meloidogyne incognita* on tomato. *Journal of Nematology* 20, 362–365.

Cabos, R.Y.M., Wang, K.-H., Sipes, B.S., Heller, W.P. and Matsumoto, T.K. (2013) Detection of plant-parasitic nematode DNA in the gut of predatory and omnivorous nematodes. *Nematropica* 43, 44–48.

Cabrera, J.A., Kiewnick, S., Grimm, C., Dababat, A.A. and Sikora, R.A. (2009) Efficacy of abamectin seed treatment on *Pratylenchus zeae*, *Meloidogyne incognita* and *Heterodera schachtii*. *Journal of Plant Diseases & Protection* 116, 124–128.

Campbell, R. and Ephgrave, J.M. (1983) Effect of bentonite clay on the growth of *Gaeumannomyces graminis* var. *tritici* and on its interaction with antagonistic bacteria. *Journal of General Microbiology* 129, 771–777.

Campos-Herrera, R., El-Borai, F.E. and Duncan, L.W. (2012a) Wild interguild relationships among entomopathogenic and free-living nematodes in soil as measured by real time qPCR. *Journal of Invertebrate Pathology* 111, 126–135.

Campos-Herrera, R., Barbercheck, M., Hoy, C.W. and Stock, S.P. (2012b) Entomopathogenic nematodes as a model system for advancing the frontiers of ecology. *Journal of Nematology* 44, 162–176.

Cannayane, I. and Sivikumar, C.V. (2001) Nematode-egg parasitic fungus 1: *Paecilomyces lilacinus*- a review. *Agricultural Reviews* 22, 79–86.

Cardon, Z.G. and Whitbeck, J.L. (2007) *The Rhizosphere. An Ecological Perspective*. Elsevier Academic Press, Burlington, Massachusetts.

Carey, J., D'Amico, R., Sutton, D.A. and Rinaldi, M.G. (2003) *Paecilomyces lilacinus* vaginitis in an immunocompetent patient. *Emerging Infectious Diseases* 9, 1155–1158.

Caride, C., Piñeiro, G. and Paruelo, J.M. (2012) How does agricultural management modify ecosystem services in the argentine Pampas? The effects on soil C dynamics. *Agriculture Ecosystems & Environment* 154, 23–33.

Carneiro, R.M.D.G., Hidalgo-Díaz, L., Martins, I., Ayres de Souza Silva, K.F., Guimarães de Sousa, M. and Tigano, M.S. (2011) Effect of nematophagous fungi on reproduction of *Meloidogyne enterolobii* on guava (*Psidium guajava*) plants. *Nematology* 13, 721–728.

Carris, L.M., Glawe, D.A., Smyth, C.A. and Edwards, D.I. (1989) Fungi associated with populations of *Heterodera glycines* in two Illinois soybean fields. *Mycologia* 81, 66–75.

Cass, A. (1999) Interpretation of some soil physical indicators for assessing soil physical fertility. In: Peverill, K.I., Sparrow, L.A. and Reuter, D.J. (eds) *Soil Analysis: An Interpretation Manual*. CSIRO Publishing, Collingwood, VIC, Australia, pp. 95–102.

Cassman, K.G. (2007) Editorial response: Can organic agriculture feed the world- science to the rescue? *Renewable Agriculture & Food Systems* 22, 83–84.

Castillo, J.D. and Lawrence, K.S. (2011) First report of *Catenaria auxiliaris* parasitizing reniform nematode *Rotylenchulus reniformis* on cotton in Alabama. *Plant Disease* 95, 490.

Castillo, J.D., Lawrence, K.S. and Kloepper, J.W. (2011) Evaluation of *Bacillus firmus* strain GB-126 seed treatment for the biocontrol of the reniform nematode on cotton plants. *Phytopathology* 101, S29.

Cayrol, J.C. (1983) Lutte biologique contre les *Meloidogyne* au moyen d'*Arthrobotrys irregularis*. *Revue de Nématologie* 6, 265–273.

Cayrol, J.C. and Frankowski, J.P. (1979) Une methode de lutte biologique contre les nematodes a galles des racines appartenant au genre *Meloidogyne*. *Pepinieristes, Horticulteurs, Maraichers. Revue Horticole* 193, 15–23.

Cayrol, J.C, Frankowski, J.P., Laniece, A., D'Hardemare, G. and Talon, J.P. (1978) Contre les nematodes en champigonniere. Mise au point d'une methode de lute biologique a l'aide d'un Hyphomycete predateur: *Arthrobotrys robusta* souche 'antipolis' (Royal 300). *Pepinieristes, Horticulteurs, Maraichers. Revue Horticole* 184, 23–30.

Cetintas, R. and Dickson, D.W. (2004) Persistence and suppressiveness of *Pasteuria penetrans* to *Meloidogyne arenaria* Race 1. *Journal of Nematology* 36, 540–549.

Cetintas, R. and Dickson, D.W. (2005) Distribution and downward movement of *Pasteuria penetrans* in field soil. *Journal of Nematology* 37, 155–160.

Cetintas, R., Lima, R.D., Mendes, M.L., Brito, J.A. and Dickson, D.W. (2003) *Meloidogyne javanica* on peanut in Florida. *Journal of Nematology* 35, 433–436.

Chan, K.Y., Conyers, M.K., Li, G.D., Helyar, K.R., Poile, G., Oates, A. and Barchia, I.M. (2011) Soil carbon dynamics under different cropping and pasture management in temperate Australia: results of three long-term experiments. *Soil Research* 49, 320–328.

Chandler, D., Bailey, A.S., Tatchell, G.M., Davidson, G., Greaves, J. and Grant, W.P. (2011) The development, regulation and use of biopesticides for integrated pest management. *Philosophical Transactions of the Royal Society B* 366, 1987–1998.

Channer, A.G. and Gowen, S.R. (1992) Selection for increased host resistance and increased pathogen specificity in the *Meloidogyne-Pasteuria* interaction. *Fundamental & Applied Nematology* 15, 331–339.

Chantanao, A. and Jensen, H.J. (1969) Saprozoic nematodes as carriers and disseminators of plant pathogenic bacteria. *Journal of Nematology* 1, 216–218.

Chao, W.L., Nelson, E.B., Harman, G.E. and Hoch, H.C. (1986) Colonization of the rhizosphere by biological control agents applied to seeds *Phytopathology* 76, 60–65.

Chaparro, J.M., Sheflin, A.M., Manter, D.K. and Vivanco, J.M. (2012) Manipulating the soil microbiome to increase soil health and plant fertility. *Biology & Fertility of Soils* 48, 489–499.

Charles, L., Carbone, I., Davies, K.G., Bird, D., Burke, M., Kerry, B.R. and Opperman, C.H. (2005) Phylogenetic analysis of *Pasteuria penetrans* by use of multiple genetic loci. *Journal of Bacteriology* 187, 5700–5708.

Chatskikh, D., Hansen, S., Olesen, J.E. and Petersen, B.M. (2009) A simplified modelling approach for quantifying tillage effects on soil carbon stocks. *European Journal of Soil Science* 60, 924–934.

Chavarriá-Carvajal, J.A. and Rodriquez-Kábana, R. (1998) Changes in soil enzymatic activity and control of *Meloidogyne incognita* using four amendments. *Nematropica* 28, 7–18.

Chavas, J.-P., Posner, J.L. and Hedtcke, J.L. (2009) Organic and conventional production systems in the Wisconsin Integrated Cropping Systems trial: II. Economic risk analysis 1993–2006. *Agronomy Journal* 101, 288–295.

Chaverri, P., Samuels, G.J. and Hodge, K.T. (2005) The genus *Podocrella* and its nematode-killing anamorph *Harposporium*. *Mycologia* 97, 433–443.

Chaves, N.P., Pocasangre, L.E., Elango, F., Rosales, F.E. and Sikora, R. (2009) Combining endophytic fungi and bacteria for the biocontrol of *Radopholus similis* (Cobb) Thorne and for effects on plant growth. *Scientia Horticulturae* 122, 472–478.

Chellemi, D.O. (2002) Nonchemical management of soilborne pests in fresh market vegetable production systems. *Phytopathology* 92, 1367–1372.

Chen, F. and Chen, S. (2002) Mycofloras in cysts, females, and eggs of the soybean cyst nematode in Minnesota. *Applied Soil Ecology* 19, 35–50.

Chen, G. and Weil, R.R. (2011) Root growth and yield of maize as affected by soil compaction and cover crops. *Soil & Tillage Research* 117, 17–27.

Chen, J. and Ferris, H. (1999) The effects of nematode grazing on nitrogen mineralization during fungal decomposition of organic matter. *Soil Biology & Biochemistry* 31, 1265–1279.

Chen, J. and Ferris, H. (2000) Growth and nitrogen mineralization of selected fungi and fungal-feeding nematodes on sand amended with organic matter. *Plant & Soil* 218, 91–101.

Chen, J., Moore, W.H., Yuen, G.Y., Kobayashi, D. and Caswell-Chen, E.P. (2006) Influence of *Lysobacter enzymogenes* strain C3 on nematodes. *Journal of Nematology* 38, 233–239.

Chen, J., Xu, L.-L., Liu, B. and Liu, X.-Z. (2007a) Taxonomy of *Dactylella* complex and *Vermispora*. I. Generic concepts based on morphology and ITS sequences data. *Fungal Diversity* 26, 73–83.

Chen, J., Xu, L.-L., Liu, B. and Liu, X.-Z. (2007b) Taxonomy of *Dactylella* complex and *Vermispora*. II. The genus *Dactylella*. *Fungal Diversity* 26, 85–126.

Chen, J., Xu, L.-L., Liu, B. and Liu, X.-Z. (2007c) Taxonomy of *Dactylella* complex and *Vermispora*. III. A new genus *Brachyphoris* and revision of Vermispora. *Fungal Diversity* 26, 127–142.

Chen, S. (1997) Infection of *Heterodera glycines* by *Hirsutella rhossiliensis* in a Minnesota soybean field. *Journal of Nematology* 29, 573.

Chen, S. (2007) Suppression of *Heterodera glycines* in soils from fields with long-term soybean monoculture. *Biocontrol Science & Technology* 17, 125–134.

Chen, S. and Dickson, D.W. (2004a) Biological control of nematodes by fungal antagonists. In: Chen, Z.X., Chen, S.Y. and Dickson, D.W. (eds) *Nematology Advances and Perspectives, Volume 2, Nematode Management and Utilization*. CAB International, Wallingford, UK, pp. 979–1039.

Chen, S. and Dickson, D.W. (2004b) Biological control of nematodes by bacterial antagonists. In: Chen, Z.X., Chen, S.Y. and Dickson, D.W. (eds) *Nematology Advances and Perspectives, Volume 2, Nematode Management and Utilization*. CAB International, Wallingford, UK, pp. 1041–1082.

Chen, S. and Liu, X. (2005) Control of the soybean cyst nematodes by the fungi *Hirsutella rhossiliensis* and *Hirsutella minnesotensis* in greenhouse studies. *Biological Control* 32, 208–219.

Chen, S. and Reese, C.D. (1999) Parasitism of the nematode *Heterodera glycines* by the fungus *Hirsutella rhossiliensis* as influenced by crop sequence. *Journal of Nematology* 31, 437–444.

Chen, S., Dickson, D.W., Kimbrough, J.W., McSorley, R. and Mitchell, D.J. (1994a) Fungi associated with females and cysts of *Heterodera glycines* in a Florida soybean field. *Journal of Nematology* 26, 296–303.

Chen, S., Dickson, D.W. and Whitty, E.B. (1994b) Response of *Meloidogyne* spp. to *Pasteuria penetrans*, fungi, and cultural practices in tobacco. *Journal of Nematology* 26, 620–625.

Chen, S., Dickson, D.W. and Mitchell, D.J. (1996a) Population development of *Heterodera glycines* in response to mycoflora in soil from Florida. *Biological Control* 6, 226–231.

Chen, S., Liu, X.Z. and Chen, F.J. (2000) *Hirsutella minnesotensis* sp. nov., a new pathogen of the soybean cyst nematode. *Mycologia* 92, 819–824.

Chen, S.Y., Dickson, D.W. and Mitchell, D.J. (1996b) Pathogenicity of fungi to eggs of *Heterodera glycines*. *Journal of Nematology* 28, 148–158.

Chen, Z.X. and Dickson, D.W. (1997a) Effect of ammonium nitrate and time of harvest on mass production of *Pasteuria penetrans*. *Nematropica* 27, 53–60.

Chen, Z.X. and Dickson, D.W. (1997b) Minimal growth temperature of *Pasteuria penetrans*. *Journal of Nematology* 29, 635–639.

Chen, Z.X. and Dickson, D.W. (1998) Review of *Pasteuria penetrans*: biology, ecology and biological control potential. *Journal of Nematology* 30, 313–340.

Chen, Z.X., Dickson, D.W. and Hewlett, T.E. (1996c) Quantification of endospore concentrations of *Pasteuria penetrans* in tomato root material. *Journal of Nematology* 28, 50–55.

Chen, Z.X., Dickson, D.W., McSorley, R., Mitchell, D.J. and Hewlett, T.E. (1996d) Suppression of *Meloidogyne arenaria* race 1 by soil application of endospores of *Pasteuria penetrans*. *Journal of Nematology* 28, 159–168.

Chen, Z.X., Dickson, D.W., Freitas, L.G. and Preston, J.F. (1997a) Ultrastructure, morphology, and sporogenesis of *Pasteuria penetrans*. *Phytopathology* 87, 273–283.

Chen, Z.X., Dickson, D.W., Mitchell, D.J., McSorley, R. and Hewlett, T.E. (1997b) Suppression mechanisms of *Meloidogyne arenaria* race 1 by *Pasteuria penetrans*. *Journal of Nematology* 29, 1–8.

Chen, Z.X., Chen, S.Y. and Dickson, D.W. (eds) (2004) *Nematology Advances and Perspectives, Volumes 1 & 2*. CAB International, Wallingford, UK.

Cherr, C.M., Scholberg, J.M.S. and McSorley, R. (2001) Green manure approaches to crop production: a synthesis. *Agronomy Journal* 98, 302–319.

Chesworth, W. (ed.) (2008) *Encyclopedia of Soil Science*. Springer, Dordrecht, the Netherlands.

Chitwood, D.J. (2002) Phytochemical based strategies for nematode control. *Annual Review of Phytopathology* 40, 221–249.

Cho, M.-R., Dickson, D.W. and Hewlett, T.E. (2005) Comparison of inoculation methods, *Meloidogyne* spp. and different host plants for production of *Pasteuria penetrans*. *Journal Asia-Pacific Entomology* 8, 297–300.

Choudhary, D.K., Prakash, A. and Johri, B.N. (2007) Induced systemic resistance (ISR) in plants: mechanism of action. *Indian Journal of Microbiology* 47, 289–297.

Chung, H., Ngo, K.J., Plante, A.F. and Six, J. (2010) Evidence for carbon saturation in a highly structured and organic matter-rich soil. *Soil Science Society of America Journal* 74, 130–138.

Ciancio, A. (1995a) Phenotypic adaptations in *Pasteuria* spp. nematode parasites. *Journal of Nematology* 27, 328–338.

Ciancio, A. (1995b) Density-dependent parasitism of *Xiphinema diversicaudatum* by *Pasteuria penetrans* in a naturally infested field. *Phytopathology* 85, 144–149.

Ciancio, A. and Bourijate, M. (1995) Relationship between *Pasteuria penetrans* infection levels and density of *Meloidogyne javanica*. *Nematologia Mediterranea* 23, 43–49.

Ciancio, A. and Mukerji, K.G. (eds) (2008) *Integrated management and biocontrol of vegetable and grain crops nematodes*. Springer, Dordrecht, the Netherlands.

Ciancio, A. and Quénéhervé, P. (2000) Population dynamics of *Meloidogyne incognita* and infestation levels by *Pasteuria penetrans* in a naturally infested field in Martinique. *Nematropica* 30, 77–86.

Ciancio, A., Mankau, R. and Mundo-Ocampo, M. (1992) Parasitism of *Helicotylenchus lobus* by *Pasteuria penetrans* in naturally infested soil. *Journal of Nematology* 24, 29–35.

Ciancio, A., Bonsignore, R., Vovlas, N. and Lamberti, F. (1994) Host records and spore morphometrics of *Pasteuria penetrans* group parasites of nematodes. *Journal of Invertebrate Pathology* 63, 260–267.

Cipriani, M.G., Stea, G., Moretti, A., Altomare, C., Mulè, G. and Vurro, M. (2009) Development of a PCR-based assay for the detection of *Fusarium oxysporum* strain FT2, a potential mycoherbicide of *Orobanche ramosa*. *Biological Control* 50, 78–84.

Cobb, N.A. (1920) Transference of nematodes (Mononchs) from place to place for economic purposes. *Science* 51, 640–641.

Cole, L., Bardgett, R.D., Ineson, P. and Adamson, J. (2002) Relationships between enchytraeid worms (Oligochaeta), climate change, and the release of dissolved organic carbon from blanket peat in northern England. *Soil Biology & Biochemistry* 34, 599–607.

Coleman, D.C. and Crossley Jr, D.A. (2003) *Fundamentals of Soil Ecology*. Academic Press, Amsterdam, the Netherlands.

Coll, P., Le Cadre, E. and Villenave, C. (2012) How are nematode communities affected during a conversion from conventional to organic farming in southern French vineyards? *Nematology* 14, 665–676.

Collange, B., Navarrete, M., Peyre, G., Mateille, T. and Tchamitchian, M. (2011) Root-knot nematode (*Meloidogyne*) management in vegetable crop production: the challenge of an agronomic system analysis. *Crop Protection* 30, 1251–1262.

Compant, S., Duffy, B., Nowak, J., Clément, C. and Barka, E.A. (2005) Use of plant growth-promoting bacteria for biocontrol of plant diseases: principles, mechanisms of action, and future prospects. *Applied and Environmental Microbiology* 71, 4951–4959.

Compant, S., Clément, C. and Sessitsch, A. (2010) Plant growth- promoting bacteria in the rhizo- and endosphere of plants: their role in colonization, mechanisms involved and prospects for utilization. *Soil Biology & Biochemistry* 42, 669–678.

Conant, R.T., Paustian, K. and Elliott, E.T. (2001) Grassland management and conversion into grassland: effects on soil carbon. *Ecological Applications* 11, 343–355.

Conn, K.L. and Lazarovits, G. (1999) Impact of animal manures on Verticillium wilt, potato scab, and soil microbial populations. *Canadian Journal of Plant Pathology* 21, 81–92.

Cook, R. and Yeates, G.W. (1993) Nematode pests of grassland and forage crops. In: Evans, K., Trudgill, D.L. and Webster, J.M. (eds) *Plant Parasitic Nematodes in Temperate Agriculture*. CAB International, Wallingford, UK, pp. 305–350.

Cook, R., Lewis, G.C. and Mizen, K.A. (1991) Effects on plant-parasitic nematodes of infection of perennial ryegrass, *Lolium perenne*, by the endophytic fungus *Acremonium lolii*. *Crop Protection* 10, 403–407.

Cook, R.J. (1993) Making greater use of introduced microorganisms for biological control of plant pathogens. *Annual Review of Phytopathology* 31, 53–80.

Cook, R.J. (2006) Toward cropping systems that enhance productivity and sustainability. *Proceedings of the National Academy of Sciences* 103, 18389–18394.

Cook, R.J. and Baker, K.F. (1983) *The Nature and Practice of Biological Control of Plant Pathogens*. American Phytopathological Society, St Paul, Minnesota.

Cook, R.J. and Papendick, R.I. (1972) Influence of water potential of soils and plants on root diseases. *Annual Review of Phytopathology* 10, 349–374.

Cook, R.J. and Weller, D.M. (2004) In defense of crop monoculture. In: *New Directions for a Diverse Planet*. Proceedings of the 4th International Crop Science Congress, Brisbane, QLD, Australia. www.cropscience.org.au/icsc2004, accessed 4 October 2013.

Cook, R.J., Bruckart, W.L., Coulson, J.R., Goettel, M.S., Humber, R.A., Lumsden, R.D., Maddox, J.V., McManus, M.L., Moore, L., Meyer, S.F., Quimby Jr, P.C., Stack, J.P. and Vaughn, J.L. (1996) Safety of microorganisms intended for pest and plant disease control: a framework for scientific evaluation. *Biological Control* 7, 333–351.

Cooke, R.C. (1962a) The ecology of nematode-trapping fungi in the soil. *Annals of Applied Biology* 50, 507–513.

Cooke, R.C. (1962b) Behaviour of nematode-trapping fungi during decomposition of organic matter in the soil. *Transactions of the British Mycological Society* 45, 314–320.

Cooke, R.C. (1963) Ecological characteristics of nematode-trapping hyphomycetes. I. Preliminary studies. *Annals of Applied Biology* 52, 431–437.

Cooke, R.C. (1968) Relationships between nematode-destroying fungi and soil-borne phytonematodes. *Phytopathology* 58, 909–913.

Cooke, R.C. and Godfrey, B.E.S. (1964) A key to the nematode-destroying fungi. *Transactions of the British Mycological Society* 47, 61–74.

Cooke, R.C. and Satchuthananthavale, V. (1968) Sensitivity to mycostasis of nematode-trapping hyphomycetes. *Transactions of the British Mycological Society* 51, 555–561.

Cordier, C. and Alabouvette, C. (2009) Effects of the introduction of a biocontrol strain of *Trichoderma atroviride* on non target soil micro-organisms. *European Journal of Soil Biology* 45, 267–274.

Costa, R., Götz, M., Mrotzek, N., Lotmann, J., Berg, G. and Smalla, K. (2006a) Effects of site and plant species on rhizosphere community structure as revealed by molecular analysis of microbial guilds. *FEMS Microbial Ecology* 56, 236–249.

Costa, S.R., Kerry, B.R., Bardgett, R.D. and Davies, K.G. (2006b) Exploitation of immunofluorescence for the quantification and characterization of small numbers of *Pasteuria* endospores. *FEMS Microbiology Ecology* 58, 593–600.

Costa, S.R., Kerry, B.R., Bardgett, R.D. and Davies, K.G. (2012) Interactions between nematodes and their microbial enemies in coastal sand dunes. *Oecologia* 170, 1053–1066.

Cox, T.S., Glover, J.D., van Tassel, D.L., Cox, C.M. and DeHaan, L.R. (2006) Prospects for developing perennial grain crops. *BioScience* 56, 649–659.

Cresswell, H.P. and Kirkegaard, J.A. (1995) Subsoil amelioration by plant roots – the process and the evidence. *Australian Journal of Soil Research* 33, 221–239.

Cronin, D., Moënne-Loccoz, Y., Fenton, A., Dunne, C., Dowling, D.N. and O'Gara, F. (1997a) Role of 2,4-diacetylphloroglucinol in the interactions of the biocontrol pseudomonad strain F113 with the potato cyst nematode *Globodera rostochiensis. Applied & Environmental Microbiology* 63, 1357–1361.

Cronin, D., Moënne-Loccoz, Y., Dunne, C. and O'Gara, F. (1997b) Inhibition of egg hatch of the potato cyst nematode *Globodera rostochiensis* by chitinase-producing bacteria. *European Journal of Plant Pathology* 103, 433–440.

Cross, J.V. and Polonenko, D.R. (1996) An industry perspective on registration and commercialization of biocontrol products in Canada. *Canadian Journal of Plant Pathology* 18, 446–454.

Crow, W.T. (2013) Nematode management in golf courses in Florida. ENY-008, Entomology and Nematology Department, Florida Cooperative Extension Service, Institute of Food and Agricultural Sciences, Gainesville, Florida.

Crow, W.T. and Brammer, A.S. (2008) Sting nematode, *Belonolaimus longicaudatus* Rau (Nematoda: Secernentea: Tylenchida: Tylenchina: Belonolaimidae: Belonolaiminae). EENY239, Entomology and Nematology Department, Florida Cooperative Extension Service, Institute of Food and Agricultural Sciences, Gainesville, Florida.

Crow, W.T., Porazinska, D.L., Giblin-Davis, R.M. and Grewal, P.S. (2006) Entomopathogenic nematodes are not an alternative to fenamiphos for management of plant-parasitic nematodes on golf courses in Florida. *Journal of Nematology* 38, 52–58.

Crow, W.T., Luc, J.E. and Giblin-Davis, R.M. (2011) Evaluation of Econem™, a formulated *Pasteuria* sp. bionematicide, for management of *Belonolaimus longicaudatus* on golf course turf. *Journal of Nematology* 43, 101–109.

Crump, D. (2004) Biocontrol – a route to market. *Nematology Monographs & Perspectives* 2, 167–174.

Crump, D.A. and Irving, F. (1992) Selection of isolates and methods of culturing *Verticillium chlamydosporium* and its efficacy as a biological control agent of beet and potato cyst nematodes. *Nematologica* 38, 367–374.

Crump, D.H. and Kerry, B.R. (1977) Maturation of females of the cereal cyst-nematode on oat roots and infection by an *Entomophthora*-like fungus in observation chambers. *Nematologica* 23, 398–402.

Crump, D.H. and Kerry, B.R. (1981) A quantitative method for extracting resting spores of two nematode parasitic fungi, *Nematophthora gynophila* and *Verticillium chlamydosporium* from soil. *Nematologica* 27, 330–339.

Crump, D.H., Sayre, R.M. and Young, L.D. (1983) Occurrence of nematophagous fungi in cyst nematode populations. *Plant Disease* 67, 63–64.

Culbreath, A.K., Rodriguez-Kábana, R. and Morgan-Jones, G. (1985) The use of hemicellulosic waste matter for reduction of the phytotoxic effects of chitin and control of root-knot nematodes. *Nematropica* 15, 49–75.

Culbreath, A.K., Rodriguez-Kábana, R. and Morgan-Jones, G. (1986) Chitin and *Paecilomyces lilacinus* for control of *Meloidogyne arenaria. Nematropica* 16, 153–166.

Curl, E.A. and Truelove, B. (1986) *The Rhizosphere.* Springer Verlag, Berlin.

Dababat, A.A. and Sikora, R.A. (2007a) Importance of application time and inoculum density of *Fusarium oxysporum* 162 for biological control of *Meloidogyne incognita* on tomato. *Nematropica* 37, 267–275.

Dababat, A.A. and Sikora, R.A. (2007b) Influence of the mutualistic endophyte *Fusarium oxysporum* 162 on *Meloidogyne incognita* attraction and invasion. *Nematology* 9, 771–776.

Dababat, A.A. and Sikora, R.A. (2007c) Induced resistance by the mutualistic endophyte, *Fusarium oxysporum* strain 162, toward *Meloidogyne incognita* on tomato. *Biocontrol Science & Technology* 17, 969–975.

Dababat, A.A., Selim, M.E., Saleh, A.A. and Sikora, R.A. (2008) Influence of Fusarium wilt resistant tomato cultivars on root colonization of the mutualistic endophyte *Fusarium oxysporum* strain 162 and its biological control efficacy toward the root-knot nematode *Meloidogyne incognita. Journal of Plant Diseases & Protection* 115, 273–278.

Dabiré, R.K. and Mateille, T. (2004) Soil texture and irrigation influence the transport and the development of *Pasteuria penetrans*, a bacterial parasite of root-knot nematodes. *Soil Biology & Biochemistry* 36, 539–543.

Dabiré, K.R., Chotte, J.-L., Fardoux, J. and Mateille, T. (2001a) New developments in the estimation of spores of *Pasteuria penetrans*. *Biology & Fertility of Soils* 33, 340–343.

Dabiré, K.R., Duponnois, R. and Mateille, T. (2001b) Indirect effects of the bacterial soil aggregation on the distribution of *Pasteuria penetrans*, an obligate bacterial parasite of plant-parasitic nematodes. *Geoderma* 102, 139–152.

Dabiré, K.R., Ndiaye, S., Chotte, J.-L., Fould, S., Diop, M.T. and Mateille, T. (2005a) Influence of irrigation on the distribution and control of the nematode *Meloidogyne javanica* by the biocontrol bacterium *Pasteuria penetrans* in the field. *Biology & Fertility of Soils* 41, 205–211.

Dabiré, K.R., Chotte, J.-L., Fould, S. and Mateille, T. (2005b) Distribution of *Pasteuria penetrans* in size fractions of a *Meloidogyne javanica*-infested field soil. *Pedobiologia* 49, 335–343.

Dabiré, R.K., Ndiaye, S., Mounport, D. and Mateille, T. (2007) Relationships between abiotic factors and epidemiology of the biocontrol bacterium *Pasteuria penetrans* in a root-knot nematode *Meloidogyne javanica*-infested field. *Biological Control* 40, 22–29.

Dackman, C., Olsson, S., Jansson, H.-B., Lundgren, B. and Nordbring-Hertz, B. (1987) Quantification of predatory and endoparasitic nematophagous fungi in soil. *Microbial Ecology* 13, 89–93.

Dalal, R.C., Strong, W.M., Weston, E.J., Cahill, M.J., Cooper, J.E., Lehane, K.J., King, A.J. and Gaffney, J. (1994) Evaluation of forage and grain legumes, no-till and fertilisers to restore fertility degraded soils. *Transactions of the 15th World Congress of Soil Science, Volume 5a*, pp. 62–74. International Society of Soil Science and Mexican Society of Soil Science, Acapulaco, Mexico.

Dalal, R.C., Strong, W.M., Weston, E.J., Cooper, J.E., Lehane, K.J., King, A.J. and Chicken, C.J. (1995) Sustaining productivity of a vertisol at Warra, Queensland, with fertilisers, no-till and legumes. I. Organic matter status. *Australian Journal of Experimental Agriculture* 35, 903–913.

Dalal, R.C., Wang, W., Robertson, G.P. and Parton, W.J. (2003) Nitrous oxide emission from Australian agricultural lands and mitigation options: a review. *Australian Journal of Soil Research* 41, 165–195.

Dallemole-Giaretta, R., de Freitas, L.G., Lopes, E.A., Ferraz, S., de Podesta, G.S. and Agnes, E.L. (2011) Cover crops and *Pochonia chlamydosporia* for the control of *Meloidogyne javanica*. *Nematology* 13, 919–926.

Dallemole-Giaretta, R., Freitas, L.G., Lopes, E.A, Pereira, O.L., Zooca, R.J.F. and Ferraz, S. (2012) Screening of *Pochonia chlamydosporia* Brazilian isolates as biocontrol agents of *Meloidogyne javanica*. *Crop Protection* 42, 102–107.

Dalpé, Y. and Monreal, M. (2004) Arbuscular mycorrhiza inoculum to support sustainable cropping systems. *Crop Management*. DOI: 10.1094/CM-2004–0301–09-RV.

Daniel, R. (2005) The metagenomics of soil. *Nature Reviews Microbiology* 3, 470–478.

Darban, D.A., Pembroke, B. and Gowen, S.R. (2004) The relationships of time and temperature to body weight and numbers of endospores of *Pasteuria penetrans*-infected *Meloidogyne javanica* females. *Nematology* 6, 33–36.

Datta, A. and Knezevic, S.Z. (2013) Flaming as an alternative weed control method for conventional and organic agronomic crop production systems: a review. *Advances in Agronomy* 118, 339–428.

Davet, P. (2004) *Microbial Ecology of the Soil and Plant Growth*. Science Publishers Inc., Enfield, Connecticut.

Davies, K.G. (2009) Understanding the interaction between an obligate hyperparasitic bacterium, *Pasteuria penetrans* and its obligate plant-parasitic nematode host, *Meloidogyne* spp. *Advances in Parasitology* 68, 211–245.

Davies, K.G. and Whitbread, R. (1989) Factors affecting the colonization of a root system by fluorescent pseudomonads: the effects of water, temperature and soil microflora. *Plant and Soil* 116, 247–256.

Davies, K.G., Kerry, B.R. and Flynn, C.A. (1988) Observations on the pathogenicity of *Pasteuria penetrans*, a parasite of root-knot nematodes. *Annals of Applied Biology* 112, 491–501.

Davies, K.G., Flynn, C.A., Laird, V. and Kerry, B.R. (1990) The life cycle, population dynamics and host specificity of a parasite of *Heterodera avenae*, similar to *Pasteuria penetrans*. *Revue de Nématologie* 13, 303–309.

Davies, K.G., Laird, V. and Kerry, B.R. (1991) The motility, development and infection of *Meloidogyne incognita* encumbered with spores of the obligate hyperparasite *Pasteuria penetrans*. *Revue de Nématologie* 14, 611–618.

Davies, K.G., Redden, M. and Pearson, T.K. (1994) Endospore heterogeneity in *Pasteuria penetrans* related to adhesion to plant-parasitic nematodes. *Letters in Applied Microbiology* 19, 370–373.

Davies, K.G., Rowe, J.A. and Williamson, V.M. (2008) Inter- and intra-specific cuticle variation between amphimictic and parthenogenetic species of root-knot nematode (*Meloidogyne* spp.) as revealed by a bacterial parasite (*Pasteuria penetrans*). *International Journal for Parasitology* 38, 851–859.

Davies, K.G., Rowe, J., Manzanilla-López, R. and Opperman, C.H. (2011) Re-evaluation of the life cycle of the nematode-parasitic bacterium *Pasteuria penetrans* in root-knot nematodes, *Meloidogyne* spp. *Nematology* 13, 825–835.

de Almeida, G.L., Santurio, J.M., Filho, J.O.J., Zanette, R.A., Camillo, G., Flores, A.G., da Silva, J.H.S. and de la Rue, M.L. (2012) Predatory activity of the fungus *Duddingtonia flagrans* in equine strongyle infective larvae on natural pasture in the southern region of Brazil. *Parasitology Research* 110, 657–662.

de Boer, W. and van Veen, J.A. (2001) Are chitinolytic rhizosphere bacteria really beneficial to plants? In: Jeger, M.J. and Spence, N.J. (eds) *Biotic Interactions in Plant-Pathogen Associations*. CAB International, Wallingford, UK, pp. 121–130.

de Boer, W., Verheggen, P., Gunnewiek, P.J.A.K., Kowalchuk, G.A. and van Veen, J.A. (2003) Microbial community composition affects soil fungistasis. *Applied & Environmental Microbiology* 69, 835–844.

de Boer, W., Wagenaar, A.-M., Klein Gunnewiek, P.J.A. and van Veen, J.A. (2007) *In vitro* suppression of fungi caused by combinations of apparently non-antagonistic soil bacteria. *FEMS Microbiology Ecology* 59, 177–185.

de Leij, F.A.A.M. (1992) The significance of ecology in the development of *Verticillium chlamydosporium* as a biological control agent against root-knot nematodes (*Meloidogyne* spp.). PhD thesis, Wageningen University, the Netherlands.

de Leij, F.A.A.M. and Kerry, B.R. (1991) The nematophagous fungus *Verticillium chlamydosporium* Goddard, as a potential biological control agent for *Meloidogyne arenaria*. *Revue de Nématologie* 14, 157–164.

de Leij, F.A.A.M., Davies, K.D. and Kerry, B.R. (1992a) The use of *Verticillium chlamydosporium* Goddard and *Pasteuria penetrans* (Thorne) Sayre and Starr alone and in combination to control *Meloidogyne incognita* on tomato plants. *Fundamental & Applied Nematology* 15, 235–242.

de Leij, F.A.A.M., Dennehy, J.A. and Kerry, B.R. (1992b) The effect of temperature and nematode species on the interactions between the nematophagous fungus *Verticillium chlamydosporium* and root-knot nematodes (*Meloidogyne* spp.). *Nematologica* 38, 65–79.

de Leij, F.A.A.M., Kerry, B.R. and Dennehy, J.A. (1993) *Verticillium chlamydosporium* as a biological control agent for *Meloidogyne incognita* and *M. hapla* in pot and microplot tests. *Nematologica* 39, 115–126.

de Ruiter, P.C., Moore, J.C., Zwart, K.B., Bouwman, L.A., Hassink, J., Bloem, J., de Vos, J.A., Marinissen, J.C.Y., Didden, W.A.M., Lebbink, G. and Brussard, L. (1993) Simulation of nitrogen mineralization in belowground food webs of two winter wheat fields. *Journal of Applied Ecology* 30, 95–106.

De Sanctis, G., Roggero, P.P., Seddaiu, G., Orsini, R., Porter, C.H. and Jones, J.W. (2012) Long-term no tillage increased soil organic carbon content of rain-fed cereal systems in a Mediterranean area. *European Journal of Agronomy* 40, 18–27.

de Weert, S., Vermeiren, H., Mulders, I.H.M., Kuiper, I., Hendrickx, N., Bloemberg, G.V., Vanderleyden, J., De Mot, R.D. and Lugtenberg B.J.J. (2002) Flagella-driven chemotaxis towards exudate components is an important trait for tomato root colonization by *Pseudomonas fluorescens*. *Molecular Plant-Microbe Interactions* 15, 1173–1180.

de Weger, L.A., van der Vlugt, C.I.M., Wijfjes, A.H.M., Bakker, P.A.H.M., Schippers, B. and Lugtenberg, B. (1987) Flagella of a plant growth stimulating *Pseudomonas fluorescens* strain are required for colonization of potato roots. *Journal of Bacteriology* 169, 2769–2773.

Deacon, J.W. (1991) Significance of ecology in the development of biocontrol agents against soil-borne plant pathogens. *Biocontrol Science & Technology* 1, 5–20.

DeAngelis, K.M., Brodie, E.L., DeSantis, T.Z., Anderson, G.L., Lindow, S.E. and Firestone, M.K. (2009) Selective progressive response of soil microbial community to wild oat roots. *International Society for Microbial Ecology Journal* 3, 168–178.

Degens, B.P. (1997) Macro-aggregation of soils by biological bonding and binding mechanisms and the factors affecting these: a review. *Australian Journal of Soil Research* 35, 431–459.

Dekkers, L.C., van der Bij, A.J., Mulders, I.H.M., Phoelich, C.C., Wentwoord, R.A., Glandorf, D.C., Wijffelman, C.A. and Lugtenberg, B.J.J. (1998) Role of the O-antigen of lipopolysaccharide, and possible roles of growth rate and NADH:ubiquinone oxidoreductase (nuo) in competitive tomato root-tip colonization by *Pseudomonas fluorescens* WCS365. *Molecular Plant-Microbe Interactions* 11, 763–771.

Den Belder, E. and Jansen, E. (1994) Capture of plant-parasitic nematodes by an adhesive hyphae forming isolate of *Arthrobotrys oligospora* and some other nematode-trapping fungi. *Nematologica* 40, 423–437.

Dennis, P.G., Miller, A.J. and Hirsch, P.R. (2010) Are root exudates more important than other sources of rhizodeposits in structuring rhizosphere bacterial communities? *FEMS Microbiology Ecology* 72, 313–327.

Diacono, M. and Montemurro, F. (2010) Long-term effects of organic amendments on soil fertility. A review. *Agronomy for Sustainable Development* 30, 401–422.

Dick, M.W. (1997) The Myzocytiopsidaceae. *Mycological Research* 101, 878–882.

Dicklow, M.B., Acosta, N. and Zuckerman, B.M. (1993) A novel *Streptomyces* species for controlling plant-parasitic nematodes. *Journal of Chemical Ecology* 19, 159–173.

Dickson, D.W. (1998) Peanut. In: Barker, K.R., Pederson, G.A. and Windham, G.L. (eds) *Plant and Nematode Interactions*. Agronomy Monographs no. 36. American Society of Agronomy Inc., Madison, Wisconsin, pp. 523–566.

Dickson, D.W., Mitchell, D.J., Hewlett, T.E., Oostendorp, M. and Kannwischer-Mitchell, M.E. (1991) Nematode-suppressive soil from a peanut field. *Journal of Nematology* 23, 526.

Dickson, D.W., Preston III, J.F., Giblin-Davis, R.M., Noel, G.R., Dieter Ebert, G.R.D. and Bird, G.W. (2009) Family Pasteuriaceae Laurent 1890. In: De Vos, P., Garrity, G.M., Jones, D., Krieg, N.R., Ludwig, W., Rainey, F.A., Schleifer, K.-H. and Whitman, W.B. (eds) *Bergey's Manual of Systematic Bacteriology, The Firmicutes, Volume 3*. Springer, New York, pp. 328–347.

Diedhiou, P.M., Hallmann, J., Oerke, E.-C. and Dehne, H.-W. (2003) Effects of arbuscular mycorrhizal fungi and non-pathogenic *Fusarium oxysporum* on *Meloidogyne incognita* infestation of tomato. *Mycorrhiza* 13, 199–204.

Dijksterhuis, J., Veenhuis, M. and Harder, W. (1990) Ultrastructural study of adhesion and initial stages of infection of nematodes by conidia of *Drechmeria coniospora*. *Mycological Research* 94, 1–8.

Dindal, D.L. (1990) *Soil Biology Guide*. John Wiley & Son, New York.

Dion, P. (ed.) (2010) *Soil Biology and Agriculture in the Tropics*. Springer, Dordrecht, the Netherlands.

Dixon, S.M. (1952) Predaceous fungi from rotten wood. *Transactions of the British Mycological Society* 35, 144–148.

Doane, T.A., Horwath, W.R., Mitchell, J.P., Jackson, J., Miyao, G. and Brittan, K. (2009) Nitrogen supply from fertilizer and legume cover crop in the transition to no-tillage for irrigated row crops. *Nutrient Cycling in Agroecosystems* 85, 253–262.

Dobbs, C.G. and Hinson, W.H. (1953) A widespread fungistasis in soil. *Nature* 172, 197–199.

Domsch, K.H., Gams, W. and Anderson, T.-H. (1980) *Compendium of soil fungi* Vol. 1. Academic Press, London.

Donald, P.A. and Hewlett, T.E. (2009) *Pasteuria nishizawae* studies in Tennessee. *Journal of Nematology* 41, 325.

Donn, S., Neilson, R., Griffiths, B.S. and Daniell, T.J. (2011) Greater coverage of the phylum Nematoda in SSU rDNA studies. *Biology & Fertility of Soils* 47, 333–339.

Donn, S., Neilson, R., Griffiths, B.S. and Daniell, T.J. (2012) A novel molecular approach for rapid assessment of soil nematode assemblages – variation, validation and potential applications. *Methods in Ecology and Evolution* 3, 12–23.

Doran, J.W. and Parkin, T.B. (1994) Defining and assessing soil quality. In: Doran, J.W., Coleman, D.C., Bezdicek, D.F. and Stewart, B.A. (eds) *Defining Soil Quality for a Sustainable Environment*. SSSA Special Publication No. 35, Soil Science Society of America, Madison, Wisconsin, pp. 3–21.

Doran, J.W. and Safley, M. (1997) Defining and assessing soil health and sustainable productivity. In: Pankhurst, C., Doube, B.M. and Gupta, V.V.S.R. (eds) *Biological Indicators of Soil Health*. CAB International, Wallingford, UK, pp. 1–28.

Doran, J.W. and Zeiss, M.R. (2000) Soil health and sustainability: managing the biotic component of soil quality. *Applied Soil Ecology* 15, 3–11.

Doran, J.W., Coleman, D.C., Bezdicek, D.F. and Stewart, B.A. (eds) (1994) *Defining Soil Quality for a Sustainable Environment*. SSSA Special Publication No. 35, Soil Science Society of America, Madison, Wisconsin.

Doran, J.W., Sarrantonio, M. and Liebig, M.A. (1996) Soil health and sustainability. *Advances in Agronomy* 56, 1–54.

Douds Jr, D.D. and Seidel, R. (2013) The contribution of arbuscular mycorrhizal fungi to the success or failure of agricultural practices. In: Cheeke, T.E., Coleman, D.C. and Wall, D.H. (eds) *Microbial Ecology in Sustainable Agroecosystems*. CRC Press, Taylor & Francis Group, Boca Raton, Florida, pp. 133–152.

Downer, A.J., Menge, J.A. and Pond, E. (2001) Association of cellulytic enzyme activities in *Eucalyptus* mulches with biological control of *Phytophthora cinnamomi*. *Phytopathology* 91, 847–855.

Drechsler, C. (1936) A fusarium-like species of *Dactylella* capturing and consuming testaceous rhizopods. *Journal Washington Academy of Sciences* 26, 397–404.

Drechsler, C. (1941) Some hyphomycetes parasitic on free-living terricolous nematodes. *Phytopathology* 31, 773–802.

Drechsler, C. (1944) A species of *Arthrobotrys* that captures springtails. *Mycologia* 36, 382–399.

Drechsler, C. (1952) Another nematode-strangulating *Dactylella* and some related hyphomycetes. *Mycologia* 44, 533–556.

Drechsler, C. (1962) Two additional species of *Dactylella* parasitic on *Pythium* oospores. *Sydowia* 15, 94–96.

Drechsler, C. (1963) A slender-spored *Dactylella* parasitic on *Pythium* oospores. *Phytopathology* 53, 993–994.

Drenovsky, R.E., Duncan, R.A. and Scow, K.M. (2005) Soil sterilization and organic carbon, but not microbial inoculants, change microbial communities in replanted peach orchards. *California Agriculture* 59, 176–181.

Drinkwater, L.E. and Snapp, S.S. (2007) Understanding and managing the rhizosphere in agroecosystems. In: Cardon, Z.G. and Whitbeck, J.L. (eds) *The Rhizosphere*. Elsevier Academic Press, New York, pp. 127–153.

Duan, W., Yang, E., Xiang, M. and Liu, X. (2008) Effect of storage conditions on the survival of two potential biocontrol agents of nematodes, the fungi *Paecilomyces lilacinus* and *Pochonia chlamydosporia*. *Biocontrol Science & Technology* 18, 605–612.

Duan, Y.P., Castro, H.F., Hewlett, T.E., White, J.H. and Ogram, A.V. (2003) Detection and characterization of *Pasteuria* 16S rRNA gene sequences from nematodes and soils. *International Journal of Systematic & Evolutionary Microbiology* 53, 105–112.

Duke, S.O. and Powles, S.B. (2008) Glyphosate: a once-in-a-century herbicide. *Pest Management Science* 64, 319–325.

Duncan, L.W. (1991) Current options for nematode management. *Annual Review of Phytopathology* 29, 469–490.

Duncan, L.W. and Noling, J.W. (1998) Agricultural sustainability and nematode integrated pest management. In: Barker, K.R., Pederson, G.A. and Windham, G.L. (eds) *Plant and Nematode Interactions*. Agronomy Monographs no. 36, American Society of Agronomy Inc., Madison, Wisconsin, pp. 251–287.

Duncan, L.W., Graham, J.H., Zellers, J., Bright, D., Dunn, D.C., El-Borai, F.E. and Porazinska, D.L. (2007) Food web responses to augmenting the entomopathogenic nematodes in bare and animal manure-mulched soil. *Journal of Nematology* 39, 176–189.

Duncan, L.W., Stuart, R.J., El-Borai, F.E., Campos-Herrera, R., Pathak, E., Giurcanu, M. and Graham, J.H. (2013) Modifying orchard planting sites conserves entomopathogenic nematodes, reduces weevil herbivory and increases citrus tree growth, survival and fruit yield. *Biological Control* 64, 26–36.

Dunn, M.T., Sayre, R.M., Carrell, A. and Wergin, W.P. (1982) Colonization of nematode eggs by *Paecilomyces lilacinus* (Thom) Samson as observed with scanning electron microscope. *Scanning Electron Microscopy* III, 1351–1357.

Duponnois, R., Ba, A.M. and Mateille, T. (1998) Effects of some rhizosphere bacteria for the biocontrol of nematodes of the genus *Meloidogyne* with *Arthrobotrys oligospora*. *Fundamental & Applied Nematology* 21, 157–163.

Duponnois, R., Ba, A.M. and Mateille, T. (1999) Beneficial effects of *Enterobacter cloacae* and *Pseudomonas mendocina* for biocontrol of *Meloidogyne incognita* with the endospore-forming bacterium *Pasteuria penetrans*. *Nematology* 1, 95–101.

DuPont, S.T., Ferris, H. and Van Horn, M. (2009) Effects of cover crop quality and quantity on nematode-based soil food webs and nutrient cycling. *Applied Soil Ecology* 41, 157–167.

Eayre, C.G., Jaffee, B.A. and Zehr, E.I. (1983) Suppression of *Criconemella xenoplax* by the fungus *Hirsutella rhossiliensis*. *Phytopathology* 73, 500.

Ebert, D., Rainey, P., Embley, T.M. and Scholz, D. (1996) Development, life cycle, ultrastructure and phylogenetic position of *Pasteuria ramosa* Metchnikoff 1888: Rediscovery of an obligate endoparasite of *Daphnia magna* Straus. *Philosophical Transactions of the Royal Society B* 351, 1689–1701.

Eddaoudi, M. and Bourijate, M. (1998) Comparative assessment of *Pasteuria penetrans* and three nematicides for the control of *Meloidogyne javanica* and their effect on yields of successive crops of tomato and melon. *Fundamental & Applied Nematology* 21, 113–118.

Edwards, C.A. and Bohlen, P. (1996) *Biology and Ecology of Earthworms*. 3rd edn. Chapman & Hall, London.

Edwards, C.A., Lal, R., Madden, P., Miller, R.H. and House, G. (eds) (1990) *Sustainable Agricultural Systems*. Soil Water Conservation Society, Ankeny, Iowa.

Elad, Y., David, D.R., Harel, Y.M., Borenshtein, M., Kalifa, H.B., Silber, A. and Graber, E.R. (2010) Induction of systemic resistance in plants by biochar, a soil-applied carbon sequestering agent. *Phytopathology* 100, 913–921.

El-Borai, F.E., Brentu, C.F. and Duncan, L.W. (2007) Augmenting entomopathogenic nematodes in soil from a Florida citrus orchard: non-target effects of a trophic cascade. *Journal of Nematology* 39, 203–210.

El-Borai, F.E., Bright, D.B., Graham, J.H., Stuart, R.J., Cubero, J. and Duncan, L.W. (2009) Differential susceptibility of entomopathogenic nematodes to nematophagous fungi from Florida citrus orchards. *Nematology* 11, 231–241.

El-Borai, F.E., Campos-Herrera, R., Stuart, R.J. and Duncan, L.W. (2011) Substrate modulation, group effects and behavioural responses of entomopathogenic nematodes to nematophagous fungi. *Journal of Invertebrate Pathology* 106, 347–356.

Elekcioglu, I.H. (1995) Occurrence of *Pasteuria* bacteria as parasites of plant-parasitic nematodes in the east Mediterranean region of Turkey. *Nematologia Mediterranea* 23, 213–215.

Elmi, A.A., West, C.P., Robbins, R.T. and Kirkpatrick, T.L. (2000) Endophyte effects on reproduction of a root-knot nematode (*Meloidogyne marylandi*) and osmotic adjustment in tall fescue. *Grass & Forage Science*, 55, 166–172.

Elsen, A., Gervacio, D., Swennen, R. and De Waele, D. (2008) AMF-induced biocontrol against plant parasitic nematodes in *Musa* sp.: a systemic effect. *Mycorrhiza* 18, 251–256.

El-Tarabily, K.A., Hardy, G.E.St.J., Sivasithamparam, K., Hussein, A.M. and Kurtböke, D.I. (1997) The potential for the biological control of cavity-spot disease of carrots, caused by *Pythium coloratum*, by streptomycete and non-streptomycete actinomycetes. *New Phytologist* 137, 495–507.

Eno, C.F., Blue, W.G. and Good, J.M. Jr (1955) The effect of anhydrous ammonia on nematodes, fungi, bacteria, and nitrification in some Florida soils. *Proceedings of the Soil Science Society of America* 19, 55–58.

Epsky, N.D., Walter, D.E. and Capinera, J.L. (1988) Potential role of nematophagous microarthropods as biotic mortality factors of entomogenous nematodes (Rhabditida: Steinernematidae, Heterorhabditidae). *Journal of Economic Entomology* 81, 821–825.

Eren, J. and Pramer, D. (1965) The most probable number of nematode-trapping fungi in soil. *Soil Science* 99, 285.

Escudero, N. and Lopez-Llorca, L.V. (2012) Effects on plant growth and root-knot nematode infection of an endophytic GFP tansformant of the nematophagous fungus *Pochonia chlamydosporia*. *Symbiosis* 57, 33–42.

Espanol, M., Verdejo-Lucas, S., Davies, K.G. and Kerry, B.R. (1997) Compatibility between *Pasteuria penetrans* and *Meloidogyne* populations from Spain. *Biocontrol Science & Technology* 7, 219–230.

Esser, R.P. (1987) Biological control of nematodes by nematodes. II. Seinura (Nematoda: Aphelenchoididae). *Florida Department of Agriculture & Consumer Services, Nematology Circular* 147, 1–3.

Esser, R.P. and El-Gholi, N.E. (1992) *Harposporium*, a fungus that parasitizes and kills nematodes utilizing conidia swallowed or sticking to its prey. *Florida Department of Agriculture & Consumer Services, Nematology Circular* 200, 1–6.

Esteves, I., Peteira, B., Atkins, S.D., Magan, N. and Kerry, B. (2009) Production of extracellular enzymes by different isolates of *Pochonia chlamydosporia*. *Mycological Research* 113, 867–876.

Ettema, C. and Bongers, T. (1993) Characterization of nematode colonization and succession in disturbed soil using the Maturity Index. *Biology & Fertility of Soils* 16, 79–85.

Evans, K. (1969) Changes in a *Heterodera rostochiensis* population through the growing season. *Annals of Applied Biology* 64, 31–41.

Evans, K., Trudgill, D.L. and Webster, J.M. (eds) (1993) *Plant Parasitic Nematodes in Temperate Agriculture*. CAB International, Wallingford, UK.

Fageria, N.K. (2012) Role of soil organic matter in maintaining sustainability of cropping systems. *Communications in Soil Science & Plant Analysis* 43, 2063–2113.

FAO (2003) *Biological management of soil ecosystems for sustainable agriculture*. World Soil Resources Reports 101, FAO, Rome, Italy.

Farrell, F.C., Jaffee, B.A. and Strong, D.R. (2006) The nematode-trapping fungus *Arthrobotrys oligospora* in soil of the Bodega marine reserve: distribution and dependence on nematode-parasitized moth larvae. *Soil Biology & Biochemistry* 38, 1422–1429.

Faske, T.R. and Starr, J.L. (2006) Sensitivity of *Meloidogyne incognita* and *Rotylenchulus reniformis* to abamectin. *Journal of Nematology* 38, 240–244.

Fattah, F.A., Saleh, H.M. and Aboud, H.M. (1989) Parasitism of the citrus nematode, *Tylenchulus semipenetrans*, by *Pasteuria penetrans* in Iraq. *Journal of Nematology* 21, 431–433.

Fegan, N. (1993) The interaction between chitinolytic bacteria and root-knot nematode eggs. PhD thesis. University of Queensland, QLD, Australia.

Félix, M.-A., Ashe, A., Piffaretti, J., Wu, G., Nuez, I., Bélicard, T., Jiang, Y, Zhao, G., Franz, C.J., Goldstein, L.D., Sanroman, M., Miska, E.A., Wang, D. (2011) Natural and experimental infection of *Caenorhabditus* nematodes by novel viruses related to nodaviruses. *PLoS Biology*, 9, 1–14.

Ferris, H. (2010) Contribution of nematodes to the structure and function of the soil food web. *Journal of Nematology* 42, 63–67.

Ferris, H. and Bongers, T. (2006) Nematode indicators of organic enrichment. *Journal of Nematology* 38, 3–12.

Ferris, H. and Bongers, T. (2009) Indices developed specifically for analysis of nematode assemblages. In: Wilson, M.J. and Kakouli-Duarte, T. (eds) *Nematodes as Environmental Indicators*. CAB International, Wallingford, UK, pp. 124–145.

Ferris, H. and Matute, M.M. (2003) Structural and functional succession in the nematode fauna of a soil food web. *Applied Soil Ecology* 23, 93–110.

Ferris, H. and Noling, J.W. (1987) Analysis and prediction as a basis for management decisions. In: Brown, R.H. and Kerry, B.R. (eds) *Principles and Practice of Nematode Control in Crops*. Academic Press, Sydney, NSW, Australia, pp. 49–85.

Ferris, H., McKenry, M.V. and McKinney, H.E. (1976) Spatial distribution of nematodes in peach orchards. *Plant Disease Reporter* 60, 18–22.

Ferris, H., Venette, R.C. and Lau, S.S. (1996) Dynamics of nematode communities in tomatoes grown in conventional and organic farming systems, and their impact on soil fertility. *Applied Soil Ecology* 3, 161–175.

Ferris, H., Venette, R.C. and Lau, S.S. (1997) Population energetics of bacterial-feeding nematodes: carbon and nitrogen budgets. *Soil Biology & Biochemistry* 29, 1183–1194.

Ferris, H., Venette, R.C., van der Meulen, H.R. and Lau, S.S. (1998) Nitrogen mineralization by bacterial-feeding nematodes: verification and measurement. *Plant & Soil* 203, 159–171.

Ferris, H., Bongers, T. and de Goede, R.G.M. (2001) A framework for soil food web diagnostics: extension of the nematode faunal concept. *Applied Soil Ecology* 18, 13–29.

Ferris, H., Venette, R.C. and Scow, K.M. (2004) Soil management to enhance bacterivore and fungivore nematode populations and their nitrogen mineralisation function. *Applied Soil Ecology* 25, 19–35.

Ferris, H., Pocasangre, L.E., Serrano, E., Muñoz, J., Garcia, S., Perichi, G. and Martinez, G. (2012a) Diversity and complexity complement apparent competition: nematode assemblages in banana plantations. *Acta Oecologica* 40, 11–18.

Ferris, H., Griffiths, B.S., Porazinska, D.L., Powers, T.O., Wang, K.-H. and Tenuta, M. (2012b) Reflections on plant and soil nematode ecology: past, present and future. *Journal of Nematology* 44, 115–126.

Ferris, H., Sánchez-Moreno, S. and Brennan, E.B. (2012c) Structure, functions and interguild relationships of the soil nematode assemblage in organic vegetable production. *Applied Soil Ecology* 61, 16–25.

Fierer, N., Grandy, A.S., Six, J. and Paul, E.A. (2009) Searching for unifying principles in soil ecology. *Soil Biology & Biochemistry* 41, 2249–2256.

Fiscus, D.A. and Neher, D.A. (2002) Distinguishing sensitivity of free-living soil nematode genera to physical and chemical disturbances. *Ecological Applications* 12, 565–575.

Fisher, J., Tozer, P. and Abrecht, D. (2012) Livestock in no-till cropping systems – a story of trade-offs. *Animal Production Science* 52, 197–214.

Fog, K. (1988) The effect of added nitrogen on the rate of decomposition of organic matter. *Biological Reviews* 63, 433–462.

Foster, R.C. (1986) The ultrastructure of the rhizoplane and rhizosphere. *Annual Review of Phytopathology* 24, 211–234.

Fould, S., Dieng, A.L., Davies, K.G., Normand, P. and Mateille, T. (2001) Immunological quantification of the nematode parasitic bacterium *Pasteuria penetrans* in soil. *FEMS Microbiology Ecology* 37, 187–195.

Franchini, J.C., Crispino, C.C., Souzsa, R.A., Torres, E. and Hungria, M. (2007) Microbiological parameters as indicators of soil quality under various soil management and crop rotation systems in southern Brazil. *Soil & Tillage Research* 92, 18–29.

Franco, C., Michelsen, P., Percy, N., Conn, V., Listiana, E., Moll, S., Loria, R. and Coombs, J. (2007) Actinobacterial endophytes for improved crop performance. *Australasian Plant Pathology* 36, 524–531.

Franco-Navarro, F. and Godinez-Vidal, D. (2008) Occurrence of *Pasteuria* forms from a biosphere reserve in Mexico. *Nematropica* 38, 187–194.

Franken, P. (2012) The plant strengthening root endophyte *Piriformospora indica*: potential application and the biology behind. *Applied Microbiology & Biotechnology* 96, 1455–1464.

Franzluebbers, A.J. (2004) Tillage and residue management effects on soil organic matter. In: Magdoff, F. and Weil, R.R. (eds) *Soil Organic Matter in Sustainable Agriculture*. CRC Press, Boca Raton, Florida, pp. 227–268.

Fravel, D. (1988) Role of antibiosis in the biocontrol of plant diseases. *Annual Review of Phytopathology* 26, 75–91.

Fravel, D.R. (2005) Commercialization and implementation of biocontrol. *Annual Review of Phytopathology* 43, 337–359.

Fravel, D., Olivain, C. and Alabouvette, C. (2003) *Fusarium oxysporum* and its biocontrol. *New Phytologist* 157, 493–502.

Freckman, D.W. and Ettema, C.H. (1993) Assessing nematode communities in agroecosystems of varying human intervention. *Agriculture, Ecosystems & Environment* 45, 239–261.

Fu, S., Coleman, D.C., Hendrix, P.F. and Crossley Jr, D.A. (2000) Responses of trophic groups of soil nematodes to residue application under conventional tillage and no-till regimes. *Soil Biology & Biochemistry* 32, 1731–1741.

Fu, S., Ferris, H., Brown, D. and Plant, R. (2005) Does the positive feedback effect of nematodes on the biomass and activity of their bacteria prey vary with nematode species and population size? *Soil Biology & Biochemistry* 37, 1979–1987.

Fulthorpe, R.R., Roesch, L.F.W., Riva, A. and Triplett, E.W. (2008) Distantly sampled soils carry few species in common. *International Society for Microbial Ecology Journal* 2, 901–910.

Gair, R., Mathias, P.L. and Harvey, P.N. (1969) Studies of cereal nematode populations and cereal yields under continuous or intensive culture. *Annals of Applied Biology* 63, 503–512.

Galeano, M., Verdejo-Lucas, S. and Ciancio, A. (2003) Morphology and ultrastructure of a *Pasteuria* form parasitic in *Tylenchorhynchus cylindricus* (Nematoda). *Journal of Invertebrate Pathology* 83, 83–85.

Galper, S., Eden, L.M., Stirling, G.R. and Smith, L.J. (1995) Simple screening methods for assessing the predacious activity of nematode-trapping fungi. *Nematologica* 41, 130–140.

Gams, W. (1988) A contribution to the knowledge of nematophagous species of *Verticillium*. *Netherlands Journal of Plant Pathology* 94, 123–48.

Gams, W. and Jansson, H.-B. (1985) The nematode parasite *Meria coniospora* Drechsler in pure culture and its classification. *Mycotaxon* 22, 33–38.

Gams, W. and Zare, R. (2001) A revision of *Verticillium* sect. Prostrata. III. Generic classification. *Nova Hedwigia* 72, 329–337.

Gao, L., Sun, M.H., Liu, X.Z. and Che, Y.S. (2007) Effects of carbon concentration and carbon to nitrogen ratio on the growth and sporulation of several biocontrol fungi. *Mycological Research* 111, 87–92.

Gao, X., Yin, B., Borneman, J. and Ole Becker, J. (2008) Assessment of parasitic activity of *Fusarium* strains obtained from a *Heterodera schachtii*-suppressive soil. *Journal of Nematology* 40, 1–6.

Garabedian, S. and Van Gundy, S.D. (1983) Use of avermectins for the control of *Meloidogyne incognita* on tomatoes. *Journal of Nematology* 15, 503–510.

Garbeva, P., van Veen, J.A. and van Elsas, J.D. (2004) Assessment of the diversity and antagonism towards *Rhizoctonia solani* AG3, of *Pseudomonas* species in soil from different agricultural regimes. *FEMS Microbial Ecology* 47, 51–64.

Garbeva, P., Postma, J., van Veen, J.A. and van Elsas, J.D. (2006) Effect of above-ground plant species on microbial community structure and its impact on suppression of *Rhizoctonia solani* AG3. *Environmental Microbiology* 8, 233–246.

Garbeva, P., Hol, W.H.G, Termorshuizen, A.J., Kowalchuk, G.A. and de Boer, W. (2011a) Fungistasis and general soil biostasis – a new synthesis. *Soil Biology & Biochemistry* 43, 469–477.

Garbeva, P., Silby, M.W., Raaijmakers, J.M., Levy, S.B. and de Boer, W. (2011b) Transcriptional and antagonistic responses of *Pseudomonas fluorescens* Pf0–1 to phylogenetically different bacterial competitors. *International Society for Microbial Ecology Journal* 5, 973–985.

Garside, A.L. and Bell, M.J. (2011a) Growth and yield responses to amendments to the sugarcane monoculture: effects of crop, pasture and bare fallow breaks and soil fumigation on plant and ratoon crops. *Crop & Pasture Science* 62, 396–412.

Garside, A.L. and Bell, M.J. (2011b) Growth and yield responses to amendments to the sugarcane monoculture: towards identifying the reasons behind the response breaks. *Crop & Pasture Science* 62, 776–789.

Garside, A.L., Bell, M.J., Robotham, B.G., Magarey, R.C. and Stirling, G.R. (2005) Managing yield decline in sugarcane cropping systems. *International Sugar Journal*, 107, 16–26.

Gaspard, J.T. and Mankau, R. (1986) Nematophagous fungi associated with *Tylenchulus semipenetrans* and the citrus rhizosphere. *Nematologica* 32, 359–363.

Gaspard, J.T., Jaffee, B.A. and Ferris, H. (1990a) Association of *Verticillium chlamydosporium* and *Paecilomyces lilacinus* with root-knot nematode infested soil. *Journal of Nematology* 22, 207–213.

Gaspard, J.T., Jaffee, B.A. and Ferris, H. (1990b) *Meloidogyne incognita* survival in soil infested with *Paecilomyces lilacinus* and *Verticillium chlamydosporium*. *Journal of Nematology* 22, 176–181.

Gené, J., Verdejo-Lucas, S., Stchigel, A.M., Sorribas, F.J. and Guarro, J. (2005) Microbial parasites associated with *Tylenchulus semipenetrans* in citrus orchards of Catalonia, Spain. *Biocontrol Science & Technology* 15, 721–731.

Georgieva, S.S., McGrath, S.P., Hooper, D.J. and Chambers, B.S. (2002) Nematode communities under stress: the long-term effects of heavy metals in soil treated with sewage sludge. *Applied Soil Ecology* 20, 27–42.

Gerber, J.F., Hewlett, T.E., Smith, K.S. and White, J.H. (2006) Materials and methods for *in vitro* production of bacteria. United States Patent No. US 7,067,299 B2, 27 June 2006. http://patft.uspto.gov/netacgi/

nph-Parser?Sect1=PTO1&Sect2=HITOFF&d=PALL&p=1&u=%2Fnetahtml%2FPTO%2Fsrchnum.htm&r=1&f=G&l=50&s1=7067299.PN.&OS=PN/7067299&RS=PN/7067299, accessed 16 October 2013

Germida, J.J. and Siciliano, S.D. (2001) Taxonomic diversity of bacteria associated with the roots of modern, recent and ancient wheat cultivars. *Biology & Fertility of Soils* 33, 410–415.

Gernandt, D.S. and Stone, J.K. (1999) Phylogenetic analysis of nuclear ribosomal DNA places the nematode parasite, *Drechmeria coniospora*, in Clavicipitaceae. *Mycologia* 91, 993–1000.

Ghorbani, R., Wilcockson, S., Koocheki, A. and Leifert, C. (2008) Soil management for sustainable crop disease control: a review. *Environmental Chemistry Letters* 6, 149–162.

Giannakou, I.O. and Gowen, S.R. (2004) Factors affecting biological control effectiveness of *Pasteuria penetrans* in *Meloidogyne javanica* and the bacterial development in the nematode body. *Nematropica* 34, 153–163.

Giannakou, I.O., Pembroke, B., Gowen, S.R. and Davies, K.G. (1997) Effects of long-term storage and above normal temperatures on spore adhesion of *Pasteuria penetrans* and infection of the root-knot nematode *Meloidogyne javanica*. *Nematologica* 43, 185–192.

Giannakou, I.O., Pembroke, B., Gowen, S.R. and Douloumpaka, S. (1999) Effects of fluctuating temperatures and different host plants on development of *Pasteuria penetrans* in *Meloidogyne javanica*. *Journal of Nematology* 31, 312–318.

Giannakou, I.O., Karpouzas, D.G. and Prophetou-Athanasiadou, D. (2004) A novel non-chemical nematicide for the control of root-knot nematodes. *Applied Soil Ecology* 26, 69–74.

Giannakou, I.G., Anastasiadis, I.A., Gowen, S.R. and Prophetou-Athanasiadou, D.A. (2007) Effects of a non-chemical nematicide combined with soil solarization for the control of root-knot nematodes. *Crop Protection* 26, 1644–1654.

Giblin-Davis, R.M. (2000) *Pasteuria* sp. for biological control of the sting nematode, *Belonolaimus longicaudatus*, in turfgrass. In: Clark, J.M. and Kenna, M.P. (eds) *Fate and Management of Turfgrass Chemicals*. American Chemical Society Symposium Series No. 743. Oxford University Press, New York, pp. 408–426.

Giblin-Davis, R.M., McDaniel, L.L. and Bilz, F.G. (1990) Isolates of the *Pasteuria penetrans* group from phytoparasitic nematodes in Bermudagrass turf. *Journal of Nematology* 22, 750–762.

Giblin-Davis, R.M., Williams, D., Hewlett, T.E. and Dickson, D.W. (1995) Development and host attachment studies using *Pasteuria* from *Belonolaimus longicaudatus* from Florida. *Journal of Nematology* 27, 500.

Giblin-Davis, R.M., Center, B.J., Hewlett, T.E. and Dickson, D.W. (1998) *Pasteuria*-infested soil suppresses *Belonolaimus longicaudatus* in a bermudagrass green. *Journal of Nematology* 30, 497.

Giblin-Davis, R.M., Williams, D.S., Wergin, W.P., Dickson, D.W., Hewlett, T.E., Bekal, S. and Becker, J.O. (2001) Ultrastructure and development of *Pasteuria* sp. (S-1 strain), an obligate endoparasite of *Belonolaimus longicaudatus* (Nemata: Tylenchida). *Journal of Nematology* 33, 227–238.

Giblin-Davis, R.M., Williams, D.S., Bekal, S., Dickson, D.W., Brito, J.A., Becker, J.O. and Preston, J.F. (2003) 'Candidatus Pasteuria usgae' sp. nov., an obligate endoparasite of the phytoparasitic nematode *Belonolaimus longicaudatus*. *International Journal of Systematic & Evolutionary Microbiology* 53, 197–200.

Giblin-Davis, R., Nong, G., Preston, J.F., Williams, D.S., Center, B.J., Brito, J.A. and Dickson, D.W. (2011) 'Candidatus Pasteuria aldrichii', an obligate endoparasite of the bacterivorous nematode, *Bursilla*. *International Journal of Systematic and Evolutionary Microbiology* 61, 2073–2080.

Gilmore, S.K. (1970) Collembola predation on nematodes. *Search Agriculture* 1, 1–12.

Gintis, B.O., Morgan-Jones, G. and Rodriguez-Kabana, R. (1983) Fungi associated with several developmental stages of *Heterodera glycines* from an Alabama soybean field soil. *Nematropica* 13, 181–200.

Giuma, A.Y. and Cooke, R.C. (1974) Potential of *Nematoctonus* conidia for biological control of soil-borne phytonematodes. *Soil Biology & Biochemistry* 6, 217–220.

Glanz, J.T. (1995) *Saving Our Soil: Solutions for Sustaining Earth's Vital Resources*. Johnson Books, Boulder, Colorado.

Glare, T., Caradus, J., Gelernter, W., Jackson, T., Keyhani, N., Köhl, J., Marrone, P., Morin, L. and Stewart, A. (2012) Have biopesticides come of age? *Trends in Biotechnology* 30, 250–258.

Glick, B.R., Todorovic, B., Czarny, J., Chen, Z., Duan, J. and McConkey, B. (2007) Promotion of plant growth by bacterial ACC deaminase. *Critical Reviews in Plant Sciences* 26, 227–242.

Gliessman, S.R. (1984) An agroecological approach to sustainable agriculture. In: Jackson, W., Berry, W. and Colman, B. (eds) *Meeting the Expectations of the Land*. North Point Press, San Francisco, California, pp. 160–171.

Gliessman, S.R. (2007) *Agroecology: The Ecology of Sustainable Food Systems*. CRC Press, Boca Raton, Florida.

Glockling, S.L. (1993) Eelworm-eaters. The *Harposporium* and the host. *Mycologist* 7, 139–142.

Glockling, S.L. and Beakes, G.W. (2000a) A review of the taxonomy, biology and infection strategies of "biflagellate holocarpic" parasites of nematodes. *Fungal Diversity* 4, 1–20.

Glockling, S.L. and Beakes, G.W. (2000b) An ultrastructural study of sporidium formation during infection of a rhabditid nematode by large gun cells of *Haptoglossa heteromorpha*. *Journal of Invertebrate Pathology* 76, 208–215.

Glockling, S.L. and Beakes, G.W. (2002) Ultrastructural morphogenesis of dimorphic arcuate infection (gun) cells of *Haptoglossa erumpens* an obligate parasite of *Bunonema* nematodes. *Fungal Genetics & Biology* 37, 250–262.

Glockling, S.L. and Beakes, G.W. (2006a) Structural and developmental studies of *Chlamydomyzium ovipara-siticum* from *Rhabditis* nematodes and in culture. *Mycological Research* 110, 1119–1126.

Glockling, S.L. and Beakes, G.W. (2006b) An ultrastructural study of development and reproduction in the nematode parasite *Myzocytiopsis vermicola*. *Mycologia* 98, 1–15.

Glover, J.D., Cox, C.M. and Reganold, J.P. (2007) Future farming: a return to roots? *Scientific American* 297, 82–89.

Godfray, H.C.J., Beddington, J.R., Crute, I.R., Haddad, L., Lawrence, D., Muir, J.F., Pretty, J., Robinson, S., Thomas, S.M. and Toulmin, C. (2010) Food security: the challenge of feeding 9 billion people. *Science* 327, 812–818.

Godoy, G., Rodriguez-Kábana, R. and Morgan-Jones, G. (1982) Parasitism of eggs of *Heterodera glycines* and *Meloidogyne arenaria* by fungi isolated from cysts of *H. glycines*. *Nematropica* 12, 111–119.

Godoy, G., Rodriguez-Kabana, R., Shelby, R.A. and Morgan-Jones, G. (1983a) Chitin amendments for control of *Meloidogyne arenaria* in infested soil. II. Effects on microbial population. *Nematropica* 13, 63–74.

Godoy, G., Rodriguez-Kabana, R. and Morgan-Jones, G. (1983b) Fungal parasites of *Meloidogyne arenaria* eggs in an Alabama soil. A mycological survey and greenhouse studies. *Nematropica* 13, 201–213.

Gold, M.V. (2007) Sustainable agriculture: definitions and terms. Special Reference Briefs 99–02, USDA, Beltsville, Maryland www.nal.usda.gov/afsic/pubs/terms/srb9902.shtml, accessed 18 November 2013.

Gomes Carneiro, R.M.D. and Cayrol, J.-C. (1991) Relationship between inoculum density of the nematophagous fungus *Paecilomyces lilacinus* and control of *Meloidogyne arenaria* on tomato. *Revue de Nématologie* 14, 629–634.

Gomiero, T., Pimentel, D. and Paoletti, M.G. (2011a) Is there a need for a more sustainable agriculture? *Critical Reviews in Plant Sciences* 30, 6–23.

Gomiero, T., Pimentel, D. and Paoletti, M.G. (2011b) Environmental impact of different agricultural management practices: conventional vs. organic agriculture. *Critical Reviews in Plant Sciences* 30, 95–124.

Gonzalez-Quiñones, V., Stockdale, E.A., Banning, N.C., Hoyle, F.C., Sawada, Y., Wherrett, A.D., Jones, D.L. and Murphy, D.V. (2011) Soil microbial biomass – interpretation and consideration for soil monitoring. *Soil Research* 49, 287–304.

Goodey, J.B. (1963) *Soil and Freshwater Nematodes*. Methuen, London.

Goodman, R.M., Bintrim, S.B., Handelsman, J., Quirino, B.F., Rosas, J.C., Simon, H.M. and Smith K.P. (1998) A dirty look: soil microflora and rhizosphere microbiology. In: Flores, H.E., Lynch, J.P. and Eissenstat, D. (eds) *Radical Biology: Advances and Perspectives on the Function of Plant Roots*. American Society of Plant Physiologists, Rockville, Maryland, pp. 219–231.

Gortari, M.C. and Hours, R.A. (2008) Fungal chitinases and their biological role in the antagonism onto nematode eggs. A review. *Mycological Progress* 7, 221–238.

Goulding, K., Jarvis, S. and Whitmore, A. (2008) Optimizing nutrient management for farm systems. *Philosophical Transactions of the Royal Society B*, 363, 667–680.

Gowen, S.R., Bala, G., Madulu, J., Mwageni, W. and Triviño, C.G. (1998) Field evaluation of *Pasteuria penetrans* for the management of root-knot nematodes. *The 1998 Brighton Conference – Pests and Diseases* 8B-3, 755–760.

Gowen, S.R., Quénéhervé, P. and Fogain, R. (2005) Nematode parasites of bananas and plantains. In: Luc, M., Sikora, R.A. and Bridge, J. (eds) *Plant Parasitic Nematodes in Subtropical and Tropical Agriculture*. CAB International, Wallingford, UK, pp. 611–643.

Gowen, S., Davies, K.G. and Pembroke, B. (2008) Potential use of *Pasteuria* spp. in the management of plant parasitic nematodes. In: Ciancio, A. and Mukerji, K.G. (eds) *Integrated Management and Biocontrol of Vegetable and Grain Crops Nematodes*. Springer, Dordrecht, the Netherlands, pp. 205–219.

Graber, E.R., Harel, Y.M., Kolton, M., Cytryn, E., Silber, A., David, D.R., Tsechansky, L., Borenshtein, M. and Elad, Y. (2010) Biochar impact on development and productivity of pepper and tomato grown in fertigated soilless media. *Plant & Soil* 337, 481–496.

Graff, N.J. and Madelin, M.F. (1989) Axenic culture of the cyst-nematode parasitizing fungus *Nematophthora gynophila*. *Journal of Invertebrate Pathology* 53, 301–306.

Grandy, A.S., Kallenbach, C., Loecke, T.D., Snapp, S.S. and Smith, R.G. (2013) The biological basis for nitrogen management in agroecosystems. In: Cheeke, T.E., Coleman, D.C. and Wall, D.H. (eds) *Microbial Ecology in Sustainable Agroecosystems*. CRC Press, Taylor and Francis Group, Boca Raton, Florida, pp. 113–132.

Grant, C., Bittman, S., Montreal, M., Plenchette, C. and Morel, C. (2005) Soil and fertilizer phosphorus: effects on plant P supply and mycorrhizal development. *Canadian Journal of Plant Science* 85, 3–14.

Gray, N.F. (1985) Ecology of nematophagous fungi: effect of soil moisture, organic matter, pH and nematode density on distribution. *Soil Biology & Biochemistry* 17, 499–507.

Gray, N.F. (1987) Nematophagous fungi with particular reference to their ecology. *Biological Reviews* 62, 245–304.

Gray, N.F. (1988) Fungi attacking vermiform nematodes. In: Poinar, G.O. and Jansson, H.-B. (eds) *Diseases of Nematodes*. Vol. II. CRC Press Inc., Boca Raton, Florida, pp. 3–38.

Grayston, S.J., Campbell, C.D., Bardgett, R.D., Mawdsley, J.L., Clegg, C.D., Ritz, K., Griffiths, B.S., Rodwell, J.S., Edwards, S.J., Davies, W.J., Elston, D.J. and Millard, P. (2004) Assessing shifts in microbial community structure across a range of grasslands of differing management intensity using CLPP, PLFA and community DNA techniques. *Applied Soil Ecology* 25, 63–84.

Gregorich, E.G. and Carter, M.R. (eds) (1997) *Soil Quality for Crop Production and Ecosystem Health*. Elsevier, Amsterdam, the Netherlands.

Gregory, P.J. (2006) Roots, rhizosphere and soil: the route to a better understanding of soil science? *European Journal of Soil Science* 57, 2–12.

Gressel, J. (2011) Global advances in weed management. *Journal of Agricultural Science* 149, 47–53.

Grewal, P.S., Martin, W.R., Miller, R.W. and Lewis, E.E. (1997) Suppression of plant-parasitic nematode populations in turfgrass by application of entomopathogenic nematodes. *Biocontrol Science & Technology* 7, 393–399.

Griffin, D.M. (1972) *Ecology of Soil Fungi*. Chapman and Hall, London.

Griffin, D.M. (1981) Water potential as a selective factor in the microbial ecology of soils. In: Parr, J.F., Gardner, W.R. and Elliot, L.F. (eds) *Water Potential Relations in Soil Microbiology*. Special publication no. 9, Soil Science Society of America, Madison, Wisconsin, pp. 141–151.

Griffin, G.J. and Roth, D.A. (1979) Nutritional aspects of soil mycostasis. In: Schippers, B. and Gams, W. (eds). *Soil-borne Plant Pathogens*. Academic Press, London, pp. 79–96.

Griffiths, B.S., Ball, B.C., Daniell, T.J., Hallett, P.D., Neilson, R., Wheatley, R.E., Osler, G. and Bohanec, M. (2010) Integrating soil quality changes to arable agricultural systems following organic matter addition, or adoption of ley-arable rotation. *Applied Soil Ecology* 46, 43–53.

Griffiths, B.S., Daniell, T.J., Donn, S. and Neilson, R. (2012) Bioindication potential of using molecular characterisation of the nematode community: response to soil tillage. *European Journal of Soil Biology* 49, 92–97.

Grootaert, P. and Maertens, D. (1976) Cultivation and life cycle of *Mononchus aquaticus*. *Nematologica* 22, 173–181.

Grootaert, P., Jaques, A. and Small, R.W. (1977) Prey Selection in *Butlerius* sp. (Rhabditida: Diplogasteridae). *Mededelingen Faculteit Landbouwwetenschappen Rijksuniversiteit Gent* 42/2, 1559–1563.

Gugino, B.K., Idowu, O.J., Schindelbeck, R.R., van Es, H.M., Wolfe, D.W., Moebius-Clone, B.N., Thies, J.E. and Abawi, G.S. (2009) *Cornell Soil Health Assessment Training Manual*. 2nd edn. http://soilhealth.cals.cornell.edu/extension/manual.htm. Cornell University, Geneva, New York.

Gupta, V.V.S.R. and Knox, O.G.G. (2010) How best can we design rhizosphere plant-microbe interactions for the benefit of plant growth? In: Gupta, V.V.S.R., Ryder, M. and Radcliffe, J. (eds) *The Rovira Rhizosphere Symposium*. The Crawford Fund, Deacon, ACT, Australia, pp. 11–24.

Gupta, V.V.S.R., Rovira, A.D. and Roget, D.K. (2011) Principles and management of soil biological factors for sustainable rainfed farming systems. In: Tow, P.G., Cooper, I., Partridge, I. and Birch, C. (eds) *Rainfed Farming Systems*. Springer, Dordrecht, the Netherlands, pp. 149–184.

Haas, D. and Defago, G. (2005) Biological control of soil-borne pathogens by fluorescent pseudomonads. *Nature Reviews Microbiology* 3, 307–319.

Haas, D. and Keel, C. (2003) Regulation of antibiotic production in root-colonising *Pseudomonas* spp. and relevance for biological control of plant disease. *Annual Review of Phytopathology* 41, 117–153.

Hackenberg, C., Muehlchen, A., Forge, T. and Vrain, T. (2000) *Pseudomonas chlororaphis* strain Sm3, bacterial antagonist of *Pratylenchus penetrans*. *Journal of Nematology* 32, 183–189.

Hagedorn, G. and Scholler, M. (1999) A reevaluation of predatory orbiliaceous fungi. 1. Phylogenetic analysis using rDNA sequence data. *Sydowia* 51, 27–48.

Halbrendt, J.M. (1996) Allelopathy in the management of plant-parasitic nematodes. *Journal of Nematology* 28, 8–14.

Hallmann, J. (2001) Plant interactions with endophytic bacteria. In: Jeger, M.J. and Spence, N.J. (eds) *Biotic Interactions in Plant-Pathogen Associations*. CAB International, Wallingford, UK, 87–119.

Hallmann, J. and Sikora, R.A. (1994) Occurrence of plant parasitic nematodes and non-pathogenic species of *Fusarium* in tomato plants in Kenya and their role as mutualistic synergists for biological control of root-knot nematodes. *International Journal of Pest Management* 40, 321–325.

Hallmann, J. and Sikora, R.A. (1996) Toxicity of fungal endophyte secondary metabolites to plant-parasitic nematodes and soil-borne plant pathogenic fungi. *European Journal of Plant Pathology* 102, 155–162.

Hallmann, J. and Sikora, R.A. (2011) Endophytic fungi. In: Davies, K. and Spiegel, Y. (eds) *Biological Control of Plant-Parasitic Nematodes: Building Coherence between Microbial Ecology and Molecular Mechanisms*. Springer, Dordrecht, the Netherlands, pp. 227–258.

Hallmann, J., Quadt-Hallmann, A., Mahaffee, W.F. and Kloepper, J.W. (1997) Bacterial endophytes in agricultural crops. *Canadian Journal of Microbiology* 43, 895–914.

Hallmann, J., Quadt-Hallmann, A., Rodriguez-Kábana, R. and Kloepper, J.W. (1998) Interactions between *Meloidogyne incognita* and endophytic bacteria in cotton and cucumber. *Soil Biology & Biochemistry* 30, 925–937.

Hallmann, J., Rodriguez-Kabana, R. and Kloepper, J.W. (1999) Chitin-mediated changes in bacterial communities of soil, rhizosphere and within roots of cotton in relation to nematode control. *Soil Biology & Biochemistry* 31, 551–560.

Hallmann, J., Quadt-Hallmann, A., Miller, W.G., Sikora, R.A. and Lindow, S.E. (2001) Endophytic colonization of plants by the biocontrol agent *Rhizobium etli* G12 in relation to *Meloidogyne incognita* infection. *Phytopathology* 91, 415–422.

Hallmann, J., Frankenburg A., Paffrath, A. and Schmidt, H. (2007) Occurrence and importance of plant-parasitic nematodes in organic farming in Germany. *Nematology* 9, 869–879.

Hallmann, J., Davies, K.G. and Sikora, R. (2009) Biological control using microbial pathogens, endophytes and antagonists. In: Perry, R.N., Moens, M. and Starr, J.L. (eds) *Root-Knot Nematodes*. CAB International, Wallingford, UK, pp. 380–411.

Hamblin, A. (1995) The concept of agricultural sustainability. *Advances in Plant Pathology* 11, 1–19.

Handelsman, J. and Stabb, E.V. (1996) Biocontrol of soilborne pathogens. *Plant Cell* 8, 1855–1869.

Haney, R.L., Senseman, S.A., Hons, F.M. and Zuberer, D.A. (2000) Effect of glyphosate on soil microbial activity and biomass. *Weed Science* 48, 89–93.

Harker, K.N. and O'Donovan, J.T. (2013) Recent weed control, weed management, and integrated weed management. *Weed Technology* 27, 1–11.

Harman, G.E. (2000) Myths and dogmas of biocontrol. Changes in perceptions derived from research on *Trichoderma harzianum* T-22. *Plant Disease* 84, 377–393.

Harman, G.E. and Kubicek, C.P. (eds) (1998) *Trichoderma and Gliocladium*. Vols. I and II. Taylor & Francis, London.

Harman, G.E., Howell, C.R., Viterbo, A., Chet, I. and Lorito, M (2004) *Trichoderma* species – opportunistic, avirulent plant symbionts. *Nature Reviews Microbiology* 2, 43–56.

Harris, R.F. (1981) Effect of water potential on microbial growth and activity. In: Parr, J.F., Gardner, W.R. and Elliot, L.F. (eds) *Water Potential Relations in Soil Microbiology*. Soil Science Society of America, Madison, Wisconsin, pp. 23–95.

Hartmann, A., Schmid, M., van Tuinen, D. and Berg, G. (2009) Plant-driven selection of microbes. *Plant & Soil* 321, 235–257.

Hartwig, E.E. (1981) Breeding productive soybean cultivars resistant to soybean cyst nematode for the southern United States. *Plant Disease* 65, 303–307.

Harwood, R.R. (1990) A history of sustainable agriculture. In: Edwards, C.A., Lal, R., Madden, P., Miller, R.H. and House, G. (eds) *Sustainable Agricultural Systems*. Soil Water Conservation Society, Ankeny, Iowa, pp. 3–19.

Hashem, M. and Abo-Elyousr, K.A. (2011) Management of the root-knot nematode *Meloidogyne incognita* on tomato with combinations of different biocontrol organisms. *Crop Protection* 30, 285–292.

Hasky-Günther, K., Hoffmann-Hergarten, S. and Sikora, R.A. (1998) Resistance against the potato cyst nematode *Globodera pallida* systemically induced by the rhizobacteria *Agrobacterium radiobacter* (G12) and *Bacillus sphaericus* (B43). *Fundamental & Applied Nematology* 21, 511–517.

Hatz, B. and Dickson, D.W. (1992) Effect of temperature on attachment, development, and interactions of *Pasteuria penetrans* on *Meloidogyne arenaria*. *Journal of Nematology* 24, 512–521.

Hawes, M.C., Brigham, L.A., Wen, F., Woo, H.H., Zhu, Y (1998) Function of root border cells in plant health: pioneers in the rhizosphere. *Annual Review of Phytopathology* 36, 311–327.

Hawes, M.C., Gunawardena, U., Miyasaka, S. and Zhao, X.W. (2000) The role of root border cells in plant defence. *Trends in Plant Science* 5, 128–133.

Hawkes, C.V., DeAngelis, K.M. and Firestone, M.K. (2007) Root interactions with soil microbial communities and processes. In: Cardon, Z.G. and Whitbeck, J.L. (eds) *The Rhizosphere*. Elsevier Academic Press, New York, pp. 1–29.

Hawksworth, D.L. (2012) Managing and coping with names of pleomorphic fungi in a period of transition. *Mycosphere* 3, 143–155. DOI 10.5943/mycosphere/3/2/4/.

Hay, R.J. (2002) *Scytalidium* infections. *Current Opinion in Infectious Diseases* 15, 99–100.

Hayat, R., Ali, S., Amara, U., Khalid, R. and Ahmed, I. (2010) Soil beneficial bacteria and their role in plant growth promotion: a review. *Annals of Microbiology* 60, 579–598.

Hayward, A.C., Fegan, N., Fegan, M. and Stirling G.R. (2009) *Stenotrophomonas* and *Lysobacter*: ubiquitous plant-associated *gamma*-proteobacteria of developing significance in applied microbiology. *Journal of Applied Microbiology* 108, 756–770.

Heap, I. (2013) The international survey of herbicide resistant weeds. www.weedscience.org, accessed 1 July 2013.

Hechler, H.C. (1963) Description, developmental biology, and feeding habits of *Seinura tenuicaudata* (De Man) J.B. Goodey, 1960 (Nematoda: Aphelenchoididae), a nematode predator. *Proceedings of the Helminthological Society of Washington* 30, 182–195.

Heidemann, K., Scheu, S., Ruess, L. and Maraun, M. (2011) Molecular detection of nematode predation and scavenging in oribatid mites: Laboratory and field experiments. *Soil Biology & Biochemistry* 43, 2229–2236.

Helander, M., Saloniemi, I. and Saikkonen, K. (2012) Glyphosate in northern ecosystems. *Trends in Plant Science* 17, 569–574.

Helgason, B.L., Walley, F.L. and Germida, J.J. (2009) Fungal and bacterial abundance in long-term no till and intensive-till soils of the northern great plains. *Soil Science Society of America Journal* 73, 120–127.

Helgason, B.L., Walley, F.L. and Germida, J.J. (2010) No-till management increases microbial biomass and alters community profiles in soil aggregates. *Applied Soil Ecology* 46, 390–397.

Hendrix, P.F., Parmelee, R.W., Crossley, D.A. Jr, Coleman, D.C., Odum, E.P. and Groffman, P.M. (1986) Detritus food webs in conventional and no-tillage agroecosystems. *BioScience* 36, 374–380.

Henriksen, T.M. and Breland, T.A. (1999) Nitrogen availability affects carbon mineralization, fungal and bacterial growth, and enzyme activities during decomposition of wheat straw in soil. *Soil Biology & Biochemistry* 31, 1121–1134.

Herencia, J.F., García-Galavís, P.A., Dorado, J.A.R. and Maqueda, C. (2011) Comparison of nutritional quality of the crops grown in an organic and conventional fertilized soil. *Scientia Horticulturae* 129, 882–888.

Hewlett, T.E. and Dickson, D.W. (1993) A centrifugation method for attaching endospores of *Pasteuria* spp. to nematodes. *Journal of Nematology* 25, 785–788.

Hewlett, T.E., Dickson, D.W., Mitchell, D.J. and Kannwischer-Mitchell, M.E. (1988) Evaluation of *Paecilomyces lilacinus* as a biocontrol agent of *Meloidogyne javanica* on tobacco. *Journal of Nematology* 20, 578–584.

Hewlett, T.E., Gerber, J.F. and Smith, K.S. (2004) *In vitro* culture of *Pasteuria penetrans*. *Nematology Monographs & Perspectives* 2, 175–185.

Hewlett, T.E., Griswold, S.T. and Smith, K.S. (2006) Biological control of *Meloidogyne incognita* using *in-vitro* produced *Pasteuria penetrans* in a microplot study. *Journal of Nematology* 38, 274.

Hewlett, T.E., Griswold, S.T., Waters, J.P. and Smith, K.S. (2008a) Production and efficacy of *in vitro Pasteuria* spp. parasitizing *Belonolaimus longicaudatus*. Abstracts, Fifth International Congress of Nematology, Brisbane, QLD, Australia, p. 96.

Hewlett, T.E., Griswold, S.T., Waters, J.P. and Smith, K.S. (2008b) *In vitro Pasteuria* spp. endospores rate of germination and infection on *Belonolaimus longicaudatus*. Abstracts, Fifth International Congress of Nematology, Brisbane, QLD, Australia, p. 295.

Hewlett, T.E., Waters, J.P., Luc, J.E. and Crow, W.T. (2008c) Field studies using *in vitro* produced *Pasteuria* endospores to control sting nematodes on turf. Abstracts, Fifth International Congress of Nematology, Brisbane, QLD, Australia, p.180.

Hewlett, T.E., Stetina, S.R., Schmidt, L.M., Waters, J.P., Simmons, L.J. and Rich, J.R. (2009) Identification of *Pasteuria* spp. that parasitize *Rotylenchulus reniformis*. *Journal of Nematology* 41, 338.

Hibbett, D.S., Binder, M., Bischoff, J.F., Blackwell, M., Cannon, P.F., Eriksson, O.E., Huhndorf, S., James, T., Kirk, P.M. *et al.* (2007) A higher-level phylogenetic classification of the fungi. *Mycological Research* 111, 509–547.

Hidalgo-Diaz, L., Bourne, J.M., Kerry, B.R. and Rodriguez, M.G. (2000) Nematophagous *Verticillium* spp. in soils infested with *Meloidogyne* spp. in Cuba: Isolation and screening. *International Journal of Pest Management* 46, 277–284.

Hinsinger, P., Bengough, A.G., Vetterlein, D. and Young, I.M. (2009) Rhizosphere: biophysics, biogeochemistry and ecological relevance. *Plant & Soil* 321, 117–152.

Hirsch, P.R., Mauchline, T.H., Mendum, T.A. and Kerry, B.R. (2000) Detection of the nematophagous fungus *Verticillium chlamydosporium* in nematode-infested plant roots using PCR. *Mycological Research* 104, 435–439.

Hirsch, P.R., Atkins, S.D., Mauchline, T.H., Morton, C.O., Davies, K.G. and Kerry, B.R. (2001) Methods for studying the nematophagous fungus *Verticillium chlamydosporium* in the root environment. *Plant & Soil* 232, 21–30.

Ho, W.C. and Ko, W.H. (1982) Characteristics of soil microbiostasis. *Soil Biology & Biochemistry* 14, 589–593.

Ho, W.C. and Ko, W.H. (1986) Microbiostasis by nutrient deficiency shown in natural and synthetic soils. *Journal of General Microbiology* 132, 2807–2815.

Hobbie, S.E. (2008) Nitrogen effects on decomposition: a five-year experiment in eight temperate sites. *Ecology* 89, 2633–2644.

Hobbs, P.R., Sayre, K. and Gupta, R. (2008) The role of conservation agriculture in sustainable agriculture. *Philosophical Transactions of the Royal Society B*, 363, 543–555.

Hodda, M., Peters, L. and Traunspurger, W. (2009) Nematode diversity in terrestrial, freshwater, aquatic and marine ecosystems. In: Wilson, M.J. and Kakouli-Duarte, T. (eds) *Nematodes as Environmental Indicators*. CAB International, Wallingford, UK, pp. 45–93.

Hodge, K.T., Viaene, N.M. and Gams, W. (1997) Two *Harposporium* species with *Hirsutella* synanamorphs. *Mycological Research* 101, 1377–1382.

Hoeksema, J.D., Chaudhary, V.B., Gehring, C.A., Johnson, N.C., Karst, J., Koide, R.T., Pringle, A., Zabinski, C., Bever, J.D., Moore, J.C., Wilson, G.W.T., Klironomos, J.N. and Umbanhowar, J. (2010) A meta-analysis of context-dependency in plant response to inoculation with mycorrhizal fungi. *Ecology Letters* 13, 394–407.

Hohberg, K. and Traunspurger, W. (2005) Predator-prey interaction in soil food web: functional response, size-dependent foraging efficiency, and the influence of soil texture. *Biology & Fertility of Soils* 41, 419–427.

Hoitink, H.A.J. and Boehm, M.J. (1999) Biocontrol within the context of soil microbial communities: a substrate-dependent phenomenon. *Annual Review of Phytopathology* 37, 427–446.

Hol, W.G.H. and Cook, R. (2005) An overview of arbuscular mycorrhizal fungi-nematode interactions. *Basic & Applied Ecology* 6, 498–503.

Hölker, U., Höfer, M. and Lenz, J. (2004) Biotechnological advantages of laboratory-scale solid-state fermentation with fungi. *Applied Microbiology & Biotechnology* 64, 175–186.

Holland, E.A. and Coleman, D.C. (1987) Litter placement effects and organic matter dynamics in an agroecosystem. *Ecology* 68, 425–433.

Holland, R.J., Williams, K.L. and Nevalainen, K.M.H. (2003) *Paecilomyces* strain Bioact 251 is not a plant endophyte. *Australasian Plant Pathology* 32, 473–478.

Holterman, M., van der Wurff, A., van den Elsen, S., van Megen, H., Bongers, T., Holovachov, O., Bakker, J. and Helder, J. (2006) Phylum-wide analysis of SSU rDNA reveals deep phylogenetic relationships among nematodes and accelerated evolution toward crown clades. *Molecular Biology & Evolution* 23, 1792–1800.

Holterman, M., Rybarczyk, K., Van den Elsen, S., Van Megen, H., Mooyman, P., Peña Santiago, R., Bongers, T., Bakker, J. and Helder, J. (2008) A ribosomal DNA-based framework for the detection and quantification of stress-sensitive nematode families in terrestrial habitats. *Molecular Ecology Resources* 8, 23–34.

Höper, H. and Alabouvette, C. (1996) Importance of physical and chemical soil properties in the suppressiveness of soils to plant diseases. *European Journal of Soil Biology* 32, 41–58.

Horiuchi, J.-I., Prithiviraj, B., Bais, H., Kimball, B. and Vivanco, J. (2005) Soil nematodes mediate positive interactions between legume plants and rhizobium bacteria. *Planta* 222, 848–857.

Howell, C.R. (2003) Mechanisms employed by *Trichoderma* species in the biological control of plant diseases: the history and evolution of current concepts. *Plant Disease* 87, 4–10.

Howie, W.J., Cook, R.J. and Weller, D.M. (1987) Effects of soil matric potential and cell motility on wheat root colonization by fluorescent pseudomonads suppressive to take-all. *Phytopathology* 77, 286–292.

Hue, N.V. and Silva, J.A. (2000) Organic soil amendments for sustainable agriculture: organic sources of nitrogen, phosphorus, and potassium. In: Silva, J.A. and Uchida, R. (eds) *Plant Nutrient Management in Hawaii's Soils, Approaches for Tropical and Subtropical Agriculture*. College of Tropical Agriculture and Human Resources, University of Hawaii at Manoa, Hawaii, pp. 133–144.

Huebner, R.A., Rodriguez-Kabana, R. and Patterson, R.M. (1983) Hemicellulosic waste and urea for control of plant parasitic nematodes: effect on soil enzyme activities. *Nematropica* 13, 37–54.

Hung, L.L. and Sylvia, D.M. (1988) Production of vesicular-arbuscular mycorrhizal fungus inoculum in aeroponic culture. *Applied & Environmental Microbiology* 54, 353–357.

Hunt, D.J. (1993) *Aphelenchida, Longidoridae and Trichodoridae: Their Systematics and Bionomics*. CAB International, Wallingford, UK.

Hunter-Fujita, F.R., Entwistle, P.F., Evans, H.F. and Crook, N.E. (1998) *Insect Viruses and Pest Management*. John Wiley & Sons, Chichester, New York.

Hussey, R.S. and Roncadori, R.W. (1981) Influence of *Aphelenchus avenae* on vesicular-arbuscular endomycorrhizal growth response in cotton. *Journal of Nematology* 13, 48–52.

Hussey, R.S. and Roncadori, R.W. (1982) Vesicular-arbuscular mycorrhizae may limit nematode activity and improve plant growth. *Plant Disease* 66, 9–14.

Hyvönen, R. and Persson, T. (1996) Effects of fungivorous and predatory arthropods on nematodes and tardigrades in microcosms with coniferous forest soil. *Biology & Fertility of Soils* 21, 121–127.

Hyvönen, R., Andersson, S., Clarhom, M. and Persson, T. (1994) Effects of lumbricids and enchytraeids on nematodes in limed and unlimed coniferous mor humus. *Biology & Fertility of Soils* 17, 201–205.

Idowu, O.J., van Es, H.M., Abawi, G.S., Wolfe, D.W., Schindelbeck, R.R., Moebius-Clune, B.N. and Gugino, B.K. (2009) Use of an integrative soil health test for evaluation of soil management impacts. *Renewable Agriculture & Food Systems* 24, 214–224.

Ilieva-Makulec, K. and Makulec, G. (2002) Effect of the earthworm *Lumbricus rubellus* on the nematode community in a peat meadow soil. *European Journal of Biology* 38, 59–62.

Imbriani, J.L. and Mankau, R. (1977) Ultrastructure of the nematode pathogen, *Bacillus penetrans*. *Journal of Invertebrate Pathology* 30, 337–347.

Ingham, R.E. (1988) Interactions between nematodes and vesicular-arbuscular mycorrhizae. *Agriculture, Ecosystems & Environment* 24, 169–182.

Ingham, R.E., Trofymow, J.A., Ingham, E.R. and Coleman, D.C. (1985) Interactions of bacteria, fungi, and their nematode grazers: effects on nutrient cycling and plant growth. *Ecological Monographs* 55, 119–140.

Ingram, J.S.I. and Fernandes, E.C.M. (2001) Managing carbon sequestration in soils: concepts and terminology. *Agriculture, Ecosystems & Environment* 87, 111–117.

Inserra, R.N., Oostendorp, M. and Dickson, D.W. (1992) *Pasteuria* sp. parasitizing *Trophonema okamotoi* in Florida. *Journal of Nematology* 24, 36–39.

Insunza, V., Alström, S. and Eriksson, K.B. (2002) Root bacteria from nematicidal plants and their biocontrol potential against trichodorid nematodes in potato. *Plant & Soil* 241, 271–278.

Irving, F. and Kerry, B.R. (1986) Variation between strains of the nematophagous fungus, *Verticillium chlamydosporium* Goddard. II. Factors affecting parasitism of cyst nematode eggs. *Nematologica* 32, 474–485.

Jackson, R.B., Canadell, J., Ehleringer, J.R., Mooney, H.A., Sala, O.E. and Schulze, E.D. (1996) A global analysis of root distributions for terrestrial biomes. *Oecologia* 108, 389–411.

Jacobs, H., Gray, S.N. and Crump, D.H. (2003) Interactions between nematophagous fungi and consequences for their potential as biological control agents for the control of potato cyst nematodes. *Mycological Research* 107, 47–56.

Jacobsen, B.J., Zidack, N.K. and Larson, B.J. (2004) The role of *Bacillus*-based biological control agents in integrated pest management systems: Plant diseases. *Phytopathology* 94, 1272–1275.

Jaffee, B.A. (1992) Population biology and biological control of nematodes. *Canadian Journal of Microbiology* 38, 359–364.

Jaffee, B.A. (1993) Density-dependent parasitism in biological control of soil-borne insects, nematodes, fungi and bacteria. *Biocontrol Science & Technology* 3, 235–246.

Jaffee, B.A. (1998) Susceptibility of a cyst and a root-knot nematode to three nematode-trapping fungi. *Fundamental & Applied Nematology* 21, 695–703.

Jaffee, B.A. (1999) Enchytraeids and nematophagous fungi in tomato fields and vineyards. *Phytopathology* 89, 398–406.

Jaffee, B.A. (2000) Augmentation of soil with the nematophagous fungi *Hirsutella rhossiliensis* and *Arthrobotrys haptotyla*. *Phytopathology* 90, 498–504.

Jaffee, B.A. (2002) Soil cages for studying how organic amendments affect nematode-trapping fungi. *Applied Soil Ecology* 21, 1–9.

Jaffee, B.A. (2003) Correlations between most probable number and activity of nematode-trapping fungi. *Phytopathology* 93, 1599–1605.

Jaffee, B.A. (2004a) Do organic amendments enhance the nematode-trapping fungi *Dactylellina haptotyla* and *Arthrobotrys oligospora*? *Journal of Nematology* 36: 267–275.

Jaffee, B.A. (2004b) Wood, nematodes, and the nematode-trapping fungus *Arthrobotrys oligospora*. *Soil Biology & Biochemistry* 35, 1171–1178.

Jaffee, B.A. (2006) Interactions among a soil organic amendment, nematodes and the nematode-trapping fungus *Dactylellina candidum*. *Phytopathology* 96, 1388–1396.

Jaffee, B.A. and McInnis, T.M. (1991) Sampling strategies for detection of density-dependent parasitism of soil-borne nematodes by nematophagous fungi. *Revue de Nématologie* 14, 147–150.

Jaffee, B.A. and Muldoon, A.E. (1989) Suppression of cyst nematode by a natural infestation of a nematophagous fungus. *Journal of Nematology* 21, 505–510.

Jaffee, B.A. and Muldoon, A.E. (1995a) Susceptibility of root-knot and cyst nematodes to the nematode-trapping fungi *Monacrosporium ellipsosporum* and *M. cionopagum*. *Soil Biology & Biochemistry* 27, 1083–1090.

Jaffee, B.A. and Muldoon, A.E. (1995b) Numerical responses of the nematophagous fungi *Hirsutella rhossiliensis*, *Monacrosporium cionopagum*, and *M. ellipsosporum*. *Mycologia* 87, 643–650.

Jaffee, B.A. and Muldoon, A.E. (1997) Suppression of the root-knot nematode *Meloidogyne javanica* by alginate pellets containing the nematophagous fungi *Hirsutella rhossiliensis*, *Monacrosporium cionopagum* and *M. ellipsosporum*. *Biocontrol Science & Technology* 7, 202–217.

Jaffee, B.A. and Schaffer, R.L. (1987) Parasitism of *Xiphinema americanum* and *X. rivesi* by *Catenaria anguillulae* and other zoosporic fungi in soil solution, Baermann funnels, or soil. *Nematologica* 33, 220–231.

Jaffee, B.A. and Strong, D.R. (2005) Strong bottom-up and weak top-down effects in soil: nematode-parasitized insects and nematode-trapping fungi. *Soil Biology & Biochemistry* 37, 1011–1021.

Jaffee, B.A. and Zasoski, R.J. (2001) Soil pH and the activity of a pelletized nematophagous fungus. *Phytopathology* 91, 324–330.

Jaffee, B.A. and Zehr, E.I. (1982) Parasitism of the nematode *Criconemella xenoplax* by the fungus *Hirsutella rhossiliensis*. *Phytopathology* 72, 1378–1381.

Jaffee, B.A. and Zehr, E.I. (1985) Parasitic and saprophytic abilities of the nematode-attacking fungus *Hirsutella rhossiliensis*. *Journal of Nematology* 17, 341–345.

Jaffee, B.A., Gaspard, J.T. and Ferris, H. (1989) Density-dependent parasitism of the soil-borne nematode *Criconemella xenoplax* by the nematophagous fungus *Hirsutella rhossiliensis*. *Microbial Ecology* 17, 193–200.

Jaffee, B.A., Muldoon, A.E., Phillips, R. and Mangel, M. (1990) Rates of spore transmission, mortality and production for the nematophagous fungus *Hirsutella rhossiliensis*. *Phytopathology* 80, 1083–1088.

Jaffee, B.A., Muldoon, A.E., Anderson, C.E. and Westerdahl, B.B. (1991) Detection of the nematophagous fungus *Hirsutella rhossiliensis* in California sugarbeet fields. *Biological Control* 1, 63–67.

Jaffee, B., Phillips, R., Muldoon, A. and Mangel, M. (1992a) Density-dependent host-pathogen dynamics in soil microcosms. *Ecology* 73, 495–506.

Jaffee, B.A., Muldoon, A.E. and Tedford, E.C. (1992b) Trap production by nematophagous fungi growing from parasitized nematodes. *Phytopathology* 82, 615–620.

Jaffee, B.A., Tedford, E.C. and Muldoon, A.E. (1993) Tests for density-dependent parasitism of nematodes by nematode-trapping and endoparasitic fungi. *Biological Control* 3, 329–336.

Jaffee, B.A., Ferris, H., Stapleton, J.J., Norton, M.V.K. and Muldoon, A.E. (1994) Parasitism of nematodes by the fungus *Hirsutella rhossiliensis* as affected by certain organic amendments. *Journal of Nematology* 26, 152–161.

Jaffee, B.A., Muldoon, A.E. and Westerdahl, B.B. (1996a) Failure of a mycelial formulation of the nematophagous fungus *Hirsutella rhossiliensis* to suppress the nematode *Heterodera schachtii*. *Biological Control* 6, 340–346.

Jaffee, B.A., Strong, D.R. and Muldoon, A.E. (1996b) Nematode-trapping fungi of a natural shrubland: tests for food chain involvement. *Mycologia* 88, 554–564.

Jaffee, B.A., Muldoon, A.E. and Didden, W.A.M. (1997a) Enchytraeids and nematophagous fungi in soil microcosms. *Biology & Fertility of Soils* 25, 382–388.

Jaffee, B.A., Santos, P.F. and Muldoon, A.E. (1997b) Suppression of nematophagous fungi by enchytraeid worms: a field enclosure experiment. *Oecologia* 112, 412–423.

Jaffee, B.A., Ferris, H. and Scow, K.M. (1998) Nematode-trapping fungi in organic and conventional cropping systems. *Phytopathology* 88, 344–350.

Jairajpuri, M.S. and Ahmad, W. (1992) *Dorylaimida: Free living, predaceous and plant-parasitic nematodes*. E.J. Brill, Leiden, the Netherlands.

Jakobsen, J. (1974) The importance of monocultures of various host plants for the population density of *Heterodera avenae*. *Tidsskrift for Planteavl* 78, 697–700.

James Jr, D.W., Suslow, T.V. and Steinback, K.E. (1985) Relationship between rapid firm adhesion and long-term colonization of roots by bacteria. *Applied and Environmental Microbiology* 50, 392–397.

James, E.K. (2000) Nitrogen fixation in endophytic and associative symbiosis. *Field Crops Research* 65, 197–209.

Jansa, J., Mozafar, A., Kuhn, G., Anken, T., Ruh, R., Sanders, I.R. and Frossard, E. (2003) Soil tillage affects the community structure of mycorrhizal fungi in maize roots. *Ecological Applications* 13, 1164–1176.

Jansson, H.-B. (1982) Predacity by nematophagous fungi and its relationship to the attraction of nematodes. *Microbial Ecology* 8, 233–240.

Jansson, H.-B. and Friman, E. (1999) Infection-related surface proteins on conidia of the nematopahgous fungus *Drechmeria coniospora*. *Mycological Research* 103, 249–256.

Jansson, H.-B. and Nordbring-Hertz, B. (1979) Attraction of nematodes to living mycelium of nematophagous fungi. *Journal of General Microbiology* 112, 89–93.

Jansson, H.-B. and Nordbring-Hertz, B. (1980) Interactions between nematophagous fungi and plant parasitic nematodes: attraction, induction of trap formation and capture. *Nematologica* 26, 383–389.

Jansson, H.-B. and Nordbring-Hertz, B. (1988) Infection events in the fungus nematode system. In: Poinar Jr, G.O. and Jansson, H.-B. (eds) *Diseases of Nematodes*. Volume II. CRC Press, Boca Raton, Florida, pp. 59–72.

Jansson, H.-B., Jeyaprakash, A. and Zuckerman, B.M. (1985a) Differential adhesion and infection of nematodes by the endoparasitic fungus *Meria coniospora* (Deuteromycetes). *Applied & Environmental Microbiology* 49, 552–555.

Jansson, H.-B., Jeyaprakash, A. and Zuckerman, B.M. (1985b) Control of root-knot nematodes on tomato by the endoparasitic fungus *Meria coniospora*. *Journal of Nematology* 17, 327–329.

Jansson, H.-B., Dackman, C. and Zuckerman, B.M. (1987) Adhesion and infection of plant parasitic nematodes by the fungus *Drechmeria coniospora*. *Nematologica* 33, 480–487.

Jansson, J.K., Björklöf, K., Elvang, A.M. and Jørgensen, K.S. (2000) Biomarkers for monitoring efficacy of bioremediation by microbial inoculants. *Environmental Pollution* 107, 217–223.

Janvier, C., Villeneuve, F., Alabouvette, C., Edel-Hermann, V., Mateille, T. and Steinberg, C. (2007) Soil health through soil disease suppression: which strategy from descriptors to indicators? *Soil Biology & Biochemistry* 39, 1–23.

Jatala, P. (1986) Biological control of nematodes. *Annual Review of Phytopathology* 24, 453–489.

Jatala, P., Kaltenbach, R. and Bocangel, M. (1979) Biological control of *Meloidogyne incognita acrita* and *Globodera pallida* on potatoes. *Journal of Nematology* 11, 303.

Javed, N., Gowen, S.R., Inam-ul-Haq, M. and Anwar, S.A. (2007a) Protective and curative effect of neem (*Azadirachta indica*) formulations on the development of root-knot nematode *Meloidogyne javanica* in roots of tomato plants. *Crop Protection* 26, 530–534.

Javed, N., Gowen, S.R., Inam-ul-Haq, M., Abdullah, K. and Shahina, F. (2007b) Systemic and persistent effect of neem (*Azadirachta indica*) formulations against root-knot nematodes, *Meloidogyne javanica* and their storage life. *Crop Protection* 26, 911–916.

Javed, N., Gowen, S.R., El-Hassan, S.A., Inam-ul-Haq, M., Shahina, F. and Pembroke, B. (2008) Efficacy of neem (*Azadirachta indica*) formulations on biology of root-knot nematodes (*Meloidogyne javanica*) on tomato. *Crop Protection* 27, 36–43.

Jeffries, P., Gianinazzi, S., Perotto, S., Turnau, K. and Barea, J.-M. (2003) The contribution of arbuscular mycorrhizal fungi in sustainable maintenance of plant health and soil fertility. *Biology & Fertility of Soils* 37, 1–16.

Jenny, H. (1984) The making and unmaking of fertile soil. In: Jackson, W., Berry, W. and Colman, B. (eds) *Meeting the Expectations of the Land*. North Point Press, San Francisco, California, pp. 44–52.

Jeschke, J.M. and Hohberg, K. (2008) Predicting and testing functional response: an example from a tardigrade-nematode system. *Basic & Applied Ecology* 9, 145–151.

Jetiyanon, K. and Kloepper, J.W. (2002) Mixtures of plant growth-promoting rhizobacteria for induction of systemic resistance against multiple plant diseases. *Biological Control* 24, 285–291.

Jetiyanon, K., Fowler, W.D. and Kloepper, J.W. (2003) Broad-spectrum protection against several pathogens by PGPR mixtures under field conditions in Thailand. *Plant Disease* 87, 1390–1394.

Jia, C., Ruan, W., Zhu, M., Ren, A. and Gao, Y. (2013) Potential antagonism of cultivated and wild grass-endophyte associations towards *Meloidogyne incognita*. *Biological Control* 64, 225–230.

Jin, R.D., Suh, J.W., Park, R.D., Kim, Y.W., Krishnan, H.B. and Kim, K.Y. (2005) Effect of chitin compost and broth on biological control of *Meloidogyne incognita* on tomato (*Lycopersicon esculentum* Mill.). *Nematology* 7, 125–132.

Joergensen, R.G. (2000) Ergosterol and microbial biomass in the rhizosphere of grassland soils. *Soil Biology & Biochemistry* 32, 647–652.

Johnson, A.W., Burton, G.W., Sumner, D.R. and Handoo, Z. (1997) Coastal bermudagrass rotation and fallow for management of nematodes and soilborne fungi on vegetable crops. *Journal of Nematology* 29, 710–716.

Johnson, L.F. (1971) Influence of oat straw and mineral fertilizer soil amendments on severity of tomato root-knot. *Plant Disease Reporter* 55, 1126–1129.

Jones, D.L., Nguyen, C. and Finlay, R.D. (2009) Carbon flow in the rhizosphere: carbon trading at the soil–root interface. *Plant & Soil* 321, 5–33.

Jones, F.G.W. (1978) The soil–plant environment. In: Southey, J.F. (ed.) *Plant Nematology*. Her Majesty's Stationery Office, London, pp. 46–62.

Joshi, R. and McSpadden Gardener, B.B. (2006) Identification and characterization of novel genetic markers associated with biological control activities in *Bacillus subtilis*. *Phytopathology*, 96, 145–154.

Juhl, M. (1985) The effect of *Hirsutella heteroderae* on the multiplication of the beet cyst nematode on sugar beet. *Tidsskrift for Planteavl* 89, 475–480.

Jun, O.K. and Kim, Y.H. (2004) *Aphelenchus avenae* and antagonistic fungi as biological control agents of *Pythium* spp. *Plant Pathology Journal* 20, 271–276.

Kabir, Z. (2005) Tillage or no-tillage: Impact on mycorrhizae. *Canadian Journal of Plant Science* 85, 23–29.

Kabir, Z. and Koide, R.T. (2002) Effect of autumn and winter mycorrhizal cover crops on soil properties, nutrient uptake and yield of sweet corn in Pennsylvania, USA. *Plant & Soil* 238, 205–215.

Kaneda, S., Miura, S., Yamashita, N., Ohigashi, K., Yamasaki, S., Murakami, T. and Urashima, Y. (2012) Significance of litter layer in enhancing mesofaunal abundance and microbial biomass nitrogen in sweet corn-white clover living mulch systems. *Soil Science & Plant Nutrition* 58, 424–434.

Kaplan, D.T. (1994) Partial characterization of a *Pasteuria* sp. attacking the citrus nematode, *Tylenchulus semipenetrans*, in Florida. *Fundamental & Applied Nematology* 17, 509–512.

Karagoz, M., Gulcu, B., Cakmak, I., Kaya, H.K. and Hazir, S. (2007) Predation of entomopathogenic nematodes by *Sancassania* sp. (Acari: Acaridae*). Experimental & Applied Acarology* 43, 85–95.

Karasawa, T., Kasahara, Y. and Takebe, M. (2001) Variable response of growth and arbuscular mycorrhizal colonization of maize plants to preceding crops in various types of soils. *Biology & Fertility of Soils* 33, 286–293.

Kariuki, G.M. and Dickson, D.W. (2007) Transfer and development of *Pasteuria penetrans*. *Journal of Nematology* 39, 55–61.

Karlen, D.L., Andrews, S.S. and Doran, J.W. (2001) Soil quality: current concepts and applications. *Advances in Agronomy* 74, 1–40.

Karlen, D.L., Andrews, S.S., Weinhold, B.J. and Doran, J.W. (2003a) Soil quality: Humankind's foundation for survival. *Journal of Soil and Water Conservation* 58, 171–179.

Karlen, D.L., Ditzler, C.A. and Andrews, S.S. (2003b) Soil quality: why and how? *Geoderma* 114, 145–156.

Kassam, A., Friedrich, T., Shaxson, F. and Pretty, J. (2009) The spread of conservation agriculture: justification, sustainability and uptake. *International Journal of Agricultural Sustainability* 7, 292–320.

Kerry, B. (1988) Fungal parasites of cyst nematodes. *Agriculture, Ecosystems & Environment* 24, 293–305.

Kerry, B.R. (1975) Fungi and the decrease of cereal cyst-nematode populations in cereal monoculture. *European Plant Protection Organization Bulletin* 5, 353–361.

Kerry, B.R. (1980) Biocontrol: Fungal parasites of female cyst nematodes. *Journal of Nematology* 12, 253–259.

Kerry, B.R. (1981) Progress in the use of biological agents for control of nematodes, In: Papavizas, G.C. (ed.) *Biological Control in Crop Production*, Allanheld, Osmum, Totowa, NJ, pp. 79–90.

Kerry, B.R. (1982) The decline of *Heterodera avenae* populations. *European Plant Protection Organization Bulletin* 12, 491–496.

Kerry, B.R. (1989) Fungi as biological control agents for plant parasitic nematodes. In: Whipps, J.M. and Lumsden, R.D. (eds) *Biotechnology of Fungi for Improving Plant Growth*. Press Syndicate of the University of Cambridge, Cambridge, UK, pp. 153–170.

Kerry, B.R. (1990) An assessment of progress toward microbial control of plant-parasitic nematodes. *Journal of Nematology* 22, 621–631.

Kerry, B.R. (1995) Ecological considerations for the use of the nematophagous fungus, *Verticillium chlamydo-sporium*, to control plant-parasitic nematodes. *Canadian Journal of Botany* 73, S65-S70.

Kerry, B.R. (2000) Rhizosphere interactions and the exploitation of microbial agents for the biological control of plant-parasitic nematodes. *Annual Review of Phytopathology* 38, 423–441.

Kerry, B.R. and Bourne, J.M. (1996) The importance of rhizosphere interactions in the biological control of plant parasitic nematodes – a case study using *Verticillium chlamydosporium*. *Pesticide Science* 47, 69–75.

Kerry, B.R. and Crump, D.H. (1977) Observations on fungal parasites of females and eggs of the cereal cyst-nematode, *Heterodera avenae,* and other cyst nematodes. *Nematologica* 23, 193–201.

Kerry, B.R. and Crump, D.H. (1980) Two fungi parasitic on females of cyst-nematodes (*Heterodera* spp.). *Transactions of the British Mycological Society* 74, 119–125.

Kerry, B.R. and Crump, D.H. (1998) The dynamics of the decline of the cereal cyst nematode, *Heterodera avenae,* in four soils under intensive cereal production. *Fundamental & Applied Nematology* 21, 617–625.

Kerry, B.R. and Hirsch, P.R. (2011) Ecology of *Pochonia chlamydosporia* in the rhizosphere at the population, whole organism and molecular scales. In: Davies, K. and Spiegel, Y. (eds) *Biological Control of Plant-Parasitic Nematodes: Building Coherence Between Microbial Ecology and Molecular Mechanisms.* Springer, Dordrecht, the Netherlands, pp. 171–182.

Kerry, B.R., Crump, D.H. and Mullen, L.A. (1980) Parasitic fungi, soil moisture and multiplication of the cereal cyst nematode, *Heterodera avenae. Nematologica* 26, 57–68.

Kerry, B.R., Crump, D.H. and Mullen, L.A. (1982a) Studies of the cereal cyst nematode, *Heterodera avenae* under continuous cereals, 1975–1978. II. Fungal parasitism of nematode eggs and females. *Annals of Applied Biology* 100, 489–499.

Kerry, B.R., Crump, D.H. and Mullen, L.A. (1982b) Natural control of the cereal cyst nematode, *Heterodera avenae* Woll. by soil fungi at three sites. *Crop Protection* 1, 99–109.

Kerry, B.R., Crump, D.H. and Mullen, L.A. (1982c) Studies of the cereal cyst-nematode, *Heterodera avenae* under continuous cereals, 1974–1978. 1. Plant growth and nematode multiplication. *Annals of Applied Biology* 100, 477–487.

Kerry, B.R., Simon, A. and Rovira, A.D. (1984) Observations on the introduction of *Verticillium chlamydo-sporium* and other parasitic fungi into soil for control of the cereal cyst-nematode *Heterodera avenae. Annals of Applied Biology* 105, 509–516.

Kerry, B.R., Irving, F. and Hornsey, J.C. (1986) Variation between strains of the nematophagous fungus, *Verticillium chlamydosporium* Goddard. I. Factors affecting growth *in vitro. Nematologica* 32, 461–473.

Kerry, B.R., Kirkwood, I.A., de Leij, F.A.A.M., Barba, J., Leijdens, M.B. and Brookes, P.C. (1993) Growth and survival of *Verticillium chlamydosporium* Goddard, a parasite of nematodes, in soil. *Biocontrol Science & Technology* 3, 355–365.

Khan, A., Williams, K.L., Molloy, M.P. and Nevalainen, H.K.M. (2003) Purification and characterization of a serine protease and chitinases from *Paecilomyces lilacinus* and detection of chitinase activity on 2D gels. *Protein Expression & Purification* 32, 210–220.

Khan, A., Williams, K.L. and Nevalainen, H.K.M. (2006a) Infection of plant-parasitic nematodes by *Paecilomyces lilacinus* and *Monacrosporium lysipagum. BioControl* 51, 659–678.

Khan, A., Williams, K.L. and Nevalainen, H.K.M. (2006b) Control of plant-parasitic nematodes by *Paecilomyces lilacinus* and *Monacrosporium lysipagum* in pot trials. *BioControl* 51, 643–658.

Khan, M.R., Majid, S., Mohidin, F.A. and Khan, N. (2011) A new bioprocess to produce low cost powder formulations of biocontrol bacteria and fungi to control fusarial wilt and root-knot nematode of pulses. *Biological Control* 59, 130–140.

Khan, Z. and Kim, Y.H. (2005) The predatory nematode, *Mononchoides fortidens* (Nematoda: Diplogasterida), suppresses the root-knot nematode, *Meloidogyne arenaria,* in potted field soil. *Biological Control* 35, 78–82.

Khan, Z. and Kim, Y.H. (2007) A review of the role of predatory soil nematodes in the biological control of plant parasitic nematodes. *Applied Soil Ecology* 35, 370–379.

Khan, Z., Bilgrami, A.L. and Jairajpuri, M.S. (1991) Observations on the predation ability of *Aporcelaimellus nivalis* (Altherr, 1952) Heyns, 1966 (Nematoda: Dorylaimida). *Nematologica* 37, 333–342.

Khan, Z., Ahmad, S., Al-Ghimlas, F., Al-Mutairi, S., Joseph, L., Chandy, R., Sutton, D.A. and Guarro, J. (2012) *Purpureocillium lilacinum* as a cause of cavitary pulmonary disease: a new clinical presentation and observation on atypical morphologic characteristics of the isolate. *Journal of Clinical Microbiology* 50, 1800–1804.

Kibblewhite, M.G., Ritz, K. and Swift, M.J. (2008) Soil health in agricultural systems. *Philosophical Transactions of the Royal Society B,* 363, 685–701.

Kiewnick, S. (2007) Practicalities of developing and registering microbial biological control agents. *CAB reviews: Perspectives in Agriculture, Veterinary Science, Nutrition and Natural Resources* 2007 2, No. 013, 1–11. DOI: 10.1079/PAVSNNR20072013.

Kiewnick, S. (2010) Importance of multitrophic interactions for successful biocontrol of plant parasitic nematodes with *Paecilomyces lilacinus* strain 251. In: Gisi, U., Chet, I. and Lodvica Gullino, M. (eds) *Recent Developments in Management of Plant Diseases, Plant Pathology in the 21st Century*. Springer Science + Business Media, pp. 81–92.

Kiewnick, S. and Sikora, R.A. (2003) Efficacy of *Paecilomyces lilacinus* (strain 251) for the control of root-knot nematodes. *Communications in Agriculture & Applied Biological Sciences* 68, 123–128.

Kiewnick, S. and Sikora, R.A. (2006a) Biological control of the root-knot nematode *Meloidogyne incognita* by *Paecilomyces lilacinus* strain 251. *Biological Control* 38, 179–187.

Kiewnick, S. and Sikora, R.A. (2006b) Evaluation of *Paecilomyces lilacinus* strain 251 for the biological control of the northern root-knot nematode *Meloidogyne hapla* Chitwood. *Nematology* 8, 69–78.

Kiewnick, S., Neumann, S. and Sikora, R. (2006) Importance of nematode inoculum density and antagonist dose for biocontrol efficacy of *Paecilomyces lilacinus* strain 251. *Phytopathology* 96, S60.

Kiewnick, S., Neumann, S., Sikora, R.A. and Frey, J.E. (2011) Effect of *Meloidogyne incognita* inoculum density and application rate of *Paecilomyces lilacinus* strain 251 on biocontrol efficacy and colonization of egg masses analyzed by real-time quantitative PCR. *Phytopathology* 101, 105–112.

Kim, D.G. and Riggs, R.D. (1991) Characteristics and efficacy of a sterile hyphomycete (ARF18), a new biocontrol agent for *Heterodera glycines* and other nematodes. *Journal of Nematology* 23, 275–282.

Kim, D.G. and Riggs, R.D. (1994) Techniques for isolation and evaluation of fungal parasites of *Heterodera glycines*. *Journal of Nematology* 26, 592–595.

Kim, D.G. and Riggs, R.D. (1995) Efficacy of the nematophagous fungus ARF18 in alginate-clay pellet formulations against *Heterodera glycines*. *Journal of Nematology* 27, 602–608.

Kim, D.G., Riggs, R.D. and Correll, J.C. (1998) Isolation, characterization, and distribution of a biocontrol fungus from cysts of *Heterodera glycines*. *Phytopathology* 88, 465–471.

Kimmons, C.A., Gwinn, K.D. and Bernard, E.C. (1990) Nematode reproduction on endophyte-infected and endophyte-free tall fescue. *Plant Disease* 74, 757–761.

Kimpinski, J., Gallant, C.E., Henry, R., Macleod, J.A., Sanderson, J.B. and Sturz, A.V. (2003) Effect of compost and manure soil amendments on nematodes and on yields of potato and barley: a 7-year study. *Journal of Nematology* 35, 289–293.

Kimura, M., Jia, Z.-J., Nakayama, N. and Asakawa, S. (2008) Ecology of viruses in soils: Past, present and future perspectives. *Soil Science and Plant Nutrition* 54, 1–32.

King, R.A., Read, D.S., Traugott, M. and Symondson, W.O.C. (2008) Molecular analysis of predation: a review of best practice for DNA-based approaches. *Molecular Ecology* 17, 947–963.

Kinloch, R.A. (1998) Soybean. In: Barker, K.R., Pederson, G.A. and Windham, G.L. (eds) *Plant and Nematode Interactions*. Agronomy Monographs no. 36. American Society of Agronomy Inc., Madison, Wisconsin, pp. 317–333.

Kirchmann, H. and Ryan, M.H. (2004) Nutrients in organic farming – are there advantages from the exclusive use of organic manures and untreated materials. Proceedings of the 4th International Crop Science Congress, 'New directions for a diverse planet'. www.cropscience.org.au, accessed 11 October 2013.

Kirchmann, H., Bergström, H., Kätterer, T., Mattsson, L. and Gesslein, S. (2007) Comparison of long-term organic and conventional crop-livestock systems on a previously nutrient-depleted soil in Sweden. *Agronomy Journal* 99, 960–972.

Kirkby, C.A., Richardson, A.E., Wade, L.J., Batten, G.D., Blanchard, C. and Kirkegaard, J.A. (2013) Carbon-nutrient stoichiometry to increase soil carbon sequestration. *Soil Biology & Biochemistry* 60, 77–86.

Kirkegaard, J.A., Peoples, M.B., Angus, J.F. and Unkovich, M.J. (2010) Diversity and evolution of rainfed farming systems in southern Australia. In: Tow, P.G., Partridge, I., Cooper, I. and Birch, C. (eds) *Rainfed Farming Systems*. Springer, Dordrecht, the Netherlands, pp. 715–754.

Kladivko, E.J. (2001) Tillage systems and soil ecology. *Soil & Tillage Research* 61, 61–76.

Kloepper, J.W. (1993) Plant-growth-promoting rhizobacteria as biological control agents. In: Metting Jr, F.B. (ed.) *Soil Microbial Ecology – Applications in Agricultural and Environmental Management*. Marcel Dekker, New York, pp. 255–274.

Kloepper, J.W. and Beauchamp, C.J. (1992) A review of issues related to measuring colonization of plant roots by bacteria. *Canadian Journal of Microbiology* 38, 1219–1232.

Kloepper, J.W., Rodríguez-Kábana, R., McInroy, J.A. and Collins, D.J. (1991) Analysis of populations and physiological characterization of microorganisms in rhizospheres of plants with antagonistic properties to phytopathogenic nematodes. *Plant & Soil* 136, 95–102.

Kloepper, J.W., Rodríguez-Kábana, R., McInroy, J.A. and Young, R.W. (1992a) Rhizosphere bacteria antagonistic to soybean cyst (*Heterodera glycines*) and root knot (*Meloidogyne incognita*) nematodes: identification by fatty acid analysis and frequency of biological control activity. *Plant & Soil* 139, 75–84.

Kloepper, J.W., Schippers, B. and Bakker, P.A.H.M. (1992b) Proposed elimination of the term endorhizos-
phere. *Phytopathology* 82, 726–727.

Kloepper, J.W., Rodríguez-Kábana, R., Zehnder, G.W., Murphy, J.F., Sikora, E. and Fernández, C. (1999) Plant
root-bacterial interactions in biological control of soilborne diseases and potential extension to systemic
and foliar diseases. *Australasian Plant Pathology* 28, 21–26.

Kloepper, J.W., Ryu, C.-M. and Zhang, S. (2004) Induced systemic resistance and promotion of plant growth
by *Bacillus* spp. *Phytopathology* 94, 1259–1266.

Kluepfel, D.A. (1993) The behaviour and tracking of bacteria in the rhizosphere. *Annual Review of
Phytopathology* 31, 441–472.

Knox, O.G.G., Killham, K., Mullins, C.E. and Wilson, M.J. (2003) Nematode-enhanced microbial colonization of the wheat rhizosphere. *FEMS Microbiology Letters* 225, 227–233.

Knox, O.G.G., Killham, K., Artz, R.R.E., Mullins, C. and Wilson, M. (2004) Effect of nematodes on rhizosphere
colonization by seed-applied bacteria. *Applied & Environmental Microbiology* 70, 4666–4671.

Knox, O.G.G., Gupta, V.V.S.R. and Lardner, R. (2009) Cotton cultivar selection impacts on microbial diversity
and function. *Aspects of Applied Biology* 98, 129–136.

Koehler, H.H. (1997) Mesostigmata (Gamasina, Uropodina), efficient predators in agroecosystems. *Agriculture,
Ecosystems & Environment* 62, 105–117.

Koehler, H.H. (1999) Predatory mites (Gamasina, Mesostigmata). *Agriculture, Ecosystems & Environment* 74,
395–410.

Koenning, S.R. and Barker, K.R. (2004) Influence of poultry litter applications on nematode communities in
cotton agroecosystems. *Journal of Nematology* 36, 524–533.

Kokalis-Burelle, N., Martinez-Ochoa, N., Rodriguez-Kábana, R. and Kloepper, J.W. (2002a) Development of
multi-component transplant mixes for suppression of *Meloidogyne incognita* on tomato (*Lycopersicon
esculentum*). *Journal of Nematology* 34, 362–369.

Kokalis-Burelle, N., Vavrina, C.S., Rosskopf, E.N. and Shelby, R.A. (2002b) Field evaluation of plant growth-
promoting rhizobactera amended transplant mixes and soil solarization for tomato and pepper production in Florida. *Plant & Soil* 238, 257–266.

Kokalis-Burelle, N., Chellemi, D.O. and Périès, X. (2005) Effect of soils from six management systems on root-
knot nematodes and plant growth in greenhouse assays. *Journal of Nematology* 37, 467–272.

Kolton, M., Harel, Y.M., Pasternak, Z., Graber, E.R., Elad, Y. and Cytryn, E. (2011) Impact of biochar application to soil on the root-associated bacterial community structure of fully developed greenhouse pepper
plants. *Applied & Environmental Microbiology* 77, 4924–4930.

Koning, G., Hamman, B. and Eicker, A. (1996) The efficacy of nematophagous fungi on predaceous nematodes in soil compared with saprophagous nematodes in mushroom compost. *South African Journal of
Botany* 62, 49–53.

Koppenhöfer, A.M., Jaffee, B.A., Muldoon, A.E., Strong, D.R. and Kaya, H.K. (1996) Effect of nematode-
trapping fungi on an entomopathogenic nematode originating from the same field site in California.
Journal of Invertebrate Pathology 68, 246–252.

Koppenhöfer, A.M., Jaffee, B.A., Muldoon, A.E. and Strong, D.R. (1997) Suppression of an entomopathogenic
nematode by the nematode-trapping fungi *Geniculifera paucispora* and *Monacrosporium eudermatum*
as affected by the fungus *Arthrobotrys oligospora*. *Mycologia* 89, 220–227.

Korthals, G.W., van de Ende, A., van Megen, H., Lexmond, T.M., Kammenga, J.E. and Bongers, T. (1996) Short-
term effects of cadmium, copper, nickel and zinc on soil nematodes from different feeding and life history groups. *Applied Soil Ecology* 4, 107–117.

Kosma, P., Ambang, Z., Begoude, B.A.D., Ten Hoopen, G.M., Kuate, J. and Akoa, A. (2011) Assessment of
nematicidal properties and phytochemical screening of neem seed formulations using *Radopholus similis*,
parasitic nematode of plantain in Cameroon. *Crop Protection* 30, 733–738.

Kosuta, S., Chabaud, M., Lougnon, G., Gough, C., Dénarié, J., Barker, D.G. and Bécard, G. (2003) A diffusible
factor from arbuscular mycorrhizal fungi induces symbiosis-specific *MtENDO11* expression in roots of
Medicago truncatula. *Plant Physiology* 131, 952–962.

Koziak, A.T.E., Cheng, K.C. and Thorn, R.G. (2007) Phylogenetic analyses of *Nematoctonus* and *Hohenbuehelia*
(Pleurotaceae). *Canadian Journal of Botany* 85, 762–773.

Kremer, R.J., Means, N.E. and Kim, S. (2005) Glyphosate affects soybean root exudation and rhizosphere
microorganisms. *International Journal of Environmental Analytical Chemistry* 85, 1165–1174.

Kuzyakov, Y. and Domanski, G. (2000) Carbon inputs by plants into soil. Review. *Journal of Plant Nutrition & Soil Science* 163, 421–431.

Kwok, O.C.H., Plattner, R., Weisleder, D. and Wicklow, D.T. (1992) A nematicidal toxin from *Pleurotus ostreatus* NRRL 3526. *Journal of Chemical Ecology* 18, 127–135.

Laakso, J., Setälä, H. and Palojärvi, A. (2000) Influence of decomposer food web structure and nitrogen availability on plant growth. *Plant & Soil* 225, 153–165.

Lackey, B.A., Muldoon, A.E. and Jaffee, B.A. (1993) Alginate pellet formulation of *Hirsutella rhossiliensis* for biological control of plant-parasitic nematodes. *Biological Control* 3, 155–160.

Ladygina, N., Johansson, T., Canbäck, B., Tunlid, A. and Hedlund, K. (2009) Diversity of bacteria associated with grassland soil nematodes of different feeding groups. *FEMS Microbiology Ecology* 69, 53–61.

Lagerlöf, J., Insunza, V., Lundegårdh, B. and Rämert, B. (2011) Interaction between a fungal plant disease, fungivorous nematodes and compost suppressiveness. *Acta Agriculturae Scandinavica Section B-Soil & Plant Science* 61, 372–377.

Lal, R. (1994) Sustainable land use systems and soil resilience. In: Greenland, D.J. and Szabolcs, I. (eds) *Soil Resilience and Sustainable Land Use*. CAB International, Wallingford, UK, pp. 41–67.

Lal, R. (2004) Soil carbon sequestration impacts on global climate change and food security. *Science* 304, 1623–1627.

Lal, R. (2005) *Encyclopedia of Soil Science*. 2nd edn. CRC Press, Boca Raton, Florida.

Lal, R. (2009) Challenges and opportunities in soil organic matter research. *European Journal of Soil Science* 60, 158–169.

Lal, R. and Stewart, B.A. (eds) (1995) *Soil Management. Experimental Basis of Sustainability and Environmental Quality*. CRC Lewis Publishers, Boca Raton, Florida.

Lal, R., Delgado, J.A., Groffman, P.M., Millar, N., Dell, C. and Rotz, A. (2011) Management to mitigate and adapt to climate change. *Journal of Soil & Water Conservation* 66, 276–285.

Lanfranco, L., Perotto, S. and Bonfante, P. (1998) Applications of PCR for studying the biodiversity of mycorrhizal fungi. In: Bridge, P.D., Arora, D.K., Elander, R.P. and Reddy, C.A. (eds) *Applications of PCR in Mycology*. CAB International, Wallingford, UK, pp. 107–124.

Larsen, M. (2006) Biological control of nematode parasites of sheep. *Journal of Animal Science* 84, E133-E139.

Larsen, T., Schjønning, P. and Axelsen, J. (2004) The impact of soil compaction on euedaphic Collembola. *Applied Soil Ecology* 26, 273–281.

Laurence, M.H., Burgess, L.W., Summerell, B.A. and Liew, E.C.Y. (2012) High levels of diversity in *Fusarium oxysporum* from non-cultivated ecosystems in Australia. *Fungal Biology* 116, 289–297.

Lavelle, P. and Spain, A. (2005) *Soil Ecology*. Springer, Dordrecht, the Netherlands.

Lazarovits, G., Conn, K.L. and Potter, J. (1999) Reduction of potato scab, verticillium wilt, and nematodes by soymeal and meat and bone meal in two Ontario potato fields. *Canadian Journal of Plant Pathology* 21, 345–353.

Lazarovits, G., Tenuta, M. and Conn, K.L. (2001) Organic amendments as a disease control strategy for soilborne diseases of high-value agricultural crops. *Australasian Plant Pathology* 30, 111–117.

Le, H.T.T., Padgham, J.L. and Sikora, R.A. (2009) Biological control of the rice root-knot nematode *Meloidogyne graminicola* on rice, using endophytic and rhizosphere fungi. *International Journal of Pest Management* 55, 31–36.

Lee, Y.-J. and George, E. (2005) Development of a nutrient film technique culture system for arbuscular mycorrhizal plants. *HortScience* 40, 378–380.

Lee, Y.S., Anees, M., Hyun, H.N. and Kim, K.Y. (2013) Biocontrol potential of *Lysobacter antibioticus* HS124 against the root-knot nematode, *Meloidogyne incognita*, causing disease in tomato. *Nematology* 15, 545–555.

Leff, B., Ramankutty, N. and Foley, J.A. (2004) Geographic distribution of major crops across the world. *Global Biogeochemical Cycles* 18, GB1009, DOI: 10.1029/2003GB002108.

Lehmann, J., Rillig, M.C., Thies, J., Masiello, C.A., Hockaday, W.C. and Crowley, D. (2011) Biochar effects on the soil biota – A review. *Soil Biology & Biochemistry* 43, 1812–1836.

Lehman, R.M., Taheri, W.I., Osborne, S.L., Buyer, J.S. and Douds Jr, D.D. (2012) Fall cropping can increase arbuscular mycorrhizae in soils supporting intensive agricultural production. *Applied Soil Ecology* 61, 300–304.

Leigh, J., Hodge, A. and Fitter, A.H. (2009) Arbuscular mycorrhizal fungi can transfer substantial amounts of nitrogen to their host plant from organic matter. *New Phytologist* 181, 199–207.

Lembright, H.W. (1990) Soil fumigation: principles and application technology. *Journal of Nematology* 22, 632–644.

Lenz, R. and Eisenbeis, G. (2000) Short-term effects of different tillage in a sustainable farming system on nematode community structure. *Biology & Fertility of Soils* 31, 237–244.

Leong, J. (1986) Siderophores: their biochemistry and possible role in the biocontrol of plant pathogens. *Annual Review of Phytopathology* 24, 187–209.

Letey, J., Sojka, R.E., Upchurch, D.R., Cassell, D.K., Olson, K.R., Payne, W.A., Petrie, S.E., Price, G.H., Reginato, R.J., Scott, H.D., Smethhurst, P.J. and Triplett, G.B. (2003) Deficiencies in the soil quality concept and its application. *Journal of Soil and Water Conservation* 58, 180–187.

Lewandowski, I., Hardtlein, M. and Kaltschmitt, M. (1999) Sustainable crop production: definition and methodological approach for assessing and implementing sustainability. *Crop Science* 39, 184–193.

Li, L., Ma, M., Liu, Y., Zhou, J., Qu, Q., Lu, K., Fu, D. and Zhang, K. (2011) Induction of trap formation in nematode-trapping fungi by a bacterium. *FEMS Microbiology Letters* 322, 157–165.

Li, S.-D., Miao, Z.-Q., Zhang, Y.H. and Liu, X.-Z. (2003) *Monacrosporium janus* sp. nov., a new nematode-trapping hyphomycete parasitizing sclerotia and hyphae of *Sclerotinia sclerotiorum*. *Mycological Research* 107, 888–894.

Li, Y., Hyde, K.D., Jeewon, R., Cai, L., Vijaykrishna, D. and Zhang, K. (2005) Phylogenetics and evolution of nematode-trapping fungi (Orbiliales) estimated from nuclear and protein encoding genes. *Mycologia* 97, 1034–1046.

Li, Y., Seymour, N. and Stirling, G.R. (2012) Antagonists of root-lesion nematode in vertosols from the northern grain-growing region. Proceedings 7th Australasian Soilborne Diseases Symposium, Fremantle, WA, Australia, p. 21.

Liddell, C.M. and Parke, J.L. (1989) Enhanced colonization of pea taproots by a fluorescent pseudomonad biocontrol agent by water infiltration into soil. *Phytopathology* 79, 1327–1332.

Liebman, M., Mohler, C.L. and Staver, C.P. (2001) *Ecological Management of Agricultural Weeds*. Cambridge University Press, Cambridge, UK.

Linford, M.B. (1937) Stimulated activity of natural enemies of nematodes. *Science* 85, 123–124.

Linford, M.B. and Oliveira, J.M. (1937) The feeding of hollow-spear nematodes on other nematodes. *Science* 85, 295–297.

Linford, M.B. and Yap, F. (1939) Root-knot nematode injury restricted by a fungus. *Phytopathology* 29, 596–608.

Linford, M.B., Yap, F. and Oliveira, J.M. (1938) Reduction of soil populations of the root-knot nematode during decomposition of organic matter. *Soil Science* 45, 127–141.

Liou, G.Y. and Tzean, S.S. (1997) Phylogeny of the genus *Arthrobotrys* and allied nematode-trapping fungi based on rDNA sequences. *Mycologia* 89, 876–884.

Litterick, A.M., Harrier, L., Wallace, P., Watson, C.A. and Wood, M. (2004) The role of uncomposted materials, composts, manures and compost extracts in reducing pest and disease incidence and severity in sustainable temperate agricultural and horticultural crop production – a review. *Critical Reviews in Plant Sciences* 23, 453–479.

Liu, X.Z. and Chen, S.Y. (2000) Parasitism of *Heterodera glycines* by *Hirsutella* spp. in Minnesota soybean fields. *Biological Control* 19, 161–166.

Liu, X.Z. and Chen, S.Y. (2001a) Screening isolates of *Hirsutella* species for biocontrol of *Heterodera glycines*. *Biocontrol Science & Technology* 11, 151–160.

Liu, S.F. and Chen, S.Y. (2001b) Nematode hosts of the fungus *Hirsutella minnesotensis*. *Phytopathology* 91, S138.

Liu, X.Z. and Chen, S.Y. (2002) Nutritional requirements of the nematophagous fungus *Hirsutella rhossiliensis*. *Biocontrol Science & Technology* 12, 381–193.

Liu, S. and Chen, S.Y. (2005) Efficacy of the fungi *Hirsutella minnesotensis* and *H. rhossiliensis* from liquid culture for control of the soybean cyst nematode *Heterodera glycines*. *Nematology* 7, 149–157.

Liu, S.F. and Chen, S.Y. (2009) Effectiveness of *Hirsutella minnesotensis* and *H. rhossiliensis* in control of the soybean cyst nematode in four soils with various pH, texture, and organic matter. *Biocontrol Science & Technology* 19, 595–612.

Liu, X.-Z., Gao, R.H., Zhang, K.Q. and Cao, L. (1996) *Dactylella tenuifusaria* sp. nov., a rhizopod-capturing hyphomycete. *Mycological Research* 100, 236–238.

Llopez-Llorca, L.V., Bordallo, J.J., Salinas, J., Monfort, E. and Llópez-Serna, M.L. (2002) Use of light and scanning electron microscopy to examine colonisation of barley rhizosphere by the nematophagous fungus *Verticillium chlamydosporium*. *Micron* 33, 61–67.

Lockwood, J.L. (1977) Fungistasis in soils. *Biological Reviews* 52, 1–43.

Loewenberg, J.R., Sullivan, T. and Schuster, M.L. (1959) A virus disease of *Meloidogyne incognita incognita*, the southern root-knot nematode. *Nature*, 184, 1896.

Long, S.R. (2001) Genes and signals in the rhizobium-legume symbiosis. *Plant Physiology* 125, 69–72.

López-Granados, F. (2010) Weed detection for site-specific weed management: mapping and real-time approaches. *Weed Research* 51, 1–11.

López-Pérez, J.A., Edwards, S. and Ploeg, S. (2011) Control of root-knot nematodes on tomato in stone wool substrate with biological nematicides. *Journal of Nematology* 43, 110–117.

Lorito, M., Woo, S.L., Harman, G.E. and Monte, E. (2010) Translational research on *Trichoderma*: from 'omics to the field. *Annual Review of Phytopathology* 48, 395–417.

Loveland, P. and Webb, J. (2003) Is there a critical level of organic matter in the agricultural soils of temperate regions: a review. *Soil & Tillage Research* 70, 1–18.

Loveys, B.R., Stoll, M. and Davies, W.J. (2004) Physiological approaches to enhance water use efficiency in agriculture: exploiting plant signalling in novel irrigation practice. In: Bacon, M.A (ed.) *Water Use Efficiency in Plant Biology*. Blackwell Publishing, Oxford, UK, pp. 113–141.

Lu, Y.-C., Watkins, K.B., Teasdale, J.R. and Abdul-Baki, A.A. (2000) Cover crops in sustainable food production. *Food Reviews International* 16, 121–157.

Luangsa-ard, J., Houbraken, J., van Doorn, T., Hong, S.-B., Borman, A.M., Hywel-Jones, N.L and Samson, R.A. (2011) *Purpureocillium*, a new genus for the medically important *Paecilomyces lilacinus*. *FEMS Microbiology Letters* 321, 141–149.

Luc, J.E., Crow, W.T., McSorley, R. and Giblin-Davis, R.M. (2010a) Suppression of *Belonolaimus longicaudatus* with *in vitro*-produced *Pasteuria* sp. endospores. *Nematropica* 40, 217–225.

Luc, J.E., Pang, W., Crow, W.T. and Giblin-Davis, R.M. (2010b) Effect of formulation and host nematode density on the ability of *in vitro*-produced *Pasteuria* endospores to control its host *Belonolaimus longicaudatus*. *Journal of Nematology* 42, 87–90.

Luc, J.E., Crow, W.T., Pang, W., McSorley, R. and Giblin-Davis, R.M. (2011) Effects of irrigation, thatch, and wetting agent on movement of *Pasteuria* sp. endospores in turf. *Nematropica* 41, 185–190.

Luc, M., Sikora, R.A. and Bridge, J. (eds) (2005) *Plant Parasitic Nematodes in Subtropical and Tropical Agriculture*. CAB International, Wallingford, UK.

Lucas, S.T. and Weil, R.R. (2012) Can a labile carbon test be used to predict crop responses to improve soil organic matter management? *Agronomy Journal* 104, 1160–1170.

Lugtenberg, B.J.J., Dekkers, L. and Bloemberg, G.V. (2001) Molecular determinants of rhizosphere colonization by *Pseudomonas*. *Annual Review of Phytopathology* 39, 461–490.

Lumsden, R.D. (1981) Ecology of mycoparasitism. In: Wicklow, D.T. and Carroll, G.C. (eds) *The Fungal Community*. Marcel Dekker, New York, pp. 295–318.

Lumsden, R.D., García, E.R., Lewis, J.A. and Frías, T. (1987) Suppression of damping-off caused by *Pythium* spp. in soil from the indigenous Mexican chinampa agricultural system. *Soil Biology & Biochemistry* 19, 501–508.

Luna, J.M., Mitchell, J.P. and Shrestha, A. (2012) Conservation tillage for organic agriculture: evolution toward hybrid systems in the western USA. *Renewable Agriculture & Food Systems* 27, 21–30.

Luo, H., Li, X., Li, G., Pan, Y. and Zhang, K. (2006) Acanthocytes of *Stropharia rugosoannulata* function as a nematode-attacking device. *Applied & Environmental Microbiology* 72, 2982–2987.

Lynch, J.M. (1978) Microbial interaction around imbibed seeds. *Annals of Applied Biology* 89, 165–167.

Lynch, J.M. (ed.) (1990) *The Rhizosphere*. Wiley, Chichester, UK.

Ma, L.-J., van der Does, H.C., Borkovich, K.A., Coleman, J.J., Daboussi, M.-J., Di Pietro, A., Dufresne, M., Freitag, M., Grabherr, M., Henrissat, B. *et al.* (2010) Comparative genomics reveals mobile pathogenicity chromosomes in *Fusarium*. *Nature* 464, 367–373.

Ma, R., Liu, X., Jian, H. and Li, S. (2005) Detection of *Hirsutella* spp. and *Pasteuria* sp. parasitizing second-stage juveniles of *Heterodera glycines* in soybean fields in China. *Biological Control* 33, 223–229.

Machado, P.L.O.A., Sohi, S.P. and Gaunt, J.L. (2003) Effect of no-tillage on turnover of organic matter in a Rhodic Ferralsol. *Soil Use & Management* 19, 250–256.

Maciá-Vicente, J.G., Jansson, H.-B., Talbot, N.J. and Lopez-Llorca, L.V. (2009a) Real-time PCR quantification and live-cell imaging of endophytic colonization of barley (*Hordeum vulgare*) roots by *Fusarium equiseti* and *Pochonia chlamydosporia*. *New Phytologist* 182, 213–228.

Maciá-Vicente, J.G., Rosso, L.C., Ciancio, A., Jansson, H.-B. and Lopez-Llorca, L.V. (2009b) Colonization of barley roots by endophytic *Fusarium equiseti* and *Pochonia chlamydosporia*: effects on plant growth and disease. *Annals of Applied Biology* 155, 391–401.

Madulu, J.D., Trudgill, D.L. and Phillips, M.S. (1994) Rotational management of *Meloidogyne javanica* and effects on *Pasteuria penetrans* and tomato and tobacco yields. *Nematologica* 40, 438–455.

Magdoff, F. and Weil, R.R. (2004a) (eds) *Soil Organic Matter in Sustainable Agriculture*. CRC Press, Boca Raton, Florida.

Magdoff, F. and Weil, R.R. (2004b) Soil organic matter management strategies. In: Magdoff, F. and Weil, R.R. (eds) *Soil Organic Matter in Sustainable Agriculture*. CRC Press, Boca Raton, Florida, pp. 45–65.

Magkos, F., Arvaniti, F. and Zampelas, A. (2006) Organic food: buying more safety or just peace of mind? A critical review of the literature. *Critical Reviews in Food Science & Nutrition* 46, 23–56.

Maheshwari, R. (2006) What is an endophytic fungus? *Current Science* 90, 1309.

Mahoney, C.J. and Strongman, D.B. (1994) Nematophagous fungi from cattle manure in four states of decomposition at three sites in Nova Scotia, Canada. *Mycologia* 86, 371–375.

Majdi, N., Tackx, M., Traunspurger, W. and Buffan-Dubau, E. (2012) Feeding of biofilm-dwelling nematodes examined using HPLC-analysis of gut pigment contents. *Hydrobiologia* 680, 219–232.

Maksimov, I.V., Abizgil'dina, R.R. and Pusenkova, L.I. (2011) Plant growth promoting rhizobacteria as an alternative to chemical crop protectors from pathogens (Review). *Applied Biochemistry & Microbiology* 47, 333–345.

Malajczuk, N. (1983) Microbial antagonism to *Phytophthora*. In: Erwin, D.C., Bartnicki-Garcia, S. and Tsao, P.H. (eds) *Phytophthora: Its Biology, Taxonomy, Ecology and Pathology*. American Phytopathological Society, St Paul, Minnesota, pp. 197–218.

Mani, A. and Anandam, R.J. (1989) Evaluation of plant leaves, oil cakes and agro-industrial wastes as substrates for mass multiplication of the nematophagous fungus, *Paecilomyces lilacinus*. *Journal of Biological Control* 3, 56–58.

Mankau, R. (1975a) Prokaryotic affinities of *Duboscqia penetrans* Thorne. *Journal of Protozoology* 21, 31–34.

Mankau, R. (1975b) *Bacillus penetrans* n. comb. causing a virulent disease of plant-parasitic nematodes. *Journal of Invertebrate Pathology* 26, 333–339.

Mankau, R. and Imbriani, J.L. (1975) The life cycle of an endoparasite in some tylenchid nematodes. *Nematologica* 21, 89–94.

Mankau, R. and Prasad, N. (1972) Possibilities and problems in the use of a sporozoan endoparasite for biological control of plant-parasitic nematodes. *Nematropica* 2, 7–8.

Mankau, R. and Prasad, N. (1977) Infectivity of *Bacillus penetrans* in plant-parasitic nematodes. *Journal of Nematology* 9, 40–45.

Manzanilla-López, R.H., Atkins, S.D., Clark, I.M., Kerry, B.R. and Hirsch, P.R. (2009) Measuring abundance, diversity and parasitic ability in two populations of the nematophagous fungus *Pochonia chlamydosporia* var. *chlamydosporia*. *Biocontrol Science & Technology* 19, 391–406.

Manzanilla-López, R.H., Esteves, I. and Finetti-Sialer, M. (2011a) *Pochonia chlamydosporia*: biological, ecological and physiological aspects in the host-parasite relationship of a biological control agent of nematodes. In: Boeri, F. and Chung, J.A. (eds) *Nematodes, Morphology and Management Strategies*. Nova Science Publishers, Inc., pp. 267–300.

Manzanilla-López, R.H., Esteves, I., Powers, S.J. and Kerry, B.R. (2011b) Effects of crop plants on abundance of *Pochonia chlamydosporia* and other fungal parasites of root-knot and potato cyst nematodes. *Annals of Applied Biology* 159, 118–129.

Manzanilla-López, R.H., Esteves, I., Finetti-Sialer, M.M., Hirsch, P.R., Ward, E., Devonshire, J. and Hidalgo-Díaz, L. (2013) *Pochonia chlamydosporia*: advances and challenges to improve its performance as a biological control agent of sedentary endo-parasitic nematodes. *Journal of Nematology* 45, 1–7.

Mao, X., Hu, F., Griffiths, B. and Li, H. (2006) Bacterial-feeding nematodes enhance root growth of tomato seedlings. *Soil Biology & Biochemistry* 38, 1615–1622.

Mao, X., Hu, F., Griffiths, B., Chen, X., Liu, M. and Li, H (2007) Do bacterial-feeding nematodes stimulate root proliferation through hormonal effects? *Soil Biology & Biochemistry* 39, 1816–1819.

Marahatta, S.P., Wang, K.-H., Sipes, B.S. and Hooks, C.R.R. (2010) Strip-tilled cover cropping for managing nematodes, soil mesoarthropods, and weeds in a bitter melon agroecosystem. *Journal of Nematology* 42, 111–119.

Maraun, M., Erdmann, G., Fischer, B.M., Pollierer, M.M., Norton, R.A., Schneider, K. and Scheu, S. (2011) Stable isotopes revisited: their use and limits for oribatid mite trophic ecology. *Soil Biology & Biochemistry* 43, 877–882.

Mateille, T., Duponnois, R., Dabiré, K., Ndiaye, S. and Diop, M.T. (1996) Influence of the soil on the transport of the spores of *Pasteuria penetrans*, parasite of the genus *Meloidogyne*. *European Journal of Soil Biology* 32, 81–88.

Mateille, T., Trudgill, D.L., Trivino, C., Bala, G., Sawadogo, A. and Vouyoukalou, E. (2002) Multisite survey of soil interactions with infestation of root-knot nematodes (*Meloidogyne* spp.) by *Pasteuria penetrans*. *Soil Biology & Biochemistry* 34, 1417–1424.

Mateille, T., Cadet, P. and Fargette, M. (2008) Control and management of plant-parasitic nematode communities in a soil conservation approach. In: Ciancio, A. and Mukerji. K.G. (eds) *Integrated Management and Biocontrol of Vegetable and Grain Crops Nematodes*. Springer, Dordrecht, the Netherlands, pp. 79–97.

Mateille, T., Fould, S., Dabiré, K.R., Diop, M.T. and Ndiaye, S. (2009) Spatial distribution of the nematode biocontrol agent *Pasteuria penetrans* as influenced by its soil habitat. *Soil Biology & Biochemistry* 41, 303–308.

Mateille, T., Dabiré, K.R., Fould, S. and Diop, M.T. (2010) Host-parasite soil communities and environmental constraints: modelling of soil functions involved in interactions between plant-parasitic nematodes and *Pasteuria penetrans*. *Soil Biology and Biochemistry* 42, 1193–1199.

Matthiessen, J.N. and Kirkegaard, J.A. (2006) Biofumigation and enhanced biodegradation: opportunity and challenge in soilborne pest and disease management. *Critical Reviews in Plant Sciences* 25, 235–265.

Matthiessen, J.N., Warton, B. and Shackelton, M.A. (2004) The importance of plant maceration and water addition in achieving high *Brassica*-derived isothiocyanate levels in soil. *Agroindustria* 3, 277–280.

Mauchline, T.H., Kerry, B.R. and Hirsch, P.R. (2002) Quantification in soil and the rhizosphere of the nematophagous fungus *Verticillium chlamydosporium* by competitive PCR and comparison with selective plating. *Applied & Environmental Microbiology* 68, 1846–1853.

Mauchline, T.H., Kerry, B.R. and Hirsch, P.R. (2004) The biocontrol fungus *Pochonia chlamydosporia* shows nematode host preference at the infraspecific level. *Mycological Research* 108, 161–169.

Mauchline, T.H., Mohan, S., Davies, K.G., Schaff, J.E., Opperman, C.H., Kerry, B.R. and Hirsch, P.R. (2010) A method for release and multiple strand amplification of small quantities of DNA from endospores of the fastidious bacterium *Pasteuria penetrans*. *Letters in Applied Microbiology* 50, 515–521.

Mauchline, T.H., Knox, R., Mohan, S., Powers, S.J., Kerry, B.R., Davies, K.G. and Hirsch, P.R. (2011) Identification of new single nucleotide polymorphism-based markers for inter- and intraspecies discrimination of obligate parasites (*Pasteuria* spp.) of invertebrates. *Applied & Environmental Microbiology* 77, 6388–6394.

Mavrodi, D.V., Blankenfeldt, W. and Thomashow, L.S. (2006) Phenazine compounds in fluorescent *Pseudomonas* spp. biosynthesis and regulation. *Annual Review of Phytopathology* 44, 417–445.

Mayerhofer, M.S., Kernaghan, G. and Harper, K.A. (2013) The effects of fungal root endophytes on plant growth: a meta-analysis. *Mycorrhiza* 23, 119–128.

Mazzola, M. (1998) Elucidation of the microbial complex having a causal role in the development of apple replant disease in Washington. *Phytopathology* 88, 930–938.

Mazzola, M. (2007) Manipulation of rhizosphere bacterial communities to induce suppressive soils. *Journal of Nematology* 39, 213–220.

Mazzola, M. and Mullinix, K. (2005) Comparative field efficacy of management strategies containing *Brassica napus* seed meal or green manure for the control of apple replant disease. *Plant Disease* 89, 1207–1213.

McCulloch, J.S. (1977) A survey of nematophagous fungi in Queensland. *Queensland Journal of Agricultural and Animal Sciences* 34, 25–33.

McInnis, T.M. and Jaffee, B.A. (1989) An assay for *Hirsutella rhossiliensis* spores and the importance of phialides for nematode inoculation. *Journal of Nematology* 21, 229–234.

McIntyre, B.D., Speijer, P.R., Riha, S.J. and Kizito, F. (2000) Effects of mulching on biomass, nutrients, and soil water in banana inoculated with nematodes. *Agronomy Journal* 92, 1081–1085.

McKenry, M.V. and Anwar, S.A. (2008) Nematode and grape rootstock interactions including an improved understanding of tolerance. *Journal of Nematology* 38, 312–318.

McKenry, M.V. and Kretsch, J. (1987) Peach tree and nematode responses to various soil treatments under two irrigation regimes. *Nematologica* 33, 343–354.

McLeod, R.W. and Steele, C.C. (1999) Effects of *Brassica* leaf green manures and crops on activity and reproduction of *Meloidogyne javanica*. *Nematology* 1, 613–624.

McLeod, R.W., Kirkegaard, J.A. and Steele, C.C. (2001) Invasion, development, growth and egg-laying by *Meloidogyne javanica* in Brassicaceae crops. *Nematology* 3, 463–472.

McSorley, R. (1998) Population dynamics. In: Barker, K.R., Pederson, G.A. and Windham, G.L. (eds) *Plant and Nematode Interactions*. American Society of Agronomy Inc., Madison, Wisconsin, pp. 109–133.

McSorley, R. (2001) Multiple cropping systems for nematode management: a review. *Soil & Crop Science Society of Florida Proceedings* 60, 132–142.

McSorley, R. (2011a) Trends in the Journal of Nematology, 1969–2009: authors, states, nematodes and subject matter. *Journal of Nematology* 43, 63–68.

McSorley, R. (2011b) Overview of organic amendments for management of plant-parasitic nematodes, with case studies from Florida. *Journal of Nematology* 43, 69–81.

McSorley, R. (2011c) Effect of disturbances on trophic groups in soil nematode assemblages. *Nematology* 13, 553–559.

McSorley, R. (2012) Ecology of the dorylaimid omnivore genera *Aporcelaimellus*, *Eudorylaimus* and *Mesodorylaimus*. *Nematology* 14, 645–663.

McSorley, R. and Duncan, L.W. (2004) Population dynamics. In: Chen, Z.X., Chen, S.Y. and Dickson, D.W. (eds) *Nematology Advances and Perspectives. Volume 1. Nematode Morphology, Physiology and Ecology.* CAB International, Wallingford, UK, pp. 469–492.

McSorley, R. and Frederick, J.J. (1995) Responses of some common cruciferae to root-knot nematode. *Journal of Nematology* 27, 550–554.

McSorley, R. and Gallaher, R.N. (1995) Cultural practices improve crop tolerance to nematodes. *Nematropica* 25, 53–60.

McSorley, R. and Phillips, M.S. (1993) Modelling population dynamics and yield losses and their use in nematode management. In: Evans, K., Trudgill, D.L. and Webster, J.M. (eds) *Plant Parasitic Nematodes in Temperate Agriculture.* CAB International, Wallingford, UK, pp. 61–85.

McSorley, R. and Wang, K.-H. (2009) Possibilities for biological control of root-knot nematodes by natural predators in Florida soils. *Proceedings of the Florida State Horticultural Society* 122, 421–425.

McSorley, R., Wang, K.-H., Kokalis-Burelle, N. and Church, G. (2006) Effects of soil type and steam on nematode biological control potential of the rhizosphere community. *Nematropica* 36, 197–214.

McSorley, R., Wang, K.-H. and Church, G. (2008) Suppression of root-knot nematodes in natural and agricultural soils. *Applied Soil Ecology* 39, 291–298.

McSpadden Gardener, B.B., Gutierrez, L.J., Joshi, R., Edema, R. and Lutton, E. (2005) Distribution and biocontrol potential of phlD+ pseudomonads in corn and soybean fields. *Phytopathology* 95, 715–724.

Meerman, F., van de Ven, G.W.J., van Keulen, H. and Bremen, H. (1996) Integrated crop management: an approach to sustainable agricultural development. *International Journal of Pest Management* 42, 13–24.

Mekete, T., Hallmann, J., Kiewnick, S. and Sikora, R. (2009) Endophytic bacteria from Ethiopian coffee plants and their potential to antagonise *Meloidogyne incognita*. *Nematology* 11, 117–127.

Melakeberhan, H. (2002) Embracing the emerging precision agriculture technologies for site-specific management of yield-limiting factors. *Journal of Nematology* 34, 185–188.

Melero, S., López-Garrido, R., Murillo, J.M. and Moreno, F. (2009) Conservation tillage: short- and long-term effects on soil carbon fractions and enzymatic activities under Mediterranean conditions. *Soil & Tillage Research* 104, 292–298.

Melki, K.C., Giannakou, I.O., Pembroke, B. and Gowen, S.R. (1998) The cumulative build-up of *Pasteuria* spores in root-knot nematode infested soil and the effect of soil applied fungicides on its infectivity. *Fundamental and Applied Nematology* 21, 679–683.

Mendes, R., Kruijt, M., de Bruijn, I., Dekkers, E., van der Voort, M., Schneider, J.H.M., Piceno, Y.M., DeSantis, T.Z., Anderson, G.L., Bakker, P.A.H.M. and Raaijmakers, J.M. (2011) Deciphering the rhizosphere microbiome for disease-suppressive bacteria. *Science.* 332, 1097–1100.

Mendoza, A.R. and Sikora, R.A. (2009) Biological control of *Radopholus similis* in banana by combined application of the mutualistic endophyte *Fusarium oxysporum* strain 162, the egg pathogen *Paecilomyces lilacinus* strain 251 and the antagonistic bacteria *Bacillus firmus*. *BioControl* 54, 263–272.

Mendoza, A.R., Sikora, R.A. and Kiewnick, S. (2007) Influence of *Paecilomyces lilacinus* strain 251 on the biological control of the burrowing nematode *Radopholus similis* on banana. *Nematropica* 37, 203–213.

Mendoza, A.R., Kiewnick, S. and Sikora, R.A. (2008) *In vitro* activity of *Bacillus firmus* against the burrowing nematode *Radopholus similis*, the root-knot nematode *Meloidogyne incognita* and the stem nematode *Ditylenchus dipsaci*. *Biocontrol Science & Technology* 18, 377–389.

Mennan, S., Chen, S. and Melakeberhan, H. (2006) Suppression of *Meloidogyne hapla* populations by *Hirsutella minnesotensis*. *Biocontrol Science & Technology* 16, 181–193.

Mensink, B.J.W.G. and Scheepmaker, J.W.A. (2007) How to evaluate the environmental safety of microbial plant protection products: a proposal. *Biocontrol Science & Technology* 17, 3–20.

Mertens, M.C.A. and Stirling, G.R. (1993) Parasitism of *Meloidogyne* spp. on grape and kiwifruit by the fungal egg parasites *Paecilomyces lilacinus* and *Verticillium chlamydosporium*. *Nematologica* 39, 400–410.

Metchnikoff, M.E. (1888) *Pasteuria ramosa*, un representant des bacteries a division longitudinale. *Annales de l'Institut Pasteur* 2, 165–170.

Meyer, S.L.F. (1994) Effects of a wild type strain and a mutant strain of the fungus *Verticillium lecanii* on *Meloidogyne incognita* populations in greenhouse studies. *Fundamental & Applied Nematology* 17, 563–567.

Meyer, S.L.F. (2003) United States Department of Agriculture – Agricultural Research Service research programs on microbes for management of plant-parasitic nematodes. *Pest Management Science* 59, 665–670.

Meyer, S.L.F. and Meyer, R.J. (1996) Greenhouse studies comparing strains of the fungus *Verticillium lecanii* for activity against the nematode *Heterodera glycines*. *Fundamental & Applied Nematology* 19, 305–308.

Meyer, S.L.F. and Wergin, W.P. (1998) Colonization of soybean cyst nematode females, cysts and gelatinous matrices by the fungus *Verticillium lecanii*. *Journal of Nematology* 30, 436–450.

Meyer, S.L.F., Johnson, G., Dimock, M., Fahey, J.W. and Huettel, R.N. (1997) Field efficacy of *Verticillium lecanii*, sex pheromone, and pheromone analogs as potential management agents for soybean cyst nematode. *Journal of Nematology* 29, 282–288.

Meyer, S.L.F., Roberts, D.P. and Wergin, W.P. (1998) Association of the plant-beneficial fungus *Verticillium lecanii* with soybean roots and rhizosphere. *Journal of Nematology* 30, 451–460.

Meyer, S.L.F., Massoud, S.I., Chitwood, D.J. and Roberts, D.P. (2000) Evaluation of *Trichoderma virens* and *Burkholderia cepacia* for antagonistic activity against root-knot nematode, *Meloidogyne incognita*. *Nematology* 2, 871–879.

Mian, I.H., Godoy, G., Shelby, R.A., Rodríguez-Kábana, R. and Morgan-Jones, G. (1982) Chitin amendments for control of *Meloidogyne arenaria* in infested soil. *Nematropica* 12, 71–84.

Mikola, J. and Setälä, H. (1998) No evidence of trophic cascades in an experimental microbial-based soil food web. *Ecology* 79, 153–164.

Miller, L.K. and Ball, L.A. (eds) (1998) *The Insect Viruses*. Plenum Press, New York.

Miller, P.M. (1976) Effects of some nitrogenous materials and wetting agents on survival in soil of lesion, stylet and lance nematodes. *Phytopathology* 66, 798–800.

Miller, P.M., Taylor, G.S. and Wihrheim, S.E. (1968) Effects of cellulosic soil amendments and fertilizers on *Heterodera tabacum*. *Plant Disease Reporter* 52, 441–445.

Min, Y.Y. and Toyota, K. (2013) Suppression of *Meloidogyne incognita* in different agricultural soils and possible contribution of soil fauna. *Nematology* 15, 459–468.

Minerdi, D., Moretti, M., Gilardi, G., Barberio, C., Gullino, M.L. and Garibaldi, A. (2008) Bacterial ectosymbionts and virulence silencing in a *Fusarium oxysporum* strain. *Environmental Microbiology* 10, 1725–1741.

Minor, M.A. and Norton, R.A. (2004) Effects of soil amendments on assemblages of soil mites (Acari: Oribatida, Mesostigmata) in short-rotation willow plantings in central New York. *Canadian Journal of Forest Research* 34, 1417–1425.

Minoshima, H., Jackson, L.E., Cavagnaro, T.R., Sánchez-Moreno, S., Ferris, H., Temple, S.R., Goyal, S. and Mitchell, J.P. (2007) Soil food webs and carbon dynamics in response to conservation tillage in California. *Soil Science Society of America Journal* 71, 952–963.

Minter, D.W. and Brady, B.L. (1980) Mononematous species of *Hirsutella*. *Transactions of the British Mycological Society* 74, 271–282.

Miransari, M. (2010) Contribution of arbuscular mycorrhizal symbiosis to plant growth under different types of soil stress. *Plant Biology* 12, 563–569.

Mirsky, S.B., Curran, W.S., Mortensen, D.M., Ryan, M.R. and Shumway, D.L. (2011) Timing of cover-crop management effects on weed suppression in no-till planted soybean using a roller crimper. *Weed Science* 59, 380–389.

Mitchell, D.J., Kannwischer-Mitchell, M.E. and Dickson, D.W. (1987) A semi-selective medium for the isolation of *Paecilomyces lilacinus* from soil. *Journal of Nematology* 19, 255–256.

Mitchell, D.A., Krieger, N. and Berovič, M. (eds) (2006) *Solid-state Fermentation Bioreactors. Fundamentals of Design and Operation*. Springer-Verlag, Heidelberg, Germany.

Mitchell, J.P., Klonsky, K., Shrestha, A., Fry, R., DuSault, A., Beyer, J. and Harben, R. (2007) Adoption of conservation tillage in California: current status and future perspectives. *Australian Journal of Experimental Agriculture* 47, 1383–1388.

Mitchell, J.P., Singh, P.N., Wallender, W.W., Munk, D.S., Wroble, J.F., Horwarth, W.R., Hogan, P., Roy, R. and Hanson, B.R. (2012a) No-tillage and high residue practices reduce soil water evaporation. *California Agriculture* 66, 55–61.

Mitchell, J.P., Klonsky, K.M., Miyao, E.M., Aegerter, B.J., Shrestha, A., Munk, D.S., Hembree, K., Madden, N.M. and Turini, T.A. (2012b) Evolution of conservation tillage systems for processing tomato in California's Central Valley. *HortTechnology* 22, 617–626.

Mitchell, R. and Alexander, M. (1962) Lysis of soil fungi by bacteria. *Canadian Journal of Microbiology* 9, 169–177.

Mitchell, R. and Hurwitz, E. (1965) Suppression of *Pythium debaryanum* by lytic rhizosphere bacteria. *Phytopathology* 55, 156–158.

Miura, F., Nakamoto, T., Kaneda, S., Okano, S., Nakajima, M. and Murakami, T. (2008) Dynamics of soil biota at different depths under two contrasting tillage practices. *Soil Biology & Biochemistry* 40, 406–414.

Mohan, S., Mauchline, T.H., Rowe, J., Hirsch, P.R. and Davies, K.G. (2012) *Pasteuria* endospores from *Heterodera cajani* (Nematoda: Heteroderidae) exhibit inverted attachment and altered germination in cross-infection studies with *Globodera pallida* (Nematoda: Heteroderidae). *FEMS Microbiology Ecology* 79, 675–684.

Molina, J.P., Dolinski, C., Souza, R.M. and Lewis, E.E. (2007) Effect of entomopathogenic nematodes (Rhabditida: Steinernematidae and Heterorhabditidae) on *Meloidogyne mayaguensis* Rammah and Hirschmann (Tylenchida: Meloidogynidae) infection in tomato plants. *Journal of Nematology* 39, 338–342.

Monfort, E., Lopez-Llorca, L.V., Jansson, H.-B., Salinas, J., Park, J.O. and Sivasithamparam, K. (2005) Colonisation of seminal roots of wheat and barley by egg-parasitic nematophagous fungi and their effects on *Gaeumannomyces graminis* var. *tritici* and development of root rot. *Soil Biology & Biochemistry* 37, 1229–1235.

Monfort, E., Lopez-Llorca, L.V., Jansson, H.-B., Salinas, J. (2006) *In vitro* soil receptivity assays to egg-parasitic nematophagous fungi. *Mycological Progress* 5, 18–23.

Monfort, W.S., Kirkpatrick, T.L., Rothrock, C.S. and Mauromoustakos, A. (2007) Potential for site-specific management of *Meloidogyne incognita* in cotton using soil textural zones. *Journal of Nematology* 39, 1–8.

Moody, P.W., Yo, S.A. and Aitken, R.L. (1997) Soil organic carbon, permanganate fractions, and the chemical properties of acidic soils. *Australian Journal of Soil Research* 35, 1301–1308.

Moore, J.C. (1994) Impact of agricultural practices on soil food web structure: theory and application. *Agriculture, Ecosystems & Environment* 51, 239–247.

Moore, J.C., Walter, D.E. and Hunt, H.W. (1988) Arthropod regulation of micro-and mesobiota in belowground detrital food webs. *Annual Review of Entomology* 33, 419–439.

Moosavi, M.R., Zare, R., Zamanizadeh, H.R. and Fatemy, S. (2010) Pathogenicity of *Pochonia* species on eggs of *Meloidogyne javanica*. *Journal of Invertebrate Pathology* 104, 125–133.

Mordue, A.J. and Blackwell, A. (1993) Azadirachtin: an update. *Journal of Insect Physiology* 39, 903–924.

Moretti, M., Grunau, A., Minerdi, D., Gehrig, P., Roschitzki, B., Eberl, L., Garibaldi, A., Gullino, M.L. and Riedel, K. (2010) A proteomics approach to study synergistic and antagonistic interactions of the fungal-bacterial consortium *Fusarium oxysporum* wild-type MSA 35. *Proteomics* 10, 3292–3320.

Morgan-Jones, G. and Rodriguez-Kábana, R. (1981) Fungi associated with cysts of *Heterodera glycines* in an Alabama soil. *Nematropica* 11, 69–74.

Morgan-Jones, G. and Rodriguez-Kábana, R. (1988) Fungi colonizing cysts and eggs. In: Poinar Jr, G.O. and Jansson, H.-B. (eds) *Diseases of Nematodes, Volume II*. CRC Press, Boca Raton, Florida, pp. 39–58.

Morgan-Jones, G., Ownley Gintis, B. and Rodriguez-Kábana, R. (1981) Fungal colonization of *Heterodera glycines* cysts in Arkansas, Florida, Mississippi and Missouri soils. *Nematropica* 11, 155–163.

Morgan-Jones, G., White, J.F. and Rodriguez-Kábana, R. (1983) Phytonematode pathology: Ultrastructural studies. 1. Parasitism of *Meloidogyne arenaria* eggs by *Verticillium chlamydosporium*. *Nematropica* 13, 245–260.

Morgan-Jones, G., White, J.F. and Rodriguez-Kábana, R. (1984a) Phytonematode pathology: Ultrastructural studies. II. Parasitism of *Meloidogyne arenaria* eggs and larvae by *Paecilomyces lilacinus*. *Nematropica* 14, 57–71.

Morgan-Jones, G., White, J.F. and Rodriguez-Kábana, R. (1984b) Fungal parasites of *Meloidogyne incognita* in an Alabama soybean field soil. *Nematropica* 14, 93–96.

Morison, J.I.L., Baker, N.R., Mullineaux, P.M. and Davies, W.J. (2008) Improving water use in crop production. *Philosophical Transactions of the Royal Society B*, 363, 639–658.

Morris, N.L., Miller, P.C.H., Orson, J.H. and Froud-Williams, R.J. (2010) The adoption of non-inversion tillage systems in the United Kingdom and the agronomic impact on soil, crops and the environment – a review. *Soil & Tillage Research* 108, 1–15.

Mortensen, D.A., Egan, J.F., Maxwell, B.D., Ryan, M.R. and Smith, R.G. (2012) Navigating a critical juncture for sustainable weed management. *BioScience* 62, 75–84.

Morton, C.O., Mauchline, T.H., Kerry, B.R. and Hirsch, P.R. (2003) PCR-based DNA fingerprinting indicates host-related genetic variation in the nematophagous fungus *Pochonia chlamydosporia*. *Mycological Research* 107, 198–205.

Morton, C.O., Hirsch, P.R. and Kerry, B.R. (2004) Infection of plant-parasitic nematodes by nematophagous fungi – a review of the application of molecular biology to understand infection processes and to improve biological control. *Nematology* 6, 161–170.

Mukerji, K.G., Manoharachary, C. and Singh, J. (eds) (2006) *Microbial Activity in the Rhizosphere*. Springer-Verlag, Berlin.

Mullen, J. (2007) Productivity growth and the returns from public investment in R&D in Australian broadacre agriculture. *Australian Journal of Agricultural & Resource Economics* 51, 359–384.

Muller, J. (1982) The influence of fungal parasites on the population dynamics of *Heterodera schachtii* on oil radish. *Nematologica* 28, 161.

Muller, J. (1985) The influence of two pesticides on fungal parasites of *Heterodera schachtii*. *Les Colloques de l'INRA* 31, 225–231.

Muller, R. and Gooch, P.S. (1982) Organic amendments in nematode control. An examination of the literature. *Nematropica* 12, 319–326.

Muschiol, D., Marković, M., Threis, I. and Traunspurger, W. (2008) Predator-prey relationship between the cyclopoid copepod *Diacyclops bicuspidatus* and a free-living bacterivorous nematode. *Nematology* 10, 55–62.

Nagy, P. (1999) Effects of an artificial metal pollution on nematode assemblage of a calcareous loamy chernozem soil. *Plant & Soil* 212, 35–47.

Nagy, P. (2009) Case studies using nematode assemblage analysis in terrestrial habitats. In: Wilson, M.J. and Kakouli-Duarte, T. (eds) *Nematodes as Environmental Indicators*. CABI, Wallingford, UK, pp. 172–187.

Nair, A. and Ngouajio, M. (2012) Soil microbial biomass, functional microbial diversity, and nematode community structure as affected by cover crops and compost in an organic vegetable production system. *Applied Soil Ecology* 58, 45–55.

Nakamoto, T. and Tsukamoto, M. (2006) Abundance and activity of soil organisms in fields of maize grown with a white clover living mulch. *Agriculture, Ecosystems & Environment* 115, 34–42.

Neal Jr, J.L., Larson, R.I. and Atkinson, T.G. (1973) Changes in rhizosphere populations of selected physiological groups of bacteria related to substitution of specific pairs of chromosomes in spring wheat. *Plant & Soil* 39, 209–212.

Neher, D.A. (1999a) Nematode communities in organically and conventionally managed agricultural soils. *Journal of Nematology* 31, 142–154.

Neher, D.A. (1999b) Soil community composition and ecosystem processes. Comparing agricultural ecosystems with natural ecosystems. *Agroforestry Systems* 45, 159–185.

Neher, D.A. (2001) Role of nematodes in soil health and their use as indicators. *Journal of Nematology* 33, 161–168.

Neher, D.A. (2010) Ecology of plant and free-living nematodes in natural and agricultural soil. *Annual Review of Phytopathology* 48, 371–394.

Neher, D.A. and Campbell, C.L. (1994) Nematode communities and microbial biomass in soils with annual and perennial crops. *Applied Soil Ecology* 1, 17–28.

Neher, D.A. and Darby, B.J. (2009) General community indices that can be used for analysis of nematode assemblages. In: Wilson, M.J. and Kakouli-Duarte, T. (eds) *Nematodes as Environmental Indicators*. CAB International, Wallingford, UK, pp. 107–123.

Nehl, D.B., Allen, S.J. and Brown, J.F. (1996) Deleterious rhizosphere bacteria: an integrating perspective. *Applied Soil Ecology* 5, 1–20.

Neilson, R., Hesketh, J. and Daniell, T. (2006) Novel application of T-RFLP techniques to characterise the diet of three species of predatory Mononchidae. *Journal of Nematology* 38, 285.

Neilson, R., Donn, S., Griffiths, B., Daniell, T., Rybarczyk, K., van den Elsen, S., Mooyman, P. and Helder, J. (2009) Molecular tools for analysing nematode assemblages. In: Wilson, M.J. and Kakouli-Duarte, T. (eds) *Nematodes as Environmental Indicators*. CAB International, Wallingford, UK, pp. 188–207.

Neipp, P.W. and Becker, J.O. (1999) Evaluation of biocontrol activity of rhizobacteria from *Beta vulgaris* against *Heterodera schachtii*. *Journal of Nematology* 31, 54–61.

Netscher, C. and Duponnois, R. (1998) Use of aqueous suspensions for storing and inoculating spores of *Pasteuria penetrans*, parasite of *Meloidogyne* spp. *Nematologica*, 44, 91–94.

Nguyen, V.L., Bastow, J.L., Jaffee, B.A. and Strong, D.R. (2007) Response of nematode-trapping fungi to organic substrates in a coastal grassland soil. *Mycological Research* 111, 856–862.

Nicol, J.M. and Rivoal, R. (2008) Global knowledge and its application for the integrated control and management of nematodes on wheat. In: Ciancio, A. and Mukerji. K.G. (eds) *Integrated Management and Biocontrol of Vegetable and Grain Crops Nematodes*. Springer, Dordrecht, the Netherlands, pp. 251–294.

Nichols, K.A. and Wright, S.F. (2004) Contributions of fungi to soil organic matter in agroecosystems. In: Magdoff, F. and Weil, R.R. (eds) *Soil Organic Matter in Sustainable Agriculture*. CRC Press, Boca Raton, Florida, pp. 179–198.

Nicolay, R. and Sikora, R.A. (1989) Techniques to determine the activity of fungal egg parasites of *Heterodera schachtii* in field soil. *Revue de Nématologie* 12, 97–102.

Nigh, E.A., Thomason, I.J. and van Gundy, S.D. (1980a) Identification and distribution of fungal parasites of *Heterodera schachtii* eggs in California. *Phytopathology* 70, 884–889.

Nigh, E.A., Thomason, I.J. and van Gundy, S.D. (1980b) Effect of temperature and moisture on parasitization of *Heterodera schachtii* eggs by *Acremonium strictum* and *Fusarium oxysporum*. *Phytopathology* 70, 889–891.

Nishizawa, T. (1987) A decline phenomenon in a population of upland rice cyst nematode, *Heterodera elachista*, caused by bacterial parasite, *Pasteuria penetrans*. *Journal of Nematology* 19, 546.

Noble, R. and Coventry, E. (2005) Suppression of soil-borne plant diseases with composts: a review. *Biocontrol Science &Technology* 15, 3–20.

Noel, G.R. and Stanger, B.A. (1994) First report of *Pasteuria* sp. attacking *Heterodera glycines* in North America. *Journal of Nematology* 26, 612–615.

Noel, G.R., Atibalentja, N. and Domier, L. (2005) Emended description of *Pasteuria nishizawae*. *International Journal of Systematic & Evolutionary Microbiology* 55, 1681–1685.

Noel, G.R., Atibalentja, N. and Bauer, S.J. (2010) Suppression of *Heterodera glycines* in a soybean field artificially infested with *Pasteuria nishizawae*. *Nematropica* 40, 41–52.

Nong, G., Chow, V., Schmidt, L.M., Dickson, D.W. and Preston, J.F. (2007) Multiple-strand displacement and identification of single nucleotide polymorphisms as markers of genotypic variation of *Pasteuria penetrans* biotypes infecting root-knot nematodes. *FEMS Microbiology Ecology* 61, 327–336.

Nordbring-Hertz, B., Jansson, H.-B. and Tunlid, A. (2006) Nematophagous fungi. *Encyclopedia of Life Sciences*. John Wiley & Sons, pp. 1–11.

Ntalli, N.C., Menkissoglu-Spiroudi, Y., Giannakou, I.O. and Prophetou-Athanasiadou, D.A. (2009) Efficacy evaluation of a neem (*Azadirachta indica* A. Juss) formulation against root-knot nematodes *Meloidogyne incognita*. *Crop Protection* 28, 489–494.

Nyczepir, A.P. and Meyer, S.L.F. (2010) Host status of endophyte-infected and noninfected tall fescue grass to *Meloidogyne* spp. *Journal of Nematology* 42, 151–158.

Nyczepir, A.P., Shapiro-Ilan, D.I., Lewis, E.E. and Handoo, Z.A. (2004) Effect of entomopathogenic nematodes on *Mesocriconema xenoplax* populations on peach and pecan. *Journal of Nematology* 36, 181–185.

Ogle, S.M., Swan, A. and Paustian, K. (2012) No-till management impacts on crop productivity, carbon input and soil carbon sequestration. *Agriculture, Ecosystems & Environment* 149, 37–49.

O'Hanlon, K.A., Knorr, K., Jørgensen, L.N., Nicolaisen, M. and Boelt, B. (2012) Exploring the potential of symbiotic fungal endophytes in cereal disease suppression. *Biological Control* 63, 69–78.

Ohnesorge, B., Friedel, J. and Oesterlin, U. (1974) Investigations on the distribution pattern of *Heterodera avenae* and its changes in a field under continuous cereal cultivation. *Zeitschrift fiir Pflanzenkrankheiten und Planzenschutz* 81, 356–363.

Oka, Y. (2010) Mechanisms of nematode suppression by organic soil amendments – a review. *Applied Soil Ecology* 44, 101–115.

Oka, Y. and Pivonia, S. (2002) Use of ammonia-releasing compounds for control of the root-knot nematode *Meloidogyne javanica*. *Nematology* 4, 65–71.

Oka, Y., Chet, I. and Spiegel, Y. (1993) Control of root-knot nematode *Meloidogyne javanica* by *Bacillus cereus*. *Biocontrol Science & Technology* 3, 115–126.

Oka, Y., Chet, I., Mor, M. and Spiegel, Y. (1997) A fungal parasite of *Meloidogyne javanica* eggs: evaluation of its use to control the root-knot nematode. *Biocontrol Science & Technology* 7, 489–497.

Oka, Y., Tkachi, N., Shuker, S., Rosenberg, R., Suriano, S. and Fine, P. (2006a) Laboratory studies on the enhancement of nematicidal activity of ammonia-releasing fertilisers by alkaline amendments. *Nematology* 8, 335–346.

Oka, Y., Tkachi, N., Shuker, S., Rosenberg, R., Suriano, S., Roded, L. and Fine, P. (2006b) Field studies on the enhancement of nematicidal activity of ammonia-releasing fertilisers by alkaline amendments. *Nematology* 8, 881–893.

Oka, Y., Shapira, N. and Fine, P. (2007) Control of root-knot nematodes in organic farming systems by organic amendments and soil solarization. *Crop Protection* 26, 1556–1565.

Okada, H. and Ferris, H. (2001) Effect of temperature on growth and nitrogen mineralization of fungi and fungal-feeding nematodes. *Plant & Soil* 234, 253–262.

Okada, H. and Harada, H. (2007) Effects of tillage and fertilizer on nematode communities in a Japanese soybean field. *Applied Soil Ecology* 35, 582–598.

Okada, H. and Kadota, I. (2003) Host status of 10 fungal isolates for two nematode species, *Filenchus misellus* and *Aphelenchus avenae*. *Soil Biology & Biochemistry* 35, 1601–1607.

Olatinwo, R., Yin, B., Becker, J.O. and Borneman, J. (2006a) Suppression of the plant-parasitic nematode *Heterodera schachtii* by the fungus *Dactylella oviparasitica*. *Phytopathology* 96, 111–114.

Olatinwo, R., Borneman, J. and Becker, J.O. (2006b) Induction of beet-cyst nematode suppressiveness by the fungi *Dactylella oviparasitica* and *Fusarium oxysporum* in field microplots. *Phytopathology* 96, 855–859.

Olatinwo, R., Becker, J.O. and Borneman, J. (2006c) Suppression of *Heterodera schachtii* populations by *Dactylella oviparasitica* in four soils. *Journal of Nematology* 38, 345–348.

Oliveira, A.R., de Moraes, G.J. and Ferraz, L.C.C.B. (2007) Consumption rate of phytonematodes by *Pergalumna* sp. (Acari: Oribatida: Galumnidae) under laboratory conditions determined by a new method. *Experimental & Applied Acarology* 41, 183–189.

Olsson, S. and Persson, Y. (1994) Transfer of phosphorus from *Rhizoctonia solani* to the mycoparasite *Arthrobotrys oligospora*. *Mycological Research* 98, 1065–1068.

Omay, A., Rice, C., Maddux, D. and Gordon, W. (1998) Corn yield and nitrogen uptake in monoculture and in rotation with soybean. *Soil Science Society of America Journal* 62, 1596–1603.

Oostendorp, M. and Sikora, R.A. (1989) Seed treatment with antagonistic rhizobacteria for the suppression of *Heterodera schachtii* early root infection of sugar beet. *Revue de Nématologie* 12, 77–83.

Oostendorp, M. and Sikora, R.A. (1990) *In-vitro* interrelationships between rhizosphere bacteria and *Heterodera schachtii*. *Revue de Nématologie* 13, 269–274.

Oostendorp, M., Dickson, D.W. and Mitchell, D.J. (1990) Host range and ecology of isolates of *Pasteuria* spp. from the Southeastern United States. *Journal of Nematology* 22, 525–531.

Oostendorp, M., Hewlett, T.E., Dickson, D.W. and Mitchell, D.J. (1991a) Specific gravity of spores of *Pasteuria penetrans* and extraction of spore-filled nematodes from soil. *Journal of Nematology* 23, 729–732.

Oostendorp, M., Dickson, D.W. and Mitchell, D.J. (1991b) Population development of *Pasteuria penetrans* on *Meloidogyne arenaria*. *Journal of Nematology* 23, 58–64.

Ophel-Keller, K., McKay, A., Hartley, D., Herdina, and Curran, J. (2008) Development of a routine DNA-based testing service for soilborne diseases in Australia. *Australasian Plant Pathology* 37, 243–253.

Orajay, J.I. (2009) Population dynamics of *Pasteuria penetrans* in a peanut field and description of a *Pasteuria* isolate infecting *Mesocriconema ornatum*. PhD thesis, University of Florida Graduate School, Gainesville, Florida.

Ordentlich, A., Elad, Y. and Chet, I. (1988) The role of chitinase of *Serratia marcescens* in biocontrol of *Sclerotium rolfsii*. *Phytopathology* 78, 84–88.

Orion, D., Kritzman, G., Meyer, S.L.F., Erbe, E.F. and Chitwood, D.J. (2001) A role of the gelatinous matrix in the resistance of root-knot nematode (*Meloidogyne* spp.) eggs to microorganisms. *Journal of Nematology* 33, 203–207.

Ornat, C. and Sorribas, F.J. (2008) Integrated management of root-knot nematodes in Mediterranean horticultural crops. In: Ciancio, A. and Mukerji, K.G. (eds) *Integrated Management and Biocontrol of Vegetable and Grain Crops Nematodes*. Springer, Dordrecht, the Netherlands, pp. 295–319.

Ornat, C., Verdejo-Lucas, S., Sorribas, F.J. and Tzortzakakis, E.A. (1999) Effect of fallow and root destruction on survival of root-knot and root-lesion nematodes in intensive vegetable cropping systems. *Nematropica* 29, 5–16.

Osler, G.H.R. and Murphy, D.V. (2005) Oribatid mite species richness and soil organic matter fractions in agricultural and native vegetation in Western Australia. *Applied Soil Ecology* 29, 93–98.

Ownley Gintis, B., Morgan-Jones, G. and Rodriguez-Kabana, R. (1983) Fungi associated with several developmental stages of *Heterodera glycines* from an Alabama soybean field soil. *Nematropica* 13, 181–200.

Padgham, J.L. and Sikora, R.A. (2007) Biological control potential and modes of action of *Bacillus megaterium* against *Meloidogyne graminicola* on rice. *Crop Protection* 26, 971–977.

Panaccione, D.G., Kotcon, J.B., Schardl, C.L., Johnson, R.D. and Morton, J.B. (2006) Ergot alkaloids are not essential for endophytic fungus-associated population suppression of the lesion nematode, *Pratylenchus scribneri*, on perennial ryegrass. *Nematology* 8, 583–590.

Pang, W., Luc, J.E., Crow, W.T., Kenworthy, K.E., McSorley, R. and Giblin-Davis, R.M. (2011) Screening Bermudagrass germplasm accessions for tolerance to sting nematodes. *HortScience* 46, 1503–1506.

Pankhurst, C.E., Doube, B.M. and Gupta, V.V.S.R. (eds) (1997) *Biological Indicators of Soil Health*. CABI, Wallingford, UK.

Pankhurst, C.E., McDonald, H.J., Hawke, B.G. and Kirkby, C.A. (2002) Effect of tillage and stubble management on chemical and microbiological properties and the development of suppression towards cereal
root disease in soils from two sites in NSW, Australia. *Soil Biology & Biochemistry* 34, 833–840.

Paparu, P., Dubois, T., Coyne, D. and Viljoen, A. (2009) Dual inoculation of *Fusarium oxysporum* endophytes
in banana: effect on plant colonization, growth and control of the root burrowing nematode and the
banana weevil. *Biocontrol Science & Technology* 19, 639–655.

Papavizas, G.C. (1985) *Trichoderma* and *Gliocladium*: biology, ecology and potential for biocontrol. *Annual
Review of Phytopathology* 23, 23–54.

Papendick, R.I. and Campbell, G.S. (1981) Theory and measurement of water potential. In: Parr, J.F., Gardner,
W.R. and Elliot, L.F. (eds) *Water Potential Relations in Soil Microbiology*. Soil Science Society of America,
Madison, Wisconsin, pp. 1–22.

Papert, A. and Kok, C.J. (2000) Size and community metabolic profile of the bacterial population of
Meloidogyne hapla egg masses. *Nematology* 2, 581–584.

Park, J.O., Hargreaves, J.R., McConville, E.J., Stirling, G.R., Ghisalberti, E.L. and Sivasithamparam, K. (2004)
Production of leucostatins and nematicidal activity of Australian isolates of *Paecilomyces lilacinus*
(Thom) Samson. *Letters in Applied Microbiology* 38, 271–276.

Parke, J.L., Moen, R., Rovira, A.D. and Bowen, G.D. (1986) Soil water flow affects the rhizosphere distribution
of a seed-borne biological control agent, *Pseudomonas fluorescens*. *Soil Biology & Biochemistry* 18,
583–588.

Pathak, E., El-Borai, F.E., Campos-Herrera, R., Johnson, E.G., Stuart, R.J., Graham, J.H. and Duncan, L.W.
(2012) Use of real-time PCR to discriminate parasitic and saprophagous behaviour by nematophagous
fungi. *Fungal Biology* 116, 563–573.

Patten, C.L. and Glick, B.R. (1996) Bacterial biosynthesis of indole-3-acetic acid. *Canadian Journal of
Microbiology* 42, 207–220.

Pattinson, G.S and McGee, P.A. (1997) High densities of arbuscular mycorrhizal fungi maintained during long
fallows in soils used to grow cotton except when soil is wetted periodically. *New Phytologist* 136,
571–580.

Pattison A.B., Moody, P.W., Badcock, K.A., Smith, L.J., Armour, J.A., Rasiah, V., Cobon, J.A., Gulino, L.-M. and
Mayer, R. (2008) Development of key soil health indicators for the Australian banana industry. *Applied
Soil Ecology* 40, 155–164.

Pattison, A.B., Badcock, K. and Sikora, R. (2011) Influence of soil organic amendments on suppression of the
burrowing nematode, *Radopholus similis*, on the growth of bananas. *Australasian Plant Pathology* 40,
385–396.

Pattison, T. and Lindsay, S. (2006) *Banana Root and Soil Health User's Manual: FR02023 Soil and Root Health
for Sustainable Banana Production*. Department of Primary Industries and Fisheries, Queensland,
Australia.

Paustian, K., Andren, O., Janzen, H.H., Lal, R., Smith, P., Tian, G., Tiessen, H., Van Noordwijk, M. and Woomer, P.L.
(1997) Agricultural soils as a sink to mitigate CO_2 emissions. *Soil Use & Management* 13, 230–244.

Paustian, K., Six, J., Elliott, E.T. and Hunt, H.W. (2000) Management options for reducing CO_2 emissions from
agricultural soils. *Biogeochemistry* 48, 147–163.

Pérez, E.E. and Lewis, E.E. (2004) Suppression of *Meloidogyne incognita* and *Meloidogyne hapla* with
entomopathogenic nematodes on greenhouse peanuts and tomatoes. *Biological Control* 30, 336–441.

Pérez-Ruiz, M., Slaughter, D.C., Gliever, C.J. and Upadhyaya, S.K. (2012) Automatic GPS-based intra-row
weed knife control system for transplanted row crops. *Computers & Electronics in Agriculture* 80, 41–49.

Perry, R.N. and Moens, M. (eds) (2006) *Plant Nematology*. CAB International, Wallingford, UK.

Persmark, L. and Jansson, H.-B. (1997) Nematophagous fungi in the rhizosphere of agricultural crops. *FEMS
Microbiology Ecology* 22, 303–312.

Persmark, L. and Nordbring-Hertz, B. (1997) Conidial trap formation of nematode-trapping fungi in soil and
soil extracts. *FEMS Microbiology Ecology* 22, 313–323.

Persmark, L., Marban-Mendoza, N. and Jansson, H.-B. (1995) Nematophagous fungi from agricultural soils of
central America. *Nematropica* 25, 117–124.

Persmark, L., Banck, A. and Jansson, H.-B. (1996) Population dynamics of nematophagous fungi and nematodes in an arable soil: vertical and seasonal fluctuations. *Soil Biology & Biochemistry* 28, 1005–1014.

Persson, C. and Jansson, H.-B. (1999) Rhizosphere colonization and control of *Meloidogyne* spp. by nematode-
trapping fungi. *Journal of Nematology* 31, 164–171.

Persson, C., Olsson, S. and Jansson, H.-B. (2000) Growth of *Arthrobotrys superba* from a birch wood resource
base into soil determined by radioactive tracing. *FEMS Microbiology Ecology* 31, 47–51.

Persson, Y. and Bååth, E. (1992) Quantification of mycoparasitism by the nematode-trapping fungus *Arthrobotrys oligospora* on *Rhizoctonia solani* and the influence of nutrient levels. *FEMS Microbiology Ecology* 101, 11–16.

Persson, Y., Veenhuis, M. and Nordbring-Hertz, B. (1985) Morphogenesis and significance of hyphal coiling by nematode-trapping fungi in mycoparasitic relationships. *FEMS Microbiology Ecology* 31, 283–291.

Pesek, J.C. (ed.) (1989) *Alternative Agriculture*. National Research Council Press, Washington, DC.

Peters, R.D., Sturz, A.V., Carter, M.R. and Sanderson, J.B. (2003) Developing disease-suppressive soils through crop rotation and tillage management practices. *Soil & Tillage Research* 72, 181–192.

Peterson, E.A. and Katznelson, H. (1965) Studies on the relationships between nematodes and other soil microorganisms. IV. Incidence of nematode-trapping fungi in the vicinity of plant roots. *Canadian Journal of Microbiology* 11, 491–495.

Pfister, D.H. (1994) *Orbilia fimicola*, a nematophagous discomycete and its *Arthrobotrys* anamorph. *Mycologia* 86, 451–453.

Pfister, D.H. (1997) Castor, Pollux and life histories of fungi. *Mycologia* 89, 1–23.

Pfister, D.H. and Liftik, M.E. (1995) Two *Arthrobotrys* anamorphs from *Orbilia auricolor*. *Mycologia* 87, 684–688.

Phillips, D.A., Ferris, H., Cook, D.R. and Strong, D.R. (2003) Molecular control points in rhizosphere food webs. *Ecology* 84, 816–826.

Pinochet, J., Calvet, C., Camprubí, A. and Fernández, C. (1996) Interactions between migratory endoparasitic nematodes and arbuscular mycorrhizal fungi in perennial crops: a review. *Plant & Soil* 185, 183–190.

Plenchette, C., Clermont-Dauphin, C., Meynard, J.M. and Fortin, J.A. (2005) Managing arbuscular mycorrhizal fungi in cropping systems. *Canadian Journal of Plant Science* 85, 31–40.

Pliego, C., Ramos, C., de Vicente, A. and Cazorla, F.M. (2011) Screening for candidate bacterial biocontrol agents against soilborne fungal plant pathogens. *Plant & Soil* 340, 505–520.

Ploeg, A. (1999) Greenhouse studies on the effect of marigolds (*Tagetes* spp.) on four *Meloidogyne* species. *Journal of Nematology* 31, 62–69.

Ploeg, A.T. (2000) Effects of amending soil with *Tagetes patula* cv. Single Gold on *Meloidogyne incognita* infestation of tomato. *Nematology* 2, 489–493.

Ploeg, A. (2008) Biofumigation to manage plant-parasitic nematodes. In: Ciancio, A. and Mukerji, K.G. (eds) *Integrated Management and Biocontrol of Vegetable and Grain Crops Nematodes*. Springer, Dordrecht, the Netherlands, pp. 239–248.

Pocasangre, L., Sikora, R.A., Vilich, V. and Schuster, R.P. (2000) Survey of banana endophytic fungi from central America and screening for biological control of the burrowing nematode (*Radopholus similis*). *InfoMusa* 9, 3–5.

Pompanon, F., Deagle, B.E., Symondson, W.O.C., Brown, D.S., Jarman, S.N. and Taberlet, P. (2012) Who is eating what: diet assessment using next generation sequencing. *Molecular Ecology* 21, 1931–1950.

Porazinska, D.L., Giblin-Davis, R.M., Faller L., Farmerie, W., Kanzaki, N., Morris, K., Powers, T.O., Tucker, A.E., Sung, W. and Thomas, W.K. (2009) Evaluating high-throughput sequencing as a method for metagenomic analysis of nematode diversity. *Molecular Ecology Resources* 9, 1439–1450

Porazinska, D.L., Giblin-Davis, R.M., Esquivel, A., Powers, T.O., Sung, W. and Thomas, W.K. (2010) Ecometagenomics confirms high tropical rainforest nematode diversity. *Molecular Ecology* 19, 5521–5530.

Porter, I.J., Brett, R.W. and Wiseman, B.M. (1999) Alternatives to methyl bromide: chemical fumigants or integrated pest management systems? *Australasian Plant Pathology* 28, 65–71.

Porter, T.M., Martin, W., James, T.Y., Longcore, J.E., Gleason, F.H., Adler, P.H., Letcher, P.M. and Vilgalys, R. (2011) Molecular phylogeny of the Blastocladiomycota (Fungi) based on nuclear ribosomal DNA. *Fungal Biology* 115, 381–392.

Posner, J.L., Baldock, J.O. and Hedtcke, J.L. (2008) Organic and conventional production systems in the Wisconsin integrated cropping systems trials: I. Productivity 1990–2002. *Agronomy Journal* 100, 253–260.

Postma-Blaauw, M.B., de Goede, R.G.M., Bloem, J., Faber, J.H. and Brussaard, L. (2010) Soil biota community structure and abundance under agricultural intensification and extensification. *Ecology* 91, 460–473.

Postma-Blaauw, M.B., de Goede, R.G.M., Bloem, J., Faber, J.H. and Brussaard, L. (2012) Agricultural intensification and de-intensification differentially affect taxonomic diversity of predatory mites, earthworms, enchytraeids, nematodes and bacteria. *Applied Soil Ecology* 57, 39–49.

Powell, P.A. and Jutsum, A.R. (1993) Technical and commercial aspects of biocontrol products. *Pesticide Science* 37, 315–321.

Power, A.G. (2010) Ecosystem services and agriculture: tradeoffs and synergies. *Philosophical Transactions of the Royal Society B* 365, 2959–2971.

Powlson, D.S., Gregory, P.J., Whalley, W.R., Quinton, J.N., Hopkins, D.W., Whitmore, A.P., Hirsch, P.R. and Goulding, K.W.T. (2011a) Soil management in relation to sustainable agriculture and ecosystem services. *Food Policy* 36, S73-S87.

Powlson, D.S., Whitmore, A.P. and Goulding, W.T. (2011b) Soil carbon sequestration to mitigate climate change: a critical re-examination to identify the true and the false. *European Journal of Soil Science* 62, 42–55.

Powlson, D.S., Bhogal, A., Chambers, B.J., Coleman, K., Macdonald, A.J., Goulding, K.W.T. and Whitmore, A.P. (2012) The potential to increase soil carbon stocks through reduced tillage or organic material additions in England and Wales: a case study. *Agriculture, Ecosystems & Environment* 146, 23–33.

Prado Vero, I.C. del., Ferris, H. and Nadler, S.A. (2010) Soil inhabiting nematodes of the genera *Trischistoma*, *Tripylina*, and *Tripyla* from Mexico and the USA with descriptions of new species. *Journal of Nematode Morphology & Systematics* 13, 29–49.

Preston, J.F., Dickson, D.W., Maruniak, J.E., Nong, G., Brito, J.A., Schmidt, L.M. and Giblin-Davis, R.M. (2003) *Pasteuria* spp.: systematics and phylogeny of these bacterial parasites of phytopathogenic nematodes. *Journal of Nematology* 35, 198–207.

Pretty, J. (2008) Agricultural sustainability: concepts, principles and evidence. *Philosophical Transactions of the Royal Society B*, 363, 447–465.

Prikryl, Z. and Vancura, V. (1980) Root exudates of plants. VI. Wheat exudation as dependent on growth, concentration gradient of exudates and the presence of bacteria. *Plant & Soil*, 57, 69–83.

Prot, J.C. and Netscher, C. (1979) Influence of movement of juveniles on detection of fields infested with *Meloidogyne*. In: Lamberti, F. and Taylor, C.E. (eds), *Root-knot Nematodes (Meloidogyne species). Systematics, Biology and Control*. Academic Press, London, pp. 193–203.

Pyrowolakis, A., Westphal, A., Sikora, R.A. and Becker, J.O. (2002) Identification of root-knot nematode suppressive soils. *Applied Soil Ecology* 19, 51–56.

Qiao, M., Zhang, Y., Li, S.-F., Baral, H.-O., Weber, E., Su, H.-Y., Xu, J.-P., Zhang, K.-Q. and Yu, Z.-F. (2012) *Orbilia blumenaviensis* and its *Arthrobotrys* anamorph. *Mycological Progress* 11, 255–262.

Qin, S., Xing, K., Jiang, J.-H., Xu, L.-H. and Li, W.-J. (2011) Biodiversity, bioactive natural products and biotechnological potential of plant-associated endophytic actinobacteria. *Applied Microbiology & Biotechnology* 89, 457–473.

Quilty, J.R. and Cattle, S.R. (2011) Use and understanding of organic amendments in Australian agriculture: a review. *Soil Research* 49, 1–26.

Quinn, M.A. (1987) The influence of saprophytic competition on nematode predation by nematode-trapping fungi. *Journal of Invertebrate Pathology* 49, 170–174.

Raaijmakers, J.M., Vlami, M. and de Souza, J.T. (2002) Antibiotic production by bacterial biocontrol agents. *Antonie van Leeuwenhoek* 81, 537–547.

Raaijmakers, J.M., Paulitz, T.C., Steinberg, C., Alabouvette, C. and Moënne-Loccoz, Y. (2009) The rhizosphere: a playground and battlefield for soilborne pathogens and beneficial organisms. *Plant & Soil* 321, 341–361.

Ramirez, K.S., Craine, J.M. and Fierer, N. (2010) Nitrogen fertilization inhibits soil microbial respiration regardless of the form of nitrogen applied. *Soil Biology & Biochemistry* 42, 2336–2338.

Rapport, D.J., McCullum, J. and Miller, M.H. (1997) Soil health: its relationship to ecosystem health. In: Pankhurst, C.E., Doube, B.M. and Gupta, V.V.S.R. (eds) *Biological Indicators of Soil Health*. CABI, Wallingford, UK, pp. 29–47.

Read, D.S., Sheppard, S.K., Bruford, M.W., Glen, D.M. and Symondson, W.O.C. (2006) Molecular detection of predation by soil micro-arthropods on nematodes. *Molecular Ecology* 15, 1963–1972.

Recous, S., Robin, D., Darwis, D. and Mary, B. (1995) Soil inorganic N availability: effect on maize residue decomposition. *Soil Biology & Biochemistry* 27, 1529–1538.

Reddy, S.R., Pindi, P.K. and Reddy, S.M. (2005) Molecular methods for research on arbuscular mycorrhizal fungi in India: problems and prospects. *Current Science* 89, 1699–1709.

Redecker, D. and Raab, P. (2006) Phylogeny of Glomeromycota (arbuscular mycorrhizal fungi): recent developments and new gene markers. *Mycologia* 98, 885–895.

Redecker, D., Hijri, I. and Wiemken, A. (2003) Molecular identification of arbuscular mycorrhizal fungi in roots: Perspectives and problems. *Folia Geobotanica* 38, 113–124.

Reeves, D.W. (1997) The role of organic matter in maintaining soil quality in continuous cropping systems. *Soil & Tillage Research* 43, 131–167.

Regaieg, H., Ciancio, A., Raouani, N.H. and Rosso, L. (2011) Detection and biocontrol potential of *Verticillium leptobactrum* parasitizing *Meloidogyne* spp. *World Journal of Microbiology & Biotechnology* 27, 1615–1623.

Reimann, S., Hauschild, R., Hildebrandt, U. and Sikora, R.A. (2008) Interrelationships between *Rhizobium etli* G12 and *Glomus intraradices* and multitrophic effects in the biological control of the root-knot nematode *Meloidogyne incognita* on tomato. *Journal of Plant Diseases & Protection* 115, 108–113.

Ren, L., Lou, Y., Zhang, N., Zhu, X., Hao, W., Sun, S., Shen, Q. and Xu, G. (2013) Role of arbuscular mycorrhizal network in carbon and phosphorus transfer between plants. *Biology & Fertility of Soils* 49, 3–11.

Renčo, M. (2013) Organic amendments of soil as useful tools of plant parasitic nematodes control. *Helminthologia* 50, 3–14.

Rhoades, H.L. and Linford, M.B. (1959) Control of *Pythium* root rot by the nematode *Aphelenchus avenae*. *Plant Disease Reporter* 43, 323–328.

Richardson, A.E. (2001) Prospects for using soil microorganisms to improve the acquisition of phosphorus by plants. *Australian Journal of Plant Physiology* 28, 897–906.

Riley, I.T., Hou, S. and Chen, S. (2010) Crop rotational and spatial determinants of variation in *Heterodera avenae* (cereal cyst nematode) population density at village scale in spring cereals grown at high altitude on the Tibetan Plateau, Qinghai, China. *Australasian Plant Pathology* 39, 424–430.

Rimé, D., Nazaret, S., Gourbière, F., Cadet, P. and Moënne-Loccoz, Y. (2003) Comparison of sandy soils suppressive and conducive to ectoparasitic nematode damage on sugarcane. *Phytopathology* 93, 1437–1444.

Ristaino, J.B. and Thomas, W. (1997) Agriculture, methyl bromide and the ozone hole. Can we fill the gaps? *Plant Disease* 81, 964–977.

Ritz, K. and Young, I.M. (2004) Interactions between soil structure and fungi. *Mycologist* 18, 52–59.

Ritz, K., Black, H.I.J., Campbell, C.D., Harris, J.A. and Wood, C. (2009) Selecting biological indicators for monitoring soils: a framework for balancing scientific and technical opinion to assist policy development. *Ecological Indicators* 9, 1212–1221.

Robb, E.J. and Barron, G.L. (1982) Nature's ballistic missile. *Science* 218, 1221–1222.

Roberts, P.A. (1993) The future of Nematology: integration of new and improved management strategies. *Journal of Nematology* 25, 383–394.

Robinson, A.F. and Carter, W.W. (1986) Effects of cyanide ion and hypoxia on the volumes of second-stage juveniles of *Meloidogyne incognita* in polyethylene glycol solutions. *Journal of Nematology* 18, 563–570.

Robinson, A.F., Westphal, A., Overstreet, C., Padgett, G.B., Greenberg, S.M., Wheeler, T.A. and Stetina, S.R. (2008) Detection of suppressiveness against *Rotylenchulus reniformis* in soil from cotton (*Gossypium hirsutum*) fields in Texas and Louisiana. *Journal of Nematology* 40, 35–38.

Rodríguez, H. and Fraga, R. (1999) Phosphate solubilizing bacteria and their role in plant growth promotion. *Biotechnology Advances* 17, 319–339.

Rodriguez, R.J., White Jr, J.F., Arnold, A.E. and Redman, R.S. (2009) Fungal endophytes: diversity and functional roles. *New Phytologist* 182, 314–330.

Rodriguez-Kábana, R. (1986) Organic and inorganic nitrogen amendments to soil as nematode suppressants. *Journal of Nematology* 18, 129–135.

Rodriguez-Kábana, R. and King, P.S. (1980) Use of mixtures of urea and blackstrap molasses for control of root-knot nematodes in soil. *Nematropica* 10, 38–44.

Rodriguez-Kábana, R., King, P.S. and Pope, M.H. (1981) Combinations of anyhdrous ammonia and ethylene dibromide for control of nematodes parasitic of soybeans. *Nematropica* 11, 27–41.

Rodriguez-Kábana, R., Shelby, R.A., King, P.S. and Pope, M.H. (1982) Combinations of anhydrous ammonia and 1,3-dichloropropenes for control of root-knot nematodes in soybean. *Nematropica* 12, 61–69.

Rodriguez-Kábana, R., Morgan-Jones, G. and Gintis, B.O. (1984) Effect of chitin amendments to soil on *Heterodera glycines*, microbial populations, and colonization of cysts by fungi. *Nematropica* 14, 10–25.

Rodriguez-Kábana, R., Morgan-Jones, G. and Chet, I. (1987) Biological control of nematodes: soil amendments and microbial antagonists. *Plant & Soil* 100, 237–247.

Rodríguez-Kábana, R., Boube, D. and Young, R.W. (1989) Chitinous materials from blue crab for control of root-knot nematode. I. Effect of urea and enzymatic studies. *Nematropica* 19, 53–74.

Rodríguez-Kábana, R., Kloepper, J.W., Robertson, D.G. and Wells, L.W. (1992) Velvetbean for the management of root-knot and southern blight of peanut. *Nematropica* 22, 75–80.

Rodriguez-Kábana, R., Kokalis-Burelle, N., Kiewnick, S., Schuster, R.-P. and Sikora, R.A. (1994) Alginate films for delivery of root-knot nematode inoculum and evaluation of microbial interactions. *Plant & Soil* 164, 147–154.

Roesch, L.F.W., Fulthorpe, R.R., Riva, A., Casella, G., Hadwin, A.K.M., Kent, A.D., Daroub, S.H., Camargo, F.A.O., Farmerie, W.G. and Triplett, E.W. (2007) Pyrosequencing enumerates and contrasts soil microbial diversity. *International Society for Microbial Ecology Journal* 1, 283–290.

Roger-Estrade, J., Anger, C., Bertrand, M. and Richard, G. (2010) Tillage and soil ecology: partners for sustainable agriculture. *Soil & Tillage Research* 111, 33–40.

Roget, D.K. (1995) Decline in root rot (*Rhizoctonia solani* AG8) in wheat in a tillage and rotation experiment at Avon, South Australia. *Australian Journal of Experimental Agriculture* 35, 1009–1013.

Roget, D.K., Coppi, J.A., Herdina, and Gupta, V.V.S.R. (1999) Assessment of suppression to *Rhizoctonia solani* in a range of soils across SE Australia. In: Magarey, R.C. (ed.) *Proceedings of the First Australasian Soilborne Diseases Symposium*, BSES, Brisbane, QLD, Australia, pp. 129–130.

Roper, M.M., Gupta, V.V.S.R. and Murphy, D.V. (2010) Tillage practices altered labile soil organic carbon and microbial function without affecting crop yields. *Australian Journal of Soil Research* 48, 274–285.

Rosenblueth, M. and Martínez-Romero, E. (2006) Bacterial endophytes and their interactions with hosts. *Molecular Plant Microbe Interactions* 19, 827–837.

Rosenheim, J.A., Kaya, H.K., Ehler, L.E., Marois, J.J. and Jaffee, B.A. (1995) Intraguild predation among biological control agents: theory and evidence. *Biological Control* 5, 303–335.

Rosenzweig, N., Tiedje, J.M., Quensen III, J.F., Meng, Q. and Hao, J.J. (2012) Microbial communities associated with potato common scab-suppressive soil determined by pyrosequencing analyses. *Plant Disease*, 96, 718–725.

Rotenberg, D., Wells, A.J., Chapman, E.J., Whitfield, A.E., Goodman, R.M. and Cooperband, L.R. (2007a) Soil properties associated with organic matter-mediated suppression of bean root rot in field soil amended with fresh and composted paper mill residuals. *Soil Biology & Biochemistry* 39, 2936–2948.

Rotenberg, D., Joshi, R., Benitez, M.-S., Chapin, L.G., Camp, A., Zumpetta, C., Osborne, A., Dick, W.A. and McSpadden Gardener, B.B. (2007b) Farm management effects on rhizosphere colonization by native populations of 2,4-diacetylphloroglucinol-producing *Pseudomonas* spp. and their contributions to crop health. *Phytopathology* 97, 756–766.

Roubstova, T., López-Pérez, J.-A., Edwards, S. and Ploeg, A. (2007) Effect of broccoli (*Brassica oleracea*) tissue, incorporated at different depths in a soil column, on *Meloidogyne incognita*. *Journal of Nematology* 39, 111–117.

Rovira, A.D. (1979) Biology of the root-soil interface. In: Harley, J.L. and Russell, R.S. (eds) *The Soil–Root Interface*. Academic Press, London, pp.145–160.

Rovira, A.D. (1991) Rhizosphere research: 85 years of progress and frustration. In: Keister, D.L. and Cregan, P.B. (eds) *The Rhizosphere and Plant Growth*. Kluwer Academic Publishers, Dordrecht, the Netherlands, pp. 3–13.

Rovira, A.D., Newman, E.I., Bowen, H.J. and Campbell, R. (1974) Quantitative assessment of the rhizoplane microflora by direct microscopy. *Soil Biology & Biochemistry* 6, 211–216.

Rowley, M.A., Ransom, C.V., Reeve, J.R. and Black, B.L. (2011) Mulch and organic herbicide combinations for in-row orchard weed suppression. *International Journal of Fruit Science* 11, 316–331.

Roy, K.W., Patel, M.V. and Baird, R.E. (2000) Colonization of *Heterodera glycines* cysts by *Fusarium solani* form A, the cause of sudden death syndrome, and other fusaria in soybean fields in the midwestern and southern United States. *Phytoprotection* 81, 57–67.

Royal Society (2009) Reaping the benefits: science and the sustainable intensification of global agriculture. Policy document 11/09. The Royal Society, London. ISBN: 978–0-85403–784–1.

Rubner, A. (1996) Revision of predacious hyphomycetes in the *Dactylella-Monacrosporium* complex. *Studies in Mycology* 39, 1–134.

Rucker, C.J. and Zachariah, K. (1987) The influence of bacteria on trap induction in predacious hyphomycetes. *Canadian Journal of Botany* 65, 1160–1162.

Rumbos, C., Mendoza, A., Sikora, R. and Kiewnick, S. (2008) Persistence of the nematophagous fungus *Paecilomyces lilacinus* strain 251 in soil under controlled conditions. *Biocontrol Science & Technology* 18, 1041–1050.

Rumbos, C.I. and Kiewnick, S. (2006) Effect of plant species on persistence of *Paecilomyces lilacinus* strain 252 in soil and on root colonization by the fungus. *Plant & Soil* 283, 25–31.

Russell, C.C. (1986) The feeding habits of a species of *Mesodorylaimus*. *Journal of Nematology* 18, 641.

Ryan, M.H. and Kirkegaard, J.A. (2012) The agronomic relevance of arbuscular mycorrhizae in the fertility of Australian extensive cropping systems. *Agriculture, Ecosystems & Environment* 163, 37–53.

Ryan, M.H., Kirkegaard, J.A. and Angus, J.F. (2006) *Brassica* crops stimulate soil mineral N accumulation. *Australian Journal of Soil Research* 44, 367–377.

Ryan, P.R., Delhaize, E. and Jones, D.L. (2001) Function and mechanism of organic anion exudation from plant roots. *Annual Review of Plant Physiology & Plant Molecular Biology* 52, 527–560.

Ryan, R.P., Germaine, K., Franks, A., Ryan, D.J. and Dowling, D.N. (2008) Bacterial endophytes: recent developments and applications. *FEMS Microbiology Letters* 278, 1–9.

Sagüés, M.F., Fusé, L.A., Fernández, A.S., Iglesias, L.E., Moreno, F.C. and Saumell, C.A. (2011) Efficacy of an energy block containing *Duddingtonia flagrans* in the control of gastrointestinal nematodes of sheep. *Parasitology Research* 109, 707–713.

Saharan, B.S. and Nehra, V. (2011) Plant growth promoting rhizobacteria: a critical review. *Life Sciences & Medicine Research* 21, 1–30.

Sahebani, N. and Hadavi, N. (2008) Biological control of the root-knot nematode *Meloidogyne javanica* by *Trichoderma harzianum*. *Soil Biology & Biochemistry* 40, 2016–2020.

Saikawa, M. (2011) Ultrastructural studies on zygomycotan fungi in the Zoopagaceae and Cochlonemataceae. *Mycoscience* 52, 83–90.

Saikawa, M. and Wada, N. (1986) Adhesive knobs in *Pleurotus ostreatus* (the oyster mushroom), as trapping organs for nematodes. *Transactions of the Mycological Society of Japan* 27, 113–118.

Sánchez Márquez, S., Bills, G.F., Herrero, N. and Zabalgogeazcoa, I. (2012) Non-systemic fungal endophytes of grasses. *Fungal Ecology* 5, 289–297.

Sánchez-Bayo, F. (2011) Impacts of agricultural pesticides on terrestrial ecosystems. In: Sánchez-Bayo, F., van den Brink, P.J. and Mann, R.M. (eds) *Ecological Impacts of Toxic Chemicals*. Bentham Science Publishers, pp. 63–87.

Sánchez-Moreno, S. and Ferris, H. (2007) Suppressive service of the soil food web: effects of environmental management. *Agriculture, Ecosystems & Environment* 119, 75–87.

Sánchez-Moreno, S., Minoshima, H., Ferris, H. and Jackson, L.E. (2006) Linking soil properties and nematode community composition: effects of soil management on soil food webs. *Nematology* 8, 703–715.

Sánchez-Moreno, S., Ferris, H. and Guil, N. (2008) Role of tardigrades in the suppressive service of a soil food web. *Agriculture, Ecosystems & Environment* 124, 187–192.

Sánchez-Moreno, S., Nicola, N.L., Ferris, H. and Zalom, F.G. (2009) Effects of agricultural management on nematode-mite assemblages: Soil food web indices as predictors of mite community composition. *Applied Soil Ecology* 41, 107–117.

Sánchez-Moreno, S., Jiménez, L., Alonso-Prados, J.L. and García-Baudín, J.M. (2010) Nematodes as indicators of fumigant effects on soil food webs in strawberry crops in Southern Spain. *Ecological Indicators* 10, 148–156.

Sanguin, H., Sarniguet, A., Gazengel, K., Moënne-Loccoz, Y. and Grundmann, G.L. (2009) Rhizosphere bacterial communities associated with disease suppressiveness stages of take-all decline in wheat monoculture. *New Phytologist* 184, 694–707.

Santos, P.F., Phillips, J. and Whitford, W.G. (1981) The role of mites and nematodes in early stages of buried litter decomposition in a desert. *Ecology* 62, 664–669.

Santurio, J.M., Zanette, R.A., Da Silva, A.S., Fanfa, V.R., Farret, M.H., Ragagnin, L., Hecktheuer, P.A. and Monteiro, S.G. (2011) A suitable model for the utilization of *Duddingtonia flagrans* fungus in small-flock-size sheep farms. *Experimental Parasitology* 127, 727–731.

Sanyal, S.K. and De Datta, S.K. (1991) Chemistry of phosphorus transformations in soil. *Advances in Soil Science* 16, 1–120.

Sarathchandra, S.U., Watson, R.N., Cox, N.R., di Menna, M.E., Brown, J.A. and Neville, F.J. (1996) Effects of chitin amendment of soil on microorganisms, nematodes, and growth of white clover (*Trifolium repens* L.) and perennial ryegrass (*Lolium perenne* L.). *Biology & Fertility of Soils* 22, 221–226.

Sarathchandra, S.U., Ghani, A., Yeates, G.W., Burch, G. and Cox, N.R. (2001) Effect of nitrogen and phosphate fertilisers on microbial and nematode diversity in pasture soils. *Soil Biology & Biochemistry* 33, 953–964.

Sasser, J.N. and Freckman, D.W. (1987) A world perspective on nematology: the role of the Society. In: Veech, J.A. and Dickson, D.W. (eds) *Vistas on Nematology*. Society of Nematologists, Hyattsville, Maryland, pp. 7–14.

Sasser, J.N., Kirkpatrick, T.L. and Dybas, R.A. (1982) Efficacy of avermectins for root-knot control in tobacco. *Plant Disease* 66, 691–693.

Saxena, G. (2008) Observations on the occurrence of nematophagous fungi in Scotland. *Applied Soil Ecology* 39, 352–357.

Sayre, R.M. (1993) *Pasteuria*, Metchnikoff, 1888. In: Sonenshein, A.L., Hoch, J.A. and Losick, R. (eds) *Bacillus subtilis and other Gram-Positive Bacteria: Biochemistry, Physiology and Molecular Genetics*. American Society for Microbiology, Washington, DC, pp. 101–111.

Sayre, R.M. and Keeley, L.S. (1969) Factors influencing *Catenaria anguillulae* infections in a free-living and a plant-parasitic nematode. *Nematologica* 15, 492–502.

Sayre, R.M and Powers, E.M. (1966) A predacious soil turbellarian that feeds on free-living and plant-parasitic nematodes. *Nematologica* 12, 619–629.

Sayre, R.M and Starr, M.P. (1985) *Pasteuria penetrans* (ex Thorne, 1940) nom. rev., comb. n., sp. n., a mycelial and endospore-forming bacterium parasitic in plant-parasitic nematodes. *Proceedings of the Helminthological Society of Washington* 52, 149–165.

Sayre, R.M. and Starr, M.P. (1988) Bacterial diseases and antagonisms of nematodes. In: Poinar Jr, G.O. and Jansson, H.-B. (eds) *Diseases of Nematodes*. Vol. 1. CRC Press, Boca Raton, Florida, pp. 69–101.

Sayre, R.M. and Wergin, W.P. (1977) Bacterial parasite of a plant nematode: morphology and ultrastructure. *Journal of Bacteriology* 129, 1091–1101.

Sayre, R.M., Wergin, W.P. and Davies, R.E. (1977) Occurrence in *Monia rectirostris* (Cladocera: Daphnidae) of a parasite morphologically similar to *Pasteuria ramosa* (Metchnikoff, 1888). *Canadian Journal of Microbiology* 23, 1573–1579.

Sayre, R.M., Adams, J.R. and Wergin, W.P. (1979) Bacterial parasite of a cladoceran: Morphology, development *in vivo*, and taxonomic relationships with *Pasteuria ramosa* Metchnikoff 1888. *International Journal of Systematic Bacteriology* 29, 252–262.

Sayre, R.M., Gherna, R.L. and Wergin, W.P. (1983) Morphological and taxonomic reevaluation of *Pasteuria ramosa* Metchnikoff 1888 and "*Bacillus penetrans*" Mankau 1975. *International Journal of Systematic Bacteriology* 33, 636–649.

Sayre, R.M., Starr, M.P., Golden, A.M., Wergin, W.P. and Endo, B.Y. (1988) Comparison of *Pasteuria penetrans* from *Meloidogyne incognita* with a related mycelial and endospore-forming bacterial parasite from *Pratylenchus brachyurus*. *Proceedings of the Helminthological Society of Washington* 55, 28–49.

Sayre, R.M., Wergin, W.P., Nishizawa, T. and Starr, M.P. (1991a) Light and electron microscopical study of a bacterial parasite from the cyst nematode, *Heterodera glycines*. *Journal of the Helminthological Society of Washington* 58, 69–81.

Sayre, R.M., Wergin, W.P., Schmidt, J.M. and Starr, M.P. (1991b) *Pasteuria nishizawae* sp. nov., a mycelial and endospore-forming bacterium parasitic on cyst nematodes of genera *Heterodera* and *Globodera*. *Research in Microbiology* 142, 551–564.

Scala, F., Bonanomi, G., Capodilupo, M., Cennicola, M. and Lorito, M. (2011) Effect of microbial diversity on soil fungistasis, disease suppression and colonization by biological control agents. *Phytopathology* 101, S160.

Schaff, J.E., Mauchline, T.H., Opperman, C.H. and Davies, K.G. (2011) Exploiting genomics to understand the interactions between root-knot nematodes and *Pasteuria penetrans*. In: Davies, K. and Spiegel, Y. (eds) *Biological Control of Plant-Parasitic Nematodes. Building Coherence Between Microbial Ecology and Molecular Mechanisms*. Springer, Dordrecht, the Netherlands, pp. 91–113.

Schardl, C.L. (2010) The Epichloae, symbionts of the grass subfamily Poöideae. *Annals of the Missouri Botanical Garden* 97, 646–665.

Scharroba, A., Dibbern, D., Hünninghaus, Kramer, S., Moll, J., Butenschoen, O., Bonkowski, M., Buscot, F., Kandeler, E., Koller, R., Krüger, D., Lueders, T., Scheu, S. and Ruess, L. (2012) Effects of resource availability and quality on the structure of the micro-food web of an arable soil across depth. *Soil Biology & Biochemistry* 50, 1–11.

Schenk, P.M., Carvalhais, L.C. and Kazan, K. (2012) Unravelling plant-microbe interactions: can multi-species transcriptomics help? *Trends in Biotechnology* 30, 177–184.

Schenk, S. (2001) Molasses soil amendment for crop improvement and nematode management. Hawaii Agriculture Research Center, Vegetable Report 3, HARC, Aiea, Hawaii.

Scheuerell, S.J., Sullivan, D.M. and Mahaffee, W.F. (2005) Suppression of seedling damping-off caused by *Pythium ultimum, P. irregulare*, and *Rhizoctonia solani* in container media amended with a diverse range of Pacific northwest compost sources. *Phytopathology* 95, 306–315.

Schippers, B., Bakker, A.W. and Bakker, P.A.H.M. (1987) Interactions of deleterious and beneficial rhizosphere microorganisms and the effect of cropping practices. *Annual Review of Phytopathology*, 25, 339–358.

Schisler, D.A., Slininger, P.J., Behle, R.W. and Jackson, M.A. (2004) Formulation of *Bacillus* spp. for biological control of plant diseases. *Phytopathology* 94, 1267–1271.

Schmidt, L.M., Preston, J.F., Dickson, D.W., Rice, J.D. and Hewlett, T.E. (2003) Environmental quantification of *Pasteuria penetrans* endospores using *in situ* antigen extraction and immunodetection with a monoclonal antibody. *FEMS Microbiology Ecology* 44, 17–26.

Schmidt, L.M., Hewlett, T.E., Green, A., Simmons, L.J., Kelley, K., Doroh, M. and Stetina, S.R. (2010) Molecular and morphological characterization and biological control capabilities of a *Pasteuria* ssp. parasitizing *Rotylenchulus reniformis*, the reniform nematode. *Journal of Nematology* 42, 207–217.

Schneider, K., Migge, S., Norton, R.A., Scheu, S., Langel, R., Reineking, A. and Maraun, M. (2004) Trophic niche differentiation in soil microarthropods (Oribatida, Acari): evidence from stable isotope ratios ($^{15}N/^{14}N$). *Soil Biology & Biochemistry* 36, 1769–1774.

Scholler, M and Rubner, A. (1994) Predacious activity of the nematode-destroying fungus *Arthrobotrys oligospora* in dependence of the medium composition. *Microbiological Research* 149, 145–149.

Scholler, M., Hagedorn, G. and Rubner, A. (1999) A reevaluation of predatory orbiliaceous fungi. II. A new generic concept. *Sydowia* 51, 89–113.

Schrimsher, D.W., Lawrence, K.S., Castillo, J., Moore, S.R. and Kloepper, J.W. (2011) Effects of *Bacillus firmus* GB-126 on the soybean cyst nematode mobility *in vitro*. *Phytopathology* 101, S161.

Schroth, M.N. and Becker, J.O. (1990) Concepts of ecological and physiological activities of rhizobacteria related to biological control and plant growth promotion. In: Hornby, D. (ed.) *Biological Control of Soilborne Plant Pathogens*. CAB International, Wallingford, UK, pp. 389–414.

Seifert, K.A. and Axelrood, P.E. (1998) *Cylindrocarpon destructans* var. *destructans*. *Canadian Journal of Plant Pathology* 20, 115–117.

Seinhorst, J.W. (1965) The relation between nematode density and damage to plants. *Nematologica* 11, 137–154.

Seiter, S. and Horwarth, W.R. (2004) Strategies for managing soil organic matter to supply plant nutrients. In: Magdoff, F. and Weil, R.R. (eds) *Soil Organic Matter in Sustainable Agriculture*. CRC Press, Boca Raton, Florida, pp. 269–293.

Semenov, A.M., van Bruggen, A.H.C. and Zelenev, V.V. (1999) Moving waves of bacterial populations and total organic carbon along roots of wheat. *Microbial Ecology* 37, 116–128.

Serracin, M., Schuerger, A.C., Dickson, D.W. and Weingartner, D.P. (1997) Temperature-dependent development of *Pasteuria penetrans* in *Meloidogyne arenaria*. *Journal of Nematology* 29, 228–238.

Seyb, A., Xing, L.J., Vyn, T.J. and Westphal, A. (2008) Effect of tillage on population levels of *Heterodera glycines* in a crop sequence of corn and a nematode-susceptible or -resistant cultivar of soybean (Abstr). *Phytopathology* 98, S204.

Shannon, D., Sen, A.M. and Johnson, D.B. (2002) A comparative study of the microbiology of soils managed under organic and conventional regimes. *Soil Use & Management* 18, 274–283.

Shapiro-Ilan, D.I., Nyczepir, A.P. and Lewis, E.E. (2006) Entomopathogenic nematodes and bacteria applications for control of the pecan root-knot nematode, *Meloidogyne partityla*, in the greenhouse. *Journal of Nematology* 38, 449–454.

Sharma, R.D. (1971) Studies on the plant parasitic nematode *Tylenchorhynchus dubius*. *Mededelingen Landbouwhogeschool Wageningen* 71, 1–154.

Sharma, R.D. and Stirling, G.R. (1991) *In vivo* mass production systems for *Pasteuria penetrans*. *Nematologica* 37, 483–484.

Sharma, S.B and Davies, K.G. (1996) Characterization of *Pasteuria* isolated from *Heterodera cajani* using morphology, pathology and serology of endospores. *Systematic & Applied Microbiology* 19, 106–112.

Sharon, E., Bar-Eyal, M., Chet, I., Herrera-Estrella, A., Kleifeld, O. and Spiegel, Y. (2001) Biological control of the root-knot nematode *Meloidogyne javanica* by *Trichoderma harzianum*. *Phytopathology* 91, 687–693.

Sharon, E., Chet, I., Viterbo, A., Bar-Eyal, M., Nagan, H., Samuels, G.J. and Spiegel, Y. (2007) Parasitism of *Trichoderma* on *Meloidogyne javanica* and role of the gelatinous matrix. *European Journal of Plant Pathology* 118, 247–258.

Sharon, E., Chet, I. and Spiegel, Y. (2011) *Trichoderma* as a biological control agent. In: Davies, K. and Spiegel, Y. (eds) *Biological Control of Plant-Parasitic Nematodes: Building Coherence Between Microbial Ecology and Molecular Mechanisms*. Springer, Dordrecht, the Netherlands, pp. 183–201.

Shennan, C. (2008) Biotic interactions, ecological knowledge and agriculture. *Philosophical Transactions of the Royal Society B*, 363, 717–739.

Shoresh, M., Harman, G.E. and Mastouri, F. (2010) Induced systemic resistance and plant responses to fungal biocontrol agents. *Annual Review of Phytopathology* 48, 21–43.

Siddiqi, M.R. (2000) *Tylenchida: Parasites of Plants and Insects*. 2nd edn. CAB International, Wallingford, UK.

Siddiqi, M.R., Bilgrami, A.L. and Tabassum, K. (2004) Description and biology of *Mononchoides gaugleri* sp. n. (Nematoda: Diplogasterida). *International Journal of Nematology* 14, 124–129.

Siddiqui, I.A., Atkins, S.D. and Kerry, B.R. (2009) Relationship between saprotrophic growth in soil of different biotypes of *Pochonia chlamydosporia* and the infection of nematode eggs. *Annals of Applied Biology* 155, 131–141.

Siddiqui, Z.A. and Mahmood, I. (1996) Biological control of plant-parasitic nematodes by fungi: a review. *Bioresource Technology* 58, 229–239.

Sikora, R.A. (1992) Management of the antagonistic potential in agricultural ecosystems for the biological control of plant parasitic nematodes. *Annual Review of Phytopathology* 30, 245–270.

Sikora, R.A. (1995) Vesicular-arbuscular mycorrhizae: their significance for biological control of plant-parasitic nematodes. *Biocontrol* 1, 29–33.

Sikora, R.A., Bridge, J. and Starr, J.L. (2005) Management practices: an overview of Integrated Nematode Management technologies. In: Luc, M., Sikora, R.A. and Bridge, J. (eds) *Plant Parasitic Nematodes in Subtropical and Tropical Agriculture*. CAB International, Wallingford, UK, pp. 793–825.

Sikora, R.A., Schäfer, K. and Dababat, A.A. (2007) Modes of action associated with microbially induced *in planta* suppression of plant-parasitic nematodes. *Australasian Plant Pathology* 36, 124–134.

Sikora, R.A., Pocasangre, L., zum Felde, A., Niere, B., Vu, T.T. and Dababat, A.A. (2008) Mutualistic endo-phytic fungi and *in-planta* suppressiveness to plant parasitic nematodes. *Biological Control*, 46, 15–23.

Simard, S.W. and Durall, D.M. (2004) Mycorrhizal networks: a review of their extent, function, and impor-tance. *Canadian Journal of Botany* 82, 1140–1165.

Singh, B.K., Millard, P., Whiteley, A.S. and Murrell, J.C. (2004) Unravelling rhizosphere-microbial interac-tions: opportunities and limitations. *Trends in Microbiology* 12, 386–393.

Singh, K.P., Jaiswal, R.K. and Kumar, N. (2007) *Catenaria anguillulae* Sorokin: a natural biocontrol agent of *Meloidogyne graminicola* causing root knot disease of rice (*Oryza sativa* L.). *World Journal of Microbiology & Biotechnology* 23, 291–294.

Singh, R.S. and Sitaramaiah, K. (1971) Control of root-knot through organic and inorganic amendments of soil. Effect of oil cakes and sawdust. *Indian Journal of Mycology & Plant Pathology* 1, 20–29.

Siviour, T.R. and McLeod, R.W. (1979) Redescription of *Ibipora lolii* (Siviour, 1978) comb. n. (Nematoda: Belonolaimidae) with observations on its host range and pathogenicity. *Nematologica* 25, 487–493.

Slana, L.J. and Sayre, R.M. (1981) A method for measuring incidence of *Bacillus penetrans* spore attachment to the second-stage-larvae of *Meloidogyne* spp. *Journal of Nematology* 13, 461.

Small, R.W. (1987) A review of the prey of predatory soil nematodes. *Pedobiologia* 30, 179–206.

Small, R.W. (1988) Invertebrate predators. In: Poinar Jr, G.O. and Jansson, H.-B. (eds) *Diseases of Nematodes*, Vol. II. CRC Press Inc., Boca Raton, Florida, pp. 73–92.

Small, R.W. and Grootaert, P. (1983) Observations on the predation abilities of some soil dwelling predatory nematodes. *Nematologica* 29, 109–118.

Smelt, J.H. and Leistra, M. (1974) Conversion of metham-sodium to methyl isothiocyanate and basic data on the behaviour of methyl isothiocyanate in soil. *Pesticide Science* 5, 401–407.

Smit, E., Leeflang, P., Glandorf, B., van Elsas, J.D. and Wernars, L. (1999) Analysis of fungal diversity in the wheat rhizosphere by sequencing of cloned PCR-amplified genes encoding 18S rRNA and temperature gradient electrophoresis. *Applied & Environmental Microbiology* 65, 2614–2621.

Smith, G.S. (1987) Interactions of nematodes with mycorrhizal fungi. In: Veech, J.A. and Dickson, D.W. (eds) *Vistas on Nematology*. Society of Nematologists, Hyattsville, Maryland, pp. 292–300.

Smith, J.L., Papendick, R.I., Bezdicek, D.F. and Lynch, J.M. (1993) Soil organic matter dynamics and crop resi-due management. In: Metting Jr, F.B. (ed.) *Soil Microbial Ecology: Applications in Agricultural and Environmental Management*. Marcel Dekker, New York, pp. 65–94.

Smith, M.E. and Jaffee, B.A. (2009) PCR primers with enhanced specificity for nematode-trapping fungi (Orbiliales). *Microbial Ecology* 58, 117–128.

Smith M.L., Smith J.P. and Stirling, G.R. (2011) Integration of minimum tillage, crop rotation and organic amendments into a ginger farming system: impacts on yield and soilborne disease. *Soil & Tillage Research* 114, 108–116.

Smith, R.S., Shiel, R.S., Bardgett, R.D., Millwards, D., Corkhill, P., Rolph, G., Hobbs, P.J. and Peacock, S. (2003) Soil microbial community, fertility, vegetation and diversity as targets in the restoration manage-ment of a meadow grassland. *Journal of Applied Ecology* 40, 51–64.

Smith, S.E. and Read, D.J. (2008) *Mycorrhizal Symbiosis*. Academic Press, New York.

Smith, S.E. and Smith, F.A. (2012) Fresh perspectives on the roles of arbuscular mycorrhizal fungi in plant nutrition and growth. *Mycologia* 104, 1–13.

Smitley, D.R., Warner, F.W. and Bird, G.W. (1992) Influence of irrigation and *Heterorhabditis bacteriophora* on plant-parasitic nematodes in turf. *Journal of Nematology* 24, 637–641.

Sneh, B., Humble, S.J. and Lockwood, J.L. (1977) Parasitism of oospores of *Phytophthora megasperma* var. *sojae*, *P. cactorum*, *Pythium* sp., and *Aphanomyces euteiches* in soil by oomycetes, chytridiomycetes, hyphomycetes, actinomycetes and bacteria. *Phytopathology* 67, 622–628.

Soder K.J. and Holden, L.A. (2005) Review: Use of nematode-trapping fungi as a biological control in grazing livestock. *Professional Animal Scientist* 21, 30–37.

Sojka, R.E. and Upchurch, D.R. (1999) Reservations regarding the soil quality concept. *Soil Science Society of America Journal* 63, 1039–1054.

Sojka, R.E., Upchurch, D.R. and Borlaug, N.E. (2003) Quality soil management or soil quality management: performance versus semantics. *Advances in Agronomy* 79, 1–68.

Somasekhar, N., Grewal, P.S., de Nardo, E.A.B. and Stinner, B.R. (2002) Non-target effects of entomopatho-genic nematodes on the soil nematode community. *Journal of Applied Ecology* 39, 735–744.

Somasundaram, S., Bonkowski, M. and Iijima, M. (2008) Functional role of mucilage-border cells: a complex facilitating protozoan effects on plant growth. *Plant Production Science* 11, 344–351.

Sørensen, J., Nicolaisen, M.H., Ron, E. and Simonet, P. (2009) Molecular tools in rhizosphere microbiology – from single-cell to whole-community analysis. *Plant & Soil* 321, 483–512.

Sorribas, F.J., Verdejo-Lucas, S., Forner, J.B., Alcaide, A., Pons, J. and Ornat, C. (2000) Seasonality of *Tylenchulus semipenetrans* Cobb and *Pasteuria* sp. in citrus orchards in Spain. *Journal of Nematology* 32, 622–632.

Spaull, V.W. (1984) Observations on *Bacillus penetrans* infecting *Meloidogyne* in sugar cane fields in South Africa. *Revue de Nématologie* 7, 277–282.

Spiegel, Y. and Chet, I. (1998) Evaluation of *Trichoderma* spp. as a biocontrol agent against soilborne fungi and plant-parasitic nematodes in Israel. *Integrated Pest Management Reviews* 3, 169–175.

Spiegel, Y., Chet, I. and Cohn, E. (1987) Use of chitin for controlling plant parasitic nematodes. II. Mode of action. *Plant and Soil* 98, 337–345.

Spiegel, Y., Cohn, E., Galper, S., Sharon, E. and Chet, I. (1991) Evaluation of a newly isolated bacterium, *Pseudomonas chitinolytica* sp. nov., for controlling the root-knot nematode *Meloidogyne javanica*. *Biocontrol Science & Technology* 1, 115–125.

Srinivasan, A. (ed.) (2006) *Handbook of Precision Agriculture - Principles and Applications*. The Haworth Press Inc., New York.

Starr, M.P and Sayre, R.M. (1988) *Pasteuria thornei* sp. nov. and *Pasteuria penetrans sensu stricto* emend., mycelial and endospore-forming bacteria parasitic, respectively, on plant-parasitic nematodes of the genera *Pratylenchus* and *Meloidogyne*. *Annales de l'Institut Pasteur/Microbiologie* 139, 11–31.

Stephens, P.M., O'Sullivan, M. and O'Gara, F. (1987) Effect of bacteriophage on colonization of sugarbeet roots by fluorescent *Pseudomonas* spp. *Applied & Environmental Microbiology* 53, 1164–1167.

Stewart, A., Ohkura, M. and McLean, K.L. (2010) Targeted screening for microbial bioactivity. In: Zydenbos, S.M. and Jackson, T.A. (eds) *Microbial Products: Exploiting Microbial Diversity for Sustainable Plant Production*. New Zealand Plant Protection Society Inc., pp. 11–19.

Stewart, T.M., Mercer, C.F. and Grant, J.L. (1993) Development of *Meloidogyne naasi* on endophyte-infected and endophyte-free perennial ryegrass. *Australasian Plant Pahology* 22, 40–41.

Stirling, G.R. (1979a) Techniques for detecting *Dactylella oviparasitica* and evaluating its significance in field soils. *Journal of Nematology* 11, 99–100.

Stirling, G.R. (1979b) Effect of temperature on parasitism of *Meloidogyne incognita* eggs by *Dactylella oviparasitica*. *Nematologica* 25, 104–110.

Stirling, G.R. (1981) Effect of temperature on infection of *Meloidogyne javanica* by *Bacillus penetrans*. *Nematologica* 27, 458–462.

Stirling, G.R. (1984) Biological control of *Meloidogyne javanica* with *Bacillus penetrans*. *Phytopathology* 74, 55–60.

Stirling, G.R. (1985) Host specificity of *Pasteuria penetrans* within the genus *Meloidogyne*. *Nematologica* 31, 203–209.

Stirling, G.R. (1988) Biological control of plant parasitic nematodes. In: Poinar Jr, G.O. and Jansson, H.-B. (eds) *Diseases of Nematodes, Vol. II*. CRC Press Inc., Boca Raton, Florida, pp. 93–139.

Stirling, G.R. (1989) Organic amendments for control of root-knot nematode (*Meloidogyne incognita*) on ginger. *Australasian Plant Pathology* 18, 39–44.

Stirling, G.R. (1991) *Biological Control of Plant-parasitic Nematodes*. CAB International, Wallingford, UK.

Stirling, G.R. (1999) Increasing the adoption of sustainable, integrated management strategies for soilborne diseases of high-value annual crops. *Australasian Plant Pathology* 28, 72–79.

Stirling, G.R. (2008) The impact of farming systems on soil biology and soilborne diseases: examples from the Australian sugar and vegetable industries – the case for better integration of sugarcane and vegetable production and implications for future research. *Australasian Plant Pathology* 37, 1–18.

Stirling, G.R. (2009) An improved sugarcane farming system enhances suppression of root-knot nematode, but what organisms are responsible? Program handbook, Fifth Australasian Soilborne Diseases Symposium, Thredbo, NSW, Australia, pp. 80–82.

Stirling, G.R. (2011a) Biological control of plant-parasitic nematodes: an ecological perspective, a review of progress and opportunities for further research. In: Davies, K. and Spiegel, Y. (eds) *Biological Control of Plant-parasitic Nematodes. Building Coherence between Microbial Ecology and Molecular Mechanisms*. Springer, Dordrecht, the Netherlands, pp. 1–38.

Stirling, G.R. (2011b) Suppressive biological factors influence populations of root lesion nematode (*Pratylenchus thornei*) on wheat in vertosols from the northern grain-growing region of Australia. *Australasian Plant Pathology* 40, 416–429. (Erratum: DOI 10.1007/s13313–012–0141–7).

Stirling, G.R. (2013) Integration of organic amendments, crop rotation, residue retention and minimum tillage into a subtropical vegetable farming system enhances suppressiveness to root-knot nematode (*Meloidogyne incognita*). *Australasian Plant Pathology* 42, 625–637.

Stirling, G.R. and Eden, L.M. (2008) The impact of organic amendments, mulching and tillage on plant nutrition, Pythium root rot, root-knot nematode and other pests and diseases of capsicum in a subtropical environment, and implications for the development of more sustainable vegetable farming systems. *Australasian Plant Pathology* 37, 123–131.

Stirling, G.R. and Kerry, B.R. (1983) Antagonists of the cereal cyst nematode *Heterodera avenae* Woll. in Australian soils. *Australian Journal of Experimental Agriculture & Animal Husbandry* 23, 318–324.

Stirling, G.R. and Lodge, G.M. (2005) A survey of Australian temperate pastures in summer and winter rainfall zones: soil nematodes, chemical and biochemical properties. *Australian Journal of Soil Research* 43, 887–904.

Stirling, G.R. and Mani, A. (1995) The activity of nematode-trapping fungi following their encapsulation in alginate. *Nematologica* 41, 240–250.

Stirling, G.R. and Mankau, R. (1978a) *Dactylella oviparasitica*, a new fungal parasite of *Meloidogyne* eggs. *Mycologia* 70, 774–783.

Stirling, G.R. and Mankau, R. (1978b) Parasitism of *Meloidogyne* eggs by a new fungal parasite. *Journal of Nematology* 10, 236–240.

Stirling, G.R. and Mankau, R. (1979) Mode of parasitism of *Meloidogyne* and other nematode eggs by *Dactylella oviparasitica*. *Journal of Nematology* 11, 282–288.

Stirling, G.R. and Nikulin, A. (1998) Crop rotation, organic amendments and nematicides for control of root-knot nematodes (*Meloidogyne incognita*) on ginger. *Australasian Plant Pathology* 27, 234–243.

Stirling, G.R. and Pattison, A.B. (2008) Beyond chemical dependency for managing plant-parasitic nematodes: examples from the banana, pineapple and vegetable industries of tropical and subtropical Australia. *Australasian Plant Pathology* 37, 254–267.

Stirling, G.R. and Smith, L.J. (1998) Field tests of formulated products containing either *Verticillium chlamydosporium* or *Arthrobotrys dactyloides* for biological control of root-knot nematode. *Biological Control* 11, 231–239.

Stirling, G.R. and Stirling, A.M. (2003) The potential of *Brassica* green manure crops for controlling root-knot nematode (*Meloidogyne javanica*) in a subtropical environment. *Australian Journal of Experimental Agriculture* 43, 623–630.

Stirling, G.R. and Wachtel, M.F. (1980) Mass production of *Bacillus penetrans* for biological control of root-knot nematodes. *Nematologica* 26, 308–312.

Stirling, G.R. and West, L.M. (1991) Fungal parasites of root-knot nematode eggs from tropical and subtropical regions of Australia. *Australasian Plant Pathology* 20, 149–154.

Stirling, G.R. and White, A.M. (1982) Distribution of a parasite of root-knot nematodes in South Australian vineyards. *Plant Disease* 66, 52–53.

Stirling, G.R., McKenry, M.V. and Mankau, R. (1979) Biological control of root-knot nematodes (*Meloidogyne* spp.) on peach. *Phytopathology* 69, 806–809.

Stirling, G.R., Bird, A.F. and Cakurs, A.B. (1986) Attachment of *Pasteuria penetrans* to the cuticles of root-knot nematodes. *Revue de Nématologie* 9, 251–260.

Stirling, G.R., Sharma, R.D. and Perry, J. (1990) Attachment of *Pasteuria penetrans* spores to the root-knot nematode *Meloidogyne javanica* and its effects on infectivity. *Nematologica* 36, 246–252.

Stirling, G.R., Dullahide, S.R. and Nikulin, A. (1995) Management of lesion nematodes (*Pratylenchus jordanensis*) on replanted apple trees. *Australian Journal of Experimental Agriculture* 35, 247–258.

Stirling, G.R., West, L.M., Fanton, J.A. and Stanton, J.M. (1996) Crops and their resistance to root-knot nematodes (*Meloidogyne* spp.). Queensland Department of Primary Industries, Information Series QI96085, Brisbane, QLD, Australia.

Stirling, G.R., Licastro, K.A., West, L.M. and Smith, L.J. (1998a) Development of commercially acceptable formulations of the nematophagous fungus *Verticillium chlamydosporium*. *Biological Control* 11, 217–223.

Stirling, G.R., Smith, L.J., Licastro, K.A. and Eden, L.M. (1998b) Control of root-knot nematode with formulations of the nematode-trapping fungus *Arthrobotrys dactyloides*. *Biological Control* 11, 224–230.

Stirling, G.R., Wilson, E.J., Stirling, A.M., Pankhurst, C.E., Moody, P.W. and Bell, M.J. (2003) Organic amendments enhance biological suppression of plant-parasitic nematodes in sugarcane soils. *Proceedings Australian Society of Sugarcane Technologists* 25: CD ROM

Stirling, G.R., Wilson, E.J., Stirling, A.M., Pankhurst, C.E., Moody, P.W., Bell, M.J. and Halpin, N. (2005) Amendments of sugarcane trash induce suppressiveness to plant-parasitic nematodes in sugarcane soils. *Australasian Plant Pathology* 34, 203–211.

Stirling, G.R., Berthelsen, J.E., Garside, A.L. and James, A.T. (2006) The reaction of soybean and other legume crops to root-knot nematodes (*Meloidogyne* spp.) and implications for crops grown in rotation with sugarcane. *Australasian Plant Pathology* 35, 707–714.

Stirling, G.R., Moody, P. and Stirling, A.M. (2007) The potential of nematodes as an indicator of the biological status of sugarcane soils. *Proceedings of the Australian Society of Sugar Cane Technologists* 29, 339–351.

Stirling, G.R., Moody, P.W. and Stirling, A.M. (2010) The impact of an improved sugarcane farming system on chemical, biochemical and biological properties associated with soil health. *Applied Soil Ecology* 46, 470–477.

Stirling, G.R., Halpin, N.V. and Bell, M.J. (2011a) A surface mulch of crop residues enhances suppressiveness to plant-parasitic nematodes in sugarcane soils. *Nematropica* 41, 109–121.

Stirling, G.R., Rames, E., Stirling, A.M. and Hamill, S. (2011b) Factors associated with the suppressiveness of sugarcane soils to plant-parasitic nematodes. *Journal of Nematology* 43, 135–148.

Stirling, G.R., Halpin, N.V., Bell, M.J. and Moody P.W. (2011c) Impact of tillage and residues from rotation crops on the nematode community in soil and surface mulch during the following sugarcane crop. *International Sugar Journal* 113, 56–64.

Stirling, G.R., Smith, M.K., Smith, J.P., Stirling, A.M. and Hamill, S. D. (2012) Organic inputs, tillage and rotation practices influence soil health and suppressiveness to soilborne pests and pathogens of ginger. *Australasian Plant Pathology* 41, 99–112.

Stirling, G.R., Stirling, A.M., Giblin-Davis, R.M., Ye, W., Porazinska, D.L., Nobbs, J.M. and Johnston, K.J. (2013) Distribution of southern sting nematode, *Ibipora lolii* (Nematoda: Belonolaimidae), on turfgrass in Australia and its taxonomic relationship to other belonolaimids. *Nematology* 15, 401–415.

Stockdale, E.A., Shepherd, M.A., Fortune, S. and Cuttle, S.P. (2002) Soil fertility in organic farming systems – fundamentally different? *Soil Use & Management* 18, 301–308.

Stone, A.G., Scheuerell, S.J. and Darby, H.M. (2004) Suppression of soilborne diseases in field agricultural systems: Organic matter management, cover cropping, and other cultural practices. In: Magdoff, F. and Weil, R.R. (eds) *Soil Organic Matter in Sustainable Agriculture*. CRC Press, Boca Raton, Florida, pp. 131–177.

Stopes, C., Lord, E.I., Phillips, L. and Woodward, L. (2002) Nitrate leaching from organic farms and conventional farms following best practice. *Soil Use & Management* 18, 256–263.

Stotzky, G. (2004) Persistence and biological activity in soil of the insecticidal proteins from *Bacillus thuringiensis*, especially from transgenic plants. *Plant & Soil* 266, 77–89.

Sturhan, D. (1988) New host and geographical records of nematode-parasitic bacteria of the *Pasteuria penetrans* group. *Nematologica* 34, 350–356.

Sturhan, D., Winkelheide, R., Sayre, R.M. and Wergin, W.P. (1994) Light and electron microscopical studies of the life cycle and developmental stages of a *Pasteuria* isolate parasitizing the pea cyst nematode, *Heterodera goettingiana*. *Fundamental & Applied Nematology* 17, 29–42.

Sturhan, D., Shutova, T.S., Akimov, V.N. and Subbotin, S.A. (2005) Occurrence, hosts, morphology, and molecular characterisation of *Pasteuria* bacteria parasitic in nematodes of the family Plectidae. *Journal of Invertebrate Pathology* 88, 17–26.

Sturz, A.V. (1995) The role of endophytic bacteria during seed piece decay and potato tuberization. *Plant & Soil* 175, 257–263.

Sturz, A.V., Carter, M.R. and Johnston, H.W. (1997) A review of plant disease, pathogen interactions and microbial antagonism under conservation tillage in temperate humid agriculture. *Soil & Tillage Research* 41, 169–189.

Sturz, A.V., Christie, B.R. and Nowak, J. (2000) Bacterial endophytes: Potential role in developing sustainable systems of crop production. *Critical Reviews in Plant Sciences* 19, 1–30.

Stutz, E., Kahr, G. and Defago, G. (1989) Clays involved in suppression of tobacco black root rot by a strain of *Pseudomonas fluorescens*. *Soil Biology & Biochemistry* 21, 361–366.

Subrahmanyam, P., Reddy, M.N. and Rao, A.S. (1983) Exudation of certain organic compounds from seeds of groundnut. *Seed Science & Technology* 11, 267–272.

Sun, M. and Liu, X. (2006) Carbon requirements of some nematophagous, entomopathogenic and mycoparasitic hyphomycetes as fungal biocontrol agents. *Mycopathologia* 161, 295–305.

Sun, M.-H., Gao, L., Shi, Y.-X., Li, B.-J. and Liu, X.-Z. (2006) Fungi and actinomycetes associated with *Meloidogyne* spp. eggs and females in China and their biocontrol potential. *Journal of Invertebrate Pathology* 93, 22–28.

Sundin, P., Valeur, A., Olsson, S. and Odham, G. (1990) Interactions between bacteria-feeding nematodes and bacteria in the rape rhizosphere: effects on root exudation and distribution of bacteria. *FEMS Microbiology Letters* 73, 13–22.

Sung, G.-H., Spatafora, J.W., Zare, R., Hodge, K.T. and Gams, W. (2001) A revision of *Verticillium* sect. Prostrata. II. Phylogenetic analyses of SSU and LSU nuclear rDNA sequences from anamorphs and teleomorphs of the Clavicipitaceae. *Nova Hedwigia* 72, 311–328.

Surette, M.A., Sturz, A.V., Lada, R.R. and Nowak, J. (2003) Bacterial endophytes in processing carrots (*Daucus carota* L. var. *sativus*): their localization, population density, biodiversity and their effects on plant growth. *Plant & Soil* 253, 381–390.

Suryanarayanan, T.S., Thirunavukkarasu, N., Govindarajulu, M.B., Sasse, F., Jansen, R. and Murali, T.S. (2009) Fungal endophytes and bioprospecting. *Fungal Biology Reviews* 23, 9–19.

Sutherland, E.D. and Lockwood, J.L. (1984) Hyperparasitism of oospores of some Peronosporales by *Actinoplanes missouriensis* and *Humicola fuscoatra* and other Actinomycetes and fungi. *Canadian Journal of Plant Pathology* 6, 139–145.

Sutherland, E.D. and Papavizas, G.C. (1991) Evaluation of oospore hyperparasites for the control of Phytophthora crown rot of pepper. *Journal of Phytopathology* 131, 33–39.

Swe, A., Jeewon, R., Pointing, S.B. and Hyde, K.D. (2009) Diversity and abundance of nematode-trapping fungi from decaying litter in terrestrial, freshwater and mangrove habitats. *Biodiversity & Conservation* 18, 1695–1714.

Sylvia, D.M., Fuhrmann, J.J., Hartel, P.G. and Zuberer, D.A. (2005) *Principles and Applications of Soil Microbiology*. Pearson Prentice Hall, Upper Saddle River, New Jersey.

Symondson, W.O.C. (2002) Molecular identification of prey in predator diets. *Molecular Ecology* 11, 627–641.

Symondson, W.O.C., Sunderland, K.D. and Greenstone, M.H. (2002) Can generalist predators be effective biocontrol agents? *Annual Review of Entomology* 47, 561–594.

Tabarant, P., Villenave, C., Riséde, J.-M., Roger-Estrade, J. and Dorel, M. (2011a) Effects of organic amendments on plant-parasitic nematode populations, root damage, and banana plant growth. *Biology & Fertility of Soils* 47, 341–347.

Tabarant, P., Villenave, C., Risède J.-M., Roger-Estrade, J. and Dorel, M. (2011b) Effects of four organic amendments on banana parasitic nematodes and soil nematode communities. *Applied Soil Ecology* 49, 59–67.

Talavera, M., Mizukubo, T., Ito, K. and Aiba, S. (2002) Effect of spore inoculum and agricultural practices on the vertical distribution of the biocontrol plant-growth-promoting bacterium *Pasteuria penetrans* and growth of *Meloidogyne incognita*-infected tomato. *Biology & Fertility of Soils* 35, 435–440.

Tan, K.H. (2009) *Environmental Soil Science*. 3rd edn. CRC Press, Boca Raton, Florida.

Tate, R.L. (2000) *Soil Microbiology*. 2nd edn. John Wiley and Sons, New York.

Tedford, E.C., Jaffee, B.A. and Muldoon, A.E. (1992) Effect of soil moisture and texture on transmission of the nematophagous fungus *Hirsutella rhossiliensis* to cyst and root-knot nematodes. *Phytopathology* 82, 1002–1007.

Tedford, E.C., Jaffee, B.A., Muldoon, A.E., Anderson, C.E. and Westerdahl, B.B. (1993) Parasitism of *Heterodera schachtii* and *Meloidogyne javanica* by *Hirsutella rhossiliensis* in microplots over two growing seasons. *Journal of Nematology* 25, 427–433.

Tedford, E.C., Jaffee, B.A. and Muldoon, A.E. (1994) Variability among isolates of the nematophagous fungus *Hirsutella rhossiliensis*. *Mycological Research* 98, 1127–1136.

Tedford, E.C., Jaffee, B.A. and Muldoon, A.E. (1995a) Effect of temperature on infection of the cyst nematode *Heterodera schachtii* by the nematophagous fungus *Hirsutella rhossiliensis*. *Journal of Invertebrate Pathology* 66, 6–10.

Tedford, E.C., Jaffee, B.A. and Muldoon, A.E. (1995b) Suppression of the nematode *Heterodera schachtii* by the fungus *Hirsutella rhossiliensis* as affected by fungus population density and nematode movement. *Phytopathology* 85, 613–617.

Temple, S.R., Friedman, D.B., Somasco, O., Ferris, H., Scow, K. and Klonsky, K. (1994) An interdisciplinary, experiment station-based participatory comparison of alternative crop management systems for California's Sacramento Valley. *American Journal of Alternative Agriculture* 9, 64–71.

Tenuta, M. and Ferris, H (2004) Sensitivity of nematode life-history groups to ions and osmotic tensions of nitrogenous solutions. *Journal of Nematology* 36, 85–94.

Terefe, M., Tefera, T. and Sakhuja, P.K. (2009) Effect of a formulation of *Bacillus firmus* on root-knot nematode *Meloidogyne incognita* infestation and the growth of tomato plants in the greenhouse and nursery. *Journal of Invertebrate Pathology* 100, 94–99.

Termorshuizen, A.J., van Rijn, E., van der Gaag, D.J., Alabouvette, C., Chen, Y., Lagerlöf, J., Malandrakis, A.A., Paplomatas, E.J., Rämert, B., Ryckeboer, J., Steinberg, C. and Zmora-Nahum, S. (2006) Suppressiveness of 18 composts against 7 pathosystems: variability in pathogen response. *Soil Biology & Biochemistry* 38, 2461–2477.

Thies, J.E. and Grossman, J.M. (2006) The soil habitat and soil ecology. In: Uphoff, N., Ball, A.S., Fernandes, E., Herren, H., Husson, O., Laing, M., Palm, C., Pretty, J., Sanchez, P., Sanginga, N. and Thies, J. (eds) *Biological Approaches to Sustainable Soil Systems*. CRC Press, Boca Raton, Florida, pp. 59–78.

Thoden, T.C., Korthals, G.W. and Termorshuizen, A.J. (2011) Organic amendments and their influences on plant-parasitic and free-living nematodes: a promising method for nematode management? *Nematology* 13, 133–153.

Thom, E.R., Popay, A.J., Hume, D.E. and Fletcher, L.R. (2012) Evaluating the performance of endophytes in farm systems to improve farmer outcomes – a review. *Crop & Pasture Science* 63, 927–943.

Thomashow, L.S. and Weller, D.M. (1988) Role of a phenazine antibiotic from *Pseudomonas fluorescens* in biocontrol of *Gaeumannomyces graminis* var. *tritici*. *Journal of Bacteriology* 170, 3499–3508.

Thomashow, L.S., Bonsall, R.F. and Weller, D.M. (2008) Detection of antibiotics produced by soil and rhizosphere microbes *in situ*. In: Karlovsky, P. (ed.) *Secondary Metabolites in Soil Ecology*. Soil Biology Volume 14, Springer Verlag, Berlin, pp. 23–36.

Thomason, I.J. (1987) Challenges facing nematology: environmental risks with nematicides and the need for new approaches. In: Veech, J.A. and Dickson, D.W. (eds) *Vistas on Nematology*. Society of Nematologists, Hyattsville, Maryland, pp. 469–476.

Thompson, J.P. (1987) Decline of vesicular-arbuscular mycorrhizae in long fallow disorder of field crops and its expression in phosphorus deficiency of sunflower. *Australian Journal of Agricultural Research* 38, 847–867.

Thompson, J.P., Brennan, P.S., Clewett, T.G., Sheedy, J.G. and Seymour, N.P. (1999) Progress in breeding wheat for tolerance and resistance to root lesion nematode (*Pratylenchus thornei*). *Australasian Plant Pathology* 28, 45–52.

Thompson, J.P., Owen, K.J., Stirling, G.R. and Bell, M.J. (2008) Root-lesion nematodes (*Pratylenchus thornei* and *P. neglectus*: a review of recent progress in managing a significant pest of grain crops in northern Australia. *Australasian Plant Pathology* 37, 235–242.

Thompson, J.P., Clewett, T.G. and Fiske, M.L. (2013) Field inoculation with arbuscular-mycorrhizal fungi overcomes phosphorus and zinc deficiencies of linseed (*Linum usitatissimum*) in a vertisol subject to long-fallow disorder. *Plant and Soil* 371, 117–137.

Thorn, R.G. and Barron, G.L. (1984) Carnivorous mushrooms. *Science* 224, 76–78.

Thorn, R.G. and Barron, G.L. (1986) *Nematoctonus* and tribe Resupinateae in Ontario, Canada. *Mycotaxon* 25, 321–453.

Thorn, R.G., Moncalvo, J.-M., Reddy, C.A. and Vilgalys, R. (2000) Phylogenetic analyses and the distribution of nematophagy support a monophyletic Pleurotaceae within the polyphyletic pleurotoid-lentinoid fungi. *Mycologia* 92, 241–252.

Thorne, G. (1940) *Duboscqia penetrans* n. sp. (Sporozoa: Microsporidia, Nosematidae) a parasite of the nematode *Pratylenchus pratensis* (de Man) Filipjev. *Proceedings of the Helminthological Society of Washington* 7, 51–53.

Thornton, C.R. (2008) Tracking fungi in soil with monoclonal antibodies. *European Journal of Plant Pathology* 121, 347–353.

Tian, B., Yang, J. and Zhang, K.-Q. (2007) Bacteria used in the biological control of plant-parasitic nematodes: populations, mechanisms of action, and future prospects. *FEMS Microbiology Ecology* 61, 197–213.

Tian, H., Riggs, R.D. and Crippen, D.L. (2000) Control of soybean cyst nematode by chitinolytic bacteria with chitin substrate. *Journal of Nematology* 32, 370–376.

Timm, L., Pearson, D. and Jaffee, B. (2001) Nematode-trapping fungi in conventionally and organically managed corn-tomato rotations. *Mycologia* 93, 25–29.

Timm, M. (1987) 'Biocon' controls nematodes biologically. *Biotechnology* 5, 772.

Timper, P. (2009) Population dynamics of *Meloidogyne arenaria* and *Pasteuria penetrans* in a long-term crop rotation study. *Journal of Nematology* 41, 291–299.

Timper, P. (2011) Utilization of biological control for managing plant-parasitic nematodes. In: Davies, K. and Spiegel, Y. (eds) *Biological Control of Plant-parasitic Nematodes: Building Coherence between Microbial Ecology and Molecular Mechanisms*. Springer, Dordrecht, the Netherlands, pp. 259–289.

Timper, P. and Brodie, B.B. (1993) Infection of *Pratylenchus penetrans* by nematode-pathogenic fungi. *Journal of Nematology* 25, 297–302.

Timper, P. and Brodie, B.B. (1995) Interaction of the microbivorous nematode *Teratorhabditis dentifera* and the nematode-pathogenic fungus *Hirsutella rhossiliensis*. *Biological Control* 5, 629–635.

Timper, P. and Kaya, H.K. (1989) Role of the second-stage cuticle of entomogenous nematodes in preventing infection by nematophagous fungi. *Journal of Invertebrate Pathology* 54, 314–321.

Timper, P. and Riggs, R.D. (1998) Variation in efficacy of isolates of the fungus ARF against the soybean cyst nematode *Heterodera glycines*. *Journal of Nematology* 30, 461–467.

Timper, P., Kaya, H.K. and Jaffee, B.A. (1991) Survival of entomogenous nematodes in soil infested with the nematode-parasitic fungus *Hirsutella rhossiliensis* (Deuteromycotina: Hyphomycetes). *Biological Control* 1, 42–50.

Timper, P., Riggs, R.D. and Crippen, D.L. (1999) Parasitism of sedentary stages of *Heterodera glycines* by isolates of a sterile nematophagous fungus. *Phytopathology* 89, 1193–1199.

Timper, P., Koné, D., Yin, J., Ji, P. and McSpadden Gardener, B.B. (2009) Evaluation of an antibiotic-producing strain of *Pseudomona fluorescens* for suppression of plant-parasitic nematodes. *Journal of Nematology* 41, 234–240.

Timper, P., Davis, R., Jagdale, G. and Herbert, J. (2012) Resiliency of a nematode community and suppressive service to tillage and nematicide application. *Applied Soil Ecology* 59, 48–59.

Tobin, J.D., Haydock, P.P.J., Hare, M.C., Woods, S.R. and Crump, D.H. (2008) Effect of the fungus *Pochonia chlamydosporia* and fosthiazate on the multiplication rate of potato cyst nematodes (*Globodera pallida* and *G. rostochiensis*) in potato crops grown under UK field conditions. *Biological Control* 46, 194–201.

Tomich, T.P., Brodt, S., Ferris, H., Galt, R., Horwath, W.R., Kebreab, E., Leveau, J.H.J., Liptzin, D., Lubell., M., Merel, P., Michelmore, R., Rosenstock, T., Scow, K., Six, J., Williams, N. and Yang, L. (2011) Agroecology: a review from a global-change perspective. *Annual Review of Environment & Resources* 36, 193–222.

Tow, P.G., Cooper, I., Partridge, I. and Birch, C. (eds) (2011) *Rainfed Farming Systems*. Springer, Dordrecht, the Netherlands.

Treonis, A.M., Austin, E.E., Buyer, J.S., Maul, J.E., Spicer, L. and Zasada, I.A. (2010a) Effects of organic amendment and tillage on soil microorganisms and microfauna. *Applied Soil Ecology* 46, 103–110.

Treonis, A.M., Michelle, E.H., O'Leary, C.A., Austin, E.E. and Marks, C.B. (2010b) Identification and localization of food-source microbial nucleic acids inside soil nematodes. *Soil Biology & Biochemistry* 42, 2005–2011.

Treseder, K.K. (2008) Nitrogen additions and microbial biomass: a meta-analysis of ecosystem studies. *Ecology Letters* 11, 1111–1120.

Trewavas, A. (2001) Urban myths of organic farming. *Nature*, 410, 409–410.

Trewavas, A. (2004) A critical assessment of organic farming-and-food assertions with particular respect to the UK and potential environmental benefits of no-till agriculture. *Crop Protection* 23, 757–781.

Tribe, H.T. (1977) Pathology of cyst nematodes. *Biological Reviews* 52, 477–507.

Trotter, J.R. and Bishop, A.H. (2003) Phylogenetic analysis and confirmation of the endospore-forming nature of *Pasteuria penetrans* based on the spo0A gene. *FEMS Microbiology Letters* 225, 249–256.

Trudgill, D.L. (1991) Resistance to and tolerance of plant-parasitic nematodes in plants. *Annual Review of Phytopathology* 29, 167–192.

Trudgill, D.L., Bala, G., Blok, V.C., Daudi, A., Davies, K.G., Gowen, S.R., Fargette, M., Madulu, J.D., Mateille, T., Mwageni, W., Netscher, C., Philips, M.S., Sawadogo, A., Trivino, C.G. and Voyoukallou, E. (2000) The importance of tropical root-knot nematodes (*Meloidogyne* spp.) and factors affecting the utility of *Pasteuria penetrans* as a biocontrol agent. *Nematology* 2, 823–845.

Tu, J.C. (1988) Antibiosis of *Streptomyces griseus* against *Colletotrichum lindemuthianum*. *Journal of Phytopathology* 121, 97–102.

Tucker, S.L. and Talbot, N.J. (2001) Surface attachment and pre-penetration stage development by plant pathogenic fungi. *Annual Review of Phytopathology* 39, 385–417.

Turaganivalu, U., Stirling, G.R. and Smith, M.K. (2013) Burrowing nematode (*Radopholus similis*): a severe pathogen of ginger in Fiji. *Australasian Plant Pathology* 42, 431–436.

Turner, N.C. (2004) Agronomic options for improving rainfall-use efficiency of crops in dryland farming systems. *Journal of Experimental Botany* 55, 2413–2425.

Tzean, S.S. and Estey, R.H. (1978) Nematode-trapping fungi as mycopathogens. *Phytopathology* 68, 1266–1270.

Tzortzakakis, E.A. and Gowen, S.R. (1994) Resistance of a population of *Meloidogyne* spp. to parasitism by the obligate parasite *Pasteuria penetrans*. *Nematologica* 40, 258–266.

Tzortzakakis, E.A., Gowen, S.R. and Goumas, D.E. (1996) Decreased ability of *Pasteuria penetrans* spores to attack to successive generations of *Meloidogyne javanica*. *Fundamental & Applied Nematology* 19, 201–204.

Uhlenbroek, J.H. and Bijloo, J.D. (1958) Investigations on nematicides. I. Isolation and structure of a nematicidal principle occurring in *Tagetes* roots. *Recueil des Travaux Chimiques des Pays-Bas et de la Belgique* 77, 1004–1009.

Uhlenbroek, J.H. and Bijloo, J.D. (1959) Investigations on nematicides. II. Structure of a second nematicidal principle isolated from *Tagetes* roots. *Recueil des Travaux Chimiques des Pays-Bas et de la Belgique* 78, 382–390.

Uphoff, N., Ball, A.S., Fernandes, E., Herren, H., Husson, O., Laing, M., Palm, C., Pretty, J., Sanchez, P., Sanginga, N. and Thies, J. (2006) *Biological Approaches to Sustainable Soil Systems*. CRC Press, Boca Raton, Florida.

Uri, N.D. (1999) Factors affecting the use of conservation tillage in the United States. *Water, Air & Soil Pollution* 116, 621–638.

Vagelas, I., Pembroke, B. and Gowen, S.R. (2011) Techniques for image analysis of movement of juveniles of root-knot nematodes encumbered with *Pasteuria penetrans* spores. *Biocontrol Science & Technology* 21, 239–250.

Vagelas, I., Dennett, M.D., Pembroke, B., Ipsilandis, P. and Gowen, S.R. (2013) Understanding the movement of root-knot nematodes encumbered with or without *Pasteuria penetrans*. *Biocontrol Science & Technology* 23, 92–100.

van Bruggen, A.H.C. and Semenov, A.M. (2000) In search of biological indicators for soil health and disease suppression. *Applied Soil Ecology* 15, 13–24.

van Bruggen, A.H.C. and Termorshuizen, A.J. (2003) Integrated approaches to root disease management in organic farming systems. *Australasian Plant Pathology* 32, 141–156.

van Bruggen, A.H.C., Semenov, A.M., van Diepeningen, A.D., de Vos, O.J. and Blok, W.J. (2006) Relation between soil health, wave-like fluctuations in microbial populations, and soil-borne disease management. *European Journal of Plant Pathology* 115, 105–122.

van Capelle, C., Schrader, S. and Brunotte, J. (2012) Tillage-induced changes in the functional diversity of soil biota – a review with a focus on German data. *European Journal of Soil Biology* 50, 165–181.

Van Damme, V., Hoedekie, A., Viaene, N. (2005) Long-term efficacy of *Pochonia chlamydosporia* for management of *Meloidogyne javanica* in glasshouse crops. *Nematology* 7, 727–736.

Van den Boogert, P.H.J.F., Velvis, H., Ettema, C.H. and Bouwman, L.A. (1994) The role of organic matter in the population dynamics of the endoparasitic nematophagous fungus *Drechmeria coniospora* in microcosms. *Nematologica*, 40, 249–257.

van der Heijden, M.G.A. and Horton, T.R. (2009) Socialism in soil? The importance of mycorrhizal fungal networks for facilitation in natural ecosystems. *Journal of Ecology* 97, 1139–1150.

van der Putten, W.H., Cook, R., Costa, S., Davies, K.G., Fargette, M., Freitas, H., Hol, W.H.G., Kerry, B.R., Maher, N., Mateille, T., Moens, M., de la Peña, E., Piśkiewicz, A.M., Raeymaekers, A.D.W., Rodríguez-Echeverría, S. and van der Wurff, A. W. G. (2006) Nematode interactions in nature: models for sustainable control of nematode pests of crop plants? *Advances in Agronomy* 89, 227–60.

Van Elsas, J.D., Jansson, J.K. and Trevors, J.T. (eds) (2007a) *Modern Soil Microbiology*. CRC Press/Taylor Francis, Boca Raton, Florida.

Van Elsas, J.D., Torsvik, V., Hartmann, A., Øvreås, L. and Jansson, J.K. (2007b) The bacteria and archaea in soil. In: Van Elsas, J.D., Jansson, J.K. and Trevors, J.T. (eds) *Modern Soil Microbiology*. CRC Press/Taylor Francis, Boca Raton, Florida.

van Gundy, S.D., Stolzy, L.H., Szuszkiewicz, T.E. and Rackham, R.L. (1962) Influence of oxygen supply on survival of plant parasitic nematodes in soil. *Phytopathology* 52, 628–632.

van Gundy, S.D., Bird, A.F. and Wallace, H.R. (1967) Aging and starvation in larvae of *Meloidogyne javanica* and *Tylenchulus semipenetrans*. *Phytopathology* 57, 559–571.

van Loon, L.C., Bakker, P.A.H.M. and Pieterse, C.M.J. (1998) Systemic resistance induced by rhizosphere bacteria. *Annual Review of Phytopathology* 36, 453–483.

van Megen, H., van den Elsen, S., Holterman, M., Karssen, G., Mooyman, P., Bongers, T., Holovachov, O., Bakker, J. and Helder, J. (2009) A phylogenetic tree of nematodes based on about 1200 full-length small subunit ribosomal DNA sequences. *Nematology* 11, 927–950.

van Oorschot, C.A.N. (1985) Taxonomy of the *Dactylaria* complex, V. A review of *Arthrobotrys* and allied genera. *Studies in Mycology* 26, 61–96.

Vandermeer, J.H. (2011) *The Ecology of Agroecosystems*. Jones & Bartlett, Sudbury, Massachusetts.

Vanstone, V.A., Hollaway, G.J. and Stirling, G.R. (2008) Managing nematode pests in the southern and western regions of the Australian cereal industry: continuing progress in a challenging environment. *Australasian Plant Pathology* 37, 220–234.

Vawdrey, L.L. and Stirling, G.R. (1997) Control of root-knot nematode (*Meloidogyne javanica*) on tomato with molasses and other organic amendments. *Australasian Plant Pathology* 26: 179–187.

Velvis, H. and Kamp, P. (1996) Suppression of potato cyst nematode root penetration by the endoparasitic nematophagous fungus *Hirsutella rhossiliensis*. *European Journal of Plant Pathology* 102, 115–122.

Venette, R.C., Mostafa, F.A.M. and Ferris, H. (1997) Trophic interactions between bacterial-feeding nematodes in plant rhizospheres and the nematophagous fungus *Hirsutella rhossiliensis* to suppress *Heterodera schachtii. Plant & Soil* 191, 213–223.

Verbruggen, E., Röling, W.F.M., Gamper, H.A., Kowalchuk, G.A., Verhoef, H.A. and van der Heijden, M.G.A. (2010) Positive effects of organic farming on below-ground mutualists: large-scale comparison of mycorrhizal fungal communities in agricultural soils. *New Phytologist* 186, 968–979.

Vessey, J.K. (2003) Plant growth promoting rhizobacteria as biofertilizers. *Plant & Soil* 255, 571–586.

Villenave, C., Bongers, T., Ekschmitt, K., Djigal, D. and Chotte, J.L. (2001) Changes in nematode communities following cultivation of soils after fallow periods of different length. *Applied Soil Ecology* 17, 43–52.

Villenave, C., Saj, S., Pablo, A.-L., Sall, S., Djigal, D. Chotte, J.-L. and Bonzi, M. (2010) Influence of long-term organic and mineral fertilization on soil nematofauna when growing *Sorghum bicolor* in Burkina Faso. *Biology & Fertility of Soils* 46, 659–670.

Vos, C.M., Tesfahun, A.N., Panis, B., De Waele, D. and Elsen, A. (2012a) Arbuscular mycorrhizal fungi induce systemic resistance in tomato against the sedentary nematode *Meloidogyne incognita* and the migratory nematode *Pratylenchus penetrans. Applied Soil Ecology* 61, 1–6.

Vos, C., Claerhout, S., Mkandawire, R., Panis, B., De Waele, D. and Elsen, A. (2012b) Arbuscular mycorrhizal fungi reduce root-knot-nematode penetration through altered root exudation of their host. *Plant & Soil* 354, 335–345.

Vos, C., Geerinckx, K., Mkandawire, R., Panis, B., De Waele, D. and Elsen, A. (2012c) Arbuscular mycorrhizal fungi affect both penetration and further life stage development of root-knot-nematodes in tomato. *Mycorrhiza* 22, 157–163.

Vos, P., Garrity, G., Jones, D., Krieg, N.R., Ludwig, W., Rainey, F.A., Schleifer, K.-H. and Whitman, W.B. (eds) (2009) *Bergey's Manual of Systematic Bacteriology Volume 3. The Firmicutes*. Springer, New York.

Vovlas, N. and Frisullo, S. (1983) *Cylindrocarpon destructans* as a parasite of *Heterodera mediterranea* eggs. *Nematologia Mediterranea* 11, 193–196.

Vreeken-Buijs, M.J., Hassink, J. and Brussaard, L. (1998) Relationships of soil microarthropod biomass with organic matter and pore size distribution in soils under different land use. *Soil Biology & Biochemistry* 30, 97–106.

Vu, T., Hauschild, R. and Sikora, R.A. (2006) *Fusarium oxysporum* endophytes induced systemic resistance against *Radopholus similis* on banana. *Nematology* 8, 847–852.

Walker, G.E. and Wachtel, M.F. (1988) The influence of soil solarisation and non-fumigant nematicides on infection of *Meloidogyne javanica* by *Pasteuria penetrans. Nematologica* 34, 477–483.

Wall, D.H., Bardgett, R.D. and Kelly, E.F. (2010) Biodiversity in the dark. *Nature Geoscience* 3, 297–298.

Wallace, H.R. (1963) *The Biology of Plant Parasitic Nematodes*. Edward Arnold, London.

Wallace, H.R. (1968) The dynamics of nematode movement. *Annual Review of Phytopathology* 6, 91–114.

Waller, P.J. (2006) Sustainable nematode parasite control strategies for ruminant livestock by grazing management and biological control. *Animal Feed Science & Technology* 126, 277–289.

Waller, P.J., Knox, M.R. and Faedo, M. (2001) The potential of nematophagous fungi to control the free-living stages of nematode parasites of sheep: feeding and block studies with *Duddingtonia flagrans. Veterinary Parasitology* 102, 321–330.

Walter, D.E. and Ikonen, E.K. (1989) Species, guilds, and functional groups: taxonomy and behaviour in nematophagous arthropods. *Journal of Nematology* 21, 315–327.

Walter, D.E. and Kaplan, D.T. (1990) Antagonists of plant-parasitic nematodes in Florida citrus. *Journal of Nematology* 22, 567–573.

Walter, D.E., Kaplan, D.T. and Davis, E.L. (1993) Colonization of greenhouse nematode cultures by nematophagous mites and fungi. *Journal of Nematology* 25, 789–794.

Wang, E.L.H. and Bergeson, G.B. (1974) Biochemical changes in root exudate and xylem sap of tomato plants infected with *Meloidogyne incognita. Journal of Nematology* 6, 194–202.

Wang, K.-H., Sipes, B.S. and Schmitt, D.P. (2001) Suppression of *Rotylenchulus reniformis* by *Crotalaria juncea, Brassica napus*, and *Tagetes erecta. Nematropica* 31, 237–251.

Wang, K.-H., Sipes, B.S. and Schmitt, D.P. (2003a) Enhancement of *Rotylenchulus reniformis* suppressiveness by *Crotalaria juncea* amendment in pineapple soils. *Agriculture, Ecosystems & Environment* 94, 197–203.

Wang, K.-H., McSorley, R. and Gallaher, R.N. (2003b) Effect of *Crotalaria juncea* amendment on nematode communities in soil with different agricultural histories. *Journal of Nematology* 35, 294–301.

Wang, K.-H., McSorley, R., Marshall, A.J. and Gallaher, R.N. (2004a) Nematode community changes associated with decomposition of *Crotalaria juncea* amendment in litterbags. *Applied Soil Ecology* 27, 31–45.

Wang, K.-H., McSorley, R. and Gallaher, R.N. (2004b) Effect of *Crotalaria juncea* amendment on squash infected with *Meloidogyne incognita*. *Journal of Nematology* 36, 290–296.

Wang, K., Riggs, R.D. and Crippen, D. (2004c) Decomposition of plant debris by the nematophagous fungus ARF. *Journal of Nematology* 36, 263–266.

Wang, K., Riggs, R.D. and Crippen, D. (2004d) Competition between two phenotypes of the nematophagous fungus ARF in infecting eggs of *Heterodera glycines* and effect of soil depth on parasitism by ARF. *Nematology* 6, 329–334.

Wang, K., Riggs, R.D. and Crippen, D. (2004e) Suppression of *Rotylenchulus reniformis* on cotton by the nematophagous fungus ARF. *Journal of Nematology* 36, 186–191.

Wang, K., Riggs, R.D. and Crippen, D. (2005) Isolation, selection, and efficacy of *Pochonia chlamydosporia* for control of *Rotylenchulus reniformis* on cotton. *Phytopathology* 95, 890–893.

Wang, K.-H., McSorley, R. and Kokalis-Burelle, N. (2006) Effects of cover cropping, solarization, and soil fumigation on nematode communities. *Plant & Soil* 286, 229–243.

Wang, K.-H., McSorley, R., Gallaher, R.N. and Kokalis-Burelle, N. (2008) Cover crops and organic mulches for nematode, weed and plant health management. *Nematology* 10, 231–242.

Wang, K.-H., Hooks, C.R.R. and Marahatta, S.P. (2011) Can using a strip-tilled cover cropping system followed by surface mulch practice enhance organisms higher up in the soil food web hierarchy? *Applied Soil Ecology* 49, 107–117.

Wang, Y., Xu, J., Shen, J., Luo, Y., Scheu, S. and Ke, X. (2010) Tillage, residue burning and crop rotation alter soil fungal community and water-stable aggregation in arable fields. *Soil & Tillage Research* 107, 71–79.

Wang, W., Dalal, R.C., Reeves, S.H., Butterbach-Bahl, K. and Kiese, R. (2011) Greenhouse gas fluxes from an Australian subtropical cropland under long-term contrasting management regimes. *Global Change Biology* 17, 3089–3101.

Wardle, D.A. (1995) Impacts of disturbance on detritus food webs in agro-ecosystems of contrasting tillage and weed management practices. *Advances in Ecological Research* 26,105–183.

Wardle, D.A. (2002) *Communities and Ecosystems: Linking the Aboveground and Belowground Components*. Monographs in Population Biology 34, Princeton University Press, Princeton, New Jersey.

Wasilewska, L. (1994) The effect of age of meadows on succession and diversity in soil nematode communities. *Pedobiologia* 38, 1–11.

Waters, C.M. and Bassler B.L. (2005) Quorum sensing: cell-to-cell communication in bacteria. *Annual Review of Cell & Developmental Biology* 21, 319–346.

Watson, A.E. and Ford, E.J. (1972) Soil fungistasis – a reappraisal. *Annual Review of Phytopathology* 10, 327–348.

Watt, M., Kirkegaard, J.A. and Passioura, J.B. (2006) Rhizosphere biology and crop productivity – a review. *Australian Journal of Soil Research* 44, 299–317.

Waweru, B.W., Losenge, T., Kahangi, E.M., Dubois, T. and Coyne, D. (2013) Potential biological control of lesion nematodes on banana using Kenyan strains of endophytic *Fusarium oxysporum*. *Nematology* 15, 101–107.

Weaver, M.A., Krutz, L.J., Zablotowicz, R.M. and Reddy, K.N. (2007) Effects of glyphosate on soil microbial communities and its mineralization in a Mississippi soil. *Pest Management Science* 63, 388–393.

Webster, J., Henrici, A. and Spooner, B. (1998) *Orbilia fimicoloides* sp. nov., the teleomorph of *Dactylella* cf. *oxyspora*. *Mycological Research* 102, 99–102.

Weibelzahl-Fulton, E., Dickson, D.W. and Whitty, E.B. (1996) Suppression of *Meloidogyne incognita* and *M. javanica* by *Pasteuria penetrans* in field soil. *Journal of Nematology* 28, 43–49.

Weil, R.R. and Magdoff, F. (2004) Significance of soil organic matter to soil quality and health. In: Magdoff, F. and Weil, R.R. (eds) *Soil Organic Matter in Sustainable Agriculture*. CRC Press, Boca Raton, Florida, pp. 1–43.

Weil, R.R., Islam, K.R., Stine, M.A., Gruver, J.B. and Samson-Liebig, S.E. (2003) Estimating active carbon for soil quality assessment: A simplified method for laboratory and field use. *American Journal of Alternative Agriculture* 18, 3–17.

Weindling, R. (1932) *Trichoderma lignorum* as a parasite of other soil fungi. *Phytopathology* 22, 837–845.

Weiß, M., Sýkorová, Z., Garnica, S., Riess, K., Martos, F., Krause, C., Oberwinkler, F., Bauer, R. and Redecker, D. (2011) Sebacinales everywhere: previously overlooked ubiquitous fungal endophytes. *PLoS One* 6, e16793.

Welbaum, G., Sturz, A.V., Dong, Z. and Nowak, J. (2004) Managing soil microorganisms to improve productivity of agroecosystems. *Critical Reviews in Plant Science* 23, 175–193.

Weller, D.M. (1988) Biological control of soil-borne plant pathogens in the rhizosphere with bacteria. *Annual Review of Phytopathology* 26, 379–407.

Weller, D.M., Raaijmakers, J.M., McSpadden Gardner, B.B. and Thomashow, L.S. (2002) Microbial populations responsible for specific soil suppressiveness to plant pathogens. *Annual Review of Phytopathology* 40, 309–348.

West, C.P., Izekor, E., Oosterhuis, D.M. and Robbins, R.T. (1988) The effect of *Acremonium coenophialum* on the growth and nematode infestation of tall fescue. *Plant & Soil* 112, 3–6.

Westphal, A. (2005) Detection and description of soils with specific nematode suppressiveness. *Journal of Nematology* 37, 121–130.

Westphal, A. (2011) Sustainable approaches to the management of plant-parasitic nematodes and disease complexes. *Journal of Nematology* 43, 122–125.

Westphal, A. and Becker, J.O. (1999) Biological suppression and natural population decline of *Heterodera schachtii* in a California field. *Phytopathology* 89, 434–40.

Westphal, A. and Becker, J.O. (2000) Transfer of biological soil suppressiveness against *Heterodera schachtii*. *Phytopathology* 90, 401–406.

Westphal, A. and Becker, J.O. (2001a) Components of soil suppressiveness against *Heterodera schachtii*. *Soil Biology & Biochemistry* 33, 9–16.

Westphal, A. and Becker, J.O. (2001b) Impact of suppressiveness on various population densities of *Heterodera schachtii*. *Annals of Applied Biology* 138, 371–376.

Westphal, A. and Becker, J.O. (2001c) Soil suppressiveness to *Heterodera schachtii* under different cropping sequences. *Nematology* 3, 551–558.

Westphal, A. and Xing, L. (2011) Soil suppressiveness against the disease complex of the soybean cyst nematode and sudden death syndrome of soybean. *Phytopathology* 101, 878–886.

Westphal, A., Snyder, N.L., Xing, L. and Camberato, J.J. (2008a) Effects of inoculations with mycorrhizal fungi of soilless potting mixes during transplant production on watermelon growth and early fruit yield. *HortScience* 43, 354–360.

Westphal, A., Mehl, H., Seyb, A. and Vyn, T.J. (2008b) Consequences of tillage intensity on population densities of *Heterodera glycines* and severity of sudden death syndrome in corn-soybean sequence. *Phytopathology* 98, S169.

Westphal, A., Xing, L.J., Pillsbury, R. and Vyn, T.J. (2009) Effect of tillage intensity on population densities of *Heterodera glycines* in intensive soybean production systems. *Field Crops Research* 113, 218–226.

Whipps, J.M. (2001) Microbial interactions and biocontrol in the rhizosphere. *Journal of Experimental Botany* 52, 487–511.

Whipps, J.M. (2004) Prospects and limitations for mycorrhizas in biocontrol of plant pathogens. *Canadian Journal of Botany* 82, 1198–1227.

Whipps, J.M. and Davies, K.G. (2000) Success in biological control of plant pathogens and nematodes by microorganisms. In: Gurr, G. and Wratten, S. (eds) *Biological Control: Measures of Success*. Kluwer Academic Publishers, Dordrecht, the Netherlands, pp. 231–269.

Whish, J.P.M., Price, L. and Castor, P.A. (2009) Do spring cover crops rob water and so reduce wheat yields in the northern grain zone of eastern Australia? *Crop & Pasture Science* 60, 517–525.

White, J.F. and Bacon, C.W. (2012) The secret world of endophytes in perspective. *Fungal Ecology* 5, 287–288.

White, R.E. (2006) *Principles and Practice of Soil Science: the Soil as a Natural Resource*, 4th edn. Blackwell Publishing, Malden, Massachusetts.

Whitehead, A.G. (1998) *Plant Nematode Control*. CAB International, Wallingford, UK.

Widmer, T.L. and Abawi, G.S. (2000) Mechanism of suppression of *Meloidogyne hapla* and its damage by a green manure of Sudan grass. *Plant Disease* 84, 562–568.

Wilkins, R.J. (2008) Eco-efficient approaches to land management: a case for increased integration of crop and animal production systems. *Philosophical Transactions of the Royal Society B*, 363, 517–525.

Williams, A.B., Stirling, G.R., Hayward, A.C. and Perry, J. (1989) Properties and attempted culture of *Pasteuria penetrans*, a bacterial parasite of root-knot nematode (*Meloidogyne javanica*). *Journal of Applied Bacteriology* 67, 145–156.

Williams, J.R. (1960) Studies on the nematode soil fauna of sugarcane fields in Mauritius. 5. Notes upon a parasite of root-knot nematodes. *Nematologica* 5, 37–42.

Williams, S.M. and Weil, R.R. (2004) Cover crop root channels may alleviate soil compaction effects on soybean crop. *Soil Science Society of America Journal* 68, 1403–1409.

Williams, S.T. (1982) Are antibiotics produced in soil? *Pedobiologia* 23, 427–435.

Williams, T.D. (1969) The effects of formalin, nabam, irrigation and nitrogen on *Heterodera avenae* Woll., *Ophiobolus graminis* Sacc. and the growth of spring wheat. *Annals of Applied Biology* 64, 325–334.

Willis, A., Rodrigues, B.F. and Harris, P.J.C. (2013) The ecology of arbuscular mycorrhizal fungi. *Critical Reviews in Plant Sciences* 32, 1–20.

Wilson, J.M. and Griffin, D.M. (1975) Water potential and the respiration of microorganisms in the soil. *Soil Biology and Biochemistry* 7, 199–204.

Windham, G.L., Windham, M.T. and Williams, W.P. (1989) Effects of *Trichoderma* spp. on maize growth and *Meloidogyne arenaria* reproduction. *Plant Disease* 73, 493–495.

Wishart, J., Blok, V.C., Phillips, M.S. and Davies, K.G. (2004) *Pasteuria penetrans* and *P. nishizawae* attachment to *Meloidogyne chitwoodi, M. fallax* and *M. hapla. Nematology* 6, 507–510.

Wolf, B. and Snyder, G.H. (2003) *Sustainable Soils. The Place of Organic Matter in Sustaining Soils and their Productivity.* The Haworth Press Inc., Binghamton, New York.

Wood, F.H. (1973a) Biology of *Aporcelaimellus* sp. (Nematoda: Aporcelaimidae). *Nematologica* 19, 528–537.

Wood, F.H. (1973b) Feeding relationships of soil-dwelling nematodes. *Soil Biology & Biochemistry* 5, 593–601.

Wood, F.H. (1974) Biology of *Seinura demani* (Nematoda: Aphelenchoididae). *Nematologica* 20, 347–353.

Worthington, M. and Reberg-Horton, C. (2013) Breeding cereal crops for enhanced weed suppression: optimizing allelopathy and competitive ability. *Journal of Chemical Ecology* 39, 213–231.

Wright, S.F. and Upadhyaya, A. (1998) A survey of soils for aggregate stability and glomalin, a glycoprotein produced by hyphae of arbuscular mycorrhizal fungi. *Plant & Soil* 198, 97–107.

Wright, S.F. and Upadhyaya, A. (1999) Quantification of arbuscular mycorrhizal fungi activity by the glomalin concentration on hyphal traps. *Mycorrhiza* 8, 283–285.

Wu, M., Zhang, H., Li, X., Zhang, Y., Su, Z. and Zhang, C. (2008) Soil fungistasis and its relations to soil microbial composition and diversity: A case study of a series of soils with different fungistasis. *Journal of Environmental Sciences* 20, 871–877.

Wyss, U., Voss, B. and Jansson, H.-B. (1992) *In vitro* observations on the infection of *Meloidogyne incognita* eggs by the zoosporic fungus *Catenaria anguillulae* Sorokin. *Fundamental & Applied Nematology* 15, 133–139.

Xiang, M., Xiang, P., Liu, X. and Zhang, L. (2010a) Effect of environment on the abundance and activity of the nematophagous fungus *Hirsutella minnesotensis* in soil. *FEMS Microbiology Ecology* 71, 413–417.

Xiang, M.C., Xiang, P.A., Jiang, X.Z., Duan, W.J. and Liu, X.Z. (2010b) Detection and quantification of the nematophagous fungus *Hirsutella minnesotensis* in soil with real-time PCR. *Applied Soil Ecology* 44, 170–175.

Xing, L. and Westphal, A. (2009) Effects of crop rotation of soybean with corn on severity of sudden death syndrome and population densities of *Heterodera glycines* in naturally infested soil. *Field Crops Research* 112, 107–117.

Xu, C., Mo, M., Zhang, L. and Zhang, K. (2004) Soil volatile fungistasis and volatile fungistatic compounds. *Soil Biology & Biochemistry* 36, 1997–2004.

Yang, E., Xu, L., Yang, Y., Zhang, X., Xiang, M., Wang, C., An, Z. and Liu, X. (2012c) Origin and evolution of carnivorism in the Ascomycota (fungi). *Proceedings National Academy Science* 109, 10960–10965.

Yang, J.-I., Loffredo, A., Borneman, J. and Becker, J.O. (2012a) Biocontrol efficacy among strains of *Pochonia chlamydosporia* obtained from a root-knot suppressive soil. *Journal of Nematology* 44, 67–71.

Yang, J.-I., Benecke, S., Jeske, D.R., Rocha, F.S., Smith Becker, J., Timper, P., Becker, J.O. and Borneman, J. (2012b) Population dynamics of *Dactylella oviparasitica* and *Heterodera schachtii:* toward a decision model for sugar beet planting. *Journal of Nematology* 44, 237–244.

Yang, Y. and Liu, X.Z. (2005) *Dactylella coccinella* sp. nov., an anamorphic species. *Mycotaxon* 91, 127–132.

Yeates, G.W. (1987a) Nematode feeding and activity: the importance of development stages. *Biology & Fertility of Soils* 3, 143–146.

Yeates, G.W. (1987b) Significance of developmental stages in the coexistence of three species of Mononchoidea (Nematoda) in a pasture soil. *Biology and Fertility of Soils* 5, 225–229.

Yeates, G.W. (1998) Feeding in free-living soil nematodes: a functional approach. In: Perry, R.N. and Wright, D.J. (eds) *The Physiology and Biochemistry of Free-living and Plant-parasitic Nematodes.* CAB International, Wallingford, UK, pp. 245–269.

Yeates, G.W. (2010) Nematodes in ecological webs. In: *Encyclopedia of Life Sciences.* John Wiley & Sons, Chichester, UK. DOI: 10.1002/9780470015902.a0021913.

Yeates, G.W. and Foissner, W. (1995) Testate amoebae as predators of nematodes. *Biology & Fertility of Soils* 20, 1–7.

Yeates, G.W. and van der Meulen, H. (1996) Recolonization of methyl-bromide sterilized soils by plant and soil nematodes over 52 months. *Biology & Fertility of Soils* 21, 1–6.

Yeates, G.W. and Wardle, D.A. (1996) Nematodes as predators and prey: relationships to biological control and soil processes. *Pedobiologia* 40, 43–50.

Yeates, G.W., Bamforth, S.S., Ross, D.J., Tate, K.R. and Sparling, G.P. (1991) Recolonization of methyl bromide sterilized soils under four different field conditions. *Biology & Fertility of Soils* 11, 181–189.

Yeates, G.W., Bongers, T., de Goede, R.G.M., Freckman, D.W. and Georgieva, S.S. (1993a) Feeding habits in soil nematode families and genera – an outline for soil ecologists. *Journal of Nematology* 25, 315–331.

Yeates, G.W., Wardle, D.A. and Watson, R.N. (1993b) Relationships between nematodes, soil microbial biomass and weed-management strategies in maize and asparagus cropping systems. *Soil Biology & Biochemistry* 25, 869–876.

Yeates, G.W., Wardle, D.A. and Watson, R.N. (1999) Responses of soil nematode populations, community structure, diversity and temporal variability to agricultural intensification over a seven-year period. *Soil Biology & Biochemistry* 31, 1721–1733.

Yin, B., Valinsky, L., Gao, X., Becker, J.O. and Borneman, J. (2003a) Identification of fungal rDNA associated with soil suppressiveness against *Heterodera schachtii* using oligonucleotide fingerprinting. *Phytopathology* 93,1006–1013.

Yin, B., Valinsky, L., Gao, X., Becker, J.O. and Borneman, J. (2003b) Bacterial rRNA genes associated with soil suppressiveness against the plant-parasitic nematode *Heterodera schachtii*. *Applied & Environmental Microbiology* 69, 1573–1580.

You, M.P. and Sivasithamparam, K. (1994) Hydrolysis of fluorescein diacetate in an avocado plantation mulch suppressive to *Phytophthora cinnamomi* and its relationship with certain biotic and abiotic factors. *Soil Biology & Biochemistry* 26, 1355–1361.

You, M.P. and Sivasithamparam, K. (1995) Changes in microbial populations of an avocado plantation mulch suppressive to *Phytophthora cinnamomi*. *Applied Soil Ecology* 2, 33–43.

Youngberg, G. and DeMuth, S.P. (2013) Organic agriculture in the United States: a 30 year retrospective. *Renewable Agriculture & Food Systems*. DOI:10.1017/S1742170513000173.

Yu, O. and Coosemans, J. (1998) Fungi associated with cysts of *Globodera rostochiensis*, *G. pallida*, and *Heterodera schachtii*; and egg masses and females of *Meloidogyne hapla* in Belgium. *Phytoprotection* 79, 63–69.

Yu, Z., Zhang, Y., Qiao, M., Baral, H.-O., Weber, E. and Zhang, K. (2006) *Drechslerella brochopaga*, the anamorph of *Orbilia* (*Hyalinia*) *orientalis*. *Mycotaxon* 96, 163–168.

Yu, Z.F., Zhang, Y., Qiao, M. and Zhang, K.Q. (2007) *Orbilia dorsalia* sp. nov., the teleomorph of *Dactylella dorsalia* sp. nov. *Cryptogamie, Mycologie* 28, 55–63.

Yu, Z.-Q., Wang, Q.-L., Liu, B., Zou, X., Yu, Z.-N. and Sun, M. (2008) *Bacillus thuringiensis* crystal protein toxicity against plant-parasitic nematodes. *Chinese Journal of Agricultural Biotechnology* 5, 13–17.

Yuen, G.Y., Broderick, K., Moore, W.H. and Caswell-Chen, E.P. (2006) Effects of *Lysobacter enzymogenes* C3 and its antibiotic dihydromaltophilin on nematodes. *Phytopathology* 96, S128.

Zabaloy, M.C., Gómez, E., Garland, J.L. and Gómez, M.A. (2012) Assessment of microbial community function and structure in soil microcosms exposed to glyphosate. *Applied Soil Ecology* 61, 333–339.

Zare, R. and Gams, W. (2001) A revision of Verticillium section Prostrata. VI. The genus *Haptocillium*. *Nova Hedwigia* 73, 271–292.

Zare, R., Gams, W. and Culham, A. (2001a) A revision of *Verticillium* sect. Prostrata. I. Phylogenetic studies using ITS sequences. *Nova Hedwigia* 71, 465–480.

Zare, R., Gams, W. and Evans, H.C. (2001b) A revision of *Verticillium* section Prostrata. V. The genus *Pochonia*, with notes on *Rotiferophthora*. *Nova Hedwigia* 73, 51–86.

Zasada, I.A. (2005) Factors affecting the suppression of *Heterodera glycines* by N-Viro soil. *Journal of Nematology* 37, 220–225.

Zasada, I.A. and Ferris, H. (2003) Sensitivity of *Meloidogyne javanica* and *Tylenchulus semipenetrans* to isothiocyanates in laboratory assays. *Phytopathology* 93, 747–750.

Zasada, I.A. and Ferris, H. (2004) Nematode suppression with brassicaceous amendments: application based on glucsinolate profiles. *Soil Biology & Biochemistry* 36, 1017–1024.

Zasada, I.A. and Tenuta, M. (2004) Chemical-mediated toxicity of N-Viro soil to *Heterodera glycines* and *Meloidogyne incognita*. *Journal of Nematology* 36, 297–302.

Zasada, I.A., Halbrendt, J.M., Kokalis-Burelle, N., LaMondia, J., McKenry, M.V. and Noling, J.W. (2010) Managing nematodes without methyl bromide. *Annual Review of Phytopathology* 48, 311–328.

Zehr, E.I. (1985) Evaluation of parasites and predators of plant parasitic nematodes. *Journal of Agricultural Entomology* 2, 130–134.

Zelenev, V.V., Berkelmans, R., van Bruggen, A.H.C., Bongers, T. and Semenov, A.M. (2004) Daily changes in bacterial-feeding nematode populations oscillate with similar periods as bacterial populations after a nutrient impulse in soil. *Applied Soil Ecology* 26, 93–106.

Zhang, J.X., Howell, C.R. and Starr, J.L. (1996) Suppression of *Fusarium* colonization of cotton roots and Fusarium wilt by seed treatments with *Gliocladium virens* and *Bacillus subtilis*. *Biocontrol Science & Technology* 6, 175–187.

Zhang, L., Liu, X., Zhu, S. and Chen, S. (2006) Detection of the nematophagous fungus *Hirsutella rhossiliensis* in soil by real-time PCR and parasitism bioassay. *Biological Control* 36, 316–323.

Zhang, L., Yang, E., Xiang, M., Liu, X. and Chen, S. (2008) Population dynamics and biocontrol efficacy of the nematophagous fungus *Hirsutella rhossiliensis* as affected by stage of the soybean cyst nematode. *Biological Control* 47, 244–249.

Zhu, M.L., Mo, M.H., Xia, Z.Y., Li, Y.H., Yang, S.J., Li, T.F. and Zhang, K.Q. (2006) Detection of two fungal biocontrol agents against root-knot nematodes by RAPD markers. *Mycopathologia* 161, 307–316.

Zou, C.-S., Mo, M.-H., Gu, Y.-Q., Zhou, J.-P. and Zhang, K.-Q. (2007) Possible contributions of volatile-producing bacteria to soil fungistasis. *Soil Biology & Biochemistry* 39, 2371–2379.

Zuckerman, B.M., Dicklow, M.B. and Acosta, N. (1993) A strain of *Bacillus thuringiensis* for the control of plant-parasitic nematodes. *Biocontrol Science & Technology* 3, 41–46.

zum Felde, A., Pocasangre, L.E., Carñizares Monteros, C.A., Sikora, R.A., Rosales, F.E. and Riveros, A.S. (2006) Effect of combined inoculations of endophytic fungi on the biocontrol of *Radopholus similis*. *Infomusa* 15, 12–18.

Index of Soil Organisms
by Genus and Species

General Index

Organisms commonly referred to in the text are included in this index. A comprehensive list can be found in the Index of Soil Organisms by Genus and Species.